Lecture Notes in Computer Science 13810

The series Lecture Notes in Computer Science (LNCS), including its subseries Lecture Notes in Artificial Intelligence (LNAI) and Lecture Notes in Bioinformatics (LNBI), has established itself as a medium for the publication of new developments in computer science and information technology research, teaching, and education.

LNCS enjoys close cooperation with the computer science R & D community, the series counts many renowned academics among its volume editors and paper authors, and collaborates with prestigious societies. Its mission is to serve this international community by providing an invaluable service, mainly focused on the publication of conference and workshop proceedings and postproceedings. LNCS commenced publication in 1973.

Giuseppe Nicosia · Varun Ojha ·
Emanuele La Malfa · Gabriele La Malfa ·
Panos Pardalos · Giuseppe Di Fatta ·
Giovanni Giuffrida · Renato Umeton
Editors

Machine Learning, Optimization, and Data Science

8th International Conference, LOD 2022
Certosa di Pontignano, Italy, September 18–22, 2022
Revised Selected Papers, Part I

Editors
Giuseppe Nicosia
University of Catania
Catania, Italy

Varun Ojha
University of Reading
Reading, UK

Emanuele La Malfa
University of Oxford
Oxford, UK

Gabriele La Malfa
University of Cambridge
Cambridge, UK

Panos Pardalos
University of Florida
Gainesville, FL, USA

Giuseppe Di Fatta
Free University of Bozen-Bolzano
Bolzano, Italy

Giovanni Giuffrida
University of Catania
Catania, Italy

Renato Umeton
Dana-Farber Cancer Institute
Boston, MA, USA

ISSN 0302-9743 ISSN 1611-3349 (electronic)
Lecture Notes in Computer Science
ISBN 978-3-031-25598-4 ISBN 978-3-031-25599-1 (eBook)
https://doi.org/10.1007/978-3-031-25599-1

This Springer imprint is published by the registered company Springer Nature Switzerland AG
The registered company address is: Gewerbestrasse 11, 6330 Cham, Switzerland

Preface

LOD is the international conference embracing the fields of machine learning, deep learning, optimization, and data science. The eighth edition, LOD 2022, was organized on September 18–22, 2022, in Castelnuovo Berardenga, Italy. LOD 2022 was held successfully online and onsite to meet challenges posed by the worldwide outbreak of COVID-19. As in the previous edition, LOD 2022 hosted the second edition of the Advanced Course and Symposium on Artificial Intelligence & Neuroscience – ACAIN 2022. In fact, this year, in the LOD proceedings we decided to also include the papers of the second edition of the Symposium on Artificial Intelligence and Neuroscience (ACAIN 2022). The ACAIN 2022 chairs were:

Giuseppe Nicosia, University of Catania, Italy, and University of Cambridge, UK
Varun Ojha, University of Reading, UK
Panos Pardalos, University of Florida, USA

The review process of papers submitted to ACAIN 2022 was double blind, performed rigorously by an international program committee consisting of leading experts in the field. Therefore, the last thirteen articles in the Table of Contents are the articles accepted at ACAIN 2022.

Since 2015, the LOD conference has brought together academics, researchers and industrial researchers in a unique *pandisciplinary community* to discuss the state of the art and the latest advances in the integration of machine learning, deep learning, nonlinear optimization and data science to provide and support the scientific and technological foundations for interpretable, explainable and trustworthy AI. Since 2017, LOD has adopted the *Asilomar AI Principles*.

The annual conference on machine Learning, Optimization and Data science (LOD) is an international conference on machine learning, computational optimization and big data that includes invited talks, tutorial talks, special sessions, industrial tracks, demonstrations and oral and poster presentations of refereed papers.

LOD has established itself as a premier multidisciplinary conference in machine learning, computational optimization and data science. It provides an international forum for presentation of original multidisciplinary research results, as well as exchange and dissemination of innovative and practical development experiences.

The manifesto of the LOD conference is:

"The problem of understanding intelligence is said to be the greatest problem in science today and "the" problem for this century – as deciphering the genetic code was for the second half of the last one. Arguably, the problem of learning represents a gateway to understanding intelligence in brains and machines, to discovering how the human brain works, and to making intelligent machines that learn from experience and improve their competences as children do. In engineering, learning techniques would make it possible to develop software that can be quickly customized to deal with the increasing amount of information and the flood of data around us."

The Mathematics of Learning: Dealing with Data
Tomaso Poggio (MOD 2015 & LOD 2020 Keynote Speaker) and Steve Smale

"Artificial Intelligence has already provided beneficial tools that are used every day by people around the world. Its continued development, guided by the Asilomar principles of AI, will offer amazing opportunities to help and empower people in the decades and centuries ahead."
The Asilomar AI Principles

LOD 2022 attracted leading experts from industry and the academic world with the aim of strengthening the connection between these institutions. The 2022 edition of LOD represented a great opportunity for professors, scientists, industry experts, and research students to learn about recent developments in their own research areas and to learn about research in contiguous research areas, with the aim of creating an environment to share ideas and trigger new collaborations.

As chairs, it was an honour to organize a premier conference in these areas and to have received a large variety of innovative and original scientific contributions.

During LOD 2022, 11 plenary talks were presented by leading experts:

LOD 2022 Keynote Speakers:

Jürgen Bajorath, University of Bonn, Germany
Pierre Baldi, University of California Irvine, USA
Ross King, University of Cambridge, UK and The Alan Turing Institute, UK
Rema Padman, Carnegie Mellon University, USA
Panos Pardalos, University of Florida, USA

ACAIN 2022 Keynote Lecturers:

Karl Friston, University College London, UK
Wulfram Gerstner, EPFL, Switzerland
Máté Lengyel, Cambridge University, UK
Christos Papadimitriou, Columbia Engineering, Columbia University, USA
Panos Pardalos, University of Florida, USA
Michail Tsodyks, Institute for Advanced Study, USA

LOD 2022 received 226 submissions from 69 countries in five continents, and each manuscript was independently reviewed by a committee formed by at least five members. These proceedings contain 85 research articles written by leading scientists in the fields of machine learning, artificial intelligence, reinforcement learning, computational optimization, neuroscience, and data science, presenting a substantial array of ideas, technologies, algorithms, methods, and applications.

At LOD 2022, Springer LNCS generously sponsored the LOD Best Paper Award. This year, the paper by *Pedro Henrique da Costa Avelar, Roman Laddach, Sophia Karagiannis, Min Wu and Sophia Tsoka* titled "*Multi-Omic Data Integration and Feature Selection for Survival-based Patient Stratification via Supervised Concrete Autoencoders*", received the LOD 2022 Best Paper Award.

This conference could not have been organized without the contributions of exceptional researchers and visionary industry experts, so we thank them all for participating.

A sincere thank you goes also to the 35 sub-reviewers and to the Program Committee, of more than 170 scientists from academia and industry, for their valuable and essential work in selecting the scientific contributions.

Finally, we would like to express our appreciation to the keynote speakers who accepted our invitation, and to all the authors who submitted their research papers to LOD 2022.

October 2022

Giuseppe Nicosia
Varun Ojha
Emanuele La Malfa
Gabriele La Malfa
Panos Pardalos
Giuseppe Di Fatta
Giovanni Giuffrida
Renato Umeton

Organization

General Chairs

Giovanni Giuffrida	University of Catania and NeoData Group, Italy
Renato Umeton	Dana-Farber Cancer Institute, MIT, Harvard T.H. Chan School of Public Health & Weill Cornell Medicine, USA

Conference and Technical Program Committee Co-chairs

Giuseppe Di Fatta	Free University of Bozen-Bolzano, Bolzano, Italy
Varun Ojha	University of Reading, UK
Panos Pardalos	University of Florida, USA

Special Sessions Chairs

Gabriele La Malfa	University of Cambridge, UK
Emanuele La Malfa	University of Oxford, UK

Tutorial Chair

Giorgio Jansen	University of Cambridge, UK

Steering Committee

Giuseppe Nicosia	University of Catania, Italy
Panos Pardalos	University of Florida, USA

Program Committee Members

Jason Adair	University of Stirling, UK
Agostinho Agra	Universidade de Aveiro, Portugal
Richard Allmendinger	University of Manchester, UK

Luca Di Gaspero	University of Udine, Italy
Mario Di Raimondo	University of Catania, Italy
Ciprian Dobre	University Politehnica of Bucharest, Romania
Andrii Dobroshynskyi	New York University, USA
Stephan Doerfel	Kiel University of Applied Sciences, Germany
Rafal Drezewski	University of Science and Technology, Krakow, Poland
Juan J. Durillo	Leibniz Supercomputing Centre, Germany
Nelson F. F. Ebecken	University of Rio de Janeiro, Brazil
Michael Elberfeld	RWTH Aachen University, Germany
Michael T. M. Emmerich	Leiden University, The Netherlands
Roberto Esposito	University of Turin, Italy
Giovanni Fasano	University Ca' Foscari of Venice, Italy
Lionel Fillatre	Université Côte d'Azur, France
Steffen Finck	FH Vorarlberg University of Applied Sciences, Austria
Enrico Formenti	Université Côte d'Azur, France
Giuditta Franco	University of Verona, Italy
Piero Fraternali	Politecnico di Milano, Italy
Valerio Freschi	University of Urbino, Italy
Nikolaus Frohner	TU Wien, Austria
Carola Gajek	University of Augsburg, Germany
Claudio Gallicchio	University of Pisa, Italy
Alfredo García Hernández-Díaz	Pablo de Olavide University (Seville), Spain
Paolo Garza	Politecnico di Torino, Italy
Romaric Gaudel	ENSAI, France
Michel Gendreau	Polytechnique Montréal, Canada
Kyriakos Giannakoglou	National Technical University of Athens, Greece
Giorgio Stefano Gnecco	IMT School for Advanced Studies, Lucca, Italy
Teresa Gonçalves	University of Évora, Portugal
Michael Granitzerv	University of Passau, Germany
Vladimir Grishagin	Lobachevsky State University of Nizhni Novgorod, Russia
Vijay K. Gurbani	Vail Systems, Inc. & Illinois Institute of Technology, USA
Jin-Kao Hao	University of Angers, France
Verena Heidrich-Meisnerv	Kiel University, Germany
Carlos Henggeler Antunes	University of Coimbra, Portugal
J. Michael Herrmann	University of Edinburgh, UK
Giorgio Jansenv	University of Cambridge, UK
Laetitia Jourdan	University of Lille/CNRS, France

Valery Kalyagin	National Research University Higher School of Economics, Russia
George Karakostas	McMaster University, Canada
Branko Kavšek	University of Primorska, Slovenia
Michael Khachay	Ural Federal University Ekaterinburg, Russia
Zeynep Kiziltan	University of Bologna, Italy
Yury Kochetov	Novosibirsk State University, Russia
Hennie Kruger	North-West University, South Africa
Vinod Kumar Chauhan	University of Cambridge, UK
Dmitri Kvasov	University of Calabria, Italy
Niklas Lavesson	Jönköping University, Sweden
Eva K. Lee	Georgia Tech, USA
Carson Leung	University of Manitoba, Canada
Gianfranco Lombardo	University of Parma, Italy
Paul Lu	University of Alberta, Canada
Angelo Lucia	University of Rhode Island, USA
Pasi Luukka	Lappeenranta-Lahti University of Technology, Finland
Anthony Man-Cho So	The Chinese University of Hong Kong, Hong Kong, China
Luca Manzoni	University of Trieste, Italy
Marco Maratea	University of Genova, Italy
Magdalene Marinaki	Technical University of Crete, Greece
Yannis Marinakis	Technical University of Crete, Greece
Ivan Martino	Royal Institute of Technology, Sweden
Petr Masa	University of Economics, Prague, Czech Republic
Joana Matos Dias	Universidade de Coimbra, Portugal
Nikolaos Matsatsinis	Technical University of Crete, Greece
Stefano Mauceri	University College Dublin, Ireland
George Michailidis	University of Florida, USA
Kaisa Miettinen	University of Jyväskylä, Finland
Shokoufeh Monjezi Kouchak	Arizona State University, USA
Rafael M. Moraes	Viasat Inc., USA
Mohamed Nadif	University of Paris, France
Mirco Nanni	CNR - ISTI, Italy
Giuseppe Nicosia	University of Catania, Italy
Wieslaw Nowak	Nicolaus Copernicus University, Poland
Varun Ojha	University of Reading, UK
Marcin Orchel	AGH University of Science and Technology, Poland
Mathias Pacher	Goethe Universität Frankfurt am Main, Germany
Pramudita Satria Palar	Bandung Institute of Technology, Indonesia

Panos Pardalos	University of Florida, USA
Konstantinos Parsopoulos	University of Ioannina, Greece
Andrea Patanè	University of Oxford, UK
David A. Pelta	Universidad de Granada, Spain
Dimitri Perrin	Queensland University of Technology, Australia
Milena Petkovic	Zuse Institute Berlin, Germany
Koumoutsakos Petros	ETH, Switzerland
Nikolaos Ploskas	University of Western Macedonia, Greece
Alessandro S. Podda	University of Cagliari, Italy
Buyue Qian	IBM T. J. Watson, USA
Michela Quadrini	University of Padova, Italy
Günther Raidl	TU Wien, Austria
Jan Rauch	University of Economics, Prague, Czech Republic
Steffen Rebennack	Karlsruhe Institute of Technology (KIT), Germany
Wolfgang Reif	University of Augsburg, Germany
Cristina Requejo	Universidade de Aveiro, Portugal
Francesco Rinaldi	University of Padova, Italy
Laura Anna Ripamonti	Università degli Studi di Milano, Italy
Humberto Rocha	University of Coimbra, Portugal
Vittorio Romano	University of Catania, Italy
Andre Rosendo	University of Cambridge, UK
Arnab Roy	Fujitsu Laboratories of America, USA
Andrea Santoro	Queen Mary University London, UK
Giorgio Sartor	SINTEF, Norway
Claudio Sartori	University of Bologna, Italy
Fréderic Saubion	Université d'Angers, France
Robert Schaefer	AGH University of Science and Technology, Kraków, Poland
Andrea Schaerf	University of Udine, Italy
Oliver Schuetze	CINVESTAV-IPN, Mexico
Bryan Scotney	Ulster University, UK
Andrea Serani	CNR-INM, National Research Council-Institute of Marine Engineering, Rome, Italy
Marc Sevaux	Université de Bretagne-Sud, France
Vladimir Shenmaier	Sobolev Institute of Mathematics, Russia
Zeren Shui	University of Minnesota, USA
Patrick Siarry	Université Paris-Est Creteil, France
Konstantina Skouri	University of Ioannina, Greece
Cole Smith	New York University, USA
Antoine Spicher	LACL University Paris Est Creteil, France
Catalin Stoean	University of Craiova, Romania

Johan Suykens	KU Leuven, Belgium
Tatiana Tchemisova Cordeiro	University of Aveiro, Portugal
Gabriele Tolomei	Sapienza University of Rome, Italy
Elio Tuci	University of Namur, Belgium
Gabriel Turinici	Université Paris Dauphine - PSL, France
Gregor Ulm	Fraunhofer-Chalmers Research Centre for Industrial Mathematics, Sweden
Renato Umeton	Dana-Farber Cancer Institute, MIT & Harvard T.H. Chan School of Public Health, USA
Werner Van Geit	Blue Brain Project, EPFL, Switzerland
Carlos Varela	Rensselaer Polytechnic Institute, USA
Herna L. Viktor	University of Ottawa, Canada
Marco Villani	University of Modena and Reggio Emilia, Italy
Mirko Viroli	Università di Bologna, Italy
Dachuan Xu	Beijing University of Technology, China
Xin-She Yang	Middlesex University London, UK
Shiu Yin Yuen	City University of Hong Kong, China
Qi Yu	Rochester Institute of Technology, USA
Zelda Zabinsky	University of Washington, USA

Best Paper Awards

LOD 2022 Best Paper Award

"Multi-Omic Data Integration and Feature Selection for Survival-based Patient Stratification via Supervised Concrete Autoencoders"
Pedro Henrique da Costa Avelar[1], Roman Laddach[2], Sophia Karagiannis[2], Min Wu[3] and Sophia Tsoka[1]
[1] Department of Informatics, Faculty of Natural, Mathematical and Engineering Sciences, King's College London, UK
[2] St John's Institute of Dermatology, School of Basic and Medical Biosciences, King's College London, UK
[3] Machine Intellection Department, Institute for Infocomm Research, A*STAR, Singapore
Springer sponsored the LOD 2022 Best Paper Award with a cash prize of EUR 1,000.

Special Mention

"A Two-Country Study of Default Risk Prediction using Bayesian Machine-Learning"
Fabio Incerti[1], Falco Joannes Bargagli-Stoffi[2] and Massimo Riccaboni[1]
[1] IMT School for Advanced Studies of Lucca, Italy
[2] Harvard University, USA

"Helping the Oracle: Vector Sign Constraints in Model Shrinkage Methodologies"
Ana Boskovic[1] and Marco Gross[2]
[1] ETH Zurich, Switzerland
[2] International Monetary Fund, USA

"Parallel Bayesian Optimization of Agent-based Transportation Simulation"
Kiran Chhatre[1,2], Sidney Feygin[3], Colin Sheppard[1] and Rashid Waraich[1]
[1] Lawrence Berkeley National Laboratory, USA
[2] KTH Royal Institute of Technolgy, Sweden
[3] Marain Inc., Palo Alto, USA

"Source Attribution and Leak Quantification for Methane Emissions"
Mirco Milletarí[1], Sara Malvar[2], Yagna Oruganti[3], Leonardo Nunes[2], Yazeed Alaudah[3] and Anirudh Badam[3]
[1] Microsoft, Singapore
[2] Microsoft, Brazil
[3] Microsoft, USA

ACAIN 2022 Special Mention

"*Brain-like combination of feedforward and recurrent network components achieves prototype extraction and robust pattern recognition*"
Naresh Balaji Ravichandran, Anders Lansner and Pawel Herman
KTH Royal Institute of Technology, Sweden

"*Topology-based Comparison of Neural Population Responses via Persistent Homology and p-Wasserstein Distance*"
Liu Zhang[1], Fei Han[2] and Kelin Xia[3]
[1] Princeton University, USA
[2] National University of Singapore, Singapore
[3] Nanyang Technological University, Singapore

Contents – Part I

Contents – Part II

Detection of Morality in Tweets Based on the Moral Foundation Theory

Luana Bulla[1,2(✉)], Stefano De Giorgis[3], Aldo Gangemi[1,2,3], Ludovica Marinucci[1], and Misael Mongiovì[1,2]

[1] ISTC - Consiglio Nazionale delle Ricerche, Rome, Italy
{luana.bulla,aldo.gangemi,ludovica.marinucci,misael.mongiovi}@istc.cnr.it
[2] ISTC - Consiglio Nazionale delle Ricerche, Catania, Italy
[3] Università degli Studi di Bologna, Bologns, Italy
stefano.giorgis@unibo.it

Abstract. Moral Foundations Theory is a socio-cognitive psychological theory that constitutes a general framework aimed at explaining the origin and evolution of human moral reasoning. Due to its dyadic structure of values and their violations, it can be used as a theoretical background for discerning moral values from natural language text as it captures a user's perspective on a specific topic. In this paper, we focus on the automatic detection of moral content in sentences or short paragraphs by means of machine learning techniques. We leverage on a corpus of tweets previously labeled as containing values or violations, according to the Moral Foundations Theory. We double evaluate the result of our work: (i) we compare the results of our model with the state of the art and (ii) we assess the proposed model in detecting the moral values with their polarity. The final outcome shows both an overall improvement in detecting moral content compared to the state of the art and adequate performances in detecting moral values with their sentiment polarity.

Keywords: Text classification · Moral foundation theory · Transformers

1 Introduction

Morality can be described as a set of social and acceptable behavioral norms [13] part of our every day commonsense knowledge. It underlies many situations in which social agents are requested to participate in the dynamics of actions in domains like societal interaction [12], political ideology [10] and commitment [1], individual conception of rightness and wrongness [27], public figure credibility [11], and plausible narratives to explain causal dependence of events or processes [6]. Therefore, moral values can be seen as parameters that allow humans to assess personal and other people's actions. Understanding this pervasive moral layer in both in person and *onlife* interaction occurrences [5] constitutes a pillar

G. Nicosia et al. (Eds.): LOD 2022, LNCS 13810, pp. 1–13, 2023.
https://doi.org/10.1007/978-3-031-25599-1_1

for a good integration of AI systems in human societal communication and cultural environment. Moral values detection from natural language text passages might help us better understand the cultural currents to which they belong. However, the difficulties in identifying data with a latent moral content, as well as cultural differences, political orientation, personal interpretation and the inherent subjectivity of the annotation work, make this an especially tough undertaking.

In our work, we aim at addressing these critical issues by fine-tuning a BERT-based model [4], a well-known architecture that has achieved cutting-edge performance in a variety of NLP tasks, including classification tasks [21]. We apply the BERT-based model on the dataset developed by Hoover and colleagues [14], which contains 35,000 Tweets tagged with Graham and Haidt's Moral Foundation Theory (MFT) [9]. Based on the work of Hoover and colleagues [14], we calculate the agreement between the annotators to estimate the moral values associated with each tweet in the dataset, and then validate the classification results in two distinct ways. The first one is based on the comparison of our system with the state of the art on the presence or absence in the text of the five MFT's dyads [9]. Each of the Moral Foundations consists in the union of the moral value and its violation (e.g. "Care" *and* "Harm"). The results of the classification show a noticeable improvement. The second assessment considers the polarity of the value evoked by the evaluated text, revealing with notable precision the value or its opposition (i.e. distinguishing "Care" from "Harm"). Our approach expands the set of labels, hence making the process harder. Finally, we propose an analysis of the results that separates "moral" passages from "non-moral" ones.

Our main contributions are:

- we propose a BERT-based model for the automatic detection of latent moral values in short text passages;
- we perform an extensive comparison of the proposed model with the state of the art in the task of inferring the MFT's dyadic dimensions;
- we perform an assessment of the model in detecting moral values and their violations, thus highlighting the polarity of moral sentiment, and discuss the results.

The paper is organized as follows. In Sect. 2 we provide a brief introduction to the theoretical background we rely on, i.e. Moral Foundations Theory (MFT) [9], which is adopted to perform moral value detection by previous work we refer to in Sect. 3, thus highlighting similarities and differences with our contribution. Section 4 details the description of the Moral Foundation Twitter Corpus (MFTC) [14], and our BERT-based approach. In Sect. 5, we describe our two evaluation methodologies, provide the results concerning precision, recall and F1 score for our approach in comparison with the state of the art and present and discuss a confusion matrix; Sect. 6 discusses the above mentioned results comparing them with those described in the work of Hoover and colleagues [14]. Section 7 concludes the paper envisaging possible future developments of our approach.

2 Theoretical Framework

Our work is framed on Haidt's Moral Foundation Theory (MFT). MFT is grounded on the idea that morality could vary widely in its *extension*, namely in what is considered a harmful or caring behavior, according to cultural, geographical, temporal, societal and other factors [10], but not in its *intension*, showing recurring patterns that allow to delineate a psychological system of "intuitive ethics" [9]. MFT is also considered a "nativist, cultural-developmentalist, intuitionist, and pluralist approach to the study of morality" [9]. "Nativist" due to its neurophysiological grounding; "cultural-developmentalist" because it includes environmental variables in the morality-building process [11]; "intuitionist" in asserting that there is no unique moral or non-moral trigger, but rather many co-occurring patterns resulting in a rationalized judgment [12]; "pluralist" in considering that more than one narrative could fit the moral explanation process [13].

MFT is built around a core of six dyads of values and violations:

- *Care/Harm:* caring versus harming behavior, it is grounded in mammals attachment system and cognitive appraisal mechanism of dislike of pain. It grounds virtues of kindness, gentleness and nurturance.
- *Fairness/Cheating:* it is based on social cooperation and typical nonzero-sum game theoretical situations based on reciprocal altruism. It underlies ideas of justice, rights and autonomy.
- *Loyalty/Betrayal:* it is based on tribalism tradition and the positive outcome coming from cohesive coalition, as well as the ostracism towards traitors.
- *Authority/Subversion:* social interactions in terms of societal hierarchies, it underlies ideas of leadership and deference to authority, as well as respect for tradition.
- *Purity/Degradation:* derived from psychology of disgust, it implies the idea of a more elevated spiritual life; it is expressed via metaphors like "the body as a temple", and includes the more spiritual side of religious beliefs.
- *Liberty/Oppression:* it expresses the desire of freedom and the feeling of oppression when it is negated. This last dyad is listed here for a complete overview of the MFT [13]. However, it is not considered in the Moral Foundation Twitter Corpus (MFTC), and thus it is not employed in our classification process, as explained in Sect. 4.

MFT's dyadic structure for defining values and their violations, which coincides with a positive vs negative polarity, and is applied to an extended labeled corpus (i.e. MFTC), offers a sound theoretical framework for the latent moral content detection task we aim to perform.

3 Related Works

Previous work on detecting MFT's moral values in text have relied on word counts [7] or have employed features based on word and sequence embeddings [8,

18]. More broadly, we observe that the most common methodological approaches in this field are divided into unsupervised and supervised methods.

The non-supervised methods rely on systems not backed by external framing annotations. This approach includes architectures based on Frame Axis technique [19], such as those by Mokhberian and colleagues [23] and Priniski and colleagues [25]. This type of approach projects the words on micro-frame dimensions characterized by two sets of opposing words. A Moral Foundations framing score captures the ideological and moral inclination of the text examined. Part of the studies take as a reference point the extended version of the Moral Foundation Dictionary (eMFD) [15], which consists of words concerning virtues and vices of the five MFT's dyads and a sixth dimension relating to the terms of general morality.

The supervised methods aim at creating and optimizing frameworks based on external knowledge databases. The main datasets in this field are: (i) the textual corpus [17], which contains 93,000 tweets from US politicians in the years 2016 and 2017, and (ii) the Moral Foundation Twitter Corpus (MFTC) [14], which consists of 35,000 Tweets from 7 distinct domains. In this context, the work of Roy and colleagues [26] extends the data set created by Johnson and Goldwasser [17] and applies a methodology for identifying moral values based on DRaiL, a declarative framework for deep structured prediction proposed by Pacheco and Goldwasser [24]. The approach adopted is mainly based on the text and information available with the unlabeled corpus such as topics, authors' political affiliations and time of the tweets.

Our research focuses on the use of supervised methods. Specifically, we are close to the work of Hoover and colleagues [14] in terms of the final goal and dataset used. Therefore, due to these similarities, in Sect. 6 we compare our results with those described in [14] following the same data processing procedures. Unlike the authors' methodology, which implement a multi-task Long Short-Term Memory (LSTM) neural network to test the MFTC dataset, we employ a more recent technology based on a transformer language model called BERT [4]. This architecture is pre-trained from unlabeled data extracted from BooksCorpus with 800 million words and English Wikipedia with 2.500 million words. BERT learns contextual embeds for words in an unsupervised way as a result of the training process. After pre-training, the model can be fine-tuned with fewer resources to optimize its performance on specific tasks. This allows it to outperform the state of the art in multiple NLP tasks.

4 Methodology

In the following we detail our approach of detecting moral foundations by applying a BERT-based model to the Moral Foundation Twitter Corpus (MFTC), which has been annotated according to the moral dyads of Haidt's MFT (cf. Sect. 2).

4.1 The Moral Foundation Twitter Corpus

The Moral Foundation Twitter Corpus (MFTC), developed by Hoover and colleagues [14], and consisting of 35k tweets, is organized into seven distinct thematic topics covering a wide range of moral concerns, relevant to current social science problems. In summary, the seven topics are:

- *All Lives Matter* (ALM), which aggregates all tweets using the hashtags #BlueLivesMatter and #AllLivesMatter published between 2015 and 2016. These materials refer to the American social movement that arose in response to the African-American community's Black Lives Matter movement.
- *Baltimore Protests* (Baltimore), which collects tweets from cities where protests over the murder of Freddie Gray took place during the 2015 Baltimore protests.
- *Black Lives Matter* (BLM), which groups all tweets using the hashtags #BLM and #BlackLivesMatter posted between 2015–2016. These data refer to the African-American community's movement born in reaction to the murders of Black people by the police forces and against discriminatory policies against the Black community.
- *2016 Presidential Election* (Election), which are scraped tweets from the followers of @HillaryClinton, @realDonaldTrump, @NYTimes, @washingtonpost and @WSJ during the 2016 Presidential election season.
- *Davidson*, whose tweets are taken from the corpus of hate speech and offensive language by Davidson and colleagues [3].
- *Hurricane Sandy* (Sandy), which includes all tweets posted before, during, and immediately after Hurricane Sandy (10/16/2012-11/05/2012).
- *#MeToo*, whose tweets contain data from 200 individuals involved in the #MeToo movement, a social movement against sexual abuse and sexual harassment.

Unlike previous datasets in this field (cf. [17,26] described in Sect. 3), the MFTC includes both issues with no connection to politics (i.e. Hurricane Sandy) and topics with political meaning (i.e. the Presidential election). Furthermore, the latter have no ideological inclinations because the facts pertain to issues that both liberals (i.e. BLM) and democrats (i.e. ALM) care about.

The corpus heterogeneity corresponds to our study goal, which is to recognize moral values that are not registered in a single area. Each tweet is labeled from three to eight different annotators trained to detect and categorize text passages following the guidelines outlined by Haidt's MFT [9]. In particular, the dataset includes ten different moral value categories, as well as a label for textual material that does not evoke a morally meaningful response. To account for their semantic independence, each tweet in the corpus was annotated with both values and violations. Furthermore, to set performance baselines, tweet annotations are processed by calculating the majority vote for each moral value, where the majority is considered 50%.

4.2 A BERT-Based Method for Detecting Moral Values

To identify moral values in short text we adopt a pre-trained linguistic model for the English language called SqueezeBERT [16]. This model is very similar to the BERT-based architecture [4], which is a self-attention network able to remove the unidirectionality constraint by using a masked language model (MLM) pre-training objective. This task masks some tokens from the input at random, with the goal of predicting the masked word's original vocabulary ID based solely on its context. Furthermore, BERT uses a next sentence prediction task that jointly pre-trains sentence-pair representations.

Bert-derived models have been utilized in a range of applications, with several of them demonstrating significant improvements in natural language task performance, including classification tasks [22]. In the latter, most BERT-derived networks generally consist of three stages: (i) the *embedding*, (ii) the *encoder*, and (iii) the *classifier*. The *embedding* translates pre-processed words into learned feature-vectors of floating-point values; the *encoder* consists of multiple blocks, each of them formed by a self-attention module followed by three fully connected layers, known as feed-forward network layers (FFN); and the *classifier* generates the network's final output. SqueezeBERT model architecture is comparable to BERT-base, except grouped convolutions replace the point-wise fully-connected layers. This change optimizes the BERT-based structure by making the SqueezeBERT model faster in executing the task.

To detect moral foundations in tweets, we fine-tuned SqueezeBERT by using labeled data from the MFTC. Each tweet went through a lemmatization and cleaning procedure before being categorized, removing references to links and Retweets (RT). Furthermore, all of the tweets were truncated to a length of 40 tokens before passing through BERT. We use a learning rate of 2e−5, batch size of 64, drop-out of 0.4, and AdamW as optimizer. To measure the distance of the model predictions from the real values, we adopted a commonly used loss function for multi-label classification tasks:

$$loss(\boldsymbol{x}, \boldsymbol{y}) = -\frac{1}{N} \cdot \sum_{i=1}^{N} \left(y_i \cdot \log \frac{1}{1+e^{-x_i}} + (1-y_i) \cdot \log \frac{e^{-x_i}}{1+e^{-x_i}} \right)$$

where N is the number of labels, x is the N-length output vector of the classifier and y is the N-length binary vector representing the real labels (1 for labels associated with the text, 0 in the other cases). The loss function compares the predicted vector x with the actual annotation y and calculates a score that penalizes probabilities that are distant from the expected value. In other words, the metric establish a criterion that optimizes a multi-label one-versus-all loss based on max-entropy, between the classifier output x and the target y.

To make the prediction, we first normalize the classifier output to return values between 0 and 1 by means of a sigmoid function. Then we set a 0.9 threshold and chose the labels whose predicted values are above or attained to the threshold.

5 Results and Evaluation

We perform an extensive experimental analysis to evaluate the performances of the proposed approach. We first compare the performances of our approach in detecting the five dyads of the MFT and the polarity of tweets (i.e. positive versus negative) with the state-of-the-art method from Hoover and colleagues [14]. Then, we present the results of detecting all moral values and their negation. All experiments were performed in the MFTC corpus introduced in Sect. 4.1.

5.1 Classification of Tweets Based on the MFT Dimentions

To examine the effectiveness of our approach in detecting moral dyadic dimensions, we compare our classifier with the results obtained by Hoover and colleagues [14] by means of employing a Long-short Term Memory (LSTM) model [2,20].

Tables 1 to 5 show the results obtained both by our model (i.e. BERT-Model) and the Hoover and colleagues' model (i.e. LSTM) on 10-fold cross-validation. We got the precision, recall, and F1 score for each of the moral dyads from Graham and Haidt MFT's taxonomy. For BERT-Model, each of the moral dimensions has been defined as the union of the two moral values that compose it (e.g. for the Care moral value, we considered as positive tweets the ones labeled either with Care or Harm). For the 10-fold cross-validation, each fold was obtained by splitting the set of tweets of each subtopic (i.e. ALM, #MeToo, Sandy) in 10 parts, after shuffling the tweets randomly. For LSTM we report the values from Hoover and colleagues [14].

We also evaluated the performances in detecting text with moral content in comparison with the scores reported by Hoover and colleagues [14] (cf. Table 6). We distinguished positive and negative tweets in the following way: items labeled with at least one of the ten moral values are considered as part of the positive set, while non-moral tweets were considered as part of the negative set (Table 2, 3 and 4).

With an F1 ranging from 87% to 81%, the data show a noticeable performance improvement over LSTM in all situations, with a few exceptions. For the whole dataset (column All), precision, recall and F1 are above or attained to 80% for all labels while for LSTM these values are usually between 28% and

Table 1. Classification results for the Care dimension of the MFT.

Model	Metric	All	ALM	Baltimore	BLM	Election	Davidson	#MeToo	Sandy
LSTM	F1	.63	.65	.26	.77	.61	.06	.36	.78
	Precision	.81	.80	.76	.86	.78	.64	.69	.81
	Recall	.52	.55	.16	.70	.50	.03	.25	.75
BERT-Model	F1	**.82**	**.86**	**.65**	**.91**	**.81**	**.23**	**.83**	**.81**
	Precision	**.86**	**.88**	**.81**	**.92**	**.85**	**.79**	**.87**	**.86**
	Recall	**.82**	**.86**	**.63**	**.91**	**.81**	**.35**	**.82**	**.81**

Table 2. Classification results for the Fairness dimension of the MFT.

Model	Metric	All	ALM	Baltimore	BLM	Election	Davidson	#MeToo	Sandy
LSTM	F1	.70	.75	.43	.88	.75	.02	.55	.10
	Precision	.81	.84	**.81**	**.91**	.85	.35	.76	.06
	Recall	.61	.68	.30	.86	.68	.01	.43	**.87**
BERT-Model	F1	**.81**	**.87**	**.60**	**.89**	**.83**	**.21**	**.77**	**.81**
	Precision	**.85**	**.89**	.74	**.91**	**.87**	**.86**	**.79**	**.87**
	Recall	**.80**	**.87**	**.60**	**.88**	**.83**	**.26**	**.77**	.79

Table 3. Classification results for the Loyalty dimension of the MFT.

Model	Metric	All	ALM	Baltimore	BLM	Election	Davidson	#Me Too	Sandy
LSTM	F1	.70	.75	.43	.88	.75	.02	.55	.10
	Precision	.81	.84	**.81**	.91	.85	.35	.76	.06
	Recall	.61	.68	.30	.86	.68	.01	.43	**.87**
BERT-Model	F1	**.85**	**.92**	**.46**	**.95**	**.85**	**.27**	**.86**	**.79**
	Precision	**.88**	**.93**	.66	**.95**	**.89**	**.81**	**.87**	**.88**
	Recall	**.83**	**.91**	**.55**	**.95**	**.84**	**.35**	**.86**	.75

Table 4. Classification results for the Authority dimension of the MFT.

Model	Metric	All	ALM	Baltimore	BLM	Election	Davidson	#Me Too	Sandy
LSTM	F1	.47	.57	.19	.83	.33	.01	.47	.59
	Precision	.80	.85	.77	.91	.80	.24	.67	.80
	Recall	.34	.43	.11	.76	.21	.01	.36	.46
BERT-Model	F1	**.82**	**.92**	**.59**	**.96**	**.87**	**.19**	**.68**	**.80**
	Precision	**.87**	**.93**	**.87**	**.96**	**.92**	**.89**	**.72**	**.86**
	Recall	**.80**	**.91**	**.50**	**.96**	**.85**	**.22**	**.67**	**.79**

Table 5. Classification results for the Purity dimension of the MFT.

Model	Metric	All	ALM	Baltimore	BLM	Election	Davidson	#MeToo	Sandy
LSTM	F1	.41	.57	.07	.48	.47	.04	.53	.15
	Precision	.80	.85	.81	.81	.79	.48	.71	.72
	Recall	.28	.43	.03	.34	.33	.02	.43	.09
BERT-Model	F1	**.87**	**.95**	**.89**	**.97**	**.84**	**.16**	**.71**	**.91**
	Precision	**.91**	**.97**	**.94**	**.97**	**.86**	**.83**	**.77**	**.96**
	Recall	**.86**	**.94**	**.86**	**.97**	**.83**	**.26**	**.71**	**.88**

81%. For both models, performances varied across the discourse domain. This is notably true in the sub-corpus "Davidson" that includes hate messages from Davidson and colleagues's corpus of hate speech and offensive language [3]. In

Table 6. Classification results for the Moral Sentiment

Model	Metric	All	ALM	Baltimore	BLM	Election	Davidson	#MeToo	Sandy
LSTM	F1	.80	.76	.69	**.89**	.77	.14	.81	.86
	Precision	.81	.77	.81	.86	.78	.49	.78	**.97**
	Recall	.79	.76	.61	**.92**	.76	.08	**.84**	.77
BERT-Model	F1	**.85**	**.83**	**.86**	.88	**.78**	**.89**	**.83**	**.88**
	Precision	**.87**	**.85**	**.89**	**.90**	**.80**	**.89**	.84	.90
	Recall	**.86**	**.85**	**.86**	.89	**.79**	**.88**	**.84**	**.90**

this scenario, the majority of tweets have several flaws and are labeled with a high numbers of 'non-moral' labels that amplify the misprediction outcomes.

5.2 Detection of Moral Values with Polarity

To evaluate the performance of the model in detecting both the moral values and their violations, we verify the results of the prediction by highlighting the moral sentiment polarity. This leads to an 11-class classification task (we added a "Non-moral" class to the 10 moral classes given by MFT).

Table 7 shows the results obtained by testing the model on 6,125 items of the MFTC test set. Each label is evaluated in terms of precision, recall and F1 score. The overall results (All in the bottom) are calculated by averaging over all labels weighted by their support (i.e. the number of elements in the ground truth with each specific label).

Table 7. Model F1, Precision, and Recall Scores for Moral Values classification

Moral value	Precision	Recall	F1
Care	0.76	0.73	0.75
Harm	0.67	0.69	0.68
Purity	0.63	0.52	0.57
Degradation	0.59	0.43	0.50
Non-moral	0.90	0.82	0.86
Loyalty	0.66	0.66	0.66
Betrayal	0.56	0.54	0.55
Fairness	0.83	0.72	0.77
Cheating	0.69	0.66	0.67
Authority	0.62	0.58	0.60
Subversion	0.44	0.51	0.47
All	0.76	0.71	0.73

The results reveal an F1 of 73% overall. As expected, performance varied significantly depending on the predicted moral value, with an F1 score ranging from 47% for "subversion" to 86% for tweets classified as "non-moral". Specifically, the best results recorded for positive moral values in terms of F1 score are related to "care" (75%) and "fairness" (77%) while for negative values we have "harm" (68%) and "cheating" (67%).

Furthermore, Table 8 shows the confusion matrix generated from the above predictions. The results demonstrate how the classifier often swaps values for "non-moral". Furthermore, the greater ambiguity is given by confusing "subversion" or "betrayal" with "cheating", and "authority" with "loyalty" or "subversion".

Table 8. Confusion Matrix for predicted Moral Values

	Care	Purity	Non-moral	Loyalty	Cheating	Fairness	Subversion	Betrayal	Degradation	Harm	Authority
Care	367	12	23	34	8	7	9	1	3	26	10
Purity	14	74	29	4	0	7	0	2	3	1	2
Non-moral	43	36	2373	59	82	27	71	14	19	140	42
Loyalty	19	3	25	215	11	5	8	4	0	7	20
Cheating	2	1	46	7	411	19	50	40	10	34	4
Fairness	7	2	17	5	21	268	15	2	1	7	11
Subversion	4	0	34	7	16	0	156	32	14	24	15
Betrayal	2	0	30	10	16	0	18	70	3	11	0
Degradation	0	1	20	1	10	1	18	2	65	18	1
Harm	17	2	41	1	31	1	21	8	3	342	4
Authority	5	1	13	11	0	4	14	0	0	2	112

6 Discussion

As expected, in classifying moral values in tweets (cf. Sect. 5.2) the performance varied substantially across labels. Although the tool performed reasonably well overall, some labels appear to be interpreted inconsistently, resulting in ambiguities that relate to the message expressed in the text. This is visible where components of subversion have been mixed together with moral betrayal feelings (i.e. the tweet "Trump Isn't Hitler? Really? #DonaldTrump is another Hitler! I can't stand Dictators and Traitors! #FDT #Resist #NoH8 #EndRacism #LoveWINS" is tagged with "subversion" by annotators and classified as "betrayal" by the model).

Furthermore, a text can simultaneously communicate multiple moral values that are not identified by the annotators but recognized by the classifier (i.e. the tweet "I'm continually shocked by the stupidity of people. Support our country, support your local police, respect authority. #AllLivesMatter" is labeled only as "authority" by annotators but it is classified thought "authority" and "loyalty" by the model). The results revealed that concepts like "Degradation" or "Subversion" have shades of meaning that are more difficult to detect. This criticality, along with the annotation task's great subjectivity, led to the display activity anomalies in the human value labeling task and increase the task detection's difficulty.

The task of classifying dimensions of the Moral Foundation (cf. Sect. 5.1) turned out to be simpler and led to better results. With an overall F1 of 83%, the proposed model outperforms the state-of-the-art architecture represented by the Hoover and colleagues' work [14], which implemented and trained a multi-task Long Short-Term Memory (LSTM) neural network [2,20] to predict moral sentiment. The likelihood of misclassification is decreased in this circumstance since the number of labels to predict is substantially lower. Indeed, only 5 dyads and a value designated as non-moral are tracked, compared to the 11 labels supplied for the classification of the single moral values in text. The most common anomalies in the classification are highlighted at the sub-corpora level and sometimes worse performances of the model are related to the low cleanliness of the data.

7 Conclusion and Future Work

In our work, we detect latent moral content from natural language text by means of a BERT-based model. The approach considers Haidt's Moral Foundation Theory as a reference for moral values and employ a BERT-based classifier fine-tuned and tested on the Moral Foundation Twitter Corpus (MFTC). The results show an advancement of the state of the art presented by Hoover and colleagues [14] in detecting the dyads that compound Moral Foundations. Furthermore, we present the results of the multi-label classifier taking into account the polarity of moral sentiment and expanding the set of reference labels.

We plan to build an implementation of our model that gives greater weight to the most significant aspects of the sentence, in order to improve the detection of the prevailing moral value, especially when associated to specific emotions evocative of a certain value. Additionally, further experiments on different datasets would help to verify the performance of the model across different domains and on text with different characteristics.

Acknowledgements. We gratefully acknowledge Morteza Dehghani for supporting us in setting up the MFTC dataset and the testset for comparison. This work is supported by the H2020 projects TAILOR: Foundations of Trustworthy AI - Integrating Reasoning, Learning and Optimization – EC Grant Agreement number 952215 – and SPICE: Social Cohesion, Participation and Inclusion through Cultural Engagement – EC Grant Agreement number 870811.

References

1. Clifford, S., Jerit, J.: How words do the work of politics: moral foundations theory and the debate over stem cell research. J. Politics **75**(3), 659–671 (2013)
2. Collobert, R., Weston, J.: A unified architecture for natural language processing: Deep neural networks with multitask learning. In: Proceedings of the 25th International Conference on Machine Learning, pp. 160–167 (2008)
3. Davidson, T., Warmsley, D., Macy, M., Weber, I.: Automated hate speech detection and the problem of offensive language. In: Proceedings of the International AAAI Conference on Web and Social Media, vol. 11, pp. 512–515 (2017)

4. Devlin, J., Chang, M.W., Lee, K., Toutanova, K.: Bert: pre-training of deep bidirectional transformers for language understanding (2019)
5. Floridi, L.: The Onlife Manifesto: Being Human in a Hyperconnected Era. Springer, Heidelberg (2015). https://doi.org/10.1007/978-3-319-04093-6
6. Forbes, M., Hwang, J.D., Shwartz, V., Sap, M., Choi, Y.: Social chemistry 101: learning to reason about social and moral norms. arXiv preprint arXiv:2011.00620 (2020)
7. Fulgoni, D., Carpenter, J., Ungar, L., Preoţiuc-Pietro, D.: An empirical exploration of moral foundations theory in partisan news sources. In: Proceedings of the Tenth International Conference on Language Resources and Evaluation (LREC 2016), pp. 3730–3736 (2016)
8. Garten, J., Boghrati, R., Hoover, J., Johnson, K.M., Dehghani, M.: Morality between the lines: detecting moral sentiment in text. In: Proceedings of IJCAI 2016 Workshop on Computational Modeling of Attitudes (2016)
9. Graham, J., et al.: Moral foundations theory: the pragmatic validity of moral pluralism. In: Advances in Experimental Social Psychology, vol. 47, pp. 55–130. Elsevier (2013)
10. Graham, J., Haidt, J., Nosek, B.A.: Liberals and conservatives rely on different sets of moral foundations. J. Pers. Social Psychol. **96**(5), 1029 (2009)
11. Graham, J., Nosek, B.A., Haidt, J.: The moral stereotypes of liberals and conservatives: exaggeration of differences across the political spectrum. PloS One **7**(12), e50092 (2012)
12. Haidt, J.: The emotional dog and its rational tail: a social intuitionist approach to moral judgment. Psychol. Rev. **108**(4), 814 (2001)
13. Haidt, J.: The righteous mind: why good people are divided by politics and religion. Vintage (2012)
14. Hoover, J., Portillo-Wightman, G., Yeh, L., Havaldar, S., Davani, A.M., Lin, Y., Kennedy, B., Atari, M., Kamel, Z., Mendlen, M., et al.: Moral foundations twitter corpus: a collection of 35k tweets annotated for moral sentiment. Social Psychol. Pers. Sci. **11**(8), 1057–1071 (2020)
15. Hopp, F.R., Fisher, J.T., Cornell, D., Huskey, R., Weber, R.: The extended moral foundations dictionary (emfd): development and applications of a crowd-sourced approach to extracting moral intuitions from text. Behav. Res. Methods **53**(1), 232–246 (2021)
16. Iandola, F.N., Shaw, A.E., Krishna, R., Keutzer, K.W.: SqueezeBERT: what can computer vision teach nlp about efficient neural networks? arXiv:2006.11316 (2020)
17. Johnson, K., Goldwasser, D.: Classification of moral foundations in microblog political discourse. In: Proceedings of the 56th Annual Meeting of the Association for Computational Linguistics, vol. 1: Long Papers, pp. 720–730 (2018)
18. Kennedy, B., et al.: Moral concerns are differentially observable in language. Cognition **212**, 104696 (2021)
19. Kwak, H., An, J., Jing, E., Ahn, Y.Y.: Frameaxis: characterizing microframe bias and intensity with word embedding. PeerJ Comput. Sci. **7**, e644 (2021)
20. Luong, M.T., Le, Q.V., Sutskever, I., Vinyals, O., Kaiser, L.: Multi-task sequence to sequence learning. arXiv preprint arXiv:1511.06114 (2015)
21. Minaee, S., Kalchbrenner, N., Cambria, E., Nikzad, N., Chenaghlu, M., Gao, J.: Deep learning-based text classification: a comprehensive review. ACM Comput. Surv. (CSUR) **54**(3), 1–40 (2021)
22. Mohammed, A.H., Ali, A.H.: Survey of bert (bidirectional encoder representation transformer) types. In: Journal of Physics: Conference Series, vol. 1963, p. 012173. IOP Publishing (2021)

23. Mokhberian, N., Abeliuk, A., Cummings, P., Lerman, K.: Moral framing and ideological bias of news. In: Aref, S., Bontcheva, K., Braghieri, M., Dignum, F., Giannotti, F., Grisolia, F., Pedreschi, D. (eds.) SocInfo 2020. LNCS, vol. 12467, pp. 206–219. Springer, Cham (2020). https://doi.org/10.1007/978-3-030-60975-7_16
24. Pacheco, M.L., Goldwasser, D.: Modeling content and context with deep relational learning. Trans. Assoc. Comput. Linguist. **9**, 100–119 (2021)
25. Priniski, J.H., et al.: Mapping moral valence of tweets following the killing of george floyd. arXiv preprint arXiv:2104.09578 (2021)
26. Roy, S., Goldwasser, D.: Analysis of nuanced stances and sentiment towards entities of us politicians through the lens of moral foundation theory. In: Proceedings of the Ninth International Workshop on Natural Language Processing for Social Media, pp. 1–13 (2021)
27. Young, L., Saxe, R.: When ignorance is no excuse: different roles for intent across moral domains. Cognition **120**(2), 202–214 (2011)

Matrix Completion for the Prediction of Yearly Country and Industry-Level CO_2 Emissions

Francesco Biancalani[1], Giorgio Gnecco[1](✉), Rodolfo Metulini[2], and Massimo Riccaboni[1]

[1] IMT School for Advanced Studies, Lucca, Italy
{francesco.biancalani,giorgio.gnecco,massimo.riccaboni}@imtlucca.it
[2] University of Bergamo, Bergamo, Italy
rodolfo.metulini@unibg.it

Abstract. In the recent past, yearly CO_2 emissions at the international level were studied from different points of view, due to their importance with respect to concerns about climate change. Nevertheless, related data (available at country-industry level and referred to the last two decades) often suffer from missingness and unreliability. To the best of our knowledge, the problem of solving the potential inaccuracy/missingness of such data related to certain countries has been overlooked. Thereby, with this work we contribute to the academic debate by analyzing yearly CO_2 emissions data using Matrix Completion (MC), a Statistical Machine Learning (SML) technique whose main idea relies on the minimization of a suitable trade-off between the approximation error on a set of observed entries of a matrix (training set) and a proxy of the rank of the reconstructed matrix, e.g., its nuclear norm. In the work, we apply MC to the prediction of (artificially) missing entries of a country-specific matrix whose elements derive (after a suitable pre-processing at the industry level) from yearly CO_2 emission levels related to different industries. The results show a better performance of MC when compared to a simple baseline. Possible directions of future research are pointed out.

Keywords: Matrix completion · Counterfactual analysis · Causal inference · Green economy · Pollution

1 Introduction

The concern about climate change is increasing day by day; media and public opinion dedicate a large room to this issue. Surely, one cause of this phenomenon is represented by the yearly emission of a huge amount of CO_2, largely produced by anthropogenic activities such as transportation, heavy industries, and electricity production from fossil fuels. For these reasons, at the international level, some countries and supranational organizations are planning strategies to reduce the use of hydrocarbons, a famous example being represented by the Paris Agreement in 2016, whose main goal is to reduce the yearly CO_2 emission levels by at least 55% by 2030 (compared to their levels achieved in 1990).

G. Nicosia et al. (Eds.): LOD 2022, LNCS 13810, pp. 14–19, 2023.
https://doi.org/10.1007/978-3-031-25599-1_2

In the last years, some academic literature studied the behavior of CO_2 emissions from different points of view. Nevertheless, to the best of our knowledge, none investigated how to solve the potential inaccuracy/missingness of data in certain countries. Thereby, in this work we would like to contribute to the academic debate by studying CO_2 emissions by applying a Statistical Machine Learning (SML) method to a country-industry level database spanning several years in the recent past. Specifically, we propose the application of a method called Matrix Completion (MC, see Hastie et al., 2015 [1]), whose main idea relies on the minimization of a suitable trade-off between the approximation error on a set of observed entries of a matrix (training set) and a proxy of the rank of the reconstructed matrix, e.g., its nuclear norm. A recent successful example of application of MC - in the field of international trade - is represented by the work Metulini et al. (2022) [2], in which MC is applied for the reconstruction of World Input-Output Database (WIOD) subtables. With the final goal of providing effective CO_2 emissions data completion, in the present work we propose a specific adjustment (i.e., a suitable pre-processing of the available dataset) that may improve MC performance for its specific use with data related to yearly CO_2 emission levels. More in details, we apply MC to country-specific subsets of the available pre-processed database of CO_2 emissions, comparing its predictive performance with the one of a simple baseline, namely, the industry-specific average over the training set. Results and their statistical significance analysis show that MC has a better performance than the baseline.

2 Description of the Dataset

The database, available at https://joint-research-centre.ec.europa.eu/scientific-activities_en and fully described in Corsatea et al. (2019) [3], refers to yearly CO_2 emission levels from 56 industries[1] and from households, for 12 energy commodities. The database covers 30 European countries and 13 other major countries in the world, in the period from 2000 to 2016 (one observation for each country, industry, and year). A significant feature of the available database is that its adopted classification of industries matches the one adopted in the World Input-Output Database (WIOD) tables (https://www.rug.nl/ggdc/valuechain/wiod/wiod-2016-release), whose elements represent yearly trade flows from input country-specific industries to output country-specific industries/final consumption sectors. In this preliminary study we restrict the application of MC to country-specific matrices of yearly CO_2 emission levels. A preliminary descriptive analysis of the dataset shows that, for each country, yearly CO_2 emission levels of each industry vary typically quite smoothly from one year to the successive one (whereas yearly CO_2 emission levels of different industries are quite different at the cross-sectional level).

[1] Data related to the total yearly CO_2 emission levels over all the 56 industries were removed from the database, likewise data related to 2 industries for which the yearly CO_2 emission levels were typically 0. Hence, a total of 54 industries was considered in our analysis.

3 Method

In this work, we apply the next formulation for the Matrix Completion (MC) optimization problem, which was studied theoretically in Mazumder et al. (2010) [4]:

$$\underset{\hat{\mathbf{M}} \in \mathbb{R}^{m \times n}}{\text{minimize}} \left(\frac{1}{2} \sum_{(i,j) \in \Omega^{\text{tr}}} \left(M_{i,j} - \hat{M}_{i,j} \right)^2 + \lambda \|\hat{\mathbf{M}}\|_* \right), \tag{1}$$

where Ω^{tr} (which, using a machine-learning expression, may be called training set) is a subset of pairs of indices (i, j) corresponding to positions of known entries of a matrix $\mathbf{M} \in \mathbb{R}^{m \times n}$, $\hat{\mathbf{M}}$ is the completed matrix (to be optimized by solving the optimization problem above), $\lambda \geq 0$ is a regularization constant, and $\|\hat{\mathbf{M}}\|_*$ is the nuclear norm of the matrix $\hat{\mathbf{M}}$, i.e., the summation of all its singular values. The regularization constant λ controls the trade-off between fitting the known entries of the matrix $\mathbf{M}$ and achieving a small nuclear norm of its reconstruction $\hat{\mathbf{M}}$. The latter plays a similar role as the well-known l_1 regularization term used in the LASSO (Hastie et al., 2015 [1]). In this work, the optimization problem (1) is solved numerically by applying the Soft Impute algorithm, developed in Mazumder et al. (2010) [4], to which we refer for its full description. In our application its tolerance parameter (which is part of the termination criterion used by that algorithm) is chosen as $\varepsilon = 10^{-10}$. Moreover, when convergence is not achieved, in order to reduce the computational time, the algorithm is stopped after $N^{\text{it}} = 10^6$ iterations. In our application an additional post-processing step is included, thresholding to 0 any negative element (when present) of the completed matrix. This step is motivated by the fact that the original matrix of yearly CO_2 emission levels is non-negative. For each λ, the resulting (country-specific) completed and thresholded matrix is denoted as $\hat{\mathbf{M}}_\lambda$.

Since MC usually performs better when the elements of the matrix to which it is applied have similar orders of magnitude, for each country, the original matrix of yearly CO_2 emission levels is pre-processed by dividing each row by the l_1 norm of that row restricted to the training set, and multiplying it by the fraction of observed entries in that row (this pre-processing step is not performed when the row contains only zero elements in the training set). Then, MC is applied to the resulting matrix. In this preliminary study, for comparison purposes, for each country the performance of MC is contrasted with that of a simple baseline, namely, the industry-specific average over the training set.

In the present application, the union of the validation and test sets corresponds to positions of matrix elements that are artificially obscured (but that are still available as a ground truth), whereas the training set corresponds to the positions of all the remaining entries of the matrix considered. More specifically, for each country, MC is applied 20 times, each time choosing the training set in the following way:

i) 50% of randomly chosen rows (industries) are observed entirely;
ii) the remaining rows are observed over all the years, except the last 2 years.

Then, in order to avoid overfitting, the regularization constant λ is selected via the following validation method. First, the set of positions of unobserved entries of the matrix $\mathbf{M}$ is divided randomly into a validation set Ω^{val} (about 25% of the positions of the unobserved entries) and a test set Ω^{test} (the positions of the remaining entries). In order to ease the comparison of the MC results when considering different countries, the random choices of the training, validation, and test sets are the same for every country, in each of the 20 repetitions of the MC application (nevertheless, different repetitions turn out to have different realizations of the training, validation, and test sets). It is worth observing that, by the construction above, the training, validation, and test sets do not overlap.

Finally, the optimization problem (1) is solved for several choices λ_k for λ. In order to explore different scales, these are exponentially distributed as $\lambda_k = 2^{k/2-25}$, for $k = 1, \ldots, 100$. For each λ_k, the Root Mean Square Error (RMSE) of matrix reconstruction on the validation set is computed as $RMSE_{\lambda_k}^{\text{val}} := \sqrt{\frac{1}{|\Omega^{\text{val}}|} \sum_{(i,j)\in\Omega^{\text{val}}} \left(M_{i,j} - \hat{M}_{\lambda_k,i,j}\right)^2}$, then the choice $\lambda = \lambda_k^{\circ}$ that minimizes $RMSE_{\lambda_k}^{\text{val}}$ for $k = 1, \ldots, 100$ is found. For each λ_k, the RMSEs of matrix reconstruction on the training and test sets ($RMSE_{\lambda_k}^{\text{tr}}$ and $RMSE_{\lambda_k}^{\text{test}}$) are defined in a similar way. In particular, focus is given to their values computed for $\lambda = \lambda_k^{\circ}$.

4 Results

Figure 1 details the results obtained in one repetition of our analysis, for a representative country (Italy), taken as case study (similar results are obtained for other repetitions and other countries). The figure shows, for various choices of λ, the RMSEs achieved by the MC method on the training, validation and test sets, and compares them with the respective RMSEs obtained by the baseline. Then, Table 1 reports, for all the 20 repetitions, the RMSE on the test set (in correspondence of the optimal choice of the regularization constant λ) for the representative country, and the RMSE on the same test set obtained by the baseline. In this case, the MC method achieves a statistically significant better performance than the baseline. Indeed, the application of a one-sided Wilcoxon matched-pairs signed-rank test (adopted for a similar purpose in Gnecco et al., 2021 [5]) rejects the null hypothesis that the difference $RMSE^{\text{test}}(\text{baseline}) - RMSE_{\lambda_k^{\circ}}^{\text{test}}(\text{MC})$ has a symmetric distribution around its median and this median is smaller than or equal to 0 (p-value $= 4.7846 \cdot 10^{-5}$, significance level $\alpha = 0.05$). Similarly, the null hypothesis is rejected for 37 among the other 42 countries.

5 Possible Developments

Potential applications of our analysis arise in the construction of counterfactuals, useful to predict the effects of policy changes able to influence the yearly CO_2 emission levels of specific industries in selected countries. As a further step, MC could be applied to matrices obtained by combining the information available from different countries, or by merging the information available on yearly trade

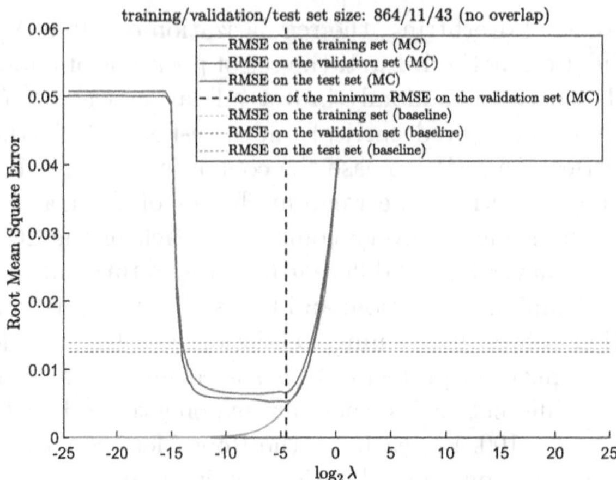

Fig. 1. Results of the application of one repetition of the MC method and of the baseline, for the case of a representative country (Italy).

Table 1. RMSEs on the test set for the two methods considered in the work, for the case of a representative country (Italy).

	Repetition number									
	1	2	3	4	5	6	7	8	9	10
$RMSE^{test}_{\lambda_k^{\circ}}$ (MC)	0.0061	0.0065	0.0057	0.0044	0.0046	0.0122	0.0059	0.0105	0.0116	0.0118
$RMSE^{test}$ (basel.)	0.0134	0.0139	0.0150	0.0182	0.0178	0.0211	0.0166	0.0208	0.0229	0.0213
	Repetition number									
	11	12	13	14	15	16	17	18	19	20
$RMSE^{test}_{\lambda_k^{\circ}}$ (MC)	0.0156	0.0127	0.0124	0.0078	0.0116	0.0130	0.0080	0.0064	0.0169	0.0060
$RMSE^{test}$ (basel.)	0.0259	0.0201	0.0195	0.0157	0.0222	0.0186	0.0118	0.0134	0.0256	0.0154

flows and yearly CO_2 emission levels. Finally, using an ensemble machine learning approach, MC could be combined with a refined baseline (e.g., an industry-specific moving average over a subset of past observations) to achieve possibly a better performance with respect to the baseline alone.

References

1. Hastie, T., Tibshirani, R., Wainwright, M.: Statistical Learning with Sparsity: The Lasso and its Generalizations. CRC Press, Boca Raton (2015)
2. Metulini, R., Gnecco, G., Biancalani, F., Riccaboni, M.: Hierarchical clustering and matrix completion for the reconstruction of world input-output tables. AStA - Adv. Stat. Anal. (2022). https://doi.org/10.1007/s10182-022-00448-6
3. Corsatea, T.D., et al.: World input-output database environmental accounts. Update 2000–2016. Publications Office of the European Union, Luxembourg (2019). https://doi.org/10.2791/947252

4. Mazumder, R., Hastie, T., Tibshirani, R.: Spectral regularization algorithms for learning large incomplete matrices. J. Mach. Learn. Res. **11**, 2287–2322 (2010)
5. Gnecco, G., Nutarelli, F., Selvi, D.: Optimal data collection design in machine learning: the case of the fixed effects generalized least squares panel data model. Mach. Learn. **110**, 1549–1584 (2021)

A Benchmark for Real-Time Anomaly Detection Algorithms Applied in Industry 4.0

Philip Stahmann(✉) and Bodo Rieger

University of Osnabrueck, Osnabrueck, Germany
{pstahmann,brieger}@uni-osnabrueck.de

Abstract. Industry 4.0 describes flexibly combinable production machines enabling efficient fulfillment of individual requirements. Timely and automated anomaly recognition by means of machine self-diagnosis might support efficiency. Various algorithms have been developed in recent years to detect anomalies in data streams. Due to their diverse functionality, the application of different real-time anomaly detection algorithms to the same data stream may lead to different results. Existing algorithms as well as mechanisms for their evaluation and selection are context-independent and not suited to industry 4.0 settings. In this research paper, an industry 4.0 specific benchmark for real-time anomaly detection algorithms is developed on the basis of six design principles in the categories timeliness, threshold setting and qualitative classification. Given context-specific input parameters, the benchmark ranks algorithms according to their suitability for real-time anomaly detection in production datasets. The application of the benchmark is demonstrated and evaluated on the basis of two case studies.

Keywords: Anomaly detection · Streaming analytics · Algorithm benchmark · Industry 4.0 · Real-time analysis

1 Introduction

Among other terms, "industry 4.0" describes the revolutionary impact that digitalization has on production. In industry 4.0, production setups shall be able to efficiently serve individual customer needs through flexible, automated machine configuration (Kagermann et al. 2013). The failure rate as well as the time necessary for countermeasures to production anomalies shall be kept low up to a zero-failure ideal. To this end, machine self-diagnosis for anomalies that may cause production failure is a key competency to fulfill the industry 4.0 vision (Cohen and Singer 2021). As concretization of generic self-X competencies, self-diagnosis is at the heart of the automation of production (Cohen and Singer 2021). Self-diagnosis builds on the integration of sensor technology and machine learning capability with production components, which creates opportunities for advanced production monitoring and control (Schütze et al. 2018).

In practice, various methods are used for timely, autonomous anomaly detection in industry 4.0 production environments (Stahmann and Rieger 2021). The goal is to find methodologies to evaluate sensor data in real-time, so that machines can respond to

G. Nicosia et al. (Eds.): LOD 2022, LNCS 13810, pp. 20–34, 2023.
https://doi.org/10.1007/978-3-031-25599-1_3

unexpected sensor values as quickly as possible (Schütze et al. 2018). Several openly available algorithms have been implemented to assess real-time data streams for anomalies (Numenta Anomaly Benchmark 2020). However, the algorithms achieve different results when applied to the same data streams due to their fundamentally different functionality. In real production scenarios, contradictory results may make it difficult to judge whether to react if only few algorithms indicate unexpected machine behavior. On the other hand, the decision to react only when there is indication by numerous algorithms bears the risk of missing subtle anomalies. Supporting the adequate decision is important especially in light of potential costs and production delays wrong decisions can entail. Figure 1 exemplarily shows the different results of five openly available state-of-the-art real-time anomaly detection algorithms applied to a real-world machine temperature dataset (Numenta Anomaly Benchmark 2020). As discussed in the previous chapter, the different results complicate the decision whether to assume the presence of an anomaly and act on it.

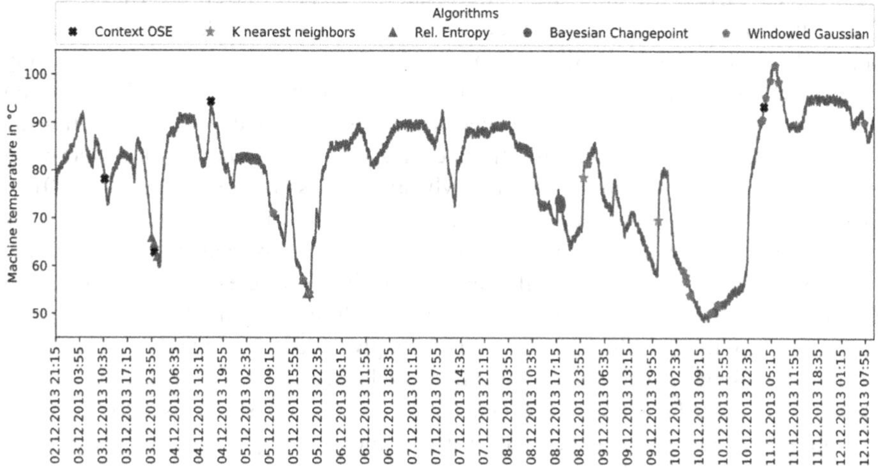

Fig. 1. Real-time anomaly detection algorithms applied to real machine temperature data set.

Usual metrics to assess algorithmic suitability, such as the F1-score or false positive rate, do not cover streaming characteristics, such as timeliness and are therefore inadequate (Lavin and Ahmad 2015). The Numenta Anomaly Benchmark (NAB) supports ranking openly available real-time algorithms to increase reliability of their results (Lavin and Ahmad 2015). However, the benchmark is generic and context-independent and does not address industry 4.0 specificities. Stahmann and Rieger (2022a) identified design principles to rank real-time anomaly detection algorithms in the context of digitalized production. Building on this, the following research question arises: *How can an industry 4.0 specific benchmark to rank real-time anomaly detection algorithms be implemented?*

This paper aims to contribute to research and practice with a formalization, prototypical implementation and evaluation of a benchmark to facilitate decision-making in algorithmic real-time anomaly detection. The research question is answered in seven subsequent sections. Section two outlines the current state of related research. Section three details the design science approach taken to develop the benchmark. Section four introduces the design principles (DPs) applied for benchmark implementation. Subsequently, the functionality of the prototypical implementation is explained and also formalized to enhance comprehensibility and reproducibility of the solution. Chapter six covers demonstration and evaluation of the prototypical implementation including two data sets from real-world case studies. Sections seven and eight include limitations, future research as well as a conclusion.

2 Algorithm Selection and Anomaly Detection

The requirement to find the right algorithm for a defined problem exists in a wide variety of domains (Bischl et al. 2016). Algorithm selection is a subjective and explorative process. The evaluator's knowledge and ability about the ranking process, data and algorithms can have a decisive influence on the result (Ho and Pepyne 2002). The suitability of algorithms is also contingent as it depends on the data to which the algorithms are applied (Brazdil and Soares 2000). Consequently, there can be no universal ranking (Wolpert and Macready 1997). Generally, two different methods for algorithm selection are distinguishable in literature, namely per-instance selection and models supporting the selection of algorithms for data sets (Bischl et al. 2016). The former method focuses on determining the best algorithm for each newly arriving instance (Kerschke et al. 2018). Applied to industrial production contexts, per-instance algorithm selection approaches may for example support the optimization of production job sequencing on multiple machines (Pavelski et al. 2018). Different from per-instance selection, other algorithm selection models require entire data sets or subsets with sufficient information to support decision making towards the best algorithm for a given task (Lavin and Ahmad 2015). Benchmark rankings are common results of related research (Bischl et al. 2016). In the domain of digitalized manufacturing, benchmarking was for instance used to determine algorithms that can reduce unnecessary waiting times during production on sequenced machines (Cheng et al. 2021).

In the context of our contribution, we refer to algorithm ranking based on an aggregated benchmark score. Accordingly, algorithms are systematically evaluated and prioritized regarding their suitability to detect anomalies in real-time (Brazdil and Soares 2000). Anomalies in data streams characterize through their variation from expectation that is created by previous data analysis (Chandola et al. 2009). Point anomalies can be differentiated from collective anomalies. The former refers to single data points that deviate from other data points. Collective anomalies are groups of related data points that in combination are anomalous.

3 Methodology

The development of our benchmark follows the steps of design science research according to Peffers et al. (2007). Figure 2 shows the five steps carried out, where the second step covers the adoption of design principles (Stahmann and Rieger 2022a).

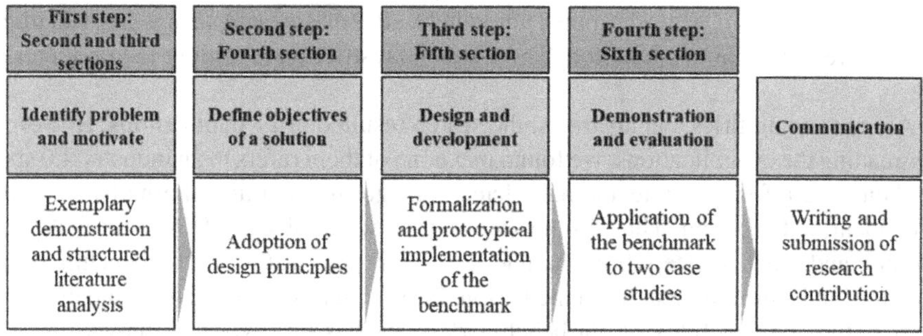

Fig. 2. Design science research approach following Peffers et al. (2007).

In the first step, the problem was identified and demonstrated by using a real data set as an example to show how different the analysis results of state-of-the-art real-time anomaly detection algorithms can be. Furthermore, we conducted a structured literature analysis according to vom Brocke et al. (2015) to examine the state of research regarding answers to our research question. We searched for contributions from science and practice in the databases *Google Scholar, IEEE Xplore, Scopus, Science Direct, Springer Link* and *Web of Science*. These databases were selected as they cover a wide range of sources on business analytics including those hosting publications with high impact. Table 1 shows the combined search strings. The column on the left in Table 1 includes search strings regarding algorithms to detect unexpected values in real-time. The search strings in the center column refer to algorithm selection methodologies that include algorithm ranking. The column on the right includes search strings that refer to the digitalization of production. We initiated the search with the term "industry 4.0" and its German equivalent as both spellings diffused in literature (Bueno et al. 2020). We subsequently added the remaining search strings of the right column with terms we encountered during the literature review.

Table 1. Search string combinations.

OR		OR		OR
Real-time anomaly detection algorithms Real-time outlier detection algorithms	**AND**	Benchmark Prioritization Ranking	**AND**	Industrie 4.0 Industry 4.0 Intelligent manufacturing Smart factory Smart manufacturing Industrial Internet Reference Architecture Intelligent Manufacturing System Architecture

First, two columns were combined alternately, after that, search strings from all three columns were used in combination. There was no restriction regarding the publication date.

After scanning titles and abstracts, the search resulted in 27 publications. However, after reading these publications, we found that none of them refers to an industry 4.0 specific benchmark for real-time anomaly detection algorithms or any alternative solution to the identified problem. Thus, in the second step, we outline six DPs identified from literature and practice. Step three includes a formal specification of the benchmarking procedure and a prototypical implementation. In step four, we demonstrate and evaluate functionality and feasibility of our benchmark on the basis of two case studies.

4 Design Principles as Solution Objectives

Stahmann and Rieger (2022a, p. 5) have formulated six DPs for an industry 4.0 specific benchmark for real-time anomaly detection based on a qualitative survey with industry experts. The interviewees responded to four questions in the categories timeliness, threshold setting and qualitative classification. According to the majority of interviewed experts, timeliness is defined as time span between anomaly occurrence, detection and notification. The interviewees responded that detection and notification as early as possible increase the risk of false alarms. On the other hand, too late detection and notification may hinder anomaly elimination in time. Two DPs could be identified:

DP1: "*Real-time anomaly detection evaluation needs to consider whether anomalies were detected and notified early or late after occurrence.*"

DP2: "*Real-time anomaly detection evaluation needs to consider that raising early anomaly alarms might increase false alarm rates. Therefore, real-time anomaly detection evaluation needs to consider a certain robustness.*"

The second category refers to the setting of upper and lower thresholds production values shall not trespass. In Stahmann and Rieger (2022a, p. 5), thresholds resulted as most frequent measure to delineate normal from anomalous data in production environments. There are two design principles regarding threshold setting:

DP3: "*Real-time anomaly detection evaluation needs to consider thresholds, as these are the most frequent mechanism for anomaly detection in production.*"

DP4: "*Real-time anomaly detection evaluation needs to consider personal experience from production step specific experts as well as simulation and statistics on materials to set fixed and dynamic thresholds.*"

Thirdly, qualitative classification refers to the assignment of qualitative criteria to each anomalous instance in data sets used for algorithm ranking (Stahmann and Rieger, 2022a, p. 5). Two further design principles are formulated:

DP5: "*Real-time anomaly detection evaluation needs to consider systematic classification of anomalies according to their impact.*"

DP6: "*Real-time anomaly detection evaluation needs to consider time, cost and intensity requirements of countermeasures to determine anomalies' impact and thus classification.*"

5 Formalization and Prototypical Implementation

We propose a benchmark solution consisting of the three categories presented in the previous section. The benchmarking procedure extends over the three phases preparation, analysis procedure as well as score calculation and ranking. Figure 3 shows the solution, which initially requires a data set with ground truth labels indicating which data points are anomalous. In a real-scenario, such data might result from previous production or simulation (Stahmann and Rieger 2021). The benchmark logic shown in Fig. 3 has been prototypically implemented in an iterative process. The preparation phase requires user input to tailor the application of the benchmark to a specific production context. Addressed users are mainly machine operators and engineers. The preparation and analysis phases consider timeliness, threshold setting and qualitative assessment individually. The last phase integrates the analytic results to an overall score eligible for ranking real-time anomaly detection algorithms. After ranking with labelled training data, the best algorithms can be selected for real-time anomaly detection during production.

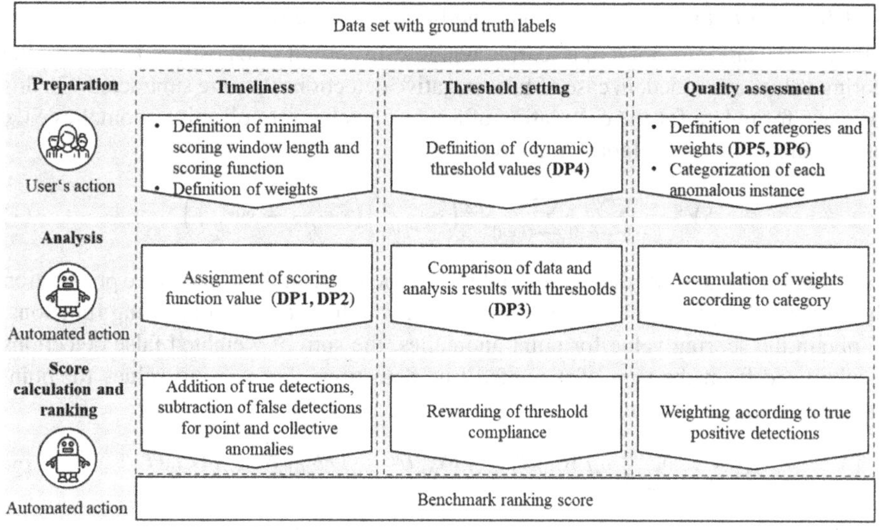

Fig. 3. Three-step solution to obtain a benchmarking score.

Timeliness: To introduce timeliness in their benchmark for real-time anomaly detection evaluation, Lavin and Ahmad (2015) use a scoring function in predefined scoring windows. Scoring windows are finite subsequences of the data. Their scoring function is supposed to reward early and penalize late anomaly detections in data sequences. The earlier an anomaly is recognized, the higher the value on the function. The function value is then used to compute a benchmark score. We adapt the idea of using a scoring function to evaluate timeliness in our solution. However, the first two DPs show the need for robustness in terms of timeliness. Real-time anomaly detection evaluation should consider that the delayed detection and notification of collective anomalies in a data sequence can be more valuable than immediate detection and notification (Singh and Olinsky 2017).

For this purpose, our solution provides for the definition of scoring windows and a scoring function f^S by the user. The scoring windows represent collective anomalies. All anomalies outside scoring windows are point anomalies. Also different from Lavin and Ahmad (2015), for each algorithmic result we define for each data point outside of scoring windows $\left(d_j, d_{j+1}, \ldots, d_m\right) = D$ whether it is a true or false positive or negative detection. We formulate these alternatives as binary variables $d_j{}^{TP}$, $d_j{}^{FP}$, $d_j{}^{TN}$and$d_j{}^{FN}$. As these alternatives are mutually exclusive, their sum must equal 1 for each data point. Additionally, we define scoring windows as $\left(sw^i, sw^{i+1}, \ldots, sw^n\right) = SW$ and their data points as $(sw_{d_k}, sw_{d_{k+1}}, \ldots, sw_{d_z}) = S$. As all data points in a scoring window are anomalous, algorithmic detections are either true positive or false negative, which are specified as $sw_{d_k^{TP}}$ and $sw_{d_k^{FN}}$. The sum of $sw_{d_k^{TP}}$ and $sw_{d_k^{FN}}$ must also equal 1 as both options are mutually exclusive. To calculate the scoring value on f^S, we use the relative position $sw_{d_{p(k)}}$ of each data point in a scoring window. For the scoring window's position, we define that $f^S\left(sw_{d_{p(z)}}\right) = 0$. This condition forces the scoring function to take the value 0 at the last data point of a scoring window. In case of true positive detections, the scoring values are added, in case of false negative detections they are subtracted. On this basis, we formulate for the calculation of a scoring value for collective anomalies SV_S for the detections of an algorithm $a \in A$:

$$SV_S^a = \sum\nolimits_{i=0}^{n} \sum\nolimits_{k=0}^{z} f^S\left(sw^i_{d_{p(k)}}\right)\left(sw^i_{d_k^{TP}} - sw^i_{d_k^{FN}}\right) \tag{1}$$

Furthermore, the user defines a weight w for each type of detection in the preparation phase. These weights apply to point anomalies, which lie outside of scoring functions. To obtain the scoring value for point anomalies, the sum of weighted false detections is subtracted from the sum of weighted true detections. For scoring values for point anomalies SV_D, we formulate:

$$SV_D^a = \sum\nolimits_{j=0}^{m} d_j{}^{TP} w^{TP} + d_j{}^{TN} w^{TN} - d_j{}^{FN} w^{FN} - d_j{}^{FP} w^{FP} \tag{2}$$

The overall scoring value defines as sum of the collective and point anomalies' scoring values in a dataset X, with$\{S, D\} \epsilon X$:

$$SV_X^a = SV_S^a + SV_D^a \tag{3}$$

Threshold Setting: According to DP3, the benchmark needs to consider whether anomaly detection algorithms reliably declare points outside threshold values anomalous. Threshold values are therefore defined in the preparation step by experts (cf. DP4). Algorithmic recognition of threshold trespassing shall lead to higher ranking scores. To this end, threshold compliance is determined and rewarded in the steps analysis procedure and score calculation and ranking. We use $sw^i_{d_k^{lower}}$ and $sw^i_{d_k^{upper}}$ as binary indications whether a data point violates a lower or an upper threshold. $SW_{S_{threshold}}$ is the sum of threshold violations in the ground truth labels of the scoring windows. We formulate threshold value compliance for collective anomalies TV_S as follows:

$$TV_S^a = 1 + \frac{\sum_{i=0}^{n} \sum_{k=0}^{z} f^S\left(sw_{d_{p(k)}}^i\right)\left(sw_{d_k^{lower}}^i + sw_{d_k^{upper}}^i\right) sw_{d_k^{TP}}^i}{SW_{S threshold}} \tag{4}$$

For point anomalies, $d_j{}^{lower}$ and $d_j{}^{upper}$ make binary indications for threshold violations. $SW_{D threshold}$ is the maximum number of threshold violations detectable outside scoring windows. We formulate:

$$TV_D^a = 1 + \frac{\sum_{j=0}^{m}\left(d_j{}^{lower} + d_j{}^{upper}\right) d_j^{TP}}{SW_{D threshold}} \tag{5}$$

The product of threshold violations of point and collective anomalies constitutes the overall threshold value:

$$TV_X^a = TV_S^a TV_D^a \tag{6}$$

Qualitative Assignment: The last two DPs show the need of classifying anomalies according to their impact. Emmott et al. (2013) propose to classify anomalies into four categories according to how difficult they are to detect. In this classification, each anomaly is either easy, medium, hard or very hard to detect. Each of these categories is assigned with a difficulty score. DP6 indicates that a qualitative assessment needs to consider time, cost and intensity of countermeasures to solve anomalies. For the purpose of quantifiability, each category must be assigned a weight in the preparation phase. During analysis, weights of correctly identified anomalies are accumulated. To this end, we define $sw_{d_k^q}$ as qualitative weight of a data point in a scoring window. The weights are multiplied with the scoring function value at each data point's position. The maximum sum of weights reachable in scoring windows denotes as SW_{SQ}. To obtain the qualitative value for collective anomalies QV_S that is reachable by an algorithm, we formulate:

$$QV_S^a = 1 + \frac{\sum_{i=0}^{n} \sum_{k=0}^{z} f^S\left(sw_{d_{p(k)}}^i\right) sw_{d_k^q}^i sw_{d_k^{TP}}^i}{SW_{SQ}} \tag{7}$$

Further, for point anomalies we formulate the qualitative weight as $\boldsymbol{q_{d_j}}$ with the maximum reachable sum of all weights of point anomalies $\boldsymbol{Q^D}$. The qualitative value outside scoring windows $\boldsymbol{QV_D}$ can be formulated as:

$$QV_D^a = 1 + \frac{\sum_{j=0}^{m} q_{d_j} d_j^{TP}}{Q^D} \tag{8}$$

The product of qualitative weight assignments of collective and point anomalies defines as overall qualitative value:

$$QV_X^a = QV_S^a QV_D^a \tag{9}$$

The final score calculates as product of the three outlined values:

$$FS_X^a = SV_X^a TV_X^a QV_X^a \tag{10}$$

Lastly, we normalize the overall score each algorithm yields to enhance comparability in the ranking. Due to normalization, the best algorithm obtains a ranking score RFS_X^{best} of 100, the worst performing algorithm yields a score of 0. We formulate ranking score calculation as follows, where FS_X^A includes the overall scores of all ranked algorithms $a \in A$:

$$RFS_X^a = 100\frac{FS_X^a - \min(FS_X^A)}{\max(FS_X^A) - \min(FS_X^A)} \tag{11}$$

The presented formalization has been implemented as a Python script in such a way that potential users only need to specify the input parameters to get results for the evaluation of their algorithms (cf. Fig. 3). The implementation proceeded in an iterative process based on the algorithms and the dataset of the first case study presented in the next section.

6 Demonstration and Evaluation

The benchmark is context-sensitive and the main design risk is user-oriented (Venable et al. 2016). Therefore, we aim to evaluate the benchmark using two demonstrative case studies in which we apply algorithms to data streams from real scenarios and subsequently rank the algorithms (Stahmann et al. 2022b). In both case studies, we rank five state-of-the-art real-time anomaly detection algorithms. The evaluation focuses on functionality and feasibility (Hevner et al. 2004). Regarding functionality, the aim is to show the benchmark's capability to rank the algorithms. With respect to feasibility, the aim is to get an idea how easy it is for potential users to understand and make use of the benchmark and its results. In both case studies, we worked with industry experts. The construction of each case study follows three steps. First, data is simulated. The data should correspond to sensor-generated machine data from the experts' practical activities. In the second step, the benchmark is explained in-depth to the experts. The focus of the explanation is on the required user input, i.e. the first phase presented in section five. The last step includes the application of the benchmark and the interpretation of the output. Also, in the last step we asked the experts for feedback on the benchmarking procedure with regards to feasibility. To determine the data relevant to the first case study, there were two face-to-face meetings that lasted a total of three hours and forty minutes. Data elicitation for the second case study required three face-to-face meetings that lasted a total of three and a half hours. Contents of the meetings were recorded with the consent of the experts.

Case Study 1. The first case study focuses on a process to manufacture metal gears. In the observed case, gear production is highly automated and shall cater individual business customer requirements. As part of the manufacturing process, gears are placed in a water bath containing various chemicals for hardening and, above all, for cleaning.

Water temperature must be at least 60 °C for effective cleaning. Gears subjected to lower temperatures for cleaning may not be used properly, as dirt may hinder their mechanical function. Water temperature must not exceed 70 °C in order not to damage sensitive materials and delicate elements. Water is heated by permanently installed immersion heaters. If water temperature reaches the minimum of 60 °C, the immersion heaters are switched on automatically. If the water temperature reaches 70 °C, the immersion heaters are switched off so that the water can cool down. The cooling process is accelerated with cold water that is added in a controlled manner to compensate for water evaporation. A sensor measures water temperature every second. The system runs all day and is regularly interrupted after about two days for changing the water and checking functionality. In an iterative process, temperature data behavior was modelled with the following function:

$$f(x) = 5sin\left(x + \frac{sin\left(x + \frac{1}{2}sin(x)\right)}{1.7}\right) + 65 \tag{12}$$

For one period, i.e. until the water is replaced, 191673 data instances were created. Overall, errors occur very rarely in the modeled process step; it is estimated that only about 1% of all measurements are anomalous. In our modeled data, a total of 1972 measurements were created that deviated from the expected temperature data behavior. Table 2 shows the distribution of the kinds of anomalies as well as their qualitative weights. 1886 of these anomalies belong to 46 *SWs* that consist of at least five data instances. Algorithms were supposed to detect anomalies soon after their occurrence, but not immediately as not all small changes in the slope of the temperature curve should lead to an immediate alarm. In an iterative process, we formulated the scoring function as:

$$f^S_{gear}(x) = sin(2x + 0.9) \tag{13}$$

Case study 2: The second case study focuses on an automated pressing process, in which metal components of automobiles are formed. For the adequate pressing of a metal component at least 3,847.5 kN and at most 3,852.5 kN must be exerted. Sensors measure the pressure every second, so that press movements can be tracked as closely as possible. If the maximum pressure falls below the lower limit, the deformation of the metal component is not sufficient for further processing. If the maximum permitted pressure is exceeded, the metal component is usually damaged to such an extent that it can no longer be used. About 5% of the measured data is anomalous. The sensor measurements can be formulated as:

$$f(x) = 1,925sin\left(x + \frac{\sin(2x)}{2}\right) + 1,925 \tag{14}$$

Using this function, we synthesized 200,000 datapoints, 10384 of these are anomalous. 10,115 anomalous data points belong to 222 SWs. Table 2 contains the most frequent causes of errors, including frequency and qualitative classification. Anomalies should be detected and reported immediately after they occur, as they lead to defective components or process delays. Due to the standardization of production, it can be assumed that an anomaly will be repeated, so that it needs to be corrected immediately. We modelled the following scoring function in an iterative process:

$$f^S_{press}(x) = sin(1.5x + 7.8) \tag{15}$$

Table 2. Details on anomalies in both case studies' datasets.

	Anomaly description	Cause description	Qualitative classification	Estimated frequency
Case study 1	Noncompliance with thresholds	Malfunction of the immersion heater	Severe	10%
	Unexpected jumps of data that are inside threshold margins	Measurement error due to dirt contamination	Relevant	0–5%
	Changed slope of temperature measurements	Maladjustment of the immersion heaters or sensors	Relevant	80%
	Monotonous measurements	Measurement error due to dirt contamination	Critical	5–10%
Case study 2	Press unexpectedly comes to a stop at top or bottom	Metal components inserted incorrectly	Critical	40%
	Noncompliance with thresholds	Maladjustment of machine parameters	Severe	45%
	Press moves too slowly or too quickly	Maladjustment of machine parameters	Critical	12.5%
	Sudden unexpected spikes	Measurement errors occur	Critical	2.5%
	Weights: Relevant: 1; Critical: 1.5; Severe: 2			

Comparable to the feedback in case study 1, the benchmark procedure was perceived as comprehensive, but time-consuming. As most manufacturing processes in the second case's company are suited to standardized mass production, measurement data are not subject to frequent changes. Therefore, the benchmarking procedure would not have to be repeated frequently, but only in rare cases of changes in manufacturing processes or for ranking new algorithms. Again, it was suggested that a GUI would simplify data preparation and understanding of the benchmarking process.

We used five algorithms for demonstration and evaluation, which are Bayesian Changepoint Detection (BCD) (Adams and MacKay, 2007), Context Anomaly Detection Open Source Edition (CAD OSE) (Ahmad et al. 2017; Numenta Anomaly Benchmark 2020) K Nearest Neighbors Conformal Anomaly Detection (KNN CAD) (Burnaev, Ishimtsev, 2016), relative entropy and expected similarity estimation (REXPOSE)

(Schneider et al. 2016) and Windowed Gaussian (WG) (Ahmad et al. 2017; Numenta Anomaly Benchmark 2020). Further information on the algorithms can be found in Ahmad et al. (2017).

Table 3 shows the result of ranking the algorithms applied to the prepared datasets. We decided not to compare our ranking with common metrics for evaluating algorithms, since no metric encompasses the evaluation aspects from the six DPs. The results show how strong the influence of SV_X^a can be compared to the values of TV_X^a and QV_X^a. This strong influence justifies by the fact that SV_X^a considers the user-weighted type of detection of each data point. However, the comparison of the results of BCD and CAD OSE in the first case study shows that the consideration of the other factors, in this case TV_X^a, may also be decisive for the ranking. In the second case study, TV_X^a and QV_X^a equal 1 as four of five algorithms' detection rate of true positives and negatives was low. This is due to the different definitions of thresholds in the two case studies. In the first case study, there is a clear upper and lower threshold that no measured value may trespass. In the second case study, the lower limit refers to the least maximum pressure. In the remaining data, which are not intended to represent the maximum of the press force, values are below the lower thresholds, but are not violations. In QV_X^a, this effect is reinforced by the fact that threshold violations were classified as severe, because they have strong effects on the outcome of the pressing process.

Table 3. Algorithm ranking.

	Rank	Algorithm	SV_X^a	TV_X^a	QV_X^a	RFS_X^a
Case study 1	1	KNN CAD	18753.29	1.01	1.07	100
	2	WG	18781.45	1.04	1.01	43.67
	3	CADOSE	18777.48	1.02	1	6.40
	4	RE	18783.23	1.01	1.01	2.31
	5	BCD	18775.51	1.01	1	0
Case study 2	1	KNN CAD	17998.64	1.01	1.01	100
	2	RE	17930.46	1	1	11.79
	3	CAD OSE	17936.2	1	1	6.58
	4	BCD	17920.2	1	1	2.38
	5	WG	17923	1	1	0

7 Limitations and Future Research

The results as well as their evaluation are not free from limitations. The premise that the benchmark requires labelled data has two drawbacks. Firstly, data preparation was perceived as time-consuming. Secondly, real production data must not differ significantly from those from the labelled datasets, otherwise benchmarking might have to be applied to new data. Therefore, the application is more suitable for standardized production processes with low variation. However, the experts from the evaluation identified as an advantage of the labeling process that the data must be dealt with intensively and context-specifically, since the labelling process itself might lead to new insights on the data. Furthermore, due to its domain-specificity, the resulting benchmark is not comparable to other benchmarks that are context-independent or specific to other domains. Also, common algorithm evaluation metrics, such as the F1-score or false positive rate, are not suitable for comparison to the resulting benchmark as they do not consider problem specificities such as timeliness or thresholding either.

The choice of case studies seemed adequate to show first results (Venable et al. 2016). In the future, the dependencies between the three mechanisms of the benchmark might be investigated in more detail. A quantitative investigation with more real, labelled production datasets would be suitable for this purpose. Based on the findings of the evaluation, our future research will also focus on extending the current prototype with a GUI to facilitate data preparation.

The results of the evaluation are subjective and therefore not generalizable. However, production processes as focused problem-solving domain are highly context-dependent by themselves.

8 Conclusion

The ability of machines to detect anomalies in production in real-time is a core component of industry 4.0 automation. Various algorithms for the real-time detection of anomalies have been developed. For their concrete application, however, it is problematic that they may output different results when applied to the same data streams. A domain-specific benchmark for algorithm evaluation can support selecting the most suitable algorithms. The research question how such a benchmark can be formalized, implemented and evaluated on the basis of six DPs was answered based on a design science research approach according to Peffers et al. (2007). The evaluation shows that the prototypical implementation may yield algorithm rankings in context-dependent scenarios. Consequently, the benchmark can support decision making towards real-time anomaly detection and algorithm reliability. There are two main points of criticism. Firstly, the effort of data preparation is high. Yet, the effort seems worthwhile in the case of standardized production processes according to production experts. Secondly, the resulting benchmark is not comparable to existing benchmarks or evaluation metrics, such as F1-score or false positive rate. This implies the need for testing the benchmark in practice to evaluate its suitability to solve the problem of reliably determining best real-time anomaly detection algorithms for a given data set.

Overall, this paper contributes to both research and practice. Future research can extend the prototypical implementation and quantitatively investigate dependencies among the used mechanisms. Practitioners may appropriate the benchmark for their own use to find the most suitable anomaly detection algorithm in production data streams.

References

Adams, E.P., MacKay, D.J.C. Bayesian Online Changepoint Detection (2007)

Ahmad, S., Lavin, A., Purdy, S., et al.: Unsupervised real-time anomaly detection for streaming data. Neurocomputing **262**, 134–147 (2017)

Bischl, B., et al.: ASlib: a benchmark library for algorithm selection. Artif. Intell. **237**, 41–58 (2016)

Brazdil, P.B., Soares, C.: A comparison of ranking methods for classification algorithm selection. In: LópezdeMántaras, R., Plaza, E. (eds.) ECML 2000. LNCS (LNAI), vol. 1810, pp. 63–75. Springer, Heidelberg (2000). https://doi.org/10.1007/3-540-45164-1_8

Bueno, A., Godinho Filho, M., Frank, A.G.: 'Smart production planning and control in the Industry 4.0 context: a systematic literature review, Comput. Industrial Eng. **149**, 106774 (2020)

Burnaev, E., Ishimtsev, V.: Conformalized density- and distance-based anomaly detection in time-series data (2016). https://arxiv.org/abs/1608.04585. Accessed 16 Nov 2021

Chandola, V., Banerjee, A., Kumar, V.: Anomaly detection. ACM Comput. Surv. **41**(3), 1–58 (2009)

Cheng, C.-Y., Pourhejazy, P., Ying, K.-C., Liao, Y.-H.: New benchmark algorithms for No-wait flowshop group scheduling problem with sequence-dependent setup times. Appl. Soft Comput. **111**, 107705 (2021)

Cohen, Y., Singer, G.: A smart process controller framework for Industry 4.0 settings. J. Intell. Manuf. **32**(7), 1975–1995 (2021). https://doi.org/10.1007/s10845-021-01748-5

Emmott, A.F., Das, S., Dietterich, T., Fern, A., Wong, W.-K.: Systematic construction of anomaly detection benchmarks from real data. In: Proceedings of the ACM SIGKDD Workshop on Outlier Detection and Description - ODD 2013. Chicago, Illinois, New York, New York, USA, 16–21 (2013)

Hevner, A.R., March, S., Park, J., Ram, S.: Design science in information systems research. MIS Q. **28**(1), 75 (2004)

Ho, Y.C., Pepyne, D.L.: Simple explanation of the no-free-lunch theorem and its implications. J. Optim. Theory Appl. **115**(3), 549–570 (2002)

Kagermann, H., Wahlster, W., Helbig, J.: Umsetzungsempfehlungen für das Zukunftsprojekt Industrie 4.0: Abschlussbericht des Arbeitskreises Industrie 4.0 (2013). https://www.acatech.de/wp-content/uploads/2018/03/Abschlussbericht_Industrie4.0_barrierefrei.pdf. Accessed 16 Nov 2021

Kerschke, P., Kotthoff, L., Bossek, J., et al.: Leveraging TSP solver complementarity through machine learning. Evol. Comput. **26**(4), 597–620 (2018)

Lavin, A., Ahmad, S.: Evaluating Real-Time Anomaly Detection Algorithms - The Numenta Anomaly Benchmark (2015). https://arxiv.org/abs/1510.03336. Accessed 16 Nov 2021

Lehmann, E.L., Romano, J.P.: Testing Statistical Hypotheses, 3rd edn. Springer, New York (2005). https://doi.org/10.1007/0-387-27605-X

Mayring, P.: Qualitative content analysis: theoretical foundation, basic procedures and software solution. SSOAR, Klagenfurt (2014)

Numenta Anomaly Benchmark (2020). Numenta Anomaly Benchmark. URL: https://github.com/numenta/NAB (visited on November 16, 2021)

Pavelski, L., Delgado, M., Kessaci, M.-E.: Meta-Learning for Optimization: A Case Study on the Flowshop Problem Using Decision Trees. IEEE Congress on Evolutionary Computation, pp. 1–8, Rio de Janeiro, Brazil (2018)

Peffers, K., Tuunanen, T., Rothenberger, M.A., Chatterjee, S.: A design science research methodology for information systems research. J. Manag. Inf. Syst. **24**(3), 45–77 (2007)

Schütze, A., Helwig, N., Schneider, T.: Sensors 4.0 – smart sensors and measurement technology enable Industry 4.0. J. Sens. Sens. Syst. **7**(1), 359–371 (2018)

Schneider, M., Ertel, W., Ramos, F.: Expected similarity estimation for large-scale batch and streaming anomaly detection. Mach. Learn. **105**(3), 305–333 (2016). https://doi.org/10.1007/s10994-016-5567-7

Singh, N., Olinsky, C.: Demystifying Numenta anomaly benchmark. In: 2017 International Joint Conference on Neural Networks (IJCNN), 1570–1577), IEEE (2017)

Stahmann, P., Rieger, B.: Requirements Identification for Real-Time Anomaly Detection in Industrie 4.0 Machine Groups: A Structured Literature Review', Proceedings of the 54th Hawaii International Conference on System Sciences (2021)

Stahmann, P., Rieger, B.: Towards design principles for a real-time anomaly detection algorithm benchmark suited to Industrie 4.0 streaming data, Proceedings of the 55th Hawaii International Conference of System Sciences (2022a)

Stahmann, P., Oodes, J., Rieger, B.: Improving machine self-diagnosis with an instance-based selector for real-time anomaly detection algorithms. In: 8th International Conference on Decision Support System Technology 2022b. Decision Support addressing modern Industry, Business and Societal needs, Lecture Notes in Business Information Processing (LNBIP) (2022b)

Tukey, J.W.: Exploratory Data Analysis. Addison-Wesley Series in Behavioral Science Quantitative Methods. Addison-Wesley, Reading, Mass. (1977)

Venable, J., PriesHeje, J., Baskerville, R.: FEDS: a framework for evaluation in design science research. Eur. J. Inf. Syst. **25**(1), 77–89 (2016)

Vom Brocke, J., Simons, A., Riemer, K., Niehaves, B., Plattfaut, R., Cleven, A.: Standing on the shoulders of giants: challenges and recommendations of literature search in information systems research. Commun. Assoc. Inf. Syst. **37** (2015)

Wolpert, D.H., Macready, W.G.: No free lunch theorems for optimization. IEEE Trans. Evol. Comput. **1**(1), 67–82 (1997)

A Matrix Factorization-Based Drug-Virus Link Prediction Method for SARS-CoV-2 Drug Prioritization

Yutong Li, Xiaorui Xu, and Sophia Tsoka(✉)

King's College London, London WC2R 2LS, UK
{yutong.li,xiaorui.xu,sophia.tsoka}@kcl.ac.uk

Abstract. Matrix factorization (MF) has been widely used in drug discovery for link prediction, which aims to reveal new drug-target links by integrating drug-drug and target-target similarity information with a drug-target interaction matrix. The MF method is based on the assumption that similar drugs share similar targets and *vice versa*. However, one major disadvantage is that only one similarity metric is used in MF models, which is not enough to represent the similarity between drugs or targets. In this work, we develop a similarity fusion enhanced MF model to incorporate different types of similarity for novel drug-target link prediction. We apply the proposed model on a drug-virus association dataset for anti-COVID drug prioritization, and compare the performance with other existing MF models developed for COVID. The results show that the similarity fusion method can provide more useful information for drug-drug and virus-virus similarity and hence improve the performance of MF models. The top 10 drugs as prioritized by our model are provided, together with supporting evidence from literature.

Keywords: COVID · drug discovery · Matrix factorization · Machine learning

1 Introduction

The COVID-19 pandemic [7], caused by Severe Acute Respiratory Syndrome (SARS)-associated coronavirus, SARS-CoV-2, was first reported in Wuhan, China in December 2019, and has spread to most countries in the world. The pandemic is considered as a serious global health issue due to its high transmittability and risk of death. By April 2022, the World Health Organization (WHO) [1] has reported more than 510 million COVID cases and over 6 million deaths worldwide. Research has shown that a significant proportion of COVID patients can continue to experience long-term symptoms. [3]. Today quite a few COVID-19 vaccines have been approved for emergency to prevent the transmission of COVID, however the effectiveness is still under investigation [8]. There is

Y. Li—Supported by China Scholarship Council.

G. Nicosia et al. (Eds.): LOD 2022, LNCS 13810, pp. 35–47, 2023.
https://doi.org/10.1007/978-3-031-25599-1_4

no approved drug for treatment of COVID currently [4,9] as drug development is a long term process which involves screening hits, compound optimization, and many phases of clinical trials. Computational methods are critical in accelerating this process [23].

Recently, various computational approaches have been employed in COVID-19 drug discovery [26]. Among them, matrix factorization (MF) [18] aims to decompose a matrix into a product of two latent matrices in lower dimension. The principle of applying MF methods in the development of drugs, is to consider the drug target interaction prediction problem as a recommendation system problem, based on the assumption that similar drugs tend to share similar targets, and *vice versa* [6].

MF methods have been used widely in drug discovery. Zhang et al. proposed a similarity constrained matrix factorization network (SCMFDD) [36] for drug-disease association prediction. The results suggested that SCMFDD outperforms the logistic regression model PREDICT [11], achieving 0.33 in AUPR and 0.92 in AUC. Zhang et al. developed a feature-derived graph regularized matrix factorization (FGRMF) [35] for drug side-effect prediction and reported that FGRMF outperforms restricted Boltzmann machine and collaborative filtering method [37] 0.08 in AUPR and 0.003 in AUC. Zhang et al. introduced a manifold regularized matrix factorization (MRMF) [34] for drug-drug interaction prediction. MRMF was compared with neighbor recommender and random walk [27], and outperformed the two models 0.03 in AUPR and 0.01 in AUC.

MF methods have been widely employed in COVID drug prioritization as well [22,24,30,30]. Mongia et al. [25] developed a drug-virus interaction dataset, and applied different MF models on it, including Matrix Completion (MC), MF, Deep Matrix Factorization (DMF), Graph Regularize Matrix Factorization (GRMF), Graph Regularized Matrix Completion (GRMC), and Graph Regularized Binary Matrix Completion (GRBMC). Results showed that GRMF, GRMC and GRBMC have achieved the best performance.

In this work, we report a similarity fusion network enhanced GRMF model, aiming to discover novel Drug-Virus Interactions (DVI) for COVID-19. Employing various evaluation and validation metrics, we demonstrate that the similarity fusion enhanced GRMF model can improve the performance of MF models, and performs better than other MF models that have been applied for COVID. We then select the top 10 prioritized drugs from each model that can be potentially used for COVID treatment, and provide supporting evidence from the literature.

2 Materials

2.1 DVA Dataset

The model to prioritize drugs against SARS-CoV-2 is applied to the Drug Virus Association (DVA) [25] dataset, which is a manually curated dataset for COVID containing 216 anti-virual drug-virus associations, comprising 121 drugs and 39 viruses.

2.2 Similarity Measurement for Drugs and Viruses

As a single similarity method may be limited in describing the biological information of drugs and viruses [16], we compute the similarity of drugs and targets from multiple perspectives, i.e. besides intrinsic biological properties such as structure-based similarity, to provide a more comprehensive description. We also consider the associated biological entities for measuring the similarity of drugs and viruses, such as pathway similarity, gene expression profile similarity, etc.

Structure-Based Similarity. Following the key concept that chemically similar drugs tend to interact with similar targets [32], two strategies are chosen to measure the structure-based drug-drug similarity, i.e. the SIMCOMP score [14] and fingerprint similarity, respectively. The SIMCOMP score is a graph-based method for comparing chemical structures, where pairwise similarity between two molecular structures is computed by converting structures into graphs and then finding the maximal common substructures. In our work SIMCOMP is computed through the KEGG API [17]. Fingerprint similarity describes the presence or absence of substructures in a molecule, and is computed by the Tanimoto coefficient (T_c similarity) [31] (in Eq. 1) between the Extended-connectivity fingerprints (ECFP) [29] of the two molecules:

$$T_c = \frac{|A \cap B|}{|A \cup |B} = \frac{|A \cap B|}{|A| + |B| - |A \cap B|} \tag{1}$$

Gene Expression Similarity. Gene expression patterns indicate the interaction profile of a drug [15]. We obtain the druggable genome data from DGIdb [12], and generate gene expression vectors describing the presence of gene names for each drug. The similarity of gene expression is assessed by Pearson's correlation coefficient:

$$r = \frac{\sum(x_i - \overline{x})(y_i - \overline{y})}{\sqrt{\sum(x_i - \overline{x})^2 \sum(y_i - \overline{y})^2}} \tag{2}$$

Sequence-Based Similarity. Virus sequences are obtained from the KEGG database, and the virus sequence similarity matrix is measured by computing the d2* Oligonucleotide frequency (ONF) [2] dissimilarity (at $k = 6$) between pairs of viral genome sequences through VirHostMatcher [2].

Pathway Similarity. Viruses that share similar infectious pathways may be inhibited by similar drugs. We download the pathways for each virus, and convert each pathway into a set of involved genes through KEGG, before computing the T_c similarity for each gene set.

3 Methodology

3.1 Problem Description

Given a set of drugs $D = \{d_1, d_2, ..., d_m\}$ and viruses $V = \{v_1, v_2, ..., v_n\}$, where m and n represent the number of drugs and viruses, their interaction Y is a $m \times n$ binary matrix where $y_{ij} = 1$ indicates interaction between drug d_i and virus v_j while $y_{ij} = 0$ means no interaction. S_d is a $m \times m$ matrix representing the similarity between each drug-drug pair, while S_v is a $n \times n$ matrix representing the similarity between each virus-virus pair. Similarity scores in each matrix are normalized in the range [0, 1]. The objective is to decompose matrix Y, and recompose it with the similarity information, to obtain novel (i.e. unknown) DVIs.

3.2 Similarity Fusion

Using a nonlinear equation [20] for similarity fusion is one of the common methods for similarity integration:

$$S_{ij} = 1 - \prod_{t=1}^{n}(1 - s_{ij}^t) \tag{3}$$

where n denotes the number of similarity metrics; s_{ij}^t denotes the the similarity between the i-th and j-th drugs or viruses in the t-th measurement. Here, we use this concept to integrate the metrics outlined above.

3.3 GRMF Model for Novel Drug-Virus Link Prediction

To predict novel drug-virus links, we adopt GRMF [10] and WGRMF [10] on the DVA dataset and the drug-drug and virus-virus similarity fusion matrices. An overview of the workflow is shown in Fig. 1. Multiple similarity matrices of drug-drug and virus-virus pairs are computed and integrated as described in Similarity measurement for drugs and viruses and similarity fusion. The similarity fusion matrices are then passed through a pre-processing PNN [10] algorithm which sparsifies the resulting similarity matrix by only keeping the p most similar neighbors for each drug or target and discarding the rest.

The interaction matrix is also pre-processed through a WKNKN [10] algorithm where the DVI values are re-calculated with respect to the interaction profile of a similar drug/virus, in order to reduce the sparsity of the DVI matrix. The pre-processed DVI matrix is then decomposed into two latent matrices in lower dimension through Singular Value Decomposition (SVD) [5], as described below:

$$Y_{m\times n} = u_{n\times n} s_{n\times m} v_{m\times m}^T \tag{4}$$

$$U = us^{\frac{1}{2}} \tag{5}$$

$$V = vs^{\frac{1}{2}} \tag{6}$$

The objective of GRMF is described as follows:

$$\begin{aligned}\min_{U,V} \quad & \|Y - UV^T\|_F^2 \\ & + \lambda_l(\|U\|_F^2 + \|V\|_F^2) \\ & + \lambda_d Tr(U^T \tilde{\mathcal{L}}_d U) \\ & + \lambda_v Tr(V^T \tilde{\mathcal{L}}_v V)\end{aligned} \tag{7}$$

where $\|.\|_F^2$ is the Frobenius norm; λ_l, λ_d, and λ_v are positive parameters; $Tr(\cdot)$ is the trace of a matrix, $\tilde{\mathcal{L}}_d$ and $\tilde{\mathcal{L}}_v$ are the normalized graph Laplacians for S_d and S_v. By setting $\frac{\partial L}{\partial U} = 0$ and $\frac{\partial L}{\partial V} = 0$, the latent matrices U and V are the updated as follows:

$$U = (YV - \lambda_d \tilde{\mathcal{L}}_d U)(V^T V + \lambda_l I_k)^{-1} \tag{8}$$

$$V = (Y^T U - \lambda_v \tilde{\mathcal{L}}_v V)(U^T U + \lambda_l I_k)^{-1} \tag{9}$$

where k is the number of latent features in U and V. The new DVI matrix is formalized as follows:

$$Y' = UV^T \tag{10}$$

Finally, the new DVI matrix Y' is utilized to derive novel drug-virus links. Based on the matlab code from [10], we have developed the python versions of GRMF and WGRMF. The software together with the integrated features implemented here are available at https://github.com/yutongLi1997/Similarity-Fusion-GRMF.

3.4 Experimental Setting and Evaluation Metrics

The evaluation steps described previously [25] are employed to compare the performance with three different variants of 10-fold cross-validation setting for different purposes of discovering new drug-virus pair, discover new drugs, or discover new viruses:

cv_p: to select 10% of the associations between drugs and viruses;
cv_d: to select 10% of the drugs;
cv_v: to select 10% of the viruses.

A train/test mask is then applied to Y where the selected set is set to be 0 in training and 1 in testing.

Area Under the Receiver Operating Characteristic Curve (AUC) [13] and Area Under the Precision-Recall curve (AUPR) are selected as evaluation metrics as they are both robust for evaluating imbalanced classes. Following Eq. 11–14, the Receiver Operating Characteristic Curve (ROC) is created by plotting the True Positive Rate (TPR) against the False Positive Rate (FPR) at various threshold settings. Similarly, the precision-recall curves plot the positive predictive value (precision) against the true positive rate (recall). The AUC score is

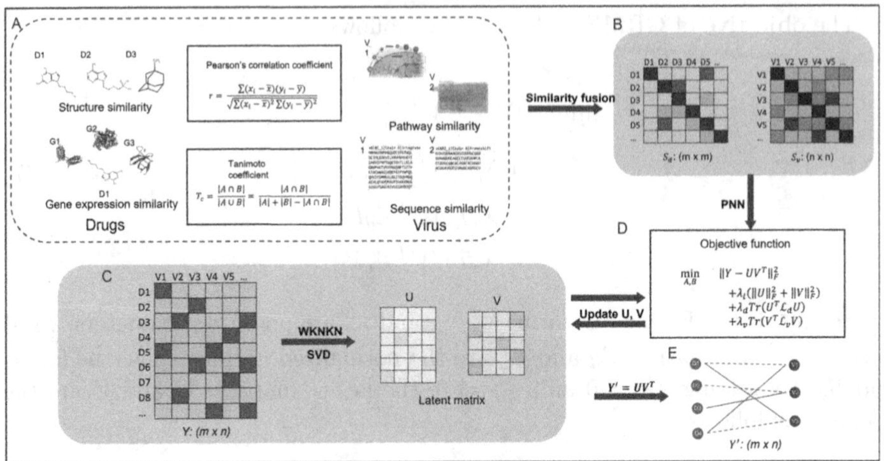

Fig. 1. The flowchart of a similarity fusion network enhanced GRMF model. (A) Different similarity measurements for drug-drug pairs and virus-virus pairs are computed, including structure similarity, gene expression similarity, sequence similarity and pathway similarity. (B) Multiple similarity matrices are integrated into one matrix through a nonlinear equation for similarity fusion. (C) The DVI matrix is pre-processed and decomposed into two lower dimension matrices U and V. (D) The pre-processed drug-drug and virus-virus similarity matrices are used to update the latent matrices U and V. (E) After the algorithm converges, U and V are used to compose the predicted DVI matrix.

obtained by computing the area under the ROC curve, while the AUPR score is computed as the area under the precision-recall curve. Both AUC and AUPR can range from 0 to 1. AUC over 0.5 suggests that the model performs better than a random classifier. The baseline of AUPR is equal to the fraction of positives, and thus AUPR over 0.04 suggests better performance than random classifier.

$$TPR = \frac{TP}{TP + FN} \tag{11}$$

$$FPR = \frac{FP}{FP + TN} \tag{12}$$

$$Precision = \frac{TP}{TP + FP} \tag{13}$$

$$Recall = \frac{TP}{TP + FN} \tag{14}$$

4 Results

4.1 Model Tuning

For hyper-parameter tuning, we run grid search to determine the best set of hyper-parameters. The detailed parameter setting is as follows:

$k \in \{5, 10, 15, 20, 25, 30, 35, 40\}$;
$\lambda_l \in \{0.25, 0.5, 1, 2\}$;
$\lambda_d, \lambda_v \in \{0, 0.0001, 0.001, 0.01, 0.1\}$.

GRMF and WGRMF achieve best performance at $k = 20$, $\lambda_l = 0.25$, and $\lambda_d, \lambda_v = 0.1$, which maximizes the contribution of similarity matrices in the loss function, and allows the model to exploit the contribution to its largest extent. This supports our research aim to explore how similarity fusion network can enhance MF model performance. For the other comparison models, we choose the optimal value recommended in the relevant publication.

4.2 Effectiveness of Similarity Fusion

An example demonstrating the change of similarity values through similarity fusion is given in Fig. 2. The integrated similarity matrix has larger values and is more populated than the single similarity matrix, showing that the similarity fusion method can reduce the sparsity of a matrix and thus provide a richer representation of the interaction network.

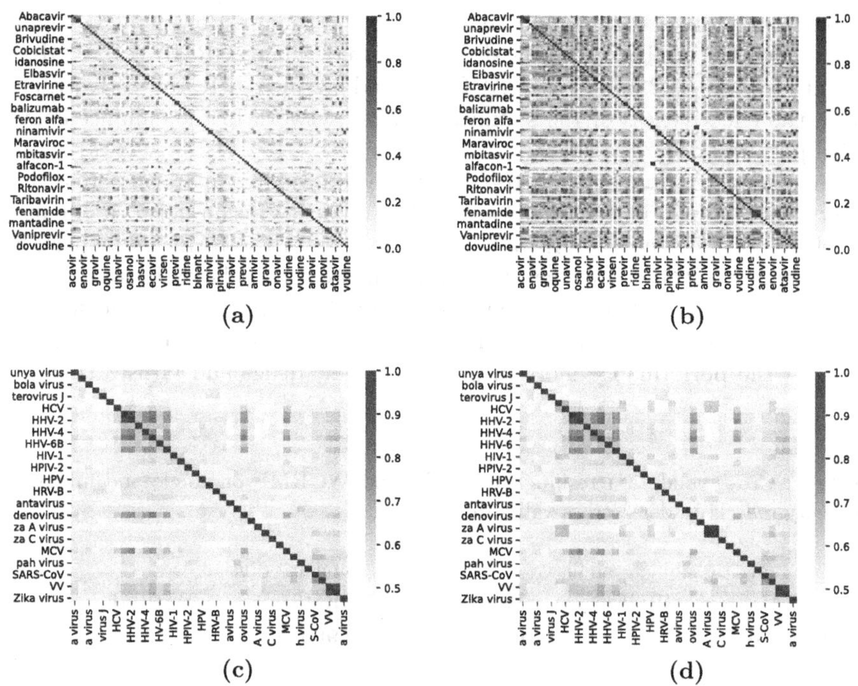

Fig. 2. The heatmap of similarity matrix before and after similarity fusion. (a) and (c) show the single similarity matrix, (b) and (d) show the integrated similarity matrix.

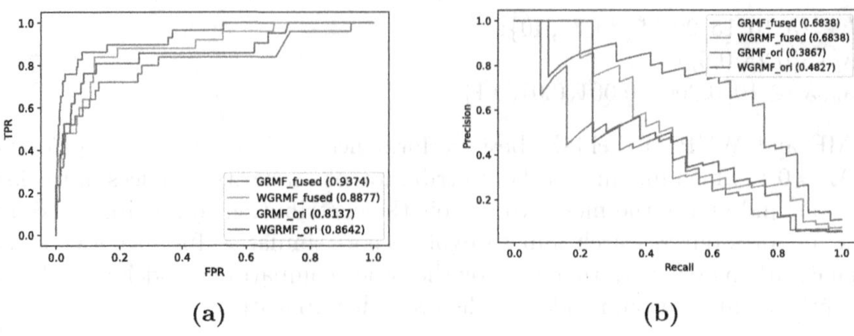

Fig. 3. The drug-virus pairwise performance comparison of GRMF and WGRMF using multiple and single similarity matrix on test set. (a) The AUC curve of the comparison models. (b) The AUPR curve of the comparison models

We then validate the effectiveness of using the similarity fusion method by comparing the performance of GRMF and WGRMF trained with multiple similarity matrices and a single similarity matrix. Figure 3 shows the AUC and AUPR curve of GRMF and WGRMF trained with single and integrated similarity matrices in one test set predicting new drug-virus pair. Both models perform better when the similarity fusion step is followed, where GRMF and WGRMF are improved by 0.12 and 0.02 in AUC, 0.3 and 0.2 in AUPR, respectively, indicating the effectiveness of the similarity fusion network in the enhancement of model performance.

The complete summary of a 10-fold cross-validation comparison across three cv settings is shown in Table 1. Overall, the similarity fusion network enhanced models perform better than the baseline models, and the GRMF model performs better than WGRMF. Moreover, the models based on similarity fusion networks are relatively more stable than the baseline models, according to the standard deviation across the 10-fold cross validation. In particular, the integrated matrix improves the performance more in predicting new drugs and viruses than new DVI.

Table 1. Cross validation results for GRMF and WGRMF on single and integrated similarity matrix under 3 cross-validation settings

	Metrics	Single similarity		Similarity fusion	
		GRMF	WGRMF	GRMF	WGRMF
cv_p	AUC	0.859 (±0.057)	0.856 (±0.041)	**0.900** (±0.033)	0.893 (±0.029)
	AUPR	0.440 (±0.110)	0.451 (±0.097)	**0.503** (±0.123)	0.466 (±0.087)
cv_d	AUC	0.776 (±0.078)	0.776 (±0.091)	**0.831** (±0.062)	0.810 (±0.052)
	AUPR	0.422 (±0.165)	0.441 (±0.144)	0.513 (±0.142)	**0.520** (±0.139)
cv_v	AUC	0.734 (±0.145)	0.680 (±0.183)	**0.751** (±0.130)	0.734 (±0.144)
	AUPR	0.241 (±0.153)	**0.322** (±0.178)	0.270 (±0.220)	0.275 (±0.198)

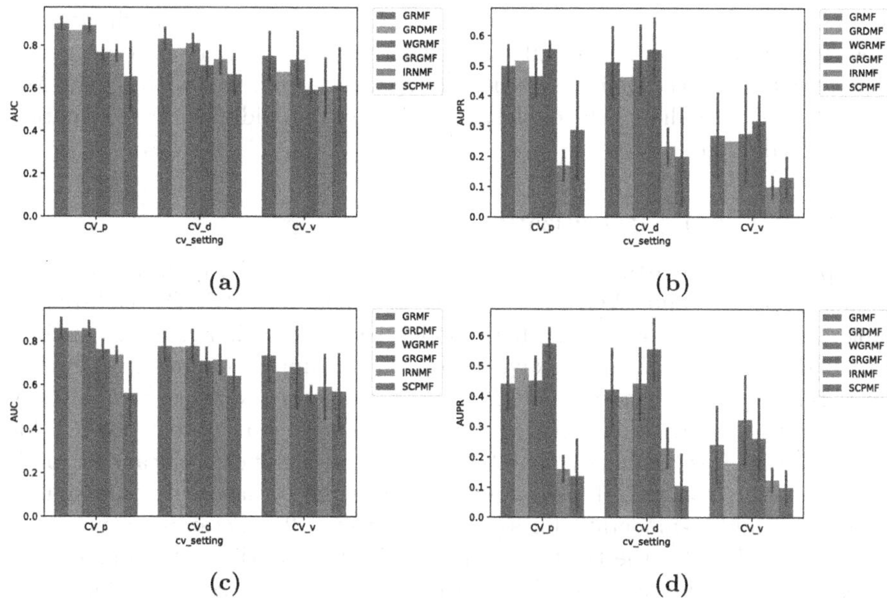

Fig. 4. Cross-validation results for model comparison on three cv settings. (a) and (b) show the mean and standard deviation of AUC and AUPR for similarity fusion enhanced GRMF, GRDMF, WGRMF, GRGMF, IRNMF, and SCPMF. (c) and (d) show the mean and standard deviation of AUC and AUPR for the baseline model of the above models.

4.3 Performance Comparison

We compare the performance of similarity fusion enhanced GRMF and WGRMF with another improved GRMF model, as well as other MF methods that have been used in drug prioritization for SARS-CoV-2, namely, GRDMF [24], GRGMF [38], IRNMF [30], and SCPMF [22]. We also perform the above models on both integrated similarity matrix and single similarity matrix, to further validate the effectiveness of similarity fusion method. The results are shown in Fig. 4.

Overall, the similarity fusion network enhanced GRMF achieves the best performance in AUC score across all three cv settings. For the task of predicting new DVI, GRMF outperforms GRDMF, WGRMF, GRGMF, IRNMF, and SCPMF 0.03, 0.01, 0.14, 0.14, and 0.25, respectively. For predicting new drugs, it outperforms the above five models 0.05, 0.02, 0.13, 0.1, and 0.17, respectively. For predicting new viruses, GRMF outperforms the five models 0.08, 0.02, 0.16, 0.15, and 0.14, respectively. GRGMF performs the best in AUPR.

Generally, most models are improved by the similarity fusion method in AUC score. For predicting new DVI, GRMF, GRDMF, WGRMF, GRGMF, IRNMF, and SCPMF are improved by 0.05, 0.03, 0.04, 0.001, 0.03, and 0.09, respectively. For predicting new drugs, the above models, excluding GRGMF, are improved by 0.06, 0.02, 0.04, 0.02, and 0.02, respectively. For predicting new viruses, the above models are improved by 0.02, 0.01, 0.05, 0.04, 0.01, and 0.05, respectively. This indicates that the similarity fusion network also works well with general MF methods, and can enhance the model performance by providing more information for link prediction.

4.4 COVID Drug Prioritization

We repurpose the existing anti-viral drugs with the similarity fusion network enhanced GRMF and WGRMF for the treatment of COVID, as listed in Table 2. The drugs prioritized by both models show a certain agreement, including Ribavirin, Umifenovir, Baloxavir marboxil, Zanamivir, Lamivudine, Favipiravir, Tenofovir alafenamide, and Triazavirin. Some of the drugs have already been investigated by other research groups, and have shown great potency for COVID treatments. Umifenovir exhibits statistically significant efficacy for mild-asymptomatic patients [28]. Favipiravir is considered to have strong possibility for treating mild-to-moderate illness COVID patients [21]. Peramivir is proposed to be a candidate for the treatment of COVID-19 and other infections related cytokine storm syndrome [33]. Triazavirin is suggested to effectively block both the entry of the pathogen into a host cell and its replication [19]. The prioritized drugs can either be investigated directly as COVID treatment, or be used to design novel drugs for treating COVID.

Table 2. Drug Prioritization for SARS-CoV-2

GRMF	WGRMF
Ribavirin	Umifenovir
Umifenovir	Baloxavir marboxil
Baloxavir marboxil	Ribavirin
Zanamivir	Lamivudine
Favipiravir	Tenofovir alafenamide
Triazavirin	Elvucitabine
Oseltamivir	Tenofovir disoproxil fumarate
Laninamivir	Zanamivir
Peramivir	Favipiravir
Tenofovir alafenamide	Triazavirin

5 Conclusion

In this study, we report a similarity fusion enhanced GRMF model that incorporates more data in relation to drugs and viruses in order to predict new drug-virus links. To evaluate the performance of the model in repurposing drugs for

COVID-19 treatment, we apply the proposed model, as well as other MF models that have been developed for COVID, on the DVA dataset and run a 10-fold cross-validation on three different cv settings. The results show that the similarity fusion method is able to improve the performance of MF models in general, and the proposed model achieves the best performance among all the MF models that have been developed for COVID. Our results include drug prioritization and we discuss the top 10 drugs that our model predicts for the treatment of COVID, some of which have also been independently shown as potent in the literature.

Some future work can be developed as follows. This work uses the original SARS-CoV-2 virus as target for anti-viral drugs repurposing. This can be replaced by one or some variants of SARS-CoV-2 to prioritize drugs for a specific variant, or compare and assess the average highly-ranked drugs. Moreover, this work can be extended to incorporate more similarity metrics, such as side effects, symptoms, etc. A feature importance inspection can also be applied to the reported algorithm to select the best set of similarity metric combination.

References

1. Who coronavirus (covid-19) dashboard. https://covid19.who.int/
2. Ahlgren, N.A., Ren, J., Lu, Y.Y., Fuhrman, J.A., Sun, F.: Alignment-free oligonucleotide frequency dissimilarity measure improves prediction of hosts from metagenomically-derived viral sequences. Nucleic Acids Res. **45**(1), 39–53 (2017)
3. Aiyegbusi, O.L., et al.: Symptoms, complications and management of long covid: a review. J. R. Soc. Med. **114**(9), 428–442 (2021)
4. Basu, D., Chavda, V.P., Mehta, A.A.: Therapeutics for covid-19 and post covid-19 complications: an update. Current Res. Pharmacol. Drug Discovery, 100086 (2022)
5. Björnsson, H., Venegas, S.: A manual for EOF and SVD analyses of climatic data. CCGCR Report **97**(1), 112–134 (1997)
6. Chen, R., Liu, X., Jin, S., Lin, J., Liu, J.: Machine learning for drug-target interaction prediction. Molecules **23**(9), 2208 (2018)
7. Ciotti, M., Ciccozzi, M., Terrinoni, A., Jiang, W.C., Wang, C.B., Bernardini, S.: The covid-19 pandemic. Crit. Rev. Clin. Lab. Sci. **57**(6), 365–388 (2020)
8. Dolgin, E.: Omicron is supercharging the covid vaccine booster debate. Nature 10 (2021)
9. Elmorsy, M.A., El-Baz, A.M., Mohamed, N.H., Almeer, R., Abdel-Daim, M.M., Yahya, G.: In silico screening of potent inhibitors against covid-19 key targets from a library of FDA-approved drugs. Environ. Sci. Pollut. Res. **29**(8), 12336–12346 (2022)
10. Ezzat, A., Zhao, P., Wu, M., Li, X.L., Kwoh, C.K.: Drug-target interaction prediction with graph regularized matrix factorization. IEEE/ACM Trans. Comput. Biol. Bioinf. **14**(3), 646–656 (2016)
11. Gottlieb, A., Stein, G.Y., Ruppin, E., Sharan, R.: Predict: a method for inferring novel drug indications with application to personalized medicine. Mol. Syst. Biol. **7**(1), 496 (2011)
12. Griffith, M., et al.: Dgidb: mining the druggable genome. Nat. Methods **10**(12), 1209–1210 (2013)

13. Hanley, J.A., McNeil, B.J.: The meaning and use of the area under a receiver operating characteristic (ROC) curve. Radiology **143**(1), 29–36 (1982)
14. Hattori, M., Tanaka, N., Kanehisa, M., Goto, S.: Simcomp/subcomp: chemical structure search servers for network analyses. Nucleic Acids Res. **38**(Suppl-2), W652–W656 (2010)
15. Hizukuri, Y., Sawada, R., Yamanishi, Y.: Predicting target proteins for drug candidate compounds based on drug-induced gene expression data in a chemical structure-independent manner. BMC Med. Genomics **8**(1), 1–10 (2015)
16. Huang, L., Luo, H., Li, S., Wu, F.X., Wang, J.: Drug-drug similarity measure and its applications. Briefings Bioinform. **22**(4), bbaa265 (2021)
17. Kanehisa, M., et al.: KEGG for linking genomes to life and the environment. Nucleic Acids Res. **36**(suppl-1), 480–484 (2007)
18. Koren, Y., Bell, R., Volinsky, C.: Matrix factorization techniques for recommender systems. Computer **42**(8), 30–37 (2009)
19. Kováč, I.M.J.Č.G., Hudecová, M.P.L.: Triazavirin might be the new hope to fight severe acute respiratory syndrome coronavirus 2 (sars-cov-2). Ceska a Slovenska farmacie: casopis Ceske farmaceuticke spolecnosti a Slovenske farmaceuticke spolecnosti **70**(1), 18–25 (2021)
20. Liu, H., Sun, J., Guan, J., Zheng, J., Zhou, S.: Improving compound-protein interaction prediction by building up highly credible negative samples. Bioinformatics **31**(12), i221–i229 (2015)
21. Manabe, T., Kambayashi, D., Akatsu, H., Kudo, K.: Favipiravir for the treatment of patients with covid-19: a systematic review and meta-analysis. BMC Infect. Dis. **21**(1), 1–13 (2021)
22. Meng, Y., Jin, M., Tang, X., Xu, J.: Drug repositioning based on similarity constrained probabilistic matrix factorization: covid-19 as a case study. Appl. Soft Comput. **103**, 107135 (2021)
23. Mohamed, K., Yazdanpanah, N., Saghazadeh, A., Rezaei, N.: Computational drug discovery and repurposing for the treatment of covid-19: a systematic review. Bioorg. Chem. **106**, 104490 (2021)
24. Mongia, A., Jain, S., Chouzenoux, E., Majumdar, A.: Deepvir: graphical deep matrix factorization for in silico antiviral repositioning-application to covid-19. J. Comput. Biol. (2022)
25. Mongia, A., Saha, S.K., Chouzenoux, E., Majumdar, A.: A computational approach to aid clinicians in selecting anti-viral drugs for covid-19 trials. Sci. Rep. **11**(1), 1–12 (2021)
26. Muratov, E.N., et al.: A critical overview of computational approaches employed for covid-19 drug discovery. Chem. Soc. Rev. (2021)
27. Park, K., Kim, D., Ha, S., Lee, D.: Predicting pharmacodynamic drug-drug interactions through signaling propagation interference on protein-protein interaction networks. PLoS ONE **10**(10), e0140816 (2015)
28. Ramachandran, R., et al.: Phase iii, randomized, double-blind, placebo controlled trial of efficacy, safety and tolerability of antiviral drug umifenovir vs standard care of therapy in non-severe covid-19 patients. Int. J. Infect. Dis. **115**, 62–69 (2022)
29. Rogers, D., Hahn, M.: Extended-connectivity fingerprints. J. Chem. Inf. Model. **50**(5), 742–754 (2010)
30. Tang, X., Cai, L., Meng, Y., Xu, J., Lu, C., Yang, J.: Indicator regularized nonnegative matrix factorization method-based drug repurposing for covid-19. Front. Immunol. **11**, 3824 (2021)
31. Tanimoto, T.T.: Elementary mathematical theory of classification and prediction (1958)

32. Vilar, S., Hripcsak, G.: The role of drug profiles as similarity metrics: applications to repurposing, adverse effects detection and drug-drug interactions. Brief. Bioinform. **18**(4), 670–681 (2017)
33. Zhang, C.x., et al.: Peramivir, an anti-influenza virus drug, exhibits potential anti-cytokine storm effects. bioRxiv (2020)
34. Zhang, W., Chen, Y., Li, D., Yue, X.: Manifold regularized matrix factorization for drug-drug interaction prediction. J. Biomed. Inform. **88**, 90–97 (2018)
35. Zhang, W., Liu, X., Chen, Y., Wu, W., Wang, W., Li, X.: Feature-derived graph regularized matrix factorization for predicting drug side effects. Neurocomputing **287**, 154–162 (2018)
36. Zhang, W., et al.: Predicting drug-disease associations by using similarity constrained matrix factorization. BMC Bioinform. **19**(1), 1–12 (2018)
37. Zhang, W., Zou, H., Luo, L., Liu, Q., Wu, W., Xiao, W.: Predicting potential side effects of drugs by recommender methods and ensemble learning. Neurocomputing **173**, 979–987 (2016)
38. Zhang, Z.C., Zhang, X.F., Wu, M., Ou-Yang, L., Zhao, X.M., Li, X.L.: A graph regularized generalized matrix factorization model for predicting links in biomedical bipartite networks. Bioinformatics **36**(11), 3474–3481 (2020)

Hyperbolic Graph Codebooks

Pascal Mettes(✉)

University of Amsterdam, Amsterdam, The Netherlands
p.s.m.mettes@uva.nl

Abstract. This work proposes codebook encodings for graph networks that operate on hyperbolic manifolds. Where graph networks commonly learn node representations in Euclidean space, recent work has provided a generalization to Riemannian manifolds, with a particular focus on the hyperbolic space. Expressive node representations are obtained by repeatedly performing a logarithmic map, followed by message passing in the tangent space and an exponential map back to the manifold at hand. Where current hyperbolic graph approaches predominantly focus on node representation, we propose a way to aggregate over nodes for graph-level inference. We introduce Hyperbolic Graph Codebooks, a family of graph encodings where a shared codebook is learned and used to aggregate over nodes. The resulting representations are permutation invariant and fixed-size, yet expressive. We show how to obtain zeroth-order codebook encodings through soft assignments over hyperbolic distances, first-order encodings with anchored logarithmic mappings, and second-order encodings by computing variance information in the tangent space. Empirically, we highlight the effectiveness of our approach, especially when few examples and embedding dimensions are available.

Keywords: Graph networks · Hyperbolic geometry · Graph encoding

1 Introduction

This work addresses the problem of performing inference over entire graphs using neural networks. Graph networks learn embedded node representations through iterative message passing based on known edge connections between nodes [33]. The permutation invariant nature of graph networks makes them applicable to a wide range of domains, from chemical structures [10] to social networks [24] and knowledge graphs [7]. Graph networks commonly embed nodes in Euclidean space, addressing tasks such as node classification, edge prediction, and graph classification. The Euclidean space might however not always be the best choice for embedding graph data, for example when data is hierarchical in nature [31,46]. Moreover, graph-level representations are typically limited to pooling over nodes, which entails a significant loss of information [42]. In this work, we propose graph-level representations that are both expressive and applicable beyond Euclidean spaces.

Building on advances in hyperbolic embeddings for trees [31], Chami *et al.* [9] and Liu *et al.* [26] propose graph networks that generalize to Riemannian

G. Nicosia et al. (Eds.): LOD 2022, LNCS 13810, pp. 48–61, 2023.
https://doi.org/10.1007/978-3-031-25599-1_5

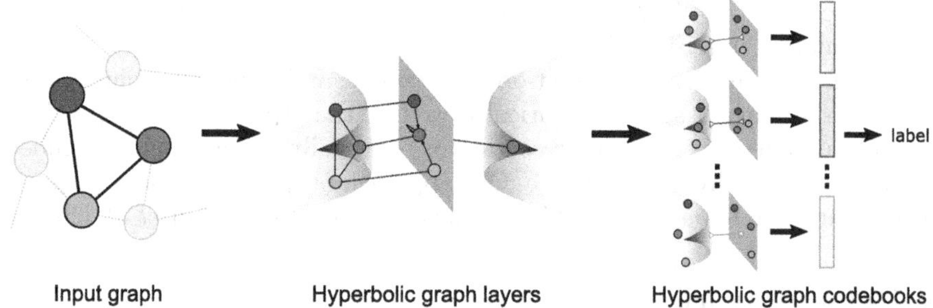

Fig. 1. Hyperbolic Graph Codebooks. For a graph, node representations are first learned through hyperbolic graph layers. Afterwards, we aggregate the nodes into a graph representation using anchored logarithmic maps and hyperbolic distances with a codebook that is learned on-the-fly. This results in an expressive representation applicable to both Euclidean and hyperbolic manifolds.

manifolds, with a focus on the hyperbolic space. The main idea behind their approach is to project graph representations in each layer to the Euclidean tangent space (logarithmic map), where the message passing is performed, followed by a mapping back to the hyperbolic manifold (exponential map). Combined with manifold-preserving non-linearities and optimization, this enables learning node-level representations for graphs on hyperbolic manifolds. Here, we take the step from node-level to graph-level hyperbolic representations.

The main contribution of this work is a family of codebook encodings for hyperbolic graph networks. We take inspiration from codebook encodings for visual data. In such encodings, the parts of an input (*e.g.* patches of an image) are compared to and aggregated over a codebook (typically given as a clustering over all training parts). The aggregated representation is used for tasks such as classification or regression. In zeroth-order approaches, inputs are assigned to nearest codebook mixtures, akin to bag-of-words [37]. First-order approaches additionally store per-dimension differences between inputs and mixtures [2], while second-order approaches also take into account the variances of the mixtures rather than only the means [32]. These approaches generally operate on visual data in Euclidean space with pre-defined codebooks. Here, we seek to aggregate over nodes in graphs on hyperbolic manifolds, with mixtures learned on-the-fly during network training. The main pipeline of the proposed encoding is shown in Fig. 1.

We introduce three codebook encodings: *(i)* We propose a zeroth-order encoding akin to bag-of-words in text and soft assignment in images [25,37]. Each node at the last graph level is soft assigned to a set of codebook mixtures using its respective hyperbolic distance metric. The soft assignments are mean pooled to obtain a graph-level encoding for classification. *(ii)* We propose a codebook that encodes first-order dimensionality information, akin to VLAD for images [2]. To encode first-order information on hyperbolic manifolds, we obtain an element-wise subtraction between a node and a codebook element through an anchored logarithmic map. *(iii)* We propose a codebook that also retains second-order variance information in a Fisher vector formulation [32]. The idea behind this graph encoding is to start from the first-order encoding and capture variance information in the tangent space. Empirically, we find that Hyperbolic Graph Codebooks obtain effective graph representations, especially when using few embedding dimensions in hyperbolic space. Our approach outperforms standard graph networks and the hyperbolic graph classification approach of Liu *et al.* [26].

2 Related Work

2.1 Graph Networks

Graph neural networks learn representations on graph-structured data, which is typically given as a set of nodes with input features with an adjacency matrix denoting edges between nodes. Graph layers learn node-level representations through iterative message passing. Each additional graph layer extends the scope of the neighbourhood used for representing nodes in latent space. Examples of graph neural networks include Graph Neural Networks [33], Graph Convolutional Networks [24,43], Graph Attention Networks [39], and Pointer Graph Networks [38]. On top of one or more graph layers, graph-level representations are obtained by pooling or attention [3,42]. Rather than pooling nodes, we propose an expressive graph-level representation using a codebook that is shared over all classes. Our approach relates to work on hierarchical graph pooling [21,45]. Rather than pooling sets of nodes, we use a shared codebook to obtain a fixed-size representation that can operate in Euclidean space and beyond.

Where conventional graph networks operate in Euclidean space, a growing body of work has shown the potential of general representation learning in non-Euclidean spaces, *e.g.* for supervised learning [14,17,18,23,27,28], unsupervised learning [8,20,40], and embedding tree-like structures [5,16,30,31]. In similar spirit, a number of recent works have proposed graph networks in non-Euclidean spaces. In [9,26], a Riemannian graph layer is defined as a logarithmic map to the Euclidean tangent space, followed by message passing in the tangent space and an exponential map back to the relevant Riemannian manifold. Extensions include constant curvature graph networks [4] and spherical graph convolutional networks [44]. Where these works focus predominantly on node-level representations, we propose a way to aggregate over nodes for graph classification on commonly used spaces such as the Euclidean space and the Poincaré ball.

Several recent works have investigated graph networks that directly operate on hyperbolic manifolds, rather than mapping back and forth between the tangent space. Lou *et al.* [29] show how to differentiate through the Fréchet mean to replace projected aggregations, while Chen *et al.* [11] propose fully hyperbolic graph networks by adapting Lorentz transformations for network operations. Similar in spirit, Shimizu *et al.* [34] propose fully hyperbolic graph layers in the Poincaré ball model and Dai *et al.* [12] do the same by switching between different hyperbolic models for different network operations. In this work, we focus on the generic Riemannian formulation of hyperbolic graph networks rather than model-specific hyperbolic networks, as this allows us to also be directly applicable to other manifolds. Specifically, we show that our codebook-based approach is also beneficial for conventional Euclidean graph networks.

2.2 Codebook Encodings

For our graph encoding, we take inspiration from codebook-based encodings for text documents, images, and videos. Encoding data using codebooks has multiple benefits relevant to our problem as they (i) provide fixed-size outputs from variable-size inputs, (ii) are permutation-invariant, and (iii) share information across categories. Codebooks require a vocabulary, for example by taking all English words for text representations or by clustering local visual features for image representations. Classical approaches include tf-idf [47] and soft-assigned bag of visual words [37]. In the visual domain, a large body of literature has investigated retaining first-order and second-order information in the comparison between inputs and codebook mixtures. The VLAD encoding aggregates per-dimension residuals, effectively storing first-order information in the encoding [2], while the Fisher vector additionally captures second-order variance [32]. In this work, we seek to take codebook encodings to Riemannian graph networks.

Where foundational work in visual codebook encodings assume hand-crafted features, a number of works have investigated codebooks in deep networks. Examples include NetVLAD [1], ActionVLAD [19], and Deep Fisher Networks [35]. These works operate on visual data in Euclidean spaces only and we generalize codebook encodings to graphs on Riemannian manifolds. A few works have investigated fixed visual codebooks on Riemannian manifolds for VLAD [15] and Fisher vectors [22]. We build upon these works and propose a family of codebooks that can operate on graphs, while jointly enabling representation learning and codebook learning as part of the network optimization, instead of using fixed codebooks.

3 Background

3.1 Graph Representation Layers

We consider the problem of graph classification, where we are given a graph $G = (V, \mathbf{A})$, consisting of a set V of n nodes with pair-wise adjacency matrix $\mathbf{A}$. Optionally, each node i is represented by a feature vector $\mathbf{h}_i^0$. We seek to predict a graph-level label Y. In the context of graph networks, global labels are obtained by first learning node-level representations, followed by a permutation-invariant pooling. For convenience, let $\widetilde{\mathbf{A}} = \mathbf{D}^{-\frac{1}{2}}(\mathbf{A} + \mathbf{I})\mathbf{D}^{-\frac{1}{2}}$ denote the normalized adjacency matrix, with $\mathbf{I}$ the identity matrix, and $\mathbf{D}$ the diagonal degree matrix of $\mathbf{A} + \mathbf{I}$ [24]. In conventional graphs, operating in Euclidean space, the representation of node i for layer $k + 1$ is obtained as follows:

$$\mathbf{h}_i^{k+1} = \sigma\Big(\sum_{j \in I(i)} \widetilde{\mathbf{A}}_{ij} \mathbf{W}^k \mathbf{h}_j^k \Big). \tag{1}$$

with $\mathbf{W}^k$ the learnable parameters in layer k, $I(i)$ the set of neighbours of i, $\mathbf{h}_j^k$ the representation of node j in the kth layer, and σ a non-linear operation. Graph-level representations are commonly obtained by pooling the node representations through summing, averaging, or using attention as shown in recent graph network surveys [3,42].

Akin to conventional graph networks, hyperbolic graph layers operate through iterative message passing of neighbours. To make the node representation learning applicable to any Riemannian manifold, several works propose a generalized graph layer with differentiable exponential and logarithmic mappings [9,26]. The representation for node i in layer $k+1$ is given as [26]:

$$\mathbf{h}_i^{k+1} = \sigma\Bigg(\exp_{\mathbf{x}'} \Big(\sum_{j \in I(i)} \widetilde{\mathbf{A}}_{ij} \mathbf{W}^k \log_{\mathbf{x}'}(\mathbf{h}_j^k) \Big) \Bigg). \tag{2}$$

In this formulation, node representation $\mathbf{h}_j^k \in \mathcal{M}^d$ is projected to the Euclidean tangent space with a logarithmic map $\log_{\mathbf{x}'}(\mathbf{h}_j^k) \in \mathcal{T}_{\mathbf{x}'}\mathcal{M}^d$ using anchor point $\mathbf{x}'$ of dimensionality d. Transformations and message passing are in turn performed on the tangent space, followed by an exponential map to the original manifold. On top, manifold preserving non-linearities are applied to allow for stacking of network layers.

3.2 Mapping to and Form the Tangent Space

In this work, we consider two common employed Riemannian manifolds: the Euclidean and the hyperbolic space. For each, we need a corresponding distance metric and logarithmic map function for our graph-level encoding, as well as an exponential map function for the node representation layers. For the Euclidean space, the distance metric is the standard Euclidean distance and the logarithmic and exponential maps are given as:

$$\log_{\mathbf{x}}(\mathbf{v}) = \mathbf{v} - \mathbf{x}, \quad \exp_{\mathbf{x}}(\mathbf{v}) = \mathbf{x} + \mathbf{v}. \tag{3}$$

Note that with $\mathbf{x}$ defined as the origin the generalized graph layer in Eq. 2 becomes the original formulation in Eq. 1.

For hyperbolic spaces, several models can be employed, such as the Poincaré ball and the Lorentz model. Following [29,31], we use the Poincaré ball $(\mathcal{B}, g_{\mathbf{x}}^{\mathcal{B}})$, where $\mathcal{B}$ denotes the open unit ball. The distance metric is as follows:

$$d(\mathbf{x}, \mathbf{v}) = \text{arcosh}\Bigg(1 + 2 \frac{||\mathbf{x} - \mathbf{v}||^2}{(1 - ||\mathbf{x}||^2)(1 - ||\mathbf{v}||^2)} \Bigg). \tag{4}$$

The logarithmic and exponential maps are given as follows:

$$\begin{aligned} \log_{\mathbf{x}}(\mathbf{v}) &= \frac{2}{\lambda_{\mathbf{x}}} \text{arctanh}(|| - \mathbf{x} \oplus \mathbf{v}||) \frac{-\mathbf{x} \oplus \mathbf{v}}{|| - \mathbf{x} \oplus \mathbf{v}||}, \\ \exp_{\mathbf{x}}(\mathbf{v}) &= \mathbf{x} \oplus \big(\tanh(\frac{\lambda_{\mathbf{x}} ||\mathbf{v}||}{2}) \frac{\mathbf{v}}{||\mathbf{v}||}\big), \end{aligned} \tag{5}$$

where $\oplus$ denotes the Möbius addition and $\lambda_{\mathbf{x}} = \frac{2}{1 - ||\mathbf{x}||^2}$.

4 Encoding Hyperbolic Graph Networks

The result of applying K hyperbolic graph layers is a set of node representations $\mathcal{H} = \{\mathbf{h}_1^K, \cdot, \mathbf{h}_n^K\}$, with $\mathbf{h}_i^K \in \mathcal{M}^d$. Since the number of nodes for each graph is different, a graph-level aggregation is required to obtain fixed-sized representations for classification. Here, we propose graph-level encodings that are permutation invariant, expressive, and fixed-sized. Let $\mathcal{C} = \{(\mu_i, \Sigma_i)\}_{i=1}^C$ denote a C-component mixture model with means μ_i and covariance vectors Σ_i. The main idea behind our approach is to encode higher-order graph information using the mixtures. We start from a zeroth-order graph representation, which aggregates the weighted contribution of the mixture components for each node, followed by a first-order representation and ultimately a Fisher-based (first- and second-order) representation of graphs.

4.1 Zeroth-Order Graph Encoding

For a zeroth-order encoding of nodes embedded on a hyperbolic manifold with respect to a set of mixture components, we only consider the component means. For a node $\mathbf{h}_i^K$, we compute a soft assignment for mixture c using the inverse distance between the node and component:

$$\phi_c(\mathbf{h}_i^K) = \frac{\exp\big(-\alpha \cdot d_{\mathcal{M}}(\mathbf{h}_i^K, \mu_c)\big)}{\sum_{c' \in \mathcal{C}} \exp\big(-\alpha \cdot d_{\mathcal{M}}(\mathbf{h}_i^K, \mu_{c'})\big)}, \tag{6}$$

with $d_{\mathcal{M}} : (\mathcal{M}^d, \mathcal{M}^d) \mapsto \mathbb{R}$ the distance function for manifold $\mathcal{M}$ and α a fixed temperature parameter. Let $\phi(\mathbf{h}_i^K) = [\phi_1(\mathbf{h}_i^K), \cdots, \phi_C(\mathbf{h}_i^K)] \in \mathbb{R}^C$ denote the assignment distribution of node i to all mixture components. We average the distributions over all nodes to obtain a zeroth-order graph encoding:

$$\Omega^0(\mathcal{H}) = \frac{1}{n} \sum_{i=1}^{n} \phi(\mathbf{h}_i^K), \tag{7}$$

The encoding is ℓ_2-normalized before feeding it to a final linear layer for graph classification.

4.2 First-Order Graph Encoding

In a first-order encoding, we go beyond soft component assignments and also consider the per dimension differences of nodes for each mixture component, see Fig. 2. This requires vector subtraction on manifold $\mathcal{M}$ to obtain a local difference vector per component. We obtain a local difference vector for each mixture component through tangent space projections using the component mean as tangent point. For component c, the encoding is as follows:

$$\Psi_c(\mathcal{H}) = \frac{1}{n} \sum_{i=1}^{n} \phi_c(\mathbf{h}_i^K) \log_{\mu_c}(\mathbf{h}_i^K). \tag{8}$$

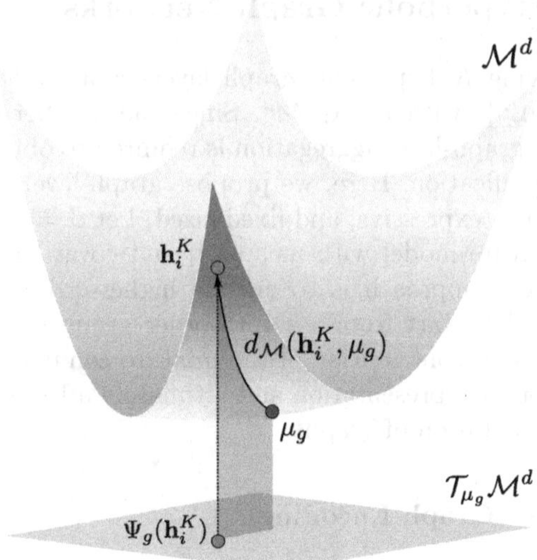

Fig. 2. First order encoding. Assignments are determined using distances $d_{\mathcal{M}}$ on $\mathcal{M}$, while the first-order difference vectors are obtained through a logarithmic map of node $\mathbf{h}_i^K$ using μ_g as anchor.

For a graph encoding, we concatenate the local difference vectors of all components to obtain a Gd-dimensional representation:

$$\Omega^1(\mathcal{H}) = [\Psi_1(\mathcal{H}), \cdots, \Psi_C(\mathcal{H})]. \tag{9}$$

We perform ℓ_2-normalization per mixture component, followed by a global ℓ_2-normalization (Fig. 3).

4.3 Second-Order Graph Encoding

Lastly, we consider a second-order encoding of graphs. This is obtained by using a Fisher kernel when modelling the nodes with respect to the mixture components [32]. Given nodes $\mathcal{H}$ and a probability distribution u_λ that governs the node distribution in the respective manifold, the score function is given by the gradient of the log-likelihood of the data on the model, $S_\lambda^{\mathcal{H}} = \nabla_\lambda \log u_\lambda(\mathcal{H})$. Subsequently, the Fisher Information Matrix is given as $F_\lambda = E_{x \sim u_\lambda}[S_\lambda^{\mathcal{H}} S_\lambda^{\mathcal{H}T}]$. Let $F_\lambda^{-1} = L_\lambda^T L_\lambda$ denote the Cholesky decomposition of F_λ, then the normalized gradient vector of X is given as: $N_\lambda^{\mathcal{H}} = L_\lambda S_\lambda^{\mathcal{H}}$. This gradient vector, the Fisher Vector, can be used to represent $\mathcal{H}$ using u_λ.

In the context of this work, we use a Gaussian mixture model C as probability distribution u_λ. Different from standard Fisher Vectors, which assume fixed observations and fixed mixtures in Euclidean space, our observations are continuously updated, while we also learn the mixture components as part of our network optimization in both Euclidean and hyperbolic spaces. We assume

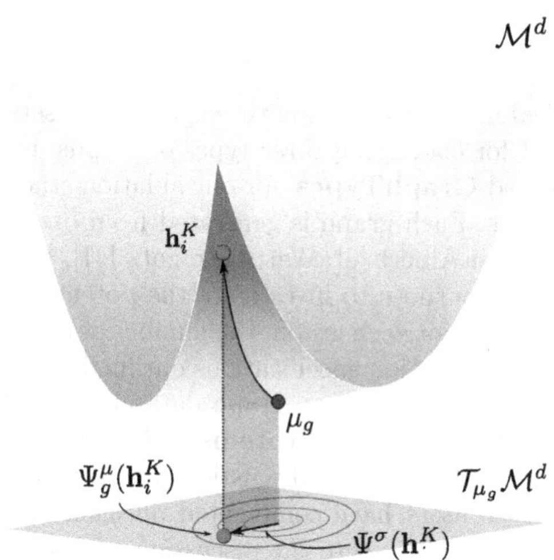

Fig. 3. Second order encoding. First-order difference vectors are again obtained using the codebook-anchored map, while second-order information is obtained in the tangent space using the learned variance of the codebook element.

diagonal covariance matrices and represent each mixture component with variance vectors, rather than a covariance matrix [35,36]. The encoding with respect to the mixture means is then given as:

$$\Phi_c^\mu(\mathcal{H}) = \frac{1}{n}\sum_{i=1}^{n}\hat{\phi}_c(\mathbf{h}_i^K)\Big(\frac{\log_{\mu_c}(\mathbf{h}_i^K)}{\Sigma_c^2}\Big), \tag{10}$$

with soft assignment function:

$$\hat{\phi}_c(\mathbf{h}_i^K) = \frac{\exp\big(-\frac{1}{2}\alpha\cdot\big(\frac{\log_{\mu_c}(\mathbf{h}_i^K)}{\Sigma_c^2}\big)^T\big(\frac{\log_{\mu_c}(\mathbf{h}_i^K)}{\Sigma_c^2}\big)\big)}{\sum_{c'\in C}\exp\big(-\frac{1}{2}\alpha\cdot\big(\frac{\log_{\mu_{c'}}(\mathbf{h}_i^K)}{\Sigma_{c'}^2}\big)^T\big(\frac{\log_{\mu_{c'}}(\mathbf{h}_i^K)}{\Sigma_{c'}^2}\big)\big)}. \tag{11}$$

with α again the temperature. The encoding with respect to the mixture variances is given as:

$$\Phi_c^\sigma(\mathcal{H}) = \frac{1}{n\sqrt{2}}\sum_{i=1}^{n}\hat{\phi}_c(\mathbf{h}_i^K)\Big(\frac{(\log_{\mu_c}(\mathbf{h}_i^K))^2}{\Sigma_c^2} - 1\Big), \tag{12}$$

A graph-level representation is obtained by concatenating the two first- and second-order representations for all mixture components:

$$\Omega^2(\mathcal{H}) = [\Phi_1^\mu(\mathcal{H}), \Phi_1^\sigma(\mathcal{H}), \cdots, \Phi_C^\mu(\mathcal{H}), \Phi_C^\sigma(\mathcal{H})]. \tag{13}$$

The graph-level encoding is of dimensionality $2Cd$ and is normalized using ℓ_2-normalization [32].

5 Experimental Setup

5.1 Datasets

For the empirical evaluation, we employ three graph datasets. The first dataset is a synthetic dataset for classifying three types of graphs, proposed in [26]. We made a variant, dubbed **GraphTypes**, for our ablation studies, with 300 training and 900 test graphs. Each graph is generated from one of three graph generators, namely Barabási-Albert [6], Watts-Strogatz [41], and Erdős-Rényi [13]. The dataset is small on purpose to investigate the potential of our approach in the few-example scenario. For each graph, the number of nodes are sampled from $U(100, 500)$, using the generation specifications outlined in [26]. We furthermore perform a comparison on two common graph datasets, namely **Reddit-multi-12k** (11,929 graphs, 11 classes) and **Proteins** (1,113 graphs, 2 classes)[1] In the Reddit dataset, each graph denotes a discussion thread with users as nodes and edges denoting whether users have commented on each other. The goal is to predict the specific sub-reddit where the discussion took place. In the Proteins dataset, a graph denotes a protein and the task is to classify whether the protein is an enzyme or not.

5.2 Implementation Details

We build our implementation on top of the codebase of [26]. The default learning rate is set to 0.001. For GraphTypes, we train for a fixed 80 epochs and always use 5 graph layers. For Reddit and Proteins, we fix the epochs and number of graph layers to respectively (20, 4) and (100, 4). For all encodings, temperature α is fixed to 10 as a balance between soft and hard assignment [1]. One training epoch takes roughly 48s (GraphTypes), 36m16s (Reddit), and 15s (Proteins) on a single 1080Ti GPU. For GraphTypes, we report each result using the mean and standard deviation over three runs. For the Reddit and Proteins dataset, we use the 10-fold cross-validation performance as given by [26].

6 Experimental Results

For our empirical evaluation, we first perform a series of ablation studies to gain insight into the Hyperbolic Graph Codebooks. We focus the ablation studies on the effect of the codebook size, the effect of embedding dimensionality, and the choice of manifold. All ablations are performed on the GraphTypes dataset. We finalize the experiments with a comparative evaluation to the approach of [26] on all three datasets.

[1] Datasets: https://ls11-www.cs.tu-dortmund.de/staff/morris/graphkerneldatasets..

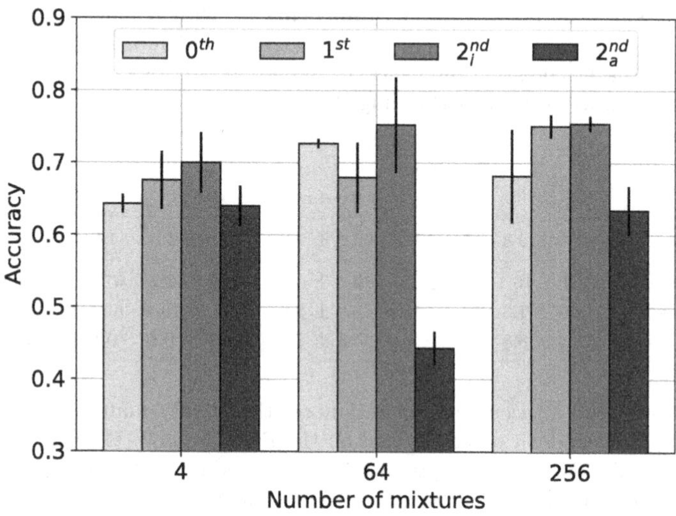

Fig. 4. Classification accuracy on GraphTypes for all graph encodings across four codebook sizes. Across all codebook sizes, the second-order graph encoding performs best. Rather than learning additional variance parameters in this encoding (2^{nd}_a), higher accuracies are obtained when sticking to an isotropic variant with identity variance (2^{nd}_i).

6.1 Ablation Studies

The Effect of Encoding and Codebook Size. First, we compare the different codebooks as a function of the codebook size. We use the Euclidean space as choice of manifold on GraphTypes, with an embedding size of 10 dimensions. We provide results for two variants of the second-order encoding, an anisotropic variant with learnable parameters per embedding dimension and an isotropic variant with identity variance. The results for 4, 64, and 256 mixtures is shown in Fig. 4.

Overall, there is a modest upwards trend when increasing the number of mixtures in the codebooks. Where the zeroth-order encoding performs best with 64 mixtures, the first-order encoding benefits from 256 mixtures. For the second-order encoding, there is a clear difference between the isotropic and anisotropic variant; across all mixture sizes, the isotropic variant performs better. The addition of learnable parameters for the anisotropic variant is not paired with improved performance. The best results are obtained with the isotropic second-order graph encoding using 256 mixtures, with a mean accuracy of 75.4%. For each selection of codebook size, we recommend the second-order graph encoding when excluding additional learnable variance parameters. For the rest of the experiments, we refer to the isotropic variant when using the second-order encoding.

Table 1. Classification accuracies on GraphTypes for three embedding sizes in both Euclidean space ($\mathbb{R}$) and on the Poincare ball ($\mathbb{H}$). When embedding dimensions are limited, all approaches benefit from using the hyperbolic space. The hyperbolic second-order encoding works best, outperforming the approach of [26].

	Euclidean			Hyperbolic		
	3-dim	5-dim	10-dim	3-dim	5-dim	10-dim
HGNN [26]	49.5 ± 0.8	57.8 ± 3.3	66.3 ± 7.6	49.6 ± 5.8	54.5 ± 11.8	72.2 ± 2.9
Ours (0th-order)	47.9 ± 10.3	63.8 ± 2.7	72.7 ± 0.7	64.3 ± 2.2	67.6 ± 3.4	69.7 ± 4.3
Ours (1st-order)	48.0 ± 1.3	57.8 ± 4.0	67.9 ± 4.8	56.0 ± 10.4	66.6 ± 1.2	63.2 ± 9.3
Ours (2nd-order)	57.2 ± 8.8	63.3 ± 7.9	75.3 ± 2.4	**64.3 ± 0.9**	**70.0 ± 5.2**	**77.4 ± 3.6**

Table 2. Comparative accuracy evaluation on the Reddit-multi-12k and Proteins datasets in both Euclidean space ($\mathbb{R}$) and on the Poincare ball ($\mathbb{H}$). On both datasets and on both manifolds, our zeroth- and first-order encodings obtain comparable results to [26], while our second-order encoding outperforms all.

	Reddit		Proteins	
	Euclidean	Hyperbolic	Euclidean	Hyperbolic
HGNN [26]	37.7 ± 1.8	43.9 ± 1.5	64.8 ± 8.9	65.9 ± 5.2
Ours (0th-order)	38.4 ± 1.2	39.4 ± 8.3	65.6 ± 3.9	60.8 ± 4.8
Ours (1^{sr}-order)	37.8 ± 2.8	44.7 ± 0.9	69.6 ± 6.2	62.5 ± 6.5
Ours (2nd-order)	38.5 ± 1.7	**44.8 ± 0.9**	**69.9 ± 5.7**	66.6 ± 5.4

The Effect of Embedding Size and Manifold. Second, we evaluate the effect of the number of embedding dimensions for the nodes and the choice of manifold. We investigate using 3, 5, and 10 dimensions in both Euclidean space and on the Poincare ball. In Table 1 we show the results for our three encodings. We include a comparison to [26], which also proposes a Riemannian graph aggregation approach based on Radial Basis Functions.

All three encodings benefit from using a non-Euclidean manifold for graph classification, especially when using few dimensions. For three dimensions, our second-order encoding improves by 7.1 percent point (from 57.2% to 64.3% mean accuracy) when using a hyperbolic space, while the improvement is 2.1 percent point (from 75.3% to 77.4%) when using 10 dimensions. We find that in both manifolds, our second-order encoding improves over the baseline across all dimensions, highlighting the effectiveness of our approach.

6.2 Comparative Evaluation

To evaluate the effectiveness of our approach, we draw a comparison to the Riemmanian graph classification approach [26] on two common graph datasets:

Reddit-multi-12k and Proteins. We run the baseline and our three encodings over all 10 folds with the same network configurations and hyperparameters.

The mean and standard deviations are reported in Table 2. For Reddit, the hyperbolic space is clearly preferred over the Euclidean space; all approaches report higher mean accuracies. On Proteins, the baseline obtains a small boost in hyperbolic space, while our encodings favor the Euclidean space. We note that the mean accuracies for the baseline are lower than reported in the original paper of the baseline, even though the same settings are used. The zeroth-order encoding has the same running time as the baseline, while the first- and second-order encodings result in a small increase in computation time since the final classification layer has more input dimensions. An explanation for the discrepancy is the use of a fixed number of iterations per dataset rather than using early stopping and the best test score over all epochs. For both datasets, the second-order encoding performs better, in hyperbolic and Euclidean space for respectively the Reddit and Proteins datasets. We conclude that the proposed Riemmanian Graph Codebooks provide an effective and general solution to graph classification, regardless of the selected manifold.

7 Conclusions

In this work, we have investigated the problem of classifying graphs using neural networks operating on hyperbolic manifolds. We propose three codebook encodings that aggregate node representation over graphs. Our Hyperbolic Graph Codebooks result in fixed-sized, permutation-invariant, and expressive graph representations that are applicable to well known manifolds such as Euclidean and hyperbolic spaces. Experimentally, we find that the second-order encoding provides a consistent boost over the other encodings, especially in hyperbolic space with few dimensions. While second-order information is beneficial, learning per-dimension variance is best omitted. From the comparisons, we conclude that Hyperbolic Graph Codebooks enable effective graph classification results, regardless of the choice of manifold.

References

1. Arandjelovic, R., Gronat, P., Torii, A., Pajdla, T., Sivic, J.: NetVLAD: CNN architecture for weakly supervised place recognition. In: CVPR (2016)
2. Arandjelovic, R., Zisserman, A.: All about VLAD. In: CVPR (2013)
3. Bacciu, D., Errica, F., Micheli, A., Podda, M.: A gentle introduction to deep learning for graphs. NN (2020)
4. Bachmann, G., Bécigneul, G., Ganea, O.: Constant curvature graph convolutional networks. In: ICML (2020)
5. Balazevic, I., Allen, C., Hospedales, T.: Multi-relational poincaré graph embeddings. In: NeurIPS (2019)
6. Barabási, A.L., Albert, R.: Emergence of scaling in random networks. Science **286**(5439), 509–512 (1999)

7. Berrendorf, M., Faerman, E., Melnychuk, V., Tresp, V., Seidl, T.: Knowledge graph entity alignment with graph convolutional networks: lessons learned. In: ECIR (2020)
8. Bojanowski, P., Joulin, A.: Unsupervised learning by predicting noise. ICML (2017)
9. Chami, I., Ying, Z., Ré, C., Leskovec, J.: Hyperbolic graph convolutional neural networks. In: NeurIPS (2019)
10. Chen, C., Ye, W., Zuo, Y., Zheng, C., Ong, S.P.: Graph networks as a universal machine learning framework for molecules and crystals. Chem. Mater. **31**(9), 3564–3572 (2019)
11. Chen, W., et al.: Fully hyperbolic neural networks. CoRR (2021)
12. Dai, J., Wu, Y., Gao, Z., Jia, Y.: A hyperbolic-to-hyperbolic graph convolutional network. In: CVPR (2021)
13. Erdős, P., Rényi, A.: On random graphs i. Publicationes Mathematicae Debrecen **6**(290–297), 18 (1959)
14. Fang, P., Harandi, M., Petersson, L.: Kernel methods in hyperbolic spaces. In: ICCV (2021)
15. Faraki, M., Harandi, M.T., Porikli, F.: More about VLAD: a leap from Euclidean to Riemannian manifolds. In: CVPR (2015)
16. Ganea, O.E., Bécigneul, G., Hofmann, T.: Hyperbolic entailment cones for learning hierarchical embeddings. ICML (2018)
17. Ghadimi Atigh, M., Keller-Ressel, M., Mettes, P.: Hyperbolic Busemann learning with ideal prototypes. In: NeurIPS (2021)
18. Ghadimi Atigh, M., Schoep, J., Acar, E., van Noord, N., Mettes, P.: Hyperbolic image segmentation. In: CVPR (2022)
19. Girdhar, R., Ramanan, D., Gupta, A., Sivic, J., Russell, B.: ActionVLAD: learning spatio-temporal aggregation for action classification. In: CVPR (2017)
20. Hsu, J., Gu, J., Wu, G., Chiu, W., Yeung, S.: Capturing implicit hierarchical structure in 3D biomedical images with self-supervised hyperbolic representations. In: NeurIPS (2021)
21. Hu, F., Zhu, Y., Wu, S., Wang, L., Tan, T.: Hierarchical graph convolutional networks for semi-supervised node classification. In: IJCAI (2019)
22. Ilea, I., Bombrun, L., Germain, C., Terebes, R., Borda, M., Berthoumieu, Y.: Texture image classification with Riemannian fisher vectors. In: ICIP (2016)
23. Khrulkov, V., Mirvakhabova, L., Ustinova, E., Oseledets, I., Lempitsky, V.: Hyperbolic image embeddings. In: CVPR (2020)
24. Kipf, T.N., Welling, M.: Semi-supervised classification with graph convolutional networks. In: ICLR (2017)
25. Liu, L., Wang, L., Liu, X.: In defense of soft-assignment coding. In: ICCV (2011)
26. Liu, Q., Nickel, M., Kiela, D.: Hyperbolic graph neural networks. In: NeurIPS (2019)
27. Liu, W., et al.: Deep hyperspherical learning. In: NeurIPS (2017)
28. Long, T., Mettes, P., Shen, H.T., Snoek, C.G.: Searching for actions on the hyperbole. In: CVPR (2020)
29. Lou, A., Katsman, I., Jiang, Q., Belongie, S., Lim, S.N., De Sa, C.: Differentiating through the fréechet mean. In: ICML (2020)
30. Nickel, M., Kiela, D.: Learning continuous hierarchies in the Lorentz model of hyperbolic geometry. In: ICML (2018)
31. Nickel, M., Kiela, D.: Poincaré embeddings for learning hierarchical representations. In: NeurIPS (2017)
32. Sánchez, J., Perronnin, F., Mensink, T., Verbeek, J.: Image classification with the fisher vector: theory and practice. IJCV **105**(3), 222–245 (2013)

33. Scarselli, F., Gori, M., Tsoi, A.C., Hagenbuchner, M., Monfardini, G.: The graph neural network model. NN **20**(1), 61–80 (2008)
34. Shimizu, R., Mukuta, Y., Harada, T.: Hyperbolic neural networks++. In: ICLR (2021)
35. Simonyan, K., Vedaldi, A., Zisserman, A.: Deep fisher networks for large-scale image classification. In: NeurIPS (2013)
36. Sydorov, V., Sakurada, M., Lampert, C.H.: Deep fisher kernels-end to end learning of the fisher kernel GMM parameters. In: CVPR (2014)
37. Van Gemert, J.C., Veenman, C.J., Smeulders, A.W.M., Geusebroek, J.M.: Visual word ambiguity. TPAMI **32**(7), 1271–1283 (2009)
38. Veličković, P., Buesing, L., Overlan, M.C., Pascanu, R., Vinyals, O., Blundell, C.: Pointer graph networks. In: NeurIPS (2020)
39. Veličković, P., Cucurull, G., Casanova, A., Romero, A., Lio, P., Bengio, Y.: Graph attention networks. In: ICLR (2018)
40. Wang, T., Isola, P.: Understanding contrastive representation learning through alignment and uniformity on the hypersphere. In: ICML (2020)
41. Watts, D.J., Strogatz, S.H.: Collective dynamics of 'small-world' networks. Nature **393**(6684), 440–442 (1998)
42. Wu, Z., Pan, S., Chen, F., Long, G., Zhang, C., Philip, S.Y.: A comprehensive survey on graph neural networks. TNNLS **32**, 4–24 (2020)
43. Yang, L., Wu, F., Wang, Y., Gu, J., Guo, Y.: Masked graph convolutional network. In: IJCAI (2019)
44. Yang, Q., Li, C., Dai, W., Zou, J., Qi, G.J., Xiong, H.: Rotation equivariant graph convolutional network for spherical image classification. In: CVPR (2020)
45. Ying, Z., You, J., Morris, C., Ren, X., Hamilton, W., Leskovec, J.: Hierarchical graph representation learning with differentiable pooling. In: NeurIPS (2018)
46. Zhang, C., Gao, J.: Hype-HAN: hyperbolic hierarchical attention network for semantic embedding. In: IJCAI (2020)
47. Zhang, W., Yoshida, T., Tang, X.: A comparative study of TF* IDF, LSI and multi-words for text classification. Expert Syst. Appl. **38**(3), 2758–2765 (2011)

A Kernel-Based Multilayer Perceptron Framework to Identify Pathways Related to Cancer Stages

Marzieh Soleimanpoor[1], Milad Mokhtaridoost[2(✉)], and Mehmet Gönen[3,4]

[1] Graduate School of Sciences and Engineering, Koç University, Istanbul, Turkey
msoleimanpoor17@ku.edu.tr

[2] Genetics and Genome Biology, The Hospital for Sick Children, Toronto ON, Canada
milad.mokhtaridoost@sickkids.ca

[3] Department of Industrial Engineering, College of Engineering, Koç University, Istanbul, Turkey
mehmetgonen@ku.edu.tr

[4] School of Medicine, Koç University, İstanbul 34450, Turkey

Abstract. Standard machine learning algorithms have limited knowledge extraction capability in discriminating cancer stages based on genomic characterizations, due to the strongly correlated nature of high-dimensional genomic data. Moreover, activation of pathways plays a crucial role in the growth and progression of cancer from early-stage to late-stage. That is why we implemented a kernel-based neural network framework that integrates pathways and gene expression data using multiple kernels and discriminates early- and late-stages of cancers. Our goal is to identify the relevant molecular mechanisms of the biological processes which might be driving cancer progression. As the input of developed multilayer perceptron (MLP), we constructed kernel matrices on multiple views of expression profiles of primary tumors extracted from pathways. We used Hallmark and Pathway Interaction Database (PID) datasets to restrict the search area to interpretable solutions. We applied our algorithm to 12 cancer cohorts from the Cancer Genome Atlas (TCGA), including more than 5100 primary tumors. The results showed that our algorithm could extract meaningful and disease-specific mechanisms of cancers. We tested the predictive performance of our MLP algorithm and compared it against three existing classification algorithms, namely, random forests, support vector machines, and multiple kernel learning. Our MLP method obtained better or comparable predictive performance against these algorithms.

Keywords: Machine learning · Multilayer perceptron · Big data · Neural network · Cancer stages · Genomic data

Our implementation of proposed algorithm in R is available at https://github.com/MSoleimanpoor/Neural-Network.git.

G. Nicosia et al. (Eds.): LOD 2022, LNCS 13810, pp. 62–77, 2023.
https://doi.org/10.1007/978-3-031-25599-1_6

1 Introduction

Cancer has become one of the leading causes of death. Mortality rates increase from early-stages to late-stages of cancer, and its diagnosis in earlier stages can significantly assist the treatment procedure. It is demonstrated that genomic changes play a crucial role in cancer progression from early-stage to late-stage [1]. This is why the genomic characterization extracted from tumor biopsies of cancer patients is used to develop new approaches to diagnose, treat, and prevent cancer diseases.

Applying machine learning methods can facilitate extracting meaningful information about the biological mechanisms. However, extraction of informative data is not easily achievable from high-dimensional and correlated genomic datasets [2]. Moreover, the limited number of profiled tumors is another significant restriction. Thus far, the evidence clearly demonstrates the need for better strategies to identify pathways and gene expression changes that drive cancer progression from early- to late-stages.

In the past decades, several studies have addressed the problem of separating cancer stages from each other using genomic measurements by applying score-based [3] and correlation-based [4] algorithms. To categorize cancer patients, various studies developed threshold-based methods using different algorithms, such as random forests (RFs) and SVMs [1,5]. Although scoring- and thresholding-based metrics determine predictive gene expression signatures with a limited number of features, they have two significant limitations: (i) difficulty in interpretation arising from high-dimensional and correlated features and (ii) for a given prediction task, if machine learning classifiers use various subsets of the same patient cohort; they might choose diverse biomarkers as predictive in high-dimensional space [6].

Unlike the above studies, a large amount of literature has been published to improve the interpretability of the result in problems with high-dimensional and correlated input data using embedded methods. Authors in [7,8] used Hallmark gene sets to build a unified model on a subset of gene sets using genomic information of profiled tumors. Authors in [9] and [10] designed embedded classification models for differentiation of early- and late-stages of prostate cancer.

ANNs have been confirmed as robust and powerful methods in cancer biology [11]. Depending on the type of input-output data and the application, ANNs can have different structures. Among various ANN structures, multilayer perceptrons (MLPs) have been more widely used to analyze high-throughput medical data. In many of these applications, algorithms based on MLPs have surpassed the performance of previous state-of-art algorithms. Gene subset selection [12,13] and gene expression prediction [14,15] are two critical active research problems that have been addressed by MLPs on genomic data with wrapper and embedded feature selection methods. One of the main advantages of MLPs over conventional methods is the ability to capture the complex and nonlinear interactions between pathways. Moreover, MLPs consider the inter-relation between pathways, which can significantly improve the model performance in extracting informative oncology knowledge [16]. However, no previous study has investigated the discrimination of cancer stages using MLPs on gene expression data.

In this study, we addressed the problem of extracting disease-related information from an extensive amount of redundant data and noise in high dimensional feature space using gene expression profiles, as well as, Pathway Interaction Database (PID) [17] and Hallmark gene sets [18] in a unified model. PID and Hallmark collections provide information about sets of genes that are similar or dependent in terms of their functions. We trained the classifier on the discovered relevant biological processes. We built kernel matrices on multiple views generated from gene expression profiles of tumors using input pathways and gene sets.

The significant contributions of our study are summarized as follows: (1) To the best of our knowledge, this is the first study using both Hallmark gene sets and PID pathways, along with gene expression data that restricts the genes to the ones related to the progression of cancer from early to late stages. (2) We developed a ℓ_1-regularized multilayer perceptron framework that is able to choose the predictive pathways and train a classification model simultaneously. This property increases the interpretability of the result. (3) Our results demonstrated our algorithm's capacity to obtain biologically meaningful and disease-specific information for the prognosis of cancers. (4) We implemented our approach on 12 different cancer cohorts and compared the result with three other existing algorithms in the literature, namely, random forests, support vector machines, and multiple kernel learning. The result indicated the better or comparable predictive performance of our MLP against other algorithms.

2 Materials

In this study, we used clinical information and genomic characterizations obtained from the Cancer Genome Atlas (TCGA) project and pathways databases from MSigDB collections.

2.1 Constructing Datasets Using TCGA Data

TCGA is a comprehensive consortium that offers genomic characterizations and clinical and histopathological information of more than 10,000 cancer patients across 33 different cancer types. We used preprocessed gene expression profiles of tumors generated by the unified RNA-Seq pipeline of the TCGA project. For each cancer type, we downloaded HTSeq-FPKM files of all primary tumors, as well as `pathologic_stage` annotation, which included stage information of patients, from the most recent data freeze (i.e., `Data Release 32 - March 29, 2022`). Metastatic tumors were not included considering that their fundamental biology would be completely different from primary tumors.

Localized cancers are considered early-stage because they are only found in the tissue where cancer originated before it spreads to the other areas of the body. Regional and distant metastasis cancers are often considered late-stage due to the fact that cancer has spread to nearby lymph nodes and/or to a different part of the body, respectively. We labeled primary tumors with `Stage I` annotation to be early-stage, which is also known as localized cancers, and the

Table 1. Summary of 12 TCGA cohorts used in our experiment

Cohort	Disease name	Stage I	Stage II	Stage III	Stage IV	Early	Late	Total
BRCA	Breast invasive carcinoma	181	619	247	20	181	886	1,067
COAD	Colon adenocarcinoma	75	176	128	64	75	368	443
HNSC	Head and neck squamous cell carcinoma	25	70	78	259	25	407	432
KIRC	Kidney renal clear cell carcinoma	265	57	123	82	265	262	527
KIRP	Kidney renal papillary cell carcinoma	172	21	51	15	172	87	259
LIHC	Liver hepatocellular carcinoma	171	86	85	5	171	176	347
LUAD	Lung adenocarcinoma	274	121	84	26	274	231	505
LUSC	Lung squamous cell carcinoma	244	162	84	7	244	253	497
READ	Rectum adenocarcinoma	30	51	51	24	30	126	156
STAD	Stomach adenocarcinoma	53	111	150	38	53	299	352
TGCT	Testicular germ cell tumors	55	12	14	0	55	26	81
THCA	Thyroid carcinoma	281	52	112	55	281	219	500
					Total	1,826	3,340	5,163

remaining tumors with Stage II, III, or IV annotations to be late-stage, which are regional or distant spreads.

We included cancer types with at least 25 samples in each stage, and that is why 12 cancer types were included in our experiment. In total, we used 5,163 primary tumors. The detailed information about included datasets in this study is represented in Table 1.

2.2 Pathway Databases

While predicting the stage of a tumor, we used PID pathways and Hallmark gene sets to restrict the gene expressions and understand the biological mechanisms that separate the cancer stages from each other. Hallmark gene sets were created through a computational procedure to determine overlaps among the gene sets in other MSigDB collections and keep genes that display coherent expression in different cancers. However, PID collection includes curated and peer-reviewed pathways related to molecular signaling, regulatory events, and key cellular processes. Hallmark and PID databases consist of 50 gene sets and 196 pathways, respectively, and the sizes of these gene sets change between 32–200 and 10–137, respectively.

3 Method

We implemented an MLP algorithm for binary classification to determine the biological mechanisms that discriminate early-stage from late-stage tumors. We used a distinct kernel matrix for each pathway, which calculated the expression features of genes included in this pathway. These Kernel matrices measured the similarity between patients for the biological process captured in the underlying pathway.

3.1 Problem Formulation

We addressed the problem of discriminating the stages of primary tumors using the gene expression data as a binary classification problem. Gene expression data is indicated as $\mathcal{X}$, and the related phenotype of these profiles tumors, which are early- and late-stages is indicated as $\mathcal{Y}$. We arbitrarily labeled early-stage tumors as class 1 and late-stage tumors as class 0 in our MLP algorithm. We represented the training dataset for a given problem as $\{(\boldsymbol{x}_i, y_i)\}_{i=1}^{N}$, where N indicates the number of tumors, $\boldsymbol{x}_i$ indicates the gene expression profile of primary tumor i, and $y_i \in \{0, 1\}$. This binary classification problem can be solved by learning a discriminant function to predict the phenotype from the gene expression profiles. The discriminant function can be shown as $f\colon \mathcal{X} \to \mathcal{Y}$. This function is learned over training data, which is a subset of gene expression data. Training data are used to make predictions for out-of-sample primary tumors, which are unseen during training of the discriminant function.

3.2 Structure of the Proposed MLP

We designed a multi-input MLP algorithm for predicting the pathological stages of primary tumors (see Fig. 1). In our network, subscripts i and m index the units in the input and hidden layers, respectively.

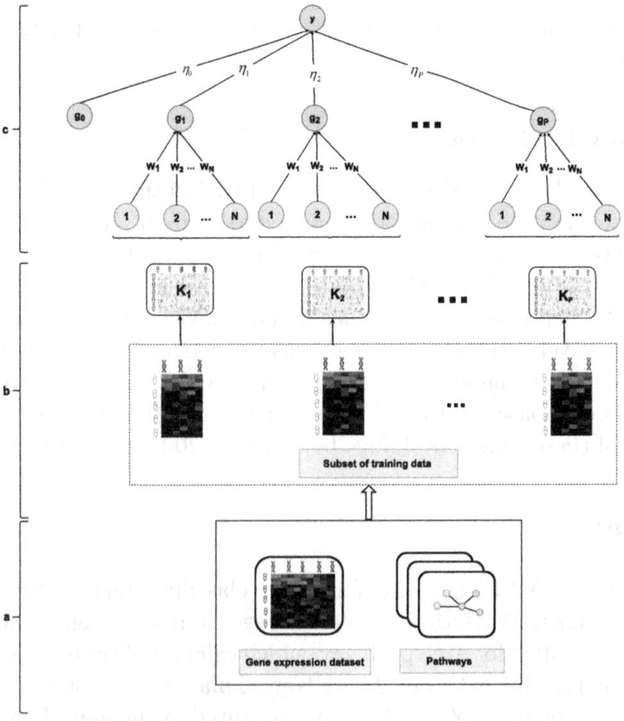

Fig. 1. The overview of our proposed MLP algorithm; Part (a) shows the inputs of our framework. Part (b) kernel matrices were calculated on the gene expression profiles of primary tumors for different pathways. Kernel matrices ($\mathbf{K}_m \ \forall m = 1, \ldots, P$) integrated pathways information into the constructed MLP model. Part (c) demonstrates the structure of our MLP network, including P input layers with N nodes, one hidden layer with $P + 1$ nodes ($g_m \ \forall m = 0, \ldots, P$), and one output layer with one node.

We calculated the kernel matrices on gene expression profiles of tumors for distinct pathways. Kernels measure the similarity between the genomic information for each pair of samples. Integrating kernels into the MLP algorithm leads to narrow gene expression features and overcoming the curse of dimensionality by discarding the ineffective and redundant gene sets. Constructed kernels are the inputs to our proposed MLP network that can be represented as $\{k_m(\boldsymbol{x}_i, \boldsymbol{x}_j)\}_{m=1}^{P}$, where P is the number of kernels and the dimension of each kernel matrix is $N \times N$. We integrated each kernel matrix into a separate input layer; thus, the constructed MLP has P different input layers for each input kernel (i.e., $\mathbf{K}_1, \ldots, \mathbf{K}_P$), and each input layer has N nodes. Every node in each input layer is connected to the corresponding kernel node in the hidden layer. Input layers have shared weights $w_1, \ldots, w_N$ because we want to use the same weights for different kernels to decrease the number of parameters and, subsequently, model complexity.

Our MLP network has one hidden node for each input kernel (i.e. $g_1, \ldots, g_P$) and one bias node (i.e. g_0). The value of each hidden node is equal to the weighted sum of its inputs:

$$g_{i,m} = \boldsymbol{w}^\top \boldsymbol{x}_{i,m} \qquad \forall m = 1, \ldots, P,$$

where $\boldsymbol{x}_{i,m}$ is the vector of the input values, which is the kernel values for $\boldsymbol{x}_i$ using kernel m. We defined the input entries for each input layer as:

$$\boldsymbol{x}_{i,m} = \left[k_m(\boldsymbol{x}_1, \boldsymbol{x}_i) \ldots k_m(\boldsymbol{x}_N, \boldsymbol{x}_i)\right]^\top \qquad \forall m = 1, \ldots, P,$$

where $k_m(\cdot, \cdot)$ measures the similarity between samples for pathway m. We applied the linear activation function to the weighted sum of the input values for each hidden unit. Then we converted this weighted sum to the output by a linear function such that the output value consists of the transformed weighted sum from the previous layer. The linear activation function applied to the weighted sum of the hidden layer produces the output of the hidden layer as follows:

$$\sum_{m=1}^{P} \eta_m g_{i,m} + \eta_0,$$

where $\eta_1, \ldots, \eta_P$ are the connection weights of the hidden layer and the output layer for every hidden unit. There is also a bias node in the hidden layer, which is indicated by g_0, where η_0 is the bias weight.

Note that in our MLP network, there is one output node, which is sufficient for two classes. We applied a nonlinear sigmoid activation function to the input values of the output neuron for binary classification. The output layer is fully connected. We enforced the Lasso regularization on top of all the connection weights between hidden layer nodes and the output layer node. Using the Lasso regularization leads to sparsity in the kernel weights. Therefore, we mostly obtained zero weights due to ℓ_1-norm and reduced overfitting risk by eliminating irrelevant kernels. Since negative kernel weights are not meaningful, we also imposed non-negativity constraints on the connection weights of the last layer to improve the interpretability of our MLP model. We obtained the output of the applied MLP model, which is the prediction value for primary tumors, by applying the sigmoid activation function in the last layer:

$$\hat{y}_i = \text{sigmoid}\left(\sum_{m=1}^{P} \eta_m g_{i,m} + \eta_0\right),$$

where $\hat{y}_i$ can be written as follows:

$$\hat{y}_i = \frac{1}{1 + \exp\left(-\left[\sum_{m=1}^{P} \eta_m g_{i,m} + \eta_0\right]\right)}.$$

The objective of MLPs is to minimize the loss function, which penalizes the difference between the desired output and the predicted output. We used binary cross-entropy loss function to find model parameters, which is the default loss function for binary classification problems and defined as

$$\mathcal{L}(\boldsymbol{y}, \hat{\boldsymbol{y}}) = -\sum_{i=1}^{N} [y_i \log(\hat{y}_i) + (1 - y_i) \log(1 - \hat{y}_i)],$$

where the $\boldsymbol{y}$ is the vector of actual class labels, and $\hat{\boldsymbol{y}}$ is the predicted values. The back-propagation algorithm uses the chain rule to calculate the gradient of connection weights from hidden to the output layer. The update equation for η_m can be written as

$$\Delta\eta_m = \mu \sum_{i=1}^{N} (y_i - \hat{y}_i) g_{i,m} \qquad \forall m = 1, \ldots, P,$$

where μ is the learning rate, which represents the step size to update the parameters. The update equations for the weights between the input layer and hidden layer can be formulated as

$$\Delta w_j = \mu \sum_{i=1}^{N} \sum_{m=1}^{P} (y_i - \hat{y}_i) g_{i,m} \boldsymbol{x}_{i,m} \qquad \forall j = 1, \ldots, N.$$

Since we implemented our MLP network in TensorFlow [19], the entire back-propagation technique was performed automatically.

4 Numerical Study

4.1 Experimental Settings

We divided the data points into two sets by randomly selecting 80% as the training set and the remaining 20% as the test set. We kept the same proportion of the early- and late-stages in training and test sets. The four-fold inner cross-validation method picks the hyper-parameters of our MLP model. The gene expression profiles of primary tumors are count data and can take only

Table 2. Parameters of our MLP network

Parameters	Value	Status
Number of input layers	P	Fixed
Number of input neurons in each input layer	N	Fixed
Number of hidden layers	1	Fixed
Number of hidden neurons	P	Fixed
Number of output neurons	1	Fixed
Activation function (hidden layer)	Linear	Fixed
Activation function (output layer)	Sigmoid	Fixed
Loss function	Binary cross-entropy	Free
Optimizer	Adam	Free
ℓ_1 regularization parameter	0.01	Free
Number of epochs	10,000	Free
Batch size	(25, 50, 75, 100)	Free

non-negative values, so we applied $\log_2$-transformation on them before building classifiers. Each feature in the training dataset was standardized to have a zero mean and a standard deviation of one. We normalized each feature in the test set by the mean and standard deviation parameters calculated over the original training set. We repeated the whole cross-validation procedure with stratification 100 times to measure the reliability of machine learning model's performance.

We constructed our multi-input MLP network in R using the TensorFlow framework, which is open-source software for deep learning. We used Keras high-level library that can be used on top of TensorFlow, and it is appropriate for conducting complex networks such as multi-input models and networks with a shared layer [20]. We used random initialization to break symmetry and provide better accuracy by employing Adam optimizer as our optimization algorithm with a learning rate of 0.001. Parameters of our MLP network are listed in Table 2. Some hyper-parameters are predefined, and we did not change them during the experiment. The batch size was selected from the set $\{25, 50, 75, 100\}$ for each replication using our four-fold inner cross-validation. We used ℓ_1-norm regularizer on kernel weights to reduce the complexity of the network and avoid overfitting. To guarantee the convergence of our MLP model, we used 10,000 epochs, which corresponds to the number of passes over the training dataset.

4.2 MLP Identifies Meaningful Biological Mechanisms

To show the biological relevance of our MLP algorithm, we checked the pathways chosen for each dataset. This process helped us analyze our MLP algorithm's capability to determine related gene sets based on weights of input kernels calculated during training. We counted the number of replications for each cohort and gene set pair in which the kernel weight is nonzero. Figures 2 and 3 show the selection frequencies of 50 gene sets using Hallmark collection and the top 50 pathways in the PID collection, respectively, over 100 replications. The column sums in Figs. 2 and 3 indicate that, in `BRCA`, `COAD`, `LIHC`, and `LUSC` cohorts, our

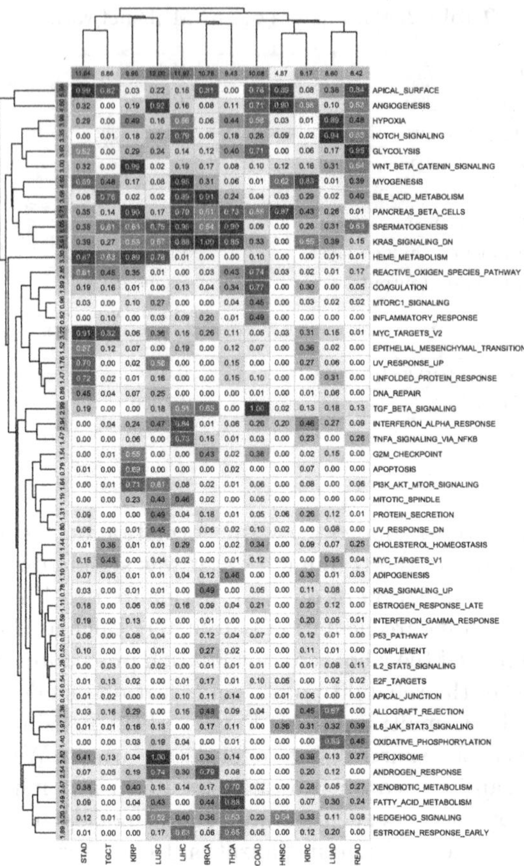

Fig. 2. The selection frequencies of 50 gene sets in the Hallmark collection for 12 datasets. We used a hierarchical clustering algorithm with Euclidean distance and full linkage functions to cluster the rows and columns of the selection frequencies matrix. The summation of each row indicates the selection frequencies of gene sets over the cohorts. The summation of each column indicates the selection frequencies of cohorts over the gene sets.

MLP used more than 10 gene sets of Hallmark and PID collections on average, whereas, in TGCT cohort, MLP used less than 7 gene sets on average to separate the cancer stages from each other. Therefore, this shows us separating early- and late-stage cancers is inherently more difficult on some cohorts.

To determine informative (uninformative) pathways for cancer stage prediction, we looked at the row sums of the selection frequency matrix. We obtained supporting evidence for the selected pathways with high (low) frequencies by matching them with the categories which are reported to be relevant (irrelevant) for cancer progression in PubMed repository at https://www.ncbi.nlm.nih.gov/pubmed. These results are summarized in Table 3.

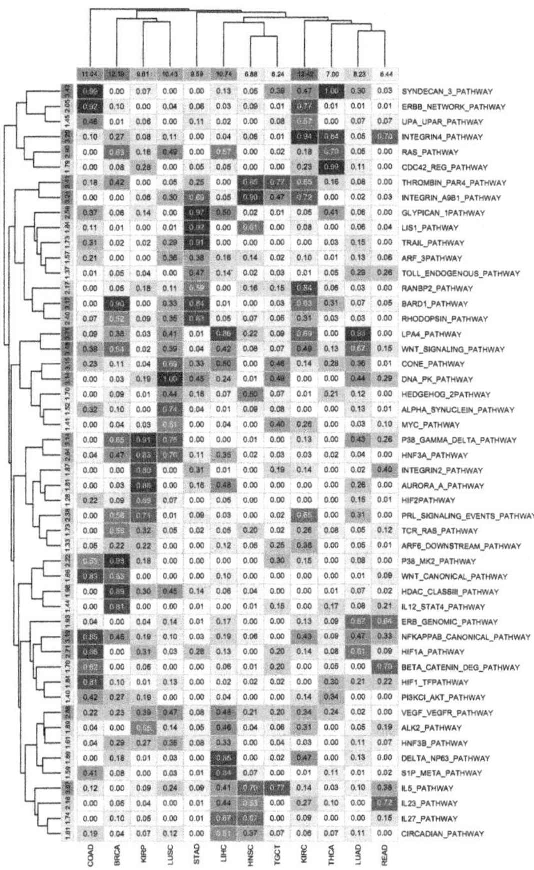

Fig. 3. The selection frequencies of the top 50 pathways in the PID collection for 12 datasets. We used a hierarchical clustering algorithm with Euclidean distance and full linkage functions to cluster the rows and columns of the selection frequencies matrix. The summation of each row indicates the selection frequencies of gene sets over the cohorts. The summation of each column indicates the selection frequencies of cohorts over the gene sets.

Some pathways are known to be relevant to cancer formation. For instance, `ANGIOGENESIS`, `KRAS_SIGNALING_DN`, `MYOGENESIS`, and `SPERMATOGENESIS` are gene sets related to cancer formation, and they were selected with high frequencies by our algorithm (see Fig. 2). In PID collection `VEGF_VEGFR_PATHWAY`, `IL5_PATHWAY`, and `GLYPICAN_1PATHWAY` correspond to cancer formation, and MLP selected them with high frequencies (see Fig. 3).

Each gene set carries information about a particular biological process. In the Hallmark collection, some gene sets are related to metabolisms, such as `BILE_ACID_METABOLISM` and `HEME_METABOLISM` in the final classifier; our MLP has selected these gene sets with high frequency in all cohorts on average. Equiv-

Table 3. Literature-based validation of the Hallmark and PID gene sets/pathways

Gene set/pathway name	Category	Article reference
KRAS_SIGNALING_DN	Cancer formation in early stage	[21]
SPERMATOGENESIS	Cancer formation in early stage	[21]
ANGIOGENESIS	Cancer formation in early stage	[21]
MYOGENESIS	Cancer formation in early stage	[21]
BILE_ACID_METABOLISM	Metabolism	[22]
HEME_METABOLISM	Metabolism	[18]
APICAL_SURFACE	Epithelial cell	[23]
HYPOXIA	Epithelial cell	[24]
DNA_REPAIR	Cell cycle	[25]
E2F_TARGETS	Cell cycle	[26]
MYC_TARGETS_V1	Cell cycle	[27]
MTORC1_SIGNALING	Cell cycle	[28]
VEGF_VEGFR_PATHWAY	Cancer formation in early stage	[21]
GLYPICAN_1PATHWAY	Cancer formation in early stage	[29]
IL5_PATHWAY	Cancer formation in early stage	[30]
NFKAPPAB_CANONICAL_PATHWAY	Cancer differentiation and development	[31]
INTEGRIN_A9B1_PATHWAY	Cancer differentiation and development	[32]
RAS_PATHWAY	Metabolism	[33]
DNA_PK_PATHWAY	Metabolism	[34]
WNT_SIGNALING_PATHWAY	Metabolism	[35]
HIF1A_PATHWAY	Metabolism	[36]
VEGF_VEGFR_PATHWAY	Epithelial cell	[37]
INTEGRIN4_PATHWAY	Epithelial cell	[38]
FOXM1_PATHWAY	Cell cycle	[39]
PLK1_PATHWAY	Cell cycle	[40]

alently, DNA_PK, WNT_SIGNALING, RAS, and HIF1A pathways in PID collection, carry information about the metabolism-related process, and all of them were used frequently in different datasets. The most important role of some organs in the body is keeping the metabolism of the body under control, such as liver and thyroid gland organs and their corresponding diseases are LIHC and THCA, respectively. We noticed that, for these diseases, our MLP selected the metabolism-related pathways with higher frequencies.

Epithelial cells are widespread in many organs such as kidney, breast, colon, lung, and stomach, and their corresponding disease abbreviations are KIRC, BRCA, COAD, LUAD, and STAD. On Hallmark collection, there are diverse gene sets that have an important effect on epithelial cells functions, such as APICAL_SURFACE and HYPOXIA. Figure 2 shows that, for most of the disease associated with epithelial cells, our MLP selected the APICAL_SURFACE and HYPOXIA gene sets with higher frequencies. On PID collection, some pathways correspond to epithelial cells, such as VEGF_VEGFR_PATHWAY and INTEGRIN4_PATHWAY. Figure 3 shows that these pathways are picked regularly by our MLP for diseases with epithelial cell-related tissues.

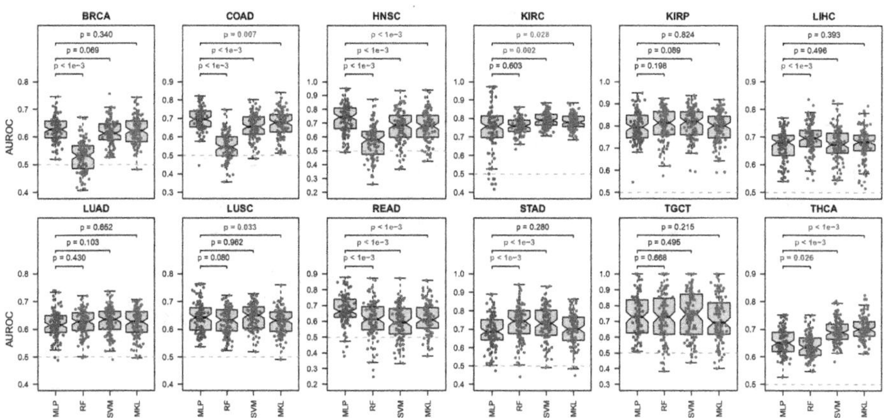

Fig. 4. Comparison of predictive performances of RF, SVM, and MKL with our MLP on 12 datasets constructed from TCGA cohorts on the Hallmark gene set. P-values determine whether there is a significant difference between the results of RF, SVM, and MKL algorithms, with MLP algorithm separately. P-value results are shown in different colors, and each of these colors indicates that corresponded algorithm has significantly better performance. Also, the black color means that there is no significant difference among the compared algorithms. (Color figure online)

Cell cycle-related pathways control the growth and division of the cell and are not related to cancer stage discrimination. We observed that on Hallmark collection and in different cohorts MLP picked cell cycle gene sets, namely, `DNA_REPAIR`, `E2F_TARGETS`, `MYC_TARGETS_V1`, and `MTORC1_SIGNALING` with lower frequencies. Similarly, most of the cell cycle-related pathways, such as `FOXM1` and `PLK1` are not selected by MLP among the top 50 pathways with higher average selection frequencies.

4.3 Predictive Performance Comparison

We compared our MLP algorithm with three other machine learning algorithms in our computational experiment in terms of their predictive performance i) random forest (RF), (ii) support vector machine (SVM), and (iii) multiple kernel learning (MKL). We considered RFs algorithm because they have been commonly used in phenotype classification problems in the bioinformatics field. Furthermore, SVMs used as a competitor method since they can capture more complex relationships. MKL algorithm integrates pathways into the model by computing kernels over them. We used the same cross-validation scheme (described in Sect. 4.1) to select hyper-parameters. For RF algorithm, we used `randomForestSRC` R package version 2.5.1 [41]. We selected the number of trees parameter (`ntree`) from the set $\{500, 1000, \ldots, 2500\}$ using the aforementioned four-fold inner cross-validation. For SVM and MKL on gene sets algorithms, we used the implementation proposed by [7] in R with Gaussian kernel. We used

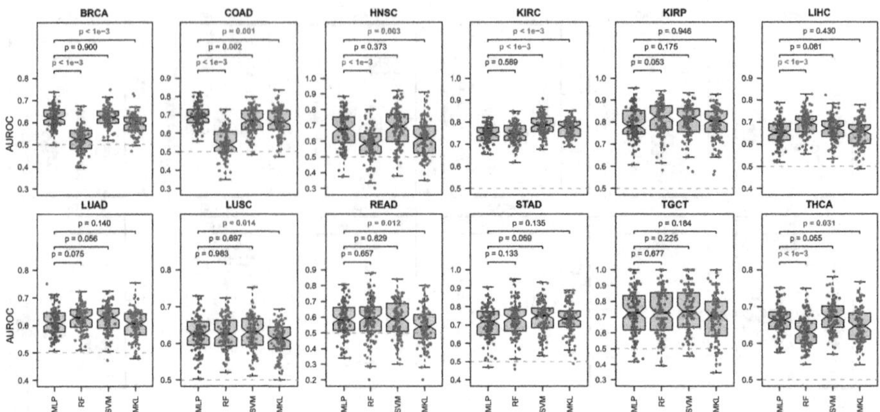

Fig. 5. Comparison of predictive performances of RF, SVM, and MKL with our MLP on 12 datasets constructed from TCGA cohorts on the PID collection. *P*-values determine whether there is a significant difference between the results of RF, SVM, and MKL algorithms, with MLP algorithm separately. *P*-value results are shown in different colors, and each of these colors indicates that corresponded algorithm has significantly better performance. Also, the black color means that there is no significant difference among the compared algorithms. (Color figure online)

the area under the receiver operating characteristic curve (AUROC) to measure the predictive performance of four machine learning algorithms.

On 12 datasets we created from TCGA cohorts to predict the stages of tumors, we measured and compared the predictive performance of MLP, RF, SVM, and MKL on the Hallmark and PID collections. The predictive performance comparisons of four algorithms on the Hallmark and PID collections have been shown in Fig. 4 and 5 for all cohorts. The baseline performance is shown as a dashed line, and we observed in both Figs. 4 and 5 that the median performance of all four classifiers is superior to the baseline performance. This is a significant reason to conclude that meaningful and valuable information can be extracted from gene expression profiles of primary tumors.

By looking at Fig. 4, we observed that MLP significantly outperformed all other algorithms in three cohorts (i.e. `COAD`, `HNSC`, and `READ`), whereas there is not a single cohort that all other three algorithms outperformed MLP. Moreover, MLP significantly outperformed RF on 5 out of 12 datasets (i.e., `BRCA`, `COAD`, `HNSC`, `READ`, and `THCA`), in contrast RF outperformed MLP on 2 cohorts, (i.e., `LIHC` and `STAD`). By comparing MLP and SVM, we found that each algorithm outperformed the other in 3 datasets. Similarly, MLP significantly outperformed MKL in 4 datasets, whereas MKL outperformed MLP in 2 cohorts. Since MLP used a smaller subset of features, we can conclude that our algorithm outperformed all three algorithms in most of the datasets on the Hallmark gene set.

The predictive performance of four algorithms on the PID collection in Fig. 5 shows that each competitor algorithm outperformed MLP in only 1 out of 12 datasets (RF in LIHC, SVM and MKL in KIRC). In contrast, MLP significantly outperformed RF, SVM, and MKL in 4, 1, and 6 datasets, respectively. Moreover, MLP significantly outperformed all other algorithms in COAD cohort. As a conclusion, MLP was able to extract biologically meaningful result with a limited number of pathways, while showed better or comparable predictive performance.

5 Conclusions

Understanding specific molecular mechanisms that cause cancer progression from early-stage to late-stage is crucial in therapeutic strategies to prevent tumor growth. In this study, we developed a kernel-based neural network model to discriminate the stages of cancers from each other based on their gene expression profiles and pathways. The major contribution of our empirical approach is implementing cancer stage classification and pathway selection conjointly by training an ℓ_1-regularized MLP. We improved the performance of classifiers by removing noisy and irrelevant gene sets during training. Therefore, the robustness and the predictive performance of the classifiers were increased by eliminating redundant pathways.

We tested our proposed MLP network (Fig. 1) on 12 different cancer datasets constructed from TCGA, which are represented in Table 1, with Hallmark gene set and PID pathway collections. Then, we found literature validation for the frequently selected pathways by our MLP model. We also observed that our MLP algorithm chose irrelevant gene sets for separating the cancer stages with relatively low frequencies. Thus, we concluded that our framework efficiently restricted the search area and extracted the biologically relevant pathways to cancer progression.

We then compared the empirical results of the MLP network on gene sets against three widely used machine learning models, namely, RF, SVM, and MKL. Our MLP model significantly outperformed RF and MKL and obtained better or comparable predictive performance values than SVM, in terms of AUROC, using fewer gene expression features.

A possible future direction that would increase the accuracy of the classifier is modeling different types of cancer that have similar mechanisms together, which would increase the sample size for training, leading to more robust classifiers (i.e., multitask learning). As another future direction, we can address the cancer staging problem as a multi-class classification using an MLP algorithm to classify the pathological stages of cancers.

References

1. Kaur, H., Bhalla, S., Raghava, G.P.: Classification of early and late stage liver hepatocellular carcinoma patients from their genomics and epigenomics profiles. PLoS ONE **14**(9), e0221476 (2019)

2. Mokhtaridoost, M., Gönen, M.: Identifying key miRNA–mRNA regulatory modules in cancer using sparse multivariate factor regression. In: Nicosia, G., et al. (eds.) LOD 2020. LNCS, vol. 12565, pp. 422–433. Springer, Cham (2020). https://doi.org/10.1007/978-3-030-64583-0_38
3. Broët, P., et al.: Identifying gene expression changes in breast cancer that distinguish early and late relapse among uncured patients. Bioinformatics **22**(12), 1477–1485 (2006)
4. Bhalla, S., et al.: Expression based biomarkers and models to classify early and late-stage samples of Papillary Thyroid Carcinoma. PLoS ONE **15**(4), e0231629 (2020)
5. Bhalla, S., et al.: Gene expression-based biomarkers for discriminating early and late stage of clear cell renal cancer. Sci. Rep. **7**(1), 1–13 (2017)
6. Ein-Dor, L., Zuk, O., Domany, E.: Thousands of samples are needed to generate a robust gene list for predicting outcome in cancer. Proc. Natl. Acad. Sci. U.S.A. **103**(15), 5923–5928 (2006)
7. Rahimi, A., Gönen, M.: Discriminating early-and late-stage cancers using multiple kernel learning on gene sets. Bioinformatics **34**(13), i412–i421 (2018)
8. Rahimi, A., Gönen, M.: A multitask multiple kernel learning formulation for discriminating early-and late-stage cancers. Bioinformatics (2020)
9. Kumar, R., et al.: Gene expression-based supervised classification models for discriminating early-and late-stage prostate cancer. Proc. Natl. Acad. Sci. India Sect. B Biol. Sci., 1–2 (2019)
10. Ma, B., et al.: Diagnostic classification of cancers using extreme gradient boosting algorithm and multi-omics data. Comput. Biol. Med. **121**, 103761 (2020)
11. Saabith, A.L.S., Sundararajan, E., Bakar, A.A.: Comparative study on different classification techniques for breast cancer dataset. Int. J. Comput. Sci. Mob. Comput. **3**(10), 185–191 (2014)
12. Seo, H., Cho, D.-H.: Cancer-related gene signature selection based on boosted regression for multilayer perceptron. IEEE Access **8**, 64 992–65 004 (2020)
13. Pati, J.: Gene expression analysis for early lung cancer prediction using machine learning techniques: an eco-genomics approach. IEEE Access **7**, 4232–4238 (2018)
14. Chaubey, V., Nair, M.S., Pillai, G.: Gene expression prediction using a deep 1D convolution neural network. In: 2019 IEEE Symposium Series on Computational Intelligence (SSCI), pp. 1383–1389. IEEE (2019)
15. Xie, R., et al.: A predictive model of gene expression using a deep learning framework. In: 2016 IEEE International Conference on Bioinformatics and Biomedicine (BIBM), pp. 676–681. IEEE (2016)
16. Mojarad, S.A., et al.: Breast cancer prediction and cross validation using multilayer perceptron neural networks. In: Communication Systems, Networks and Digital Signal Processing (CSNDSP), pp. 760–764. IEEE (2010)
17. Schaefer, C.F., et al.: PID: the pathway interaction database. Nucleic Acids Res. **37**(suppl_1), D674–D679 (2009)
18. Liberzon, A., et al.: The molecular signatures database hallmark gene set collection. Cell Syst. **1**(6), 417–425 (2015)
19. Abadi, M., et al.: TensorFlow: large-scale machine learning on heterogeneous systems (2015). http://tensorflow.org/
20. Chollet, F., et al.: Keras (2015). https://keras.io
21. Bergers, G., Benjamin, L.E.: Tumorigenesis and the angiogenic switch. Nat. Rev. Cancer **3**(6), 401–410 (2003)
22. Chiang, J.Y.L.: Bile acid metabolism and signaling. Compr. Physiol. **3**(3), 1191–1212 (2013)

23. Kotha, P.L., et al.: Adenovirus entry from the apical surface of polarized epithelia is facilitated by the host innate immune response. PLoS Pathog. **11**(3), e1004696 (2015)
24. Zeitouni, N.E., Chotikatum, S., von Köckritz-Blickwede, M., Naim, H.Y.: The impact of hypoxia on intestinal epithelial cell functions: consequences for invasion by bacterial pathogens. Mol. Cellular Pediatr. **3**(1), 1–9 (2016). https://doi.org/10.1186/s40348-016-0041-y
25. Branzei, D., Foiani, M.: Regulation of DNA repair throughout the cell cycle. Nat. Rev. Mol. Cell Biol. **9**(4), 297–308 (2008)
26. Blais, A., Dynlacht, B.D.: E2F-associated chromatin modifiers and cell cycle control. Curr. Opin. Cell Biol. **19**(6), 658–662 (2007)
27. Zhao, L., et al.: Identification of a novel cell cycle-related gene signature predicting survival in patients with gastric cancer. J. Cell. Physiol. **234**(5), 6350–6360 (2019)
28. Palaniappan, M., Menon, B., Menon, K.: Stimulatory effect of insulin on theca-interstitial cell proliferation and cell cycle regulatory proteins through MTORC1 dependent pathway. Mol. Cell. Endocrinol. **366**(1), 81–89 (2013)
29. Melo, S.A., et al.: Glypican1 identifies cancer exosomes and facilitates early detection of cancer. Nature **523**(7559), 177 (2015)
30. Kandikattu, H.K., Venkateshaiah, S.U., Mishra, A.: Synergy of interleukin (IL)-5 and IL-18 in eosinophil mediated pathogenesis of allergic diseases. Cytokine Growth Factor Rev. **47**, 83–98 (2019)
31. Xia, L., et al.: Role of the NFκB-signaling pathway in cancer. OncoTargets Ther. **11**, 2063 (2018)
32. Gupta, S.K., et al.: Integrin $\alpha9\beta1$ promotes malignant tumor growth and metastasis by potentiating epithelial-mesenchymal transition. Oncogene **32**(2), 141–150 (2013)
33. Dard, L., et al.: RAS signalling in energy metabolism and rare human diseases. Biochim. Biophys. Acta, Bioenerg. **1859**(9), 845–867 (2018)
34. Chung, J.H.: The role of DNA-PK in aging and energy metabolism. FEBS J. **285**(11), 1959–1972 (2018)
35. Mo, Y., et al.: The role of WNT signaling pathway in tumor metabolic reprogramming. J. Cancer **10**(16), 3789 (2019)
36. Xia, Y., Jiang, L., Zhong, T.: The role of HIF-1α in chemo-/radioresistant tumors. OncoTargets Ther. **11**, 3003 (2018)
37. Wild, J.R., et al.: Neuropilins: expression and roles in the epithelium. Int. J. Exp. Pathol. **93**(2), 81–103 (2012)
38. Joly, D., et al.: $\beta4$ integrin and laminin 5 are aberrantly expressed in polycystic kidney disease: role in increased cell adhesion and migration. Am. J. Pathol. **163**(5), 1791–1800 (2003)
39. Bai, H., et al.: Integrated genomic characterization of idh1-mutant glioma malignant progression. NAT **48**(1), 59–66 (2016)
40. Chen, C., et al.: Analysis of the expression of cell division cycle-associated genes and its prognostic significance in human lung carcinoma: a review of the literature databases. Biomed Res. Int. **2020** (2020)
41. Ishwaran, H., Kogalur, U.: Random forests for survival, regression, and classification (RF-SRC), R package version 2.5. 1 (2017)

Loss Function with Memory for Trustworthiness Threshold Learning: Case of Face and Facial Expression Recognition

Stanislav Selitskiy[1(✉)] and Natalya Selitskaya[2]

[1] School of Computer Science and Technology, University of Bedfordshire, Park Square, Luton LU1 3JU, UK
stanislav.selitskiy@study.beds.ac.uk
[2] IEEE Member, Georgia, USA
https://www.beds.ac.uk/computing

Abstract. We compare accuracy metrics of the supervisor meta-learning artificial neural networks (ANN) that learn the trustworthiness of the Inception v.3 convolutional neural networks (CNN) ensemble prediction *a priori* of the "ground truth" verification on the face and facial expression recognition. One of the compared meta-learning ANN modes uses a simple majority of the ensemble votes and its predictions. In contrast, another uses dynamically learned "trusted" ensemble vote count and its *a priori* prediction to decide on the trustworthiness of the underlying CNN ensemble prediction. A custom loss function with memory is introduced to collect trustworthiness predictions and their errors during training. Based on the collected statistics, learning gradients for the "trusted" ensemble vote count parameter is calculated, and the "trusted" ensemble vote count threshold is dynamically determined. A facial data set with makeup and occlusion is used for computational experiments in the partition that ensures high out of the training data distribution conditions, where only non-makeup and non-occluded images are used for CNN model ensemble training, while the test set contains only makeup and occluded images.

Keywords: Meta-learning · Statistical loss function · Trustworthiness · Uncertainty calibration · Explainability · Continuous learning

1 Introduction

The unquestionable progress of the Deep Learning (DL) trained on the Big Data (BD) in many practical tasks, such as Natural Language Processing, Image and Speech and other Pattern Recognition, Generative Modeling, Conversational Bots and other Agent Systems, have inspired not only admiration and fury of development activity [8,13,43], but also sparked discussions on limitations of such an approach and feasibility of the alternatives. Fusion of the really big models with

G. Nicosia et al. (Eds.): LOD 2022, LNCS 13810, pp. 78–92, 2023.
https://doi.org/10.1007/978-3-031-25599-1_7

the really big and diverse data, led even to the replacement of the "old-school" DL and BD terminology with a new concept of Foundation Models (FM) [14], which could encompass the representation of a big chunk of the world within, and then be utilized and tuned to achieve narrower specific goals [2,17,53].

However, such an approach of the brute force unstructured use of all available data to train huge models raises the question of whether the belief that FM "is all you need" and their ubiquitous dissemination via cloud services make an estimation of their uncertainty obsolete because they count in "everything". Despite all possible resource use, which also becomes an issue raised by the "Green AI" discussion [33,35,55], these models have questionable performance in general or adversarial settings [47,56], and can be not only not beneficial enough, but also harmful [6] due to ethical, and other built-in biases due to lack of the real-world understanding [3,7,37].

Of course, such a large and complicated task of providing quantitative estimates of the uncertainty of the ML algorithm predictions, based on that, assessing the trustworthiness of the predictions, and giving explanations to the former assessments, and doing that for a variety of applications, is a huge task far surpassing limits of this work. Here we address a limited number of aspects in a limited scope. We ask such questions as to whether an external Artificial Neural Network (ANN) can learn uncertainty in the operation of another ANN? Can such information be used to produce the trustworthiness of the prediction status? Can we explain why the trusted or non-trusted decision was made? We limit applications in our research to Face Recognition (FR) and Facial Expression Recognition (FER) tasks. Out of the number of Convolutional Neural Network (CNN) models used for the mentioned tasks, we present results for the best performing model and the data set we see best fitting our goals.

The paper is organized as follows. Section 2 briefly introduces ML and uncertainty concepts used in developing the proposed solution and reasons related to choosing FR and FER application and experimental data set. Section 4 describes the data set used for experiments; Sect. 5 outlines experimental algorithms in detail; Sect. 3 presents the obtained results, and Sect. 7 discusses the results, draws practical conclusions, and states directions of the research of not yet answered questions.

2 Existing Work

This contribution is a continuation of the authors' research in building (via meta-learning) elements of the self-awareness about their uncertainty into CNN models applied to the FR and FER tasks in conditions of the extreme Out of (training data) Distribution (OOD) caused by the makeup and occlusions [50,51]. A more detailed review of the existing research on meta-learning, FR and FER in conditions of OOD with the use of makeup and occlusions and other types of attacks, probabilistic ML frameworks for uncertainty estimation can be found in the mentioned above works. Here, due to lack of space, we briefly highlight key points giving necessary context to the novel part of the research.

2.1 Meta-learning

The idea of learning the ML processes was introduced in the 90's [57], and recently gained traction in various flavours of meta-learning [58]. There exist multiple flavours of "learning to learn" or meta-learning targeting narrower and specific tasks such as either as an extension of the transfer learning [1,15,41], or model hyper-parameter optimization [4,42], or a wider horizon "learning about learning" approach conjoint to the explainable learning [28,31].

We propose to use a broader view of meta-learning. An underlying ANN is treated as a "natural" process, the behaviour of which (including uncertainty patterns) is learned with another supervisor ANN.

2.2 Uncertainty

Using the Bayesian learning rule with ANNs, which allows capturing the latter uncertainty, was introduced in the mid-1990s [34,40] and remains a popular probabilistic framework for uncertainty representation.

The first term of the model's uncertainty can be viewed as responsible for so-called *aleatoric* uncertainty, or uncertainty related to handling data by the model. In contrast, the second term - *epistemic* uncertainty or uncertainty related to the model stability [25].

While Bayesian formulas are straightforward, in practice, their analytical solution is unrealistic due to the multi-dimensionality, multi-modality, and non-linearity of the activation functions. Therefore, practical solutions for Bayesian Neural Networks (BNN) include approximation via various sampling methods of either *aleatoric* or *epistemic* uncertainties or both, such as Variational Inference [18], Markov Chain Monte Carlo [40], Monte Carlo Dropout [16], Deep Ensembles [11,29].

Bayesian concepts of uncertainty estimation have been efficiently used to evaluate brain development [21,23], trauma severity [22,44,45], and collision avoidance [46].

2.3 Face and Facial Expression Recognition in Out-of-Distribution Settings

Using makeup and occlusions in FR experiments, when non-makeup only photo-sessions compound the training set and makeup and occlusion sessions are used for testing, such a data set is suited very well for benchmarking uncertainty estimation for the real-life conditions of OOD when the training data are not well-representing test data.

Research in the makeup and occlusion influence on the face recognition ML algorithms, as well as a collection of the benchmark data sets, has been conducted since early face recognition algorithms development [38]. Before CNN architectures made ANN, especially in DL architectures, feasible on the commonly available hardware, the engineered visual features algorithms were popular and achieved decent accuracy.

Experiments with more successful CNN models have also shown that such architectures are still vulnerable to makeup and occlusion additions to test sets. Their accuracy significantly drops in such cases. It was even noted that the targeted spoofing of another person with means of makeup [10,20].

Adjacent research areas in the presentation attack detection [24] and deep fake detection [26,36,60] have parallels to the actual face variation detection and is an active area of research using modern ANN architectures.

Facial Expression Recognition (FER) significantly aggravate OOD conditions, especially if the task is accompanied by makeup and occlusions. The earlier works on facial and emotion expression recognition used the engineered feature methods [5,52,59]. CNN were later employed for solving the facial expression recognition problem [27,30,32,39]. Both engineered features algorithms and head-on CNN models are poorly generalized on the unseen in the training set example.

The idea that the whole spectre of emotion expressions can be reduced to six basic facial feature complexes [12] was challenged in the sense that human emotion recognition is context-based. The same facial feature complexes may be interpreted differently depending on the situational context [9,19].

Therefore we have chosen FR and FER tasks as a test-cases for our trustworthiness estimation experiments.

3 Proposed Solution

In [51], two approaches to assigning a trustworthiness flag to the FR prediction were proposed: statistical analysis of the distributions of the maximal softmax activation value for correct and wrong verdicts and use of the meta-learning supervisor ANN (SNN) that uses the whole set of softmax activations for all FR classes (sorted into the "uncertainty shape descriptor" (USD) to provide class-invariant generalization) as an input and generates trusted or not-trusted flag.

This contribution "marries" these two approaches by collecting statistical information about training results in the memory of the loss layer (LL) of the meta-learning supervisor ANN. The information in the LL's memory holds prediction result y_t, training label result l_t, and trustworthiness threshold TT. The latter parameter is the learnable one, and the derivative of the loss error, calculated from these statistical data, is used to auto-configure the TT to optimize the sum of square errors loss: $SSE_{TT} = \sum_{t=1}^{K} SE_t$, where K is a number of entries in the memory table:

$$SE_{TTt} = \begin{cases} (y_t - TT)^2, & (l_t < TT \wedge y_t > TT) \vee (l_t > TT \wedge y_t < TT) \\ 0, & (l_t > TT \wedge y_t > TT) \vee (l_t < TT \wedge y_t < TT) \end{cases} \quad (1)$$

The input of the meta-learning supervisor ANN ("uncertainty shape descriptor" (USD)) was built from the softmax activations of the ensemble of the underlying CNN models. The USD is built by sorting softmax activations inside each model vector, ordering model vectors by the highest softmax activation, flattening the list of vectors, and rearranging the order of activations in each vector to the order of activations in the vector with the highest softmax activation. The detailed algorithm description can be found in [51].

Examples of the descriptor for the $M = 7$ CNN models in the underlying FR or FER ensemble, for the cases when none of the models, 4, and 6 detected the face correctly, are presented in Fig. 2. It could be seen that shapes of the distribution of the softmax activations are quite distinct and, therefore, can be subject to the pattern recognition task performed by the meta-learning SNN.

However, unlike in the mentioned above publication, for simplification reasons, supervisor ANN was not categorizing the predicted number of the correct members of the underlying ensemble but instead was performing the regression task of the transformation. On the high level (ANN layer details are given in Sect. 5), the transformation can be seen as Eq. 2, where $n = |C| * M$ is the dimensionality of the $\forall USD \in \mathcal{X}$, $|C|$ - cardinality of the set of FR or FER categories (subjects or emotions), and M - size of the CNN ensemble Fig. 1.

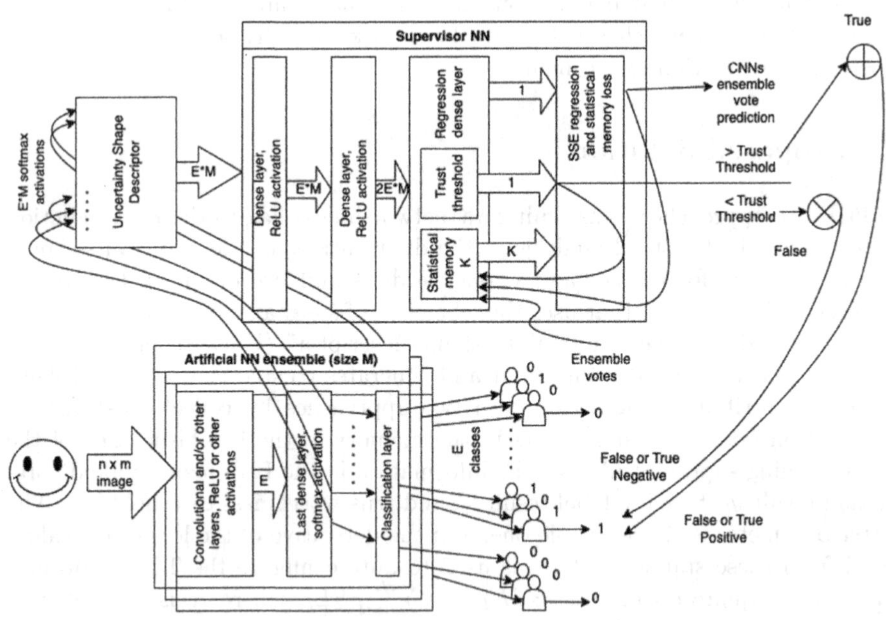

Fig. 1. Meta-learning Supervisor ANN over underlying CNN ensemble.

$$reg : \mathcal{X} \subset \mathbb{R}^n \mapsto \mathcal{Y} \subset \mathbb{R} \tag{2}$$

where $\forall \mathbf{x} \in \mathcal{X}, \mathbf{x} \in (0 \ldots 1)^n, \forall y \in \mathcal{Y}, E(y) \in [0 \ldots M]$.

The loss function used for y is the usual for regression tasks, sum of squared error: $SSE_y = \sum_{t=1}^{N_{mb}} (y_j - e_j)^2$, where e is the label (actual number of the members of CNN ensemble with correctly prediction), and N_{mb} - minbatch size.

From the trustworthiness categorization and ensemble vote point of view, the high-level transformation of the combined CNN ensemble together with the meta-learning supervisor ANN can be represented as Eq. 3:

$$cat : \mathcal{I} \subset \mathbb{I}^l \mapsto \mathcal{C} \subset \mathbb{C} \times \mathbb{B} \tag{3}$$

where $\mathbf{i}$ are images, l - mage size, c - classifications, and b - binary trustworthy flags, such as $\forall \mathbf{i} \in \mathcal{I}, \mathbf{i} \in (0 \dots 255)^l, \forall c \in \mathcal{C}, c \in \{c_1, \dots, c_{|C|}\}, \forall b \in \mathcal{B}, b \in \{1, 0\}$.

$$b_i = \begin{cases} 1, & (y_i > TT_t) \\ 0, & (y_i < TT_t) \end{cases} \tag{4}$$

where i is an index of the image at the moment t of the state of the loss function memory.

$$c_i = argmin(|y_i - e_i(c_i)|) \tag{5}$$

Equations above describe the ensemble vote that chooses category c_i, which received the closest number of votes e_i to the predicted regression number y_i.

4 Data Set

The BookClub artistic makeup data set contains images of $E = |C| = 21$ subjects. Each subject's data may contain a photo-session series of photos with no makeup, various makeup, and images with other obstacles for facial recognition, such as wigs, glasses, jewellery, face masks, or various headdresses. The data set features 37 photo sessions without makeup or occlusions, 40 makeup sessions, and 17 sessions with occlusions. Each photo session contains circa 168 JPEG images of the 1072×712 resolution of six basic emotional expressions (sadness, happiness, surprise, fear, anger, disgust), a neutral expression, and the closed eyes photoshoots taken with seven head rotations at three exposure times on the off-white background. The subjects' age varies from their twenties to sixties. The race of the subjects is predominately Caucasian and some Asian. Gender is approximately evenly split between sessions.

The photos were taken over two months, and several subjects were posed at multiple sessions over several weeks in various clothing with changed hairstyles, downloadable from https://data.mendeley.com/datasets/yfx9h649wz/3. All subjects gave consent to use their anonymous images in public scientific research.

For FR experiments, the data set was partitioned into the subject, makeup, and time-centred photo sessions. For FER experiments, the former sessions were additionally partitioned by the emotion expressions. Only images without makeup and occlusions were selected for training sub-sets, while for the test subset, only makeup and occluded sessions (11125 images) were used. Training subsets comprise two parts: one is used for CNN ensemble training (4508 images), and another - is for meta-learning supervisor ANN training (1653 images).

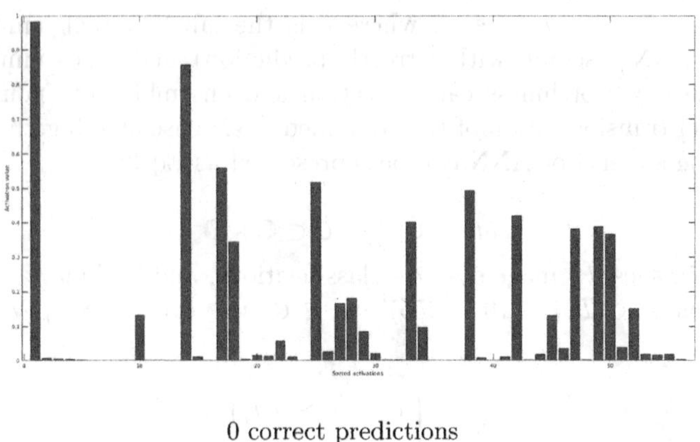

0 correct predictions

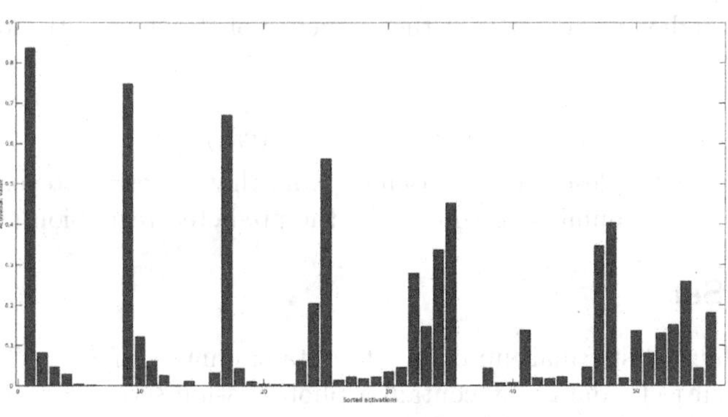

4 correct predictions

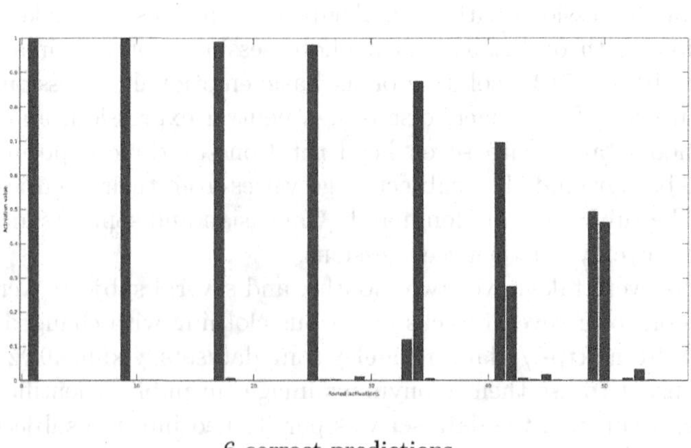

6 correct predictions

Fig. 2. Uncertainty shape descriptors for 0, 4, and 6 correct CNN ensemble FER predictions.

5 Experiments

The experiments were run on the Linux (Ubuntu 20.04.3 LTS) operating system with two dual Tesla K80 GPUs (with 2 × 12GB GDDR5 memory each) and one QuadroPro K6000 (with 12GB GDDR5 memory, as well), X299 chipset motherboard, 256 GB DDR4 RAM, and i9-10900X CPU. Experiments were run using MATLAB 2022a.

The experiments were done using MATLAB with Deep Learning Toolbox. For FR and FER experiments, the Inception v.3 CNN model was used. Out of the other State-of-the-Art (SOTA) models applied to FR and FER tasks on the BookClub data set (AlexNet, GoogLeNet, ResNet50, InceptionResnet v.2), Inception v.3 demonstrated overall the best result over such accuracy metrics as trusted accuracy, precision, and recall [50,51]. Therefore, the Inception v.3 model, which contains 315 elementary layers, was used as an underlying CNN. Its last two layers were resized to match the number of classes in the BookClub data set (21), and re-trained using "adam" learning algorithm with 0.001 initial learning coefficient, "piecewise" learning rate drop schedule with 5 iterations drop interval, and 0.9 drop coefficient, mini-batch size 128, and 10 epochs parameters to ensure at least 95% learning accuracy. The Inception v.3 CNN models were used as part of the ensemble with a number of models $N = 7$ trained in parallel.

Meta-learning supervisor ANN models were trained using the "adam" learning algorithm with 0.01 initial learning coefficient, mini-batch size 64, and 200 epochs. For online learning experiments, naturally, batch size was set to 1, as each consecutive prediction was used to update meta-learning model parameters. The memory buffer length, which collects statistics about previous training iterations, was set to $K = 8192$.

The *reg* meta-learning supervisor ANN transformation represented in the Eq. 2 implemented with two hidden layers with $n + 1$ and $2n + 1$ neurons in the first and second hidden layer, and ReLU activation function. All source code and detailed results are publicly available on GitHub (https://github.com/Selitskiy/StatLoss).

5.1 Trusted Accuracy Metrics

Although the accuracy metrics such as accuracy itself, precision, recall, and others, are defined unambiguously via true or false and negative or positive parameters, the very meaning of what they are may be subjective or, at least, depending on the task. For example, for an ANN trained to recognize image classes, the prediction with the highest softmax activation will be positive, true or false depending on the prediction correctness, but what would be negative? All other classes with lower softmax? Or only with 0 softmax? Or something in the middle? In our problem formulation, negative would be CNN's positive with SNN negative (non-trusted) flag, and the true or false qualifiers would depend on the prediction correctness.

Suppose only the classification verdict is used as a final result of the ANN model. In that case, the accuracy of the target CNN model can be calculated only as the ratio of the number of correctly identified test images by the CNN model to the number of all test images:

$$Accuracy = \frac{N_{correct}}{N_{all}} \tag{6}$$

When additional dimension in classification is used, for example amending verdict of the meta-learning supervisor ANN, (see Formula 3), and $cat(\mathbf{i}) = c \times b$, where $\forall \mathbf{i} \in \mathcal{I}$, $\forall c \times b \in \mathcal{C} \times \mathcal{B} = \{(c_1, b_1), \ldots (c_p, b_p)\}$, $\forall b \in \mathbb{B} = \{True, False\}$, then the trusted accuracy and other trusted quality metrics can be calculated as:

$$Accuracy_t = \frac{N_{correct:f=T} + N_{wrong:f \neq T}}{N_{all}} \tag{7}$$

As a mapping to a more usual notations, $N_{correct:f=T}$ can be as the True Positive (TP) number, $N_{wrong:f \neq T}$ - True Negative (TN), $N_{wrong:f=T}$ - False Positive (FP), and $N_{correct:f \neq T}$ - False Negative (FN).

Trusted precision, as a measure of the "pollution" of the true positive verdicts by the false-positive errors:

$$Precision_t = \frac{N_{correct:f=T}}{N_{correct:f=T} + N_{wrong:f=T}} \tag{8}$$

Trusted recall, as a measure of the true positive verdicts "loss" due to false-negative errors:

$$Recall_t = \frac{N_{correct:f=T}}{N_{correct:f=T} + N_{correct:f \neq T}} \tag{9}$$

Alternatively, trusted specificity, as a measure of the true-negative verdicts "loss" by false-positive errors, or, in the context of A/B testing, is equal to the confidence level percentage of the wrongly identified images recognized as such.

$$Specificity_t = \frac{N_{wrong:f \neq T}}{N_{wrong:f \neq T} + N_{wrong:f=T}} \tag{10}$$

where $N_{correct}$ and N_{wrong} number of correctly and incorrectly identified test images by the CNN model.

As well as the trusted F1 score, the harmonic mean of trusted Precision and Recall:

$$F1_t = 2\frac{Precision_t * Recall_t}{Precision_t + Recall_t} \tag{11}$$

To obtain statistical metrics, k-fold ($k = 5$) validation was used. Because mixing even a small amount of test data into training data, with makeup and occlusions, or even alternative non-makeup sessions leads to interruption of the OOD conditions [48,49], validation folds were created only by pair-wise exchange of the non-makeup sessions between CNN-training and SNN-training data set, with no makeup or occlusion test sessions swapped with training sessions. Paired samples Wilcoxon test [54] was used to compare distributions of the accuracy metrics between the original expert-selected trust threshold and the proposed learned trust threshold from the historically collected accuracy data during SNN training.

6 Results

Results of the FR experiments are presented in Table 1. The first column holds metrics calculated using Formulae 6–11 using ensemble maximum vote. The second column - using the ensemble vote closest to the meta-learning supervisor ANN prediction and trustworthiness threshold learned only on the test set, see Formulae 4, 5. The mean and standard deviation values are calculated using the 5-fold validation method for CNN ensembles and SNN models retrained on the folded training data sets. The next column presents the p-value of the paired samples Wilcoxon (signed-rank) test applied to the two previous distributions.

Table 1. Accuracy metrics (mean and standard deviation) for FR task. Maximal ensemble vote with static expert threshold, meta-learning predicted vote with learned threshold, Wilcoxon p-value of the null-hypothesis.

Metric	Maximal/Static	Predicted/Learned	Wilcoxon p-value
Untrusted accuracy	0.686389 ± 0.038693	0.636878 ± 0.055959	0.0625
Trusted accuracy	0.776869 ± 0.039128	0.819381 ± 0.014560	0.125
Trusted precision	0.862624 ± 0.040948	0.900145 ± 0.043690	0.0625
Trusted recall	0.799961 ± 0.044728	0.804489 ± 0.013531	1.0000
Trusted F1 score	0.829739 ± 0.038440	0.849174 ± 0.020737	0.1875
Trusted specificity	0.724704 ± 0.065074	0.849943 ± 0.041706	0.0625

Results of the FER experiments are presented in Table 2. The column content is the same as for FR experiments.

Except for the trusted recall for FR and trusted precision for FER, the null hypothesis that both distributions belong to the same population can be rejected, and therefore, the proposed SNN architecture offers a statistically significant improvement over the previous architecture. Depending on the tasks and goals of the FR/FER application: maximize or minimize false-positive or false-negative errors, or find an optimal balance between them, and what kind of the input (structured or unstructured) is expected, the stationary training time or online learning approach would be better for the meta-learning supervisor ANN.

Table 2. Accuracy metrics (mean and standard deviation) for FER task. Maximal ensemble vote with static expert threshold, meta-learning predicted vote with learned threshold, Wilcoxon p-value of the null-hypothesis.

Metric	Maximal/Static	Predicted/Learned	Wilcoxon p-value
Untrusted accuracy	0.388962 ± 0.008462	0.351335 ± 0.005896	0.0625
Trusted accuracy	0.686005 ± 0.006285	0.724297 ± 0.011069	0.0625
Trusted precision	0.608469 ± 0.021387	0.608983 ± 0.019473	0.8125
Trusted recall	0.544722 ± 0.025094	0.605274 ± 0.020582	0.0625
Trusted F1 score	0.574152 ± 0.008712	0.606691 ± 0.009127	0.0625
Trusted specificity	0.776247 ± 0.023406	0.788541 ± 0.024681	0.0625

7 Discussion, Conclusions, and Future Work

For the experimentation with CNN model ensemble based on Inception v.3 architecture and data set with significant Out of (training data) Distribution in the form of makeup and occlusions, using meta-learning supervisor ANN, noticeably increases selected accuracy metrics for FR tasks (by ten plus of per cent) and significantly (by tens of per cent) - for FER task. The proposed novel loss layer with memory architecture, which allows to learn the trustworthy threshold instead of a static expert setting, increases selected accuracy metrics by an additional (up to 5) percentage. The trustworthiness threshold learned with the loss layer with memory offers a simple explanation of why prediction for a given image was categorized as trusted or non-trusted.

As mentioned in the beginning, true or false positive or negative predictions, and metrics that use them, may have significance depending on the task. For example, in the legal setting, we do not want to allow false convictions (positives), but conversely, in the security setting, we do not want missed threats (false negatives). Also, in some settings, the "I do not know" answer (equivalent to the SNN's untrusted flag) may not be accepted from the ML model or, on the opposite, maybe more preferable than the false prediction. Therefore, depending on the task, this contribution's results may be valuable or not.

Using the training-history-preserving loss function presented here is not the algorithm's most exciting and effective use. Dynamic threshold adjustment during the test phase, after each successful or unsuccessful prediction, especially on the structured input data, is a more intriguing task for future research. A more detailed investigation of the continuous, lifetime learning mode, improvement of the meta-learning supervisor ANN itself, and loss layer with statistical memory are natural areas for future work, as well as testing hypothesis that SNN implements implicit Bayesian integration without assumption on priors.

References

1. Andrychowicz, M., et al.: Learning to learn by gradient descent by gradient descent. In: Proceedings of the 30th International Conference on Neural Information Processing Systems, NIPS 2016, pp. 3988–3996. Curran Associates Inc., Red Hook (2016)
2. Baevski, A., Hsu, W.N., Xu, Q., Babu, A., Gu, J., Auli, M.: data2vec: a general framework for self-supervised learning in speech, vision and language (2022)
3. Bender, E.M., Gebru, T., McMillan-Major, A., Shmitchell, S.: On the dangers of stochastic parrots: Can language models be too big? In: Proceedings of the 2021 ACM Conference on Fairness, Accountability, and Transparency, FAccT 2021, pp. 610–623. Association for Computing Machinery, New York (2021). https://doi.org/10.1145/3442188.3445922
4. Bergstra, J., Bardenet, R., Bengio, Y., Kégl, B.: Algorithms for hyper-parameter optimization. In: Shawe-Taylor, J., Zemel, R., Bartlett, P., Pereira, F., Weinberger, K.Q. (eds.) Advances in Neural Information Processing Systems, vol. 24. Curran Associates, Inc. (2011). https://proceedings.neurips.cc/paper/2011/file/86e8f7ab32cfd12577bc2619bc635690-Paper.pdf
5. Berretti, S., Del Bimbo, A., Pala, P., Amor, B.B., Daoudi, M.: A set of selected sift features for 3d facial expression recognition. In: 2010 20th International Conference on Pattern Recognition, pp. 4125–4128. IEEE (2010)
6. Blodgett, S.L., Madaio, M.: Risks of AI foundation models in education. CoRR abs/2110.10024 (2021). https://arxiv.org/abs/2110.10024
7. Bommasani, R., et al.: On the opportunities and risks of foundation models. CoRR abs/2108.07258 (2021). https://arxiv.org/abs/2108.07258
8. Brown, T.B., et al.: Language models are few-shot learners (2020)
9. Cacioppo, J.T., Berntson, G.G., Larsen, J.T., Poehlmann, K.M., Ito, T.A., et al.: The psychophysiology of emotion. Handbook of emotions **2**(01), 2000 (2000)
10. Chen, C., Dantcheva, A., Swearingen, T., Ross, A.: Spoofing faces using makeup: an investigative study. In: 2017 IEEE International Conference on Identity, Security and Behavior Analysis, pp. 1–8 (Feb 2017)
11. Dietterich, T.G.: Ensemble methods in machine learning. In: Kittler, J., Roli, F. (eds.) MCS 2000. LNCS, vol. 1857, pp. 1–15. Springer, Heidelberg (2000). https://doi.org/10.1007/3-540-45014-9_1
12. Ekman, P., Friesen, W.V.: Constants across cultures in the face and emotion. J. Pers. Soc. Psychol. **17**(2), 124 (1971)
13. Fedus, W., Zoph, B., Shazeer, N.: Switch transformers: scaling to trillion parameter models with simple and efficient sparsity. CoRR abs/2101.03961 (2021). https://arxiv.org/abs/2101.03961
14. Field, H.: At Stanford's "foundation models" workshop, large language model debate resurfaces. Morning Brew, August 2021. https://www.morningbrew.com/emerging-tech/stories/2021/08/30/stanfords-foundation-models-workshop-large-language-model-debate-resurfaces
15. Finn, C., Abbeel, P., Levine, S.: Model-agnostic meta-learning for fast adaptation of deep networks. In: Precup, D., Teh, Y.W. (eds.) Proceedings of the 34th International Conference on Machine Learning. Proceedings of Machine Learning Research, vol. 70, pp. 1126–1135. PMLR, 06–11 Aug 2017. http://proceedings.mlr.press/v70/finn17a.html

16. Gal, Y., Ghahramani, Z.: Dropout as a bayesian approximation: representing model uncertainty in deep learning. In: Proceedings of the 33rd International Conference on International Conference on Machine Learning - Volume 48, ICML2016, pp. 1050–1059. JMLR.org (2016)
17. Girdhar, R., Singh, M., Ravi, N., van der Maaten, L., Joulin, A., Misra, I.: Omnivore: A single model for many visual modalities (2022). 10.48550/ARXIV.2201.08377. https://arxiv.org/abs/2201.08377
18. Graves, A.: Practical variational inference for neural networks. In: Proceedings of the 24th International Conference on Neural Information Processing Systems, NIPS 2011, pp. 2348–2356. Curran Associates Inc., Red Hook (2011)
19. Gross, C.T., Canteras, N.S.: The many paths to fear. Nat. Rev. Neurosci. **13**(9), 651–658 (2012)
20. Huang, G.B., Mattar, M., Berg, T., Learned-Miller, E.: Labeled faces in the wild: a database for studying face recognition in unconstrained environments. In: 'Real-Life' Images: Detection, Alignment, and Recognition. Erik Learned-Miller and Andras Ferencz and Frédéric Jurie, Marseille, France (2008)
21. Jakaite, L., Schetinin, V., Maple, C.: Bayesian assessment of newborn brain maturity from two-channel sleep electroencephalograms. Computational and Mathematical Methods in Medicine, pp. 1–7 (2012). https://doi.org/10.1155/2012/629654
22. Jakaite, L., Schetinin, V., Maple, C., Schult, J.: Bayesian decision trees for EEG assessment of newborn brain maturity. In: The 10th Annual Workshop on Computational Intelligence UKCI 2010 (2010). https://doi.org/10.1109/UKCI.2010.5625584
23. Jakaite, L., Schetinin, V., Schult, J.: Feature extraction from electroencephalograms for Bayesian assessment of newborn brain maturity. In: 24th International Symposium on Computer-Based Medical Systems (CBMS), pp. 1–6 (2011). https://doi.org/10.1109/CBMS.2011.5999109
24. Jia, S., Li, X., Hu, C., Guo, G., Xu, Z.: 3d face anti-spoofing with factorized bilinear coding (2020)
25. Kendall, A., Gal, Y.: What uncertainties do we need in bayesian deep learning for computer vision? (2017). http://arxiv.org/abs/1703.04977
26. Khodabakhsh, A., Busch, C.: A generalizable deepfake detector based on neural conditional distribution modelling. In: 2020 International Conference of the Biometrics Special Interest Group (BIOSIG), pp. 1–5 (2020)
27. Kim, B.-K., Roh, J., Dong, S.-Y., Lee, S.-Y.: Hierarchical committee of deep convolutional neural networks for robust facial expression recognition. J. Multimod. User Interfaces **10**(2), 173–189 (2016). https://doi.org/10.1007/s12193-015-0209-0
28. Lake, B.M., Ullman, T.D., Tenenbaum, J.B., Gershman, S.J.: Building machines that learn and think like people. Behav. Brain Sci. **40**, e253 (2017). https://doi.org/10.1017/S0140525X16001837
29. Lakshminarayanan, B., Pritzel, A., Blundell, C.: Simple and scalable predictive uncertainty estimation using deep ensembles. In: Proceedings of the 31st International Conference on Neural Information Processing Systems, NIPS 2017, pp. 6405–6416. Curran Associates Inc., Red Hook (2017)
30. Liu, M., Li, S., Shan, S., Chen, X.: Au-inspired deep networks for facial expression feature learning. Neurocomputing **159**, 126–136 (2015)
31. Liu, X., Wang, X., Matwin, S.: Interpretable deep convolutional neural networks via meta-learning. In: 2018 International Joint Conference on Neural Networks (IJCNN), pp. 1–9 (2018). https://doi.org/10.1109/IJCNN.2018.8489172

32. Lopes, A.T., De Aguiar, E., De Souza, A.F., Oliveira-Santos, T.: Facial expression recognition with convolutional neural networks: coping with few data and the training sample order. Pattern Recogn. **61**, 610–628 (2017)
33. Lottick, K., Susai, S., Friedler, S.A., Wilson, J.P.: Energy usage reports: environmental awareness as part of algorithmic accountability. CoRR abs/1911.08354 (2019). http://arxiv.org/abs/1911.08354
34. MacKay, D.J.C.: A practical bayesian framework for backpropagation networks. Neural Comput. **4**(3), 448–472 (1992). https://doi.org/10.1162/neco.1992.4.3.448
35. Mai, F., Pannatier, A., Fehr, F., Chen, H., Marelli, F., Fleuret, F., Henderson, J.: Hypermixer: An mlp-based green ai alternative to transformers. arXiv preprint arXiv:2203.03691 (2022)
36. Mansourifar, H., Shi, W.: One-shot gan generated fake face detection (2020)
37. Marcus, G.: Deep learning: A critical appraisal. CoRR abs/1801.00631 (2018). http://arxiv.org/abs/1801.00631
38. Martinez, A., Benavente, R.: The ar face database. Technical report 24, Computer Vision Center, Bellatera, June 1998
39. Mollahosseini, A., Chan, D., Mahoor, M.H.: Going deeper in facial expression recognition using deep neural networks. In: 2016 IEEE Winter Conference on Applications of Computer Vision (WACV), pp. 1–10. IEEE (2016)
40. Neal, R.M.: Bayesian learning for neural networks, Lecture Notes in Statistics, vol. 118. Springer-Verlag New York, Inc. (1996). https://doi.org/10.1007/978-1-4612-0745-0
41. Nichol, A., Achiam, J., Schulman, J.: On first-order meta-learning algorithms. ArXiv abs/1803.02999 (2018)
42. Ram, R., Müller, S., Pfreundt, F., Gauger, N., Keuper, J.: Scalable hyperparameter optimization with lazy gaussian processes. 2019 IEEE/ACM Workshop on Machine Learning in High Performance Computing Environments (MLHPC), pp. 56–65 (2019)
43. Rosset, C.: Turing-NLG: A 17-billion-parameter language model by Microsoft - Microsoft Research, February 2020. https://www.microsoft.com/en-us/research/blog/turing-nlg-a-17-billion-parameter-language-model-by-microsoft. Accessed 16 Jan 2022
44. Schetinin, V., Jakaite, L., Krzanowski, W.: Bayesian averaging over decision tree models: an application for estimating uncertainty in trauma severity scoring. Int. J. Med. Informatics **112**, 6–14 (2018). https://doi.org/10.1016/j.ijmedinf.2018.01.009
45. Schetinin, V., Jakaite, L., Krzanowski, W.: Bayesian averaging over decision tree models for trauma severity scoring. Artif. Intell. Med. **84**, 139–145 (2018). https://doi.org/10.1016/j.artmed.2017.12.003
46. Schetinin, V., Jakaite, L., Krzanowski, W.: Bayesian learning of models for estimating uncertainty in alert systems: application to air traffic conflict avoidance. Integrated Comput.-Aided Eng. **26**, 1–17 (2018). https://doi.org/10.3233/ICA-180567
47. Schick, T., Schütze, H.: It's not just size that matters: small language models are also few-shot learners. CoRR abs/2009.07118 (2020). https://arxiv.org/abs/2009.07118
48. Selitskaya, N., Sielicki, S., Christou, N.: Challenges in real-life face recognition with heavy makeup and occlusions using deep learning algorithms. In: Nicosia, G., Ojha, V., La Malfa, E., Jansen, G., Sciacca, V., Pardalos, P., Giuffrida, G., Umeton, R. (eds.) LOD 2020. LNCS, vol. 12566, pp. 600–611. Springer, Cham (2020). https://doi.org/10.1007/978-3-030-64580-9_49

49. Selitskaya, N., Sielicki, S., Christou, N.: Challenges in face recognition using machine learning algorithms: case of makeup and occlusions. In: Arai, K., Kapoor, S., Bhatia, R. (eds.) IntelliSys 2020. AISC, vol. 1251, pp. 86–102. Springer, Cham (2021). https://doi.org/10.1007/978-3-030-55187-2_9
50. Selitskiy, S., Christou, N., Selitskaya, N.: Isolating Uncertainty of the Face Expression Recognition with the Meta-Learning Supervisor Neural Network, pp. 104–112. Association for Computing Machinery, New York (2021). https://doi.org/10.1145/3480433.3480447
51. Selitskiy, S., Christou, N., Selitskaya, N.: Using statistical and artificial neural networks meta-learning approaches for uncertainty isolation in face recognition by the established convolutional models. In: Nicosia, G., Ojha, V., La Malfa, E., La Malfa, G., Jansen, G., Pardalos, P.M., Giuffrida, G., Umeton, R. (eds.) Machine Learning, Optimization, and Data Science, pp. 338–352. Springer International Publishing, Cham (2022). https://doi.org/10.1007/978-3-030-95470-3_26
52. Shan, C., Gong, S., McOwan, P.W.: Facial expression recognition based on local binary patterns: a comprehensive study. Image Vis. Comput. **27**(6), 803–816 (2009)
53. Singh, A., Hu, R., Goswami, V., Couairon, G., Galuba, W., Rohrbach, M., Kiela, D.: Flava: A foundational language and vision alignment model (2021). https://doi.org/10.48550/ARXIV.2112.04482. https://arxiv.org/abs/2112.04482
54. Sprent, P.: Applied Nonparametric Statistical Methods. Springer, Dordrecht (1989)
55. Strubell, E., Ganesh, A., McCallum, A.: Energy and policy considerations for deep learning in NLP. CoRR abs/1906.02243 (2019). http://arxiv.org/abs/1906.02243
56. Szegedy, C., et al.: Intriguing properties of neural networks. arXiv preprint arXiv:1312.6199 (2013)
57. Thrun S., P.L.: Learning To Learn. Springer, Boston, MA (1998). https://doi.org/10.1007/978-1-4615-5529-2
58. Vanschoren, J.: Meta-learning: a survey. ArXiv abs/1810.03548 (2018)
59. Whitehill, J., Omlin, C.W.: Haar features for facs au recognition. In: 7th International Conference on Automatic Face and Gesture Recognition (FGR06), pp. 5-pp. IEEE (2006)
60. Zhao, T., Xu, X., Xu, M., Ding, H., Xiong, Y., Xia, W.: Learning to recognize patch-wise consistency for deepfake detection (2020)

Machine Learning Approaches for Predicting Crystal Systems: A Brief Review and a Case Study

Gaetano Settembre[1], Nicola Corriero[3], Nicoletta Del Buono[1,2(✉)], Flavia Esposito[1,2], and Rosanna Rizzi[3]

[1] Department of Mathematics, University of Bari Aldo Moro, Bari, Italy
g.settembre@studenti.uniba.it,
{nicoletta.delbuono,flavia.esposito}@uniba.it
[2] Members of INDAM-GNCS Research Group, Rome, Italy
[3] CNR – Institute of Crystallography, Bari, Italy
{nicola.corriero,rosanna.rizzi}@ic.cnr.it

Abstract. Determining the crystal system and space group for a compound from its powder X-ray diffraction data represents the initial step of a crystal structure analysis. This task can constitute a bottleneck in the material science workflow and often requires manual interventions by the user. The fast development of Machine Learning algorithms has strongly impacted crystallographic data analysis. It offered new opportunities to develop novel strategies for accelerating the crystal structure discovery processes. This paper aims to provide an overview of approaches recently proposed for crystal system prediction, grouped according to the input features they use to construct the prediction model. It also presents the results obtained in predicting the crystal system of polycrystalline compounds, by using the lattice parameters to train some learning models.

Keywords: Machine learning · Powder X-ray diffraction · Crystal systems · Data-driven models

1 Introduction

Determining the crystal system and space group of a compound from its A *crystal* is a solid in which atoms (molecules or ions) are arranged in an orderly, repeating, three-dimensional pattern. The three-dimensional arrangement of a solid crystal is referred to as the *crystal lattice*. The smallest repeating element in the crystal lattice, the *unit cell*, can be defined by some lattice parameters: the length of three crystal axes a, b and c (expressed in Angström), and the angles between the axes at which the faces intersect α, β and γ (expressed in degrees). Figure 1 illustrates an example of a unit cell. Seven crystal systems can be identified based on the relationship between the angles between the sides of the unit cell and the distances between points in the unit cell. The seven crystal systems are triclinic, monoclinic, orthorhombic, trigonal, tetragonal, hexagonal,

G. Nicosia et al. (Eds.): LOD 2022, LNCS 13810, pp. 93–107, 2023.
https://doi.org/10.1007/978-3-031-25599-1_8

and cubic. The characterization of a crystalline substance, available as a single crystal or powder, is fundamental to establishing its physicochemical properties. The X-ray diffraction (XRD) is the most appropriate technique for investigating crystalline compounds, currently used in many scientific fields such as chemistry, pharmacology, materials science, mineralogy, etc., to address a wide range of scientific tasks: qualitative and quantitative analysis of a powder sample; structure solution; microstructural analysis; investigation of highly complex materials; etc. In the case of polycrystalline compounds, the structural characterization needs a long sequence of nontrivial steps: the identification of the lattice parameters (indexation), space-group determination, and the decomposition of the diffraction profile into integrated intensities, crystal structure determination at the atomic level, and structural refinement. The structure solution requires that the cell parameters are known, so the primary step in the solution process is a correct indexation, which often constitutes a bottleneck in the workflow and frequently requires manual-based interventions.

In the last decades, however, the large development of Machine Learning (ML) algorithms and Artificial Intelligence (AI) approaches had a great impact on the analysis of crystallographic data and offered new opportunities for developing novel strategies to construct data-driven models designed to accelerate the crystal structures discovery processes. A widespread literature demonstrates how machine and deep learning models are able to make a prediction based on correlations found in measured or calculated diffraction data [3,13,17], and the number of applications in chemistry and in manufacturing propose specialized models performing crystallographic data analysis [1,6,22]. These ever-increasing number of models use different input features in their training phase which often depend on the specific materials (inorganic or organic) under studies. This work aims to provide a brief, clearly non-comprehensive, review of some papers using ML approaches in a crystallographic context, accordingly to input data features. The work also discusses a preliminary case study on which some general ML algorithms are applied to derive data-driven models performing crystal system prediction.

The paper is organized as follows. In Sect. 2 some papers proposing ML approaches to handle some crystallographic problems, one above all the crystal system classification, are discussed and grouped according to the input data features they use to construct the prediction model. Section 3 illustrates the results obtained by some learning models to predict the crystal system of inorganic, organic, metal-organic compounds and minerals, using only the lattice constant values as input features. Finally, a section of discussions closes the paper.

2 Methods for Crystal System Prediction

In the literary panorama, many different ML strategies appeared to tackle the crystal system prediction problem when different characteristics of powder diffraction data are available. Concerning the type of features used to train the predictive model, the ML models and methods can be grouped into approaches using:

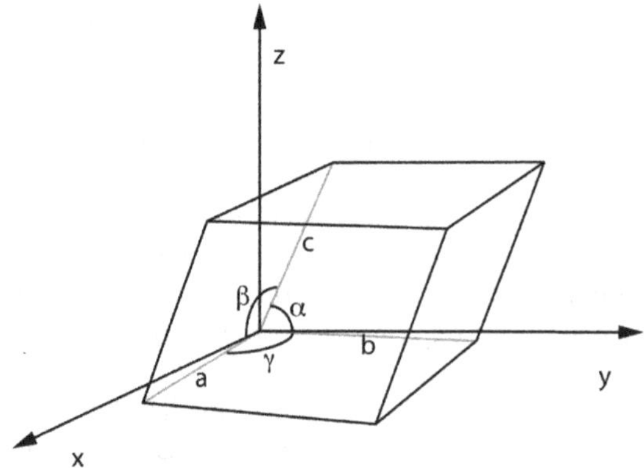

Fig. 1. Unit cell and lattice constants: a, b, c and α, β, γ.

- the entire (or a part) of the X-ray powder diffraction (XRPD) pattern;
- extracted features directly derived from XRPD pattern;
- other input crystal features;
- useful patterns extracted from XRD data through unsupervised learning approaches (including clustering and dimensionality reduction mechanisms).

In the following, some recent works for crystal system prediction are briefly reviewed.

2.1 ML Methods Using XRPD Patterns

One of the most successful learning approaches for screening crystalline materials applies the Convolutional Neural Networks (CNNs) model (a specialized type of Artificial Neural Networks using convolution in place of general matrix multiplication in at least one of their layers) trained with powder diffraction data. When the entire pattern is used to feed ML methods, it is treated as a string of real values representing the profile intensities of the spectral signal. In [13], CNNs were developed for crystal system, extinction group, and space group classifications. The model used 150000 simulated XRPD patterns derived from the crystal structures stored in the Inorganic Crystal Structure Database (ICSD)[1] and reached 94.99% test accuracy. Moreover, the proposed approach, which does not incorporate human expertise or needs human assistance, was not able to properly classify experimental patterns of novel structures not included in the ICSD.

In [19], a deep dense Neural Network solves the crystal system and space group determination problems. It was trained with 128404 diffraction patterns of inorganic non-magnetic materials, calculated from the CIF files in the ICSD and reaching a test accuracy of 73%. This deep dense network demonstrated a

[1] http://www.fiz-karlsruhe.de/icsd.html.

higher classification accuracy (54%) on experimental data when compared with the CNN model proposed in [13].

In [11], an all CNN (a-CNN) was trained starting from a small set of thin-film perovskite structures (115 patterns) and performing some data augmentation pre-processing technique (i.e. peak scaling, peak elimination, pattern shifting). This approach constructs the most accurate and interpretable classifier for dimensionality and space group classification. Via Class Activation Map, it also provides a possible explanation of the pattern regions used by the network to identify the correct class, and explanations of the possible causes of misclassification.

Differently from previous studies, generic raw XRPD data have been used in [3] to train a series of one-dimensional CNN (1D-CNNs) which predict the lattice parameters. In [20], some theoretic and very limited experimental XRPD patterns were used to train a CNN devoted to a one-by-one material classification of metal-organic frameworks.

2.2 ML Methods Using Features Derived from XRPD Patterns

An alternative approach is to train ML models by using as input some features characterizing the experimental pattern. In [24], the crystal structure is represented as an RGB image, in which each color channel shows the diffraction patterns obtained by rotating the crystal around a given axis (i.e., red (R) for the x-axis, green (G) for the y-axis, and blue (B) for z-axis). Each two-dimensional pixel image is then used as an input feature (called a two-dimensional diffraction fingerprint) to train a deep NN to solve the classification task.

In [18], the image-pixel representations of raw shaped DPs are used to train a combined approach adopting multi-stream dense NN for crystal symmetry identification, reaching high accuracy performances also when several classes of symmetry are considered.

Other approaches identify and extract from the XRPD pattern, useful input descriptors of the peaks they contain, such as shape and width, intensity, full-width at half-maximum, and position. For example, in [1] a modular and flexible NN hierarchical-based model was proposed. It combines the chemical data with the diffraction profile of the compound to obtain input features useful to perform crystal structure classification. The diffraction profiles are first converted into a single feature vector parsed based on peak positions, which are successively divided into 900 bins uniformly partitioning the range from 0.5 to 6 Angstroms in the reciprocal lattice spacing.

In [17], the positions of the first 10 peaks in the lower-angle range and the number of peaks extracted from the entire diffraction pattern (range 2θ of the interval [0,90]) are used as derived features. These are then used to train the tree-ensemble-based ML model reaching over 90% accuracy for crystal system classification, except for triclinic case, and the 88% accuracy for space group classification. This study is the first approach which demonstrates a high interpretability acting as an alternative to the use of deep learning and CNN techniques which are still considered black-box algorithms.

2.3 Methods Using Other Features

Some ML algorithms trained using composition information which do not require the XRPD pattern have also been proposed for crystal system and space group prediction. For example, in [22] a method which uses only some information, such as Magpie, atom vector and atom frequency as input features, was proposed for inorganic materials. Particularly, the Magpie descriptor set few statistics for each property of the elements in a given compound, integrating into a one-dimensional vector physical, chemical, electronic, ionic and basic properties of the material as features.

In [8], the tool CRYSPNet (Crystal Structure Prediction Network) based on a series of NN models, was proposed to predict the space group of inorganic materials. This tool uses 132 Magpie predictors to describe the properties of the elements which constitutes the compound. It further aggregates and transforms them to obtain additional features, for a total of 228.

In [7], an ensemble of ML algorithms (including Random Forest, Gradient Boosting, and Deep NN) was proposed for multi-label classification. The models work based on 22 kinds of features in Magpie element property statistics (including atomic number, Mendeleev number, atomic weight, melting temperature, etc.) and reached the 81% of accuracy on the crystal system prediction task using the Material Project Database.

In [14] input data in the form of multiperspective atomic fingerprints are used to train deep neural networks able to analyse a large collection of crystallographic data in the crystal structure repositories.

A different input feature representation is adopted in [23], where CNN are trained using the periodic table representation as an input feature.

2.4 Other Approaches

Other crystal structure prediction algorithms present in the literary panorama are based on reliable and efficient exploratory algorithms which can make predictions with little or no pre-existing knowledge. These algorithms do not belong to the class of data-driven approaches, which can face prediction problems when trained on prototypes available in the training dataset (XRPD patterns or features extracted from them or other physical and chemical properties of the crystals). On the contrary, these approaches combine some theoretical calculations with global optimization algorithms [21], including among others: evolutionary algorithms [5,9], simulated annealing [4] and random searching. Black-box optimization methods have also been used to tune the parameters in the Rietveld method[2] [12].

Going beyond the direct solution of the crystal system classification task, unsupervised approaches (such as clustering and low-rank dimensionality reduction methods) can be used to extract structural similarity information embedded in the XRD data representation. This information often describes the link

[2] The Rietveld method is an approach for refining structure and lattice parameters directly from the whole X-ray or neutron powder diffraction patterns.

between composition, structure, and property of crystals, acting as novel input descriptors of any supervised learning approach to solving the crystal systems prediction problem.

A promising low rank dimensionality technique able to properly analyzes XRD data is the Non-negative Matrix Factorization (NMF). The basic idea of NMF is to deconvolve a large number of non-negative spectral patterns into a smaller number of non-negative basis patterns to be used as new descriptors. In [10], NMF was firstly used to analyze hundreds of micro-diffraction XRD patterns extracted from a combinatorial materials library. Preserving the non negativeness of data, NMF produced basis patterns which can be interpreted as diffraction patterns, and then used as interpretable descriptors for the crystal systems prediction task.

NMF approach handling XRD pattern has been expanded in [16] by combining it with custom clustering and cross-correlation algorithms. This improved NMF model was capable of a robust determination of the number of basis patterns present in the data allowing the identification of any possible peak-shifted patterns. Since the peak-shifting in diffraction patterns is caused by slight variations of the lattice parameter, the robustness of this approach allowed a proper quantification and classification of the basis XRD patterns and the extracting of a salient structural representation.

Using Scopus as a bibliographical source (without any loss of generalization), we conclude the bibliometric analysis extracting some bibliographical data which can be visualized using the VOS viewer[3]. Through the "Advanced Search" tool provided by Scopus, we ran the general query[4] excluding work published in year 2022, TITLE-ABS-KEY ("crystal" AND "structure" AND "prediction" AND "machine" AND "learning") AND (EXCLUDE (PUBYEAR, 2022)). These terms are required to be present in the title of the retrieved document or in the associated keywords (provided by authors or automatically indexed), or in the abstract. As a result, we found out 402 documents which can provide a global picture of the research field under examination. Figure 2 shows the citation map obtained by the VOS viewer restricting the visualization to the largest set of connected items (83 papers). Size of nodes is proportional to the number of citations, colors represents nodes belonging to the same cluster, while link weights come from the co-citation index. Five clusters can be identified, they contain articles connected to the paper [15]. This latter provides a comprehensive overview and analysis of the most recent research in machine learning to solid-state systems. The obtained clusters mainly group papers accordingly to the type of learning algorithms is used to make prediction (deep neural networks [14,19,24], convolutional neural network [13,23], random forest and neural networks [17,22]).

Figure 3 shows a graph with the most popular author keywords in the articles under consideration. Each node is associated with a keyword with a proportional

[3] This tool constructs and visualizes bibliometric networks related to citation, co-citation, bibliographic coupling, co-authorship or co-occurrence of words relations.

[4] On May 1st, 2022.

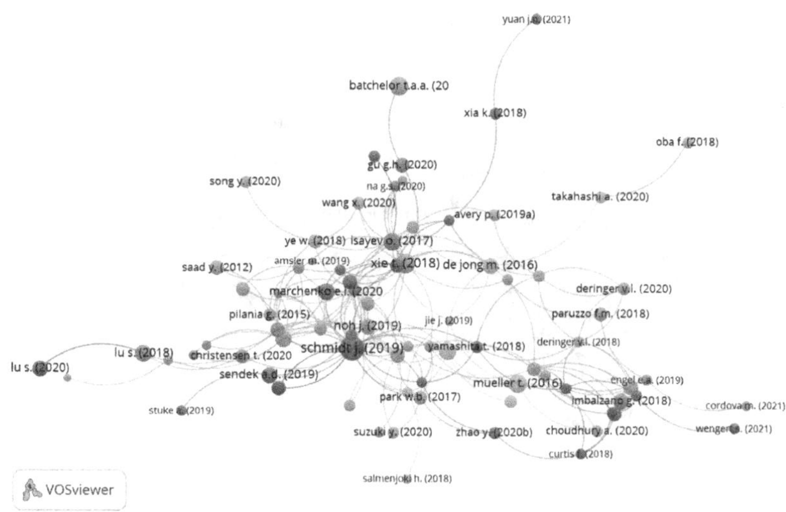

Fig. 2. Graph of co-citation map. Node size is proportional to the number of citations, colors represents nodes belonging to the same cluster, link weights come from the co-citation index.

size to the number of documents where the keyword appears with groups of closely related nodes depicted in the same color. Machine Learning is the main keyword since it is associated with the larger node. Keywords related to organic topics appear on the left (either red or blue clusters points), while on the right part of the graph, some words related to computer science real are highlighted. The keyword x-ray diffraction is also quite evident.

3 A Case Study: ML Approach Using Lattice Features

This section presents as a case study the construction of classification data-driven models to predict the crystal systems, inputting only the lattice constants. Particularly, dataset extraction and its preprocessing are reported, together with some notes about the training phase and the evaluation of the obtained classification models.

3.1 Data Preparation

Diffraction data used in this paper have been extracted from POW_COD [2], a non-commercial relational database developed by some researchers from the Institute of Crystallography (CNR) of Bari, Italy. It contains a collection of crystal structures of organic, inorganic, metal-organic compounds and minerals, it is built from the structure information in CIF format stored in the Crystallographic Open Database (COD)[5]. POW_COD is continuously growing, and when

[5] http://www.crystallography.net/cod.

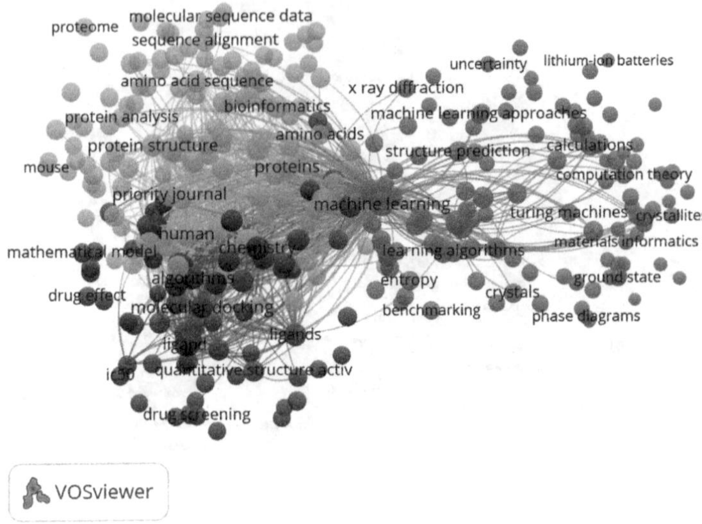

Fig. 3. Graph of most popular author keywords in the considered articles. Colors represent groups of nodes closely related to the same keyword. Size of each node is proportional to the occurrence value of the keyword.

we started this work, it contained 400144 compounds. The dataset $\{(x, l_x)\}$ is a labeled dataset, where x is the input vector whose elements are the cell parameters and l_x is the class label representing the crystal system the sample belongs. Table 1 summarizes the distribution of compounds included in the Database, and Table 2 reports some basic statistics on the six features in the dataset (i.e., for each cell parameter, the minimum and maximum value, the mean and the standard deviation).

Table 1. Distribution of samples for each crystal system as it appeared in the dataset

Crystal system	No. of samples
Monoclinic	186325
Triclinic	92938
Orthorhombic	67120
Cubic	16613
Tetragonal	16087
Trigonal	12881
Hexagonal	8180

Before the learning process, a data pre-processing phase has been performed. Particularly, samples with huge lattice parameters (>40 Å) or small

Table 2. Some basic statistics on the six numerical features describing the dataset.

Feature	Min	Max	Mean	Std. dev.
a	2.015	189.800	12.40323	6.77363
b	2.015	150.000	13.25367	6.56444
c	2.166	475.978	16.14686	8.01951
α	13.930	150.172	89.77909	6.71217
β	11.930	173.895	95.48034	10.45269
γ	13.930	149.900	91.24751	9.81564

ones (<2.5 Å) were discarded. This is because these values indicate particularly complex structures or compounds which are under high pressure, infrequent in the real cases and whose presence in the training data could compromise the performance of the ML models. Compounds with a weighted profile R-factor (R_{wp}) value larger than 10% were also discarded in agreement with domain experts. Large discrepancy index values imply poor quality of the structure model refinement. The reduced dataset (241758 XRPD patterns) was used in the present study. The distribution of samples among the seven crystal systems in the complete (POW_COD data) and reduced dataset is shown in Fig. 4; a strong imbalance between the classes is evident, as it appears also in Table 1.

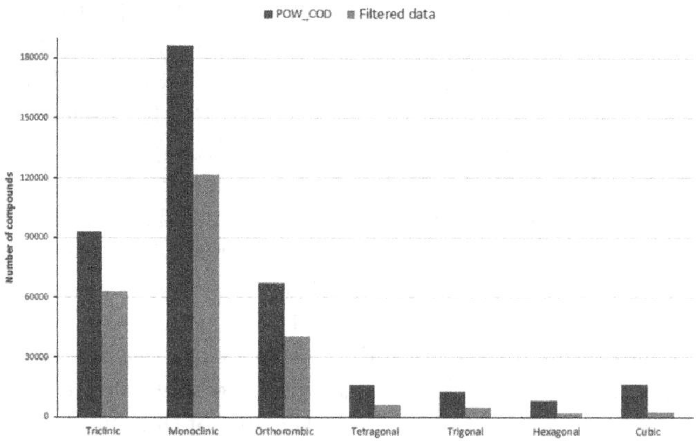

Fig. 4. Distribution of samples among the seven crystal systems in the complete POW_COD (blue) and reduced (orange) dataset. (Color figure online)

A data dimensionality reduction was also performed via Principal Component Analysis (PCA) (based on the data covariance matrix); however the poor significance and low informative results obtained and illustrated in Fig. 5 suggested the inappropriateness of this approach. To construct classification models able to predict the crystal systems from the six lattice constant input features,

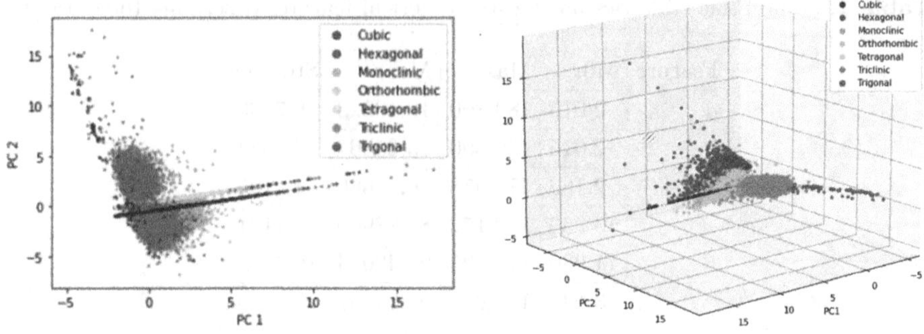

Fig. 5. Compound projections after computing the first two (left) and three (right) principal components (based on the analysis of data covariance matrix).

two datasets were considered: the complete and a balanced one. The first contains all records stored in the reduced dataset; the second is obtained using a random sub-sampling based on the less represented class (the Hexagonal class). The balanced dataset contained 57260 samples.

3.2 Learning Models Using Lattice Values

We trained some classic ML models using a 70/30 stratified split between training and test data with a StratifiedKFold (K = 5) Cross-Validation. Stratification is necessary to generate a subset of data having examples of all classes with the same aspect ratio as the complete dataset. The used methods are Random Forest (RF), Decision Tree (DT), k-nearest neighbors (K-NN), Support Vector Machine (SVM), Naive Bayes, MultiLayer Perceptron (MLP), Extremely Randomized Trees (ExRT) and eXtreme Gradient Boosting (XGB). The algorithms were used with the default options for their hyper-parameters.

Table 3 reports the accuracy of the test set (balanced set of data) for the ML models trained for the crystal system classification task. Random Forest[6] got the best result. Table 4 summarizes the values of the classification metrics (Precision, Recall, and F1-score) obtained by RF model, while Fig. 6 shows the related confusion matrices on balanced (a) and unbalances (b) test set, respectively.

The classification results obtained on the balanced dataset are quite good. For completeness, we also trained the models on the 70% of the entire reduced dataset (strongly unbalanced as reported in Table 1). Table 5 reports the accuracy results obtained on the test set.

The RF model represents the best classifier in terms of its performance on the unbalanced dataset, presenting an accuracy value of over 98%. The three performance measures (Precision, Recall, and F1-score) calculated for each class are comparable with those obtained in the previous test (balanced dataset case).

RF and DT performed very similarly in terms of the accuracy (0.9804 and 0.9743 as shown in Table 5), so that we conducted an analysis on DT. Following

[6] This model uses 200 trees and Gini impurity.

Table 3. Accuracy results reported by each ML model trained on the balanced dataset.

Trained model	Accuracy on test set
Random Forest	0.9239
K-NN (k = 5)	0.8923
SVM	0.6849
DT	0.9089
Naive Bayes	0.5826
MLP	0.8872
ExRT	0.9191
XGB	0.8938

Table 4. Classification report for the Random Forest model on the balanced dataset. The performance measures reported are related to Precision, Recall, and F1-score.

Class	Precision	Recall	F1-score	Support
Cubic	0.969	0.998	0.983	2454
Hexagonal	0.820	0.807	0.813	2454
Monoclinic	1.000	0.993	0.996	2454
Orthorhombic	0.972	0.902	0.936	2454
Tetragonal	0.908	0.953	0.930	2454
Triclinic	0.994	0.998	0.996	2454
Trigonal	0.809	0.817	0.813	2454
Accuracy			**0.924**	**17178**

Table 5. Accuracy results reported by each ML model on test set (120,044 samples) when trained on the unbalanced dataset.

Trained model	Accuracy on test set
Random Forest	0.9804
K-NN (k = 5)	0.9495
SVM	0.8686
DT	0.9743
Naive Bayes	0.7322
MLP	0.9437
ExRT	0.9768
XGB	0.9602

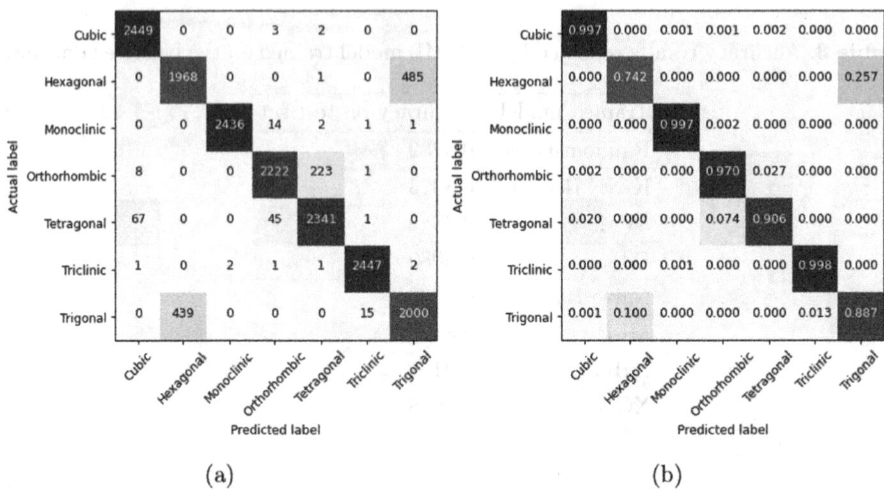

Fig. 6. Confusion matrices. a) Balanced test set. b) Unbalanced test set.

the Occam's razor principle (one should choose the simplest model with the best performance), among the many possible solutions, the simplest hypothesis that explains the data was preferred to get the better interpretable data-driven model. Accuracy values, higher than 90%, can already be reached with a single decision tree with depth parameter set to 5, while with a depth larger than level 13, 95% total accuracy is exceeded. Moreover, the accuracy value over 97% reported in Table 5 is obtained with parameter $max\_depth = None$, i.e., the nodes of the decision tree are expanded until all the leaves are pure or until all the leaves contain less than min_samples_split. An interesting characteristic of all decision tree-based models (DT, RF and ExRT) is their capability to provide some kind of explanation about the features importance in the prediction model going towards "white box" and explainable learning models. Figure 7 reports the

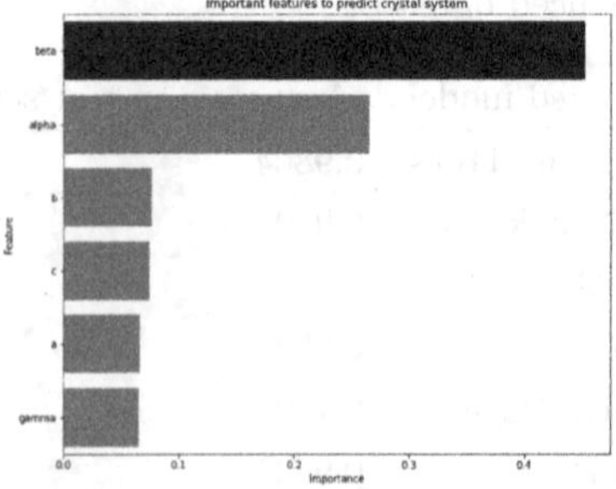

Fig. 7. Features importance for a Decision Tree with depth equals to 13. For each feature $\beta, \alpha, \gamma, b, c, a$, the importance values are $0.4830, 0.2838, 0.0696, 0.0603, 0.0587, 0.0445$.

```
|-- beta <= 90.00
|   |-- beta <= 90.00
|   |   |-- alpha <= 90.00
|   |   |   |-- beta <= 60.73
|   |   |   |   |-- gamma <= 61.23
|   |   |   |   |   |-- a <= 21.91
|   |   |   |   |   |   |-- class: Trigonal
|   |   |   |   |   |-- a >  21.91
|   |   |   |   |   |   |-- class: Triclinic
|   |   |   |   |-- gamma >  61.23
|   |   |   |   |   |-- class: Triclinic
|   |   |   |-- beta >  60.73
|   |   |   |   |-- b <= 4.68
|   |   |   |   |   |-- a <= 2.89
|   |   |   |   |   |   |-- class: Triclinic
|   |   |   |   |   |-- a >  2.89
|   |   |   |   |   |   |-- class: Trigonal
|   |   |   |   |-- b >  4.68
|   |   |   |   |   |-- c <= 6.87
|   |   |   |   |   |   |-- class: Triclinic
|   |   |   |   |   |-- c >  6.87
|   |   |   |   |   |   |-- class: Triclinic
|   |   |-- alpha >  90.00
|   |   |   |-- alpha <= 90.00
|   |   |   |   |-- gamma <= 89.50
|   |   |   |   |   |-- a <= 11.17
|   |   |   |   |   |   |-- class: Triclinic
|   |   |   |   |   |-- a >  11.17
|   |   |   |   |   |   |-- class: Orthorhombic
|   |   |   |   |-- gamma >  89.50
|   |   |   |   |   |-- a <= 5.62
|   |   |   |   |   |   |-- class: Monoclinic
|   |   |   |   |   |-- a >  5.62
|   |   |   |   |   |   |-- class: Monoclinic
|   |   |   |-- alpha >  90.00
|   |   |   |   |-- c <= 51.67
|   |   |   |   |   |-- b <= 32.42
|   |   |   |   |   |   |-- class: Triclinic
|   |   |   |   |   |-- b >  32.42
|   |   |   |   |   |   |-- class: Orthorhombic
|   |   |   |   |-- c >  51.67
|   |   |   |   |   |-- class: Orthorhombic
```

Fig. 8. Left branch of the root extracted from textual representation of a decision tree with max_depth = 6.

scores assigned to the input features in the RF predictive model trained on our data. These results indicate the relative importance of each feature when the learner makes a prediction.

Another possibility of a trained tree-model is to represent the entire decision tree in text form. This category of models is among the most interpretable in ML applications. Decision trees are like a rule system: starting from the root node, one follows the path for a record to a leaf node where the prediction can be seen. It is harder to interpret when the tree has a higher depth but still doable. Figure 8 shows an example of a textual representation.

4 Discussions

The direct prediction or generation of a crystal structure is still a challenging problem. The success that ML algorithms had in several fields of science and technology, stimulated the research interest in the real of crystal studies. In this study, we proposed an overview on approaches recently proposed for crystal system prediction accordingly to the type of the input features the ML uses to construct the prediction model. We also described a simple case study. We derive models predicting the crystal system from the six lattice parameters of the crystal unit cell. However, in a real crystallographic context, these features are generally not immediately available, and frequently real data need manual trial-and-error operations on the acquired diffraction patterns (as discussed in Sect. 2.1). Therefore, ML methods are recognized to be very important to improving this process. The future challenges of research in computational crystallography are to provide domain experts with learning systems that support them with interpretable and transparent algorithms taking into considerations the optimal use of available datasets and the specific research goal.

Acknowledgments. This work was supported in part by the GNCS-INDAM (Gruppo Nazionale per il Calcolo Scientifico of Istituto Nazionale di Alta Matematica) Francesco Severi, P.le Aldo Moro, Roma, Italy. The author F.E. was funded by REFIN Project, grant number 363BB1F4, Reference project idea UNIBA027 "Un modello numerico-matematico basato su metodologie di algebra lineare e multilineare per l'analisi di dati genomici".

References

1. Aguiar, J.A., Gong, M.L., Tasdizen, T.: Crystallographic prediction from diffraction and chemistry data for higher throughput classification using machine learning. Comput. Mater. Sci. **173**, 109409 (2020)
2. Altomare, A., Corriero, N., Cuocci, C., Falcicchio, A., Moliterni, A., Rizzi, R.: Qualx2.0: a qualitative phase analysis software using the freely available database pow-cod. J. Appl. Crystallogr. **48**, 04 2015
3. Chitturi, S.R., et al.: Automated prediction of lattice parameters from x-ray powder diffraction patterns. J. Appl. Cryst. **54**, 1799–1810 (2021)
4. Doll, K., Schoen, J.C., Jansen, M.: Structure prediction based on ab initio simulated annealing for boron nitride. Phys. Rev. B **78**, 144110 (2008)
5. Falls, Z., Avery, P., Wang, X., Hilleke, K.P., Zurek, E.: The xtalopt evolutionary algorithm for crystal structure prediction. J. Phys. Chem. C (2020)
6. Frade, A.P., McCabe, P., Cooper, R.I.: Increasing the performance, trustworthiness and practical value of machine learning models: a case study predicting hydrogen bond network dimensionalities from molecular diagrams. CrystEngComm **22**, 7186–7192 (2020)
7. Li, Y., Dong, R., Yang, W., Hu, J.: Composition based crystal materials symmetry prediction using machine learning with enhanced descriptors. Comput. Mater. Sci. **198**, 110686 (2021)
8. Liang, H., Stanev, V.G., Kusne, A.G., Takeuchi, I.: Cryspnet: crystal structure predictions via neural networks. Mater. Sci. (2020). arXiv

9. Liu, X., Niu, H., Oganov, A.R.: Copex: co-evolutionary crystal structure prediction algorithm for complex systems. NPJ Comput. Mater. **7**, 1–11 (2021)
10. Long, C.J., Bunker, D.T., Li, X., Karen, V.L., Takeuchi, I.: Rapid identification of structural phases in combinatorial thin-film libraries using x-ray diffraction and non-negative matrix factorization. Rev. Sci. Instrum. **80**, 10:103902 (2009)
11. Oviedo, F., et al.: Fast and interpretable classification of small x-ray diffraction datasets using data augmentation and deep neural networks. NPJ Comput. Mater. **5**, 1–9 (2019)
12. Ozaki, Y., Suzuki, Y., Hawai, T., Saito, K., Onishi, M., Ono, K.: Automated crystal structure analysis based on blackbox optimisation. NPJ Comput. Mater. **6**, 1–7 (2020)
13. Park, W.B., et al.: Classification of crystal structure using a convolutional neural network. IUCrJ **4**(4), 486–494 (2017)
14. Ryan, K., Lengyel, J., Shatruk, M.: Crystal structure prediction via deep learning. J. Am. Chem. Soc. **140**(32), 10158–10168 (2018)
15. Schmidt, J., Marques, M.R.G., Botti, S., Marques, M.A.L.: Recent advances and applications of machine learning in solid-state materials science. NPJ Comput. Mater. **5**, 1–36 (2019)
16. Stanev, V.G., Vesselinov, V.V., Kusne, A.G., Antoszewski, G., Takeuchi, I., Alexandrov, B.S.: Unsupervised phase mapping of x-ray diffraction data by non-negative matrix factorization integrated with custom clustering. NPJ Comput. Mater. **4**, 1–10 (2018)
17. Suzuki, Y., Hino, H., Hawai, T., Saito, K., Kotsugi, M., Ono, K.: Symmetry prediction and knowledge discovery from x-ray diffraction patterns using an interpretable machine learning approach. Sci. Rep. **10**(1), 1–11 (2020)
18. Tiong, L.C.O., Kim, J., Han, S.S., Kim, D.: Identification of crystal symmetry from noisy diffraction patterns by a shape analysis and deep learning. NPJ Comput. Mater. **6**, 1–11 (2020)
19. Vecsei, P.M., Choo, K., Chang, J., Neupert, T.: Neural network based classification of crystal symmetries from x-ray diffraction patterns. Phys. Rev. B (2019)
20. Wang, H., et al.: Rapid identification of x-ray diffraction patterns based on very limited data by interpretable convolutional neural networks. J. Chem. Inf. Model. (2020)
21. Yin, X., Gounaris, C.E.: Search methods for inorganic materials crystal structure prediction. Curr. Opin. Chem. Eng. **35**, 100726 (2022)
22. Zhao, Y., et al.: Machine learning-based prediction of crystal systems and space groups from inorganic materials compositions. ACS Omega **5**, 3596–3606 (2020)
23. Zheng, X., Zheng, P., Zhang, R.-Z.: Machine learning material properties from the periodic table using convolutional neural networks. Chem. Sc. **9**, 8426–8432 (2018)
24. Ziletti, A., Kumar, D., Scheffler, M., Ghiringhelli, L.M.: Insightful classification of crystal structures using deep learning. Nat. Commun. **9** (2018)

LS-PON: A Prediction-Based Local Search for Neural Architecture Search

Meyssa Zouambi(✉), Julie Jacques, and Clarisse Dhaenens

Univ. Lille, CNRS, Centrale Lille, UMR 9189 CRIStAL, 59000 Lille, France
{meyssa.zouambi,julie.jacques,clarisse.dhaenens}@univ-lille.fr

Abstract. Neural architecture search (NAS) is a subdomain of AutoML that consists of automating the design of neural networks. NAS has become a hot topic in the last few years. As a result, many methods are being developed in this area. Local search (LS), on the other hand, is a famous heuristic that has been around for many years. It is extensively used for optimization problems due to its simplicity and efficiency. LS has a lot of advantages in the world of NAS; it can naturally exploit methods that accelerate the global search time such as weight inheritance and network morphism. LS is also easy to implement and does not require a complex encoding or any parameter tuning. In the present work, we aim at making LS faster by guiding the exploration of the neighborhood. Our objective is to limit the number of solution evaluations, which are particularly time-consuming in NAS. We propose the method LS-PON (Local Search with a Predicted Order of Neighbors) that uses linear regression models to order the exploration of neighbors during the search. LS-PON, unlike other prediction-based NAS methods, requires neither pre-sampling nor tuning. Our experiments on popular NAS benchmarks show that LS-PON keeps the simplicity and advantages of LS while being as efficient in quality as state-of-the-art methods and can be more than twice as fast as classical LS.

Keywords: Neural architecture search · Local search · Optimization · Machine learning

1 Introduction

Neural architecture search has attracted a lot of attention in recent years. Researchers aim to develop the most efficient algorithms to automate the time-consuming task of architecture design. Although many complex methods were proposed in the past few years, the necessity of having such complex algorithms is sometimes questioned [28]. Local search is a simple heuristic that proved its efficiency in the field of NAS. Recent works that studied local search for NAS proved that it is competitive with state-of-the-art methods [3,24]. On top of its easy implementation, LS has a lot of advantages in this area. First, it does not require a complex encoding or any parameter tuning. It can also easily exploit

G. Nicosia et al. (Eds.): LOD 2022, LNCS 13810, pp. 108–122, 2023.
https://doi.org/10.1007/978-3-031-25599-1_9

strategies that accelerate global search time like weight inheritance [20] and network morphism [23].

Our work focuses on using machine learning to speed up local search for NAS. The objective is to keep all of the benefits that LS offers while being more time-efficient. We achieve this speed-up by ordering the exploration of neighbors of a solution using performance predictors. Evaluating architectures in NAS is the most time-consuming step of the process, so to be efficient, it is worth visiting the most promising neighbors first. We rely on machine learning to predict the ranking of a solution based on its accuracy and determine the neighborhood order. Unlike most prediction-based NAS methods, this one does not require any pre-sampling or parameter-tuning. The method starts with a random ordering and progressively gives more accurate rankings. The experimental results show that LS-PON can be more than two times faster than classical LS while achieving similar state-of-the-art performances.

The contributions of our work are summarized as follows:

- We create a simple and parameter-free method for NAS based on a local search and machine learning. This method is easy to implement and requires neither pre-sampling nor parameter-tuning.
- We use three different NAS benchmarks (with small-scale and large-scale search spaces) to validate the performance of our method. We confirm the competitiveness of LS and show that our method gives similar results while being significantly faster in almost all cases.

The remaining of this paper is organized as follows: related works are presented in Sect. 2. Section 3 provides a brief description of neural architecture search and local search. Section 4 details the proposed approach. Section 5 is dedicated to the experiments. It presents the benchmarks, the experimental protocol, and the results obtained. Section 6 gives the conclusion of our findings.

2 Related Works

Very few works investigated the efficiency of local search for NAS. In [24], authors explore the theoretical characterization of the landscape and its effect on the performance of local search. They proved that LS is a very competitive method when the noise in the evaluation pipeline is reduced to a minimum. Within this setting, they demonstrate that hill-climbing -the most known form of LS- can outperform many popular state-of-the-art algorithms on the popular NAS benchmark datasets. These results are confirmed in [3], where LS is employed in a multi-objective context. It shows that local search competes with state-of-the-art evolutionary algorithms, even up to thousands of evaluated architectures in the multi-objective setting. LS is therefore a method that is very easy to implement yet yields competitive performances against more complex algorithms.

Local search was also used in conjunction with network morphism as seen in [6,12]. Network morphism is a popular method to rapidly search efficient convolutional neural networks [23]. This technique allows the expansion of the network using function preserving operations and prevents training the resulting architectures from scratch. LS is a natural way of exploiting this method since its resulting architecture generates "neighbors" of the current solution.

Another way to speed up the search is to avoid the computationally expensive network training by using performance predictors [7]. Performance predictors help to identify good architectures using their characteristics only. They vary from simple decision trees to deep neural networks [1,15,18]. Recent works however, opt for using complex models to accurately represent the huge number of possible architectures of the search space [14,22]. These methods usually require a pre-sampling step to gather enough architecture-performance pairs for building the prediction model. To evaluate these NAS approaches, it is necessary to consider the training time required to sample and train architectures for the predictor, as well as the design and tuning of its hyperparameters. The latter task can be considered as counterproductive in this context, especially if the prediction models used are neural networks.

In [26], an interesting perspective is given on performance predictors. Authors state that using multiple simple prediction models at different stages of the search can be more efficient than using a single complex predictor. They emphasize that the goal of NAS is to sample the best architectures, and most of the solutions in the search space will not be evaluated at all. So it is not necessary for a predictor to accurately estimate the performance of all of them. This work iteratively creates weak predictors to determine which architecture is the best in the current subset of the search space. This aspect of locality is important for prediction. As this work demonstrates, solutions close to each other in a search space are more likely to fit well using a simple predictor.

3 Background

Before proceeding to the proposed method, we provide in this section a brief overview of neural architecture search and local search.

3.1 Neural Architecture Search

To design an efficient neural network, many parameters need to be taken into account, such as the number of layers, the type of operations to use, the hyperparameters linked to each type of operation, etc. This leaves us with a lot of potential models that perform differently based on their architecture.

The goal of neural architecture search is to automatically find an architecture a among a set of architectures $\mathcal{A}$, that achieves the best objective value. Usually, the purpose is to minimize the error on the validation dataset after training the network on the training dataset.

Formally, we can express the NAS problem as follows:

$$\arg\min_{a \in \mathcal{A}} = \mathcal{L}\left(a, \mathcal{D}_{\mathrm{t}}, \mathcal{D}_{\mathrm{v}}\right)$$

$\mathcal{A}$ defines the search space, it contains all potential neural architectures. $\mathcal{L}(\cdot)$ is the cost function that measures the error of the architecture a on the validation dataset $\mathcal{D}_\mathrm{v}$ after being trained on the training dataset $\mathcal{D}_\mathrm{t}$.

Due to the large search space of possible architectures in NAS, many search strategies were developed to explore it more efficiently. The most popular NAS strategies are gradient-based approaches [16], evolutionary algorithms [17], and reinforcement learning strategies [9]. During the search, architectures are sampled and evaluated. The classical way of evaluating an architecture is to train it using data from a training set and then assess its performance on a validation set. This task is time-consuming and is considered to be the bottleneck of most NAS algorithms. For this reason, many techniques were proposed for estimating the performance of neural networks without having to fully train them. Such methods are weight sharing [27], network morphism [6,12], weight inheritance [20], and neural predictors [25].

3.2 Local Search

Local search is a popular heuristic used to approximately solve NP-hard optimization problems. It starts from an initial solution s_0, chosen at random or by using another heuristic. It then generates neighbors of this solution by applying a *neighborhood* function N. This function applies small changes to the current solution to create neighboring ones that are close to it. Different ways of exploring the neighborhood lead to different variants of local search. The heuristic evaluates these neighbors using a cost function f that assesses their quality. It substitutes the current solution s with a better one from $N(s)$. After this update, it reiterates the process until convergence. The search stops when no neighbor is better than the current solution, so we can no longer improve it (the heuristic reaches a local optimum).

The most popular form of LS is called *hill-climbing*. In this form, the search updates the current solution with a better one from the neighborhood. Several exploration strategies can be chosen: the *first-improve* updates the current solution with the first improving neighbor found. The *best-improve* strategy updates the solution after evaluating all neighbors and picking the one that improves it the most. The *worst-improve* strategy evaluates all neighbors and chooses the one that improves the current solution the least.

In our work, we will be using hill-climbing, with the first-improve strategy, which allows exploring only a subset of the neighbors of each solution, as our purpose is to evaluate the least number of architectures for a faster search.

4 Proposed Approach

In the following, we define the important components of the proposed method and its process. It is important to note that the solution encoding, neighborhood function, and solution evaluation highly depend on the NAS task. In this paper, we focus on image classification (using NAS benchmarks) but the global idea of the proposed approach can be adapted to other types of tasks as well.

4.1 Solution Encoding

In the context of NAS, a solution s defines an architecture of a neural network. A representation that can be translated to a neural network can be used to encode it. In this work, the used encoding is a list of categorical, discrete, and/or continuous values for representing an architecture. This list can specify the type of operations of each node in the network, the connections between these nodes, the hyperparameters used for each operation, etc. This representation in the form of a list of values is similar to an entry of tabular data. It allows to directly use it as a dataset for building machine learning predictors later in this work.

Figure 1, shows a simplified example of a CNN encoding. On the top, the encoding is represented, which is the list of operations applied to the data and their hyperparameters (note that other hyperparameters not mentioned in the encoding are fixed and not optimized during the search). On the bottom, there is the corresponding CNN with its operations. In our work, we use three different NAS benchmarks, NAS-Bench-201 [4], MacroNasBenchmark [3], and NAS-Bench-301 [21]. Each benchmark defines its own set of operations and their corresponding parameters. This gives a different number of possible solutions for each of them, which defines the size of the search space.

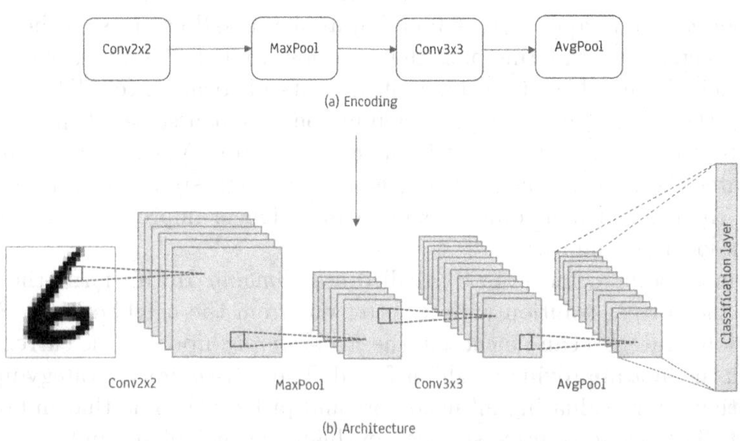

Fig. 1. Encoding of an architecture.

4.2 Neighborhood Function

The neighborhood function N generates a neighborhood $N(s)$ that contains the architectures close to s. These architectures can defer by a single element such as an operation or an edge from the initial solution. In our work, we use the *one exchange neighborhood* as defined in [8].

In the studied NAS benchmarks, variables are all categorical. Hence, a neighbor is obtained by modifying a single variable. The neighborhood of a solution is the set of solutions obtained by selecting one by one each variable and enumerating all the possible values. The number of neighbors for each solution in NAS-Bench-201, MacroNasBenchmark, and NAS-Bench-301 are respectively 24, 28, and 136 neighbors. The size of the neighborhood is relatively small, but let us recall that the evaluation of a solution is very costly.

Figure 2 shows an example of a neighbor generation. On top, there is the current solution. Generating one neighbor consists in choosing one operation and changing it by another, for example here the max pooling operation is replaced with a convolution of 3×3 kernel.

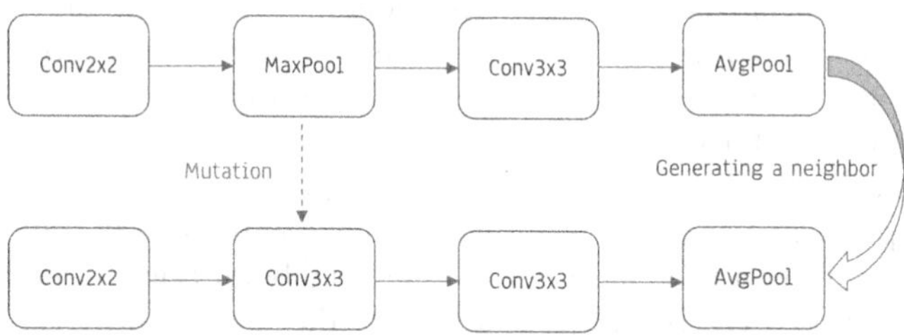

Fig. 2. Neighbor generation with *One exchange neighborhood* function.

4.3 Solution Evaluation

Evaluating an architecture is the most time-consuming step in NAS. It requires training the architecture using a training set and assessing its performance using a validation set.

Depending on the task, architecture, and hardware used, this step can take several hours for a single evaluation. Therefore, in this work, we use NAS benchmarks that provide surrogate performance metrics on both the training set and validation set for all possible architectures. Since these benchmarks deal with image classification, their evaluation is based on the classification accuracy and is calculated as the sum of well-classified images divided by the total number of images of the set.

4.4 Performance Prediction

In a normal setting, the local search engine evaluates neighbors in a random order. The proposed method, however, orders these neighbors to evaluate the most promising ones first. This order relies on performance predictions made with linear regression models. We choose to work with linear regressions, as they are simple, easy to implement, and do not require parameter tuning. This choice is also based on the work presented in [26], which states that using simple predictors is sufficient to estimate the performances of architectures that are close to each other.

As a reminder, linear regression is a method that assumes a linear relationship between a set of variables $X = (x_1, \ldots, x_p)$ and an output variable y as follows:

$$y = \beta_0 + \beta_1 x_1 + \beta_2 x_2 + \ldots + \beta_p x_p$$

Coefficients $\beta = (\beta_0, \ldots, \beta_p)$ are learned by minimizing the residual sum of squares between the observed targets in the dataset, and the targets predicted by the linear approximation.

In this work, the linear models are trained on architecture-performance pairs. The y variable is the performance we want to predict using the x_i variables which are the list of values describing an architecture. Note that in this work, the predicted performance is a score corresponding to the ranking of a solution (the normalized value of its rank). Since the encoding contains categorical data, we use one-hot-encoding to create a binary column for each category and use it as a numerical value for the linear regression.

Our method does not require any pre-sampling to build the dataset for architecture-performance pairs. The first models created can be random, which is equivalent to using a random order for the neighbors. Each architecture evaluated during the search, is added to a history database. They will be used in future iterations for creating more accurate prediction models.

4.5 LS-PON Process

Our LS-PON algorithm is an improved version of hill-climbing specifically designed for NAS. As we mentioned earlier, hill-climbing updates the current solution as soon as it finds a better one in the neighborhood. For this reason, the speed of this method relies on the order in which the neighbors are evaluated. The algorithm progresses more quickly if we assess the performance of the most promising neighbors first. Then, the method is more likely to immediately improve the current solution and move on with the search.

To determine the best order in which to evaluate neighborhoods, we use linear regression models. The models do not need to be sophisticated or accurate to yield good performances, but just good enough to provide an approximate ranking of the best neighbors quickly.

The process of LS-PON is illustrated in Fig. 3 and works as follows: after choosing an initial solution, the method generates the neighborhood of this architecture in step one (1). In step two (2), the method creates a linear regression model to predict the performance of architectures based on their parameters. It predicts the ranking of the neighbors and evaluates them in that order in the third step (3). Each evaluated architecture gets added to the database of architecture-performance pairs (step 4). This database will later be used to create a new (more accurate) linear model in the next iteration. If the solution is not better than the current one, it moves to the next one in the neighborhood (step 5a), else, it updates the current solution and reiterates the process (step 5b). If the search can no longer improve the current solution and the max budget of evaluations is not reached, it samples a new random solution and restarts the search. Algorithm 1 gives a detailed description of the proposed method.

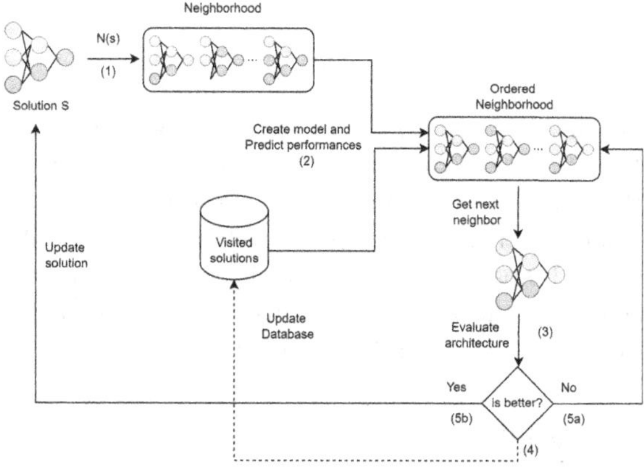

Fig. 3. Main steps of LS-PON.

Algorithm 1. Local search with a predicted order of neighbors - LS-PON

Input: $\mathcal{A}$: search space
N: neighborhood function
$maxBudget$: maximum number of evaluations
Initialization
Randomly pick an architecture $a \in \mathcal{A}$
$acc\_a, best\_acc \leftarrow Evaluate(a)$
$D \leftarrow \cup\{(a, acc\_a)\}$
$nbEval \leftarrow 1$
while $nbEval \leq maxBudget$ **do**
  Create prediction model M and train it using D
  Predict rank of each $u \in N(a)$ using M
  Order $N(a)$ based on the predicted ranks
  for $u \in$ Ordered $N(a)$ **do**
    $acc \leftarrow Evaluate(u)$
    $D \leftarrow D \cup \{(u, acc)\}$
    $nbEval \leftarrow nbEval + 1$
    if $acc \geq best\_acc$ **then**
      Update current solution: $a \leftarrow u$
      Update best score: $best\_acc \leftarrow acc$
      Exit for loop
    end if
    if $nbEval > maxBudget$ **then**
      return best architecture
    end if
  end for
  if best solution not updated, randomly sample a new one and continue searching
end while
Output: Best architecture found

5 Experiments

To test the performance of the proposed algorithm, three different benchmarks are used: NAS-Bench-201 [4], NAS-Bench-301 [21], and MacroNASBenchmark [3]. The choice of the first two benchmarks is to test the algorithm in a cell-based search space, with a small-scale and a large-scale setting. The last benchmark is for assessing the performance in a macro search space. The difference between the two types of search spaces can be found in [5]. In the following, a detailed description of each benchmark is given. This section also presents the experimentation protocol as well as the results and their discussion.

5.1 Benchmark Details

NAS-Bench-201 [4]. The NAS-Bench-201 benchmark consists of 15,625 unique architectures, representing the search space. An architecture of the NAS-Bench-201 search space consists of one repeated cell. This cell is a complete directed acyclic graph (DAG) of 4 nodes and six edges. Each edge can take one of the five available operations (1×1 convolution, 3×3 convolution, 3×3 avg-pooling, skip, no connection), which leads to 15,625 possible architectures. For each of these architectures, precomputed training, validation, and test accuracies are provided for CIFAR-10, CIFAR-100 [10], and ImageNet-16-120 [2]. CIFAR and ImageNet [11] are the most popular datasets for image classification. Two solutions from this benchmark are neighbors if they differ by exactly one operation in one of their edges. Thus the number of neighbors is 24.

MacroNASBenchmark [3]. The MacroNASBenchmark is a relatively large-scale benchmark with more than 200k unique architectures. It is based on a macro-level search consisting of 14 unrepeated modules. Each module can take one of three options (Identity, MBConv with expansion rate 6 and kernel size 3×3, MBConv with expansion rate 6 and kernel size of 5×5). For each architecture, precomputed validation and test accuracies are provided for CIFAR-10 and CIFAR-100. A neighboring solution in this benchmark is a solution that differs by one module from the current solution.

NAS-Bench-301 [21]. The NAS-Bench-301 is a surrogate benchmark based on the DARTS [16] search space for the CIFAR-10 dataset. This popular search space for large-scale NAS experiments contains 10^{18} architectures. An architecture of the DARTS search space consists of two repeated cells, a convolutional cell and a reduction cell, each with six nodes. The first two nodes are input from previous layers, and the last four nodes can take any DAG structure such that each node has degree two. Each edge can take one of the eight possible operations. In this benchmark, the classification accuracies of the architectures on the CIFAR-10 datasets are estimated through a surrogate model, removing the constraint to evaluate the entire search space. A neighboring solution is a solution that differs by an edge connection or an operation from the current solution.

5.2 Experimentation Protocol

The experimentation protocol used to evaluate our algorithm goes as follows:

For NAS-Bench-201 and MacroNasBenchmark, we use the validation accuracy as the search metric. Hence, the reference will be the best validation accuracy (Optimal) provided by the benchmark. We compare the algorithm to a standard local search, and a random search. For NAS-Bench-201, we further compare the results to Regularized Evolution Algorithm (REA) [19], which is the best-achieving algorithm reported on the NAS-Bench-201 paper [4]; results on REA are taken as-is from this paper. We set the maximum number of evaluations for these two benchmarks to 1500 evaluated architectures.

For NAS-Bench-301 [21], we report the validation accuracy given by the surrogate model provided in the benchmark. In the literature, experiments conducted on this search space suggest that the best architectures perform around 95% of validation accuracy [21]. As previously, we compare our algorithm against a local search and a random search. Since it has a larger search space than the previous two benchmarks, we set the maximum number of evaluations to 3000 architectures.

For the prediction models, we use the linear regression model from the Scikit-learn library v0.23 (*sklearn.linear_model.LinearRegression* with default parameters).

For each benchmark, all experiments are averaged over 150 runs. LS and LS-PON start with the same initial solution in every run. Note that the algorithms will restart after converging as long as they have not exhausted the maximum number of evaluations budget. We compare the algorithms from different standpoints: the mean accuracy, the convergence speed, and the dynamic of each method.

5.3 Results

Table 1 reports the validation accuracy for random search (RS), local search (LS), and LS-PON on the three benchmarks. It also reports the REA [19] method for NAS-Bench-201 for comparison. We notice that LS and LS-PON give state-of-the-art results in all benchmarks. Finding either optimal or close to optimal accuracy in all cases. With LS-PON slightly surpassing LS in some cases. Both methods significantly surpass random search. For REA, despite being the best-reported algorithm on the NAS-Bench-201 paper, random search still outperforms it in all available datasets. This could suggest that the hyperparameters of REA are too specific to the NAS problem it was designed for, and did not generalize well. This also confirms that random search is a strong baseline in NAS [13].

Since both algorithms, LS ans LS-PON, give similar performances on data quality, to further compare them, we need to analyze their speed. Indeed, in a none-benchmark setting, each sampled architecture needs to be trained. Training an architecture takes significantly more time than running a local search. For

Table 1. Performance evaluation on the three benchmarks. Each algorithm uses the validation accuracy as a search signal. *Optimal* indicates the highest validation accuracy provided by the benchmark. We report the mean and std deviation of 150 runs for Random search, LS, LS-PON. † taken from [4].

Method	NAS-Bench-201			MacroNasBench		NAS-Bench-301
	Cifar10	Cifar100	ImageNet	Cifar10	Cifar100	Cifar10
REA†	91.19 ± 0.31	71.81 ± 1.12	45.15 ± 0.89	–	–	–
RS	91.48 ± 0.09	72.93 ± 0.38	46.39 ± 0.23	92.22 ± 0.06	70.22 ± 0.08	94.52 ± 0.08
LS	$\mathbf{91.61 \pm 0.00}$	$\mathbf{73.49 \pm 0.00}$	46.72 ± 0.03	92.46 ± 0.04	70.45 ± 0.02	$\mathbf{95.11 \pm 0.05}$
LS-PON	$\mathbf{91.61 \pm 0.00}$	$\mathbf{73.49 \pm 0.00}$	$\mathbf{46.73 \pm 0.00}$	$92.48 \pm \mathbf{0.02}$	$\mathbf{70.47 \pm 0.02}$	$\mathbf{95.11 \pm 0.06}$
Optimal	**91.61**	**73.49**	**46.73**	**92.49**	**70.48**	**≈95**

Table 2. Speed evaluation on the three benchmarks. The table indicates the mean and std deviation of the number of evaluated architectures before convergence. Values in bold mean statistically better (Wilcoxon's test).

Method	NAS-Bench-201			MacroNasBench		NAS-Bench-301
	Cifar10	Cifar100	ImageNet	Cifar10	Cifar100	Cifar10
LS	76 ± 54	70 ± 48	617 ± 389	566 ± 414	628 ± 393	$\mathbf{1499 \pm 859}$
LS-PON	$\mathbf{64 \pm 54}$	$\mathbf{44 \pm 23}$	$\mathbf{276 \pm 159}$	$\mathbf{426 \pm 346}$	$\mathbf{483 \pm 417}$	1588 ± 857
Speedup	**18%**	**59%**	**123%**	**32%**	**30%**	**-6%**

this reason, the number of sampled architectures is a strong indicator of the speed of each method.

Table 2 reports the mean and standard deviation of the number of evaluated architectures during the search. Both LS and LS-PON restart after convergence as long as the evaluation budget is not reached. Since both algorithms end with almost similar accuracies, we use a *leveled* convergence for a fair speed comparison. We calculate how many evaluations were required for each algorithm to reach the same final accuracy. Since random search never surpasses these two methods after exhausting all of its evaluated budget (1500 for the first two benchmarks and 3000 for the last one) it is omitted from this table. The acceleration obtained using LS-PON is reported in the last line of the table. It shows that in both NAS-Bench-201 and MacroNasBenchmark, there is a significant speedup ranging from 18% to up to 123%. In NAS-Bench-301, however, LS-PON shows less efficiency with a slower convergence time. Figure 4 gives examples that illustrate the evolution of the validation accuracy across the number of evaluated architectures for a different dataset in each benchmark. It shows how LS-PON is faster than LS in NAS-Bench-201 and MacroNasBenchmark for ImageNet16-120 (IMNT) and Cifar100 respectively, and how it is less slightly less efficient in NAS-Bench-301 for its Cifar10 dataset.

Table 3 further shows the dynamic of these methods. It reports the number of evaluated neighbors before improving on the current solution. In NAS-Bench-201 and MacroNasBenchmark, it requires from 2 times to 3 times the number

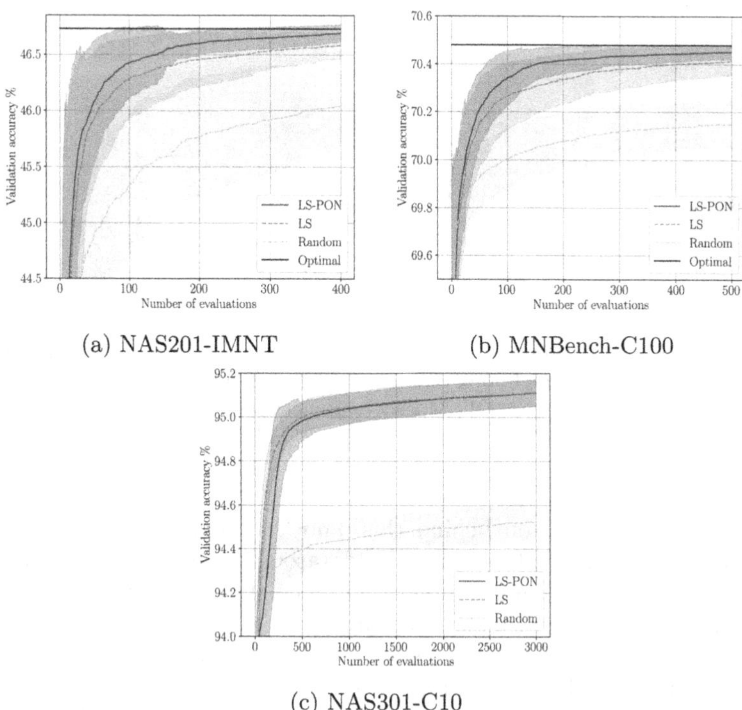

(a) NAS201-IMNT (b) MNBench-C100

(c) NAS301-C10

Fig. 4. Example of the evolution of the validation accuracy across the number of evaluated architectures for a dataset of each benchmark. NAS201-IMNT, MNBench-C100, and NAS301-C10 correspond to NAS-Bench-201 for ImageNet16-120, MacroNasBenchmark for Cifar100 and NAS-Bench-301 for Cifar10. Results are averaged for 150 runs for all algorithms. The shaded region indicates standard deviation of each search method.

of evaluations for LS to find a better neighbor compared to LS-PON. For this reason, LS-PON progresses more quickly during the search. In NAS-Bench-301, however, it takes around 11% more neighbors evaluation for LS-PON to find a better neighbor compared to LS, which makes LS-PON slightly slower in this case.

Table 3. Results on all benchmarks. Table indicates the mean and std deviation of number of required evaluations before improving on the current solution during the search. Values in bold mean statistically better (Wilcoxon's test).

Method	NAS-Bench-201			MacroNasBench		NAS-Bench-301
	Cifar10	Cifar100	ImageNet	Cifar10	Cifar100	Cifar10
LS	4.28 ± 4.14	4.29 ± 4.10	4.20 ± 4.10	5.07 ± 5.54	5.20 ± 5.68	$\mathbf{17.33 \pm 25.31}$
LS-PON	$\mathbf{1.87 \pm 2.32}$	$\mathbf{1.69 \pm 1.99}$	$\mathbf{1.61 \pm 1.71}$	$\mathbf{2.35 \pm 3.02}$	$\mathbf{2.28 \pm 2.77}$	$19.33 \pm \pm 29.60$

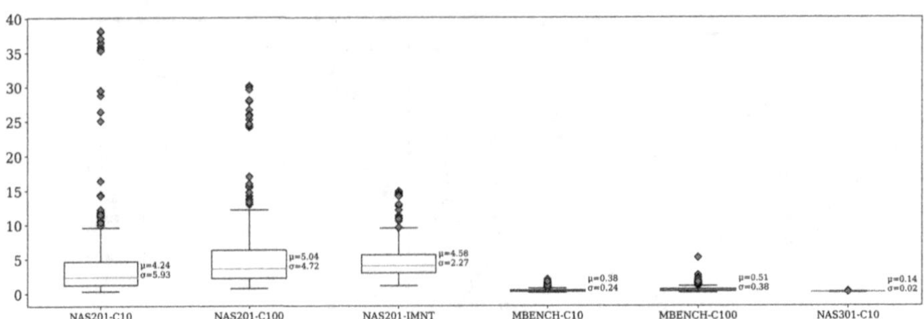

Fig. 5. Box plots representing the distribution of distances between the validation accuracy of a solution and its neighbors, diamonds represent outliers. NAS201, MNBench, NAS301 stand for NAS-Bench-201, MacroNasBenchmark and NAS-Bench-301. C10, C100 and IMNT are respectively Cifar10, Cifar100 and ImageNet16-120.

To investigate the reason behind this, and understand the effectiveness of this method in the different benchmarks/datasets, an analysis on the distance between the current solution's accuracy and its neighbors accuracy is conducted. This analysis seeks to determine if the difference between the neighbors accuracies affects the method's effectiveness. To do this, the mean absolute error between the current solution accuracy, and its neighbors accuracy is calculated for a sample of solutions. Results of this are represented in Fig. 5. We see that for the three datasets in NAS-Bench-201, there is a notable difference between the neighbors performances. Hence, the predictor can more easily classify them and identify the best ones. This difference is less visible on MacroNasBenchmark but the benchmark still has a certain number of outliers (represented by diamonds in the Figure) that can be easily recognized by the predictor.

On the other hand, in NAS-Bench-301's dataset, the difference between the neighbors is negligible (there is a mean of 0.14% difference in their accuracy) and there are also not many noticeable outliers to recognize during the search. The search space of this benchmark is mostly composed of good architectures with very close performances as presented in their paper [21].

This shows that the method is more efficient if the improvement to make is relatively observable. Which is expected in problems where the search space contains a diverse set of solutions.

6 Conclusion

In this paper, we introduced LS-PON, an improved local search based on performance predictors for neighborhood ordering. This method is fast, does not require any hyper-parameter tuning, is easy to implement, and yields state-of-the-art results on three popular NAS benchmarks.

LS-PON proved that it can be more than twice as fast as the LS without the ordering mechanism. Its effectiveness relies on the diversity of solutions in the search space, and works better if there is an observable difference between neighbors.

In the future, we will test other types of predictors and analyze their impact on the results. We also aim at making this method more robust to search spaces that mostly contain solutions with very close performances.

Another interesting addition would be to apply this method in a multi-objective context, and see if it scales well with the growing objectives of architecture design, such as the size of the network, the energy consumption, the inference times, etc.

References

1. Baker, B., Gupta, O., Raskar, R., Naik, N.: Accelerating neural architecture search using performance prediction. arXiv:1705.10823 (2017)
2. Chrabaszcz, P., Loshchilov, I., Hutter, F.: A downsampled variant of imagenet as an alternative to the cifar datasets. arXiv:1707.08819 (2017)
3. Den Ottelander, T., Dushatskiy, A., Virgolin, M., Bosman, P.A.N.: Local search is a remarkably strong baseline for neural architecture search. In: Ishibuchi, H., Zhang, Q., Cheng, R., Li, K., Li, H., Wang, H., Zhou, A. (eds.) EMO 2021. LNCS, vol. 12654, pp. 465–479. Springer, Cham (2021). https://doi.org/10.1007/978-3-030-72062-9_37
4. Dong, X., Yang, Y.: Nas-bench-201: Extending the scope of reproducible neural architecture search. arXiv:2001.00326 (2020)
5. Elsken, T., Metzen, J.H., Hutter, F.: Neural architecture search, pp. 69–86
6. Elsken, T., Metzen, J.-H., Hutter, F.: Simple and efficient architecture search for convolutional neural networks. arXiv:1711.04528 (2017)
7. Elsken, T., Metzen, J.H., Hutter, F.: Neural architecture search: a survey. J. Mach. Learn. Res., 1997–2017 (2019)
8. Hutter, F., Hoos, H.H., Leyton-Brown, K.: Sequential model-based optimization for general algorithm configuration. In: Coello, C.A.C. (ed.) LION 2011. LNCS, vol. 6683, pp. 507–523. Springer, Heidelberg (2011). https://doi.org/10.1007/978-3-642-25566-3_40
9. Jaafra, Y., Laurent, J.L., Deruyver, A., Naceur, M.S.: Reinforcement learning for neural architecture search: a review. Image and Vision Computing, pp. 57–66 (2019)
10. Krizhevsky, A., Hinton, G., et al.: Learning multiple layers of features from tiny images (2009)
11. Krizhevsky, A., Sutskever, I., Hinton, G.E.: Imagenet classification with deep convolutional neural networks. In: Advances in Neural Information Processing Systems, pp. 1097–1105 (2012)
12. Kwasigroch, A., Grochowski, M., Mikolajczyk, M.: Deep neural network architecture search using network morphism. In: International Conference on Methods and Models in Automation and Robotics, pp. 30–35. IEEE (2019)
13. Li, L., Talwalkar, A.: Random search and reproducibility for neural architecture search. In: Uncertainty in Artificial Intelligence, pp. 367–377. PMLR (2020)
14. Li, Y., Dong, M., Wang, Y., Xu, C.: Neural architecture search in a proxy validation loss landscape. In: International Conference on Machine Learning, pp. 5853–5862. PMLR (2020)

15. Liu, C., et al.: Progressive neural architecture search. In: Ferrari, V., Hebert, M., Sminchisescu, C., Weiss, Y. (eds.) ECCV 2018. LNCS, vol. 11205, pp. 19–35. Springer, Cham (2018). https://doi.org/10.1007/978-3-030-01246-5_2
16. Liu, H., Simonyan, K., Yang, Y.: Darts: differentiable architecture search. arXiv:1806.09055 (2018)
17. Liu, Y., Sun, Y., Xue, B., Zhang, M., Yen, G.G., Tan, K.C.: A survey on evolutionary neural architecture search. IEEE Trans. Neural Networks Learn. Syst. (2021)
18. Luo, R., Tan, X., Wang, R., Qin, T., Chen, E., Liu, T.-Y.: Accuracy prediction with non-neural model for neural architecture search. arXiv:2007.04785 (2020)
19. Real, E., Aggarwal, A., Huang, Y., Le, Q.V.: Regularized evolution for image classifier architecture search. In: Proceedings of the AAAI Conference on Artificial Intelligence, pp. 4780–4789 (2019)
20. Real, E., et al.: Large-scale evolution of image classifiers. In: International Conference on Machine Learning, pp. 2902–2911. PMLR (2017)
21. Siems, J., Zimmer, L., Zela, A., Lukasik, J., Keuper, M., Hutter, F.: Nas-bench-301 and the case for surrogate benchmarks for neural architecture search. arXiv:2008.09777 (2020)
22. Wei, C., Niu, C., Tang, Y., Wang, Y., Hu, H., Liang, J.: Npenas: neural predictor guided evolution for neural architecture search. arXiv:2003.12857 (2020)
23. Wei, T., Wang, C., Rui, Y., Chen, C.W.: Network morphism. In: International Conference on Machine Learning, pp. 564–572. PMLR (2016)
24. White, C., Nolen, S., Savani, Y.: Exploring the loss landscape in neural architecture search. arXiv:2005.02960 (2020)
25. White, C., Zela, A., Ru, B., Liu, Y., Hutter, F.: How powerful are performance predictors in neural architecture search? arXiv:2104.01177 (2021)
26. Wu, J., et al.: Weak nas predictors are all you need. arXiv:2102.10490 (2021)
27. Xie, L., et al.: Weight-sharing neural architecture search: a battle to shrink the optimization gap. ACM Computing Surveys (CSUR), pp. 1–37 (2021)
28. Yu, K., Sciuto, C., Jaggi, M., Musat, C., Salzmann, M.: Evaluating the search phase of neural architecture search. arXiv:1902.08142 (2019)

Local Optimisation of Nyström Samples Through Stochastic Gradient Descent

Matthew Hutchings(✉) and Bertrand Gauthier

School of Mathematics, Cardiff University Abacws, Senghennydd Road, Cardiff CF24 4AG, UK
{hutchingsm1,gauthierb}@cardiff.ac.uk

Abstract. We study a relaxed version of the column-sampling problem for the Nyström approximation of kernel matrices, where approximations are defined from multisets of landmark points in the ambient space; such multisets are referred to as Nyström samples. We consider an unweighted variation of the radial squared-kernel discrepancy (SKD) criterion as a surrogate for the classical criteria used to assess the Nyström approximation accuracy; in this setting, we discuss how Nyström samples can be efficiently optimised through stochastic gradient descent. We perform numerical experiments which demonstrate that the local minimisation of the radial SKD yields Nyström samples with improved Nyström approximation accuracy in terms of trace, Frobenius and spectral norms.

Keywords: Low-rank matrix approximation · Nyström method · Reproducing kernel Hilbert spaces · Stochastic gradient descent

1 Introduction

In Data Science, the Nyström method refers to a specific technique for the low-rank approximation of symmetric positive-semidefinite (SPSD) matrices; see e.g. [4,5,10,11,18]. Given an $N \times N$ SPSD matrix $\mathbf{K}$, with $N \in \mathbb{N}$, the Nyström method consists of selecting a sample of $n \in \mathbb{N}$ columns of $\mathbf{K}$, generally with $n \ll N$, and next defining a low-rank approximation $\hat{\mathbf{K}}$ of $\mathbf{K}$ based on this sample of columns. More precisely, let $\boldsymbol{c}_1, \cdots, \boldsymbol{c}_N \in \mathbb{R}^N$ be the columns of $\mathbf{K}$, so that $\mathbf{K} = (\boldsymbol{c}_1|\cdots|\boldsymbol{c}_N)$, and let $I = \{i_1, \cdots, i_n\} \subseteq \{1, \cdots, N\}$ denote the indices of a sample of n columns of $\mathbf{K}$ (note that I is a multiset, i.e. the indices of some columns might potentially be repeated). Let $\mathbf{C} = (\boldsymbol{c}_{i_1}|\cdots|\boldsymbol{c}_{i_n})$ be the $N \times n$ matrix defined from the considered sample of columns of $\mathbf{K}$, and let $\mathbf{W}$ be the $n \times n$ principal submatrix of $\mathbf{K}$ defined by the indices in I, i.e. the k, l entry of $\mathbf{W}$ is $[\mathbf{K}]_{i_k,i_l}$, the i_k, i_l entry of $\mathbf{K}$. The Nyström approximation of $\mathbf{K}$ defined from the sample of columns indexed by I is given by

$$\hat{\mathbf{K}} = \mathbf{C}\mathbf{W}^{\dagger}\mathbf{C}^T, \tag{1}$$

with $\mathbf{W}^{\dagger}$ the Moore-Penrose pseudoinverse of $\mathbf{W}$. The column-sampling problem for Nyström approximation consists of designing samples of columns such that the induced approximations are as accurate as possible (see Sect. 1.2 for more details).

G. Nicosia et al. (Eds.): LOD 2022, LNCS 13810, pp. 123–140, 2023.
https://doi.org/10.1007/978-3-031-25599-1_10

1.1 Kernel Matrix Approximation

If the initial SPSD matrix $\mathbf{K}$ is a kernel matrix, defined from a SPSD kernel K and a set or multiset of points $\mathcal{D} = \{x_1, \cdots, x_N\} \subseteq \mathscr{X}$ (and with $\mathscr{X}$ a general ambient space), i.e. the i,j entry of $\mathbf{K}$ is $K(x_i, x_j)$, then a sample of columns of $\mathbf{K}$ is naturally associated with a subset of $\mathcal{D}$; more precisely, a sample of columns $\{\boldsymbol{c}_{i_1}, \cdots, \boldsymbol{c}_{i_n}\}$, indexed by I, naturally defines a multiset $\{x_{i_1}, \cdots, x_{i_n}\} \subseteq \mathcal{D}$, so that the induced Nyström approximation can in this case be regarded as an approximation induced by a subset of points in $\mathcal{D}$. Consequently, in the kernel-matrix framework, instead of relying only on subsets of columns, we may more generally consider Nyström approximations defined from a multiset $\mathcal{S} \subseteq \mathscr{X}$. Using matrix notation, the Nyström approximation of $\mathbf{K}$ defined by a subset $\mathcal{S} = \{s_1, \cdots, s_n\}$ is the $N \times N$ SPSD matrix $\hat{\mathbf{K}}(\mathcal{S})$, with i,j entry

$$\big[\hat{\mathbf{K}}(\mathcal{S})\big]_{i,j} = \mathbf{k}_{\mathcal{S}}^T(x_i)\mathbf{K}_{\mathcal{S}}^{\dagger}\mathbf{k}_{\mathcal{S}}(x_j), \tag{2}$$

where $\mathbf{K}_{\mathcal{S}}$ is the $n \times n$ kernel matrix defined by the kernel K and the subset $\mathcal{S}$, and where

$$\mathbf{k}_{\mathcal{S}}(x) = \big(K(x, s_1), \cdots, K(x, s_n)\big)^T \in \mathbb{R}^n.$$

We refer to such a set or multiset $\mathcal{S}$ as a *Nyström sample*, and to the elements of $\mathcal{S}$ as *landmark points* (the terminology *inducing points* can also be found in the literature); the notation $\hat{\mathbf{K}}(\mathcal{S})$ emphasises that the considered Nyström approximation of $\mathbf{K}$ is induced by $\mathcal{S}$. As in the column-sampling case, the landmark-point-based framework naturally raises questions related to the characterisation and the design of efficient Nyström samples (i.e. samples leading to accurate approximations of $\mathbf{K}$). In this work, for a fixed $n \in \mathbb{N}$, we interpret Nyström samples of size n as elements of $\mathscr{X}^n$, and we investigate the possibility of directly optimising Nyström samples over $\mathscr{X}^n$. We consider the case $\mathscr{X} = \mathbb{R}^d$, with $d \in \mathbb{N}$, but $\mathscr{X}$ may more generally be a differentiable manifold.

Remark 1. Denoting by $\mathcal{H}$ the reproducing kernel Hilbert space (RKHS, see e.g. [1,14]) of real-valued functions on $\mathscr{X}$ associated with K, we may note that the matrix $\hat{\mathbf{K}}(\mathcal{S})$ is the kernel matrix defined by $K_{\mathcal{S}}$ and $\mathcal{D}$, with $K_{\mathcal{S}}$ the reproducing kernel of the closed linear subspace

$$\mathcal{H}_{\mathcal{S}} = \operatorname{span}\{k_{s_1}, \cdots, k_{s_n}\} \subseteq \mathcal{H},$$

where, for $t \in \mathscr{X}$, the function $k_t \in \mathcal{H}$ is defined as $k_t(x) = K(x, t)$, for all $x \in \mathscr{X}$. ◁

1.2 Assessing the Accuracy of Nyström Approximations

In the classical literature on the Nyström approximation of SPSD matrices, the accuracy of the approximation induced by a Nyström sample $\mathcal{S}$ is often assessed through the following criteria:

(C.1) $C_{\mathrm{tr}}(\mathcal{S}) = \|\mathbf{K} - \hat{\mathbf{K}}(\mathcal{S})\|_*$, with $\|.\|_*$ the trace norm;

(C.2) $C_{\mathrm{F}}(\mathcal{S}) = \|\mathbf{K} - \hat{\mathbf{K}}(\mathcal{S})\|_{\mathrm{F}}$, with $\|.\|_{\mathrm{F}}$ the Frobenius norm;
(C.3) $C_{\mathrm{sp}}(\mathcal{S}) = \|\mathbf{K} - \hat{\mathbf{K}}(\mathcal{S})\|_2$, with $\|.\|_2$ the spectral norm.

Although defining relevant and easily interpretable measures of the approximation error, these criteria are relatively costly to evaluate. Indeed, each of them involves the inversion or pseudoinversion of the kernel matrix $\mathbf{K}_{\mathcal{S}}$, with complexity $\mathcal{O}(n^3)$. The evaluation of the criterion (C.1) also involves the computation of the N diagonal entries of $\hat{\mathbf{K}}(\mathcal{S})$, leading to an overall complexity of $\mathcal{O}(n^3+Nn^2)$. The evaluation of (C.2) involves the full construction of the matrix $\hat{\mathbf{K}}(\mathcal{S})$, with an overall complexity of $\mathcal{O}(n^3 + n^2N^2)$, and the evaluation of (C.3) in addition requires the computation of the largest eigenvalue of an $N \times N$ SPSD matrix, leading to an overall complexity of $\mathcal{O}(n^3 + n^2N^2 + N^3)$. For $\mathscr{X} = \mathbb{R}^d$, the evaluation of the partial derivatives of these criteria (regarded as maps from $\mathscr{X}^n$ to $\mathbb{R}$) with respect to a single coordinate of a landmark point has a complexity similar to the complexity of evaluating the criteria themselves (and there are in this case nd such partial derivatives). Consequently, a direct optimisation of these criteria over $\mathscr{X}^n$ is intractable in most practical applications.

1.3 Radial Squared-Kernel Discrepancy

As a surrogate for the criteria (C.1)–(C.3), and following the connections between the Nyström approximation of SPSD matrices, the approximation of integral operators with SPSD kernels and the kernel embedding of measures, we consider the following *radial squared-kernel discrepancy* criterion (radial SKD, see [7,9]), denoted by R and given by, for $\mathcal{S} = \{s_1, \cdots, s_n\}$,

$$R(\mathcal{S}) = \begin{cases} \|\mathbf{K}\|_{\mathrm{F}}^2 - \frac{1}{\|\mathbf{K}_{\mathcal{S}}\|_{\mathrm{F}}^2} \left(\sum_{i=1}^{N} \sum_{j=1}^{n} K^2(x_i, s_j) \right)^2, & \text{if } \|\mathbf{K}_{\mathcal{S}}\|_{\mathrm{F}} > 0, \\ \|\mathbf{K}\|_{\mathrm{F}}^2, & \text{otherwise,} \end{cases} \tag{3}$$

where $K^2(x_i, s_j)$ stands for $\big(K(x_i, s_j)\big)^2$. We may note that $0 \leqslant R(\mathcal{S}) \leqslant \|\mathbf{K}\|_{\mathrm{F}}^2$. In (3), the evaluation of the term $\|\mathbf{K}\|_{\mathrm{F}}^2$ has complexity $\mathcal{O}(N^2)$; nevertheless, this term does not depend on the Nyström sample $\mathcal{S}$, and may thus be regarded as a constant. The complexity of the evaluation of the term $R(\mathcal{S}) - \|\mathbf{K}\|_{\mathrm{F}}^2$, i.e. of the radial SKD up to the constant $\|\mathbf{K}\|_{\mathrm{F}}^2$, is $\mathcal{O}(n^2 + nN)$; for $\mathscr{X} = \mathbb{R}^d$, the same holds for the complexity of the evaluation of the partial derivative of $R(\mathcal{S})$ with respect to a coordinate of a landmark point, see Eq. (5) below. Importantly, and in contrast to the criteria discussed in Sect. 1.2, the evaluation of the radial SKD criterion or of its partial derivatives does not involve the inversion or pseudoinversion of the $n \times n$ matrix $\mathbf{K}_{\mathcal{S}}$.

Remark 2. From a theoretical standpoint, the radial SKD criterion measures the distance, in the Hilbert space of all Hilbert-Schmidt operators on $\mathcal{H}$, between the integral operator corresponding to the initial matrix $\mathbf{K}$ (i.e. the integral operator defined from the kernel K and a uniform measure on $\mathcal{D}$), and the projection of this operator onto the subspace spanned by an integral operator defined from the kernel K and a uniform measure on $\mathcal{S}$. The radial SKD may also be defined for non-uniform measures, and the criterion in this case depends not only on $\mathcal{S}$, but also on a set of relative weights associated with each landmark point in $\mathcal{S}$; in this work, we only focus on the uniform-weight case. See [7,9] for more details. ◁

The following inequalities hold:

$$\big\|\mathbf{K}-\hat{\mathbf{K}}(\mathcal{S})\big\|_2^2 \leqslant \big\|\mathbf{K}-\hat{\mathbf{K}}(\mathcal{S})\big\|_{\mathrm{F}}^2 \leqslant R(\mathcal{S}) \leqslant \|\mathbf{K}\|_{\mathrm{F}}^2, \text{ and } \frac{1}{N}\big\|\mathbf{K}-\hat{\mathbf{K}}(\mathcal{S})\big\|_*^2 \leqslant \big\|\mathbf{K}-\hat{\mathbf{K}}(\mathcal{S})\big\|_{\mathrm{F}}^2,$$

which, in complement to the theoretical properties enjoyed by the radial SKD, further support the use of the radial SKD as a numerically-affordable surrogate for (C.1)–(C.3) (see also the numerical experiments in Sect. 4).

From now on, we assume that $\mathscr{X} = \mathbb{R}^d$. Let $[s]_l$, with $l \in \{1, \cdots, d\}$, be the l-th coordinate of s in the canonical basis of $\mathbb{R}^d$. For $x \in \mathscr{X}$, we denote by (assuming they exist)

$$\partial^{[\mathrm{l}]}_{[s]_l} K^2(s,x) \quad \text{and} \quad \partial^{[\mathrm{d}]}_{[s]_l} K^2(s,s) \tag{4}$$

the partial derivatives of the maps $s \mapsto K^2(s,x)$ and $s \mapsto K^2(s,s)$ at s and with respect to the l-th coordinate of s, respectively; the notation $\partial^{[\mathrm{l}]}$ indicates that the left entry of the kernel is considered, while $\partial^{[\mathrm{d}]}$ refers to the diagonal of the kernel; we use similar notations for any kernel function on $\mathscr{X} \times \mathscr{X}$.

For a fixed number of landmark points $n \in \mathbb{N}$, the radial SKD criterion can be regarded as a function from $\mathscr{X}^n$ to $\mathbb{R}$. For a Nyström sample $\mathcal{S} = \{s_1, \cdots, s_n\} \in \mathscr{X}^n$, and for $k \in \{1, \cdots, n\}$ and $l \in \{1, \cdots, d\}$, we denote by $\partial_{[s_k]_l} R(\mathcal{S})$ the partial derivative of the map $R : \mathscr{X}^n \to \mathbb{R}$ at $\mathcal{S}$ with respect to the l-th coordinate of the k-th landmark point $s_k \in \mathscr{X}$. We have

$$\begin{aligned}\partial_{[s_k]_l} R(\mathcal{S}) = \frac{1}{\|\mathbf{K}_{\mathcal{S}}\|_{\mathrm{F}}^4}\Big(\sum_{i=1}^{N}\sum_{j=1}^{n} K^2(s_j,x_i)\Big)^2 \Big(\partial^{[\mathrm{d}]}_{[s_k]_l} K^2(s_k,s_k) + 2\sum_{\substack{j=1,\\ j\neq k}}^{n} \partial^{[\mathrm{l}]}_{[s_k]_l} K^2(s_k,s_j)\Big) \\ - \frac{2}{\|\mathbf{K}_{\mathcal{S}}\|_{\mathrm{F}}^2}\Big(\sum_{i=1}^{N}\sum_{j=1}^{n} K^2(s_j,x_i)\Big)\Big(\sum_{i=1}^{N} \partial^{[\mathrm{l}]}_{[s_k]_l} K^2(s_k,x_i)\Big).\end{aligned} \tag{5}$$

By mutualising the evaluation of the terms in (5) that do not depend on k and l, the evaluation of the nd partial derivatives of R at $\mathcal{S}$ has a complexity of $\mathcal{O}\big((d+1)(n^2+nN)\big)$; by contrast (and although the pseudoinversion of $\mathbf{K}_{\mathcal{S}}$ can be mutualised), evaluating the nd partial derivatives of the trace criterion has a complexity of $\mathcal{O}\big(d(n^4+n^3N)\big)$.

In this work, we investigate the possibility to use the partial derivatives (5), or stochastic approximations of these derivatives, to directly optimise the radial SKD criterion R over $\mathscr{X}^n$ via gradient or stochastic gradient descent; the stochastic approximation schemes we consider aim at reducing the burden of the numerical cost induced by the evaluation of the partial derivatives of R when N is large.

The document is organised as follows. In Sect. 2, we discuss the convergence of a gradient descent with fixed step size for the minimisation of R over $\mathscr{X}^n$. The stochastic approximation of the gradient of the radial SKD criterion (3) is discussed in Sect. 3, and some numerical experiments are carried out in Sect. 4. Section 5 consists of a concluding discussion, and the Appendix contains a proof of Theorem 1.

2 A Convergence Result

We use the same notation as in Sect. 1.3 (in particular, we still assume that $\mathscr{X} = \mathbb{R}^d$), and by analogy with (4), for s and $x \in \mathscr{X}$, and for $l \in \{1, \cdots, d\}$, we denote by $\partial^{[r]}_{[s]_l} K^2(x, s)$ the partial derivative of the map $s \mapsto K^2(x, s)$ with respect to the l-th coordinate of s. Also, for a fixed $n \in \mathbb{N}$, we denote by $\nabla R(\mathcal{S}) \in \mathscr{X}^n = \mathbb{R}^{nd}$ the gradient of $R : \mathscr{X}^n \to \mathbb{R}$ at $\mathcal{S}$; in matrix notation, we have

$$\nabla R(\mathcal{S}) = \left(\left(\nabla_{s_1} R(\mathcal{S})\right)^T, \cdots, \left(\nabla_{s_1} R(\mathcal{S})\right)^T \right)^T,$$

with $\nabla_{s_k} R(\mathcal{S}) = \left(\partial_{[s_k]_1} R(\mathcal{S}), \cdots, \partial_{[s_k]_d} R(\mathcal{S})\right)^T \in \mathbb{R}^d$ for $k \in \{1, \cdots, n\}$.

Theorem 1. *We make the following assumptions on the squared-kernel K^2, which we assume hold for all x and $y \in \mathscr{X} = \mathbb{R}^d$, and all l and $l' \in \{1, \cdots, d\}$, uniformly:*

(A.1) there exists $\alpha > 0$ such that $K^2(x, x) \geqslant \alpha$;
(A.2) there exists $M_1 > 0$ such that $\left|\partial^{[d]}_{[x]_l} K^2(x, x)\right| \leqslant M_1$ and $\left|\partial^{[l]}_{[x]_l} K^2(x, y)\right| \leqslant M_1$;

(A.3) there exists $M_2 > 0$ such that $\left|\partial^{[d]}_{[x]_l} \partial^{[d]}_{[x]_{l'}} K^2(x, x)\right| \leqslant M_2$, $\left|\partial^{[l]}_{[x]_l} \partial^{[l]}_{[x]_{l'}} K^2 (x, y)\right| \leqslant M_2$ and $\left|\partial^{[l]}_{[x]_l} \partial^{[r]}_{[y]_{l'}} K^2(x, y)\right| \leqslant M_2$.

Let $\mathcal{S}$ and $\mathcal{S}' \in \mathbb{R}^{nd}$ be two Nyström samples; under the above assumptions, there exists $L > 0$ such that

$$\left\|\nabla R(\mathcal{S}) - \nabla R(\mathcal{S}')\right\| \leqslant L \left\|\mathcal{S} - \mathcal{S}'\right\|$$

with $\|.\|$ the Euclidean norm of $\mathbb{R}^{nd}$; in other words, the gradient of $R : \mathbb{R}^{nd} \to \mathbb{R}$ is Lipschitz-continuous with Lipschitz constant L.

Since R is bounded from below, for $0 < \gamma \leqslant 1/L$ and independently of the considered initial Nyström sample $\mathcal{S}^{(0)}$, Theorem 1 entails that a gradient descent from $\mathcal{S}^{(0)}$, with fixed stepsize γ for the minimisation of R over $\mathscr{X}^n$, produces a sequence of iterates that converges to a critical point of R. Barring some specific and largely pathological cases, the resulting critical point is likely to be a local minimum of R, see for instance [12]. See the Appendix for a proof of Theorem 1.

The conditions considered in Theorem 1 ensure the existence of a general Lipschitz constant L for the gradient of R; they, for instance, hold for all sufficiently regular Matérn kernels (thus including the Gaussian, or squared-exponential, kernel). These conditions are only sufficient conditions for the convergence of a gradient descent for the minimisation of R. By introducing additional problem-dependent conditions, some convergence results might be obtained for more general squared kernels K^2 and adequate initial Nyström samples $\mathcal{S}^{(0)}$. For instance, the condition (A.1) simply aims at ensuring that $\|\mathbf{K}_{\mathcal{S}}\|_{\mathrm{F}}^2 \geqslant n\alpha > 0$ for all $\mathcal{S} \in \mathscr{X}^n$; this condition might be relaxed to account for kernels with vanishing diagonal, but one might then need to introduce ad hoc conditions to ensure that $\|\mathbf{K}_{\mathcal{S}}\|_{\mathrm{F}}^2$ remains large enough during the minimisation process.

3 Stochastic Approximation of the Radial-SKD Gradient

The complexity of evaluating a partial derivative of $R : \mathscr{X}^n \to \mathbb{R}$ is $\mathcal{O}(n^2 + nN)$, which might become prohibitive for large values of N. To overcome this limitation, stochastic approximations of the gradient of R might be considered (see e.g. [2]).

The evaluation of (5) involves, for instance, terms of the form $\sum_{i=1}^N K^2(s, x_i)$, with $s \in \mathscr{X}$ and $\mathcal{D} = \{x_1, \cdots, x_N\}$. Introducing a random variable X with uniform distribution on $\mathcal{D}$, we can note that

$$\sum_{i=1}^{N} K^2(s, x_i) = N\mathbb{E}\big[K^2(s, X)\big],$$

and the mean $\mathbb{E}[K^2(s, X)]$ may then, classically, be approximated by random sampling. More precisely, if $X_1, \cdots, X_b$ are $b \in \mathbb{N}$ copies of X, we have

$$\mathbb{E}[K^2(s, X)] = \frac{1}{b}\sum_{j=1}^{b} \mathbb{E}[K^2(s, X_j)] \text{ and } \mathbb{E}[\partial^{[1]}_{[s]_l} K^2(s, X)] = \frac{1}{b}\sum_{j=1}^{b} \mathbb{E}[\partial^{[1]}_{[s]_l} K^2(s, X_j)],$$

so that we can easily define unbiased estimators of the various terms appearing in (5). We refer to the sample size b as the *batch size*.

Let $k \in \{1, \ldots, n\}$ and $l \in \{1, \ldots, d\}$; the partial derivative (5) can be rewritten as

$$\partial_{[s_k]_l} R(\mathcal{S}) = \frac{T_1^2}{\|\mathbf{K}_{\mathcal{S}}\|_{\mathrm{F}}^4} \Upsilon(\mathcal{S}) - \frac{2T_1 T_2^{k,l}}{\|\mathbf{K}_{\mathcal{S}}\|_{\mathrm{F}}^2},$$

with $T_1 = \sum_{i=1}^N \sum_{j=1}^n K^2(s_j, x_i)$ and $T_2^{k,l} = \sum_{i=1}^N \partial^{[1]}_{[s_k]_l} K^2(s_k, x_i)$, and

$$\Upsilon(\mathcal{S}) = \partial^{[\mathrm{d}]}_{[s_k]_l} K^2(s_k, s_k) + 2 \sum_{\substack{j=1,\\ j\neq k}}^{n} \partial^{[1]}_{[s_k]_l} K^2(s_k, s_j).$$

The terms T_1 and $T_2^{k,l}$ are the only terms in (5) that depend on $\mathcal{D}$. From a random sample $\mathbf{X} = \{X_1, \cdots, X_b\}$, we define the unbiased estimators $\hat{T}_1(\mathbf{X})$ of T_1, and $\hat{T}_2^{k,l}(\mathbf{X})$ of $T_2^{k,l}$, as

$$\hat{T}_1(\mathbf{X}) = \frac{N}{b}\sum_{i=1}^{n}\sum_{j=1}^{b} K^2(s_i, X_j), \quad \text{and} \quad \hat{T}_2^{k,l}(\mathbf{X}) = \frac{N}{b}\sum_{j=1}^{b} \partial_{[s_k]_l}^{[1]} K^2(s_k, X_j).$$

In what follows, we discuss the properties of some stochastic approximations of the gradient of R that can be defined from such estimators.

One-Sample Approximation. Using a single random sample $\mathbf{X} = \{X_1, \cdots, X_b\}$ of size b, we can define the following stochastic approximation of the partial derivative (5):

$$\hat{\partial}_{[s_k]_l} R(\mathcal{S}; \mathbf{X}) = \frac{\hat{T}_1(\mathbf{X})^2}{\|\mathbf{K}_\mathcal{S}\|_F^4} \Upsilon(\mathcal{S}) - \frac{2\hat{T}_1(\mathbf{X})\hat{T}_2^{k,l}(\mathbf{X})}{\|\mathbf{K}_\mathcal{S}\|_F^2}. \tag{6}$$

An evaluation of $\hat{\partial}_{[s_k]_l} R(\mathcal{S}; \mathbf{X})$ has complexity $\mathcal{O}(n^2 + nb)$, as opposed to $\mathcal{O}(n^2 + nN)$ for the corresponding exact partial derivative. However, due to the dependence between $\hat{T}_1(\mathbf{X})$ and $\hat{T}_2^{k,l}(\mathbf{X})$, and to the fact that $\hat{\partial}_{[s_k]_l} R(\mathcal{S}; \mathbf{X})$ involves the square of $\hat{T}_1(\mathbf{X})$, the stochastic partial derivative $\hat{\partial}_{[s_k]_l} R(\mathcal{S}; \mathbf{X})$ will generally be a biased estimator of $\partial_{[s_k]_l} R(\mathcal{S})$.

Two-Sample Approximation. To obtain an unbiased estimator of the partial derivative (5), instead of considering a single random sample, we may define a stochastic approximation based on two independent random samples $\mathbf{X} = \{X_1, \cdots, X_{b_\mathbf{X}}\}$ and $\mathbf{Y} = \{Y_1, \cdots, Y_{b_\mathbf{Y}}\}$, consisting of $b_\mathbf{X}$ and $b_\mathbf{Y} \in \mathbb{N}$ copies of X (i.e. consisting of uniform random variables on $\mathcal{D}$), with $b = b_\mathbf{X} + b_\mathbf{Y}$. The two-sample estimator of (5) is then given by

$$\hat{\partial}_{[s_k]_l} R(\mathcal{S}; \mathbf{X}, \mathbf{Y}) = \frac{\hat{T}_1(\mathbf{X})\hat{T}_1(\mathbf{Y})}{\|\mathbf{K}_\mathcal{S}\|_F^4} \Upsilon(\mathcal{S}) - \frac{2\hat{T}_1(\mathbf{X})\hat{T}_2^{k,l}(\mathbf{Y})}{\|\mathbf{K}_\mathcal{S}\|_F^2}, \tag{7}$$

and since $\mathbb{E}\big[\hat{T}_1(\mathbf{X})\hat{T}_1(\mathbf{Y})\big] = T_1^2$ and $\mathbb{E}\big[\hat{T}_1(\mathbf{X})\hat{T}_2^{k,l}(\mathbf{Y})\big] = T_1 T_2^{k,l}$, we have

$$\mathbb{E}\Big[\hat{\partial}_{[s_k]_l} R(\mathcal{S}; \mathbf{X}, \mathbf{Y})\Big] = \partial_{[s_k]_l} R(\mathcal{S}).$$

Although being unbiased, for a common batch size b, the variance of the two-sample estimator (7) will generally be larger than the variance of the one-sample estimator (6). In our numerical experiments, the larger variance of the unbiased estimator (7) seems to actually slow down the descent when compared to the descent obtained with the one-sample estimator (6).

Remark 3. While considering two independent samples $\mathbf{X}$ and $\mathbf{Y}$, the two terms $\hat{T}_1(\mathbf{X})\hat{T}_1(\mathbf{Y})$ and $\hat{T}_1(\mathbf{X})\hat{T}_2^{k,l}(\mathbf{Y})$ appearing in (7) are dependent. This dependence may complicate the analysis of the properties of the resulting SGD; nevertheless, this issue might be overcome by considering four independent samples instead of two. ◁

4 Numerical Experiments

Throughout this section, the matrices $\mathbf{K}$ are defined from multisets $\mathcal{D} = \{x_1, \cdots, x_N\} \subset \mathbb{R}^d$ and from kernels K of the form $K(x,t) = e^{-\rho\|x-t\|^2}$, with $\rho > 0$ and where $\|.\|$ is the Euclidean norm of $\mathbb{R}^d$ (Gaussian kernel). Except for the synthetic example of Sect. 4.1, all the multisets $\mathcal{D}$ we consider consist of the entries of data sets available on the UCI Machine Learning Repository; see [6].

Our experiments are based on the following protocol: for a given $n \in \mathbb{N}$, we consider an initial Nyström sample $\mathcal{S}^{(0)}$ consisting of n points drawn uniformly at random, without replacement, from $\mathcal{D}$. The initial sample $\mathcal{S}^{(0)}$ is regarded as an element of $\mathscr{X}^n$, and is used to initialise a SGD (except in Sect. 4.1, where GD is used), with fixed stepsize $\gamma > 0$, for the minimisation of R over $\mathscr{X}^n$, yielding, after $T \in \mathbb{N}$ iterations, a locally optimised Nyström sample $\mathcal{S}^{(T)}$. The SGDs are performed with the one-sample estimator (6) and are based on independent and identically distributed uniform random variables on $\mathcal{D}$ (i.e. i.i.d. sampling), with batch size $b \in \mathbb{N}$; see Sect. 3. We assess the accuracy of the Nyström approximations of $\mathbf{K}$ induced by $\mathcal{S}^{(0)}$ and $\mathcal{S}^{(T)}$ in terms of radial SKD and of the classical criteria (C.1)–(C.3) (for large matrices, we only consider the trace norm). We in parallel investigate the impact of the Nyström-sample size (Sects. 4.1 and 4.3) and of the kernel parameter (Sect. 4.2), and demonstrate the ability of the proposed approach to tackle problems of relatively large size (Sect. 4.4).

For a Nyström sample $\mathcal{S} \in \mathscr{X}^n$ of size $n \in \mathbb{N}$, the matrix $\hat{\mathbf{K}}(\mathcal{S})$ is of rank at most n. Following [4,10], to assess the efficiency of the approximation of $\mathbf{K}$ induced by $\mathcal{S}$, we consider the *approximation factors*

$$\mathcal{E}_{\mathrm{tr}}(\mathcal{S}) = \frac{\|\mathbf{K} - \hat{\mathbf{K}}(\mathcal{S})\|_*}{\|\mathbf{K} - \hat{\mathbf{K}}_n^*\|_*}, \quad \mathcal{E}_{\mathrm{F}}(\mathcal{S}) = \frac{\|\mathbf{K} - \hat{\mathbf{K}}(\mathcal{S})\|_{\mathrm{F}}}{\|\mathbf{K} - \hat{\mathbf{K}}_n^*\|_{\mathrm{F}}}, \quad \text{and } \mathcal{E}_{\mathrm{sp}}(\mathcal{S}) = \frac{\|\mathbf{K} - \hat{\mathbf{K}}(\mathcal{S})\|_2}{\|\mathbf{K} - \hat{\mathbf{K}}_n^*\|_2}, \tag{8}$$

where $\hat{\mathbf{K}}_n^*$ denotes an optimal rank-n approximation of $\mathbf{K}$ (i.e. the approximation of $\mathbf{K}$ obtained by truncation of a spectral expansion of $\mathbf{K}$ and based on n of the largest eigenvalues of $\mathbf{K}$). The closer $\mathcal{E}_{\mathrm{tr}}(\mathcal{S})$, $\mathcal{E}_{\mathrm{F}}(\mathcal{S})$ and $\mathcal{E}_{\mathrm{sp}}(\mathcal{S})$ are to 1, the more efficient the approximation is.

4.1 Bi-Gaussian Example

We consider a kernel matrix $\mathbf{K}$ defined by a set $\mathcal{D}$ consisting of $N = 2{,}000$ points in $[-1,1]^2 \subset \mathbb{R}^2$ (i.e. $d = 2$); for the kernel parameter, we use $\rho = 1$. A graphical representation of the set $\mathcal{D}$ is given in Fig. 1; it consists of N independent realisations of a bivariate random variable whose density is proportional to the restriction of a bi-Gaussian density to the set $[-1,1]^2$ (the two modes of the underlying distribution are located at $(-0.8, 0.8)$ and $(0.8, -0.8)$, and the covariance matrix of each Gaussian density is $\mathbb{I}_2/2$, with $\mathbb{I}_2$ the 2×2 identity matrix).

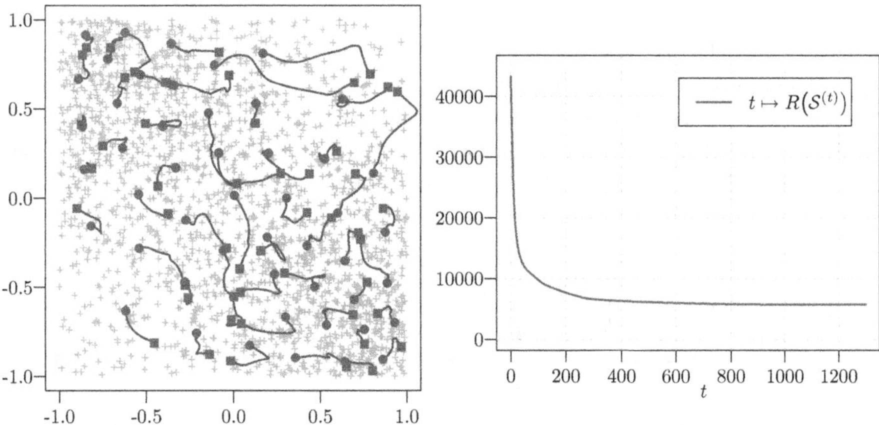

Fig. 1. For the bi-Gaussian example, graphical representation of the path $t \mapsto \mathcal{S}^{(t)}$ followed by the landmark points of a Nystrom sample during the local minimisation of R through GD, with $n = 50$, $\gamma = 10^{-6}$ and $T = 1{,}300$; the green squares are the landmark points of the initial sample $\mathcal{S}^{(0)}$, the red dots are the landmark points of the locally optimised sample $\mathcal{S}^{(T)}$, and the purple lines correspond to the paths followed by each landmark point (left). The evolution, during the GD, of the radial-SKD and trace criteria is also presented (right). (Color figure online)

The initial samples $\mathcal{S}^{(0)}$ are optimised via GD with stepsize $\gamma = 10^{-6}$ and for a fixed number of iterations T. A graphical representation of the paths followed by the landmark points during the optimisation process is given in Fig. 1 (for $n = 50$ and $T = 1{,}300$); we observe that the landmark points exhibit a relatively complex dynamic, some of them showing significant displacements from their initial positions. The optimised landmark points concentrate around the regions where the density of points in $\mathcal{D}$ is the largest, and inherit a space-filling-type property in accordance with the stationarity of the kernel K. We also observe that the minimisation of the radial-SKD criterion induces a significant decay of the trace criterion (C.1).

To assess the improvement, in terms of Nyström approximation, yielded by the optimisation of the radial-SKD, for a given number of landmark points $n \in \mathbb{N}$, we randomly draw an initial Nyström sample $\mathcal{S}^{(0)}$ from $\mathcal{D}$ (uniform sampling without replacement) and compute the corresponding locally optimised sample $\mathcal{S}^{(T)}$ (GD with $\gamma = 10^{-6}$ and $T = 1{,}000$). We then compare $R(\mathcal{S}^{(0)})$ with $R(\mathcal{S}^{(T)})$, and compute the corresponding approximation factors with respect to the trace, Frobenius and spectral norms, see (8). We consider three different values of n, namely $n = 20$, 50 and 80, and each time perform $m = 1{,}000$ repetitions of this experiment. Our results are presented in Fig. 2; we observe that, independently of n, the local optimisation produces a significant improvement of the Nyström approximation accuracy for all the criterion considered; the improvements are particularly noticeable for the trace and Frobenius norms, and slightly less for the spectral norm (which of the three, appears the coarsest measure of the approximation accuracy). Remarkably, the efficiencies of the locally optimised Nyström samples are relatively close to each other, in particular in terms of trace and Frobenius norms, suggesting that a large proportion of the local minima of the radial SKD induce approximations of comparable quality.

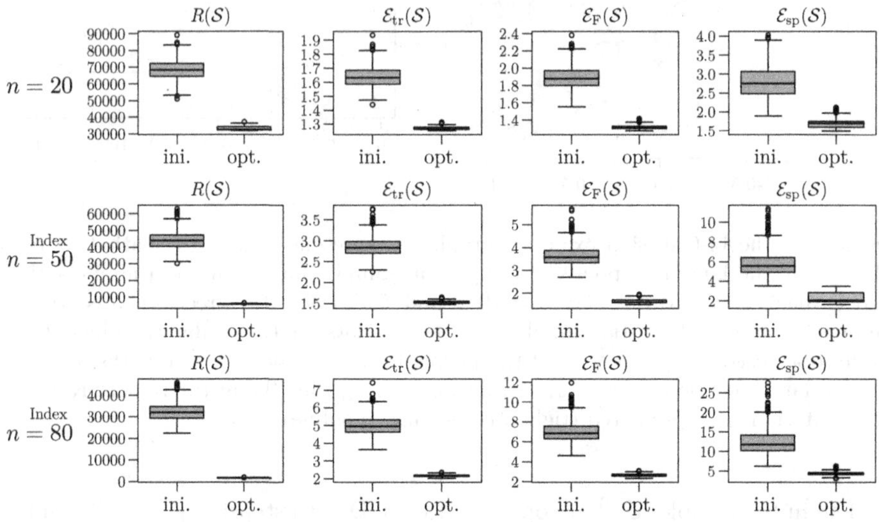

Fig. 2. For the bi-Gaussian example, comparison of the efficiency of the Nyström approximations for the initial samples $\mathcal{S}^{(0)}$ and the locally optimised samples $\mathcal{S}^{(T)}$ (optimisation through GD with $\gamma = 10^{-6}$ and $T = 1{,}000$). Each row corresponds to a given value of n; in each case $m = 1{,}000$ repetitions are performed. The first column corresponds to the radial SKD, and the following three correspond to the approximation factors defined in (8).

To further illustrate the relationship between the radial SKD and the criteria (C.1)–(C.3), for $m = 200$ random initial samples of size $n = 15$, we perform direct minimisations, through GD, of the criteria R and C_{tr} (we consider the trace norm as it is the less costly to implement). For each descent, we assess the accuracy of the locally-optimised Nyström samples in terms of radial SKD and trace norm; the results are presented in Fig. 3. We observe some strong similarities between the radial-SKD and trace-norm landscapes, further supporting the use of the radial SKD as a surrogate for the trace criterion (the minimisation of the radial SKD being, from a numerical standpoint, significantly more affordable than the minimisation of the trace norm; see Sect. 1.3).

4.2 Abalone Data Set

We now consider the $d = 8$ attributes of the Abalone data set. After removing two observations that are clear outliers, we are left with $N = 4{,}175$ entries. Each

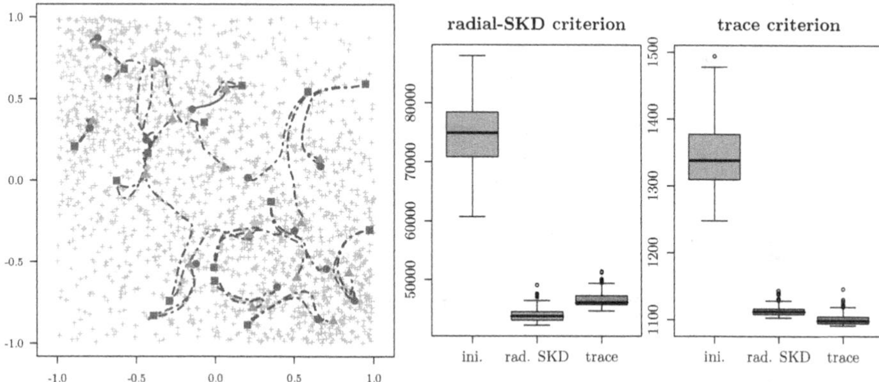

Fig. 3. For the bi-Gaussian example, graphical representation of the paths followed by the landmark points of a random initial sample of size $n = 15$ during the local minimisations of R and C_{tr} through GD; the green squares are the initial landmark points, and the red dots and orange triangles are the optimised landmark points for R and C_{tr}, respectively. The solid purple lines correspond to the paths followed by the points during the minimisation of R, and the dashed blue lines to the paths followed during the minimisation of C_{tr} (left). For $m = 200$ random initial Nyström samples of size $n = 15$, comparison of the improvements yielded by the minimisations of R and C_{tr} in terms of radial SKD (middle) and trace norm (right). Each GD uses $T = 1{,}000$ iterations, with $\gamma = 10^{-6}$ for R and $\gamma = 8 \times 10^{-5}$ for C_{tr}. (Color figure online)

of the 8 features is standardised such that it has zero mean and unit variance. We set $n = 50$ and consider three different values of the kernel paramater ρ, namely $\rho = 0.25$, 1, and 4; this values are chosen so that the eigenvalues of the kernel matrix $\mathbf{K}$ exhibit sharp, moderate and shallower decays, respectively. For the Nyström sample optimisation, we use SGD with i.i.d. sampling and batch size $b = 50$, $T = 10{,}000$ and $\gamma = 8 \times 10^{-7}$; these values were chosen to obtain relatively efficient optimisations for the whole range of values of ρ we consider. For each value of ρ, we perform $m = 200$ repetitions. The results are presented in Fig. 4.

We observe that regardless of the values of ρ and in comparison with the initial Nyström samples, the efficiencies of the locally optimised samples in terms of trace, Frobenius and spectral norms are significantly improved. As observed in Sect. 4.1, the gains yielded by the local optimisations are more evident in terms of trace and Frobenius norms, and the impact of the initialisation appears limited.

4.3 MAGIC Data Set

We consider the $d = 10$ attributes of the MAGIC Gamma Telescope data set. In pre-processing, we remove the 115 duplicated entries in the data set, leaving us with $N = 18{,}905$ data points; we then standardise each of the $d = 10$ features of the data set. For the kernel parameter, we use $\rho = 0.2$.

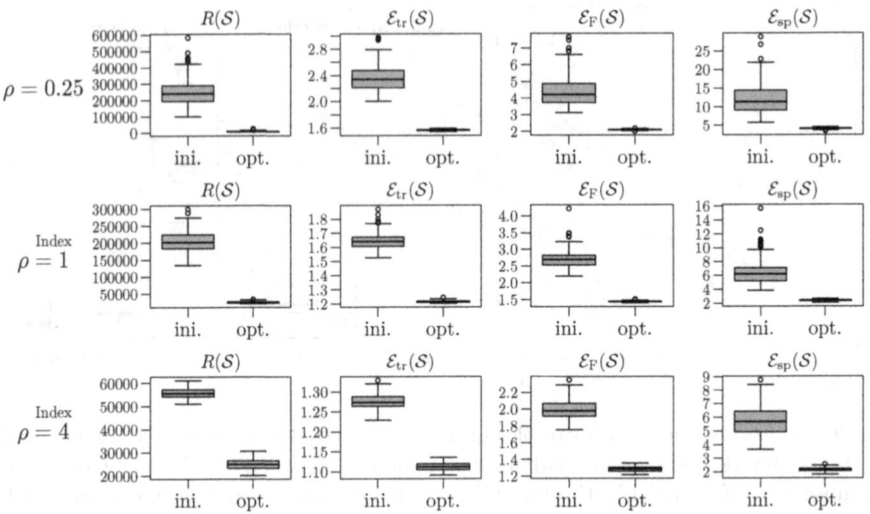

Fig. 4. For the Abalone data set with $n = 50$ and $\rho \in \{0.25, 1, 4\}$, comparison of the efficiency of the Nyström approximations for the initial Nyström samples $\mathcal{S}^{(0)}$ and the locally optimised samples $\mathcal{S}^{(T)}$ (SGD with i.i.d sampling, $b = 50$, $\gamma = 8 \times 10^{-7}$ and $T =$ 10,000). Each row corresponds to a given value of ρ; in each case, $m = 200$ repetitions are performed.

In Fig. 5, we present the results obtained after the local optimisation of $m =$ 200 random initial Nyström samples of size $n = 100$ and 200. Each optimisation was performed through SGD with i.i.d. sampling, batch size $b = 50$ and stepsize $\gamma = 5 \times 10^{-8}$; as number of iterations, for $n = 100$, we used $T =$ 3,000, and $T =$ 4,000 for $n = 200$. The optimisation parameters were chosen to obtain relatively efficient but not fully completed descents, as illustrated in Fig. 5. Alongside the radial SKD, we only compute the approximation factor corresponding to the trace norm (the trace norm is indeed the least costly to evaluate of the three matrix norms we consider, see Sect. 1.2). As in the previous experiments, we observe a significant improvement of the initial Nyström samples obtained by local optimisation of the radial SKD.

4.4 MiniBooNE Data Set

In this last experiment, we consider the $d = 50$ attributes of the MiniBooNE particle identification data set. In pre-processing, we remove the 471 entries in the data set with missing values, and 1 entry appearing as a clear outlier, leaving us with $N =$ 129,592 data points; we then standardise each of the $d = 50$ features of the data set. We use $\rho = 0.04$ (kernel parameter).

We consider a random initial Nyström sample of size $n =$ 1,000, and optimise it through SGD with i.i.d. sampling, batch size $b = 200$, stepsize $\gamma = 2 \times 10^{-7}$; the descent is stopped after $T =$ 8,000 iterations. The resulting decay of the

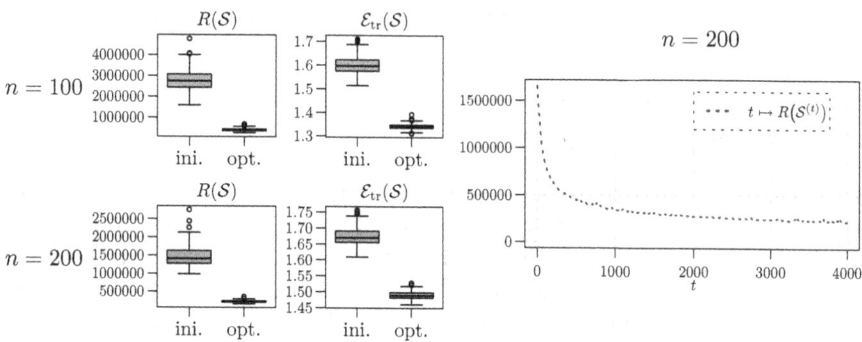

Fig. 5. For the MAGIC data set, boxplots of the radial SKD R and of the approximation factor $\mathcal{E}_{tr}$ before and after the local optimisation via SGD of random Nyström samples of size $n = 100$ and 200; for each value of n, $m = 200$ repetitions are performed. The SGD is based on i.i.d. sampling, with $b = 50$ and $\gamma = 5 \times 10^{-8}$; for $n = 100$, the descent is stopped after $T = 3{,}000$ iterations, and after $T = 4{,}000$ iterations for $n = 200$ (left). A graphical representation of the decay of the radial SKD is also presented for $n = 200$ (right).

radial SKD is presented in Fig. 6 (the cost is evaluated every 100 iterations), and the trace norm of the Nyström approximation error for the initial and locally optimised samples are reported.

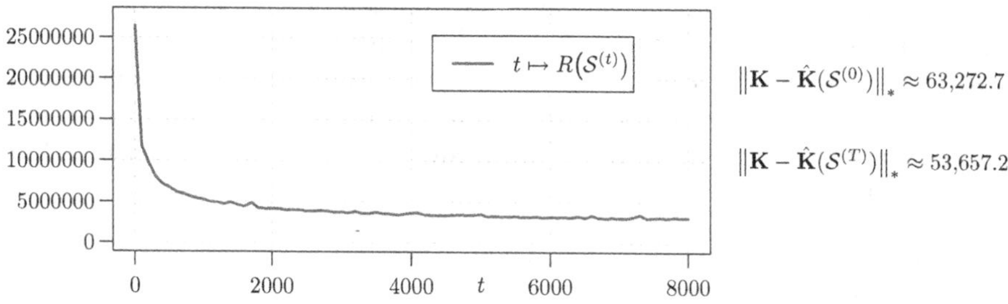

Fig. 6. For the MiniBooNE data set, decay of the radial SKD during the optimisation of a random initial Nyström sample of size $n = 1{,}000$. The SGD is based on i.i.d. sampling with batch size $b = 200$ and stepsize $\gamma = 2 \times 10^{-7}$, and the descent is stopped after $T = 8{,}000$ iterations; the cost is evaluated every 100 iterations.

In terms of computation time, on our machine (endowed with an 3.5 GHz Dual-Core Intel Core i7, and using a single-threaded C implementation interfaced with R), for $n = 1{,}000$, an evaluation of the radial SKD (up to the constant $\|\mathbf{K}\|_{\mathrm{F}}^2$) takes 6.8 s, while an evaluation of the trace criterion $\|\mathbf{K} - \hat{\mathbf{K}}(S)\|_*$ takes 6,600 s (the pseudoinverse of $\mathbf{K}_{\mathcal{S}}$ being computed in R); performing the optimisation reported in Fig. 6 without checking the decay of the cost takes 1,350 s. In this specific setting, the full radial-SKD optimisation process is thus roughly 5 times faster than a single evaluation of the trace criterion.

5 Conclusion

We demonstrated the relevance of the radial-SKD framework for the local optimisation, through SGD, of Nyström samples for SPSD kernel-matrix approximation. We studied the Lipschitz continuity of the underlying gradient and discussed its stochastic approximation. We performed numerical experiments illustrating that local optimisation of the radial SKD yields significant improvement of the Nyström approximation in terms of trace, Frobenius and spectral norms.

In our experiments, we used SGD with i.i.d. sampling, fixed stepsize and fixed number of iterations. Although already bringing satisfactory results, to improve the efficiency of the approach, the optimisation could be accelerated by considering for instance adaptive stepsizes or momentum-type techniques (see [16] for an overview), and parallelisation may be implemented. The initial Nyström samples $\mathcal{S}^{(0)}$ we considered were drawn uniformly at random without replacement; while our experiments suggest that the local minima of the radial SKD often induce approximations of comparable quality, the use of more efficient initialisation strategies may be investigated (see e.g. [3,4,11,13,18]). To evaluate the involved partial derivatives, we relied on analytical expressions of the partial derivatives of the kernel; nevertheless, in cases where implementing such analytical expressions might prove challenging, and at the cost of a loss in computational efficiency, numerical approximation of the partial derivatives (through finite differences for instance) may be considered.

As a side note, when considering the trace norm, the Nyström sampling problem is intrinsically related to the *integrated-mean-squared-error* design criterion in kernel regression (see e.g. [8,15,17]); consequently the approach considered in this paper may be used for the design of experiments for such models.

Acknowledgments. M. Hutchings thankfully acknowledges funding from the Engineering and Physical Sciences Research Council grant EP/T517951/1. All data supporting this study is openly available in the UCI Machine Learning repository at https://archive.ics.uci.edu/.

Appendix

Proof of Theorem 1. We consider a Nyström sample $\mathcal{S} \in \mathscr{X}^n$ and introduce

$$c_{\mathcal{S}} = \frac{1}{\|\mathbf{K}_{\mathcal{S}}\|_{\mathrm{F}}^2} \sum_{i=1}^{N} \sum_{j=1}^{n} K^2(x_i, s_j). \tag{9}$$

In view of (5), the partial derivative of R at $\mathcal{S}$ with respect to the l-th coordinate of the k-th landmark point s_k can be written as

$$\partial_{[s_k]_l} R(\mathcal{S}) = c_{\mathcal{S}}^2 \bigg(\partial_{[s_k]_l}^{[\mathrm{d}]} K^2(s_k, s_k) + 2 \sum_{\substack{j=1, \\ j \neq k}}^{n} \partial_{[s_k]_l}^{[\mathrm{l}]} K^2(s_k, s_j) \bigg) - 2c_{\mathcal{S}} \sum_{i=1}^{N} \partial_{[s_k]_l}^{[\mathrm{l}]} K^2(s_k, x_i). \tag{10}$$

For k and $k' \in \{1, \cdots, n\}$ with $k \neq k'$, and for l and $l' \in \{1, \cdots, d\}$, the second-order partial derivatives of R at $\mathcal{S}$, with respect to the coordinates of the landmark points in $\mathcal{S}$, verify

$$\begin{aligned} \partial_{[s_k]_l} \partial_{[s_k]_{l'}} R(\mathcal{S}) = {} & c_{\mathcal{S}}^2 \partial_{[s_k]_l}^{[\mathrm{d}]} \partial_{[s_k]_{l'}}^{[\mathrm{d}]} K^2(s_k, s_k) + 2c_{\mathcal{S}} (\partial_{[s_k]_{l'}} c_{\mathcal{S}}) \partial_{[s_k]_l}^{[\mathrm{d}]} K^2(s_k, s_k) \\ & + 2c_{\mathcal{S}}^2 \sum_{\substack{j=1, \\ j \neq k}}^{n} \partial_{[s_k]_l}^{[\mathrm{l}]} \partial_{[s_k]_{l'}}^{[\mathrm{l}]} K^2(s_k, s_j) + 4c_{\mathcal{S}} (\partial_{[s_k]_{l'}} c_{\mathcal{S}}) \sum_{\substack{j=1, \\ j \neq k}}^{n} \partial_{[s_k]_l}^{[\mathrm{l}]} K^2(s_k, s_j) \\ & - 2c_{\mathcal{S}} \sum_{i=1}^{N} \partial_{[s_k]_l}^{[\mathrm{l}]} \partial_{[s_k]_{l'}}^{[\mathrm{l}]} K^2(s_k, x_i) - 2(\partial_{[s_k]_{l'}} c_{\mathcal{S}}) \sum_{i=1}^{N} \partial_{[s_k]_l}^{[\mathrm{l}]} K^2(s_k, x_i), \text{ and} \end{aligned} \tag{11}$$

$$\begin{aligned} \partial_{[s_k]_l} \partial_{[s_{k'}]_{l'}} R(\mathcal{S}) = {} & 2c_{\mathcal{S}} (\partial_{[s_{k'}]_{l'}} c_{\mathcal{S}}) \partial_{[s_k]_l}^{[\mathrm{d}]} K^2(s_k, s_k) + 2c_{\mathcal{S}}^2 \partial_{[s_k]_l}^{[\mathrm{l}]} \partial_{[s_{k'}]_{l'}}^{[\mathrm{r}]} K^2(s_k, s_{k'}) \\ & + 4c_{\mathcal{S}} (\partial_{[s_{k'}]_{l'}} c_{\mathcal{S}}) \sum_{\substack{j=1, \\ j \neq k}}^{n} \partial_{[s_k]_l}^{[\mathrm{l}]} K^2(s_k, s_j) - 2(\partial_{[s_{k'}]_{l'}} c_{\mathcal{S}}) \sum_{i=1}^{N} \partial_{[s_k]_l}^{[\mathrm{l}]} K^2(s_k, x_i), \end{aligned} \tag{12}$$

the partial derivative of $c_{\mathcal{S}}$ with respect to the l-th coordinate of the k-th landmark point s_k is given by

$$\partial_{[s_k]_l} c_{\mathcal{S}} = \frac{1}{\|\mathbf{K}_{\mathcal{S}}\|_{\mathrm{F}}^2} \bigg(\sum_{i=1}^{N} \partial_{[s_k]_l}^{[\mathrm{l}]} K^2(s_k, x_i) - c_{\mathcal{S}} \partial_{[s_k]_l}^{[\mathrm{d}]} K^2(s_k, s_k) - 2c_{\mathcal{S}} \sum_{\substack{j=1, \\ j \neq k}}^{n} \partial_{[s_k]_l}^{[\mathrm{l}]} K^2(s_k, s_j) \bigg). \tag{13}$$

From (A.1), we have

$$\|\mathbf{K}_{\mathcal{S}}\|_{\mathrm{F}}^2 = \sum_{i=1}^{n} \sum_{j=1}^{n} K^2(s_i, s_j) \geqslant \sum_{i=1}^{n} K^2(s_i, s_i) \geqslant n\alpha. \tag{14}$$

By the Schur product theorem, the squared kernel K^2 is SPSD; we denote by $\mathcal{G}$ the RKHS of real-valued functions on $\mathscr{X}$ for which K^2 is reproducing. For x and $y \in \mathscr{X}$, we have $K^2(x,y) = \langle k_x^2, k_y^2 \rangle_{\mathcal{G}}$, with $\langle \cdot, \cdot \rangle_{\mathcal{G}}$ the inner product on $\mathcal{G}$, and where $k_x^2 \in \mathcal{G}$ is such that $k_x^2(t) = K^2(t,x)$, for all $t \in \mathscr{X}$. From the Cauchy-Schwartz inequality, we have

$$\begin{aligned}\sum_{i=1}^{N}\sum_{j=1}^{n} K^2(s_j,x_i) &= \sum_{i=1}^{N}\sum_{j=1}^{n}\langle k_{s_j}^2, k_{x_i}^2\rangle_{\mathcal{G}} = \Big\langle \sum_{j=1}^{n} k_{s_j}^2, \sum_{i=1}^{N} k_{x_i}^2 \Big\rangle_{\mathcal{G}} \\ &\leqslant \Big\| \sum_{j=1}^{n} k_{s_j}^2 \Big\|_{\mathcal{G}} \Big\| \sum_{i=1}^{N} k_{x_i}^2 \Big\|_{\mathcal{G}} = \|\mathbf{K}_{\mathcal{S}}\|_{\mathrm{F}} \|\mathbf{K}\|_{\mathrm{F}}. \end{aligned} \tag{15}$$

By combining (9) with inequalities (14) and (15), we obtain

$$0 \leqslant c_{\mathcal{S}} \leqslant \frac{\|\mathbf{K}\|_{\mathrm{F}}}{\|\mathbf{K}_{\mathcal{S}}\|_{\mathrm{F}}} \leqslant \frac{\|\mathbf{K}\|_{\mathrm{F}}}{\sqrt{n\alpha}} = C_0. \tag{16}$$

Let $k \in \{1, \dots, n\}$ and let $l \in \{1, \dots, d\}$; from Eq. (13), and using inequalities (14) and (16) together with (A.2), we obtain

$$|\partial_{[s_k]_l} c_{\mathcal{S}}| \leqslant \frac{M_1}{n\alpha}[N + (2n-1)C_0] = C_1. \tag{17}$$

In addition, let $k' \in \{1, \dots, n\} \setminus \{k\}$ and $l' \in \{1, \dots, d\}$; from Eqs. (11), (12), (16) and (17), and conditions (A.2) and (A.3), we get

$$\begin{aligned} &|\partial_{[s_k]_l}\partial_{[s_k]_{l'}} R(\mathcal{S})| \\ &\leqslant C_0^2 M_2 + 2C_0C_1M_1 + 2(n-1)C_0^2M_2 + 4(n-1)C_0C_1M_1 + 2C_0M_2N + 2C_1M_1N \\ &= (2n-1)C_0^2M_2 + (4n-2)C_0C_1M_1 + 2N(C_0M_2 + C_1M_1), \end{aligned} \tag{18}$$

and

$$\begin{aligned} |\partial_{[s_k]_l}\partial_{[s_{k'}]_{l'}} R(\mathcal{S})| &\leqslant 2C_0C_1M_1 + 2C_0^2M_2 + 4(n-1)C_0C_1M_1 + 2C_1M_1N \\ &= 2C_0^2M_2 + (4n-2)C_0C_1M_1 + 2\,NC_1M_1. \end{aligned} \tag{19}$$

For $k, k' \in \{1, \dots, n\}$, we denote by $\mathbf{B}^{k,k'}$ the $d \times d$ matrix with l, l' entry given by (11) if $k = k'$, and by (12) otherwise. The Hessian $\nabla^2 R(\mathcal{S})$ can then be represented as a block-matrix, that is

$$\nabla^2 R(\mathcal{S}) = \begin{bmatrix} \mathbf{B}^{1,1} & \dots & \mathbf{B}^{1,n} \\ \vdots & \ddots & \vdots \\ \mathbf{B}^{n,1} & \dots & \mathbf{B}^{n,n} \end{bmatrix} \in \mathbb{R}^{nd \times nd}.$$

The d^2 entries of the n diagonal blocks of $\nabla^2 R(\mathcal{S})$ are of the form (11), and the d^2 entries of the $n(n-1)$ off-diagonal blocks of $\nabla^2 R(\mathcal{S})$ are the form (12). From inequalities (18) and (19), we obtain

$$\|\nabla^2 R(\mathcal{S})\|_2^2 \leqslant \|\nabla^2 R(\mathcal{S})\|_{\mathrm{F}}^2 = \sum_{k=1}^{n}\sum_{l=1}^{d}\sum_{l'=1}^{d}[\mathbf{B}^{k,k}]_{l,l'}^2 + \sum_{k=1}^{n}\sum_{\substack{k'=1,\\ k'\neq k}}^{n}\sum_{l=1}^{d}\sum_{l'=1}^{d}[\mathbf{B}^{k,k'}]_{l,l'}^2 \leqslant L^2,$$

with

$$\begin{aligned} L = \big(&nd^2[(2n-1)C_0^2M_2 + (4n-2)C_0C_1M_1 + 2N(C_0M_2 + C_1M_1)]^2 \\ &+ 4n(n-1)d^2[C_0^2M_2 + (2n-1)C_0C_1M_1 + NC_1M_1]^2\big)^{\frac{1}{2}}. \end{aligned}$$

For all $\mathcal{S} \in \mathscr{X}^n$, the constant L is an upper bound for the spectral norm of the Hessian matrix $\nabla^2 R(\mathcal{S})$, so the gradient of R is Lipschitz continuous over $\mathscr{X}^n$, with Lipschitz constant L. □

References

1. Berlinet, A., Thomas-Agnan, C.: Reproducing Kernel Hilbert Spaces in Probability and Statistics. Springer, New York (2004). https://doi.org/10.1007/978-1-4419-9096-9
2. Bottou, L., Curtis, F.E., Nocedal, J.: Optimization methods for large-scale machine learning. SIAM Rev. **60**(2), 223–311 (2018)
3. Cai, D., Chow, E., Erlandson, L., Saad, Y., Xi, Y.: SMASH: structured matrix approximation by separation and hierarchy. Numer. Linear Algebra Appl. **25** (2018)
4. Derezinski, M., Khanna, R., Mahoney, M.W.: Improved guarantees and a multiple-descent curve for Column Subset Selection and the Nyström method. In: Advances in Neural Information Processing Systems (2020)
5. Drineas, P., Mahoney, M.W.: On the Nyström method for approximating a Gram matrix for improved kernel-based learning. J. Mach. Learn. Res. **6**, 2153–2175 (2005)
6. Dua, D., Graff, C.: UCI Machine Learning Repository (2019). http://archive.ics.uci.edu/ml
7. Gauthier, B.: Nyström approximation and reproducing kernels: embeddings, projections and squared-kernel discrepancy. Preprint (2021). https://hal.archives-ouvertes.fr/hal-03207443
8. Gauthier, B., Pronzato, L.: Convex relaxation for IMSE optimal design in random-field models. Comput. Stat. Data Anal. **113**, 375–394 (2017)
9. Gauthier, B., Suykens, J.: Optimal quadrature-sparsification for integral operator approximation. SIAM J. Sci. Comput. **40**, A3636–A3674 (2018)
10. Gittens, A., Mahoney, M.W.: Revisiting the Nyström method for improved large-scale machine learning. J. Mach. Learn. Res. **17**, 1–65 (2016)
11. Kumar, S., Mohri, M., Talwalkar, A.: Sampling methods for the Nyström method. J. Mach. Learn. Res. **13**, 981–1006 (2012)
12. Lee, J.D., Simchowitz, M., Jordan, M.I., Recht, B.: Gradient descent only converges to minimizers. In: Conference on Learning Theory, pp. 1246–1257. PMLR (2016)

13. Niederreiter, H.: Random Number Generation and Quasi-Monte Carlo Methods. SIAM (1992)
14. Paulsen, V.I., Raghupathi, M.: An Introduction to the Theory of Reproducing Kernel Hilbert Spaces. Cambridge University Press, Cambridge (2016)
15. Rasmussen, C., Williams, C.: Gaussian Processes for Machine Learning. MIT Press, Cambridge (2006)
16. Ruder, S.: An overview of gradient descent optimization algorithms. arXiv preprint arXiv:1609.04747 (2016)
17. Santner, T.J., Williams, B.J., Notz, W.I.: The Design and Analysis of Computer Experiments. Springer, New York (2018). https://doi.org/10.1007/978-1-4757-3799-8
18. Wang, S., Zhang, Z., Zhang, T.: Towards more efficient SPSD matrix approximation and CUR matrix decomposition. J. Mach. Learn. Res. **17**, 7329–7377 (2016)

Explainable Machine Learning for Drug Shortage Prediction in a Pandemic Setting

Jiye Li[1(✉)], Bruno Kinder Almentero[1], and Camille Besse[2]

[1] Thales Research and Technology Canada, Québec, QC, Canada
jiye.li@thalesgroup.com

[2] Institut intelligence et données, Université Laval, Quebec, QC, Canada
camille.besse@iid.ulaval.ca

Abstract. The COVID-19 pandemic poses new challenges on pharmaceutical supply chain including the delays and shortages of resources which lead to product backorders. Backorder is a common supply chain problem for pharmaceutical companies which affects inventory management and customer demand. A product is on backorder when the received quantity from the suppliers is less than the quantity ordered. Knowing when a product will be on backorder can help companies effectively plan their inventories, propose alternative solutions, schedule deliveries, and build trust with their customers. In this paper, we investigate two problems. One is how to use machine learning classifiers to predict product backorders for a pharmaceutical distributor. The second problem we focused on is what are the particular challenges and solutions for such task under a pandemic setting. This backorder dataset is different from existing benchmark backorder datasets with very limited features. We discuss our challenges for this task including empirical data pre-processing, feature engineering and understanding the special definitions of backorder under the pandemic situation. We discuss experimental design for predicting algorithm and comparison metrics, and demonstrate through experiments that decision tree algorithm outperforms other algorithms for our particular task. We show through explainable machine learning approaches how to interpret the prediction results.

Keywords: Backorder · Drug shortage · Inventory management · COVID-19 pandemic · Classification · Product velocity · Explainable machine learning

1 Introduction

The COVID-19 pandemic causes many products shortages and backorders in the hospital and pharmacies. It is discussed that inhaler related medications faced

B. K. Almentero—This work was performed while Mr. Kinder Almentero was a student at Université Laval performing an internship at Thales Digital Solutions Inc.

G. Nicosia et al. (Eds.): LOD 2022, LNCS 13810, pp. 141–155, 2023.
https://doi.org/10.1007/978-3-031-25599-1_11

significant shortages during the pandemic [1]. Multiple drugs have been facing shortages or discontinued from this impact [2].

Backorder happens when the quantity received from the suppliers for a product are less than the quantity ordered, or when the product is not in the inventory, but are expected to be available in a future date. This is a common problem in various domains including health care and pharmaceutical industries. There are multiple reasons that cause products to be on backorder, such as unusual demand due to pandemic, social trend, seasonality as well as manufacturer or supply chain shortage problem. When backorder happens, it may cause frustrations to the customers, lead them to look at competitors, thus causing significant financial challenges to companies. Therefore, it is pertinent to identify potential products backorders, reduce the effects of backorders by forecasting the right amount of inventories or find the alternative solutions when the backorder happens.

There exists many research efforts from management science and operations research area, using optimization techniques to plan for the inventories in order to tackle the backorder problems. Such approaches are traditionally based on stochastic approximations. However, there exists unknown and hidden information or trends in the data that can potentially impact backorders, which are not captured in the traditional optimization approaches.

Machine learning algorithms have been well known for modeling and predictive analysis in multiple domains, and have been demonstrated to be very effective on predicting related tasks. An inventory backorder prediction model [3] was proposed integrating a profit-based measure to forecast backorders for a material inventory dataset. Similar research effort on predicting backorders using classification algorithms have been discussed in [4] on the same data set. However, machine learning algorithms have been rarely applied on modeling the backorder problems in the health care domain, particularly on drug backorder predictions. Since the existing benchmark dataset from Kaggle competition[1] for backorders contains features that have already been extracted and data that has also been initially processed, current research work seldom provides insights on how to create richer features to represent the data, therefore the state-of-the-art techniques may not be easily adapted for other application domains.

In this paper, we present our investigation on pharmacy drug backorder predictions using machine learning classification algorithms, in this particular COVID-19 pandemic situation. We consider that a product is on backorder when the received quantity from the supplier is less than the quantity ordered, which also includes situations when the supplier does not send any orders. We focus on weekly predictions of whether an individual product will be on backorder.

Machine learning algorithms have rarely been applied to pharmacy drug backorder predictions. In addition, our work is different from the current state-of-the-art research work in that we work with an the empirical pharmacy data set with limited features available. We also face challenges on learning from factors which are impacted by the pandemic. Therefore, how to effectively forecast

[1] https://www.kaggle.com/c/untadta/overview.

backorders with limited features is a challenging problem. Instead of forecasting based on all the data, we design personalized model and forecast backorders for individual products according to the historical sales per product. This data is also impacted by multiple unusual external factors such as the pandemic, material shortage, competitor lockout and etc. Modeling a prediction problem containing unknown factors is not an easy task.

Our contributions lie in two aspects. First, we analyze the particular challenges on pharmacy data processing under the pandemic situation. We demonstrate data manipulating techniques on creating rich features to represent the backorder data. Secondly, we show through experiments that machine learning classifiers can be effectively adapted on predicting backorders in the pharmaceutical domain, which has rarely been studied previously. The prediction model is designed for individual product backorder predictions. Interpreting and explaining the predictions are important to demonstrate the robustness of machine learning models. The approaches discussed can be applied in other inventory management or supply chain applications for forecasting product backorders.

The paper is organized as follows. We first discuss related work on backorder predictions in Sect. 2. We describe the pharmacy backorder data set in Sect. 3. Data preprocessing and the feature creations are discussed in Sect. 4. Eexperimental design, algorithms and results comparisons with analysis are discussed in Sect. 5. Finally we conclude our paper with discussions and future works in Sect. 6.

2 Related Work

Most of the related work on predicting backorder problems in the field of machine learning and data analytics are from the Kaggle challenge[2] of predicting material product backorders.

The data has already been previously processed into training and testing data, and has been labeled as backorder or not with boolean values. Features include current inventory level, transit time, minimum recommended amount in stock and amount of stock orders overdue, quantity in transit and etc. Since each row in the data set represents an individual SKU (Stock Keeping Unit), there is no multiple backorder histories for individual product. Therefore, the machine learning algorithm is trained to forecast for any given product. This data set is highly imbalanced, so sampling techniques such as SMOTE [5] are required to balance the class distributions before feeding the data into any prediction models.

Various modeling techniques have been discussed in the field of machine learning and big data analytics on this data set. A profit function based prediction [3] model was proposed to forecast backorders using tree-based classification algorithms. Random Forest classifier, C4.5, KNN, XGBoost and other classification algorithm has been applied on the prediction task and random forest

[2] https://www.kaggle.com/c/untadta/data. Last accessed in May 2021.

classifier has been shown to outperform the others. Ensemble learning algorithms [4] have been demonstrated to be effective on predicting backorders for inventory control. Tree-based machine learning algorithms such as distributed random forest and gradient boosting machine [6] have been shown to be effective on predicting backorders. Recurrent neural network has been used to modeling the backorder prediction [7,8] and different data balancing techniques were applied on this data set simultaneously for the prediction. This kaggle data set was also used to simulate backorder predictions for Danish craft beer breweries using classification algorithms [9].

It is shown that such machine learning algorithm based classifiers are very effective on predicting backorders. However, the related works are limited on this benchmark data set, there has not been much research effort on empirical backorder data predictions using machine learning algorithms on pharmacy drug data.

Other efforts have been demonstrated on using artificial intelligent algorithms on pharmaceutical product related predictions and management field. Recent research [10] also demonstrated that different machine learning and deep learning neural network models such as SARIMA, Prophet, linear regression, random forest, XGBoost and LSTM [11], are useful to predict pharmaceutical products. A fuzzy rule-based approach [12] is demonstrated to evaluate and analysis the impact factors from the health care practitioners in order to understand how to control the COVID-19 related infections transmission.

Explainable machine learning tools such as permutation feature importance [13] and SHAP [14] are more and more used to help analyze and understand machine learning results thus can help enterprise understand better their data and plan for the upcoming changes [15].

3 Data Description

The pharmacy drug data comes from 15 months of historical backorder data from one pharmaceutical distributor from June 1, 2019 to September 30, 2020. There are more than 14,000 products available during this period, among which 7,505 products have at least one historical backorder, which is 51% of all the products. The average backorder percentage per product is 31%.

There are 7 features in the original data set:

1. Date of the transaction
2. Document ID which identifies a transaction. For a given date, there could be multiple transactions.
3. SKU, which is the product ID.
4. Client ID
5. Product descriptions
6. Quantity ordered (`Qty_PO`)
7. Quantity received (`Qty_Received`)

We are interested in the date of the transaction, the document ID, SKU, and the quantity ordered and received. In our data set, comparing to the Kaggle benchmark data, there is no inventory information nor lead time available.

4 Data Preprocessing

4.1 What is Considered as Backorder

According to our domain expert, when a product is on backorder, the company keeps the orders already made in the database, but removes this product from the company's website until the previous orders are fulfilled. So even though the dataset contains the real ordering quantities per product, the fact that the customers are not able to order the product online indicates that the true demand which would have been accumulated during the backorders is unknown. In the database, if a product has been on backorder, the daily quantities ordered to the suppliers stay the same until the product is no longer on backorder; and the quantities received remains zero, as shown in the following Table 1.

Table 1. Sample historical backorder data

Date	Document ID	Product ID	`Qty_PO`	`Qty_Received`
2020-01-05	001	P1	6	6
2020-01-05	001	P2	12	12
2020-01-05	002	P3	10	0
2020-01-05	004	P4	8	8
2020-01-06	005	P3	10	0

Table 1 demonstrates sample data containing backorder products and non-backorder products. The 3rd row shows an example of backorder product. $P3$ was ordered for 10 units on 2020-01-05, but received zero units. Therefore this product is on backorder. The next day 2020-01-06, the same product was ordered 10 units, but still received zero units. There exist also situations that the received quantity is above zero, but less than the quantity ordered. Such cases are considered as backorders as well. For those products that are not on backorder, the received quantities are equal to the ordered quantities, such as $P1$, $P2$ and $P4$.

According to such information, from the two columns `Qty_PO` and `Qty_Received` per products, we are able to identify backorder products, whose `Qty_PO` is greater than `Qty_Received` . For example, if a product has an order of 10 units, and `Qty_Received` is 3 units, even if the received quantity is not zero, we consider this product to be a backorder product.

We assume that if there are multiple orders for certain products on the same day, the sum of the `Qty_PO` represents the total demand, and the sum of the `Qty_Received` indicates the total received amount. Therefore, for each product, we first group the sales by day, and obtain the 3 features: the transaction date, `Qty_PO` and `Qty_Received`.

4.2 Features

We focus on predicting weekly backorders per product. We resample data by week, then create the following features for the predictive analysis algorithms, is_holiday, month, year, week_of_year, #_weeks_bo, week_1, week_2, week_3, week_4, is_backorder, as shown in the following Table 2.

Table 2. Features for backorder algorithm

Features	Type	Description
Qty_PO	int	Quantity ordered towards the suppliers
is_holiday	Boolean	Whether there is a holiday within the week
month	int	The month of the date
year	int	The year of the date
week_of_year	int	The week of the date
#_weeks_bo	int	Number of weeks the product has been on backorder
week_n	Boolean	Whether the product was on backorder the previous $n (1 \leq n \leq 4)$ week
is_backorder	Boolean	Whether a product is on backorder

Note that the boolean feature is_backorder represents whether the product is on backorder or not, by computing the differences between Qty_PO and Qty_Received. If Qty_Received<Qty_PO, then is_backorder= 1; otherwise, is_backorder= 0.

Qty_PO represents the forecasted quantities to be ordered for this product next week. We assume this value is known. The average of previous week's order could be an indication for next week's order. Recently, time series algorithm and machine learning algorithm have been demonstrated for their effective forecasting capabilities [10].

From the feature constructions, we prepare the input data using the above-mentioned features for our prediction algorithms. Table 3 shows a sample data set after preprocessing for a given product.

Table 3. Sample data with backorder features

Date	Qty_PO	is_holiday	Month	Year	W_of_Y	is_bo	#_weeks_bo	week1	week2	week3	week4
2019-06-16	12.0	0	6	2019	24	0	0	0	0	0	0
2019-06-23	0	0	6	2019	25	0	0	0	0	0	0
2019-06-30	0	0	6	2019	26	0	0	0	0	0	0
2019-07-07	0	0	7	2019	27	0	0	0	0	0	0
2019-07-14	6.0	0	7	2019	28	0	0	0	0	0	0

There exists no missing data in the data set. As discussed earlier, the average percentage of backorder per product is 31%. There exists many techniques to solve the imbalance class issues, however, the approaches focuses on data with class imbalance ratio ranging from 1:4 to the extreme ratio of 1:100 and above [16]. We do not consider our data as extremely imbalanced, thus we do not use any sampling techniques for imbalanced data set. Our data set includes 15 months' period which are 60 weeks. Since we group the data by week, we have maximum 60 transactions per product.

5 Experiments

We consider predicting the backorder problem as a classification problem and we use the following six machine learning algorithms in the experiment. Decision tree, random forest and XGBoost are all tree-based algorithms. Such algorithms build the classifier into a tree structure, where node represents the features, and leaves represent the decision. This tree-structured classifier provides a straightforward visualization of how predictions are made, as well as important features to separate two classes. Naïve bayes classifier is a probabilistic model which is easy to implement, but it requires the assumption that the features must be independent. Logistic regression classifier is used to predict binary outcome, such as whether an event will happen or not. Support vector machine (SVM) classifier has the advantage of classifying in high dimensional space. We model the backorder data with these algorithms and compare their performance. Among the prediction algorithms, XGBoost algorithm is from `github/dmlc`[3], and the rest of the algorithms are from scikit-learn library [17].

We use machine learning standard metrics for classifiers such as precision, recall, f1-score and ROC curve [18] to compare the performance of the algorithms.

5.1 Experimental Design

The empirical backorder history data is collected by time, thus not static. Therefore, one cannot use a more recent backorder histories to predict for a previous backorder. The traditional cross validation approach cannot be applied straightforward to our experiment.

We have considered three training periods and tested the data on the following 3 testing periods as sliding windows shown in the following Table 4.

We train the model based on the data in the training period (shown in red), then test the model on data in the testing period (shown in orange) as shown in Fig. 1. We average the results from these 3 testing periods.

[3] https://github.com/dmlc/xgboost/.

Table 4. Experiment training and testing period

Training data coverage	Testing data coverage
June 1, 2019 - June 30, 2020	July 1, 2020 - July 31, 2020
June 1, 2019 - July 31, 2020	Aug. 1, 2020 - Aug. 31, 2020
June 1, 2019 - Aug. 31, 2020	Sept. 1, 2020 - Sept. 30, 2020

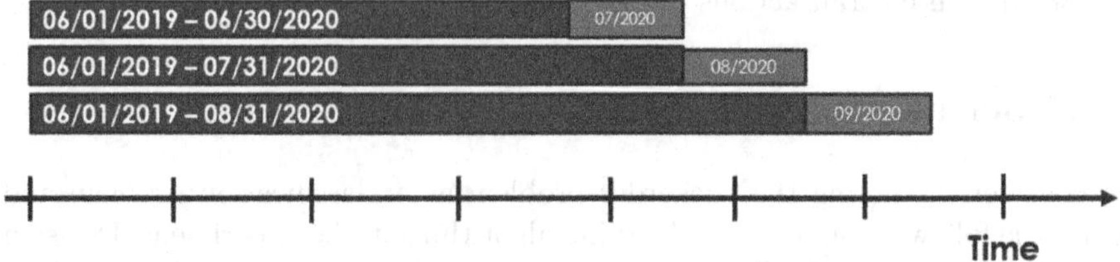

Fig. 1. Experiment training and testing period

5.2 Comparison Results

We compare six classification algorithms' performance on predicting the top 500 most backordered products. Hyperparameter tuning is applied on each of the classifiers. The following Table 5 shows the evaluation metrics for each classifier.

Table 5. Comparisons between different classifiers for Top500 most backordered products

Classifier	Accuracy	Precision	Recall	f1_score
Decision Tree	**0.786089**	0.754750	**0.621622**	**0.681747**
XGBoost	0.730514	0.755636	0.397345	0.520820
Random Forest	0.760573	**0.802126**	0.465149	0.588836
SVM	0.766515	0.775874	0.515410	0.619373
Logistic Regression	0.759523	0.799673	0.463727	0.587035
Naive Bayes	0.682628	0.581434	0.495970	0.535312

From the results, we observe that decision tree algorithm provides the highest accuracy of 78.6% and F1 score of 0.68. The following Fig. 2 shows the comparisons of ROC curve with AUC values. The higher the AUC value is, the better the classifier performs.

5.3 Results Discussions for Backorder Class

We further investigate the classifier's performance on predicting individual "backorder" and "non-backorder" classes. It is interesting to know how well

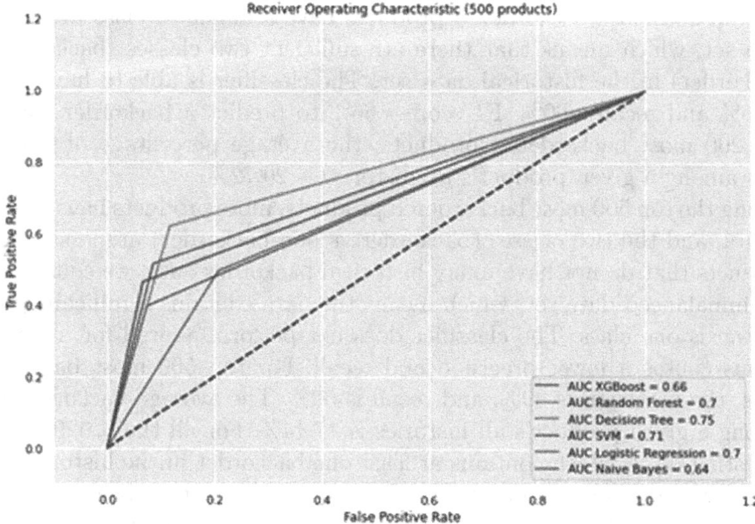

Fig. 2. Experimental results from six algorithms

the classifier can perform in general, however, it is more important to know how well the classifier can predict minority classes, which is the backorder class. For the testing periods range from July 1, 2020 to September 30, 2020, there exist 10,937 products, among which there are 6,029 products that have at least one backorder. We rank the products according to the total number of backorders and obtain the top-N most backordered products. We show in Table 6 the decision tree classifier's performance on different top-N products.

Table 6. Performance for Top-N most backordered products by decision tree classifier (Class 1 represents backorder Class)

TopN products	Decision Tree classifier performance				
	Class	Precision	R	f1-score	Accuracy
N = 200	0	0.78	0.88	0.82	0.77
	1	0.75	0.60	0.67	0.77
N = 500	0	0.79	0.86	0.81	0.78
	1	0.69	0.59	0.61	0.78
N = 1000	0	0.83	0.88	0.84	0.80
	1	0.59	0.54	0.54	0.80
N = 6029	0	0.91	0.89	0.89	0.84
	1	0.26	0.26	0.24	0.84

The top 200 most backordered products have a balanced class in the training data set, which means that there are sufficient two classes (backorder and non-backorder) in the historical data set. The classifier is able to have a precision = 75% and recall = 60%, F1 score = 66% to predict a backorder class. For the top 200 most backordered products, the average percentage of backorder histories among a given product's all histories is 20.72%.

Among the top 500 most backordered products, more products have less backorder data, and the two classes (backorder vs non-backorder) are less balanced. For products that do not have many historical backorder data, we consider such data as imbalanced data set, which means the data contains significant proportions towards one class. The classifier does not perform as well for imbalanced data, thus causes a lower precision and recall. For top 500 most backordered products, the precision is 69%, and recall is 59%. The average backorder histories among a given product's all histories is 17.14%. For all the 6,029 products in our testing data, which contains at least one backorder in the historical data, we have obtained a precision of 25% and recall of 25% for the "backorder" class. Since we are predicting for individual product, some products have very few backorder histories, thus the individual classifier's performance is low because of the imbalanced data. As we average the performance on more and more products, such products lower down the overall precision and recall.

5.4 Decision Tree Visualization

Being able to explain the machine learning model and interpret this black box are more and more in demand from both the research and the industrial community. One of the questions we ask often about a trained model is that how the predictions are generated. There exists various approaches to visualize the model. Here we demonstrate a sample approach which uses a decision tree visualization library to disclose this model.

We have plotted the following decision tree from our sample output[4] as shown in Fig. 3[5]. The depth of the tree is 5, and the root node indicates the most important features to separate two classes, which is `#weeks_bo` ≤ 0.5. If a product has not had any backorders in the past, then go to the left branch; otherwise, go to the right branch. The Gini coefficient indicates the purity of this feature. For leaf node, the Gini coefficient is always equal to 0. For non-leaf node, the Gini coefficients is between 0 and 1. "Samples" indicates the number of records in the data sets at this node; "value = [44, 20]" indicates that in the data set, there are 44 records with non-backorder; 20 backorder records. "Class" indicates the majority class label. In the example tree, the blue leaf node represents "backorder" class; the orange leaf node represents "non-backorder" class. As we can see, the most important feature is `#weeks_bo`, followed by `week_3` (whether the

[4] For confidentiality reasons, we omit the drug name and SKU.

[5] The python library for decision tree visualization used in this paper can be found here https://graphviz.readthedocs.io/en/stable/manual.html.

product was on backorder 3 weeks ago), and Qty_PO (products ordering quantities). For each product, the decision tree algorithm generates a tree to visualize how the backorder class is predicted.

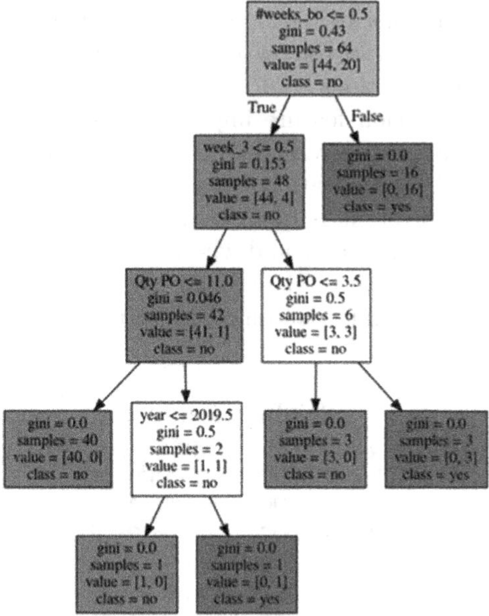

Fig. 3. Sample decision tree

5.5 Permutation Feature Importance

Knowing which features have the biggest impact on the predictions is important for both data collection and testing purposes. From the decision tree visualization, one can conclude the important features from the non-leaf node. However, features for the training data may not be as important as in the testing data which may cause the model to overfit. One way to explore the feature importance is by permutating the features and compare the prediction loss [13]. The performance deteriorates for more important features. If the performance does not change after permutating the feature, then it indicates that feature is not important.

The following figure demonstrates permutation feature importances for one particular product. Figure 4 shows the important features ranked by their importance values from a sample backorder product testing data[6]. The most important features show on the top, and the least important features are towards the bottom. The most important feature for this test data is week_1, indicating whether

[6] https://eli5.readthedocs.io/en/latest/.

the product has been on backorder last week. #_weeks_bo, the number of weeks a product has been on backorder is ranked as the second most important feature for the prediction. week_3 and Qty_PO have negative values for their permutation importances, which implies that the predictions on the permutated data happened to be more precise than the real data. This happens when the random chance causes the predictions to be more accurate, which is a common issue for smaller data set. The rest of the features have a weight of 0, which are not important towards the predictions.

Note that the permutation feature importance is only applied to the test data set, and it requires the true values to be known. This could be one of the disadvantages of using the permutation feature importance since the true values are often unknown. However, it is interesting to know how much the prediction varies when changing a feature and whether the model can be generalized well. SHAP (SHapley Additive exPlanations) [14] values can also be used to investigate the feature importance and model interpretations.

Weight	Feature
0.2800 ± 0.3200	week_1
0.0400 ± 0.1600	#weeks_bo
0 ± 0.0000	week_4
0 ± 0.0000	week_2
0 ± 0.0000	week_of_year
0 ± 0.0000	year
0 ± 0.0000	month
0 ± 0.0000	is_holiday
-0.0400 ± 0.2993	week_3
-0.0400 ± 0.1600	Qty PO

Fig. 4. Feature importance for decision tree classifier

5.6 Comparison Between Accuracy and Velocity

According to our domain experts, instead of the general performance of the classifier, it will be more interesting to know how the classifier performs for different types of products, especially by velocity.

We investigate how the velocity of the products affects the algorithm performance. Product velocity in this context is defined as the price of the product multiply by the average quantities per order. Thus, we use the formula

$$Velocity = Price * \frac{\sum_{i=1}^{n} \texttt{Qty\_PO}_i}{n}$$

to obtain the value of product velocity, where n is the total number of orders existing in the data set, $\texttt{Qty\_PO}_i (1 \leq i \leq n)$ is the quantities per order.

The price information is not available for all the products, thus in this comparison, we select top 2000 most backordered products which have the price information available to obtain their velocities. Figure 5 shows the performance of the decision tree classification prediction accuracy vs. velocity. We observe that the average accuracy for the top 2000 most backordered products is around 80%. For higher velocity products, the prediction accuracy is around 90%. Most of the products have a velocity below 2000, and the average accuracy for these low velocity products falls in the range of 60% to 80%.

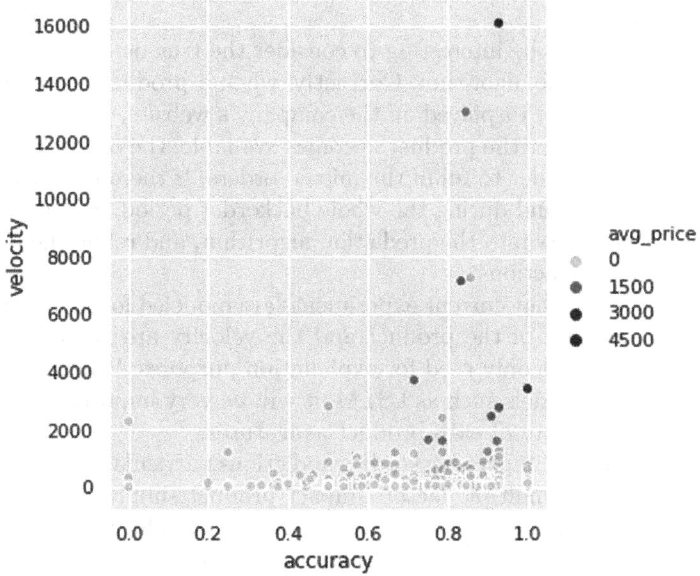

Fig. 5. Velocity vs. prediction accuracy for Top 2000 most backordered products

6 Conclusion

In this paper, we present a use case of machine learning algorithms for pharmacy drug backorder predictions under the COVID-19 pandemic setting, which posed particular challenges on data pre-processing, feature construction and the interpretations of the prediction results.

Machine learning algorithms are seldom used to predict inventory backorders in pharmaceutical domain. We demonstrate the differences of this empirical data set between existing benchmark backorder data set, the challenges on data pre-processing and feature constructions. Through comparison experiments, we show that decision tree algorithm outperforms other classification algorithms on predicting backorders. Our algorithm is designed for weekly backorder predictions for individual product. This model can easily be adapted to predict backorders in terms of days or multiple weeks. Through this study, we recognized the challenges of applying machine learning algorithms into the pharmaceutical domain. It is pertinent to understand the domain knowledge, and the data collection in this field in order to design richer features and choose the appropriate modeling algorithm. Applying machine learning algorithm may not be as straightforward in this domain as in other field. For example, the traditional 10-fold cross validation testing for the robustness of the model cannot be performed in our study since the data evolves as time. In addition, it is interesting to understand the focus from the domain expert's perspective in order to create features and explain the prediction results.

In the future, it will be interesting to consider the true demand of a product and integrate it into the algorithm. Currently, when a product is on backorder, this product is no longer displayed on the company's website. However, the true demand still exists. After the product becomes available, the orders would surge more than normal in order to fulfill the missed orders. If there is a way to know ahead of time the demand during the whole backorder period, we would be able to integrate such factors into the prediction algorithm, and refine the algorithm for more accurate predictions.

It should be noted that current experiments are modeled for individual product, therefore the price of the product and the velocity are not considered as part of the features, but only used for explanation purposes. When using multi-product predictions models such as LSTM, it will be very important to consider the price and the velocity of each product as features.

External factors can impact backorder predictions particularly under the pandemic situation, when multiple factors impact product supply chains. Without such features, the classifiers have limited performance. Often the distributors receive verbal warnings from manufacturers on the upcoming backorders, however, such information is not publicized ahead of time thus not available in the database. There also exist situations when a raw material concentrated from other countries could not be shipped to local manufacturer on time, thus the whole supply chain is affected. How to integrate such external factors into modeling and what are the substitutions knowing that a product will be on backorder, are two important subjects to investigate.

Acknowledgments. The authors would like to thank MITACS, SCALE AI and Thales Digital Solutions for the financial and scientific support for the project. This work was realized based on pharmaceutical data made available by Distribution Pharmaplus Inc. We would like to thanks Jérôme Lavoie from Group Lavoie for his advice during this work. The authors would also like to thank Pierre Gravel, Ihsen Hedhli, Camélia Dadouchi, and Freddy Lecue for their insightful comments and suggestions during the performance of this work.

References

1. Elbeddini, A., Tayefehchamani, Y., Yang, L.: Strategies to conserve salbutamol pressurized metered-dose inhaler stock levels amid COVID-19 drug shortage. Drugs Ther. Perspect. **36**(10), 451–454 (2020). https://doi.org/10.1007/s40267-020-00759-1
2. Drug Shortages Canada. Drug shortages homepage. https://www.drugshortages canada.ca. Accessed 23 Mar 2022
3. Hajek, P., Abedin, M.Z.: A profit function-maximizing inventory backorder prediction system using big data analytics. IEEE Access **8**, 58982–58994 (2020)
4. de Santis, R.B., de Aguiar, E.P., Goliatt, L.: Predicting material backorders in inventory management using machine learning. In: 2017 IEEE Latin American Conference on Computational Intelligence (LA-CCI), pp. 1–6 (2017)
5. Chawla, N.V., Bowyer, K.W., Hall, L.O., Kegelmeyer, W.P.: SMOTE: synthetic minority over-sampling technique. J. Artif. Intell. Res. **16**, 321–357 (2002)
6. Islam, S., Amin, S.H.: Prediction of probable backorder scenarios in the supply chain using distributed random forest and gradient boosting machine learning techniques. J. Big Data **7**(1), 1–22 (2020). https://doi.org/10.1186/s40537-020-00345-2
7. Shajalal, Md., Hajek, P., Abedin, MZ.: Product backorder prediction using deep neural network on imbalanced data. Int. J. Prod. Res. https://doi.org/10.1080/00207543.2021.1901153
8. Akintola, K.G., Lawal, S.O.: A product backorder predictive model using recurrent neural network. Iconic Res. Eng. J. **4**(8) (2021)
9. Li, Y.: Backorder prediction using machine learning for Danish craft beer breweries. Master thesis, Aalborg University, Denmark, September 2017
10. Almentero, B.K., Li, J., Besse, C.: Forecasting pharmacy purchases orders. In: 2021 IEEE 24th International Conference on Information Fusion (FUSION), pp. 1–8 (2021)
11. Connor, J.T., Martin, R.D., Atlas, L.E.: Recurrent neural networks and robust time series prediction. In IEEE Trans. Neural Netw. **5**(2), 240–254 (1994)
12. Asadi, S., Nilashi, M., Abumalloh, R.A., et al.: Evaluation of factors to respond to the COVID-19 pandemic using DEMATEL and fuzzy rule-based techniques. Int. J. Fuzzy Syst. **24**, 27–43 (2022)
13. Breiman, L.: Random forests. Mach. Learn. **45**(1), 5–32 (2001). https://doi.org/10.1023/A:1010933404324
14. Lundberg, S., Lee, S.: A Unified Approach to Interpreting Model Predictions: NIPS'17: Proceedings of the 31st International Conference on Neural Information Processing Systems, pp. 4768–4777 (2017)
15. Roscher, R., Bohn, B., Duarte, M.F., Garcke, J.: Explainable machine learning for scientific insights and discoveries. IEEE Access **8**, 42200–42216 (2020)
16. Krawczyk, B.: Learning from imbalanced data: open challenges and future directions. Progr. Artif. Intell. **5**(4), 221–232 (2016). https://doi.org/10.1007/s13748-016-0094-0
17. Pedregosa, F., et al.: Scikit-learn: machine learning in Python. J. Mach. Learn. Res. 2825–2830 (2011)
18. Bradley, AP.: The use of the area under the ROC curve in the evaluation of machine learning algorithms. Pattern Recogn. (1997)

Intelligent Robotic Process Automation for Supplier Document Management on E-Procurement Platforms

Angelo Impedovo(✉), Emanuele Pio Barracchia, and Giuseppe Rizzo

Niuma s.r.l., Via Giacomo Peroni, 400, 00131 Rome, Italy
{angelo.impedovo,emanuele.barracchia,giuseppe.rizzo}@niuma.it

Abstract. E-procurement platforms are marketplaces where manufacturing companies, termed buyers, frequently interact with tens of thousands of suppliers. Conversely, different suppliers compete against each other to be selected, by one or more buyers, as those to be commissioned with procuring goods and services. However, such interactions are risky because suppliers may trick buyers by issuing false information about themselves. For this reason, procurement has become a regulated environment in which rigorous supplier evaluation accounts for information from documents exchanged through e-procurement platforms. In this paper, we propose the adoption of Intelligent Robotic Process Automation based on Machine Learning for automatizing the task of supplier document management. Experimental results on suppliers' documents from a real e-procurement platform prove the effectiveness of the solution.

Keywords: Robotic process automation · Machine learning · E-procurement · Supply-chain management

1 Introduction

Recently, the global supply-chain scenario has suffered many shocks and disruptions, notably due to COVID-19, climate change, the Ukrainian campaign, and the semiconductor crisis. Consequently, on the one hand, the suppliers are struggling with material shortages while, on the other hand, manufacturing companies (*buyers* hereafter) experience a significant slowdown in their processes. Such an effect ripples throughout the whole supply chain. Buyers transform procured goods and services into manufactured ones that, in turn, are requested by third-party manufacturing companies. Therefore, a bottleneck supplier in the supply chain poses an indirect threat to the business of many other companies.

When facing worldwide-scale and local-scale events, the supply chain naturally reorganizes: buyers autonomously select the most appropriate suppliers depending on the adverse circumstances. Often, this manifests as a change in the buyer's preferences against the considered suppliers for subsequent relationships: only those performing well should be considered. For instance, a supplier that timely sends the requested goods to buyers is preferable over slower ones

G. Nicosia et al. (Eds.): LOD 2022, LNCS 13810, pp. 156–166, 2023.
https://doi.org/10.1007/978-3-031-25599-1_12

since it is less likely to slow down the production lines. Otherwise, if this does not happen, the production lines will wait for the goods to come, leading to increased costs. Similarly, a buyer may completely substitute obsolete goods for manufacturing a specific product, thus selecting a different supplier for the purchase.

To our perception: i) a medium-sized buyer requests tens of thousands of goods and services from thousands of suppliers in a year, and ii) the performance of suppliers may be affected by internal and external features which may be unknown from the buyers' side. As a result, buyers spend considerable effort collecting a large amount of data to stay updated on the suppliers' business and compliance levels. E-procurement platforms were born to promote communication between buyers and suppliers [6]. In particular, they allow buyers to request, order, and purchase goods from suppliers, while, on the other side, allowing suppliers to share their updated information with buyers. Therefore, e-procurement platforms i) store an increasing amount of data about the monitored suppliers over time and ii) orchestrate procurement execution according to shared regulations between buyers and suppliers.

Unfortunately, adopting traditional e-procurement platforms alone does not eliminate the threat of selecting harmful suppliers. In fact, although they provide regulated environments for buyers to interact with suppliers, unreliable interactions are still possible. Consider two examples: a malicious supplier that uploads an overdue or false ISO-9001 certificate and a supplier that forgets to update his due certificate. These are both sufficient reasons for a supplier to be excluded from a call for tenders. When the overdue document is left undetected, two scenarios arise depending on the procurement setting. In the public procurement setting, the overdue document may be discovered later by competitor suppliers, which in turn starts an expensive and time-consuming legal recourse against the buyer. In the private procurement setting, selecting the malicious supplier may result in low-quality procurement that directly affects the business.

Solving the problem mentioned before requires a novel class of procurement platforms. In this paper, we claim that augmenting traditional platforms with Robotic Process Automation (RPA) and document-based machine learning models helps procurement users from the buyer side. Specifically, RPA empowers the users while i) managing the document flow from their suppliers and ii) spotting suppliers not compliant with the requested documents. To this end, we introduce a framework for intelligent document recognition processes, then we propose an actual implementation accommodating different machine learning capabilities at different steps. This way, the whole automated process provides an encompassing approach to detect potentially invalid, overdue or incorrect documents, which in turn can prove to be valid predictors for spotting non-compliant suppliers. The remainder of this paper comprises the following sections: Sect. 2 introduces the recent literature on RPA in the e-procurement domain. Section 3 presents the methodological approach. Then, Sect. 4 discusses the experimental results, before concluding the manuscript in Sect. 5.

2 Related Works

The emergence of artificial intelligence (AI) and machine learning (ML) solutions in the procurement domain is relatively recent. Such a research area is concerned with employing a set of methods that should allow one to move, in principle, from traditional procurement to intelligent procurement, as described in [6]. The authors identified document management as one of four pillars of AI-based procurement. In particular, they marginally describe an approach assimilable to an *intelligent robotic process automation* process.

Robotic Process Automation (RPA) is a technology that mimics how humans interact with software to perform high-volume, repeatable tasks [5]. Recent works highlight the importance of introducing RPA processes and infrastructure in an organization, listing various advantages deriving from such technologies, such as cost savings, efficiency, and quality improvements [2,4,9]. Such works also discuss adoption barriers related to IT infrastructure, internal communication, financial resources, and supplier-related issues that could make adopting RPA unfeasible. Unfortunately, these works describe neither a general framework nor quantitative evaluations, unlike the one described in [10] where the authors aimed at assessing the efficiency of the customer documentation process via RPA. Instead, our work focuses on a supplier document management process where machine learning is crucial in enhancing RPA.

Regarding the employment of machine learning to support RPA, to the best of our knowledge, there is a shortage of literature about applications in procurement. However, some recent works showed there are some application fields where machine learning can be successfully employed. For instance, in [8], the authors exploit *Convolutional Neural Network* to dynamically detect objects in software applications interface, allowing flexibility, performance, and accuracy. A similar approach is also illustrated in [7], where the authors identify the set of challenges a company will face to move towards an intelligent RPA. In the context of RPA, an anomaly detection model has been proposed in the context of title insurance to cope with high-dimensionality data [3]. Finally, in [1], an approach combining RPA and classification model has been proposed.

3 Document Management Process

Our claim is that RPA should automatize the management of suppliers' documents at any stage in the overall procurement process, from supplier registration to purchase. It turns out that buyers typically ask suppliers to upload documents i) at the qualification stage, that is, when registering a new supplier to the platform, and ii) at the post-qualification stage, that is, when periodically reviewing already registered suppliers. Therefore, the document management process in Fig. 1, which comprises a document-agnostic and a document-specific phase, is triggered whenever a new document is uploaded.

During the document-agnostic phase, uploaded documents are i) filtered based on a set of desirable conditions (e.g., file extensions), ii) transformed into a

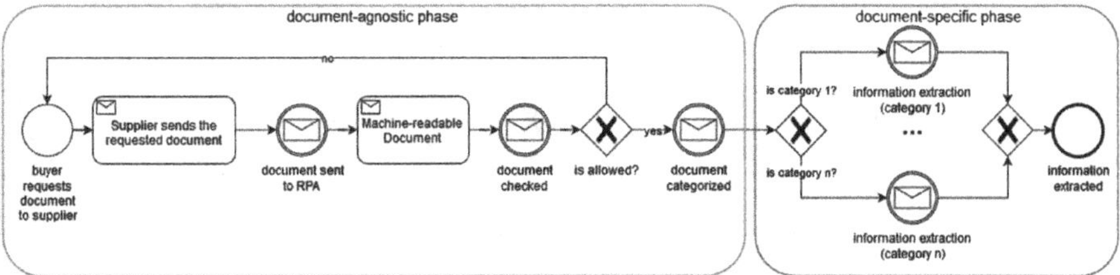

Fig. 1. Document management process

machine-readable format for further processing, iii) checked for unexpectedness (e.g., an ID card when an ISO certification is expected), and, finally, iv) categorized. In the case of non-allowed documents, an early warning is raised to buyers who request a new document upload. Often, procurement platforms only check whether supplier documents have been uploaded or not, without managing them, thus leaving the burden of validating documents on the shoulder of users. Three risks arise: i) suppliers send incorrect or unreadable files for the sole purpose of hijacking the platform or ii) suppliers send non-requested documents, and iii) users on the buyer side wrongly categorize the uploaded documents. In this perspective, it is important that the RPA manages to tackle these problems coherently, providing both early detection of undesired documents and correct categorization.

After having been filtered and categorized, documents in the document-specific phase are further analyzed, based on the category they belong to, for supporting decision-making on supplier document compliance. In particular i) documents are sent to the most appropriate information extraction pipeline, and ii) extracted information is then inspected by end-users for decision-making. Often, traditional procurement platforms are not able to automatically extract information from documents. In fact, suppliers are asked to upload the documents and yet manually fill electronic forms with information reported on the documents. Two risks arise: i) suppliers, inadvertently or not, fill the forms with false information that diverge from what is reported on documents, ii) users from the buyer side only take into account information from the electronic forms while neglecting the document content. Intelligent RPA eliminates such a redundancy by extracting information from documents in the most possible accurate way.

3.1 Intelligent Document Management RPA

We customized the process in Fig. 1 into the one reported in Fig. 2. At each step, the process accommodate different machine learning models trained on the most requested supplier documents by Italian buyers, that is i) DURC documents, ii) CCIAA documents and iii) ISO certifications.

According to the Italian regulation, a DURC ("*Documento Unico di Regolarità Contributiva*") is the document declaring whether a supplier is up to date with payment contributions in respect of INPS and INAIL authorities. The CCIAA documents, instead, provide information on any Italian company listed in the business register held by the Chamber of Commerce, Industry, Crafts

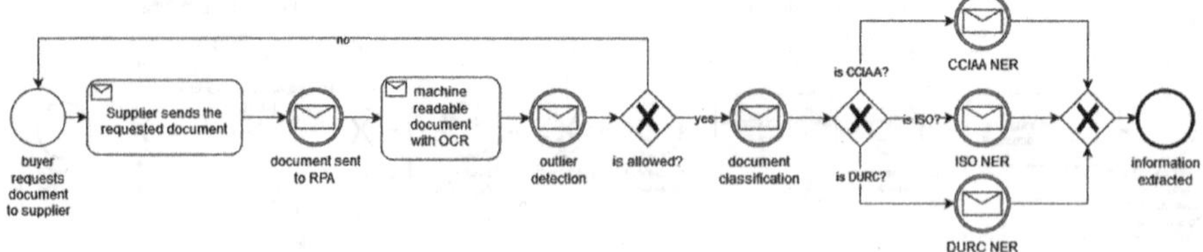

Fig. 2. Intelligent document management process

and Agriculture (*Camera di Commercio, Industria, Artigianato e Agricoltura*). Lastly, the ISO certifications are statements of approval by a third party that a company conforms to one of the international standards developed and published by the International Organization for Standardization (ISO).

Specifically, the document-agnostic phase accepts only Portable Document Files (.pdf) and converts them into machine-readable format by performing OCR, in case of scanned pdf, or by reading the string content directly, otherwise. The document-agnostic phase resorts to supervised outlier detection algorithms for spotting non-allowed documents based on their content, and supervised multi-class classifiers for categorizing them into one of the three considered document categories (DURC, CCIAA, and ISO). Consequently, the document-specific phase resorts to information extraction pipelines based on pre-trained supervised named entity recognition (NER) models for the Italian language.

Since alternative processes can be instantiated by considering alternative solutions for each step, the optimal one is that which minimizes the risks discussed in the previous section. From a machine learning viewpoint, this means that the outlier detection model must be as accurate as possible to detect non-allowed documents and hijacking attempts from the supplier side. Also, classification models should be able to assign the correct class to each provided document, hence reducing the classification error from the buyer side. Finally, NER models should be able to spot, as accurately as possible, the relevant information based solely on what is declared on supplier documents.

Concerning the classification step, we will consider supervised classification algorithms, such as Gaussian Naïve Bayes, Multinomial Naïve Bayes, and Decision Trees. As for the outlier detection step, since traditional algorithms fail on high-dimensional data, we consider a more simple threshold-based detection algorithm, *ThrCls* hereafter. The detector infers the σ probability threshold based on the α-quantile of the predicted-category probability over training data. Then, at inference time, documents with category-predicted probability lower than α are considered outliers, where α is a user-defined parameter that should be proportional to the number of expected outliers in training data.

4 Experiments

Given the 3 document categories reported in the previous section, we quantitatively evaluate the performance of both the document-agnostic and document-specific phases. Specifically, we try to answer the following research questions:

Table 1. Macro averaged (10-fold) classification and outlier detection accuracy

Model	Representation	F1-Score	Recall	Precision	MAR
ThrCls+MNB	w2v_64	0.895	0.976	0.867	**0.000**
	w2v_128	0.863	0.921	0.847	0.204
	w2v_256	0.906	0.978	0.879	**0.000**
	NMF_5	0.765	0.770	0.774	0.728
	NMF_50	0.760	0.768	0.763	0.784
ThrCls+GNB	w2v_64	0.986	0.998	0.977	**0.000**
	w2v_128	0.995	**0.999**	0.992	**0.000**
	w2v_256	**0.998**	**0.999**	**0.997**	**0.000**
	NMF_5	0.959	0.993	0.938	**0.000**
	NMF_50	0.945	0.923	0.994	0.308
ThrCls+DT	w2v_64	0.988	0.994	0.986	0.020
	w2v_128	0.990	0.994	0.988	0.020
	w2v_256	0.984	0.984	0.987	0.060
	NMF_5	0.991	0.987	0.995	0.045
	NMF_50	0.827	0.817	0.846	0.663

RQ1 How accurate is the document-agnostic phase on real-world documents?
RQ2 How accurate is the document-specific phase on real-world documents?

Note that we answer the research questions by evaluating the performance of the underlying machine learning models in an isolated way. At this stage, we conducted no end-to-end evaluation of the whole process. Instead, we prefer to focus on single models to select the most promising one, thus leaving the end-to-end evaluation to future advancements.

The experiments have been performed on a corpus made of the most recent 1764 documents from our archives, subdivided as follows: 598 DURC, 597 CCIAA, and 569 ISO certifications. Since the proposed RPA has been conceived to run on digital or scanned pdf documents, we equally represent both scanned and digital documents within each document category to force OCR in at most 50% of documents. The resulting dataset is kept private due to non-disclosure agreements with our customers.

The proposed RPA solves 3 learning problems at different stages, namely outlier detection, multi-class classification, and named entity recognition. Therefore, the same corpus has been annotated differently, depending on the task at hand. The starting dataset is already labeled for the multi-class classification problem since document categories are known in advance. While all the selected documents are inliers, outliers are missing. For this reason, in the case of the outlier detection problem, we perturbated the inliers by replacing some of them (5%) with outlier texts randomly sampled from Wikipedia pages written in Italian, French, or English language. Finally, considering that a specific NER model is needed for each known document category, we split the inliers into 3 separate corpora, on which custom-named entities have been manually annotated by human experts according to the BILOU scheme using the Doccano tool[1].

[1] https://doccano.github.io/doccano/.

Experiments have been performed on machines equipped with Intel(R) Core (TM) i7-10750H CPU @ 2.60 GHz, 16 GB RAM, Nvidia RTX 2070 Max-Q Design GPU. The code has been executed in Jupyterlab 3.0.12 environment, running with Python 3.8 on top of the Windows 11 Pro (21H2) operating system.

4.1 Accuracy of Document-Agnostic Models

Results in Table 1 depict i) the document classification accuracy (in terms of F1-Score, Precision, and Recall), and ii) the outlier detection accuracy (in terms of Missed Alarm Rate, MAR), while macro-averaging in a 10-fold cross-validation.

As for the classification accuracy, we compared approaches mixing 3 *classification algorithms*, i.e. Gaussian Naïve Bayes (*GNB*), Multinomial Naïve Bayes (*MNB*), and Decision Tree (*DT*), and 2 *data representation* schemes, namely the non-negative matrix factorization (*NMF*) of the tf-idf term-document matrix, and the min-max scaled vectors obtained by summing up the word embeddings generated by *Word2Vec* (*w2v*). As for the outlier detection accuracy, we compared approaches mixing the threshold-based detector discussed in the previous section *ThrCls* with both the 2 data representation methods (*NMF* and *w2v*): in particular, we set the user-defined parameter $\alpha = 0.05$ since we inject 5% of outliers in training data. The low-rank NMF approximation was built by tuning the following parameters: we used 5 and 50 as the number of components, 500 as the number of iterations, and we exploited the random initialization strategy using a seed equal to 42. By adopting, instead, the embedding approach we tuned only the embedding size, changing it to 64, 128, and 256. The number of epochs used for w2v has been fixed to 200.

The results show that, regardless of the classification algorithms, the embedding-based data representation is often associated with the increased classification and outlier-detection accuracy while, on the contrary, NMF data representation leads to decreased accuracy. No clear tendency emerges for increasing embedding size, leading to stable accuracy, while decreasing accuracy is observed for an increasing number of NMF components. This behavior could be due to the presence of scanned documents in the training set, and hence to noisy machine-readable documents built with OCR. Words that are not correctly recognized by OCR become additional noisy features that negatively affect the latent features discovered by the NMF. On the contrary, the embedding approach is more resilient because the resulting representations take into account the text context around discovered tokens.

We can notice the most accurate model is ThrCls + GNB trained on w2v, both in terms of classification accuracy (F1-Score, Recall, and Precision equal to 0.998, 0.999, and 0.997) and outlier detection accuracy (MAR equals to 0). In the given domain, a model with a high recall score is a further desirable outcome due to the skewness deriving from the presence of outliers in the dataset. However, it could be not enough to estimate the accuracy of the outliers. To this end, MAR provides further insight into the results. As one can observe, MAR scores are considerably higher when using NMF than w2v. In particular, we observe that MAR is non-zero in 5 out of 6 configurations on outlier detection performed with

Table 2. NER accuracy as macro-avg F1 over the named entities on the 3 corpora.

SpaCy model	Learning rate	Dropout Rate	DURC	ISO	CCIAA
it_core_news_sm	0.01	0.35	0.917	0.736	0.929
		0.50	0.924	0.734	0.934
	0.001	0.35	0.929	**0.742**	**0.943**
		0.50	**0.930**	0.741	0.941
it_core_news_md	0.01	0.35	0.922	0.713	0.899
		0.50	0.895	0.719	0.896
	0.001	0.35	0.919	0.731	0.931
		0.50	0.917	0.730	0.932

NMF. The same is not true when using w2v, where it is often equal to zero in 5 out of 9 cases. From an RPA viewpoint, integrating an outlier detection model with the lowest MAR minimizes the risk of uploading non-request documents as well as unreadable ones (such as ID cards, bills, and self-certifications that are not interesting for a buyer but we found in our systems). Additionally, a classification model with the highest F1-Score allows a buyer to properly group documents by their class, improving their retrieval and analysis (unlike traditional procurement systems whose suppliers' documents were organized badly).

4.2 Accuracy of Document-Specific Models

Results in Table 3 shows the entity-level accuracy of the most accurate (in terms of F1-Score averaged over the entities) document-specific NER models selected according to a grid search and a 5-fold cross-validation (Table 2). In particular, the best models have been found with a grid search by setting: i) two pre-trained Italian language models (`it_core_news_sm` and `it_core_news_md`), ii) two values of learning-rate (0.01 and 0.001), and iii) two values of dropout-rate (0.35 and 0.5). Each run was limited to 200 training epochs.

From the results in Table 2, we conclude that the size of the adopted pre-trained NER model does not seem to affect the accuracy of the fine-tuned models on the considered corpora. This behavior can be justified by the fact that a model trained on more instances would be more resilient to forgetting prior knowledge and, as a result, it will adapt to new data with more difficulty. The most accurate NER models are, thus, obtained by fine-tuning the `it_core_news_sm` model, with a learning rate of 0.001 and setting the dropout rate to i) 0.35 for the ISO and CCIAA corpora, and ii) 0.50 for the DURC corpus.

After identifying the best NER models for each corpus, we performed a more thorough analysis of their entity-level accuracy in terms of Precision, Recall, and F1-Score (see Table 3). The 3 document-specific models perform considerably well on almost every considered named entity, i.e., the information worth extracting. In particular, the DURC-specific model performs considerably well (F1-Score greater than 80%) on 4 out of 5 entities, the ISO-specific one on 4 out 7 entities, while the CCIAA-specific model on 6 out of 6 entities.

The DURC-specific model performs badly on the `DURC_NO` entity, mostly due to a certain amount of scanned DURC documents in the training corpus. In

Table 3. Macro-averaged (5-fold) entity-level NER accuracy.

Corpus	Named entity	NER		
		F1-score	Precision	Recall
DURC	DUE_DATE	1.000	1.000	1.000
	DURC_NO	0.666	0.667	0.666
	REQUEST_DATE	0.997	1.000	0.993
	TAX_CODE	0.948	1.000	0.917
	VALIDITY_DAYS	1.000	1.000	1.000
ISO	COMPANY_NAME	0.614	0.662	0.574
	DUE_DATE	0.867	0.870	0.869
	IAF_SUBJECT	0.430	0.373	0.506
	ISO_CERTIFICATION_NO	0.832	0.879	0.790
	ISO_STANDARD_NO	0.905	0.880	0.933
	RELEASE_DATE	0.827	0.831	0.825
	VALIDITY_DATE	0.725	0.710	0.744
CCIAA	CCIAA_LOC	0.973	0.990	0.958
	CCIAA_NO	0.887	0.895	0.878
	PEC	0.895	0.892	0.900
	REA_NO	0.972	0.979	0.967
	RELEASE_DATE	0.967	0.994	0.943
	TAX_CODE	0.968	0.989	0.950

these cases, the employed OCR cannot correctly recognize the corresponding tokens. The ISO-specific model performs poorly at recognizing the `COMPANY_NAME` and `VALIDITY_DATE` entities, and badly at recognizing the `IAF_SUBJECT` entity. Despite this consideration, it is worth mentioning that we were expecting far worse results because `it_core_news_sm` is a pre-trained model for the Italian language, while ISO documents are inherently multi-language (Italian-English). Furthermore, differently from DURC documents, the ISO certifications do not share the same document template, which is determined by each certification authority. As for the ISO documents, we expected worse results considering the CCIAA-specific NER model. In fact, the considered CCIAA corpus collects a balanced set of documents that differs in their document templates. Specifically, single-column scanned CCIAA documents and multi-column digital CCIAA documents. Despite this diversity in the training corpora, the model has an acceptable accuracy on every considered entity. Integrating NER models with the best F1-score in an RPA enables advanced capabilities in procurement platforms, such as (semi-)automatic form-filling with information that are relevant for buyers to perform some preliminary supplier evaluations (e.g. checks on ISO certification expiry date, etc.) and to avoid time-consuming and error-prone activities.

5 Conclusions

In this paper, we proposed Robotic Process Automation (RPA) supported by machine learning models, to automatize the management process of documents provided by a supplier via e-procurement platforms. We aim to pave the way toward intelligent procurement. In this regard, we claim that intelligent RPA integrating effective machine learning models provides the interesting information needed to evaluate suppliers' documents and, potentially, the suppliers themselves.

Then, we quantitatively evaluated the proposed solution on a corpus of real-world suppliers' documents. Experiments have shown promising results in terms of accuracy achieved at the different steps of the RPA. Future research directions involve i) comparative evaluation that takes into account more sophisticated machine learning algorithms, and ii) an end-to-end evaluation measuring how errors at early steps affect later ones.

Acknowledgements. The EPICS (*E-Procurement Innovation For Challenging Scenarios*) project has been co-funded by *Programma del Regolamento regionale della Puglia per gli aiuti in esenzione n. 17 del 30/09/2014 (BURP n. 139 suppl. del 06/10/2014) titolo II capo 2 del regolamento generale aiuti ai programmi integrati promossi da medie imprese ai sensi dell'articolo 26 del Regolamento.*

References

1. Baidya, A.: Document analysis and classification: a robotic process automation (rpa) and machine learning approach. In: 2021 4th International Conference on Information and Computer Technologies (ICICT), pp. 33–37 (2021). https://doi.org/10.1109/ICICT52872.2021.00013
2. Flechsig, C., Anslinger, F., Lasch, R.: Robotic process automation in purchasing and supply management: a multiple case study on potentials, barriers, and implementation. J. Purchasing Supply Manage. **28**(1), 100718 (2022). https://doi.org/10.1016/j.pursup.2021.100718. https://www.sciencedirect.com/science/article/pii/S1478409221000522
3. Guha, A., Samanta, D.: Hybrid approach to document anomaly detection: an application to facilitate RPA in title insurance. Int. J. Autom. Comput. **18**(1), 55–72 (2020). https://doi.org/10.1007/s11633-020-1247-y
4. Hartley, J.L., Sawaya, W.J.: Tortoise, not the hare: digital transformation of supply chain business processes. Bus. Horizons **62**(6), 707–715 (2019). https://doi.org/10.1016/j.bushor.2019.07. https://ideas.repec.org/a/eee/bushor/v62y2019i6p707-715.html
5. Hofmann, P., Samp, C., Urbach, N.: Robotic process automation. Electron. Markets **30**(1), 99–106 (2020). https://doi.org/10.1007/s12525-019-00365. https://ideas.repec.org/a/spr/elmark/v30y2020i1d10.1007_s12525-019-00365-8.html
6. Impedovo, A., Barracchia, E.P., Rizzo, G., Caprera, A., Landrò, E.: EPICS: pursuing the quest for smart procurement with artificial intelligence. In: Epifania, F., et al. (eds.) Proceedings of the 1st Italian Workshop on Artificial Intelligence and Applications for Business and Industries (AIABI 2021) co-located with 20th International Conference of the Italian Association for Artificial Intelligence (AI*IA 2021), Online, originally held in Milan, Italy, November 30th, 2021. CEUR Workshop Proceedings, vol. 3102. CEUR-WS.org (2021). http://ceur-ws.org/Vol-3102/paper3.pdf
7. Jha, N., Prashar, D., Nagpal, A.: Combining Artificial Intelligence with Robotic Process Automation—An Intelligent Automation Approach. In: Ahmed, K.R., Hassanien, A.E. (eds.) Deep Learning and Big Data for Intelligent Transportation. SCI, vol. 945, pp. 245–264. Springer, Cham (2021). https://doi.org/10.1007/978-3-030-65661-4_12

8. Martins, P., Sá, F., Morgado, F., Cunha, C.: Using machine learning for cognitive robotic process automation (rpa). In: 2020 15th Iberian Conference on Information Systems and Technologies (CISTI), pp. 1–6 (2020). https://doi.org/10.23919/CISTI49556.2020.9140440
9. Viale, L., Zouari, D.: Impact of digitalization on procurement: the case of robotic process automation. Supply Chain Forum: Int. J. **21**(3), 185–195 (2020). https://doi.org/10.1080/16258312.2020.1776089
10. Waiyanet, P., Madonkha, P.: A study on the implementation of robotic process automation (rpa) to decrease the time required for the documentation process: a case study of abc co., ltd. In: 2022 7th International Conference on Business and Industrial Research (ICBIR), pp. 430–434 (2022). https://doi.org/10.1109/ICBIR54589.2022.9786521

Batch Bayesian Quadrature with Batch Updating Using Future Uncertainty Sampling

Kelly Smalenberger[1(✉)] and Michael Smalenberger[2]

[1] Department of Mathematics and Physical Sciences, Belmont Abbey College, Belmont, NC, USA
kellysmalenberger@bac.edu

[2] Department of Mathematics and Statistics, The University of North Carolina at Charlotte, Charlotte, NC, USA

Abstract. We consider the approximation of unknown or intractable integrals using quadrature when the evaluation of the integrand is considered very costly. This is a central problem both within and without machine learning, including model averaging, (hyper-)parameter marginalization, and computing posterior predictive distributions. Recently, Batch Bayesian Quadrature has successfully combined the probabilistic integration techniques of Bayesian Quadrature with the parallelization techniques of Batch Bayesian Optimization, resulting in improved performance when compared to state-of-the-art Markov Chain Monte Carlo techniques, especially when parallelization is increased. While the selection of batches in Batch Bayesian Quadrature mitigates costs associated with individual point selection, every point within every batch is nevertheless chosen serially, which impedes the realization of the full potential of batch selection. We resolve this shortcoming. We have developed a novel Batch Bayesian Quadrature method that allows for the updating of points within a batch without incurring the costs traditionally associated with non-serial point selection. We show that our method efficiently reduces uncertainty, leads to lower error estimates of the integrand, and therefore results in more numerically robust estimates of the integral. We demonstrate our method and support our findings using a synthetic test function from the Batch Bayesian Quadrature literature.

Keywords: Batch bayesian quadrature · Batch updating · Future uncertainty sampling · Intractable integrals · Machine learning

1 Introduction

Problems frequently encountered in machine learning call for the approximation of intractable integrals of the form

$$Z = \int l(\mathbf{x})\pi(\mathbf{x})d\mathbf{x} \tag{1}$$

where both $l(\mathbf{x})$ (e.g., a likelihood) and $\pi(\mathbf{x})$ (e.g., a prior) are non-negative. Various approaches exist to complete this task, including Laplace approximation, variational

G. Nicosia et al. (Eds.): LOD 2022, LNCS 13810, pp. 167–180, 2023.
https://doi.org/10.1007/978-3-031-25599-1_13

inference, and Markov Chain Monte Carlo (MCMC) [1–3]. However, all these approximation methods have their drawbacks [4, 5] and are unsuitable when evaluating the desired likelihood is expensive, such as when estimates must be made on only a few evaluations [6].

The literature continues to make progress in approximating unknown or intractable integrals when the evaluation of the integrand is expensive. These include Bayesian Quadrature (BQ) [7–9], transformations when the integrand is non-negative [10–12], active sampling to reduce uncertainty [13, 14], and the selection of batches of quadrature points to reduce the cost of the evaluations [6, 15]. However, while the combination of these techniques both increase the accuracy of the approximation of the integral and mitigate costs associated with individual point selection, every point within every batch is nevertheless chosen serially [16, 17], which impedes the realization of the full potential of these techniques. We resolve this shortcoming. We have developed a novel Batch Bayesian Quadrature method that allows for the updating of points within a batch without incurring the costs traditionally associated with non-serial point selection [6]. We show that our method efficiently reduces uncertainty, leads to lower error estimates of the integrand, and therefore results in more numerically robust estimates of the integral. Since our novel contribution builds directly on the aforementioned methods, we highlight the relevant aspects of this literature and describe our implementation of these methods before detailing our novel contribution.

2 Background

2.1 Bayesian Quadrature

BQ considers the approximation of intractable integrals as a problem of inference from limited data to which probability theory can be applied. Additionally, [6] and [11] have shown that BQ techniques achieve faster convergence and smaller absolute errors when compared to state-of-the-art MCMC methods.

BQ utilizes a probabilistic model to induce both the functional form of the integrand and a probability distribution over the value of the integral. Gaussian Process (GP) is a commonly used method for placing probability distributions over functions [6, 7, 18–20], and we use it in our probabilistic model.

Given Eq. 1, we choose to place the GP on the likelihood, l, though we could place it on the entire integrand. The GP is parameterized by a mean $\mu(x)$ and a scaled Gaussian covariance

$$K(x, x') = \lambda^2 \exp\left(\frac{-(x - x')^2}{2\sigma^2}\right) \tag{2}$$

where the output length-scale λ and input length-scale σ control the standard deviation of the output and the autocorrelation range of each function evaluation, respectively, and will be jointly denoted by $\theta = \{\lambda, \sigma\}$. Conditioned on samples $x_d = \{x_1,\ldots, x_N\}$ and corresponding functional values $l(x_d)$, our GP is given by

$$l|D \sim GP(m_D, C_D) \tag{3a}$$

$$m_D(x) = \mu(x) + K(x, x_d)K^{-1}(x_d, x_d)(l(x_d) - \mu(x_d)) \quad (3b)$$

$$C_D(x, x') = K(x, x') - K(x, x_d)K^{-1}(x_d, x_d)K(x_d, x') \quad (3c)$$

where $D = \{x_d, l(x_d), \theta\}$. This leads to a Gaussian distribution of the value of Z since GPs are closed under affine transformations [21]. The mean and variance of Z are given by

$$\mathbb{E}_{l|D}D[Z] = \int mD(x)p(x)dx \quad (4a)$$

$$\mathbb{V}_{l|D}[Z] = \int\int CD(x, x')p(x)p(x')dxdx' \quad (4b)$$

These are analytic given our K and when $\pi(x)$ is Gaussian, which we use here [22]. Other combinations of K and $\pi(x)$ also lead to analytic results [11].

2.2 Warped Bayesian Quadrature

Since we place the GP on l, utilizing a standard GP prior would ignore the range and non-negativity of l leading to pathologies [20]. To address this, several prior works have investigated warping the output space of the GP [11–13]. That is, instead of modeling l as a GP, a transformed likelihood $g(l)$ is modeled, where for example $g = \log(x)$ or $g = \sqrt{x}$, such that $g^{-1}(g(l))$ is non-negative. While this leads to a posterior on l that is not a GP, it is possible to derive a GP that closely approximates it. To that end, we follow the method of [9] termed WSABI-L, which makes use of the square-root transformation $g = \sqrt{x}$. Specifically, we define

$$\tilde{l}(x) = \sqrt{2(l(x) - \alpha)} \quad (5)$$

where α is a small scalar. Prior investigations found that the performance was insensitive to the choice of this parameter and used $\alpha = 0.8 \min l(x_d)$ [11], which we also implement here.

We take a GP prior on $\tilde{l}(x)$: $\tilde{l}(x) \sim GP(0, K)$, for which the posterior is

$$p(\tilde{l}|D) = GP(\tilde{l}; \tilde{m}_D(\cdot), \tilde{C}_D(\cdot, \cdot)) \quad (6a)$$

$$\tilde{m}_D(x) = K(x, x_d)K^{-1}(x_d, x_d)\tilde{l}(x_d) \quad (6b)$$

$$\tilde{C}_D(x, x') = K(x, x') - K(x, x_d)K^{-1}(x_d, x_d)K(x_d, x') \quad (6c)$$

However, with this transformation we arrive at a GP whose marginal distribution for any $l(x)$ is a non-central χ^2 with one degree of freedom, and hence the posterior of our integral is not closed-form.

Since GPs are closed under linear transformations and given our GP on $\tilde{l}$, a local linearization of the form

$$f : \tilde{l} \mapsto l = \alpha + \frac{1}{2}\tilde{l}^2 \quad (7)$$

will give us a GP for l. That is, linearization around $\tilde{l}_0$ results in

$$l(\mathrm{x}) \approx f(\tilde{l}_0) + f'(\tilde{l}_0)(\tilde{l} - \tilde{l}_0) \tag{8}$$

We choose $\tilde{l} = \tilde{m}_\mathrm{D}$ such that

$$l(\mathrm{x}) \approx \alpha - \frac{1}{2}\tilde{m}_\mathrm{D}(\mathrm{x})^2 + \tilde{m}_\mathrm{D}(\mathrm{x})\tilde{l}(\mathrm{x}) \tag{9}$$

and therefore, l is an affine transformation of $\tilde{l}$, which results in the following posterior:

$$p(l|D) \approx GP(l; m_D^L(\cdot), C_D^L(\cdot,\cdot)) \tag{10a}$$

$$m_D^L(\mathrm{x}) = \alpha + \frac{1}{2}\tilde{m}D(\mathrm{x})^2 \tag{10b}$$

$$C_D^L(\mathrm{x},\mathrm{x}') = \tilde{m}_\mathrm{D}(\mathrm{x})\tilde{C}_\mathrm{D}(\mathrm{x},\mathrm{x}')\tilde{m}_\mathrm{D}(\mathrm{x}') \tag{10c}$$

Since $\tilde{m}_\mathrm{D}$ and $\tilde{C}_\mathrm{D}$ are mixtures of un-normalized Gaussians K, m_D^L and C_D^L are also mixtures of un-normalized Gaussians. Therefore, substituting Eqs. 10b and 10c into Eqs. 4a and 4b, respectively, yields closed-form expressions for the mean and variance of Z.

2.3 Active Sampling

One characteristic of the standard BQ model is that the posterior covariance only depends on the locations sampled, not on the functional values at those points [21]. However, given the Bayesian model of the likelihood, Bayesian decision theory can be used to define an acquisition function to guide the selection of sample locations, including the reduction in uncertainty about either the integrand or the integral.

One possibility in selecting the next sample location x_* would be to follow [10] in minimizing the expected entropy of the integral by selecting (Fig. 1).

$$\mathrm{x}_* = \underset{\mathrm{x}}{\mathrm{argmin}}\langle \mathbb{V}_{\mathrm{l}|\mathrm{D},\mathrm{l}(\mathrm{x})}[\mathrm{Z}]\rangle \tag{11}$$

where

$$\langle \mathbb{V}_{l|D,l(x)}[\mathrm{Z}]\rangle = \int \mathbb{V}_{l|D,l(x)}[\mathrm{Z}]N(l(\mathrm{x}); \mathrm{mD}(\mathrm{x}), \mathrm{CD}(\mathrm{x},\mathrm{x}))\mathrm{dl}(\mathrm{x}) \tag{12}$$

Another possibility would be what is known as *uncertainty sampling*. Here the reduction of entropy in the integrand is targeted by selecting x_* with the largest uncertainty, i.e.,

$$\mathrm{x}_* = \underset{x}{\mathrm{argmax}}\mathbb{V}_{l|D}[l(x)\pi(x)] \tag{13}$$

where for our warped integrand we have

$$\mathbb{V}_{l|D}^L[l(x)\pi(x)] = \pi(\mathrm{x})2\tilde{C}\mathrm{D}(\mathrm{x},\mathrm{x})\tilde{m}\mathrm{D}(\mathrm{x})2 \tag{14}$$

It should be noted, as the work by [11] correctly stated, that uncertainty sampling reduces the entropy of the GP to p(l) rather than the true intractable distribution, and that the computation of (11) is considerably more expensive and less numerically stable than that of (13).

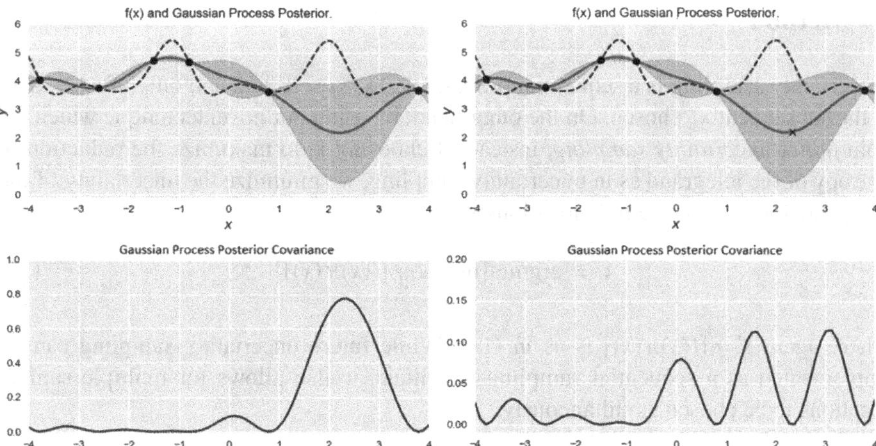

Fig. 1. Selection of one quadrature point demarcated by an x using uncertainty sampling.

2.4 Batch Bayesian Quadrature

A characteristic of the standard BQ model that is important to our paper is that the choice of sample locations is a sequential process, i.e., first x_* is chosen, and only after $x_* \in x_d$ is the subsequent x_* chosen [21]. Since each sample is expensive, this sequential process is very costly.

Recently [6] drew on the Bayesian Optimization literature to implement what they referred to as Batch Bayesian Quadrature (BBQ), where several samples are taken in parallel, i.e., each $\mathbf{x}_*$ is a batch of sample locations. To accomplish this, the authors utilized the probabilistic model of BQ to guide exploration of a search space by defining an acquisition function to be maximized. The method which we use here is termed Kriging Believer [23], although others exist [24].

In Kriging Believer, instead of updating the BQ model after every sample, the functional value of a sample location is set equal to the posterior mean, then the GP model is updated before selecting a subsequent sample location. This process is repeated until the batch of sample locations chosen is sufficiently large, and then the evaluations of those sample locations are made and the BQ model is updated.

A common characteristic of the batch selection procedures in BBQ is that sample locations within a batch are selected serially. While such parallelization procedures reduce the cost of sample selection, they do not consider how changes in sample locations within a batch impact the efficiency of the reduction in uncertainty of the integrand, *ceteris paribus*. In the next section we describe our method of taking this into consideration, and subsequently show that it results in lower absolute error estimates of the integral.

3 Method

Uncertainty sampling is a sequential process, i.e., x_* is selected, and only after $x_* \in x_d$ is the subsequent x_* chosen. On the other hand, by using our novel technique which we coin *future uncertainty sampling*, instead of choosing $x_.$ to maximize the reduction in entropy of the integrand as in uncertainty sampling, we minimize the uncertainty of the integrand by selecting sample locations x_* such that

$$\mathbf{x}_. = \underset{x_*}{\operatorname{argmin}}[\max_{x} \mathbb{V}_{l|D}[l(x)\pi(x)]] \tag{15}$$

where again $\mathbb{V}_{l|D}[l(x)\pi(x)]$ is as in (14). While future uncertainty sampling can be implemented as a sequential sampling technique, it also allows for multiple sample locations to be chosen simultaneously.

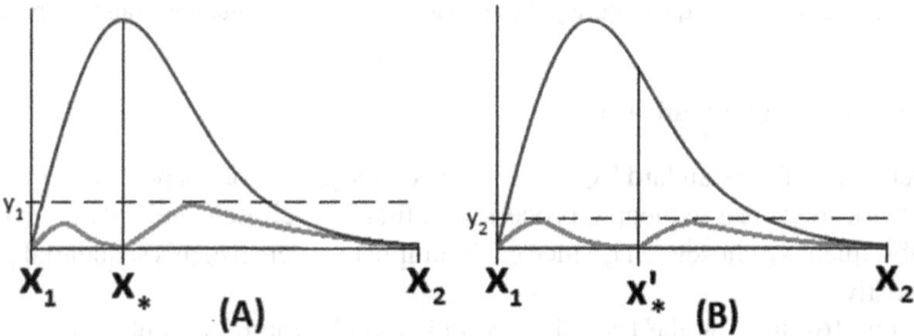

Fig. 2. Panel A: Selection of quadrature point using Uncertainty Sampling by choosing point at the largest uncertainty (blue curve). Panel B; Selection of quadrature point using Future Uncertainty Sampling by choosing point that results in smallest future uncertainty (brown curve). (Color figure online)

Figure 2 provides the intuition when comparing uncertainty sampling (panel A) and our novel method of future uncertainty sampling (panel B) given by Eq. 15. The large blue curve represents the posterior covariance of a GP, and $\mathbf{x}_1$ and $\mathbf{x}_2$ and represents previously chosen quadrature points the model has used to update the GP, and y demarcates the highest posterior covariance of the GP in this section after the next quadrature point is sampled indicated by the brown curves. As can be seen, uncertainty sampling chooses the next quadrature point where the posterior covariance is highest (choose x-value corresponding to the pinnacle of blue curve in panel A), whereas future uncertainty sampling chooses the next quadrature point so that the resulting posterior covariance of the GP is as small as possible (choose x-value corresponding to lowest possible y-value for peaks of brown curves).

Figure 3 shows this process for when batches of quadrature points are selected (denoted by x). In the left panel two quadrature points are selected sequentially using uncertainty sampling, whereas in the right panel two quadrature points are selected simultaneously using future uncertainty sampling. As can be seen in the GP posterior covariance (bottom), the maximum value is smaller for future uncertainty sampling. Since batch selection of quadrature points using uncertainty sampling is necessarily a sequential process, but batch selection of quadrature points using future uncertainty sampling can be done simultaneously, the quadrature points selected using future uncertainty sampling can work in unison to reduce the uncertainty more efficiently than using uncertainty sampling.

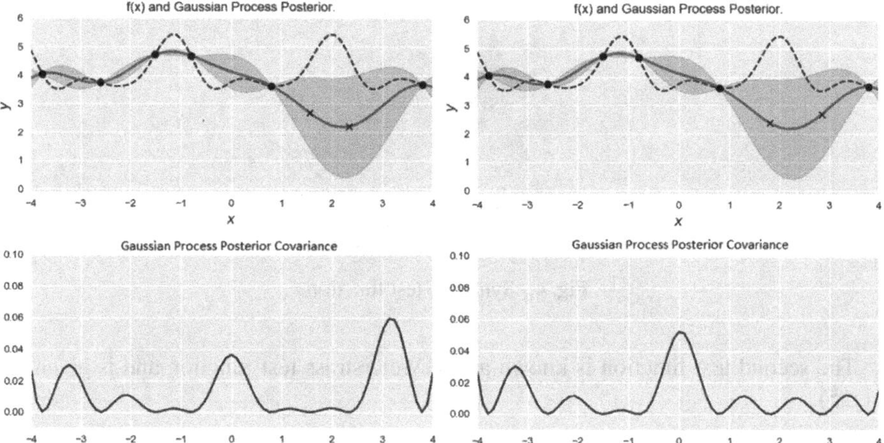

Fig. 3. Selection of batch of two quadrature points demarcated by x using uncertainty sampling (left) and future uncertainty sampling (right).

4 Experiment

In this section, we introduce, to our knowledge, the first Batch Bayesian Quadrature routine where sample locations within each batch do not need to be selected sequentially, or what we refer to as *batch updating*.

We are interested in intractable integrals of the form given by (1). We place the warped GP given in (10a) on the likelihood, and then choose sample locations $\mathbf{x}_*$ given by (15) which minimize the uncertainty of the integrand given by (14), setting the functional value of each sample location equal to the posterior mean given by (10b). Once the requisite number of points are chosen, we evaluate the integrand at these points, use these evaluations to update our model, and calculate an estimate of the integral using BQ.

4.1 Test Functions

To compare the results of our method to those of past BBQ methods, we employ two test functions well known in the optimization and quadrature literature. The first is a synthetic test function given by (Fig. 4).

$$\frac{sin(x) + \frac{1}{2}cos(3x)^2}{\left(\frac{x}{2}\right)^2 + 0.3} \tag{16}$$

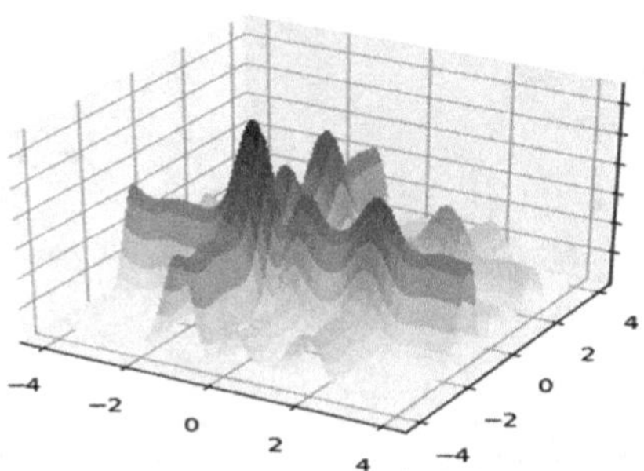

Fig. 4. Synthetic test function

The second test function is known as the Weierstrass test function and is given by (Fig. 5).

$$\sum_{i=1}^{n}[\sum_{k=0}^{kmax} a^k cos\Big(2\pi b^k(x_i + 0.5)\Big) - n\sum_{k=0}^{kmax} a^k cos\Big(\pi b^k\Big)] \tag{17}$$

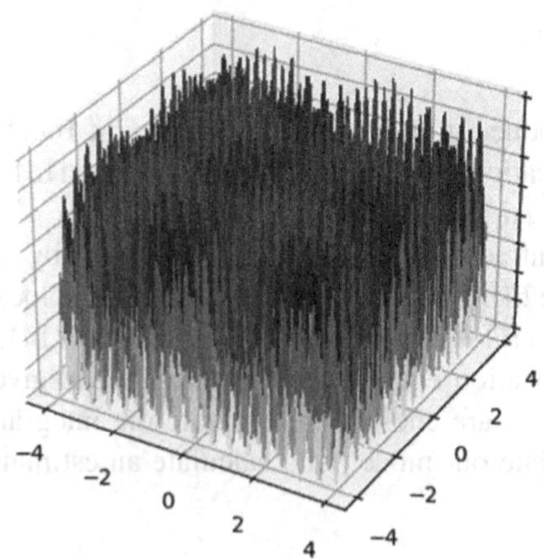

Fig. 5. Weierstrass test function

For both test functions, we conduct our experiment in two dimensions which allows us to evaluate the function on a fine grid to obtain the ground truth. Even though we only employ two test functions, given the description of our novel technique earlier, we believe our results will hold for many other functions. Exactly what class of functions this holds for is left for future research.

4.2 Dynamic Domain Decomposition

A common argument against the simultaneous selection of a set of n points is that it increases the dimensionality of the selection problem by a factor of n, and therefore in practice becomes too costly to compute. Fortunately, in quadrature, locations close to sampled points are uninformative. Therefore, to mitigate the cost of the selection problem, we implement the dynamic domain decompositions outlined in [25] and perform a discretized search over possible sample locations which does not incur the usual cost of non-serial point selection. As we will show below, the combination of future uncertainty sampling and this dynamic domain decomposition more efficiently reduces uncertainty, leads to lower error estimates of the integrand, and therefore results in more numerically robust estimates of the integral when compared to state-of-the-art BBQ techniques.

4.3 Cessation Criteria

Since standard BQ determines a variance for its integral estimate, previous applications have set a stopping criterion of $\sqrt{\mathbb{V}_D[Z]} = 0.015$ and stopped once below this threshold [26]. However, our application of the square-root transformation in the warping of the GP on the likelihood halves the dynamic range of the function we model. On one hand, this mitigates typically large variations in the likelihood, and extends the autocorrelation range of the GP yielding improved predictive power when extrapolating away from the data. On the other hand, the model is overconfident away from the data causing the BQ variance to be erroneously low, making the typical threshold unsuitable as a stopping criterion. Hence previous applications of BBQ have sidestepped this issue by setting a fixed computational budget [6]. We adhere to this precedent and fix our computational budget at 25 batches of size 4.

5 Results

We show here is that by minimizing the uncertainty of the integrand, BBQ with batch updating more efficiently explores the search space leading to better estimates of the integrand, and therefore more numerically robust estimates of the integral.

To see this, note that when multiple points are selected within a batch, future uncertainty sampling can update selected points to choose the quadrature points which better work in harmony to reduce the posterior covariance of the GP.

Pictorially, this can be seen in Fig. 6 and 7 from graphs stemming from our numerical simulations. Since in BBQ both uncertainty sampling and future uncertainty sampling use the posterior covariance of the GP as the acquisition function, we can see that the space in much less explored in uncertainty sampling in panel (A) than in future uncertainty sampling in panel (D).

This has a significant impact on the estimation of the integrand. To see this, note that panel (C) and panel (F) represent the true integrand, and are identical. Furthermore, the posterior mean of the GP is used to estimate the true integrand. Panel (B) represents the posterior mean of the GP using uncertainty sampling, and panel (E) represents the represents the posterior mean of the GP using future uncertainty sampling. It is evident that panel (E) more closely represents the true integrand than panel (B).

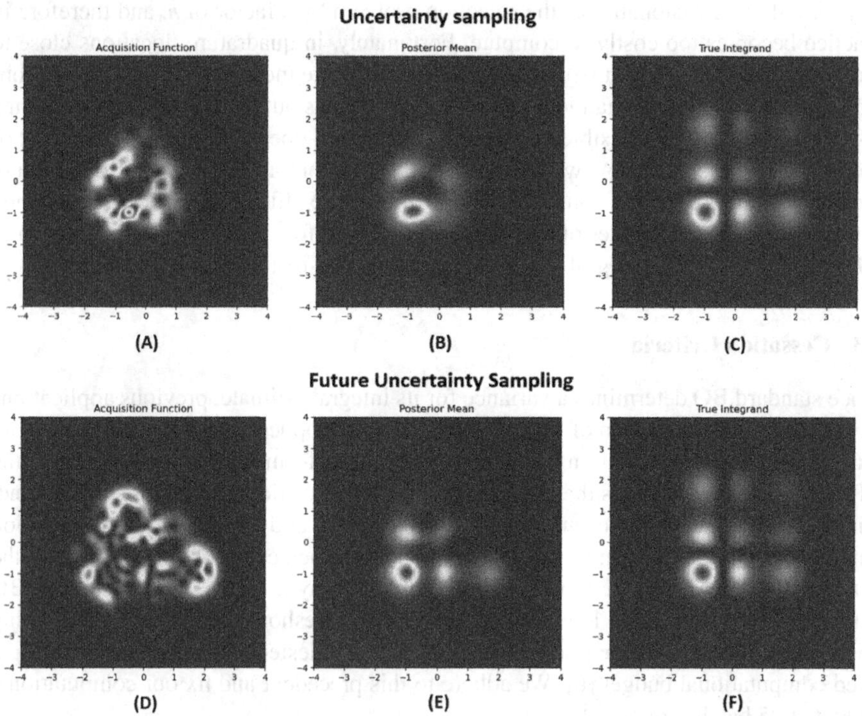

Fig. 6. Synthetic test function approximation using uncertainty sampling (top) and future uncertainty sampling (bottom).

Additionally, we can also see numerically that BBQ using our novel future uncertainty sampling more accurately estimates the integrand, and hence leads to more robust estimates of the integral. Specifically, while there is a plethora of differing evaluation techniques regarding the approximation of an unknown function, a common criterion used in estimating the accuracy of determining the function in black box scenarios is the sum of squared error (SSE) from the estimate to the true function.

It should be noted that in BBQ, using the SSE for the quadrature points would result in no information gleaned. That is, when quadrature points are selected using the full GP model, the posterior mean of the GP is used for functional values of those quadrature points in the batch selection process. Therefore, if the SSE were to be calculated before the model is updated, all the values would be equal to the posterior mean of the GP. If the SSE were to be calculated after the model is updated, all the values would be equal to the true integrand. To circumvent this obstacle, we define a grid of 2,000 equally spaced points as our evaluation points. Even though this many evaluation points will lead to rather large SSE when there are minor deviations in the estimated function, this sensitivity was conducive for our comparisons.

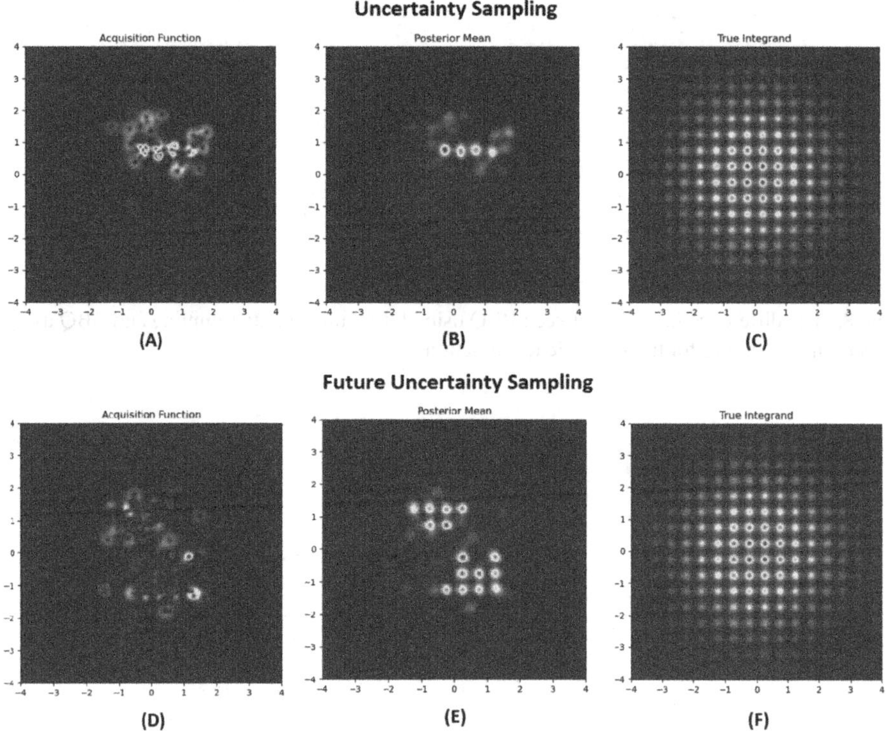

Fig. 7. Weierstrass test function approximation using uncertainty sampling (top) and future uncertainty sampling (bottom).

Another numerical comparison of the integral estimate of BBQ using future uncertainty sampling to BBQ using uncertainty sampling is to determine the absolute error the estimates are from the truth.

The way these graphs in Figs. 8 and 9 are to be understood is that it represents the difference in errors using BBQ techniques in the literature minus the errors of our novel technique. Therefore, if the value is positive, our novel technique performs better. Conversely, if the value is negative, our novel technique performs worse. As can be seen, our novel future uncertainty sampling almost consistently estimates the integrand more accurately, leading to consistently more robust estimates of the integral.

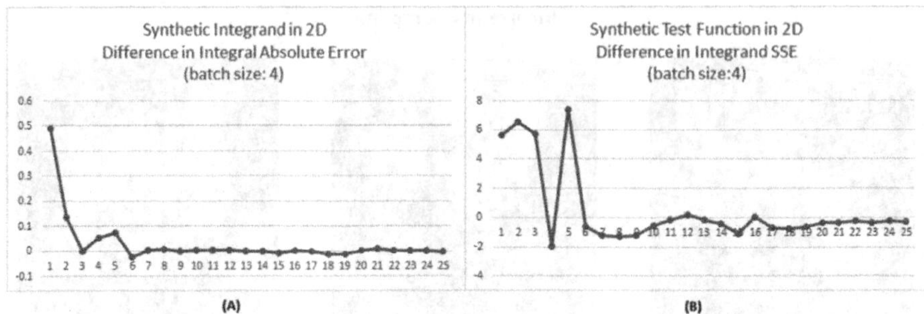

Fig. 8. The difference in errors between BBQ using future uncertainty sampling and BBQ using uncertainty sampling for the synthetic test function.

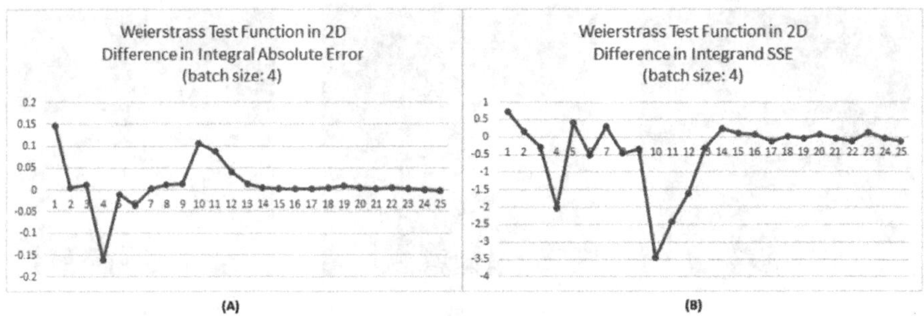

Fig. 9. The difference in errors between BBQ using future uncertainty sampling and BBQ using uncertainty sampling for the Weierstrass test function.

6 Discussion and Conclusion

We considered the approximation of unknown or intractable integrals using quadrature when the evaluation of the integrand is considered very costly. To address this, we introduced a novel method coined future uncertainty sampling which leads to the smallest posterior covariance of the GP. Importantly, this seemingly minor change allows for the tremendous ability of selecting quadrature points in a non-sequential manner while not incurring the costs traditionally assumed for non-serial point selection. We showed that our method efficiently reduces uncertainty, leads to lower error estimates of the integrand, and therefore results in more numerically robust estimates of the integral when compared to state-of-the-art BBQ techniques. We demonstrated our method and supported our findings using a synthetic test function from the Batch Bayesian Quadrature literature. This is an important contribution which builds on the recent BBQ literature addressing a central problem both within and without machine learning, including model averaging, (hyper-)parameter marginalization, and computing posterior predictive distributions.

Acknowledgments. We would like to thank Dr. Xingjie "Helen" Li, Dr. Hae-Soo Oh, Dr. Duan Chen, and Dr. Milind Khire for your generous insights and support. As always, K.H.S., J.M.S., E.M.S., and W.J.S. thank you and I l. y.

References

1. Chen, M., Shao, Q., Ibrahim, J.: Monte Carol Methods in Bayesian Computation. Springer, Heidelberg (2000)
2. Hennig, P., Osborne, M., Girolami, M.: Probabilistic numerics and uncertainty in computations. Proc. Roy. Soc. A: Math. Phys. Eng. Sci. **471**(2179) (2015). arXiv:1506.01326
3. Neal, R.: Probabilistic inference using Markov Chain Monte Carlo methods. Technical Report CRG-TR-93-1, University of Toronto (1993)
4. Blei, D., Kucukelbir, A., McAuliffe, J.: Variational inference: a review for statisticians. J. Am. Stat. Assoc. **112**(518), 859–877 (2016). arXiv:1601.00670
5. O'Hagan, A.: Monte Carlo is fundamentally unsound. J. Roy. Stat. Soc. Series D (The Stat.) **36**(2), 247–249 (1987)
6. Wagstaff, E., Hamid, S., Osborne, M.: Batch Selection for Parallelization of Bayesian Quadrature. arXiv: 1812.01553v1 (2018)
7. O'Hagan, A.: Bayes-Hermite quadrature. J. Stat. Plan. Inference **29**, 245–260 (1991)
8. Kennedy, M.: Bayesian Quadrature with non-normal approximating functions. Stat. Comput. **8**(4), 365–375 (1998)
9. Huszar, F., Duvenaud, D.: Optimally-weighted herding in bayesian quadrature. In: From Proceedings of the Twenty-Eight Conference on Uncertainty in Artificial Intelligence. AUAI Press, Arlington (2012)
10. Osborne, M., et al.: Bayesian quadrature for ratios. In: Proceedings of the Fifteenth International Conference on Artificial Intelligence and Statistics (AISTATS 2012) (2012)
11. Gunter, T., Osborne, M., Garnett, R., Hennig, P., Roberts, S.: Sampling for inference in probabilistic models with fast bayesian quadrature. In Advances in neural information processing systems (nips), pp. 1–9 (2014). arXiv: 1411.0439v1
12. Chai, H., Garnett, R.: An Improved Bayesian Framework for Quadrature. arXiv:1802.04782 (2018)
13. Osborne, M., Duvenaud, D., Garnett, R., Rasmussen, C., Roberts, S., Ghahramani, Z.: Active learning of model evidence using bayesian quadrature. Adv. Neural. Inf. Process. Syst. **1**, 46–54 (2012)
14. Garnett, R., Krishnamurthy, Y., Xiong, X., Schneider, J., Mann, R.: Bayesian optimal active search and surveying. In: Langford, J., Pineau, J. (eds.) Proceedings of the 29th International Conference on Machine Learning (ICML 2012), Omnipress, Madison, WI, USA (2012)
15. Nguyen, V., Rana, S., Gupta, S., Li, C., Venkatesh, S.: Budgeted batch bayesian optimization with unknown batch sizes. In IEEE International Conference on Data Mining, ICDM, pp. 1107–1112. arXiv:1703.04842 (2017)
16. Neal, R.: Annealed importance Sampling. Stat. Comput. **11**(2), 125–139 (2001)

17. Skilling, J.: Nested Sampling. Bayesian Inference Max. Entropy Methods Sci. Eng. **735**, 395–405 (2004)
18. Diaconis, P.: Bayesian numerical analysis. In: Statistical Decision Theory and Related Topics IV, pp. 163–175. Springer, New York (1988)
19. Minka, T.: Deriving quadrature Rules from Gaussian processes. Technical report, Statistics Department, Carnegie Mellon University, pp. 1–21 (2000)
20. Rasmussen, C.E., Ghahramani, Z., Becker, S., Obermayer, K. (eds.) Advances in Neural Information Processing Systems, vol. 15. MIT Press, Cambridge (2003)
21. Rasmussen, C.E., Williams, C.K.I.: Gaussian Processes for Machine Learning. MIT Press, Cambridge (2006)
22. Briol, F., Oates, C., Girolami, M., Osborne, M., Sejdinovic, D.: Probabilistic Integration: A Role in Statistical Computation?, pp. 1–49. arXiv:1512.00933 (2015)
23. Ginsbourger, D., Le Riche, R., Carraro, L.: Kriging is well-suited to parallelize optimization. In: Tenne, Y., Goh, C.-K. (eds.) Computational Intelligence in Expensive Optimization Problems. ALO, vol. 2, pp. 131–162. Springer, Heidelberg (2010). https://doi.org/10.1007/978-3-642-10701-6_6
24. Gonzales, J., Dai, Z., Hennig, P., Lawrence, N.: Batch Bayesian optimization via local penalization. In: Artificial intelligence and statistics, pp. 648–657 (2016)
25. Smalenberger, K., Smalenberger, M.: On the cessation criteria for batch bayesian quadrature using future uncertainty sampling. The University of North Carolina at Charlotte (2022)
26. Garnett, R., Osborne, M., Reece, S., Rogers, A., Roberts, S.: Sequential Bayesian prediction in the presence of changepoints and faults. Comput. J. **53**(9), 1430 (2010)

Sensitivity Analysis of Engineering Structures Utilizing Artificial Neural Networks and Polynomial Chaos Expansion

Lukáš Novák[(✉)], David Lehký, and Drahomír Novák

Brno University of Technology, Brno, Czech Republic
novak.l@fce.vutbr.com

Abstract. This paper is focused on sensitivity analysis of engineering structures using surrogate models. Two different techniques for surrogate modeling are utilized in order to obtain various sensitivity measures of quantity of interest. The artificial neural networks and polynomial chaos expansion are used for efficient sensitivity analysis. Each of the techniques is superior in different areas of uncertainty quantification and thus each of them is used for estimating of different sensitivity measures in two engineering examples – simplified analytical function and complex non-linear finite element model of an existing concrete bridge. On the one hand, artificial neural network is utilized for estimation of sensitivity measures based on Monte Carlo simulation and on the other hand, polynomial chaos expansion is exploited for derivation of global sensitivity measures without additional simulations. It is shown that utilization of both methods leads to efficient and complex sensitivity analysis of engineering structures, and it could be recommended to use combination of both techniques in industrial applications.

Keywords: Uncertainty quantification · Sensitivity analysis · Artificial neural network · Polynomial chaos expansion

1 Introduction

This paper is focused on two metamodeling techniques and their possibilities for sensitivity analysis of engineering structures. The first technique is Polynomial Chaos Expansion (PCE) [1, 2] – a representative from the field of uncertainty quantification (UQ). PCE is often employed in applications with limited number of samples, since it achieves high accuracy and it allows for analytical derivation of Sobol indices [3]. Moreover, it is also possible to derive the first four statistical moments directly from PCE and utilize this information for distribution-based sensitivity analysis. The second technique is Artificial Neural Network (ANN), very well-known from various fields. ANN represents powerful surrogate model, its accuracy is highly dependent on number of statistical samples which determines its common employment in big data analysis. However, this paper is focused on comparison of ANN and PCE for sensitivity analysis with low number of samples.

On the one hand, PCE offers high accuracy using low number of samples for construction together with powerful analytical post-processing of coefficients [4]. However,

G. Nicosia et al. (Eds.): LOD 2022, LNCS 13810, pp. 181–196, 2023.
https://doi.org/10.1007/978-3-031-25599-1_14

accuracy of PCE is always limited for highly non-linear functions (or models with discontinuities) due to polynomial approximation. On the other hand, ANN is general technique without any assumptions limiting its accuracy for specific functions and it achieves higher accuracy of prediction in case of non-linear functions, though it typically needs more samples for training in comparison to PCE. Moreover, it is necessary to use ANN as a surrogate model in order to obtain sensitivity indices [5], which might be time-consuming in case of global sensitivity analysis of mathematical models with many input random variables.

Since both techniques have some practical limitations, this study investigates their synergy for complex sensitivity analysis of engineering structures. Although there are various types of sensitivity measures [6–8], the most suitable for PCE and ANN were selected in order to achieve the optimal efficiency of both methods. ANN is utilized for sensitivity analysis based on results of Monte Carlo simulation (Spearman rank-order correlation, sensitivity in terms of coefficients of variations) and PCE is utilized for global sensitivity analysis (Sobol indices, distribution-based sensitivity indices). Such methodology combines utilizes the accuracy of ANN and computational efficiency of PCE for UQ. The utilization of both surrogate models is beneficial especially for engineering applications typically dealing with highly computationally expensive mathematical models. In that case, the construction of surrogate model is not a bottleneck, since most of the computational cost is required for obtaining of training data set. Here, the described methodology is applied to two engineering examples – deflection of a fixed beam (analytical function) and post-tensioned concrete bridge (non-linear finite element model).

Although a sensitivity analysis is very active research area for both methods – ANN [9–11] and PCE [3, 4], their combination for a complex analysis using various sensitivity measures has not been investigated in literature. Both techniques are typically used separately for different types of applications though they have potential to create general efficient tool for UQ. Naturally, both techniques offer different sensitivity information and thus their combination brings complex insight to behavior of the mathematical models representing physical systems. Moreover, sensitivity analysis of structures using ANN or PCE is still not very common in civil engineering. This paper aims to fill this gap by the proposed methodology utilizing advantages of both PCE and ANN for sensitivity analysis of real-life structure represented by an existing concrete bridge.

2 Theoretical Background of PCE and ANN

2.1 Polynomial Chaos Expansion

Assume a probability space $(\Omega, \mathscr{F}, \mathscr{P})$, where Ω is an event space, $\mathscr{F}$ is a σ-algebra on Ω and $\mathscr{P}$ is a probability measure on $\mathscr{F}$. Assuming random vector $\mathbf{X}(\omega)$, $\omega \in \Omega$, then the model response $Y(\omega)$ is a random variable referenced as quantity of interest (QoI). Assuming $Y = f(\mathbf{X})$ has the finite variance σ^2, i.e. $Y \in L_2(\Omega, \mathcal{F}, \mathcal{P})$, the polynomial chaos expansion can be used as an approximation of QoI in the following form [1]:

$$Y = f(\mathbf{X}) = \sum\nolimits_{\alpha \in N^M} \beta_\alpha \psi_\alpha(\boldsymbol{\xi}), \tag{1}$$

where M is the number of input random variables, β_α are unknown deterministic coefficients and ψ_α are multivariate basis functions orthonormal with respect to the joint probability density function (PDF) of $\boldsymbol{\xi}$. The basis functions must be selected in dependence to distributions of input random vector $\mathbf{X}$ which must be transformed to associated standardized variables $\boldsymbol{\xi}$. It is common to use Legendre polynomials for Uniform distributions and Hermite polynomials for Normal/Lognormal distributions in engineering applications. Another associated polynomials to specific distributions can be found in Wiener-Askey scheme [2]. In case of arbitrary distributed input random variables, it is possible to construct orthogonal polynomials numerically by Gram-Schmidt orthogonalization.

For practical computation, it is necessary to use PCE truncated to a finite number of terms P. Truncated set of basis functions $\mathcal{A}^{\mathcal{M},\surd}$ is dependent on given maximal polynomial order p and M as follows:

$$\mathcal{A}^{\mathcal{M},\surd} = \left\{ \alpha \in N^M : |\alpha| = \sum\nolimits_{i=1}^{M} \alpha_i \le p \right\} \quad (2)$$

Deterministic coefficients β_α can be obtained by intrusive and non-intrusive approaches [12]. Non-intrusive methods utilize the original mathematical model (e.g. FEM) as a black-box, which allows for their easy applications in combination with commercial software. The most popular type of the non-intrusive approach is based on linear regression. Regression is typically less expensive than spectral projection and thus it is often employed in engineering applications involving highly computationally expensive models. Experimental design (ED) contains sample points in M-dimensional space and corresponding results of the original model. Size of ED must be higher than P (recommended size is typically $3P$).

The original mathematical model must be evaluated to obtain a vector of results corresponding to generated sample points. Once the basis functions are created and experimental design is prepared, PCE coefficients can be estimated by ordinary least square (OLS) regression method. Unfortunately, truncated PCE solved by OLS is not highly efficient and cannot be employed for practical examples with large number of input random variables due to the curse of dimensionality. The solution is additional reduction of truncated basis set by any model-selection algorithm such as Least Angle Regression (LAR) [13]. LAR automatically detects the most important basis functions for given ED and create so called sparse PCE. Approximation of engineering models by sparse PCE can be justified by the sparsity-of-effects principle, which is often observed in practical applications. For further reduction of computational cost, it is beneficial to employ advanced sampling schemes for sequential enrichment of [14].

An important part of surrogate modelling is an estimation of the accuracy of the surrogate model. It is common to use the coefficient of determination R^2. Unfortunately, because R^2 might often lead to over-fitting. Therefore, more sophisticated cross-validation techniques should be employed, such as Leave-one-out error Q^2.

In this paper, an algorithm for construction of non-intrusive sparse PCE based on LAR is employed for numerical examples [15]. The algorithm is depicted in simplified Fig. 1. It consists of 3 virtual parts: pre-processing (construction of stochastic model and ED), processing (adaptive and possibly sequential construction of PCE) and post-processing including analytical part (analysis of PCE function) and numerical part (utilization of PCE as a surrogate model).

This paper is focused on analytical exploitation of PCE to perform global sensitivity analysis. Powerful post-processing is main advantage of PCE over to ANN and thus its possibilities will be investigated in two engineering examples. First of all, thanks to the explicit form of PCE it is possibility to obtain Q^2 analytically without additional computational demands which can be further used for adaptive construction of PCE approximation. Besides analytical derivation of Q^2 directly from PCE, it is possible to derive also the first four statistical moments directly from estimated coefficients. Estimated statistical moments can be further utilized for global sensitivity analysis. Specifically, variance of the approximated QoI together with its conditional variances can be efficiently used for estimation of Sobol indices. Moreover, one can use also higher statistical moments together with an approximations of probability distribution of QoI and its conditional distributions in order to perform distribution-based sensitivity analysis.

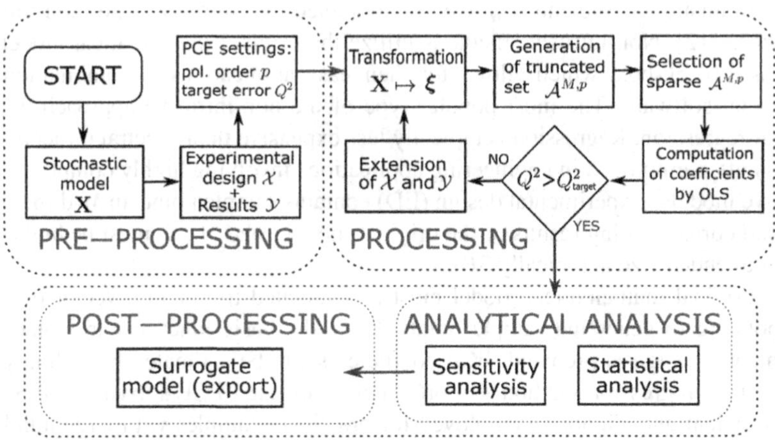

Fig. 1. Simplified algorithm for construction of PCE consisting of three parts: pre-processing, processing and post-processing (including analytical and numerical part).

2.2 Artificial Neural Network

The second universal and frequently used surrogate model is an artificial neural network (ANN) belonging to the category of machine learning models. ANN is a highly parallel system that is designed to model the relation among the input and output parameters through a training process. The typical ANN model is a cluster of multiple interconnected processing units called neurons or nodes grouped together into layers. When the input data is sorted in the forward direction, this is called the feed-forward neural network. The basic structure of an ANN consists of three types of layers: the input layer, hidden layers, and output layer (Fig. 2). The input layer receives the input data; the hidden layers compute the input data, and the output layer provides outcomes. The neighboring layers are connected by synapses through which the neuron receives signal data and consequently proceeds it to the next layer. A bias is added to the sum of the weighted signals and the resulting product is passed through the activation function, which determines the neuron output. The output is computed layer by layer starting from the input layer. The output of the kth neuron in the uth layer of the network is calculated as:

$$y_k^u = f^u\left(\left(\sum_{j=1}^{J} w_{kj}^u y_j^{u-1}\right) + b_k^u\right) = f^u\left(x_k^u\right) \tag{3}$$

where w_{kj}^u is the synaptic weight of the connection between the kth neuron in the uth layer (current) and the jth neuron in the $(u-1)$th layer (previous), y_j^{u-1} is the output of jth neuron in the previous layer, b_k^u is the bias of the kth neuron in the current layer, and f^u is the activation function of the current layer. If the current layer u is the last one, then the output y_k^u is part of the ANN output vector $\mathbf{y}^u$.

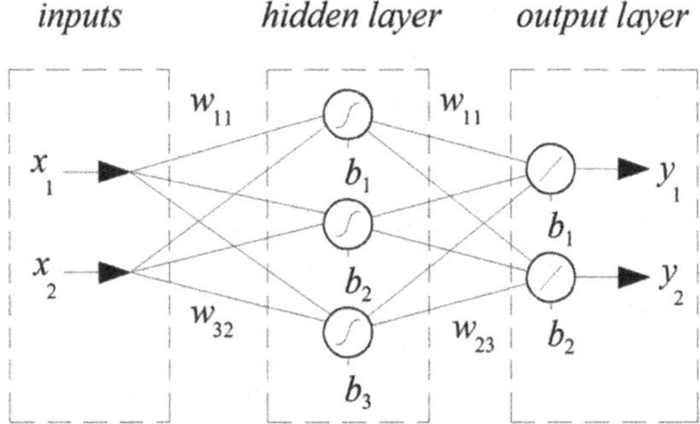

Fig. 2. Diagram of a feed-forward multi-layer network

The behavior of an ANN is determined by its structure (i.e., the number of hidden layers and the corresponding number of neurons), the types of activation functions and the values of synaptic weights and biases. The later parameters are adjusted during the training process. The choice of the activation function f^u has a key influence on the complexity and performance of the ANN. Depending on the type of the original model, both linear and nonlinear activation functions can be used. In general, the nonlinear activation function must be employed if one wishes to introduce nonlinearity into the ANN. In that case, we use the hyperbolic tangent activation function:

$$f^u(x_k) = \tanh\left(x_j\right) = \frac{2}{1 + e^{-2x_k}} - 1. \tag{4}$$

A feed-forward type network is trained using "supervised" learning, where a set of example pairs of inputs and corresponding outputs (p, y), $p \in \mathbf{P}$, $y \in \mathbf{Y}$ is introduced to the network. The aim of the subsequent optimization procedure is to find a neural network function f_{ANN}: $\mathbf{P} \rightarrow \mathbf{Y}$ in the allowed class of functions that matches the examples. This is performed by minimizing the error function E:

$$E = \frac{1}{2N}\sum_{i=1}^{N}\sum\nolimits_{k=1}^{K}\left(y_{ik}^{o} - y_{ik}^{t}\right)^{2}, \tag{5}$$

where N defines the size of the training set, y_{ik}^{t} is the desired (target) output of the kth output neuron for the ith input signal and y_{ik}^{o} is the real output of the same neuron for the same input signal, which depends on the current network parameters. ANN training is an optimization task which can be solved via an appropriate minimization method. Combining local deterministic gradient-based methods with global probabilistic methods is often advantageous in terms of accuracy and time.

For sensitivity analysis, ANN can be used in two basic modes: (1) sensitivity extracted from the ANN parameters, and (2) sensitivity with ANN as a metamodel used in combination with other sensitivity method. In this paper, the second mode is used, i.e., ANN is used primarily as a metamodel for accurate approximation of a complex nonlinear relationship between dependent and independent variables and subsequent rapid evaluation of the metamodel required by sensitivity method procedure. Here the sensitivity in the terms of Spearman rank-order correlation and in terms of coefficient of variation are used, see Sect. 3. The implementation of an ANN-based surrogate model consists of the following three steps:

1. Preparation of training data for adjusting ANN parameters. It is done by randomizing the input variables of the structural model using appropriate sampling methods and the subsequent numerical simulation of the original (FEM) model.
2. Creation of the ANN structure, its training to obtain a reasonably accurate surrogate model, and validation. The training set consists of generated random samples of input variables and the corresponding simulated responses of the structure. Validation of the surrogate model can be performed using a validation set (if available) or by performing k-fold cross-validation over the test data.
3. Utilization of the ANN surrogate model in sensitivity analysis when the evaluation of structural response for given input data is needed.

Further theoretical details on ANNs can be found, for example, in Cichocki & Unbehauen [16]. For more details on ANN-based surrogate modeling, see Lehký & Šomodíková [17].

3 Sensitivity Analysis

Sensitivity analysis is a crucial step in computational modeling and assessment. Through sensitivity analysis we gain essential insights into the behavior of computational models, their structure, and their response to changes in model inputs. Sensitivity analysis is also important to reduce a space of random variables for stochastic calculation, building the surrogate model, etc. Many sensitivity analysis methods have been developed, giving rise to a large and growing literature. Overview of available methods is given in review papers, e.g., [6–8, 18]. This paper presents sensitivity methods using surrogate models suitable for use in combination with the small sample simulation technique. This need is driven by the extreme computational burden of nonlinear structural analysis and hence the need to reduce the computational time to an acceptable level. Specifically in this paper, two groups of sensitivity methods are utilized in combination with described surrogate models: sensitivity methods based on Monte Carlo sampling in combination with ANN; and sensitivity methods which can be analytically derived directly from PCE.

Sensitivity Analysis in Terms of Spearman Rank-Order Correlation

The relative effect of each basic random variable on structural response can be measured using the partial correlation coefficient between each basic input variable and the response variable. With respect to the small-sample simulation techniques of the Monte Carlo type utilized for the reliability assessment of time-consuming nonlinear problems, the most straightforward and simplest approach uses non-parametric rank-order statistical correlation [8]. For a detailed discussion of rank-order statistical correlation see Vořechovský [19]. Non-parametric correlation is more robust than linear correlation and more resistant to defects in data. It is also independent of probability distribution. Because the model for the structural response is generally nonlinear, a non-parametric rank-order correlation is used by means of the Spearman correlation coefficient:

$$r_{XY} = \frac{\sum_{i=1}^{n}\left(\hat{x}_i - m_{\hat{x}}\right)\left(\hat{y}_i - m_{\hat{y}}\right)}{\sqrt{\sum_{i=1}^{n}\left(\hat{x}_i - m_{\hat{x}}\right)^2}\sqrt{\sum_{i=1}^{n}\left(\hat{y}_i - m_{\hat{y}}\right)^2}}, \tag{6}$$

where $\hat{x}_i, \hat{y}_i$ are the rank orders of the individual samples and $m_{\hat{x}} = \frac{1}{n}\sum_{i=1}^{n}\hat{x}_i$ is the mean of the rank orders of the variable X samples (for the variable Y the mean value of the rank orders $m_{\hat{y}}$ is determined by analogy). Equation 6 can be used even if the data set of one of the variables contains identical rank values (so-called "tied ranks"). For these values, Eq. 6 works with the average ranks.

The crude Monte Carlo simulation method can be used for the preparation of random samples. However, it is recommended that a proper sampling scheme should be used, e.g., stratified Latin hypercube sampling [20–22]. This method utilizes random permutations of the number of layers of the distribution function of the basic random variables to obtain representative values for the simulation. When using this method, the ranks $\hat{x}_i, \hat{y}_i$ in Eq. 6 are directly equivalent to the permutations used in sampling.

Sensitivity Analysis in Terms of Coefficient of Variation

Sensitivity analysis in terms of coefficient of variation is the second method used for the optimum selection of dominant random variables in the examples presented in this paper [18]. In this approach, the ratio between the partial coefficient of variation of resistance and the coefficient of variation of a selected basic variable is calculated for a case in which the selected random variable is the only one treated as random in the simulation process. This method can be seen as simplified analysis of variance.

When using a Monte Carlo type simulation, a simulated set of realizations of structural response variable R_j ($j = 1, 2, \ldots, N$), where N is the number of simulations, is statistically evaluated and its coefficient of variation COV_R can be determined. The number of variables M, representing e.g. material properties or load, can be defined as random in the simulation process. Let us designate the partial coefficient of variation COV_{Ri} ($i = 1, 2, \ldots, M$) for a case in which the variable X_i is the only one treated as random and is defined using its mean value and coefficient of variation COV_{Xi}. The other basic variables are kept as deterministic constants at their mean values. The partial sensitivity factor α_i^{COV} for the basic random variable X_i is then defined as:

$$\alpha_i^{COV} = \frac{COV_{Ri}}{COV_{Xi}}. \tag{7}$$

This procedure requires additional computational effort, since the set of values of structural response variable R_i and its coefficient of variation COV_{Ri} need to be evaluated with additional M sets of simulations. Therefore, in cases when the evaluation of an original model is time-consuming, e.g. a nonlinear FEM model is employed, a suitable type of surrogate model is needed in order to reduce the computational time to an acceptable level. Here, above mentioned ANN-based surrogate model was employed for this type of sensitivity analysis.

Sensitivity factor α_i^{COV} in Eq. 7 expresses the relative influence of individual input random variables on the variability of structural response. If M basic variables are considered as random ones, the coefficient of variation of the response variable COV_R can be calculated using an approximate formula in the form:

$$COV_R \approx \sqrt{\sum_{i=1}^{M} \left(\alpha_i^{COV} COV_{Xi}\right)^2}. \tag{8}$$

It can be seen from Eq. 8 that the actual influence (not the relative one) of random variable X_i is represented by the value $COV_{Ri}^2 = \left(\alpha_i^{COV} COV_{Xi}\right)^2$. This absolute effect is influenced by the values of the coefficients of variation COV_{Xi} and can be conveniently presented as a percentage ($COV_R = 100\%$). It shows the proportion of the influence of each input random variable on the resulting coefficient of variation of the output variable. Such sensitivity may be easily depicted using a pie chart and thus this method can be easily employed in practical engineering applications.

Sobol Sensitivity Indices

One of the most complex types of sensitivity analysis is so called analysis of variance (ANOVA), specifically Sobol indices [23]. This type of global sensitivity analysis aims at quantifying the importance of input variables on the variance of model response. First order Sobol indices can be computed as follows:

$$S_i = \frac{\sigma^2[\mu(Y|X_i)]}{\sigma^2[Y]} \tag{9}$$

where the variance of conditional expectation $\sigma^2[\mu(Y|X_i)]$ is normalized by the variance $\sigma^2[Y]$. If the random variables Y and X_i are independent, then $S_i = 0$ and if Y does depend only on X_i, then $S_i = 1$. Note that first order Sobol indices don't take into account the interactions with other variables. For this purpose, total Sobol indices S_i^T can be computed.

It was shown by Sudret [3] that Sobol indices of any order or total Sobol indices can be computed from PCE without additional computational effort due to orthogonality of the PCE terms. The PCE can be rewritten in form of Hoeffding-Sobol' decomposition by reordering of the terms:

$$f(\mathbf{x}) \approx \beta_0 + \sum_{\substack{u \subset \{1, \ldots, M\} \\ u \neq \varnothing}} f_{\mathbf{u}}(\mathbf{x_u}) = \beta_0 + \sum_{\alpha \in A_{\mathbf{u}}} \beta_{\alpha} \Psi_{\alpha}(\boldsymbol{\xi}) \tag{10}$$

where set of multivariate polynomials depend on set of input variables $\mathbf{u}$ is defined as:

$$A_{\mathbf{u}} = \left\{ \boldsymbol{\alpha} \in A^{M,p} : \alpha_k \neq 0 \leftrightarrow k \in \mathbf{u} \right\} \tag{11}$$

Therefore, the first order Sobol' indices can be analytically obtained directly from PCE as follows:

$$S_i^{PCE} = \frac{\sum_{\alpha \in A_i} \beta_{\alpha}^2}{\sigma_Y^2} \quad A_i = \left\{ \boldsymbol{\alpha} \in A^{M,p} : \alpha_i > 0, \alpha_{j \neq i} = 0 \right\} \tag{12}$$

and total order Sobol' indices as

$$S_i^{T,PCE} = \frac{\sum_{\alpha \in A_i^T} \beta_{\alpha}^2}{\sigma_Y^2} \quad A_i^T = \left\{ \boldsymbol{\alpha} \in A^{M,p} : \alpha_i > 0 \right\} \tag{13}$$

Note that from computational point of view, it is just selection of deterministic coefficients and their simple post-processing, which does not bring any additional computational demands.

Distribution-Based Sensitivity Analysis

Following the idea of Sobol' indices, it is possible to take into account also higher statistical moments of QoI. The sensitivity measure is thus not governed by variance but the whole distribution approximated by suitable method, e.g. Gram-Charlier expansion (G-C) or Kernel density estimation. In this paper, the first four statistical moments derived analytically from PCE are utilized for a construction of a cumulative distribution function (CDF) of QoI using G-C. Further similarly to Sobol' indices, conditional statistical moments are derived from PCE and used for construction of conditional CDFs by G-C. The sensitivity measure is based on difference between CDF and CCDFs associated to each input random variable as depicted in Fig. 3. Here, the relative entropy of two distributions is compared by the Kullback-Leibler divergence [24]. Obtained sensitivity indices K_i is a measure of the information lost when the original CDF of QoI is approximated by CCDF assuming i-th variable as a deterministic variable and thus the influence of i-th variable to the uncertainty of QoI. In general case, CCDF can be associated with set of input variables defined in subset $\mathbf{u}$ similarly as for Sobol indices:

$$K_{F_Y}^{\mathbf{u}} = \int F_Y^{PCE}(t) ln \frac{F_Y^{PCE}(t)}{F_{\mathbf{u}}^{PCE}(t)} dt \tag{14}$$

For the sake of simplicity of results, obtained indices are further normalized by the sum of all indices in order to obtain percentual influence of i-th variable (or set of input variables defined in $\mathbf{u}$):

$$K_{\mathbf{u}} = \frac{K^{\mathbf{u}}_{F_Y}}{\sum_{\Delta} K^{\Delta}_{F_Y}} \tag{15}$$

where Δ is the power set of $\mathbf{u}$ i.e. it contains all possible indices of the given interaction orders. More computational details can be found in recently proposed algorithm [4]. Note that the first four statistical moments used for construction of analytical G-C are obtained directly from PCE coefficients and thus this sensitivity analysis does not bring any additional computational demands.

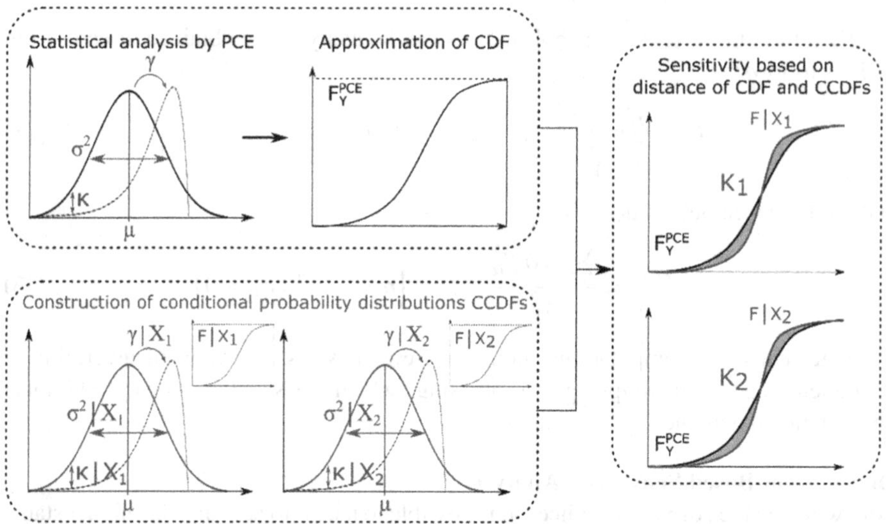

Fig. 3. Graphical interpretation of distribution-based sensitivity analysis.

4 Applications

This section presents two practical applications that demonstrate the benefits of ANN and PCE combination for complex sensitivity analysis of engineering structures.

4.1 Fixed Beam

The first example is represented by a typical engineering problem with a known analytical solution, a maximum deflection of a fixed beam loaded by a single force in mid-span:

$$Y = \frac{1}{16}\frac{FL^3}{Ebh^3}, \tag{16}$$

containing 5 lognormally distributed uncorrelated random variables according to Table 1, where b and h represent the width and height of the rectangular cross-section, E is the modulus of elasticity of concrete, F is the loading force and L is the length of the beam.

Table 1. Stochastic model of mid-span deflection of a fixed beam

Parameter (unit)	b (m)	h (m)	L (m)	E (GPa)	F (kN)
Mean μ	0.15	0.3	5	30	100
Standard deviation σ	0.0075	0.015	0.05	4.5	20

For this example, it is simple to obtain an analytically lognormal distribution of Y, and thus quality of surrogate models can be also verified by the estimated statistical moments. The product of lognormally distributed variables is a lognormal variable $Y \sim \mathcal{LN}(\lambda_Y, \zeta_Y)$, where parameters of distribution are obtained as:

$$\lambda_Y = \ln\frac{1}{16} + \lambda_F + 3\lambda_L - \lambda_E - \lambda_b - 3\lambda_h \tag{17}$$

$$\zeta_Y = \sqrt{\zeta_F^2 + (3\zeta_L)^2 + \zeta_E^2 + \zeta_b^2 + (3\zeta_h)^2} \tag{18}$$

The sparse PCE is created by the adaptive algorithm depicted in Fig. 1 but generally any algorithm might be employed for efficient construction of PCE. Using the adaptive algorithm, it was possible to build the PCE with ED containing 100 samples generated by LHS and maximal polynomial order $p = 5$. Since the function is not in a polynomial form, the PCE is an approximation of the original model with an accuracy measured by Leave-one-out cross validation $Q^2 = 0.998$. The ANN surrogate model consists of five input neurons, one output neuron with a linear activation function, and five nonlinear neurons in the hidden layer with a hyperbolic tangent activation function. The resulting superior ANN structure and associated hyperparameters emerged from initial testing. ED contains once again 100 samples generated by LHS method. Validation of the model was performed using 100 additional test samples with accuracy $R^2 = 0.998$.

The statistical moments obtained analytically, from PCE, and from ANN using 1 million Monte Carlo simulations are listed in Table 2 and compared with analytical reference solution in order to confirm accuracy of both surrogate models.

Table 2. Comparison of first four statistical moments obtained by surrogate models with analytical solution.

Method	μ	σ^2	γ	κ
Analytical	6.69	4.09	0.91	4.46
PCE	6.69	4.08	0.88	4.41
ANN	6.69	4.07	0.89	4.25

Further, both surrogate models were utilized for various types of sensitivity analysis as described in the previous section. It can be seen from results in Table 3 that global sensitivity indices in terms of CoV and Sobol indices are in good agreement, this is typical for linear mathematical models. However, calculation of COV^2_{Ri} is significantly cheaper than Sobol indices and thus the later are typically obtained from PCE. Note that r_{XY} are not comparable, since they give us different information – direction and strength of correlation between input variable and QoI. Distribution-based sensitivity indices K_i are typically comparable to Sobol indices but they emphasize the importance of input random variables, which significantly affects also higher statistical moments.

Table 3. Sensitivity analysis by ANN and PCE of deflection of a fixed beam

Variable	r_{XY}(ANN)	COV^2_{Ri}(ANN)	S_i^{PCE}(PCE)	K_i(PCE)
b	−0.156	0.026	0.026	0.004
h	−0.484	0.246	0.251	0.185
L	0.094	0.010	0.010	0.001
E	−0.494	0.246	0.250	0.177
F	0.654	0.472	0.438	0.633

4.2 Post-tensioned Concrete Bridge

The proposed methodology is applied for the existing post-tensioned concrete bridge with three spans. The super-structure of the mid-span is 19.98 m long with total width 16.60 m and it is crucial part of the bridge for assessment. In transverse direction, each span is constructed from 16 post-tensioned bridge girders KA-61 commonly used in Czech Republic. Load is applied according to national annex of Eurocode for load-bearing capacity of road bridges by exclusive loading (by a six-axial truck).

The NLFEM is created using software ATENA Science based on theory of non-linear fracture mechanics [25]. The NLFEM of each girder consists of 13,000 elements of hexahedra type in the major part of the volume and triangular 'PRISM' elements in the part with complicated geometry as depicted in Fig. 4 (top left). Reinforcement and tendons are represented by discrete 1D elements with geometry according to original documentation. As can be seen from stress field depicted in Fig. 4 (right), girders fail in bending after significant development of cracks in midspan area with tensile stresses (highlighted by red color). The numerical model is further analyzed in order to investigate the ultimate limit state (ULS) (peak of a load-deflection diagram) to determine the load-bearing capacity of the bridge and the first occurrence of cracks in super-structure (Cracking) leading to corrosion of tendons and possibly failure of structure.

The stochastic model contains 4 lognormal random material parameters of a concrete C50/60: Young's modulus E (Mean: 36 GPa, CoV: 0.16); compressive strength of concrete f_c (Mean: 56 MPa, CoV: 0.16); tensile strength of concrete f_{ct} (Mean: 3.6 MPa, CoV: 0.22) and fracture energy G_f (Mean: 195 Jm^2, CoV: 0.22). The last random variable P representing procentual prestressing losses is normally distributed with Mean: 20 and CoV: 0.3. Mean values and coefficients of variation were obtained according to Eurocode prEN 1992–1-1: 2021 (Annex A) [26] for adjustment of partial factors for materials and JCSS Probabilistic model code [27]. The ED contains 30 numerical simulations generated by LHS. Note that each simulation takes approximately 24 h and construction of the whole ED took approx. 1 week of computational time.

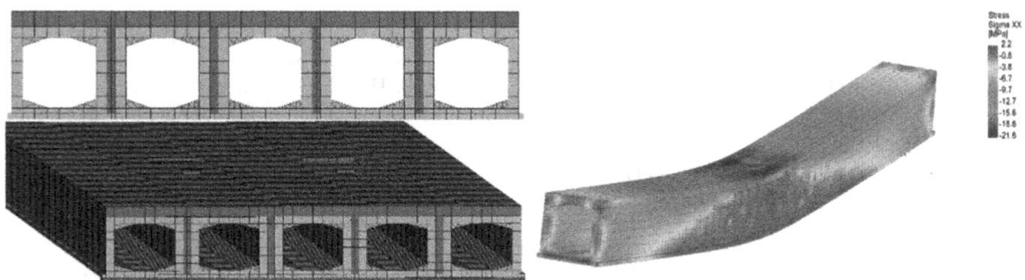

Fig. 4. NLFEM of the analyzed bridge: The cross-section (top left), FEM of the middle 4 girders (bottom left) and stress field of the single girder KA-61 (right).

The PCE is created with maximum polynomial order $p = 5$ using the adaptive algorithm based on LAR presented in Fig. 1. Obtained accuracy measured by Q^2 is higher than 0.98 for all limit states, which is sufficient for sensitivity analysis. The ANN surrogate models consist of five input neurons, one output neuron with a linear activation function, and five nonlinear neurons in the hidden layer with a hyperbolic tangent activation function. The resulting superior ANN structure and associated hyperparameters emerged from initial testing. Validation of the model was performed using 5-fold cross-validation with average accuracy $Q^2 = 0.97$. The results in Table 4 correspond to the average of five ANNs trained and validated on a different subset of the data.

From the obtained results it is clear that both limit states are highly affected by uncertainty of P and G_f. Moreover, ANN identified also substantial influence of E in Cracking limit state. Distribution-based K_i also reflects Normal distribution of P (contrasting from all other variables having Lognormal distribution) and its influence on the final probability distribution of QoI. Note that from r_{XY} we can see that P has almost linear negative influence on QoIs while G_f and E affects its variance measured by COV^2_{Ri} and $S_i^{T,PCE}$ but there is not clear linear connection to value of QoI. For further practical analysis, it can be concluded that uncertainty of f_c and f_{ct} can be neglected and they can be assumed as deterministic input variables.

Table 4. Sensitivity analysis by ANN and PCE of the concrete bridge.

Variable	r_{XY}(ANN)	COV^2_{Ri}(ANN)	$S_i^{T,PCE}$(PCE)	K_i(PCE)
ULS				
E	0.17	0.04	0.04	0.01
f_c	–0.01	0.01	0.10	0.03
f_{ct}	0.02	0.00	0.01	0.01
G_f	0.27	0.10	0.16	0.05
P	–0.92	0.86	0.85	0.90
Cracking				
E	0.38	0.15	0.01	0.01
f_c	–0.06	0.04	0.01	0.01
f_{ct}	0.10	0.02	0.01	0.01
G_f	0.19	0.09	0.17	0.05
P	–0.83	0.70	0.80	0.92

5 Conclusions

This paper presents application of artificial neural networks and polynomial chaos expansion for sensitivity analysis of engineering structures. Although ANN achieve high accuracy for sensitivity analysis, it is shown that each surrogate model is suitable for different types of sensitivity measures, and it is beneficial to use both of them to complex sensitivity analysis. Artificial neural networks are utilized for sensitivity analysis based on Monte Carlo simulation – Spearman rank-order sensitivity and sensitivity in terms of coefficients of variation. Polynomial chaos expansion is used for global sensitivity analysis based on statistical moments (Sobol indices) and distribution-based sensitivity analysis. Both surrogate models are used for sensitivity analysis of two typical engineering examples represented by analytical function and non-linear finite element model. The utilization of ANN and PCE offers efficient and accurate approach for industrial applications and obtained results from various sensitivity measures could be further used for decision making and uncertainty quantification.

Acknowledgement. This work was supported by the project No. 20-01734S, awarded by the Czech Science Foundation (GACR).

References

1. Ghanem, R.G., Spanos, P.D.: Stochastic finite elements: a spectral approach. Springer, Berlin (1991). https://doi.org/10.1007/978-1-4612-3094-6
2. Xiu, D., Karniadakis, G.E.: The Wiener-Askey polynomial chaos for stochastic differential equations. SIAM J. Sci. Comput. **24**(2), 619–644 (2002)

3. Sudret, B.: Global sensitivity analysis using polynomial chaos expansions. Reliab. Eng. Syst. Saf. **93**(7), 964–979 (2008)
4. Novák, L.: On distribution-based global sensitivity analysis by polynomial chaos expansion. Comput. Struct. **267**, 106808 (2022)
5. Pan, L., Novák, L., Novák, D., Lehký, D., Cao, M.: Neural network ensemble-based sensitivity analysis in structural engineering: Comparison of selected methods and the influence of statistical correlation. Comput. Struct. **242** (2021)
6. Kleijnen, J.P.C.: Sensitivity analysis of simulation models: an overview. Procedia Soc. Beh. Sci. **2**, 7585–7586 (2010)
7. Borgonovo, E., Plischke, E.: Sensitivity analysis: A review of recent advances. Eur. J. Oper. Res. **248**, 869–887 (2016)
8. Iman, R.L., Conover, W.J.: Small sample sensitivity analysis techniques for computer models with an application to risk assessment. Commun. Stat. Theory Methods **9**(17), 1749–1842 (1980)
9. Taylor, R., Ojha, V., Martino, I., Nicosia, G.: Sensitivity analysis for deep learning: ranking hyper-parameter influence. In: 33rd IEEE International Conference on Tools with Artificial Intelligence, ICTAI 2021, 1–3 November 2021, pp. 512–516. IEEE (2021)
10. Greco, A., Riccio, S.D., Timmis, J., Nicosia, G.: Assessing algorithm parameter importance using global sensitivity analysis. In: Kotsireas, I., Pardalos, P., Parsopoulos, K.E., Souravlias, D., Tsokas, A. (eds.) SEA 2019. LNCS, vol. 11544, pp. 392–407. Springer, Cham (2019). https://doi.org/10.1007/978-3-030-34029-2_26
11. Conca, P., Stracquadanio, G., Nicosia, G.: Automatic tuning of algorithms through sensitivity minimization. In: International Workshop on Machine Learning, Optimization and Big Data, pp. 14–25 (2015)
12. Chatzimanolakis, M., Kantarakias, K.-D., Asouti, V., Giannakoglou, K.: A painless intrusive polynomial chaos method with RANS-based applications. Comput. Methods Appl. Mech. Eng. **348**, 207–221 (2019)
13. Efron, B., Hastie, T., Johnstone, I., Tibshirani, R.: Least angle regression. Ann. Statist. **32**(2), 407–451 (2004)
14. Novák, L., Vořechovský, M., Sadílek, V., Shields, M.D.: Variance-based adaptive sequential sampling for polynomial chaos expansion. Comput. Methods Appl. Mech. Eng. **386**, 114105 (2021). https://doi.org/10.1016/j.cma.2021.114105
15. Novak, L., Novak, D.: Polynomial chaos expansion for surrogate modelling: theory and software. Beton-und Stahlbetonbau **113**, 27–32 (2018). https://doi.org/10.1002/best.201800048
16. Cichocki, A., Unbehauen, R.: Neural networks for optimization and signal processing. Wiley & B.G, Teubner, Stuttgart (1993)
17. Lehký, D., Šomodíková, M.: Reliability calculation of time-consuming problems using a small-sample artificial neural network-based response surface method. Neural Comput. Appl. **28**(6), 1249–1263 (2016). https://doi.org/10.1007/s00521-016-2485-3
18. Novák, D., Teplý, B., Shiraishi, N.: Sensitivity analysis of structures: a review. In: Proceedings of International Conference CIVIL COMP 1993, pp. 201–207. Edinburgh, Scotland (1993)
19. Vořechovský, M.: Correlation control in small sample Monte Carlo type simulations II: analysis of estimation formulas, random correlation and perfect uncorrelatedness. Probab. Eng. Mech. **29**, 105–120 (2012)
20. McKay, M.D., Conover, W.J., Beckman, R.J.: A comparison of three methods for selecting values of input variables in the analysis of output from a computer code. Technometrics **21**, 239–245 (1979)
21. Stein, M.: Large sample properties of simulations using Latin hypercube sampling. Technometrics **29**(2), 143–151 (1987)

22. Novák, D., Vořechovský, M., Teplý, B.: FReET: software for the statistical and reliability analysis of engineering problems and FReET-D: degradation module. Adv. Eng. Softw. **72**, 179–192 (2014)
23. Sobol, I.: Global sensitivity indices for nonlinear mathematical models and their Monte Carlo estimates. Math. Comput. Simul. **55**, 271–280 (2001)
24. Park, S., Rao, M., Shin, D.W.: On cumulative residual Kullback-Leibler information. Statist. Probab. Lett. **82**(11), 2025–2032 (2012). https://doi.org/10.1016/j.spl.2012.06.015
25. Červenka, J., Papanikolaou, V.K.: Three dimensional combined fracture-plastic material model for concrete. Int. J. Plasticity, 2192–2220 (2008)
26. CEN: prEN 1992-1-1 Eurocode 2: Design of concrete structures - Part 1-1: General rules - Rules for buildings, bridges and civil engineering structures (2021)
27. Joint Committee for Structural Safety. JCSS Probabilistic Model Code. Joint Committee on Structural Safety (2001)

Transformers for COVID-19 Misinformation Detection on Twitter: A South African Case Study

Irene Francesca Strydom(✉) and Jacomine Grobler

Department of Industrial Engineering, Stellenbosch University, Stellenbosch, Western Cape, South Africa
strydomfrancesca@gmail.com
https://ie.sun.ac.za/

Abstract. The aim of this paper is to investigate the use of transformer-based neural network classifiers for the detection of misinformation on South African Twitter. Twitter COVID-19 misinformation data from four publicly available datasets are used for training. Four different transformer-based embedding methods are used, namely: BERT, CT-BERT, ELECTRA, and LAMBERT. A neural network classifier is trained for each embedding method, and the architectures are optimized with the hyperband optimization algorithm. The model using the LAMBERT embedding method attains the highest F1-score (0.899) on the test data. The model does not generalize well to the South African context however, since it fails to reliably distinguish between general Tweets and COVID-19 misinformation Tweets when applied to the unlabeled South African data. The classifier does detect instances of misinformation that are consistent with known COVID-19 misinformation spread in South Africa, but these are in the minority. It is therefore recommended that misinformation datasets specific to the South African context be curated in order to facilitate future research efforts dedicated to misinformation detection on South African Twitter.

Keywords: Misinformation · Transformers · Social media · Twitter · Transfer learning · Natural language processing

1 Introduction

Nearly three years after the COVID-19 pandemic emerged, many countries are still grappling with its effects. As of 31 March 2022, more than six million deaths have been attributed to COVID-19, and although vaccines have provided much needed relief, many countries still have restrictions in place [17]. In such high-risk and stressful environments, the public relies heavily on updated information to make decisions. Governments, organizations and medical experts alike have gone to great lengths to provide the public with accurate information regarding important issues such as infection rates, prevention measures, treatment

G. Nicosia et al. (Eds.): LOD 2022, LNCS 13810, pp. 197–210, 2023.
https://doi.org/10.1007/978-3-031-25599-1_15

options and the characteristics of the virus. Social media has become a popular way for official information to be broadcast, and has proven a very useful tool when quick and efficient communication to a large audience is necessary. The pandemic, however, has also brought to light the potential of social media for spreading rumours, conspiracies and 'fake news'. COVID-19 misinformation has become so prevalent that the director-general of the World Health Organization declared that the world is not only fighting a pandemic, but also an infodemic [9]. Combatting misinformation on social media is difficult due to the large volumes of content produced daily, and therefore automatically detecting misinformation has become an important field of research.

Advances in natural language processing (NLP) and machine learning (ML) techniques have led to the increased use of these methods to detect misinformation on social media. Text embedding models are often used to learn representations of words, where words similar in meaning have similar representations. Transformer-based text embedding models improve upon early, static text embedding models by learning contextualized representations of words, where each word's representation is adapted based on the words occurring with it in the text [23]. This paper uses data from four publicly available Twitter datasets to train a neural network misinformation classifier that uses a transformer-based embedding approach to representing text. The classifier is then applied to novel datasets containing more than 180 000 South African Tweets posted between 1 March 2020 and 30 April 2020, and more than 11 000 Tweets posted between 1 March 2021 and 30 April 2021.

The paper compares the performance of four different embedding approaches for classifying COVID-19 misinformation on Twitter. The paper also applies the best-performing model to novel South African Twitter data and provides insight to the ability of these transformer-based models to generalize to the South African context. In addition, the paper provides suggestions for better characterising misinformation on South African social media.

The rest of the paper is structured as follows: Sect. 2 provides the background for the study, Sect. 3 discusses work related to misinformation detection on social media, Sect. 4 describes the methodology applied in the study, and results are discussed in Sect. 5. The study is concluded and recommendations for future work are made in Sect. 6.

2 Background: The South African Context

On the 15th of March 2020, South African president Cyril Ramaphosa declared a National State of Disaster [7] in preparation for mitigating the effects of COVID-19 in South Africa. The first of several waves of COVID-19 infections hit the country soon after, and a series of preventative measures were implemented, including lockdowns, social distancing, the wearing of masks, and a number of sanitizing practices. Similar to other parts of the world, South Africa not only experienced the effects of the pandemic, but also the effects of the infodemic. Examples of rumours spread in South Africa during the first few months of the pandemic include: "wearing masks can cause you to die from monoxide poisoning" [2], "COVID-19 vaccines cause infertility" [3], "the virus does not really exist" [22], "and 5G radio signals cause COVID-19" [16]. The South African government responded promptly by outlawing the spread of misinformation about COVID-19, and other independent organizations also joined the fight against misinformation. Real411, for example, is a South African online platform managed by Media Monitoring Africa, where any member of the public can report digital offences, which include (1) harmful false information, (2) hate speech, (3) incitement to violence, and (4) harassment of journalists. The platform seeks to ensure *"...that online content is assessed and addressed in an independent, open, transparent and accountable manner within our laws and constitutional rights"* [21]. Since March 2020, hundreds of complaints regarding the COVID-19 pandemic have been submitted to Real411. Figure 1, shows the most common tags for complaints, and it is clear to see that COVID-19 and related topics are among the most prevalent.

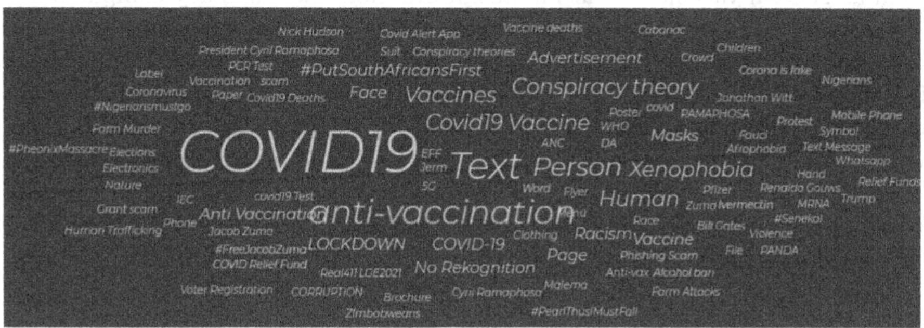

Fig. 1. Most common complaints by tag on the Real411 platform, as presented by [21].

Twitter is one of the most popular social media platforms in the country, and also a platform whose content is regularly reported to Real411. Twitter has a global reach, and is designed for easy sharing and dissemination of content. Thus, rumors that are spread on Twitter in other countries may also gain traction in South Africa. One example is the rumor that 5G signals cause COVID-19. This rumor was spread widely in the United Kingdom in April 2020, and lead to several attacks on 5G towers [1]. Similar rumors later surfaced on South African Twitter, and lead to the destruction of 5G towers in the South African province of KwaZulu Natal in July 2021 [16]. These rumors were also reported to Real411 and even addressed by the South African minister of health at the time, Dr Zweli Mkhize. Other instances of misinformation may have been overlooked by the public or not reported to Real411, however, and therefore an opportunity exists to better understand the spread of COVID-19 misinformation on South African social media, and the potential impact thereof on South Africans.

3 Related Work

Misinformation is described as false information that is spread, regardless of intent to mislead [13], while disinformation is described as false information that is spread specifically with the intent to mislead [8]. Since determining the intent of the author or publisher of a text is oftentimes difficult, even after extensive fact-checking and research, literature is mainly focused on detecting instances of misinformation. Misinformation detection is traditionally performed manually by fact-checking professionals, journalists, scientists, and even legal specialists. Manual misinformation detection is not easily scaled to the large volumes of social media data being generated daily. As such, a whole new field of automated misinformation detection has developed, relying heavily on ML and deep learning methods.

Automated misinformation detection is often cast as a supervised binary text classification task, where a machine learning model is trained to classify a text as a positive or negative example of misinformation. Hayawi *et al.* [10] presented a misinformation detection framework focusing specifically on vaccine misinformation. The authors collected more than 15 000 vaccine Tweets and annotated them as general vaccine or vaccine misinformation Tweets. They compared the performance of three classification models, the bidirectional encoder representations from encoders (BERT) model performed the best with an F1 score of 0.98. The dataset used in the study, called ANTiVax, was also made publicly available. While [10] focus on vaccine misinformation, they also suggest experimenting with a combination of general and vaccine COVID-19 misinformation. During the proceedings of the First Workshop on Combating Online Hostile Posts in Regional Languages during Emergency Situation (CONSTRAINT) [18], one of the two shared tasks presented was titled 'COVID-19 Fake News Detection in English'. The best performing classification system for the task was an ensemble of three Covid Twitter BERT (CT-BERT) models, which attained an F1-score of 0.9869. The data for the English language task were collected, annotated and released publicly by Patwa *et al.* [19], and include more than 10 000 Tweets annotated as either general or misinformation Tweets. Cui *et al.* [5] introduced a misinformation dataset called CoAID, containing real and false claims in news articles and Tweets. They applied a number of machine learning algorithms to classify the data, and a variation of the hierarchical attention network (HAN) model performed the best with an F1-score of 0.58. Shahan *et al.* [12] collected Tweets posted on 29 March, 15 June, and 24 June 2020 and manually labeled them as belonging to one of 17 categories, e.g. fake treatment or conspiracy. They publicly released their dataset, CMU-MisCOV19, to aid in future COVID-19 misinformation research.

Studies on misinformation within the South African context have also been conducted recently. De Wet and Marivate [6] trained misinformation detection models on articles from South African news websites and compared them with models trained on articles found on US news websites. In their study models trained on data from US websites did not perform well when applied to data from South African websites. Models trained on data from South African websites also did not perform well when applied to data from US websites. The South African data used in their study has been released publicly, although the focus of their paper is on general misinformation found in articles news websites, and not on shorter social media posts. Mutanga *et al.* [15] used topic modeling to identify the main themes of South African COVID-19 Tweets between 15 March and 30 April 2020. Of the nine themes identified by their study, two were labeled as misinformation topics, relating to 5G and virus origin conspiracies, respectively. Little other research has been done on COVID-19 misinformation on social media within the South African context. The lack of labeled South African COVID-19 misinformation datasets in particular means that researchers must rely on publicly available data to detect misinformation on South African social media with supervised machine learning.

4 Method

The Cross-Industry Process for Data Mining (CRISP-DM) is the guiding methodology for this study. The data understanding, data preparation, and modeling phases are discussed in Sects. 4.1, 4.2, and 4.3, respectively.

4.1 Data Understanding

The data understanding phase entails collecting labeled data for model development, and collecting unlabeled case study data to which the final classification model is applied. The rest of this section describes the collection of labeled and unlabeled data respectively.

Labeled Training Data. Several labeled datasets have been made publicly available for misinformation detection on Twitter. The scope of training data collected is limited to publicly available English language Twitter posts, or Tweets, related to the COVID-19 pandemic. Tweets from datasets released in [5,10,12,19] are used to train the misinformation detection model. The data in [10] is a set of COVID-19 vaccine Tweets posted from 1 December 2020 to 31 July 2021. The posts were manually annotated as vaccine misinformation or general vaccine Tweets and validated by medical experts. The data in [5] is a set of Tweets containing true and false claims about COVID-19 posted from 1 May 2020 to 1 November 2020, annotated in accordance with information from trusted sources such as the World Health Organisation. The data in [12] contain COVID-19 Tweets posted from 29 March 2020 to 24 June 2020, and annotated as belonging to one of 17 categories, five of which constitute misinformation. For the purposes of this study only Tweets in misinformation categories (such as "Fake Cure" and "Conspiracy") from the data in [12] are labeled as misinformation, and Tweets from other categories (such as "Sarcasm/Satire" and "Ambiguous/Difficult to Classify") are ignored. The data in [19] is a set of COVID-19 Tweets annotated as real or fake in accordance with a number of fact-checking websites. The number of Tweets collected for each dataset is specified in Table 1.

Table 1. Number of Tweets collected for each publicly available dataset.

Source	Dataset	Misinformation tweets	General tweets
[5]	CoAID	7127	130 532
[10]	ANTiVax	4 504	8 353
[12]	CMU-MisCov	348	–
[19]	Constraint	5098	5599
	Total	17 077	144 484

Unlabeled Case Study Data. The unlabeled Twitter data is collected through the Twitter API Full Archive Search endpoint, which allows for combining a number of search parameters to collect Tweets. Twitter also uses ML algorithms to categorize Tweets into broad categories with context annotations, and Tweets can also be filtered according to these annotations. The collected Tweets (1) contain either one of the keywords "covid", "coronavirus", "covid-19", "pandemic", or the Tweet context annotation "COVID-19", and (2) are geo-tagged with the "ZA" place country tag. Retweets are ignored, since they contain the same text as the original Tweet and therefore do not provide any additional information. A total of 186 520 Tweets posted from 1 March 2020 to 30 April 2020 and 11 574 Tweets posted from 1 March 2021 to 30 April 2021 are collected.

4.2 Data Preparation

The annotated training datasets are combined into a single dataset which contains 17 077 instances labeled as COVID-19 misinformation Tweets and 144 484 instances labeled as general COVID-19 Tweets. Duplicate Tweets, user mentions, hyperlinks, and any Tweets containing two or less words are removed from the dataset. No further preprocessing of Tweet text is required or recommended, since transformer models are specifically trained on naturally occurring text. Since there exists a major imbalance in the classes, undersampling of the majority general Tweet class is performed by randomly selecting instances from the class, which yields the final dataset containing 17 046 instances labeled as misinformation Tweets and 17 046 instances labeled as general COVID-19 Tweets. The text in the unlabeled dataset is subjected to the same pre-processing steps as the labeled dataset.

4.3 Modeling

A number of candidate models are optimized and trained in order to identify a final model that will be applied to the case study data. Each model consists of two submodels: (1) the embedding submodel, and (2) the classification submodel. The embedding submodel consists of a pretrained transformer model that takes raw text as input and outputs a contextualized sequence embedding. This embedding is then passed to the classification submodel, which then outputs a label for the instance.

The first embedding submodel that is used for the study is BERT, based on the multi-head attention mechanism as used by Vaswani *et al.* [23], and the model is trained using masked language modeling (MLM) and next-sentence prediction (NSP). In MLM, a random selection of tokens in a sequence are masked, and the model is tasked with predicting the correct tokens. In next NSP, the model is tasked with predicting whether one sequence follows another sequence. BERT is trained on data from Wikipedia and Google's BookCorpus. BERT has significantly advanced the state-of-the-art for the majority of NLP tasks, including text classification. The remaining three embedding submodels used in this study are CT-BERT, ELECTRA, and LAMBERT, each a version of the BERT architecture that uses different training data or a different training approach.

CT-BERT is trained on 22.5 million COVID-19 Tweets instead of the BookCorpus and Wikipedia [14]. The model attains an improvement in performance over BERT on one COVID-19 Twitter dataset, three non-COVID-19 Twitter datasets, and one general non-Twitter dataset. Since CT-BERT is optimized for COVID-19 Twitter content, the greatest relative improvement in marginal performance (25.88%) is achieved on the COVID-19 Twitter dataset. These results make CT-BERT an attractive option for classification of COVID-19 misinformation on Twitter.

ELECTRA also makes use of the original BERT architecture, but instead of masking specific tokens during training, a discriminator neural network must predict whether a specific token is real, or was replaced by a token from a trained generator neural network [4]. A major advantage of this approach is that the model learns from all input tokens in an input sequence, as opposed to learning from only the masked tokens as is the case with MLM used in pre-training BERT. Consequently, ELECTRA produces embeddings that outperform BERT embeddings on a number of NLP tasks, including text classification.

LAMBERT makes use of techniques from the LAMB optimizer and RoBERTa to train a BERT model [24]. In addition to reducing the pre-training time of the model from three days to less than 80 min, LAMBERT also achieves a higher performance than BERT on the SQuAD NLP benchmark tasks [20].

Each embedding submodel has 24 layers and 16 attention heads, and outputs a sequence embedding of size 1024, which is then passed to the classification submodel. The embedding submodel's weight parameters remain fixed, while the classification submodel's hyperparameters and parameters are learned during training. The hyperparameters of the classification submodel to be optimized are: (1) the number of hidden layers, (2) the number of neurons in each hidden layer, (3) the activation function used in the hidden layers, (4) the normalization method used, (5) the dropout rate per hidden layer, (6) the optimizer to use during training, and (7) the learning rate for the optimizer. The hyperband optimization algorithm, as implemented in [11], yields the optimal classification submodel hyperparameters to be used in conjunction with each respective embedding submodel, as shown in Table 2.

Table 2. Optimal model architecture and training configurations.

	Model 1	Model 2	Model 3	Model 4
Embedding Submodel	BERT	CT-BERT	ELECTRA	LAMBERT
# Hidden Layers	4	2	3	2
# Neurons	840, 640, 140, 140	700, 140	260, 180, 140	940, 400
Dropout Rate for Hidden Layers	0, 0.4, 0.1, 0	0, 0.2	0, 0.4, 0.2	0.1, 0.4
Activation Function	Swish	Swish	Swish	Selu
Normalization	Batch	None	Batch	Batch
Optimizer	Adamax	Nadam	Adagrad	Adagrad
Learning Rate	0.000146	0.000205	0.0052	0.00917

Each of the models summarized in Table 2 are then trained using 30 fold cross-validation of the training data (75% of the total data), and their final metrics determined on the test set (25% of the total data). Table 3 shows the average validation metrics and final test metrics for each of the trained models, and Fig. 2 shows the confusion matrices for the models on the test set. Model 2 (CT-BERT) performs slightly better than model 1 across all metrics except test precision. The better performance of model 2 is expected, since CT-BERT is pre-trained on COVID-19 Tweets. Both models 3 and 4 perform significantly better than model 2 across all metrics, however, even though they are not pre-trained on domain specific COVID-19 Tweets. Their improved performance suggests that the alternative pre-training procedures used for embedding models have a greater effect on the quality of the embeddings than training with domain specific data. Model 4 produces slightly better results than model 3 across all metrics except validation and test recall.

Depending on the use case, either recall or precision may be deemed more important for a misinformation classifier. A low recall may lead to damaging misinformation going undetected, while a low precision may lead to content being unnecessarily removed or reviewed, which in turn wastes limited resources. The F1-score, which takes the harmonic mean of accuracy and precision, is therefore used to identify the best performing model. As such, model 4 performs the best and is applied in Sect. 5.

5 Results

The application of model 4 to the unlabeled case study data and a subsequent inspection of the models output is discussed in this section. The case study data stretches over two time periods, namely 1 March to 30 April 2020, and 1 March to 30 April 2021.

Table 3. Performance of cross-validated and final models.

	Model 1	Model 2	Model 3	Model 4
Validation Accuracy	0.809	0.835	0.895	**0.901**
Validation Precision	0.833	0.835	0.896	**0.908**
Validation Recall	0.794	0.836	**0.894**	0.892
Validation F1-score	0.813	0.818	0.895	**0.900**
Test Accuracy	0.822	0.836	0.892	**0.901**
Test Precision	0.838	0.826	0.879	**0.913**
Test Recall	0.798	0.852	**0.909**	0.887
Test F1-score	0.818	0.838	0.894	**0.899**

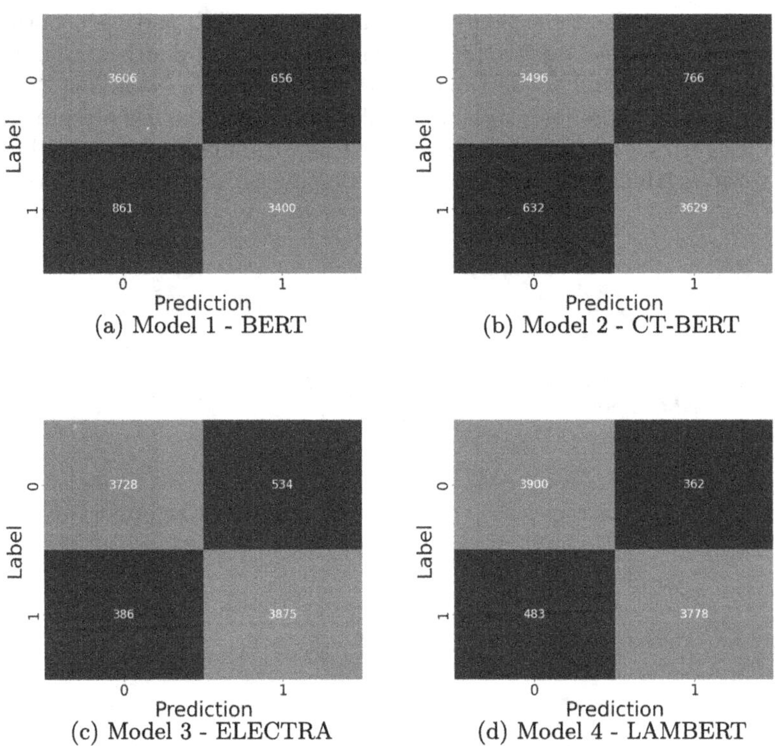

(a) Model 1 - BERT (b) Model 2 - CT-BERT (c) Model 3 - ELECTRA (d) Model 4 - LAMBERT

Fig. 2. Confusion matrices for each model on the test set. The y-axis shows the actual labels for instances in the test set, while the x-axis shows predicted labels as output by each respective model.

A sample of South African Tweets classified as misinformation by model 4 is shown in Table 4, and Fig. 3 shows the most frequent words for misinformation Tweets in the respective datasets. Some of the Tweets contain misinformation statements that are consistent with misinformation reported to the

Real411 platform, such as Tweets 3, 7, and 17. Most other Tweets labeled as misinformation however, merely contain general statements or emotive language. Although model 4 attains an F1-score of 0.899 on the test set, when applied to the unlabeled South African data, the model fails to reliably distinguish between misinformation Tweets and general Tweets.

Inspecting the most frequent terms in the South African Tweets after inference and comparing them with the most frequent terms in the training corpus sheds some light on the reason the model does not perform as well as expected on the case study data. In the training corpus for example, president and Trump are among the most common words (see Fig. 3 a)). While president Trump has been associated with misinformation in the US, this leads to the model incorrectly classifying South African Tweets about president Cyril Ramaphosa as misinformation, such as Tweets 1, 2, 11, 12, and 14 in Table 4. Common words in both time periods of the South African data are "ke" and "le", both from the Setswana language, one of the 11 official languages of South Africa. Tweets 4, 9, 10, 15, 16, 18 and 20 in Table 4 each contain words (including "ke" and "le") from two or more languages. Since the embedding models are pre-trained on only English data, the appearance of out of vocabulary words may bias the model towards labeling Tweets containing non-English words as misinformation.

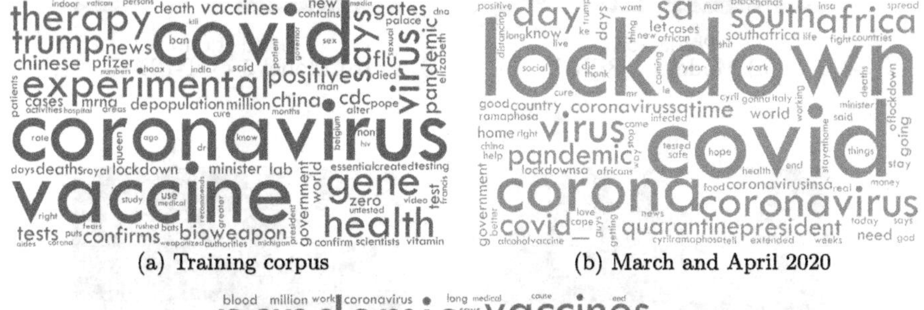

(a) Training corpus (b) March and April 2020

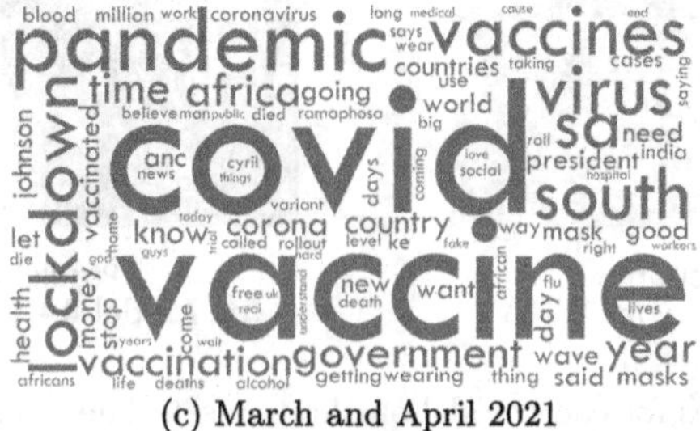

(c) March and April 2021

Fig. 3. Word clouds for Tweets labeled as misinformation.

Table 4. Examples of Tweets classified as misinformation by model 4 for the months of March and April 2020 and March and April 2021.

March, April 2020		March, April 2021	
1	God bless South Africa. Thank you Mr. President #CoronavirusInSA	11	Has the president been vaccinated yet?
2	#CoronaVirus President will address the Nation at 17:00 this evening live on SABC 2 and SABC News Channel 404 on DStv. #sabcnews	12	as a president of this country where are your priorities??
3	Is it no longer the "Chinese Virus" Mr President?	13	If I was president, I would've resigned the minute Covid got serious. Sorry folks
4	Still in Clayville, ke tlo shapa round that side once this Corona thing is under control	14	The worst thing is corona virus mot our president
5	Social distancing on the wet roads of Krugersdorp.	15	Depression le Covid ke masepa shame sies
6	Royal Eagles FC training field is closed by municipality today due to Covid. This is serious .	16	Nna ke na le corona klaar, my body is aching AF. I tick all the secondary symptoms boxes
7	An actual Dr. told my sister and a bunch or cops that" we shouldn't worry about this COVID-19 because it's a disease for white people." This man proceeded to say "it's just the flu anyway."	17	The most targeted people to be depopulated is Africa, so far Africa is saved by the corruption of their leaders; African leaders are eating this money of buying vaccines saving many sheeple from poison
8	A good read on Corona Virus. Considering our state of affairs, we might record a higher number of deaths than Italy	18	Mara how is India selling vaccine to the rest of the world and they get hit this hard....?
9	South Africans mara. A whole #coronadancechallenge	19	She took care of my cousin who had the virus, and later lost her life from it.
10	Kusiwe apha on these streets, did not know ke Mr matric, a whole honourable Steyn...nton nton	20	Haai. At least someone got a promotion during the pandemic.

6 Conclusion

This research paper combined data from four publicly available COVID-19 Twitter misinformation datasets. Four different neural network classifiers were trained on the combined dataset, making use of BERT, CT-BERT, ELECTRA, and LAMBERT embedding approaches, respectively. Each classifier's architecture was optimized and final models trained. Although the classifiers performed well on validation and test sets, the models could still not effectively distinguish misinformation Tweets from general Tweets when applied to an unlabeled South

African dataset containing COVID-19 Tweets. These findings are consistent with those of De Wet and Marivate [6], who found that misinformation detection models trained on data from US news websites did not perform well when applied to data from South African websites.

The South African context differs in terms of (1) language used, (2) the events discussed and (3) the roleplayers in the pandemic. Even though only English tweets were analyzed, these Tweets still contain words from several of South Africa's local languages, including isiXhosa, Setswana, and Afrikaans, on which the embedding submodels have not been pre-trained. South African English also contains expressions and terms unique to the country which are not reflected in the English training data. South Africa's political structure, healthcare system and socioeconomic situation affects the regulations and safety measures implemented during the pandemic, and these differences are also not reflected in the training data. The authors therefore believe that the greatest opportunity for improving the performance of a misinformation classifier on South African Tweets lies in using training data that is specific to the South African context, i.e. labeled South African COVID-19 misinformation Tweets. In addition, transformer models pre-trained on specific African languages would greatly improve the resources available to study these languages in the digital space. A multilingual transformer trained on multiple African languages would also be very useful for studying South African social media, where mixing languages in a single text is common.

Alternatively, unsupervised machine learning methods such as topic modeling or clustering may be applied to the unlabeled data to identify misinformation topics in the data and label Tweets accordingly. While inspecting the most frequent terms in the datasets provides some insight to the models performance, an in-depth statistical analysis of the classified South African Tweets and assessment of Tweets by medical and fact-checking experts would prove more useful for improving performance in future.

Furthermore, text data is only one aspect of social media data that can be utilized to detect misinformation. Other types of media, specifically images and videos, are oftentimes very effective vehicles for the dissemination of misinformation on social media. Other aspects of Twitter data, such as user characteristics, may also be helpful in detecting misinformation. Future work may, therefore, use these additional characteristics in addition to text data when training misinformation classifiers.

Acknowledgements. This work is based on the research supported in part by the National Research Foundation of South Africa (Grant number: 129340), and the school for Data Science and Computational Thinking at Stellenbosch University.

References

1. Bahja, M., Safdar, G.A.: Unlink the link between covid-19 and 5g networks: an nlp and sna based approach. IEEE Access **8**, 209127–209137 (2020)
2. Bird, W., Smith, T.: Disinformation during covid-19: weekly trends from real411 in south Africa (2020). https://www.dailymaverick.co.za/article/2020-06-08-disinformation-during-covid-19-weekly-trends-from-real411-in-south-africa/. Accessed 20 Oct 2021
3. Bird, W., Smith, T.: Disinformation during covid-19: weekly trends from real411 in south Africa (2020). https://www.dailymaverick.co.za/article/2020-06-14-disinformation-amid-covid-19-weekly-trends-in-south-africa/. Accessed 20 Oct 2021
4. Clark, K., Luong, M.T., Le, Q.V., Manning, C.D.: Electra: pre-training text encoders as discriminators rather than generators. In: International Conference on Learning Representations (2020)
5. Cui, L., Lee, D.: Coaid: COVID-19 healthcare misinformation dataset. Computing Research Repository (2020)
6. De Wet, H., Marivate, V.: Is it fake? news disinformation detection on south African news websites. In: 2021 IEEE AFRICON, pp. 1–6. IEEE (2021)
7. Department of Co-operative Governance and Traditional Affairs (South Africa): Disaster management act 57 of 2002: Regulations issued in terms of section 27(2)() of the disaster management act, 2002 (2020)
8. Disinformation (2020). https://www.merriam-webster.com/dictionary/disinformation. Accessed 20 Oct 2021
9. Ghebreyesus, T.A.: Munich security conference (2020), https://www.who.int/director-general/speeches/detail/munich-security-conference, accessed: 20/10/2021
10. Hayawi, K., Shahriar, S., Serhani, M., Taleb, I., Mathew, S.: Anti-vax: a novel twitter dataset for covid-19 vaccine misinformation detection. Public Health **203**, 23–30 (2022)
11. Li, L., Jamieson, K., DeSalvo, G., Rostamizadeh, A., Talwalkar, A.: Hyperband: A novel bandit-based approach to hyperparameter optimization. Journal of Machine Learning Research **18**(185), 1–52 (2018)
12. Memon, S.A., Carley, K.M.: Characterizing COVID-19 misinformation communities using a novel twitter dataset. Computing Research Repository (2020)
13. Misinformation (2020). https://www.merriam-webster.com/dictionary/misinformation. Accessed 20 Oct 2021
14. Müller, M., Salathé, M., Kummervold, P.E.: Covid-twitter-bert: A natural language processing model to analyse covid-19 content on Twitter. CoRR abs/2005.07503 (2020)
15. Mutanga, M.B., Abayomi, A.: Tweeting on covid-19 pandemic in South Africa: Lda-based topic modelling approach. African J. Sci. Technol. Innov. Dev. **12**, 1–10 (2020)
16. Njilo, N.: Torching of cellphone towers 'linked to 5g and covid conspiracies' (2020). https://www.sowetanlive.co.za/news/south-africa/2021-01-08-torching-of-cellphone-towers-linked-to-5g-and-covid-conspiracies/. Accessed 20 Oct 2021
17. Organization, W.H.: Who coronavirus (covid-19) dashboard: Measures (2022). https://covid19.who.int/measures. Accessed 03 May 2022

18. Patwa, P., et al.: Overview of CONSTRAINT 2021 shared tasks: detecting English COVID-19 fake news and Hindi hostile posts. In: Chakraborty, T., Shu, K., Bernard, H.R., Liu, H., Akhtar, M.S. (eds.) CONSTRAINT 2021. CCIS, vol. 1402, pp. 42–53. Springer, Cham (2021). https://doi.org/10.1007/978-3-030-73696-5_5
19. Patwa, P., et al.: Fighting an Infodemic: COVID-19 fake news dataset. In: Chakraborty, T., Shu, K., Bernard, H.R., Liu, H., Akhtar, M.S. (eds.) CONSTRAINT 2021. CCIS, vol. 1402, pp. 21–29. Springer, Cham (2021). https://doi.org/10.1007/978-3-030-73696-5_3
20. Rajpurkar, P., Zhang, J., Lopyrev, K., Liang, P.: Squad: 100,000+ questions for machine comprehension of text. In: Conference on Empirical Methods in Natural Language Processing (2016)
21. Real411: Real411 homepage (2020). https://www.real411.org/. Accessed 20 Oct 2021
22. Smith, T., Bird, W.: Disinformation during covid-19: Weekly trends from real411 in south africa (2020). https://www.dailymaverick.co.za/article/2020-06-29-disinformation-during-covid-19-weekly-trends-in-south-africa/. Accessed 20 Oct 2021
23. Vaswani, A., et al.: Attention is all you need. In: Advances in Neural Information Processing Systems, vol. 30. Curran Associates, Inc. (2017)
24. You, Y., Li, J., Hseu, J., Song, X., Demmel, J., Hsieh, C.: Large batch optimization for deep learning: training bert in 76 minutes. In: International Conference on Learning Representations (2019)

MI2AMI: Missing Data Imputation Using Mixed Deep Gaussian Mixture Models

Robin Fuchs[1], Denys Pommeret[1,2(✉)], and Samuel Stocksieker[2]

[1] Aix Marseille Univ, CNRS, Centrale Marseille, I2M, Marseille, France
{robin.fuchs,denys.pommeret}@univ-amu.fr
[2] Lyon 1 Univ, ISFA, Lab. SAF EA2429, 69366 Lyon, France
samuel.stocksieker@univ-amu.fr

Abstract. Imputing missing data is still a challenge for mixed datasets containing variables of different nature such as continuous, count, ordinal, categorical, and binary variables. The recently introduced Mixed Deep Gaussian Mixture Models (MDGMM) explicitly handle such different variable types. MDGMMs learn continuous and low dimensional representations of mixed datasets that capture the inter-variable dependence structure. We propose a model inversion that uses the learned latent representation and maps it with the observed parts of the signal. Latent areas of interest are identified for each missing value using an optimization method and synthetic imputation values are drawn. This new procedure is called MI2AMI (Missing data Imputation using MIxed deep GAussian MIxture models). The approach is tested against state-of-the-art mixed data imputation algorithms based on chained equations, Random Forests, k-Nearest Neighbours, and Generative Adversarial Networks. Two missing values designs were tested, namely the Missing Completly at Random (MCAR) and Missing at Random (MAR) designs, with missing value rates ranging from 10% to 30%.

Keywords: Missing data · Mixed data · Data augmentation

1 Introduction

The increase in data availability also comes with the missing data issue. This is particularly the case of mixed datasets; that is data containing continuous, as well as discrete, ordinal, binary, or categorical variables. Fully removing the observations containing missing values is often sub-optimal because it can impact the representativeness of the sample introducing biases and increasing the variance of the estimators. Furthermore, the number of incomplete observations tends to increase with the number of variables. Imputation methods must then be used to complete the data and use the whole sample. Several imputations methods

Granted by the Research Chair NINA under the aegis of the Risk Foundation, an initiative by BNP Cardif.

G. Nicosia et al. (Eds.): LOD 2022, LNCS 13810, pp. 211–222, 2023.
https://doi.org/10.1007/978-3-031-25599-1_16

have been proposed but most of them handle quantitative variables and cannot be satisfactorily extended to deal with mixed data.

Among the approaches to deal with mixed data, the "MICE" (Multivariate Imputations by Chained Equations) method proposed by [2] is one of the most popular. It used conditional distributions through regression models for each variable given the others. Another popular solution, proposed by [10], is based on the k-nearest neighbor (kNN) approach. In the same spirit, [4] proposed an adaptation of kNN with Grey Relational Analysis which is more appropriate to mixed data. [20] proposed an alternative method to impute missing value using Random Forest. It is considered as one of the reference methods to deal with mixed data for single imputation. [5] and [23] developed imputation methods using copula for mixed data and try to solve some drawbacks of the previous methods. Finally, a last family of models called "generative models" by [22] gathers Bayesian methods as well as machine learning models that generate pseudo-observations to perform imputation. One can cite [17] who proposed a Bayesian joint model to generate data. Adversarial networks also belong to this category. [22] proposed Generative Adversarial Imputation Networks (GAIN) based on Generative Adversarial Nets to get value imputations for continuous and binary data. [19], [11] also used GANs to propose imputation methods but they are not usable for mixed data. Independently, within the frame of imputation for images, [12] utilized two separate GANs for data and mask (a matrix for indicating missing values). In the same spirit [7] adapted GAN for machine-learned ranking problem with missing data.

The main difficulty encountered in the literature is to take into account non-continuous variables through a convenient latent space or an adequate transformation. In this work, we propose to use the recent Mixed Deep Gaussian Mixture Model (MDGMM) which extends the Deep Gaussian Mixture Model (DGMM) proposed by [21] to mixed data. The DGMM is a natural extension of the Gaussian Mixed Models (GMM) (see for instance [14] which proposes a neural network architecture (as represented in Fig. 1) where each layer contains latent mixture components. These continuous latent spaces capture the dependence structure between the variables of different types and we propose to use this latent representation to impute missing data. The MDGMM has two advantages: i) it keeps the dependence structure, even if the data are mixed; ii) it proposes a very flexible parametric distribution to capture a large diversity of non-standard distribution. Moreover, it can handle all types of data, unlike other methods that do not deal with ordinals for example (or considering them as qualitative).

In this paper, we combine MDGMM with an optimization procedure to generate pseudo-observations with a specific target: impute missing data. We call this data generation approach Missing data Imputation using MIxed deep GAussian MIxture models (MI2AMI) and we apply it on two datasets: the Contraceptive dataset from the UCI repository and on the well-known Pima Indians Diabetes dataset.

2 Background on MDGMM

The MDGMM is a multi-layer clustering model introduced by [8]. It consists in plunging mixed data into a continuous latent space using an extended version of a Generalized Linear Latent Variable Model (GLLVM) ([16]- [15]- [3]) that acts as an embedding layer. The embedded data are then clustered while going through DGMM layers [21].

More precisely let $Y = (Y_1, \cdots, Y_n)$ denote the n observations of dimension p. The MDGMM can be written as follows: for $i = 1, \cdots, n$

$$\begin{cases} Y_i \to z_i^{(1)} \text{ through GLLVM link via } (\lambda^{(0)}, \Lambda^{(0)}) \\ z_i^{(1)} = \eta_{k_1}^{(1)} + \Lambda_{k_1}^{(1)} z_i^{(2)} + u_{i,k_1}^{(1)} \text{ with probability } \pi_{k_1}^{(1)} \\ \cdots \\ z_i^{(L-1)} = \eta_{k_{L-1}}^{(L-1)} + \Lambda_{k_{L-1}}^{(L-1)} z_i^{(L)} + u_{k_{L-1}}^{(L-1)} \text{ with} \\ \text{probability } \pi_{k_L}^{(L-1)} \\ z_i^{(L)} \sim \mathcal{N}(0, I_{r_L}), \end{cases} \tag{1}$$

where Θ denotes the set of trainable model parameters (see [8] for more details). and where the factor loading matrices $\Lambda_{k_1}^{(1)}, \cdots, \Lambda_{k_{L-1}}^{(L-1)}$ have dimensions $r_0 \times r_1, \cdots, r_{L-1} \times r_L$, respectively, with the constraint $p > r_0 > r1 > \cdots > r_L$. Here the "GLLVM link" stands for the fact that the components of $f(Y_i|z_i^{(1)}, \Theta)$ belongs to an exponential family. For example if Y_{ij} is a count variable with support $[1, n_{ij}]$, one can choose a Binomial distribution as link function as

$$f(Y_{ij}|z^{(1)}, \Theta) = \binom{n_{ij}}{Y_{ij}} h(z^{(1)})^{Y_{ij}} (1 - h(z^{(1)}))^{n_{ij} - Y_{ij}}, \tag{2}$$

where h is a parametric function with parameters denoted by $(\lambda^{(0)}, \Lambda^{(0)})$. This example can be adapted for binary variables, by taking $n_{ij} = 1$ and considering a Bernoulli distribution. In the simulations, the ordered multinomial distribution is used for the ordinal variables, the unordered multinomial distribution for categorical variables, the Binomial distribution for count variables, and the Gaussian distribution for continuous variables. More examples of link functions can be found in [3].

The graphical model associated with (1) is given in Fig. 1. It corresponds to the M1DGMM architecture as introduced by [8]. Note that there exists a second architecture denoted by M2DGMM which is made of two heads (one to deal with continuous variables, one to deal with non-continuous variables). Here M1DGMM is more appropriate since it controls all the structure of dependence between the variables of Y in the first layer, through the latent variable $z^{(1)}$ and has shown to be more stable than the M2DGMM [8]. In the sequel, we use a simple one layer deep architecture with a two-dimensional latent space and with $K_1 = 4$ components. Even if this is not the purpose of this paper we also modified these parameters and the results seemed very stable.

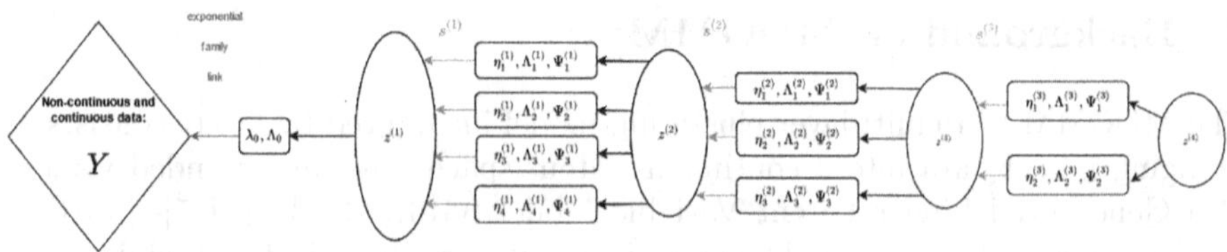

Fig. 1. Graphical model of a M1DGMM

3 MI2AMI Description

Let $Y = (Y^{obs}, Y^{mis})$ be the complete mixed data composed of observed values Y^{obs} and missing values Y^{mis}. The values estimated or imputed by MI2AMI will be denoted by a hat.

3.1 General Overview

The MI2AMI imputation method follows the following steps:

Algorithm 1. MI2AMI

Input: $\mathbf{Y} = (Y^{obs}, Y^{mis})$
1. Completion step:
Train the MDGMM on Y^{obs} and estimate Θ by $\hat{\Theta}_{init}$
Impute Y^{mis} from $\hat{\Theta}_{init}$ and obtain $\hat{Y}_{init} = (Y^{obs}, \hat{Y}^{mis}_{init})$
2. MDGMM step:
Train the MDGMM on $\hat{Y}_{init}$ and estimate Θ by $\hat{\Theta}$
2. Imputation step:
Impute Y^{mis} from $\hat{\Theta}$ and obtain $\hat{Y}^{completed} = (Y^{obs}, \hat{Y}^{mis})$
Output: $\hat{Y}^{completed}$

The first step of Algorithm 1 is a completion step which allows to have a complete dataset. In a second step we then restart the MDGMM. This second step starts with the initialization procedure included in the MDGMM algorithm which is important to stabilize the procedure. The algorithm concludes with an imputation step that we describe below.

3.2 The Imputation Step

The imputation step is based on an optimization procedure that finds the areas of the latent space that best describes the observed parts of the data. More formally for each observation i the optimization program first search the latent variables z as follows:

$$\begin{cases} \arg\min_{z_i} d(Y_i^{obs}, \hat{Y}_i^{obs}) \\ \text{such that } \min(Y_j) \leq \hat{Y}_{i,j} \leq \max(Y_j), \ \forall j \text{ s.t. } Y_j \text{ is continuous.} \end{cases} \quad (3)$$

where d denotes the Gower's distance and $\hat{Y} = (\hat{Y}_i^{obs}, \hat{Y}^{mis})$. By last constraint in (3) the pseudo-observations belong to the range of the observations. The pseudo-observation are then generated from the latent variables z_i as follows:

$$\hat{Y}_{i,j} = \begin{cases} \text{mean}(Y_{i,j}|z_i), & \text{if } Y_j \text{ is continuous} \\ m|\forall m' \neq m, P(Y_{i,j} = m|z_i) \geq P(Y_{i,j} = m'|z_i), & \text{else.} \end{cases}$$

Roughly speaking, the pseudo-observations are imputed as the conditional mean when the variable is continuous and as the conditional mode otherwise.

The optimization procedure itself is based on trust-region methods [6], with the constraint that the pseudo-observations stay in the support of the observed values for each continuous variable.

4 Numerical Illustration

The code is available at: https://github.com/RobeeF/M1DGMM.

4.1 Framework

Missingness Designs. [18] has introduced a typology of the missingness designs: Missing Completely At Random (MCAR), Missing At Random (MAR), and Missing Not At Random (MNAR). Roughly speaking, MCAR means that missing data are generated independently of the data, MAR requires that the probability of missing values depends only on the observed data and MNAR allows missing to depend on unobserved data. In the MCAR design, all marginal distributions are unchanged, in the MAR designs, observed margins can be modified.

We restrict our attention to both MCAR and MAR mechanisms. The specified percentage of missing values for these two designs is 10%, 20%, and 30%. In the following, we refer to a given dataset by its design and its missing values rate. For instance, the MAR design with 10% of missing values is called MAR10.

For the MAR process, the missing probabilities concerns the variables WR, WW and ME, and are governed by a logistic form that depends on the other variables WA, WE, HE, and NC as follows:

$$\mathbb{P}(missing) \quad = \quad \frac{\exp(\alpha_0 + \alpha_1 WA + \alpha_2 WE + \alpha_3 HE + \alpha_4 NC)}{1 + \exp(\alpha_0 + \alpha_1 WA + \alpha_2 WE + \alpha_3 HE + \alpha_4 NC)}.$$

We chose $\alpha_0, \alpha_1, \cdots, \alpha_4$ in such a way that we have 10, 20, 30 % of missing respectively.

Competitors. We propose to compare our approach to five popular competitors:

- *MICE* (in the MICE package on R), developed by [2], imputes continuous, binary, unordered categorical, and ordered categorical data using Chained Equations.
- *MissForest* (in the MissForest package on R), proposed in [20], uses Random Forest to handle missing values particularly in the case of mixed-type data: continuous and/or categorical data including complex interactions and nonlinear relations.
- *imputeFAMD* (in the MissMDA package on R), proposed in [1], which imputes the missing values of a mixed dataset, with continuous and categorical variables, using the principal component method "factorial analysis for mixed data" (FAMD).
- *kNN* (in the VIM package on R), proposed in [10], uses k-Nearest Neighbour Imputation coupled with the Gower's Distance to deal with numerical, categorical, ordered, and semi-continuous variables.
- *GAIN* (on Python), proposed in [22], which uses a generative adversarial network to impute missing values for continuous or binary variables. As mentioned, GAIN is not designed for mixed data but the algorithm can be exploited by accepting some approximations (as GAIN respects probability distribution supports, we consider that the approach is usable).

Evaluation Metrics. The quality of the latent representation learned by the model can be asserted using latent association matrices on the imputed and full datasets. The association matrix is a generalization of the correlation matrix for mixed datasets. The closest the two matrices are the less disturbed the latent representation constructed by the MDGMM is.

Concerning the comparison with the competitor models, three standard criteria are used as follows:

- The Normalized Root Mean Squared Error (NRMSE) for continuous, count and ordinal imputations. The NRMSE of the jth variable is defined by

$$NRMSE(y_j^{mis}, \hat{y}_j^{mis}) = \sqrt{\frac{mean((y_j^{mis} - \hat{y}_j^{mis})^2)}{var(y_j)}}, \tag{4}$$

  with y_j^{mis} the ground-truth value of the missing values, $\hat{y}_j^{mis}$ their estimated values, and y_j all the values of the j variable (observed and ground truth value of the missing values). "Mean" and "var" stands for the mean and variance over the observations, respectively.
- The Proportion of Falsely Classified entries (PFC) for binary and categorical imputations given by

$$PFC(y_j^{mis}, \hat{y}_j^{mis}) = mean(1_{y_j^{mis} = \hat{y}_j^{mis}})$$

where $1_{.}$ denotes the indicatrix function equal to 1 if the imputed value and its ground truth values coincide.

- The Gower's distance (see [9]) which is a combination of the PFC criteria and the NRMSE (but replacing $var(y_j)$ in (4) by the range of the observed y) and gives a global criterion for all data types.

The presented results are the medians over ten runs for each competitor.

Datasets. We test our approach on the Contraceptive dataset from the UCI repository and on the Pima Indians Diabetes dataset.

The Contraceptive dataset is a subset of the 1987 National Indonesia Contraceptive Prevalence Survey from [13]. The samples are married women who were either not pregnant or do not know if they were at the time of the interview. The dataset is composed of one continuous and one count variable, three binary variables, three ordinal variables, and two categorical variables as follows:

- WA = Wife's age (integer between 16 and 19)
- WE = Wife's education (1 = low, 2, 3, 4 = high)
- HE = Husband's education (1 = low, 2, 3, 4 = high)
- NC = Number of children ever born
- WR = Wife's religion (0 = Non-Islam, 1 = Islam)
- WW = Wife's now working? (0 = Yes, 1 = No)
- HO = Husband's occupation (1, 2, 3, 4)
- SL = Standard-of-living index (1 = low, 2, 3, 4 = high)
- ME = Media exposure (0 = Good, 1 = Not good)
- CM = Contraceptive method used (1 = No-use, 2 = Long-term, 3 = Short-term)

The Pima Diabetes data give personal information (the blood pressure, BMI, etc.) of 768 Indian individuals among which one third has diabetes. It is composed of one count variable, one binary variable, and seven continuous variables. Hence, the Contraceptive and Pima Indian Diabetes data present opposite patterns concerning continuous/non-continuous share of variables as follows:

The variables of the Pima dataset are:

- Pregnancies: Number of times pregnant
- Glucose: Plasma glucose concentration a 2 h in an oral glucose tolerance test
- BloodPressure: Diastolic blood pressure (mm Hg)
- SkinThickness:Triceps skin fold thickness (mm)
- Insulin: 2-Hour serum insulin (mu U/ml)
- BMI: Body mass index (weight in $kg/(heightinm)^2$)
- DiabetesPedigreeFunction: Diabetes pedigree function
- Age: Age (years)
- Outcome: Class variable (0 or 1) 268 of 768 are 1, the others are 0

4.2 Results

As shown in Fig. 2, the latent representation of the data is not substantially disturbed by the 30% of missing values. The main associations existing in the latent representation of the full dataset are reconstructed properly. This is the case for instance of the association between the number of Pregnancies and the Glucose rate or between the Blood Pressure and the thickness of the Skin. There is a slight tendency of the approach to strengthen already existing associations. This pattern is not surprising considering that missing values are imputed using values that are likely with respect to the estimated associations.

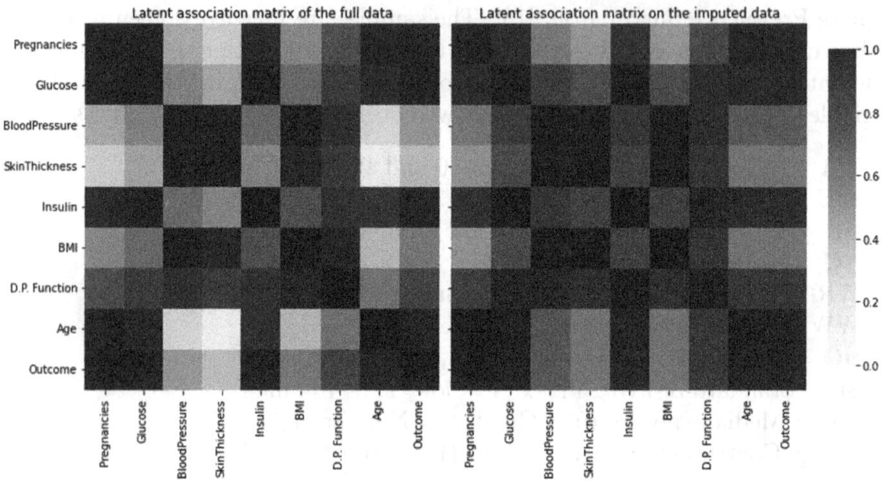

Fig. 2. Variables latent association matrices for the full data and for the completed data learned by MI2AMI on the Diabetes dataset for the MCAR30 design.

Concerning the compared performances with the competitor models on the Contraceptive dataset and for the MCAR20% design, MissMDA achieves the best performances and GAIN gets the worst as seen in Fig. 3 where the y-axis represents either the PFC for categorical as well as binary data, or the NRMSE. MI2AMI obtains above-average performances on the binary data (WifeRelig, WifeWork, Media) but under average performances on the count data especially for the number of children imputation. This can be explained because the variance of the observed number is small and our generated values are more variable which increases the NRMSE. In that case, even if the method is less efficient, it nevertheless preserves the dependence structure as shown in Fig. 2 and it seems to propose consistent imputed values (however often different from the true values). Conversely, MissForest and MICE perform well over the ordinal and continuous variables but obtain the worst performances on the binary data.

On the Pima Diabetes dataset and the MAR10 design, MissForest presents the best performance for all the variables as seen in Fig. 4. Conversely, MissMDA

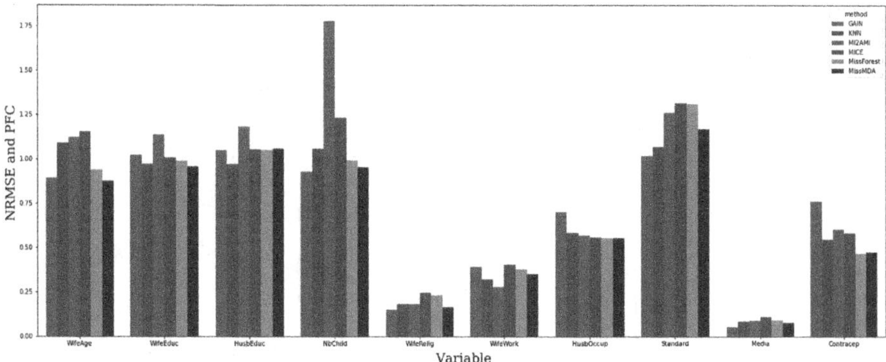

Fig. 3. Median NRMSE and PFC in MCAR scenario with 20% of missing values for the Contraceptive dataset

and MICE get the biggest error. MI2AMI obtains solid imputed values for the continuous variables "Glucose" and "Diabetes Pedigree". The non-continuous variables are the less well imputed: the number of pregnancies and the age presented the highest error rates for most methods with NRSME ≥ 1.0, *i.e.* RMSE higher than the variance of the variables to impute. The binary variable "Outcome" presented false prediction rates between 42% to 72%.

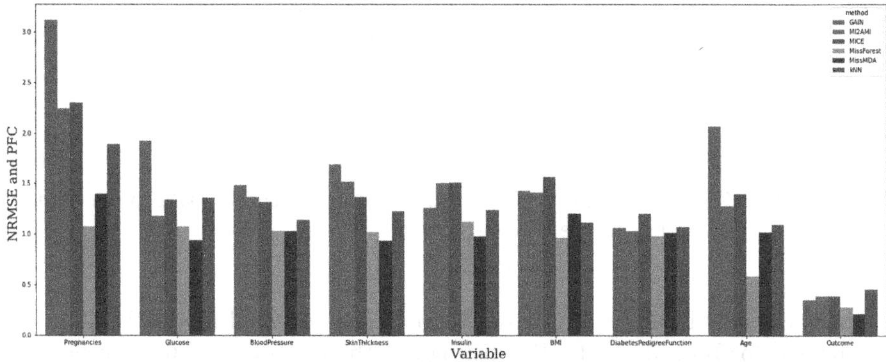

Fig. 4. Median NRMSE and PFC in MAR scenario with 10% of missing values for the Diabetes dataset

The last two designs/competitors couples are presented in Fig. 5 where the metric used is the median (over all runs) of the Gower's distance. Here kNN seems to be very efficient and stable. This is overall the best method here as well as the Missforest approach. The ranking of the method is similar to the one evoked and remains relatively unchanged with the percentage of missing values. MissMDA looks very unstable, sometimes very good, sometimes last, while MICE appears less efficient. Finally M2IAMI seems stable, but never the most efficient here.

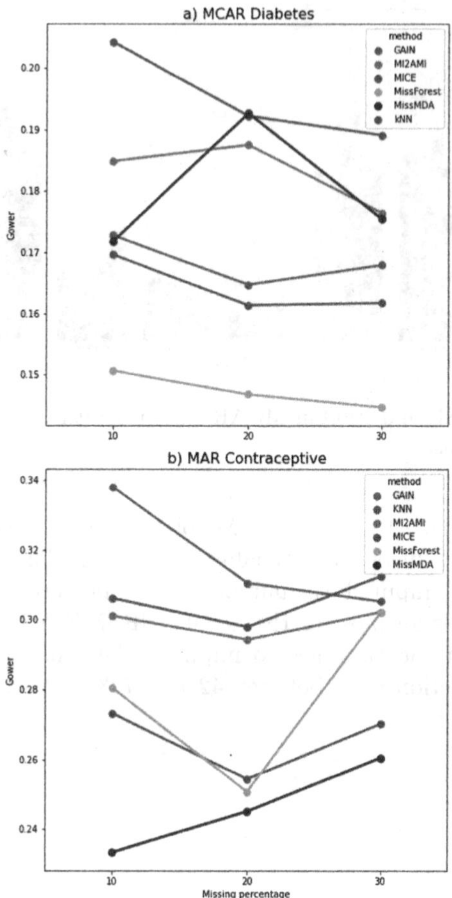

Fig. 5. Median Gower's distances for the MAR design on the Contraceptive dataset (a), for the MCAR design on the Diabetes dataset.

In conclusion, Table 1 summarizes the performance of all the methods for all designs and for both datasets. We distinguish the performance with respect to the nature of the missing data.

Table 1. Performances (from low = $---$ to high = $+++$) of the methods in terms of NRMSE, PFC, and Gower distances.

Method	All	Continuous	Discrete	Ordinal	Non ordinal
MIAMI	++	+	−	−−	++
GAIN	++	−−	−	++	+
MICE	+	−	−−	+	−−
MissForest	+++	++	+++	+	++
MissMDA	++	+++	++	++	++
kNN	+++	−	++	+++	+

5 Discussion and Perspective

In this work, an extension of the recently introduced MDGMM model has been proposed to impute missing values in mixed dataset. It takes advantage of the simple and parametric structure that links the original data with the latent space. The quality of the latent representation can easily be characterized as shown in this paper. Once the representation of the data is learned, the optimization procedure enables finding its most promising regions. As the optimization is point-wise this procedure is easily parallelizable. Numerous other improvements to the model are possible. For instance, a simple architecture with one two-dimensional layer has been chosen here to keep a low estimation variance. Using deeper architectures could enable to accept a higher variance to obtain a lower bias.

From the competitors, M2IAMI has proven to be a proper baseline model that provides reliable imputations on the two presented datasets. More generally, the ranking of the competitor models was less affected by the missing value designs (MCAR, MAR) and the percentage of the missing values than by the dataset choice. In absolute level, the imputation errors remained quite high, denoting how challenging the imputation of mixed datasets is. This shows that there is still work to optimize the number of parameters (number of layers and number of components) of the MDGMM in order to improve MI2AMI.

References

1. Audigier, V., Husson, F., Josse, J.: A principal component method to impute missing values for mixed data. Adv. Data Anal. Classif. **10**, 5–26 (2016)
2. van Buuren, S., Groothuis-Oudshoorn, K.: mice: multivariate imputation by chained equations in r. J. Stat. Softw. **45**(3), 1–67 (2011). https://doi.org/10.18637/jss.v045.i03. https://www.jstatsoft.org/index.php/jss/article/view/v045i03
3. Cagnone, S., Viroli, C.: A factor mixture model for analyzing heterogeneity and cognitive structure of dementia. AStA Adv. Stat. Anal. **98**(1), 1–20 (2014)
4. Choudhury, A., Kosorok, M.R.: Missing data imputation for classification problems (2020)

5. Christoffersen, B., Clements, M., Humphreys, K., Kjellström, H.: Asymptotically exact and fast gaussian copula models for imputation of mixed data types (2021)
6. Conn, A.R., Gould, N.I., Toint, P.L.: Trust region methods. SIAM (2000)
7. Deng, G., Han, C., Matteson, D.S.: Learning to rank with missing data via generative adversarial networks. arXiv preprint arXiv:2011.02089 (2020)
8. Fuchs, R., Pommeret, D., Viroli, C.: Mixed deep gaussian mixture model: a clustering model for mixed datasets. Advances in Data Analysis and Classification, pp. 1–23 (2021)
9. Gower, J.C.: A general coefficient of similarity and some of its properties. Biometrics, pp. 857–871 (1971)
10. Kowarik, A., Templ, M.: Imputation with the r package vim. J. Stat. Softw. **74**(7), 1–16 (2016). https://doi.org/10.18637/jss.v074.i07. https://www.jstatsoft.org/index.php/jss/article/view/v074i07
11. Lee, D., Kim, J., Moon, W.J., Ye, J.C.: Collagan : Collaborative gan for missing image data imputation (2019)
12. Li, S.C.X., Jiang, B., Marlin, B.: Learning from incomplete data with generative adversarial networks. In: International Conference on Learning Representations (2019). https://openreview.net/forum?id=S1lDV3RcKm
13. Lim, T., Loh, W., Shih, Y.: A comparison of prediction accuracy, complexity, and training time of thirty-three old and new classification algorithms. Mach. Learn. **40**(3), 203–228 (2000)
14. McLachlan, G.J., Basford, K.E.: Mixture Models: Inference and Applications to Clustering. Marcel Dekker, New York (1988)
15. Moustaki, I.: A general class of latent variable models for ordinal manifest variables with covariate effects on the manifest and latent variables. Br. J. Math. Stat. Psychol. **56**(2), 337–357 (2003)
16. Moustaki, I., Knott, M.: Generalized latent trait models. Psychometrika **65**(3), 391–411 (2000)
17. Murray, J.S., Reiter, J.P.: Multiple imputation of missing categorical and continuous values via bayesian mixture models with local dependence. J. Am. Stat. Assoc.**111**(516), 1466–1479 (2016). https://doi.org/10.1080/01621459.2016.117. https://ideas.repec.org/a/taf/jnlasa/v111y2016i516p1466-1479.html
18. Rubin, D.B.: Inference and missing data. Biometrika **63**(3), 581–592 (1976). https://doi.org/10.1093/biomet/63.3.581
19. Shang, C., Palmer, A., Sun, J., Chen, K.S., Lu, J., Bi, J.: Vigan: Missing view imputation with generative adversarial networks (2017)
20. Stekhoven, D.J., Bühlmann, P.: MissForest-non-parametric missing value imputation for mixed-type data. Bioinformatics **28**(1), 112–118 (2011). https://doi.org/10.1093/bioinformatics/btr597
21. Viroli, C., McLachlan, G.J.: Deep gaussian mixture models. Stat. Comput. **29**(1), 43–51 (2019)
22. Yoon, J., Jordon, J., van der Schaar, M.: GAIN: missing data imputation using generative adversarial nets. In: Dy, J., Krause, A. (eds.) Proceedings of the 35th International Conference on Machine Learning. Proceedings of Machine Learning Research, vol. 80, pp. 5689–5698. PMLR, 10–15 July 2018. https://proceedings.mlr.press/v80/yoon18a.html
23. Zhao, Y., Udell, M.: Missing value imputation for mixed data via gaussian copula (2020)

On the Utility and Protection of Optimization with Differential Privacy and Classic Regularization Techniques

Eugenio Lomurno(✉) and Matteo Matteucci

Politecnico di Milano, Milano, Italy
{eugenio.lomurno,matteo.matteucci}@polimi.it

Abstract. Nowadays, owners and developers of deep learning models must consider stringent privacy-preservation rules of their training data, usually crowd-sourced and retaining sensitive information. The most widely adopted method to enforce privacy guarantees of a deep learning model nowadays relies on optimization techniques enforcing differential privacy. According to the literature, this approach has proven to be a successful defence against several models' privacy attacks, but its downside is a substantial degradation of the models' performance. In this work, we compare the effectiveness of the differentially-private stochastic gradient descent (DP-SGD) algorithm against standard optimization practices with regularization techniques. We analyze the resulting models' utility, training performance, and the effectiveness of membership inference and model inversion attacks against the learned models. Finally, we discuss differential privacy's flaws and limits and empirically demonstrate the often superior privacy-preserving properties of dropout and l2-regularization.

Keywords: Differential privacy · Regularization · Membership inference · Model inversion

1 Introduction

In recent times, the number and variety of machine learning applications have noticeably grown thanks to the computational progress and to the availability of large amounts of data. In fact, given the proportionality between the performance of machine learning models and the amount of data fed into them, more and more data has been and is going to be collected to get reliable and accurate results. This data is usually crowd-sourced and may contain private information. Thus, it is necessary to guarantee high privacy levels for data owners, concerning both the sharing of their information and the machine learning models trained over it.

In fact, malicious agents can target deep learning models to exploit the information that remains within them after training is complete, even in black-box scenarios. Among the most famous and dangerous attacks, membership inference [23] and model inversion [9] techniques represent the main threat for

G. Nicosia et al. (Eds.): LOD 2022, LNCS 13810, pp. 223–238, 2023.
https://doi.org/10.1007/978-3-031-25599-1_17

machine learning models. The former aims at guessing the presence of an instance inside the training data of the attacked model, while the second has the goal of reconstructing the input data from the accessible or leaked information.

To reduce the effectiveness of these attacks against artificial neural networks, a widely spread countermeasure taken by model owners is applying differential privacy [5] in the models' training procedure. The main advantage of this solution is the guarantee that the privacy leakage at the end of models' training is limited and measurable. The main downside of this approach relies in the noise injected in the models during their training to achieve the desired privacy budget level. In fact, noise addition significantly impacts over models' training performance in terms of time and utility.

In this work, we investigate the topic of privacy preservation in deep learning. We test the effectiveness of the most known implementation of differential privacy training for deep learning models, the differentially private stochastic gradient descent (DP-SGD) [1], as a defense mechanism. We evaluate the validity of this method in terms of protection against model inversion and membership inference attacks. We also measure the impact that DP-SGD has on the model under attack, analyzing both the level of accuracy achieved and the training time required in white-box and black-box scenarios.

We perform the same analysis considering two regularization techniques, i.e., dropout and the l2-regularizer. Their effect on improving model's generalization capability is widely known, and previous works have confirmed an existing connection between privacy attacks' effectiveness and overfitting in the target model [22]. Our work empirically demonstrates how l2 regularization and dropout achieve similar or even better privacy-preserving performance with respect to DP-SGD training while preserving models utility and training time efficiency. To the best of our knowledge, this is the first work to systematically analyze the impact of different levels of differential privacy and regularization techniques in terms of models' accuracy, training time, and resistance to membership inference and model inversion attacks.

2 Privacy Threats

Privacy in machine learning is a much-debated topic in the literature. If, on the one hand, deep learning algorithms efficiently learn from data how to solve complex tasks, on the other hand, it has been demonstrated how these models exhibit vulnerabilities to malicious attacks despite their complexity [7,13,24]. The targets of these attacks are various and characterized by different risk levels from a user perspective.

We differentiate the scenarios in which an attack occurs depending on the extent of the adversarial knowledge, that is, the ensemble of information concerning the model and the data under attack at the disposal of the attacker. From this point of view, we distinguish the threat scenarios into two types: black-box and white-box. In a black-box scenario, the adversary knows only the target model elements available to the public, such as prediction vectors, but has no access to the model structure or information about the training dataset outside its format. In a white-box scenario, the adversary has complete knowledge of the target model and knows the data distribution of the training samples. Among the most famous and dangerous attacks, we focus our attention on membership inference and model inversion attack families [2].

2.1 Membership Inference Attacks

Membership inference attacks take as input a sample and try to determine if it belongs or not to the training dataset of the model under attack. The most common design paradigm involves the use of shadow models and meta-models, also known as attack models [23]. The basic idea behind this white-box approach is to train several shadow models that imitate the behavior of the target on surrogate or shadow datasets. Shadow datasets must contain samples with the same format and similar distribution to the training data of the target model. After the training of shadow models is complete, their outputs and the known labels from the shadow datasets form the attack dataset. This dataset is used to train a meta-model which learns to make membership inference based on the shadow models' results.

The main limitation of this approach is represented by the mimic capabilities of the shadow models with respect to the target model and the strong assumptions related to the adversarial knowledge of both the target model structure and the training data distribution. To overcome these issues, Salem *et al.* [22] proposed three different attacks considering scenarios with more relaxed assumptions on the adversarial knowledge. The first two approaches maintain the idea of shadow models, while the third proposal abandons the shadow model paradigm in favor of a threshold-based attack. In this approach, the attacking model is a simple binary classifier that takes the highest posterior from the prediction vector of the target model and compares it against a given threshold. If its value is greater than the threshold, the input sample obtained from that output is considered a member of the training dataset. The advantages of this approach concern the complete independence from the target model and its training data and the elimination of the overhead costs due to the design of shadow models and the creation of suitable shadow datasets. Besides, it requires no training of the attack model. A novel type of membership inference attack, named BlindMI, has recently been proposed by Hui *et al.* [14] to probe the target model and extract membership semantics via differential comparison and data generation.

2.2 Model Inversion Attacks

Model inversion attacks try to reconstruct training samples of the attacked model starting from environmental elements known by the attacker. The first designs of reconstruction attacks assumed a white-box scenario in which the adversary knows the output label, the prior distribution of features for a given sample, and has complete access to the model. With these assumptions, the attacker estimates the sensitive attributes' values that maximize the probability of observing the known model parameters. These first forms of attacks are referred to as Maximum a-posteriori (MAP) attacks [10]. However, they were soon abandoned because their performance degrades as the feature space to reconstruct grows. To overcome this limitation, Frederikson *et al.* [9] proposed to formulate the attack as an optimization problem where the objective function depends on the target model output. Starting from assumptions similar to those considered in MAP attacks, the attacker reconstructs the input sample through gradient descent in the input space. This attack can be performed in white and black box scenarios, depending on whether the attacker has access to the intermediate maps of the target model or its prediction vector respectively.

Yang *et al.* [28] proposed a black-box attack where the adversarial does not know any detail about the model and its training data. Instead, it knows the generic training data distribution and the output format of the model, i.e., the prediction vector. Zhang *et al.* [30] designed a novel solution that involves the use of a generative adversarial network (GAN) [12] to learn the training data representation, exploiting the properties of this kind of network to increase the feasibility of the attack. Zhao *et al.* [31] exploited information provided by artificial intelligence explainability tools to achieve high fidelity reconstruction of the target model input data. The authors focus on understanding which explanations are more useful for the attacker, measuring which of them leaks higher information amounts about target data.

Lim and Chan [17] exploited a privacy-breaking algorithm in a federated learning scenario to reconstruct users' input data from their leaked gradients. The idea is to generate dummy gradients from a randomly initialized dummy input and compute the loss between these and the true ones. The loss is used to update the dummy input to reduce the distance with respect to real input data according to Geiping *et al.* [11]. Finally, Yin *et al.* [29] designed GradInversion, an algorithm able to recover with great precision single images from deep networks gradients trained with large batch sizes. The first step of this approach is to convert the input reconstruction problem into an optimization task, where, starting from random noise, synthesized images are exploited to minimize the distance between the gradients of these and the real gradients provided by the environment. Then, the core of the algorithm lies in a batch-wise label restoration method, together with the use of auxiliary losses that aim to ensure fidelity and group consistency regularization to the final result.

3 Privacy Defences

With the spread of machine learning as a service, the attack surface for the world of machine learning has undergone rapid growth. Nowadays, machine learning threats and defences are involving disparate scenarios and techniques [20]. Differential privacy [5] represents the most proposed technique to guarantee protection and ensure data owners' privacy.

Differential privacy is devised as an effective privacy guarantee for algorithms that work with aggregated data. It was initially proposed in the domain of database queries, defining the concept of adjacent databases as two sets that differ in a single entry. More formally, a randomized mechanism $M\colon D \rightarrow R$ with domain D and range R satisfies ε-differential privacy if for any two adjacent inputs $d,\ d' \in D$ and for any subset of outputs $S \subseteq R$ it holds that:

$$Pr[M(d) \in S\,] \leq e^{\varepsilon} Pr[M(d') \in S\,] \tag{1}$$

The parameter ε is called privacy budget because it represents how much information leakage can be afforded in a system. The lower the ε value, the stricter the privacy guarantee. Differential privacy represents a significant development in the field of privacy-preserving techniques because it guarantees three properties that are very useful, namely: composability, group privacy, and robustness to auxiliary information.

- Composability concerns the possibility of having a composite mechanism so that, if each of its components is differentially private, then is also the overall mechanism itself. This property stays true for sequential and parallel compositions.
- Group privacy assures that if the dataset contains correlated data, like the ones provided by the same individual, the privacy guarantee degrades gracefully and not abruptly.
- Robustness to auxiliary information guarantees that the privacy level assured by the theory stands regardless of the knowledge available to the adversary.

The main theoretical issue related to this definition of differential privacy relies in its rigor. In fact, in order to make it exploitable for real uses it is necessary to relax its constraints. There exists many formulations that generalize the privacy-budget ε and provide its relaxation, for instance the *f*-differential privacy [4] or the concentrated differential privacy [6]. Among them, the most applied formulations are the (ε, δ)-differential privacy [1] and the Rényi differential privacy [19].

(ε, δ)-differential privacy is defined via a randomized mechanism $M\colon D \rightarrow R$ with domain D and range R that satisfies its constraints if for any two adjacent inputs $d,\ d' \in D$ and for any subset of outputs $S \subseteq R$ it holds that:

$$Pr[\ M(d) \in S\,] \leq e^{\varepsilon} Pr[\ M(d') \in S\,] + \delta, \tag{2}$$

where the additive factor δ represents the probability that plain ε-DP is broken. In the case of several mechanism, the composition property still holds, even if in a more complex version that keeps track of the privacy loss accumulated during the execution of each component. Starting from this composability property, Abadi *et al.* [1] designed an new function called moments accountant, that computes the privacy cost needed for each access to the data and uses this information to define the overall privacy loss of the mechanism. Then, in the same work, Abadi *et al.* defined the so called differentially-private stochastic gradient descent (DP-SGD) that is actually one of the most adopted optimizers to implement differential privacy.

The Rényi Differential Privacy is instead another form of relaxation of differential privacy based on the concept of Rényi divergence. Given that for two probability distributions P and Q defined over R, the Rényi divergence of order $\alpha > 1$ defined as:

$$D_\alpha(P \parallel Q) \triangleq \frac{1}{\alpha - 1} \log E_{x \sim Q} \left(\frac{P(x)}{Q(x)} \right)^\alpha \tag{3}$$

The relationship between the differential privacy formulation and the Rényi divergence can be defined by a randomized mechanism M: $D \rightarrow R$ that is ε-differentially private if and only if its distribution over any pair of adjacent inputs $d, d' \in D$ satisfies:

$$D_\infty(M(d) \parallel M(d')) \leq \varepsilon \tag{4}$$

Putting all together, it is possible to define the (α, ε)-Rényi differential privacy as a randomized mechanism M: $D \rightarrow R$ is said to have ε-Rényi differential privacy of order α if for any adjacent $d, d' \in D$ it holds that:

$$D_\alpha(M(d) \parallel M(d')) \leq \varepsilon \tag{5}$$

It is demonstrated that the three properties of composability, robustness to auxiliary information, and group privacy are still valid for (α, ε)-Rényi differential privacy.

Thus, differential privacy turns out to be the most adopted privacy-preserving technique for its guarantees and its quantification of the privacy-budget [16,18, 20,26]. However, Bagdasaryan *et al.* [3] argue its negative impact when applied via DP-SGD in terms of performance loss, especially for low ε values. Due to its implementation, more privacy guarantees from DP-SGD mean higher noise injections during the training procedure as well as fewer training iterations for the model. This is coherent with the results of Salem *et al.* [22] work, in which is demonstrated the proportionality between deep learning models' overfitting and their vulnerability to membership inference attacks. This makes sense, on one hand, concerning that DP-SGD noise addiction can be seen as a regularization form, and on the other hand, recalling that differential privacy can be interpreted also as a direct countermeasure to membership inference attacks.

Besides, other works have tried to mitigate these threats with different techniques. For instance Jain *et al.* [15] discussed the dropout [25] differentially-private properties and its protection against membership inference attacks. Ermis *et al.* [8] defined a form of differential privacy starting from the Bayesian definition of Gaussian dropout. Differently, Nasr *et al.* [21] defined an adversarial regularization to protect their models by changing the loss function, while Yang *et al.* [27] implemented a procedure named prediction purification to protect against both membership inference and model inversion again with adversarial learning.

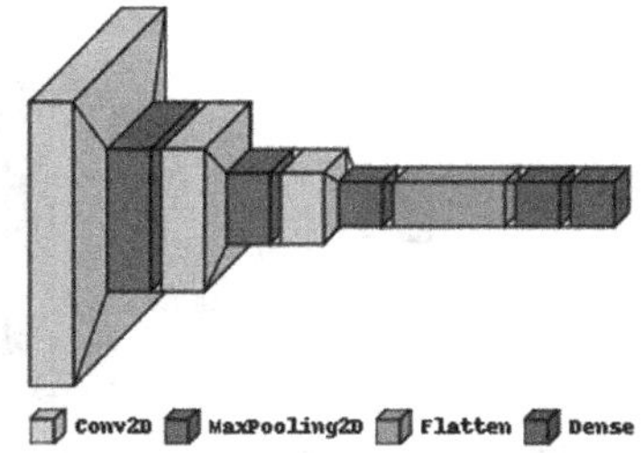

Fig. 1. The neural network architecture of the target model.

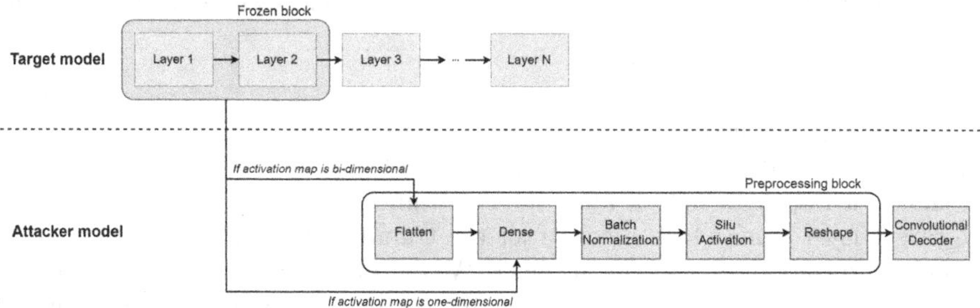

Fig. 2. The schema of the proposed model inversion attack. In a white-box scenario the attacker can reconstruct the target model input from an arbitrary activation map (in the example, from the second layer's one). In a black-box scenario, the attacker can only exploit the output of the layer N to invert the target model.

4 Method

This work aims to provide an exhaustive and detailed comparison between deep learning models with and without privacy-preservation techniques. We tested the effectiveness of both DP-SGD and regularization techniques in defending the target model, as well as their impacts over the accuracy and the training time of the target model itself. To adopt the DP-SGD we relied on the TensorFlow Privacy implementation based on Abadi *et al.* work [1] which corresponds to the (ε, δ)-differential privacy version.

The target model always has the same architecture despite the training procedure and the regularization techniques adopted, as shown in Fig. 1. For fair comparisons and simplicity, we chose a straightforward convolutional neural network with ReLU activation for hidden layers and a Softmax output activation.

To evaluate its resistance to membership inference attacks, we adopted again a TensorFlow Privacy tool implementing two black-box attacks inspired by Salem *et al.* work [22]. The first is a threshold-based attack, while the second involves the training of a single shadow model and assumes no knowledge about the data distribution. The attacking tool automatically selects the most effective technique among the aforementioned two. We measured the Area Under Curve (AUC) as a metric to evaluate the attacks' effectiveness.

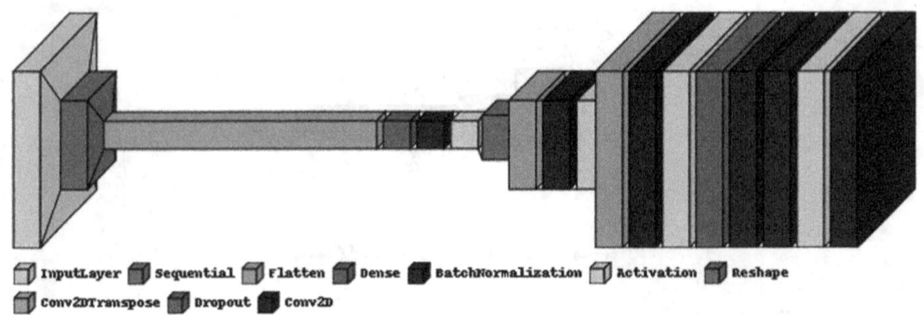

Fig. 3. The neural network architecture of the attacker model used for model inversion.

Concerning the model inversion attack, we have devised an approach that exploits the activation maps of the target model to reconstruct its training data, as shown in Fig. 2. In detail, after having trained the target model, we select the target layer from which reconstruct the training input, that in a black-box scenario corresponds to the output layer. We cut out the subsequent part of the target network and free the weights. Finally, we use this target network fraction as preprocessing layer for the attacker network to compute the model inversion attack. A detailed overview of the architecture of the adversary model is shown in Fig. 3, in which the sequential layer represents the frozen fraction of the target model. After some tuning, we selected mean squared error (MSE) as loss function, SiLU as activation function for all the hidden layers and Adam as optimizer with a learning rate of 10^{-3}. The output of the network is a convolutional layer with the same number of channels of the input data of the target and a Sigmoid activation function. To measure the reconstruction quality of this model inversion attack, we adopted the MSE as evaluation metric.

To recap, in order to introduce differential privacy in the training of the target model is sufficient to substitute its optimizer with the DP-SGD, thus preserving the same structure and number of parameters. To have a better comprehension of the impact of privacy-preserving techniques, we selected three different privacy budget levels, i.e., $\varepsilon = 2$, $\varepsilon = 4$, $\varepsilon = 8$ keeping the same privacy leak probability among the experiments, i.e., $\delta = 10^{-5}$. We trained a distinct target model with each of these privacy budget.

In parallel, we also trained target models with regularization techniques. We decided to inspect the impact of the most adopted mechanisms to avoid overfitting, i.e., dropout and weight decay or l2. In particular, we tested one target model for each of the two techniques and one with both of them together. The dropout is applied between every weighted layer, i.e., after each convolutional layer and before and after the first dense layer of the target model. The l2 is applied directly to the last dense layer, i.e., the prediction layer.

Table 1. The accuracy scores of the target model over the test set (higher is better).

Dataset	Baseline	$\varepsilon = 2$	$\varepsilon = 4$	$\varepsilon = 8$	L2	Dropout	L2+Dropout
CIFAR-10	0.664	0.518	0.538	0.536	0.649	**0.692**	0.648
MNIST	0.992	0.942	0.963	0.968	0.991	**0.996**	0.994
F-MNIST	**0.900**	0.820	0.823	0.829	0.896	0.890	0.879

5 Experiments and Results

All experiments described below have been carried out on a system equipped with an Intel(R) Xeon(R) CPU E5-2630 v4 @ 2.20GHz and an Nvidia GeForce GTX TITAN X GPU. We performed our analysis on three image datasets, i.e., CIFAR-10, MNIST and Fashion-MNIST, each of them pre-processed with a normalization step.

In order to have comparable experiments, we fixed some hyperparameters among the different target model configurations. The ε parameter is typically computed a-posteriori, to evaluate the privacy budget achieved by a model with a differentially-private optimizer. In the formulation provided by Abadi *et al.* [1], ε is function of the number of data samples, the number of micro-batches, the noise multiplier and the number of training epochs. In our setting, we have fed the differentially-private target models with the same data as the other models, with the same batch size = 200. In order to maximize the utility, we set the number of micro-batches equal to the number of mini-batches. Finally, in order to control a-priori the privacy budget, we fixed the number of training epochs to the optimal value found for the target model without privacy-preserving techniques. In this way, we could control the ε privacy budget by varying the noise multiplier parameter. We did similar reasoning concerning the regularization techniques by fixing batch size, the number of data samples, and the optimizer. After some tuning, we also fixed the dropout rate to 20% and the l2-weight to $2\text{x}10^{-2}$. As an optimizer to compare with the DP-SGD we chose the SGD. For both of them, we tuned the learning rate that converged to the same optimal value of 10^{-1}.

Table 1 shows the results of each target model over the test sets, which confirms the thesis of Bagdasaryan *et al.* [3]. According to their results, we notice a performance drop inversely proportional to the privacy budget ε. This behavior is reasonable concerning that higher differential privacy levels correspond to higher levels of noise injection during the training. However, in the most private scenario with $\varepsilon = 2$, the target model left from 5% to 14.6% of accuracy, while in the other differentially-private cases, the performance loss is slightly lower but still consistent. Concerning the regularization techniques, the performance is almost identical to the baseline for l2 regularization, while it is ever improved with the dropout application, especially over CIFAR-10. Finally, the combination of l2 and dropout brings to mildly pejorative performance. We argue that this behavior is due to the probable excess of regularization.

Table 2. The training execution times of the target model, expressed in seconds (lower is better).

Dataset	Baseline	$\varepsilon = 2$	$\varepsilon = 4$	$\varepsilon = 8$	L2	Dropout	L2+Dropout
CIFAR-10	73.3	3029.8	2923.1	2897.1	69.1	**67.4**	70.4
MNIST	39.5	1886.2	1870.2	1859.8	31.2	**29.5**	29.6
F-MNIST	**77.4**	3876.7	3952.1	3967.1	78.2	83.0	78.5

Table 3. The AUC scores of the membership inference attack against the proposed models over the test set (lower is better).

Dataset	Baseline	$\varepsilon = 2$	$\varepsilon = 4$	$\varepsilon = 8$	L2	Dropout	L2+Dropout
CIFAR-10	0.689	**0.527**	0.545	0.536	0.619	0.571	0.551
MNIST	0.738	0.585	0.598	0.583	0.567	0.614	**0.555**
F-MNIST	0.642	**0.541**	0.564	0.549	0.564	0.552	0.575

Table 2 shows the time required by each target model configuration to compute the same number of epochs on the proposed datasets. In this dimension, all the models trained via DP-SGD completed their training in a largely greater time amount than the one required by the other configurations. This behavior, which for $\varepsilon = 2$ means an increment factor between 41 and 50, is related to the implementation of the DP-SGD. In fact, to apply the differential privacy definition and compute its budget, it is necessary to expand the dimension of the tensor fed in the network to include the micro-batch channel. The number of micro-batches, constrained between one and the number of mini-batches, represents an essential trade-off between time and accuracy: increasing the number of micro-batches slows the computation and increases the performance, and vice-versa. Our configuration aimed to maximize the performance, which is already translated into a noticeable accuracy drop. Target models with regularizers behave instead very similarly with respect to the baseline.

Table 3 show the results of the membership inference attack against the proposed target models, expressed via the AUC metric. As a first observation, all the models trained via DP-SGD achieved a score closer to the random guessing attack, meaning privacy preservation is effectively guaranteed. In detail, for $\varepsilon = 2$ there is an average AUC reduction for the attacker of 13.86%. The most interesting aspect concerns the results from the target models with regularization. In fact, all of them achieved better protection from the membership inference attack with respect to the baseline. L2 works better where the DP-SGD is less effective, reaching even better performance than all the ε over the MNIST dataset, and an average AUC reduction of 10.63%. Dropout also demonstrates competitive privacy guarantees, almost like DP-SGD ones, achieving an AUC average reduction of 11.07%. Combining the two regularization techniques improves the average score, reaching an AUC reduction of 12.93%. From these results, we argue that

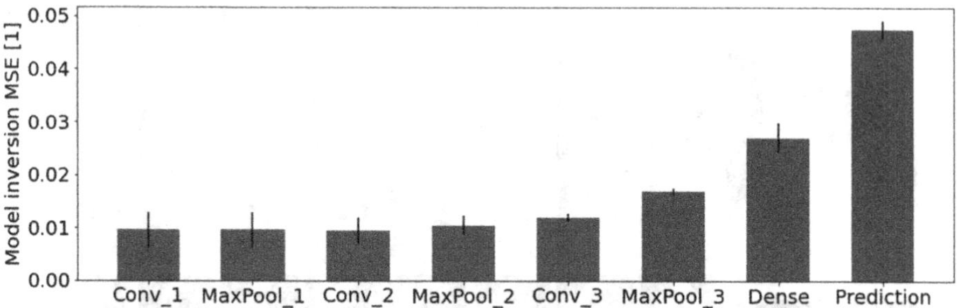

Fig. 4. The average reconstruction MSE against the baseline target model over the test set. The scores are averaged over all the datasets (lower is better).

Table 4. The variation of model inversion reconstruction MSE starting from each layer of the target model. The scores are expressed in percentages and are referred to the baseline model's score. The scores are averaged over all the datasets' test sets (higher is better).

Layer	$\varepsilon = 2$	$\varepsilon = 4$	$\varepsilon = 8$	L2	Dropout	L2+Dropout
Conv1	2.1	2.5	-1.4	-0.74	**22.0**	10.8
MaxPool1	3.1	0.6	1.7	-0.52	**19.0**	-0.2
Conv2	4.6	0.7	3.2	-2.99	**18.2**	1.6
MaxPool2	3.4	-0.5	-1.3	-3.28	**28.2**	4.2
Conv3	4.2	-9.3	-10.5	-7.13	**11.4**	2.7
MaxPool3	-14.6	-15.6	-17.2	-11.5	**8.6**	-3.6
Dense	-21.7	-25.8	-25.2	17.1	17.3	**21.3**
Prediction	-17.7	-18.2	-15.8	**73.8**	-1.8	64.2

regularization techniques empirically demonstrate a protection level comparable with the one obtained via differential privacy optimizers.

Figure 4 represents the results of the model inversion attacks against the baseline target model. Each bar corresponds to a layer and represents the reconstruction MSE averaged over all the datasets. This result shows that a model inversion attack is more effective as it is made closer to the network's input. It is reasonable given the growing amount of transformations that the attacker model would have to invert attacking from deeper layers. It is interesting to note how the best case for an attacker in a white-box setting would be around four times more effective than the same attack done in a black-box setting, i.e., from the prediction layer.

Table 4 shows the percentage variations for each layer and each approach with respect to the baseline. From these results, we notice an MSE reduction for all the final layers of the target models trained via DP-SGD. A reconstruction error reduction means an advantage for the attacker, which implies, at least in this scenario, that differential privacy tends to improve the model inversion attack's

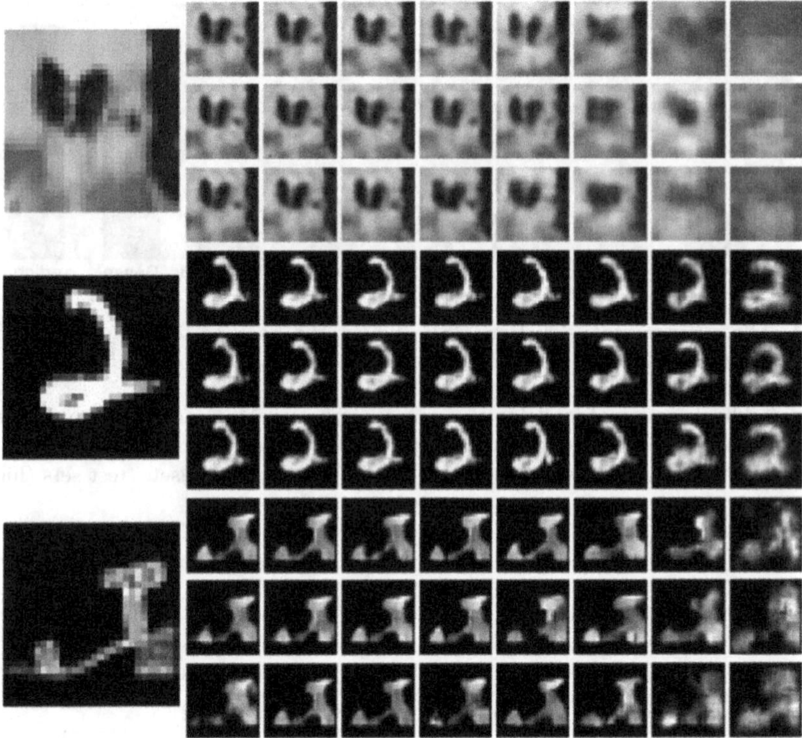

Fig. 5. Three examples of reconstructions from CIFAR-10 (top), MNIST (middle) and Fashion-MNIST (down). For each sample, the bigger image represents the ground truth to reconstruct. Each column j represents the model inversion reconstruction starting from the layer j of the target model (from left to right, Convolution, MaxPooling, Convolution, MaxPooling, Convolution, MaxPooling Flatten+Dense, Prediction). Each row corresponds to a different target model (from top to down, the baseline model, the ($\varepsilon = 2$)-DP model and the l2+dropout model).

quality rather than degrade it. Target models with regularization demonstrate completely different behavior. We notice that l2 reduces the effects of model inversion in the layer on which it is applied, with an average improvement of the reconstruction error of 73.8%. Dropout has a similar effect on the target model, improving the protection of the layers before its application. The main drawback is that it is impossible to apply dropout after the prediction layer. By putting together l2 and dropout techniques, the obtained result is an average between the two, slightly improving the protection for hidden layers and increasing the defence in the prediction layer.

Figure 5 shows three examples of a complete model inversion attack for the most significant target model configurations. The proposed images show how the attacker network degrades its reconstructions as it starts to reconstruct the image from deeper layers. In agreement with the numerical results, reconstructions from

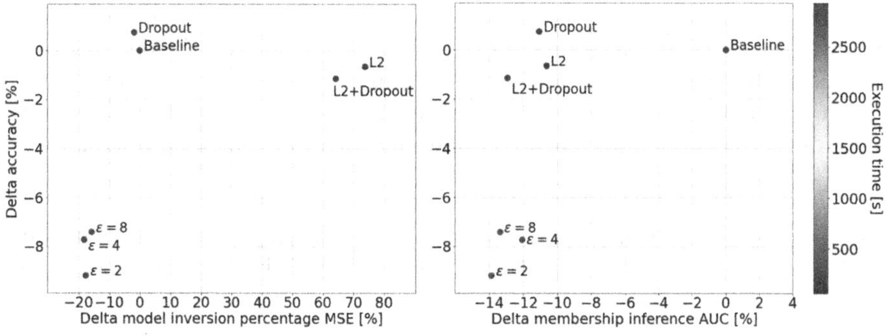

Fig. 6. The final comparison showing the performance of all the proposed models in a black-box scenario, averaged over all the datasets. On the left, the x-axis represents the percentage MSE variation for model inversion attacks with respect to the Baseline model (higher is better), the y-axis represents the classification accuracy (higher is better). On the right, the x-axis represents the percentage AUC variation for membership inference attacks with respect to the Baseline model (lower is better), the y-axis represents again the classification accuracy (higher is better).

DP-SGD target model tend to present a lower artifacts amount with respect to the ones from the baseline or the combination of l2 and dropout regularization.

Finally, Fig. 6 summarizes all the results obtained so far. The left image shows the percentage variations of the black-box model inversion MSE function of the accuracy variations. In the lower-left part are the worst results, in the upper-right the best. Instead, the right image shows the percentage variations of membership inference AUC function of the accuracy variations. In this case, the best region is the upper-left, while the worst is the lower-right. The scatter colors represent the clear difference between the training time with and without DP-SGD optimizer.

Despite the undeniable benefits against membership inference attacks and the privacy measurability property, we can conclude that differential privacy techniques are not always a suitable option. In fact, in our scenario, this mechanism demonstrated poor quality trade-offs between protection guarantees, utility and training time. Instead, the analyzed regularization techniques demonstrated promising performance. They are candidates for possible replacements in conditions where the training times and the required performances are subject to severe constraints.

6 Conclusion

This paper analyzed the benefits, drawbacks, and limitations of differentially-private optimization. In particular, we empirically showed how a differentially-private stochastic gradient descent optimizer could not always be considered a general protection paradigm for deep learning models despite its privacy budget guarantees. Despite its effectiveness in protecting them against membership inference attacks, the mechanism degrades models' utility, requires much longer training times and increases model inversion attacks' quality. We demonstrated how the combination of l2 and dropout regularization techniques is a valid protection alternative that does not degrade utility, requires contained training times, and provides a simultaneous resilience against membership inference and model inversion attacks.

Acknowledgment. The European Commission has partially funded this work under the H2020 grant N. 101016577 AI-SPRINT: AI in Secure Privacy-pReserving computINg conTinuum.

References

1. Abadi, M., et al.: Deep learning with differential privacy. In: Proceedings of the 2016 ACM SIGSAC Conference on Computer and Communications Security, pp. 308–318 (2016)
2. Al-Rubaie, M., Chang, J.M.: Privacy-preserving machine learning: threats and solutions. IEEE Secur. Priv. **17**(2), 49–58 (2019)
3. Bagdasaryan, E., Poursaeed, O., Shmatikov, V.: Differential privacy has disparate impact on model accuracy. In: Advances in Neural Information Processing Systems **32** (2019)
4. Dong, J., Roth, A., Su, W.J.: Gaussian differential privacy. arXiv preprint arXiv:1905.02383 (2019)
5. Dwork, C.: Differential Privacy: a survey of results. In: Agrawal, M., Du, D., Duan, Z., Li, A. (eds.) TAMC 2008. LNCS, vol. 4978, pp. 1–19. Springer, Heidelberg (2008). https://doi.org/10.1007/978-3-540-79228-4_1
6. Dwork, C., Rothblum, G.N.: Concentrated differential privacy. arXiv preprint arXiv:1603.01887 (2016)
7. Dwork, C., Smith, A., Steinke, T., Ullman, J., Vadhan, S.: Robust traceability from trace amounts. In: 2015 IEEE 56th Annual Symposium on Foundations of Computer Science, pp. 650–669. IEEE (2015)
8. Ermis, B., Cemgil, A.T.: Differentially private dropout. arXiv preprint arXiv:1712.01665 (2017)
9. Fredrikson, M., Jha, S., Ristenpart, T.: Model inversion attacks that exploit confidence information and basic countermeasures. In: Proceedings of the 22nd ACM SIGSAC Conference on Computer and Communications Security, pp. 1322–1333 (2015)
10. Fredrikson, M., Lantz, E., Jha, S., Lin, S., Page, D., Ristenpart, T.: Privacy in pharmacogenetics: An {End-to-End} case study of personalized warfarin dosing. In: 23rd USENIX Security Symposium (USENIX Security 14), pp. 17–32 (2014)
11. Geiping, J., Bauermeister, H., Dröge, H., Moeller, M.: Inverting gradients-how easy is it to break privacy in federated learning? Adv. Neural Inf. Process. Syst. **33**, 16937–16947 (2020)
12. Goodfellow, I., et al.: Generative adversarial nets. In: Advances in Neural Information Processing Systems **27** (2014)
13. Hu, H., Salcic, Z., Sun, L., Dobbie, G., Yu, P.S., Zhang, X.: Membership inference attacks on machine learning: a survey. In: ACM Computing Surveys (CSUR) (2021)

14. Hui, B., Yang, Y., Yuan, H., Burlina, P., Gong, N.Z., Cao, Y.: Practical blind membership inference attack via differential comparisons. arXiv preprint arXiv:2101.01341 (2021)
15. Jain, P., Kulkarni, V., Thakurta, A., Williams, O.: To drop or not to drop: Robustness, consistency and differential privacy properties of dropout. arXiv preprint arXiv:1503.02031 (2015)
16. Jordon, J., Yoon, J., Van Der Schaar, M.: PATE-GAN: generating synthetic data with differential privacy guarantees. In: International Conference on Learning Representations (2018)
17. Lim, J.Q., Chan, C.S.: From gradient leakage to adversarial attacks in federated learning. In: 2021 IEEE International Conference on Image Processing (ICIP), pp. 3602–3606. IEEE (2021)
18. Lomurno, E., Di Perna, L., Cazzella, L., Samele, S., Matteucci, M.: A generative federated learning framework for differential privacy. arXiv preprint arXiv:2109.12062 (2021)
19. Mironov, I.: Rényi differential privacy. In: 2017 IEEE 30th Computer Security Foundations Symposium (CSF), pp. 263–275. IEEE (2017)
20. Mothukuri, V., Parizi, R.M., Pouriyeh, S., Huang, Y., Dehghantanha, A., Srivastava, G.: A survey on security and privacy of federated learning. Future Gener. Comput. Syst. **115**, 619–640 (2021)
21. Nasr, M., Shokri, R., Houmansadr, A.: Machine learning with membership privacy using adversarial regularization. In: Proceedings of the 2018 ACM SIGSAC Conference on Computer and Communications Security, pp. 634–646 (2018)
22. Salem, A., Zhang, Y., Humbert, M., Berrang, P., Fritz, M., Backes, M.: ML-leaks: model and data independent membership inference attacks and defenses on machine learning models. arXiv preprint arXiv:1806.01246 (2018)
23. Shokri, R., Stronati, M., Song, C., Shmatikov, V.: Membership inference attacks against machine learning models. In: 2017 IEEE Symposium on Security and Privacy (SP), pp. 3–18. IEEE (2017)
24. Song, L., Shokri, R., Mittal, P.: Membership inference attacks against adversarially robust deep learning models. In: 2019 IEEE Security and Privacy Workshops (SPW), pp. 50–56. IEEE (2019)
25. Srivastava, N., Hinton, G., Krizhevsky, A., Sutskever, I., Salakhutdinov, R.: Dropout: a simple way to prevent neural networks from overfitting. J. Mach. Learn. Res. **15**(1), 1929–1958 (2014)
26. Wei, K., et al.: Federated learning with differential privacy: algorithms and performance analysis. IEEE Trans. Inf. Forensics Secur. **15**, 3454–3469 (2020)
27. Yang, Z., Shao, B., Xuan, B., Chang, E.C., Zhang, F.: Defending model inversion and membership inference attacks via prediction purification. arXiv preprint arXiv:2005.03915 (2020)
28. Yang, Z., Zhang, J., Chang, E.C., Liang, Z.: Neural network inversion in adversarial setting via background knowledge alignment. In: Proceedings of the 2019 ACM SIGSAC Conference on Computer and Communications Security, pp. 225–240 (2019)
29. Yin, H., Mallya, A., Vahdat, A., Alvarez, J.M., Kautz, J., Molchanov, P.: See through gradients: image batch recovery via gradinversion. In: Proceedings of the IEEE/CVF Conference on Computer Vision and Pattern Recognition, pp. 16337–16346 (2021)

30. Zhang, Y., Jia, R., Pei, H., Wang, W., Li, B., Song, D.: The secret revealer: generative model-inversion attacks against deep neural networks. In: Proceedings of the IEEE/CVF Conference on Computer Vision and Pattern Recognition, pp. 253–261 (2020)
31. Zhao, X., Zhang, W., Xiao, X., Lim, B.: Exploiting explanations for model inversion attacks. In: Proceedings of the IEEE/CVF International Conference on Computer Vision, pp. 682–692 (2021)

MicroRacer: A Didactic Environment for Deep Reinforcement Learning

Andrea Asperti(✉) and Marco Del Brutto

Department of Informatics: Science and Engineering (DISI), University of Bologna, Bologna, Italy
andrea.aaperti@unibo.it

Abstract. MicroRacer is a simple, open source environment inspired by car racing especially meant for the didactics of Deep Reinforcement Learning. The complexity of the environment has been explicitly calibrated to allow users to experiment with many different methods, networks and hyperparameters settings without requiring sophisticated software or exceedingly long training times. Baseline agents for major learning algorithms such as DDPG, PPO, SAC, TD3 and DSAC are provided too, along with a preliminary comparison in terms of training time and performance.

1 Introduction

Deep Reinforcement Learning (DRL) is the new frontier of Reinforcement Learning [31,32,34], where Deep Neural Networks are used as function approximators to address the scalability issues of traditional RL techniques. This allows agents to make decisions from high-dimensional, unstructured state descriptions without requiring manual engineering of input data. The downside of this approach is that learning may require very long trainings, depending on the acquisition of a large number of unbiased observations of the environment; in addition, since observations are dynamically collected by agents, this leads to the well known exploitation vs. exploration problem. The need of long training times, combined with the difficulty of monitoring and debugging the evolution of agents, and the difficulty to understand and explain the reasons for possible failures of the learning process, makes DRL a much harder topic than other traditional Deep Learning tasks.

This is particularly problematic from a didactic point of view. Most existing environments are either too simple and not particularly stimulating, like most of the legacy problems of OpenAIGym [8] (cart-pendulum, downhill slope simulator, ...), or far too complex, requiring hours of training (even relatively trivial problems such as those in the Atari family [6,25] may take 12–24 hours of training on a standard laptop, or Colab [7]). Even if, at the end of training, you may observe an advantage of a given technique over another, it is difficult to grasp the pros and cons of the different algorithms, and forecast their behaviour in different scenarios. The long training times make tuning or ablation studies very hard

G. Nicosia et al. (Eds.): LOD 2022, LNCS 13810, pp. 239–252, 2023.
https://doi.org/10.1007/978-3-031-25599-1_18

and expensive. In addition, complex environments are often given in the form of a black-box that essentially prevents event-based monitoring of the evolution of the agent (e.g. observe the action of the agent in response to a given environment situation). Finally, sophisticated platforms like OpenAIgym already offer state-of-the-art implementations of many existing algorithms; understanding the code is complex and time-demanding, frequently obscured by several modularization layers (good to maintain but not to understand code); as a consequence students are not really induced to put their hands on the code and try personal solutions.

For all these reasons, we created a simple environment explicitly meant for the didactics of DRL. The environment is inspired by car racing, and has a stimulating competitive nature. Its complexity has been explicitly calibrated to allow students to experiment with many different methods, networks and hyper-parameters settings without requiring sophisticated software or exceedingly long training times. Differently from most existing racing simulation frameworks that struggle in capturing realism, like Torcs, AWS Deep Racer or Learn-to-race (see Sect. 2 for a comparison) we do not care for this aspect: one of the important points of the discipline is the distinction between model-free vs. model-based approaches, and we are mostly interested in the former class. From this respect, it is important to communicate to students that model-free RL techniques are supposed to allow interaction with any environment, evolving according to unknown, unexpected and possibly unrealistic dynamics to be discovered by acquiring experience. In the case of MicroRacer, the complexity of the environment can be tuned in several different ways, changing the difficulty of tracks, adding obstacles or chicanes, modifying the acceleration or the timestep. Another important point differentiating MicroRacer from other car-racing environments is that the track is randomly generated at each episode, and unknown to the agent, preventing any form of adaptation to a given scenario (so typical of many autonomous driving competitions). In addition to the environment, we provide simple baseline implementations of several DRL algorithms, comprising DDPG [22], TD3 [16], PPO [28], SAC [19] and DSAC [13].

The environment was proposed to students of the course of Machine Learning at the University of Bologna during the past academic year as a possible project for their examination, and many students accepted the challenge obtaining interesting results and providing valuable feedback. We plan to organize a championship for the incoming year.

The code is open source, and it is available at the following github repository: https://github.com/asperti/MicroRacer. Collaboration with other universities and research groups is more than welcome.

1.1 Structure of the Article

We start with a quick review of related applications (Sect. 2), followed by an introduction to the MicroRacer environment (Sect. 3). The baseline learning models currently integrated into the systems are discussed in Sect. 4; their comparative training costs and performances are evaluated in Sect. 5. Concluding remarks and plans for future research and collaborations are given in Sect. 6.

2 Related Software

We arrived to the decision of writing a new application as a consequence of our dissatisfaction, for the didactic of Reinforcement Learning, of all environments we tested. Several thesis developed under the supervision of the first author [17,33] have been devoted to study the suitability of these environments for didactic purposes, essentially leading to negative conclusions. Here, we briefly review some of these applications, closer to the spirit of MicroRacer. Many more systems exists, such as [12,18,26], but they have a strong robotic commitment and a sym2real emphasis that is distracting from the actual topic of DRL, and quite demanding in terms of computational resources.

2.1 AWS Deep Racer

AWS Deep Racer[1] [3] is a cloud based 3D racing simulator developed by Amazon. It emulates a fully autonomous 1/18th scale race car; a global racing league is organized each year. Amazon only provides utilities to train agents remotely, and with very limited configurability: essentially, the user is only able to tune the system of rewards, that gives a wrong didactic message: manipulating rewards is a bad and easily biased way of teaching a behaviour. Moreover, at the time it was tested, the AWS DeepRacer console only supported the proximal policy optimization (PPO) algorithm [28]; the most recent release should also support Soft Actor Critic [19].

Due to this limitations, a huge effort has been done by the aws-community to pull together the different components required for DeepRacer local training (see e.g. https://github.com/aws-deepracer-community/deepracer-core.

The primary components of DeepRacer are four docker containers:

- Robomaker Container: Responsible for the robotics environment. Based on ROS + Gazebo as well as the AWS provided "Bundle";
- Sagemaker Container: Responsible for training the neural network;
- Reinforcement Learning (RL) Coach: Responsible for preparing and starting the Sagemaker environment;
- Log-Analysis: Providing a containerized Jupyter Notebook for analyzing the logfiles generated.

The resulting platform is extremely complex, computationally demanding, difficult to install and to use. See [33] for a deeper discussion of the limitations of this environment for the didactics of Reinforcement Learning.

2.2 Torcs

TORCS[2] is a portable, multi platform car racing simulation environment, originally conceived by E.Espié and C.Guionneau. It can be used as an ordinary car

[1] https://aws.amazon.com/it/deepracer/.
[2] https://sourceforge.net/projects/torcs/.

racing game, or a platform for AI research [9,23]. It runs on Linux, FreeBSD, OpenSolaris and Windows. The source code of TORCS is open source, licensed under GPL. While supporting a sophisticated and realistic physical model, it provides a sensibly simpler platform than AWS DeepRacer, and it is a definitely better choice. It does not support random generation of tracks, but many tracks, opponents and cars are available.

A gym-compliant python interface to Torcs was recently implemented in [17], under the supervision of the first author. While this environment can be a valuable testbench for experts of Deep Reinforcement Learning, its complexity and especially the difficulty of training agents is an insurmountable obstacle for neophytes.

2.3 Learn-to-race

Learn-to-Race[3] [10,20] is a recent Gym-compliant open-source framework based on a high-fidelity racing simulator developed by Arrival, able to capture complex vehicle dynamics and to render 3D photorealistic views.

Learn-to-Race provides customizable, multi-model sensory inputs giving information about the state of the vehicle (pose, speed, etc.), and comprising RGB image views with semantic segmentations. A challenge based on Learn-to-Race is organized by AICrowd (similarly to AWS): https://www.aicrowd.com/challenges/learn-to-race-autonomous-racing-virtual-challenge.

Learn-to-race is very similar, in its intents and functionalities, to Torcs (especially to the gym-compliant python interface developed in [17]). It also shares with TORCS most of the defects: learning the environment and training an agent requires a commitment far beyond the credits associated with a typical course in DRL; it can possibly be a subject for a thesis, but cannot be used as a didactic tool. Moreover, the complexity of the environment and its fancy (but onerous) observations are distracting students from the actual content of the discipline.

2.4 CarRacing-v0

This is a racing environment available in OpenAI gym. The state consists of a 96×96 pixels top-down view of the track. The action is composed of three continuous values: steering, acceleration and braking. Reward is -0.1 every frame and +1000/N for every track tile visited, where N is the total number of tiles in track. Episode finishes when all tiles are visited. The track is randomly generated at each episode. A few additional indicators at the bottom of the window provide additional information about the car: speed, four ABS sensors, steering wheel position, gyroscope. The game is considered solved when an agent consistently get 900 or more points per episode. As observed in [21], the problem is quite challenging due to the peculiar notion of state, that requires learning from pixels: this shifts the focus of the problem from the learning task to the

[3] https://learn-to-race.org/.

elaboration of the observation, adding a pointless and onerous burden. In addition, while it is a good practice to stick to a gym-compliant interface for the interaction between the agent and the environment, for the didactic reasons already explained in the introduction, we prefer to avoid a direct and extensive use of OpenAI gym libraries (while we definitely encourage students to use these libraries as a valuable source of documentation).

3 MicroRacer

MicroRacer generates new random circular tracks at each episode. The Random track is defined by CubicSplines delimiting the inner and outer border; the number of turns and the width of the track are configurable. From this description, we derive a dense matrix of points of dimension 1300×1300 providing information about positions inside the track. This is the actual definition of the track used by the environment. The basic track can be further complicated by optionally adding obstacles (similar to cars stopped along the track) and "chicanes". More details about the environment can be found in [11].

Table 1. (left) Random track generated with splines; (right) derived boolean map. The dynamic of the game is entirely based on the map. The map is unknown to agents, that merely have agent-centric sensor-based observations: speed and lidar-like view.

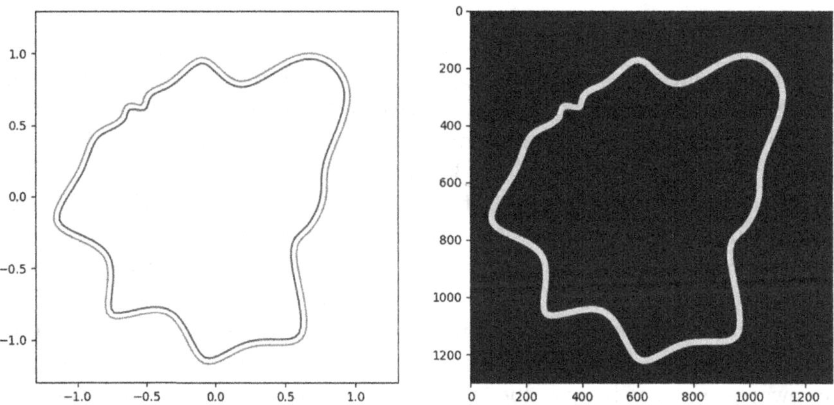

3.1 State and Actions

MicroRacer does not intend to model realistic car dynamics. The model is explicitly meant to be as simple as possible, with the minimal amount of complexity that still makes learning interesting and challenging. The maximum car acceleration, both linear and angular, are configurable. The angular acceleration is used to constraint the maximum admissible steering angle in terms of the car speed, forbidding the car to go too fast.

The state information available to actors is composed by:

- a lidar-like vision of the track from the car's frontal perspective. This is an array of 19 values, expressing the distance of the car from the track's borders along uniformly spaced angles in the range –30, +30.
- the car scalar velocity.

The actor (the car) has no global knowledge of the track, and no information about its absolute or relative position/direction w.r.t. the track[4].

The actor is supposed to answer with two actions, both in the range [–1,1]:

- acceleration/deceleration
- turning angle.

Maximum values for acceleration and turning angles can be configured. In addition, a simple law depending on a tolerated angular acceleration (configurable) limit the turning angle at high speeds. This is not meant to achieve a realistic behaviour, but merely to force agents to learn to accelerate and decelerate according to the configuration.

The lidar signal is computed by a simple iterative function written in cython [4] for the sake of efficiency.

3.2 Rewards

Differently from other software applications for autonomous driving, shaping rewards from a wide range of data relative to the distance of the car from borders, deviation from the midline, and so on, [3,14,15] MicroRacer induces the use of a simple, almost intrinsic [30], rewarding mechanism. Since the objective is to run as fast as possible, it is natural to use speed as the only reward. The cumulative reward is thus the integral of speed, namely the expected (discounted) total distance covered by the car. A negative reward is given in case of termination with failure (too slow, or out of borders). Users are free to shape different rewarding mechanisms, but the limited state information is explicitly meant to discourage this pursuit. It is important for students to realize that ad-hoc rewards may easily introduce biases in the learning process, inducing agents to behave according to possibly sub-optimal strategies.

3.3 Environment Interface

To use the environment, it is necessary to instantiate the Racer class in tracks.py. On initialization, it is possible to turn off obstacles, chicanes, turn and low-speed constraints. The Racer class has two main methods, implementing a OpenAI compliant interface with the environment:

---

[4] Our actors exploit a simplified *observation* of the state discussed in Sect. 5.

reset() − > state
: this method generates a new track and resets the racer position at the starting point. It returns the initial state.

step(action) − > state, reward, done
: this method takes an action composed by [acceleration, turn] and lets the racer perform a step in the environment according to the action. It returns the new state, the reward for the action taken and a boolean done that is true if the episode has ended.

3.4 Competitive Race

In order to graphically visualize a run it is necessary to use the function:

```
newrun(actors, obstacles=True, turn_limit=True,
        chicanes=True, low_speed_termination=True)
```

defined in tracks.py. It takes as input a list of actors (Keras models), simulating a race between them. At present, the different agents are not supposed to interfere with each other: each car is running separately and we merely superpose their trajectories.

3.5 Dependencies

The project just requires basic libraries: tensorflow, matplotlib, scipy.interpolate (for Cubic Splines) numpy, and cython. A `requirements` file is available so you can easily install all the dependencies just using the following command"pip install -r requirements.txt".

4 Learning Models

In this section, we list the learning algorithms for which a base code is currently provided, namely DDPG, TD3, PPO, SAC and DSAC. The code is meant to offer to students a starting point for further development, extending the code and implementing variants. All implementations take advantage of *target networks* [24] to stabilize training.

4.1 Deep Deterministic Policy Gradient (DDPG)

DDPG [22,29] is an off-policy algorithm that extends deep Q-learning to continuous action spaces, jointly learning a Q-function and a policy. It uses off-policy data and the Bellman equation to learn the Q-function, and uses the Q-function to learn the policy. The optimal action-value function $Q^*(s,a)$, and the optimal policy $\pi^*(s)$ should satisfy the equation

$$Q^*(s,a) = \mathbb{E}_{s \sim P}\, r(s,a) + \gamma Q(s', \pi^*(s'))$$

that allows direct training of the Q-function from transitions (s, a, s', r, T), similarly to DQN [24]; in turn, the optimal policy is trained by maximizing, over all possible states, the expected reward

$$Q(s, \pi^*(s))$$

4.2 Twin Delayed DDPG (TD3)

This is a variant of DDPG meant to overcome some shortcomings of this algorithm mostly related to a possible over-estimation of the Q-function [16]. Specifically, TD3 exploits the following tricks:

1. Clipped Double-Q Learning. Similarly to double Q-learning, two "twin" Q-functions are learned in parallel, and the smaller of the two Q-values is used in the r.h.s. of the Bellman equation for computing gradients;
2. "Delayed" Policy Updates. The policy (and its target network) is updated less frequently than the Q-function;
3. Target Policy Smoothing. Noise is added to the target action inside the Bellman equation, essentially smoothing out Q with respect to changes in action.

4.3 Proximal Policy Optimization (PPO)

A typical problem of policy-gradient techniques is that they are very sensitive to training settings: since long trajectories are into account, modifications to the policy are amplified, possibly leading to very different behaviours and numerical instabilities. Proximal Policy Optimization (PPO) [28] simply relies on ad-hoc clipping in the objective function to ensure that the deviation from the previous policy is relatively small.

4.4 Soft Actor-Critic (SAC)

Basically, this is a variant of DDPG and TD3, incorporating ideas of Entropy-regularized Reinforcement Learning [19]. The policy is trained to maximize a trade-off between expected return and entropy, a measure of randomness of the policy. Entropy is related to the exploration-exploitation trade-off: increasing entropy results in more exploration, that may prevent the policy from prematurely converging to a bad local optimum; in addition, it add a noise component to the policy producing an effect similar to Target Policy Smoothing of T3D. It also exploits the clipped double-Q trick, to prevent fast deviations.

4.5 DSAC

Distributional Soft Actor-Critic (DSAC) is an off-policy actor-critic algorithm developed by Jingliang et al. [13] that is essentially a variant of SAC where the clipped double-Q learning is substituted by a distributional action-value function [5]. The idea is that learning a distribution, instead of a single value, can help to mitigate Q-function overestimation. Furthermore, DSAC uses a single network for the action-value estimation, with a gain in efficiency.

5 Baselines Benchmarks

In this section we compare our baselines implementations in the case of an environment with a time step of 0.04 ms, and comprising obstacles, chicanes, low speed termination and turn limitations.

The different learning models are those mentioned in Sect. 4. In the case of DDPG we also considered a variant, called DDPG2 making use of parameter space noise [27] for the actor's weights. This noise is meant to improve exploration and it can be used as a surrogate for action noise.

All models work with a simplified *observation* of the environment state, where the full lidar signal is replaced by 4 values: the angle (relative to the car) of the lidar max distance, the value of this distance and the values of the distances for the two adjacent positions. In mathematical terms, if ℓ is the vector of lidar signals, $m = argmax(\ell)$ and $\alpha_m = angle(m)$ is the corresponding direction, the observation is composed by

$$\alpha_m, \ell(m-1), \ell(m), \ell(m+1)$$

The DDPG actor's neural network makes use of two towers. One of them calculates the direction, while the other calculates the acceleration. Each of them is composed of two hidden layers of 32 units, with relu activation. The output layer uses a tanh activation for each action. At the same time, the critic network uses two layers, one of 16 units and one of 32, for the state input and one layer of 32 units for the action input. The outputs of these layers are then concatenated and go through another two hidden layers composed of 64 units. All of them make use of relu activation.

In DDPG2, the actor has two hidden layers with 64 units and relu activation and one output with tanh activation . Meanwhile, the critic is the same as in DDPG.

In TD3, the actor is the same as DDPG2. The critic has two hidden layer with 64 units and relu activation.

In SAC, the actor has two hidden layer with 64 units each and relu activation and output a μ and a σ of a normal distribution for each action. The critic is equal to TD3.

In DSAC, the actor is the same as SAC. The critic has the same structure as the actor.

In PPO both the actor and the critic have two hidden layers of 64 units with tanh activation, but the actor has also tanh activation on the output layer.

All learning methods have been trained with a discount factor $\gamma = 0.99$, using Adam as optimizer. All methods except PPO share the following hyperparameters:

- Actor and Critic Learning Rate 0.001
- Buffer Size 50000
- Batch Size 64
- Target Update Rate τ 0.005

Additional methods-specific hyperparameters are listed in Table 2.

Table 2. Hyperparameters used in the various methods.

Hyperparameter	Value	Hyperparameter	Value
TD3, DDPG		*DDPG2*	
Exploration Noise	$\mathcal{N}(0, 0.1)$	Parameter Noise Std Dev	0.2
TD3		*PPO*	
Target Update Frequency	2	Actor/Critic Learning Rate	0.0003
Target Noise Clip	0.5	Mini-batch Size	64
SAC, DSAC		Epochs	10
Target Entropy	$-A$	GAE lambda	0.95
DSAC		Policy clip	0.25
Target Update Frequency	2	Target entropy	0.01
Minimum critic sigma	1	Target KL	0.01
Critic difference boundary	10		

5.1 Results

Training times have been computed as an average of ten different trainings, each one consisting of 50000 training steps. In the case of PPO, that unlike all the other methods, starts collecting a complete trajectory before executing a training step on it, we trained the agent for a fixed number of episodes (600).

The training times collected are relative to the execution on two different machines: a laptop equipped with an NVIDIA GeForce GTX 1060 GPU, Intel Core i7-8750H CPU and 16GB 2400MHz RAM, and a workstation equipped with an Asus GeForceDUALGTX1060-O6G GPU and a Intel Core i7-7700K CPU and 64GB 2400 MHz MHz RAM..

As can be observed in Table 3, the methods that train an higher number of Neural Networks require higher training times.

Table 3. Average training time (5 runs) required to perform 50000 training iterations (600 episodes for PPO) for each different method. Times are relative to two different machines: M1 is a laptop equipped with an NVIDIA GeForce GTX 1060 GPU, Intel Core i7-8750H CPU and 16GB 2400MHz RAM, M2 is a workstation equipped with an Asus GeForceDUALGTX1060-O6G GPU and a Intel Core i7-7700K CPU and 64GB 2400 MHz MHz RAM.

Machine	DDPG	DDPG2	TD3	SAC	DSAC	PPO
M1	30 m	44 m	38 m	19 m	24 m	27 m
M2	14 m	24 m	23 m	11 m	12 m	20 m

As can be seen in Fig. 1, the training process has large fluctuations, also due to frequent occurrences of catastrophic forgetting (more on it below). TD3

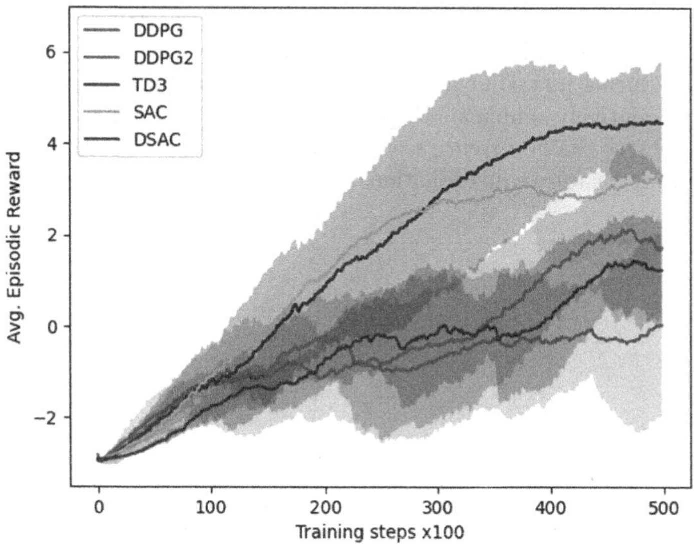

Fig. 1. Training curves of all methods except PPO. The solid lines correspond to the mean and the shaded regions correspond to 95% confidence interval over 10 trainings.

and SAC are the most stable methods, usually requiring less observations and training steps to improve. The other methods learns at a slower pace and seem to be more prone to catastrophic forgetting. However, they are occasionally able to produce reasonably performant agents.

After each training, 100 evaluation episodes has been run to collect the real performance of each trained agents. The average of these results over 10 different trainings and the best results obtained for each method can be seen in Table 4. As it can be noticed, a higher number of completed episodes usually corresponds to slower speeds. This may indicate difficulties in the process of learning the right acceleration action. Similarly to the training curves, TD3 and SAC seem to have the best performances even in evaluation, as expected.

Table 4. Average and maximum of 100 evaluation episodes executed after each training over 5 trainings of 50000 iterations (600 episodes for PPO).

Method	DDPG	DDPG2	TD3	SAC	DSAC	PPO
Average completed episodes	38	18	54	69	37	37
Average episodic reward	2.48	0.80	3.52	4.61	2.84	2.05
Average speed	0.34	0.30	0.26	0.29	0.34	0.23
Max completed episodes	90	39	80	79	75	62

In Fig. 2 we show a few examples of catastrophic forgetting, that is the tendency of a learning model to completely and abruptly forget previously learned information during its training. The phenomenon is still largely misunderstood, so having a relatively simple and highly configurable environment where we can frequently observe its occurrence seems to provide a very interesting and promising framework for future investigations.

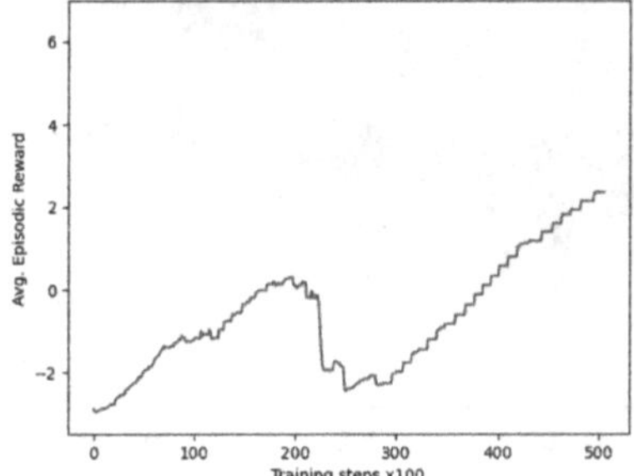

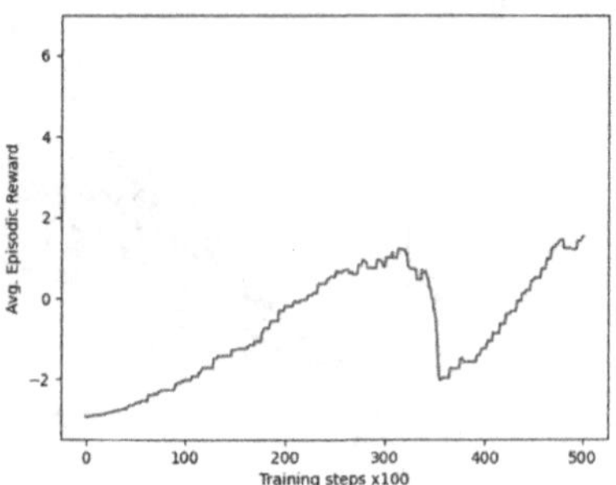

Fig. 2. Examples of catastrophic forgetting during training of DDPG (left) and DSAC (right).

6 Conclusions

In this article, we introduced the MicroRacer environment, offering a simple educational platform for the didactic of Reinforcement Learning. Similarly to our previous environment based on the old and prestigious Rogue game [1,2], we try to spare the useless burden of relying on two-dimensional state observations requiring expensive image-preprocessing, using instead more direct and synthetic state information. Moreover, differently from Rogue, that was based on a discrete action-space, MicroRacer is meant to investigate RL-algorithms with continuous actions.

On the contrary of most existing car-racing systems, MicroRacer does not make any attempt to implement realistic dynamics: autonomous driving is just a simple pretext to create a pleasant and competitive setting. This drastic simplification allows us to obtain an environment that, although far from trivial, still has acceptable training times (between 10 and 60 min depending on the learning methods and the underlying machine).

The environment was already experimented by students of the course of Machine Learning at the University of Bologna during the past academic year, that provided valuable feedback. In view of the welcome reception, we plan to organize a championship for the incoming year. The code is open source, and it is available at the following github repository: https://github.com/asperti/MicroRacer. We look forward for possible collaborations with other Universities and research institutions.

References

1. Asperti, A., Cortesi, D., De Pieri, C., Pedrini, G., Sovrano, F.: Crawling in rogue's dungeons with deep reinforcement techniques. IEEE Trans. Games **12**(2), 177–186 (2020)
2. Asperti, A., Cortesi, D., Sovrano, F.: Crawling in rogue's dungeons with (partitioned) A3C. In: Machine Learning, Optimization, and Data Science - 4th International Conference, LOD 2018, Volterra, Italy, 13–16 Sep 2018, Revised Selected Papers, vol. 11331 of Lecture Notes in Computer Science, pp. 264–275. Springer (2018). https://doi.org/10.1007/978-3-030-13709-0_22
3. Balaji, B., et al. DeepRacer: educational autonomous racing platform for experimentation with sim2real reinforcement learning. arXiv preprint arXiv:abs/1911.01562 (2019)
4. Behnel, S., Bradshaw, R., Citro, C., Dalcin, L., Seljebotn, D.S., Smith, K.: Cython: the best of both worlds. Comput. Sci. Eng. **13**(2), 31–39 (2011)
5. Bellemare, M.G., Dabney, W., Rowland, M.: Distributional Reinforcement Learning. MIT Press, Cambridge (2022). https://www.distributional-rl.org
6. Bellemare, M.G., Naddaf, Y., Veness, J., Bowling, M.: The arcade learning environment: an evaluation platform for general agents. J. Artif. Intell. Res. (JAIR) **47**, 253–279 (2013)
7. Bisong, E.: Google Colaboratory, pp. 59–64. Apress, Berkeley (2019)
8. Brockman, G., et al.: Openai gym. arXiv preprint arXiv:abs/1606.01540 (2016)
9. Cardamone, L., Loiacono, D., Lanzi, P.L., Bardelli, A.P.: Searching for the optimal racing line using genetic algorithms. In: Proceedings of the 2010 IEEE Conference on Computational Intelligence and Games, pp. 388–394 (2010)
10. Chen, B., Francis, J., Oh, J., Nyberg, E., Herbert, S.L.: Safe autonomous racing via approximate reachability on ego-vision (2021)
11. Brutto, M.D.: MicroRacer: development of a didactic environment for deep reinforcement learning. Master's thesis, University of Bologna, School of Science, Session III 2021–22
12. Dosovitskiy, A., Ros, G., Codevilla, F., López, A.M., Koltun, V.: CARLA: an open urban driving simulator. In: 1st Annual Conference on Robot Learning, CoRL 2017, Mountain View, California, USA, 13–15 Nov 2017, Proceedings, pp. 1–16. PMLR (2017)
13. Duan, J., Guan, Y., Li, S.E., Ren, Y., Sun, Q., Cheng, B.: Distributional soft actor-critic: off-policy reinforcement learning for addressing value estimation errors. In: IEEE Transactions on Neural Networks and Learning Systems, pp. 1–15 (2021)
14. Evans, B., Engelbrecht, H.A., Jordaan, H.W.: Learning the subsystem of local planning for autonomous racing. In: 20th International Conference on Advanced Robotics, ICAR 2021, Ljubljana, Slovenia, Dec 6–10 2021, pp. 601–606. IEEE (2021)
15. Evans, B., Engelbrecht, H.A., Jordaan, H.W.: Reward signal design for autonomous racing. In: 20th International Conference on Advanced Robotics, ICAR 2021, Ljubljana, Slovenia, 6–10 Dec 2021, pp. 455–460. IEEE (2021)
16. Fujimoto, S., Hoof, H.V., Meger, D.: Addressing function approximation error in actor-critic methods. In: Jennifer, G., Dy., Krause, A. (eds.) Proceedings of the 35th International Conference on Machine Learning, ICML 2018, Stockholmsmässan, Stockholm, Sweden, 10–15 Jul 2018, vol. 80 of Proceedings of Machine Learning Research, pp. 1582–1591. PMLR (2018)

17. Galletti, G.: Deep reinforcement learning nell'ambiente pytorcs. Master's thesis, University of Bologna, school of Science, Session III 2021
18. Goldfain, B., et al.: AutoRally: an open platform for aggressive autonomous driving. IEEE Control Syst. Mag. **39**(1), 26–55 (2019)
19. Haarnoja, T., Zhou, A., Abbeel, P., Levine, S.: Soft actor-critic: off-policy maximum entropy deep reinforcement learning with a stochastic actor. In: Proceedings of the 35th International Conference on Machine Learning, ICML 2018, Stockholmsmässan, Stockholm, Sweden, 10–15 Jul 2018, vol. 80 of Proceedings of Machine Learning Research, pp. 1856–1865. PMLR (2018)
20. Herman, J., et al.: Learn-to-Race: a multimodal control environment for autonomous racing. In: Proceedings of the IEEE/CVF International Conference on Computer Vision, pp. 9793–9802 (2021)
21. Li, C.: Challenging on car racing problem from openai gym. arXiv preprint arXiv:abs/1911.04868 (2019)
22. Lillicrap, T.P., et al.: Continuous control with deep reinforcement learning. In: Bengio, Y., LeCun, Y. (eds.) 4th International Conference on Learning Representations, ICLR 2016, San Juan, Puerto Rico, 2–4 May 2016, Conference Track Proceedings (2016)
23. Loiacono, D., et al.: The 2009 simulated car racing championship. IEEE Trans. Comput. Intell. AI Games **2**(2), 131–147 (2010)
24. Mnih, V., et al.: Playing atari with deep reinforcement learning. arXiv preprint arXiv:abs/1312.5602 (2013)
25. Mnih, V., et al.: Human-level control through deep reinforcement learning. Nature **518**(7540), 529–533 (2015)
26. Paull, L., et al.: Duckietown: an open, inexpensive and flexible platform for autonomy education and research. In: 2017 IEEE International Conference on Robotics and Automation, ICRA 2017, Singapore, Singapore, May 29 - June 3, 2017, pp. 1497–1504 (2017)
27. Plappert, M., et al.: Parameter space noise for exploration. In: 6th International Conference on Learning Representations, ICLR 2018, Vancouver, BC, Canada, April 30 - May 3, 2018, Conference Track Proceedings. OpenReview.net (2018)
28. Schulman, J., Wolski, F., Dhariwal, P., Radford, A., Klimov, O.: Proximal policy optimization algorithms. arXiv preprint arXiv:abs/1707.06347 (2017)
29. Silver, D., Lever, G., Heess, N., Degris, T., Wierstra, D., Riedmiller, M.A.: Deterministic policy gradient algorithms. In: Proceedings of the 31th International Conference on Machine Learning, ICML 2014, Beijing, China, 21–26 June 2014, vol. 32 of JMLR Workshop and Conference Proceedings, pp. 387–395. JMLR.org (2014)
30. Singh, S.P., Barto, A.G., Chentanez, N.: Intrinsically motivated reinforcement learning. In: Advances in Neural Information Processing Systems 17 [Neural Information Processing Systems, NIPS 2004, December 13–18, 2004, Vancouver, British Columbia, Canada], pp. 1281–1288 (2004)
31. Richard, S.: Sutton and Andrew G, 1st edn. Barto. Introduction to Reinforcement Learning. MIT Press, Cambridge, MA, USA (1998)
32. Vamvoudakis, K.G., Wan, Y., Lewis, F.L., Cansever, D. (eds.): Handbook of Reinforcement Learning and Control. SSDC, vol. 325. Springer, Cham (2021). https://doi.org/10.1007/978-3-030-60990-0
33. Vorabbi, S.: Analisi dell'ambiente aws deepracer per la sperimentazione di tecniche di reinforcement learning. Master's thesis, University of Bologna, school of Science, Session II 2021
34. Wang, H., et al.: Deep reinforcement learning: a survey. Frontiers Inf. Technol. Electron. Eng. **21**(12), 1726–1744 (2020)

A Practical Approach for Vehicle Speed Estimation in Smart Cities

Silvio Barra[1], Salvatore Carta[2], Antonello Meloni[2], Alessandro Sebastian Podda[2](✉), and Diego Reforgiato Recupero[2]

[1] Department of Electric Engineering and Information Technology, University of Naples Federico II, Naples, Italy
silvio.barra@unina.it

[2] Department of Mathematics and Computer Science, University of Cagliari, Cagliari, Italy
{salvatore,antonello.meloni,sebastianpodda,diego.reforgiato}@unica.it

Abstract. The last few decades have witnessed the increasing deployment of digital technologies in the urban environment with the goal of creating improved services to citizens especially related to their safety. This motivation, enabled by the widespread evolution of cutting edge technologies within the Artificial Intelligence, Internet of Things, and Computer Vision, has led to the creation of smart cities. One example of services that different cities are trying to provide to their citizens is represented by evolved video surveillance systems that are able to identify perpetrators of unlawful acts of vandalism against public property, or any other kind of illegal behaviour. Following this direction, in this paper, we present an approach that exploits existing video surveillance systems to detect and estimate vehicle speed. The system is currently being used by a municipality of Sardinia, an Italian region. An existing system leveraging Convolutional Neural Networks has been employed to tackle object detection and tracking tasks. An extensive experimental evaluation has been carried out on the Brno dataset and against state-of-the-art competitors showing excellent results of our approach in terms of flexibility and speed detection accuracy.

Keywords: Surveillance systems · Convolutional neural networks · Object detection · Smart cities

1 Introduction

The last few decades have witnessed the increasing deployment of digital technologies in the urban environment in order to improve services to citizens and ensure ever-greater security needs. Such a scenario, combined with the strong scientific and industrial expansion of disciplines such as Artificial Intelligence, Internet-of-Things, and Computer Vision, has led to the emergence of *smart cities* [10]. One of the most significant aspects of this evolution is the increasingly

G. Nicosia et al. (Eds.): LOD 2022, LNCS 13810, pp. 253–267, 2023.
https://doi.org/10.1007/978-3-031-25599-1_19

widespread adoption of video surveillance systems and cameras, whose main utility is considered to be that of acting as a deterrent and at the same time facilitating the identification of perpetrators of unlawful acts or vandalism against public property [5,14,21]. More recently, these infrastructures - integrated into more sophisticated software platforms - have gradually taken on more pervasive roles, from traffic monitoring and automatic access management, to crowd detection, to name a few [2,3,16]. In this context, this work aims to exploit existing city video surveillance systems to provide, in a simple and low-cost way, a solution based on Deep Learning techniques for the automatic detection and estimation of vehicle speed. Contrary to most of the scientific works in the literature, which address the problem through hand-crafted image processing methodologies, the proposed approach innovatively leverages object detection and tracking methods (based on *YOLOv3* and *Darknet*), with an extremely limited setup overhead and a fast *one-off* calibration phase. The system has been adopted by a municipality of Sardinia, an Italian region: more in detail, a couple of cameras in the main road roundabouts have been installed and continuously send videos to the main system installed within the police department. In light of the above, the main contributions of this paper are the following:

- an innovative approach for estimating the speed of vehicles to be applied to fixed-frame video surveillance cameras, which relies on object recognition and tracking methods based on Convolutional Neural Networks, and a detection dynamics built on low complexity heuristics;
- the design of an installation and calibration phase of the proposed method, for the application on new cameras, with reduced execution time;
- an extensive experimental validation of the proposed approach through the public *Brno dataset* and with an in-depth comparison against literature competitors, showing a significant performance of our method both in terms of flexibility and estimation accuracy.

The remainder of this paper is organized as follows. Section 2 explores the related works. Section 3 illustrates the proposed method. Section 4 presents the results obtained from the validation of our approach and comparisons against competitors. Finally, Sect. 5 ends the paper.

2 Related Works

In recent years, the speed estimation topic has been faced from several points of view, whether the speed is measured from the perspective of other vehicles, from the inside of the vehicle, or the outside (lampposts, traffic lights, and so on). Different sensors may be exploited for the topic and the state of the art offers different solutions for speed estimation, by involving several data sources: as an example, in [12], the authors have exploited a single multi-function magnetic sensor for estimating vehicle speed using three different methods, Vehicle Length based (VLB), Time Difference based (TDB) and Mean Value-based (MVB), so to obtain different reference speed. In [1], the authors have proposed VILDAR

(Visible Light Detection and Ranging), built upon sensing visible light variation of vehicle's headlamp, which has been proven to be able to outperform the accuracy of both RADAR (Radio Detection and Ranging) and LiDAR (Light Detection and Ranging) systems. More information about features, advantages, drawbacks and limitations of RADAR and LiDAR systems is provided in [6]. However, given the multitude of cameras that are currently used for video surveillance purposes [3], most of the approaches are deployed for working on camera streams. Using cameras as the data source for speed estimation is quite complicated, given the fact that some activities are needed before being able to detect the speed of a vehicle: in fact, the car first needs to be detected from the real scenario, then a tracking phase is necessary for understanding its direction and only at the end the speed can be estimated [2]. Also, the camera needs to be calibrated for such a goal [15], according to the context (single/multiple camera/s locations in the scenario) and its/their inner parameters (focal and sensor size). Vehicle speed estimation on camera can be achieved by following two different approaches: the first regards the estimation of the speed of the vehicles in a segment of the road, thus obtaining the average speed of the traffic in that specific section [8,11,20]. In [4] the authors have proposed a full tracking system involving vehicle detection, tracking, and speed estimation based on the displacement in the image of the vehicle concerning the actual position. The second approach is based on the analysis of the n consecutive (or non-consecutive) frames to obtain the pixel displacement of the vehicle in the video [9,13,24]. The operative difference between the two approaches is based on the moment in which the speed estimation module is triggered: while in the first case, the trigger comes from the scenario, i.e. the moment in which a car overcomes the analyzed segment as in [23], in the second case, the trigger is timed from the number of fps the video is taken with.

3 Materials and Methods

Our algorithm is fed with an input video of a road section that may contain a certain number of vehicles (i.e., cars, trucks, motorbikes, and so on). The video is split into frames and for each frame, we perform an object detection task to identify vehicles and follow them during their path. Any recognized object different from a vehicle is discarded. No pre-processing approaches have been applied neither to enhance the image nor for augmenting the local/global contrast of the image. The reason behind this choice is twofold: i) firstly, the method does not need any enhancement of the contrast augmentation approach because the speed detection is an analytical process in which image quality is a side problem; Yolo detection and tracking have not, in fact, been affected at all; ii) secondly, pre-processing stages would need a certain amount of time, which eventually could burden the overall speed detection computation time. Figure 1 shows the diagram of the proposed method. In the next paragraphs, we will give more details on our vehicle speed estimation system.

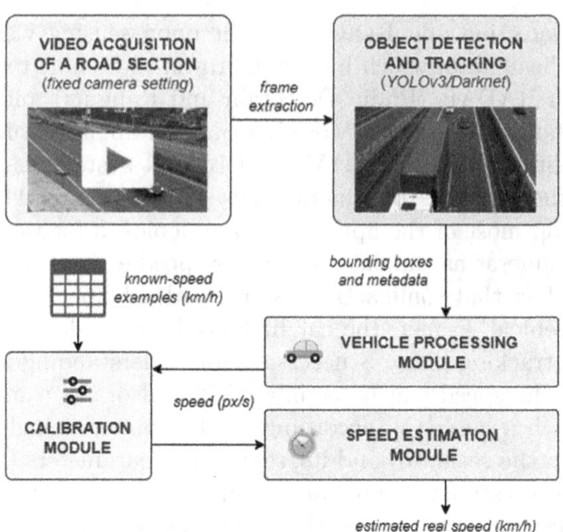

Fig. 1. From the video frames of a road section to the speed in km/h: frames are processed with $Yolo_{V3}$+*Darknet*, which returns the bounding boxes of the detected vehicles. They are used to follow the trajectory of vehicles and estimate their speed in $pixel/s$. Existing real speeds of some vehicles are used to calibrate the system. The *Speed Estimation module* applies the calculated transformation function and returns the speed in km/h.

3.1 Vehicle Speed Estimation System

Our system exploits *YOLO* (You Only Look Once - a real-time object detection system) [17] with *Darknet* [18] (an open-source neural network framework written in C and CUDA)[1] to locate vehicles in each frame. Differently from other object detection systems, like Fast R-CNN [7] and Faster R-CNN [19] which lie in the field of the Region proposal network, the YOLO family of object detectors, as the name suggests, achieves region analysis and object localization within an image or frame in just one stage. Indeed, the canonical RPN approaches divide the region proposal stage and the object detection stage in two different steps. During a detection session, the video frames are analyzed one by one. If a new vehicle is detected, it will be given an ID and its position will be tracked over the next few frames until it leaves the scene. A small amount of time passes between each frame (usually less than 50 milliseconds) so the position of the same vehicle in two successive frames cannot be too different. If multiple vehicles are identified within the same frame, the approach performs a nearest neighbor computation of the objects within the different frames. In such a case, the tracking method calculates which of the vehicles in the frame is closest to the position of each vehicle in the previous frame. The location found is used to update the one associated with that ID. For example, let us suppose

[1] https://pjreddie.com/darknet/yolo/.

that vehicles A and B have been recognized within the frame i. Then, if in frame $i+1$ we identify multiple vehicles, the one corresponding to A will be that for which the distance between its position in frame $i+1$ and the position of the identified vehicle A in frame i is the shortest. One more constraint that vehicles must satisfy is the detection area they must reside within. If they are detected out of the detection area, they are discarded. The detection area is fixed for the entire video sequence and is considered one input of the problem. An example of a detection area is shown in Fig. 2, corresponding to the green and red crossed lines. A vehicle is analyzed as soon as it passes the first green light (indicated on top of the figure) and is continuously tracked until it leaves the last green line (that at the bottom of the image). When a vehicle enters the detection area, its entry point and the current frame number are stored. When it exits the area, the distance between the exit point EX and the entry point EN is calculated in pixels and this value is divided by the elapsed time. Time is calculated as the number of frames elapsed between entering and exiting the detection area divided by the number of frames per second of the video. For example, if F_1 is the frame related to the entry point and F_2 the first frame when the vehicle left the detection area, the speed is computed as $\frac{dist(EX,EN)}{\frac{F_2-F_1}{50}}$, assuming the frame rate equal to 50 fps. The speeds returned in $pixels/s$ are converted to km/h by linear interpolation and the real speed of a few reference vehicles known as input. In particular, in the Cartesian plane we construct a graph with the calibration points $X(detectedspeed\ _{(pixel/s)},\ real\ speed\ _{(km/h)})$ (they are known and belong to the input vehicles) and, using the *Least Squares* method, we calculate the equation of the line that best approximates the position of these points. The identified equation (which is computed only once) is used to calculate the speed in km/h from that in $pixels/s$. For example, let us assume we have the following five points (903, 82.93) (910, 79.99) (856, 80.34) (935, 83.70) (810, 68.69), each of type ($pixel/s$, km/h), representing five calibration points known in advance, and the speed of 927 $pixels/s$ of a vehicle just detected. Using the *Least Squares* method we find $m = 0.0862$ and $q = 3.0157$ related to the equation $y = mx + q$. By substituting the x with 927 we obtain an estimated speed of 82.92 km/h, a value relatively close to the real one (82.81 km/h).

3.2 The BrnoCompSpeed Dataset

Authors in [22] worked on visual traffic surveillance and captured a new dataset of 18 full-HD videos, each lasted one hour, plus three more videos (of around 10 min each) recorded at seven different locations. The videos contain vehicles that are annotated with the exact speed measurements verified with different reference GPS tracks. The dataset is made up of 21 road section videos. These videos are split into six sessions, plus one (manually annotated), called session zero. More in detail, each session includes three videos, one taken from the left camera, one taken from a central camera, and the third one taken from the right camera. We will refer to each video as *Session ID-left|central|right* where *ID* $\in \{0, \ldots, 6\}$ and *left, central,* and *right* depend on whether the video was taken using the left camera, central camera, or right camera. Each session is stored in a folder that contains the videos of a different road section taken from the three points of view. For each video, there is a file containing vehicle data, a photo of the road section, and an image containing the detection area. In total, in the dataset, there are video recordings and speed data of 20,865 vehicles.

The data files were produced with the Pickle module - Python object serialization - for Python 2.7. The data contained in the Pickle files include:

- [**distanceMeasurement**]: list of measurement points and actual distances.
- [**measurementLines**]: list of the coefficients (a, b, c) of the measurement lines, perpendicular to the road axis, to be used with the canonical form of the line equation $ax + by + c = 0$. A digital image is a grid in which each point (pixel) is identified by a pair of numbers (x, y) and a straight line can be identified by a linear equation such as $ax+by+c=0$ (canonical equation). To automatically understand if a vehicle is within the detection area and in which lane, a system of linear equations corresponding to the sides of the street and the lanes and the perpendicular lines to the street must be solved.
- [**cars**]: list of data related to running vehicles. Each info includes *laneIndex* (lane number - 0 on left), *carId*, *timeIntersectionLastShifted*, *valid*, *intersections*, and *speed*. The *intersections* field is a list containing information about the exact time the vehicle crosses the measurement lines and can be useful to identify it in the video. The value of the *valid* field determines whether the measurement has to be taken into account or not.
- [**fps**]: frames per second of the video.
- [**invalidLanes**]: list of invalid lanes.
- [**laneDivLines**]: list of the coefficients (a, b, c) of the lines which divide the lanes, to be used with the canonical form of the line equation $ax + by + c = 0$.

In Fig. 2, the measurement lines (in red) and the delimitation lines (in green) of the lanes, obtained from the coefficients present in the data file of the Session 0-left of the dataset, are plotted. In Fig. 3, one of the vehicles, taken from the Session 0-left video, can be visually identified in the frame corresponding to the exact moment it crosses the measurement line.

Fig. 2. Measurement lines.

Fig. 3. Vehicle identification.

The vehicle data obtained from the videos with five different calibration systems by the authors of the dataset were included in five JSON files per video. Each JSON file includes a list of the identified vehicles with the following items:

- **[id]** -identifier of the vehicle;
- **[frames]** - the frames in which the vehicle appears;
- **[posX]** - X coordinates of the vehicle in each frame in which it appears;
- **[posY]** - Y coordinates of the vehicle in each frame in which it appears.

The BrnoCompSpeed Dataset Evaluation Code is available in the project's GitHub repository[2]. The code in the repository generates the statistics on the errors obtained in the phases of calibration, detection of distances, and estimation of speeds. These statistics include, in addition to the number of the performed measurements, mean, median, 95 percentile, and worst case (absolute and percentage values) for each video of all sessions from 1 to 6. In Table 1 we show the summary of the results obtained by the competitor methods on the Brno dataset. In the next section, we exploit such methods as the baselines for evaluating the performance of the proposed speed measurement system.

Table 1. Competitor's performance on the Brno dataset for each calibration system and for the videos of sessions 1–6. (MAE: Mean Absolute Error; RE: Relative Error)

System	MAE (km/h)	RE (%)
FullACC	8.59	10.89
OptScale	1.71	2.13
OptScaleVP2	15.66	19.83
OptCalib	1.43	1.81
OptCalibVP2	2.43	3.08

4 Experimental Analysis

A statistically significant subset of videos from the Brno dataset has been used to experimentally analyze and validate the proposed speed estimation system. For each video, the vehicles were identified and associated with the real speed measured by the authors of the dataset. The selected piece of video was then analyzed with YOLO + Darknet to identify the positions of the vehicles in each frame. The real speeds of some of the vehicles were used to calculate the transformation parameters from $pixel/s$ to km/h. The purpose of the tests is to compare the results obtained by the authors of the datasets above with those of our speed estimation system. In all the tests, the video analysis - the detection of vehicles and the calculation of their speed, returned in $pixel/s$ - were performed on *Google Colab*. The equation of the line for the linear interpolation, the vehicle speeds in km/h and the statistics were instead obtained using a Python script and offline computing.

[2] https://github.com/JakubSochor/BrnoCompSpeed.

4.1 Preliminary Quality Test

We first started the validation of our method by testing the quality of the results obtained for the measurement of the speed using the initial parameters for the detection and calculation of the position of the vehicles.

In this analysis, the center of the bounding box was used to denote the vehicle's position. Then, we estimate the speed for the first 20 vehicles of the video belonging to the *Session 0-left* of the dataset (one vehicle passed on lane 1, fourteen on lane 2, and five on lane 3), by considering the vehicles which transited on lane 2 only. The speed value obtained by the system in $pixels/s$ was multiplied by 3.6 to get $kpixel/h$. Moreover, the first five detected vehicles (see Table 2) were used for the calibration stage. At the last step, the coefficients $m = 0.0158$ and $q = 29.9095$ of the line equation $y = mx + q$ were derived with the least-squares method. The speed in km/h was obtained from the equation by substituting the speed measured in $kpixel/h$ for x. The test produced the following results (data from the first five vehicles, used for calibration, are not considered): standard deviation of 0.0325, mean relative error of 3.27%, mean absolute error of 2.36 km/h.

Table 2. Preliminary quality test: speed estimation results (MAE: Mean Absolute Error; RE: Relative Error.

Vehicle id	Detected speed (kpixel/h)	Detected speed (km/h)	Real speed (km/h)	MAE (km/h)	RE (%)
0	3252	81,27	82,93	1,67	2,01
3	3276	81,65	79,99	1,66	2,07
4	3084	78,62	80,34	1,73	2,15
5	3367	83,09	83,70	0,61	0,73
8	2967	76,77	74,42	2,35	3,15
9	2864	75,14	73,75	1,40	1,89
10	3305	82,11	81,74	0,36	0,44
11	2838	74,73	70,09	4,65	6,63
12	2916	75,96	68,69	7,27	10,59
14	3258	81,36	81,59	0,22	0,27
16	3339	82,64	82,81	0,17	0,20
17	3093	78,76	75,51	3,25	4,31
18	3100	78,87	77,11	1,75	2,27
19	3087	78,66	76,49	2,17	2,84

We can observe from Table 2 that the errors for vehicles 11 and 12 are much higher than the average. We looked at these vehicles (which we considered outliers) in the video and noticed that they have much bigger dimensions (length and height) than the others in the test. This property affects the size of the bounding box and therefore the position of its center, which is used to define the position of the vehicle itself. The bounding box assumes different sizes at the times of entry and exit from the detection area, introducing errors, especially if the entry or exit lines are close to the edge of the frame.

We also carried out a second test using a different detection point, to see whether the results were confirmed or disproved Indeed, for some vehicles, the centroid of their bounding boxes is not properly computed. As we noticed that its lower right corner is usually more stable and precise, from this test on we used it as the point of reference for the calculation of the position of the vehicle.

Table 3. Results obtained with a different detection point.

Vehicle id	Detected speed (kpixel/h)	Detected speed (km/h)	Real speed (km/h)	MAE (km/h)	RE (%)
0	3089	83,75	82,93	0,81	0,98
3	2777	79,51	79,99	0,48	0,60
4	2764	79,34	80,34	1,01	1,25
5	2958	81,97	83,70	1,73	2,06
8	2579	76,82	74,42	2,40	3,23
9	2546	76,38	73,75	2,63	3,57
10	2858	80,61	81,74	1,13	1,38
11	2439	74,92	70,09	4,84	6,90
12	2409	74,52	68,69	5,82	8,48
14	2833	80,27	81,59	1,31	1,61
16	2873	80,81	82,81	2,00	2,41
17	2665	77,99	75,51	2,48	3,29
18	2662	77,95	77,11	0,84	1,08
19	2651	77,80	76,49	1,31	1,72

The results of this test are reported in Table 3 and show a standard deviation of 0.0246, a mean relative error of 3.38%, and a mean absolute error of 2.49 km/h (the vehicle data used for calibration are excluded). They also reveal that the variation of the considered detection does not appear to significantly affect, on balance, the accuracy of the method.

4.2 Impact of the Observation Angle

We then investigate the possibility that the angle of observation may impact the goodness of measurement. In particular, with regard to the Brno dataset, the framing is the same for the different lanes considered: this implies a different angle of observation and consequent a different perspective for each lane, and such a configuration could impact the measurement result. To estimate the magnitude of this effect, the video of Session 1 - Central was chosen, with the lower right corner of the bounding box being used as the vehicle location. Consequently, the first 20 vehicles of Session 1 - central (two lanes) - were considered, while the first five vehicles for each lane were used for calibration. Moreover, to automate the calibration and test procedure, specific checks have been implemented to avoid cases of double detection (a bicycle hanging on the back of a camper detected as an additional vehicle or a vehicle with two bounding boxes almost completely overlapping) or detection in non-interesting areas of the image. The results are shown in Table 4.

Table 4. Impact of the observation angle.

	Lane 1	Lane 2	Average
Standard deviation	0,0192	0,0053	0.0122
Mean relative error (%)	2.44	1.42	1.93
Mean absolute error (km/h)	1.66	1.17	1.42

We recall that the lane 1 is lateral to the frame, while the lane 2 is central. Hence, from the results in Table 4, it emerges that a greater alignment of the camera with respect to the monitored lane (and therefore a reduced or absent angle of observation), as is the case for central lane 2, also results in less detection error, as intuitively expected.

4.3 Impact of the Detection Area Size

To clarify the possible relationship between the vertical length (i.e., the *height*, with respect to the single frame) of the detection area and the precision of the measurements, we compare the speed errors obtained by considering four different dimensions of the detection area. In this additional case scenario, the lower right corner of the bounding box was used to denote the vehicle's position, and the first 25 vehicles of the Session 1 - central (lane 1) - were considered. Four measurement sessions were thus performed with detection areas of variable height between 200 and 920 pixels (see Figures 4, 5, 6 and 7).

In such a scenario, the results highlight a significant decrease in errors as the size of the detection area increases (see Table 5 and Fig. 8).

4.4 Entry/Exit Point Improved Estimation

At last, we explore a method to improve the precision of the speed estimation by determining the exact entry and exit points in/from the detection area. In this case, the lower right corner of the bounding box was used to locate the vehicle, and the first 25 vehicles of the Session 1 - central (lane 1) were considered. Here, the detection area height was fixed at 920 pixels, while the first three vehicles

Table 5. Impact of the detection area size.

Area height (px)	MAE (km/h)	MRE (%)	Standard deviation
200	4.54	6.43	0.0437
400	2.57	2.57	0.0189
620	1.30	1.87	0.0113
920	0.79	1.06	0.0070

(other than the fastest and the slowest), the fastest vehicle, and the slowest vehicle were used for the calibration.

To implement such an improvement, we just keep the frame just before the vehicle crosses the detection area and the subsequent frame where the vehicle touches it. Similarly, when the vehicle exits from the detection area, we keep the frame where it exits it and the frame just before the exit. The moments when the vehicles enter and leave the detection area can therefore be calculated comparing the vehicle positions in the previous and current frame. If A is the point in the frame right before the vehicle enters the detection area and B is the point in the next frame (where the vehicle touches the detection area), we can compute the line between A and B. As we know the equation of the line of the detection area, we can compute the intersection point I of the two lines

Fig. 4. h = 200 pixels.

Fig. 5. h = 400 pixels.

Fig. 6. h = 620 pixels.

Fig. 7. h = 920 pixels.

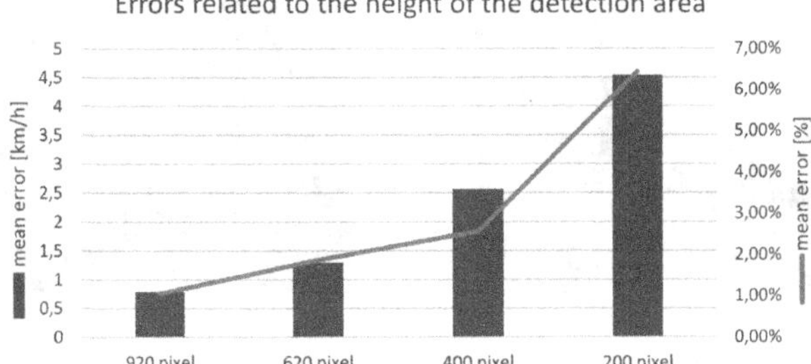

Fig. 8. Errors related to the height of the detection area.

and have the exact coordinate of the entry point. Similarly, we can compute the exact exit point.

The number of frames we take into account to compute the speed in *pixel/s* is equal to the number of frames elapsed between the frame F_1, when the vehicle enters the detection area, and the frame F_2, the last frame before the vehicle exits the detection area, plus the fraction of the frame right before F_1 and the fraction of frame right after F_2. The fraction of the frame before F_1 is equal to the distance IB divided by the distance AB. With similar calculations we compute the fraction of the frame after F_2.

Table 6. Results obtained by improving the estimation of the entry/exit points.

MAE (km/h)	MRE (%)	Standard deviation
0.24	0.35	0.0025

In the previous test scenarios, the starting position of the vehicle was considered in the first frame after the vehicle passed the starting point of the detection area and the ending position was considered in the first frame after the vehicle was out of the detection area. Indeed, this led to a minor precision as the vehicle was always some pixels over the starting/ending point of the detection area. The results are shown in Table 6, where such an improvement generates a better performance of the proposed method.

Fig. 9. Inconsistent bounding boxes for the same vehicle.

4.5 Comparison with the State-of-the-Art

Finally, we compare our results with those obtained by the authors of the dataset and the other state-of-the-art competitors on the Brno dataset. To do so, we executed a conservative configuration of our algorithm, thus excluding possible improvements identified *ex-post*, which might have suffered from overfitting due to the specific scenario related to the Brno dataset. In addition, we ran the evaluation code available in the Brno project repository for the video related to the Session 1-Central.

Table 7. Summary results of the comparison with the state-of-the-art methods.

	MAE (km/h)	**RE (%)**
FullACC	10,15	13.08
OptScale	3.13	4.12
OptScaleVP2	48.71	62.69
OptCalib	3.12	4.10
OptCalibVP2	7.02	9.06
UniCA *(proposed)*	**1.42**	**1.93**

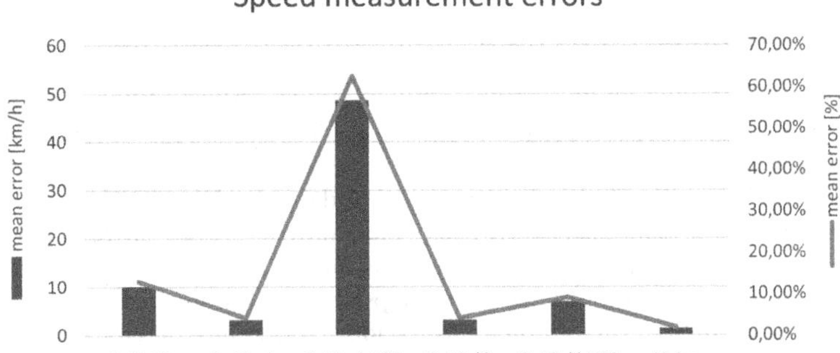

Fig. 10. Summary results of the comparison with the state-of-the-art methods.

The summary data of this comparison are illustrated in Table 7 and Fig. 10, where the proposed method is referred to as *UniCA*. The results obtained not only show a clear superiority of the proposed method in terms of lower average error, but are also overall slightly below the 2% (in relative terms) or 2km/h (in absolute terms) thresholds that are used by several world Countries to accredit automatic speed detection systems.

5 Conclusions

The developed algorithm[3], although its operating logic is relatively simple, performs well, with a relatively average error of 1.93% compared to its literature competitors, and a low-cost setup and calibration phase that allows for easy extension to existing traffic surveillance systems, provided that an affordable embedded GPU system (e.g., the *NVIDIA Jetson* board) is available to execute the real-time object detection and tracking algorithms.

Through the experiments carried out, we have also highlighted how such results can be further improved through the adoption of specific expedients, which, however, depend on the type of framing and road setting considered. However, some limitations remain. Looking at the video output generated by the system, in some cases, the size of the bounding box changes abruptly in a few frames due to particular occlusion or misdetection conditions that can deceive the YOLO + Darknet framework (see Fig. 9 for an example). If this happens in the specific frame in which the distance measurement is performed, higher errors for speed measurement can be generated. A possible way to mitigate this that we will tackle in a forthcoming enhancement of this approach, could be to consider the size of the box in each frame and a moving average of its shape to avoid inconsistent spots changes. We finally point out that the proposed approach - in a revised engineered implementation made more stable and robust - performs well enough to have enabled its usage by the *Monserrato* municipality in Sardinia, Italy, to help the local police identify hazard behaviours of drivers in selected critical road sections.

References

1. Abuella, H., Miramirkhani, F., Ekin, S., Uysal, M., Ahmed, S.: ViLDAR-visible light sensing-based speed estimation using vehicle headlamps. IEEE Trans. Veh. Technol. **68**(11), 10406–10417 (2019)
2. Atzori, A., Barra, S., Carta, S., Fenu, G., Podda, A.S.: HEIMDALL: an AI-based infrastructure for traffic monitoring and anomalies detection. In: 2021 IEEE International Conference on Pervasive Computing and Communications Workshops and other Affiliated Events (PerCom Workshops), pp. 154–159. IEEE (2021)
3. Balia, R., Barra, S., Carta, S., Fenu, G., Podda, A.S., Sansoni, N.: A deep learning solution for integrated traffic control through automatic license plate recognition. In: Gervasi, O., et al. (eds.) ICCSA 2021. LNCS, vol. 12951, pp. 211–226. Springer, Cham (2021). https://doi.org/10.1007/978-3-030-86970-0_16
4. Cheng, G., Guo, Y., Cheng, X., Wang, D., Zhao, J.: Real-time detection of vehicle speed based on video image. In: 2020 12th International Conference on Measuring Technology and Mechatronics Automation (ICMTMA), pp. 313–317 (2020). https://doi.org/10.1109/ICMTMA50254.2020.00076
5. Feldstein, S.: The global expansion of AI surveillance, vol. 17. Carnegie Endowment for International Peace Washington, DC (2019)
6. Fisher, P.: Improving on police radar. IEEE Spectr. **29**(7), 38–43 (1992). https://doi.org/10.1109/6.144510

[3] We publicly release the code at http://aibd.unica.it/speed_detection.zip.

7. Girshick, R.: Fast r-CNN. In: Proceedings of the IEEE international conference on computer vision, pp. 1440–1448 (2015)
8. Gunawan, A.A., Tanjung, D.A., Gunawan, F.E.: Detection of vehicle position and speed using camera calibration and image projection methods. Procedia Comput. Sci. **157**, 255–265 (2019)
9. Kamoji, S., Koshti, D., Dmonte, A., George, S.J., Pereira, C.S.: Image processing based vehicle identification and speed measurement. In: 2020 International Conference on Inventive Computation Technologies (ICICT), pp. 523–527. IEEE (2020)
10. Kunzmann, K.R.: Smart cities: a new paradigm of urban development. Crios **4**(1), 9–20 (2014)
11. Lee, J., Roh, S., Shin, J., Sohn, K.: Image-based learning to measure the space mean speed on a stretch of road without the need to tag images with labels. Sensors **19**(5), 1227 (2019)
12. Li, H., Dong, H., Jia, L., Xu, D., Qin, Y.: Some practical vehicle speed estimation methods by a single traffic magnetic sensor. In: 2011 14th International IEEE Conference on Intelligent Transportation Systems (ITSC), pp. 1566–1573 (2011). https://doi.org/10.1109/ITSC.2011.6083076
13. Liu, C., Huynh, D.Q., Sun, Y., Reynolds, M., Atkinson, S.: A vision-based pipeline for vehicle counting, speed estimation, and classification. IEEE Trans. Intell. Transp. Syst. **22**(12), 7547–7560 (2020)
14. Mohammed, F., Idries, A., Mohamed, N., Al-Jaroodi, J., Jawhar, I.: UAVs for smart cities: Opportunities and challenges. In: 2014 International Conference on Unmanned Aircraft Systems (ICUAS), pp. 267–273. IEEE (2014)
15. Neves, J.C., Moreno, J.C., Barra, S., Proença, H.: Acquiring high-resolution face images in outdoor environments: a master-slave calibration algorithm. In: 2015 IEEE 7th International Conference on Biometrics Theory, Applications and Systems (BTAS), pp. 1–8. IEEE (2015)
16. Radu, L.D.: Disruptive technologies in smart cities: a survey on current trends and challenges. Smart Cities **3**(3), 1022–1038 (2020)
17. Redmon, J., Divvala, S., Girshick, R., Farhadi, A.: You only look once: Unified, real-time object detection. In: Proceedings of the IEEE conference on computer vision and pattern recognition, pp. 779–788 (2016)
18. Redmon, J., Farhadi, A.: Yolov3: An incremental improvement (2018). arXiv preprint arXiv:1804.02767
19. Ren, S., He, K., Girshick, R., Sun, J.: Faster R-CNN: Towards real-time object detection with region proposal networks. In: Advances in neural Information Processing Systems, Vol. 28 (2015)
20. Schoepflin, T.N., Dailey, D.J.: Dynamic camera calibration of roadside traffic management cameras for vehicle speed estimation. IEEE Trans. Intell. Transp. Syst. **4**(2), 90–98 (2003)
21. Shorfuzzaman, M., Hossain, M.S., Alhamid, M.F.: Towards the sustainable development of smart cities through mass video surveillance: A response to the COVID-19 pandemic. Sustain. Urban Areas **64**, 102582 (2021)
22. Sochor, J., et al.: Comprehensive data set for automatic single camera visual speed measurement. IEEE Trans. Intell. Transp. Syst. **20**(5), 1633–1643 (2019). https://doi.org/10.1109/TITS.2018.2825609
23. Tourani, A., Shahbahrami, A., Akoushideh, A., Khazaee, S., Suen, C.Y.: Motion-based vehicle speed measurement for intelligent transportation systems. Int. J. Image Graph. Sig. Process. **10**(4), 42 (2019)
24. Vakili, E., Shoaran, M., Sarmadi, M.R.: Single-camera vehicle speed measurement using the geometry of the imaging system. Multimedia Tools Appl. **79**(27), 19307–19327 (2020)

Corporate Network Analysis Based on Graph Learning

Emre Atan[1], Ali Duymaz[1], Funda Sarısözen[1], Uğur Aydın[1], Murat Koraş[1], Barış Akgün[2], and Mehmet Gönen[2](✉)

[1] QNB Finansbank R&D Center, İstanbul, Turkey
{emre.atan,ali.duymaz,funda.sarisozen,ugur.aydin, murat.koras}@qnbfinansbank.com

[2] Koç University, İstanbul, Turkey
{baakgun,mehmetgonen}@ku.edu.tr

Abstract. We constructed a financial network based on the relationships of the customers in our database with our other customers or other bank customers using our large-scale data set of money transactions. There are two main aims in this study. Our first aim is to identify the most profitable customers by prioritizing companies in terms of centrality based on the volume of money transfers between companies. This requires acquiring new customers, deepening existing customers and activating inactive customers. Our second aim is to determine the effect of customers on related customers as a result of the financial deterioration in this network. In this study, while creating the network, a data set was created over money transfers between companies. Here, text similarity algorithms were used while trying to match the company title in the database with the title during the transfer. For customers who are not customers of our bank, information such as IBAN numbers are assigned as unique identifiers. We showed that the average profitability of the top 30% customers in terms of centrality is five times higher than the remaining customers. Besides, the variables we created to examine the effect of financial disruptions on other customers contributed an additional 1% Gini coefficient to the model that the bank is currently using even if it is difficult to contribute to a strong model that already works with a high Gini coefficient.

Keywords: Graph learning · Centrality metrics · Corporate network

1 Introduction

Networks can be found almost everywhere today, and they are frequently used in the analysis of large-scale data. Examples of networks that can be encountered in many parts of life including financial networks, social networks, and traffic networks. All examples of networks are becoming more complex and larger every day. Therefore, understanding large complex networks has become much more important due to the increase in networked data and information.

G. Nicosia et al. (Eds.): LOD 2022, LNCS 13810, pp. 268–278, 2023.
https://doi.org/10.1007/978-3-031-25599-1_20

Networks are usually irregular and have complex structures, and they are not easy to understand or analyze. However, graph representations offer an ideal tool to understand and analyze massive amounts of network data. Graphs have an important place in the visualization and understanding of these networks. A graph is simply a collection of nodes (i.e., vertices) and lines (i.e., edges) connecting these nodes, and do not provide geometric information, but only show the relationship between nodes.

Due to the complexity of network data, analyzing the data requires graph metrics such as centrality, betweenness, and closeness. The most frequently used one of these graph metrics is centrality. The idea of "centrality", primarily applied to human communication for small groups, was introduced by Bavelas in 1948 [1]. The centrality metric assigns a centrality (importance) value to each node in a network. Numerous measures of centrality have been designed, each capturing slightly different aspects of what it means for a node to be important. The simplest measure of centrality is "in-degree centrality", where the centrality of a node is the number of edges pointing to it. Another measure of centrality is "eigenvector centrality", which is a measure of the influence a node has on a network. If a node is pointed to by many nodes, then that node will have a high eigenvector centrality.

The current study is performed in Finansbank A/S, which was established as a deposit bank in 1987 with a paid-in capital of 4 million USD. Later, Finansbank A/S was acquired by Qatar National Bank S.A.Q (QNB Group), and the Bank's name was changed to QNB Finansbank in 2016. As of June 30, 2021, QNB Finansbank provides a wide range of financial services to its customers with more than 10,000 employees through an extensive distribution network of more than 460 domestic branches and 1 foreign branch. QNB Finansbank is Turkey's fifth largest private bank with unconsolidated assets of 258 billion Turkish Lira (TRY) as of June 30, 2021. Its wide range of product variability has a significant role in reaching 157 billion TRY net loans and 150 billion TRY customer deposits

In this study, we built a network of QNB customers (named as Corporate Network Analysis), which illustrates not only QNB customers but also their suppliers and customers. It provides an opportunity of analyzing customer needs through their volume of transactions in the financial sector. Knowledge of customers and their relations helps us to create more successful marketing strategies and to improve decision making procedures in risk management.

1.1 Related Work

Many studies in the literature emphasize the power of network databases and graph metrics such as centrality, betweenness, and closeness. Examples of graph metrics and network databases used in many different fields can be listed as traffic modeling, healthcare applications, criminal investigations, social media analysis, and financial systems.

In the studies that represent world cities as networks, the intersections are illustrated as nodes, and the streets are illustrated as edges. Comparative analyses of different centrality metrics were performed in these studies for investigating the spatial networks of urban streets. Each centrality metric captures a different aspect of one place's "being central" in geographic space [2].

Network analyses have a very diverse set of applications in health studies. For example, networks were used to represent neural activities in the brain, and centrality measures showed which parts of the brain are important when specific activities are happening [3].

In criminal investigation scenarios, networks and centrality metrics are used to detect key persons in drug traffic. Results showed that network representations are also useful for these kinds of investigations [4].

It is also possible to find influencer persons in social media using networks and centrality metrics. Some studies investigated financial decisions while others investigated political decisions in social networks [5].

There are few studies in the financial sector. One of them investigated the impact of global financial integration on liquidity risk using the network approach. Results in that study suggest that the degree of connectedness between banks is inversely related to funding stability [6]. Also, recently there are studies on annual financial reports (10-Ks) of US companies. One of them has 10-K based network named Lazy Network, that aims to capture textual changes derived from financial or economic changes on the equity market [7]. Another one measured modifications and semantic changes from 10-Ks and showed that firms that do not change their 10-Ks in a semantically important way from the previous year tend to have large and statistically significant future risk-adjusted abnormal returns [8]. Besides, one another study analyzed product differentiation and its effect on competition in their market, which was based on text-based analysis of product descriptions from 50,673 firm 10-K statements [9]. In another study, which used financial networks in the global banking sector, investigated social networks among top banks. Study claimed that connections between banks within social networks may facilitate valuable information flows. Their findings showed that connected banks are more likely to partner together [10].

1.2 Our Contributions

According to our literature review, there are few studies using networks and graph metrics in the financial sector. The main contributions of this work can be detailed as follows:

- This work discusses the data cleaning challenges as well as performance of text similarity functions in big data.
- A very big data set is collected from customer's money transactions that includes 22.3 million financial entities (nodes) and 286.2 million transactions (transfers between nodes). A real-life big data set with this size is rarely used in network studies.
- It is possible to make several analyses from graph data that illustrate customer relations from money transfers between them.
- Eigenvector centrality is used in order to find influencers and key customers in the customer network.

- The results will be used to create more successful marketing strategies based on this graph analysis.
- Network data is also important from a risk management perspective. An economic disruption in the network will affect strongly connected customers. Consequently, the network is used to extract features about customers, and these features are fed into a risk model based on machine learning.

2 Data Sources

There are billions of money transactions happening every day. We were able to collect data from one bank to understand and analyze relations between customers at the corporate level. Our data consists of 286.2 million transactions in the years between 2017 and 2019. The total transaction amount included in the data set is 3.3 trillion TRY. Money transactions with other banks, inward money transactions and Direct Debiting System Loans are collected from the bank database. Table 1 shows the total volumes of transfer types.

Table 1. Summary statistics of transactions included in the study.

Transfer type	Count	Amount (Million TRY)
Direct debiting system	891,802	37,991
Transactions with other banks	226,561,492	2,935,737
Transactions within the bank	58,820,597	399,244

Direct Debiting System and Transactions within the bank are transactions between customers of the bank with unique customer IDs. On the other hand, most of the data is from transactions with other banks, and this data does not include a unique customer ID.

Incoming transactions to the bank have a sender title written in free format text and sender account number while outgoing transactions have receiver title and receiver account number as well. That is why we need to use data cleaning techniques for assigning a customer id to accounts in other banks. An example for raw data is illustrated in Table 2.

Data cleaning steps for data from transactions with other banks are below.

1. Character cleaning – converting to capital letters: In this step, all characters except A to Z and 0 to 9 are removed. Moreover, all letters are converted into capital letters. A single <<space>> character is allowed between two valid characters.
2. Identifying corporations or households: In this step, the sender and receiver are identified as a company or a household. For example, customers with names including specific keywords such as company, real estate, software, etc. are identified as a company.

Table 2. Raw data for sample transactions.

SENDER	SENDER NAME	SENDER ACCOUNT NO	RECIPIENT	RECIPIENT NAME	RECIPIENT ACCOUNT NO	DATE	AMOUNT
22334455	XYA-Z CONS. A/S	TR41001122334455	0	Aaa Software. Company	TR41001122334461	08/12/2020	5000
22334456	ABC TRADE COMPANY	TR41001122334456	0	abc Trade Company	TR41001122334462	08/12/2020	5000
0	General Real Estate	TR41001122334457	22334461	Food and Bev. A/S	TR41001122334463	08/12/2020	5000
0	abc trad comp	TR41001122334462	22334464	Electronics Tra. And Man.	TR41001122334464	08/12/2020	5000
22334459	EDS MECHANIC	TR41001122334459	22334465	EDS Automotive	TR41001122334465	08/12/2020	5000
22334460	AAA SOFTWARE	TR41001122334460	22334466	HAPPY INSURANCE	TR41001122334466	08/12/2020	5000

3. Assigning customer ID to other bank accounts from transfers: When we find a perfect match between the sender's name and recipient name, we assume another bank account should be assigned to a customer ID. Therefore, we have started creating a table with account numbers and assigned customer IDs.
4. Similarity between unknown customer name and bank's customer names: To assign an ID to an unknown account number, "Jaro Winkler" text similarity function is applied between free format written customer name and bank database names. If similarity is over 90%, we assign customer ID from the bank database to other bank accounts.
5. Creating network data: Finally, after all data cleaning processes are done network data is created with transfer type, sender id, recipient id, transaction date and transaction amount columns. In Table 3, an example of final network data is shown after all data cleaning processes are done.

Table 3. Sample of final network data.

SENDER	RECIPIENT	DATE	AMOUNT
22334455	22334461	08/12/2020	5000
22334456	22334456	08/12/2020	5000
TR41001122334457	22334461	08/12/2020	5000
22334462	22334464	08/12/2020	5000
22334459	22334465	08/12/2020	5000
22334460	22334466	08/12/2020	5000

In conclusion, we identified 99 million transactions with 2.8 billion TRY done by corporate customers over 286 million total transactions with 3.3 billion TRY. These 99 million transactions are from 1.2 million corporate customers among which 868,000 customers lack unique customer IDs before data cleaning and text mining operations. After these steps, we assigned a bank ID to 42% of customers, a tax number to 19%, and the remaining 39% used their account number as their unique ID.

3 Customer Acquisition Application

After the data cleaning process, we tried to find out key customers from the network. Finding key customers will provide a successful marketing strategy to the bank. Eigenvector centrality metric is used for identifying the key customers in a network.

In financial sector, corporates are tended to create a micro network between their suppliers and customers. Also, these micro networks are usually led by one or two bigger sized corporate and connected corporates are always take some value from these central customers. In centrality measures, a high eigenvector centrality score means that a node is connected to many nodes who themselves have high scores. On the other hand, closeness centrality score depends on the sum of the distances to all other nodes, and betweenness centrality score is depends on the total number of shortest paths between nodes [11]. Eigenvector centrality is used to determine the importance of a node in a network. Unlike other types of centralities, when calculating the score of the relevant node in eigenvector centrality, attention is paid to the importance levels of that node's neighbors. For example, in the corporate network of our study, it does not only use the number / amount of connections, but also determines the importance level of the money receivers in the network and provides an extra score contribution to the relevant customer. Because of its calculation methodology and its convenience to financial sector, eigenvector centrality is used for identifying the key customers in a network.

Eigenvector centrality of ith node corresponds to the ith element of the eigenvector $\boldsymbol{v}$ associated with the maximum eigenvalue (i.e., λ) of the adjacency matrix $\mathbf{A} \in \mathbb{R}^{n \times n}$:

$$\mathbf{A}\boldsymbol{v} = \lambda \boldsymbol{v},$$

where elements of the adjacency matrix $\mathbf{A}$ are the weighted edges between nodes [11].

Before calculating the centrality metric, we eliminated the transactions between households, so at least one corporate customer remained in each transaction. Also, we have eliminated the customer transactions with amounts less than 1000 TRY and the customers with fewer than two transactions in total. Furthermore, customer transactions to their own accounts were also eliminated from the data set. After these eliminations, we created the graph data with 1,242 billion TRY amount and 73.7 million transactions from 3.6 million customers (nodes). These eliminations are done for creating a corporate network which includes only strong connections.

After creating the graph, we calculated the eigenvector centrality metric for each node. The calculations were done in RStudio using `igraph` package. Calculation took 22 s for 3.6 million nodes and 6.2 million edges. To determine the key nodes in the graph data, percentile analyses were done. First 30 percent of the nodes with the highest score were chosen as key nodes according to percentile analyses. Results show that customers with high centrality cover 88.9%

of the total number of transactions and even more they cover 94.7% of the total amount of transactions in the graph data. Results are summarized in Table 4.

Table 4. Results of percentile analyses.

	Nodes with high centrality	**Nodes without high centrality**	**Total**
Number of customers	1,091,217	2,546,172	3,637,389
Number of customers (%)	30%	70%	100%
Number of transactions	65,557,415	8,206,359	73,763,774
Number of transactions (%)	88.9%	11.1%	100.0%
Amount (Million TRY)	1,176,281	66,345	1,242,625
Amount (%)	94.7%	5.3%	100.0%

To analyze how the centrality effects on profitability and finding key customers in the graph data with 3.6 million nodes, we compared the customers with a high centrality score to the rest of customers. Firstly, we divided customers into three groups according to their centrality scores. Top 30% was determined as high centrality customers before, next 30% as group 2, and the remaining 40% as group 3. We compared the active customers in the bank with their monthly average profits and we calculated monthly average profit for each group based on customers' average profits. Profit of customers is calculated by summing interest revenue from credits and commissions from noncredit products and deducting the costs on that customer. Usage of credits and other products makes a higher revenue while a good credit score decreases costs.

According to these results summarized in Fig. 1, the centrality metric is highly affecting the customer profitability. Average profit of customers with high centrality is almost five times higher than other groups. Moreover, we observed that the profitability is decreasing with the decreasing centrality score.

The contribution of the centrality metric on the current customer acquisition methods in the bank is quite promising. The customer acquisition strategy before this analysis was based on random customer selection. Randomly selected customers' average profit for the bank would be the average of all customers, which is 389 TRY/month. Therefore, building a customer acquisition strategy targeting high centrality customers will contribute around 380 TRY/month for each new customer.

4 Credit Risk Modeling Application for Manufacturing Industry

Network data contains the edges of firms to their customers and their suppliers. It is also an informal cash flow statement. With this point of view, any financial stress in the network will influence the firms that are linked. Therefore, we made another application for credit risk models and created features from network data.

We trained a multilayer perceptron model with these new features to improve the existing credit risk models. Our extracted features are based on attributes of linked suppliers or customers to target firms. The list of created features are given in Table 5.

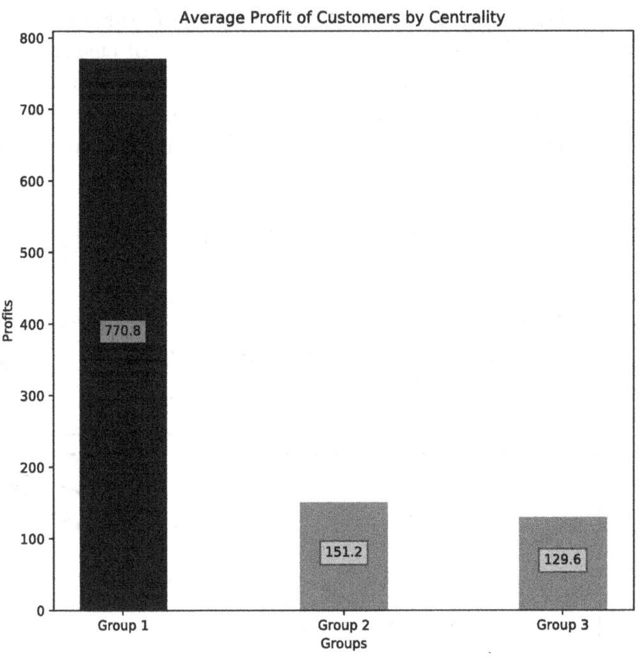

Fig. 1. Average profit of customers by centrality.

Table 5. Created features for credit risk model.

Features
Total number of related customers
Total transfer amount
Total number of transfers
Average probability of default (PD) of related customers
Number of sectors of related customers
Average number of bad check in related customers
With/without bad check in related customers
Amount weighted average PD of related customers
With/without defaulted related customer
Count weighted average PD of related customers
Count weighted average class note of related customers
Amount weighted average class note of related customers
Average class note of related customers

There was already a very strong model for the manufacturing sector which has been used in the bank. It is named Micro Policy Scorecard Model and contains internal and external data sources such as demographic features and credit payment performance features. These data sets were formed from customers who had revenue less than 6 million Turkish Liras and development period between 01/01/2017 – 31/12/2017 (12 snapshots of last day of consecutive months). As a result of this model, it is estimated what percentage of customers will default or not. The default definition is Non-Performing Loans in 24 months.

Gini coefficient is used as a performance indicator in our machine learning models. Gini coefficient has a linear relationship with AUC, and it is calculated with the formula below:

$$\text{Gini} = 2 \times \text{AUC} - 1.$$

At this point, AUC is expressed as the area under the ROC curve obtained by using the confusion matrix. In Fig. 2, the x-axis represents the False Positive Rate and y-axis represents the True Positive Rate [12].

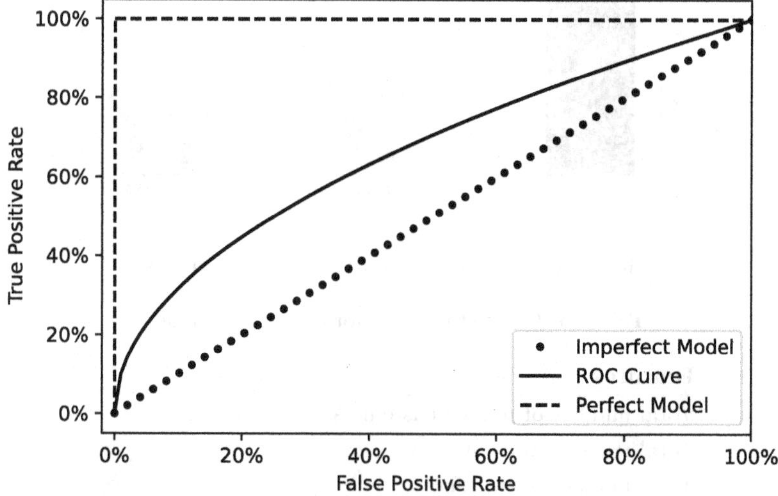

Fig. 2. ROC curve.

Data sample was split into three subgroups as train and validation sample with proportions of 60% and 40% randomly. As seen in Table 6, the train and validation Gini coefficients without network features for the manufacturing sector are 0.798 and 0.735, respectively.

Table 6. Credit risk model results for 24 months performance window.

Manufacturing sector model	Train (Gini)	Validation (Gini)
Model without network features	0.798	0.735
Model with all features	0.817	0.741
Difference	0.019	0.006

Although it is very hard to contribute to this successful model, network features made a consistent contribution to the Gini coefficient with 1% in 2017 and 2018. Depending on credit volumes, 1% addition to the bank decision system for credit approvals will make a very significant difference over time.

5 Discussion

There are some main points to be emphasized after this study. One key discussion topic is data cleaning challenges. There were lots of free format written titles in the large-scale data. Without successful data cleaning procedures, a single customer might be considered as different customers. In this study, we spent our maximum effort to make customers unique to prevent missing connections between subnetworks.

The second discussion topic is the performance of the centrality metric calculated from the constructed graph. Due to good performance of the centrality metric, we were able to select 30% of key customers covering 94% of transfer volume. Moreover, our results showed that customers with high centrality led to higher profitability.

The third discussion topic is how the network data is contributing to machine learning based risk models. Our hypothesis that financial stress is contagious through the network has been empirically shown by observing that credit risk models were improved using the features created from the constructed graph.

As a final observation, additional data may be extracted from digital invoices if firms stored their invoices in the database. Since every invoice will lead to money transfer, this addition may increase the data quality. In this study, we only focus on money transfers due to very limited access to digital invoices of firms.

6 Conclusion

Network analysis provides us with the opportunity to determine the trade relations between companies through money transfers, to understand the networks of companies, and to use the acquired information for marketing and risk management purposes. To construct a customer network and perform the analysis, we created a data set from money transactions in a specific period and ran a very detailed data cleaning process. The data cleaning step was performed using similarity algorithms between free format written transaction logs and customer titles in the bank database. The biggest question in marketing strategies is determining the most profitable customers. Using the network representation, it is easier to measure the importance of customers from the network, and the eigenvector centrality metric was used for this analysis. Working together with the metric calculated as the outcome of the study showed that customers with high centrality cover a high portion of the total transaction volume and have significantly higher profitability. Another analysis we performed was concentrated on checking the contribution of the features extracted from network data on the credit risk models. Thanks to the extracted network features, it can identify customers with financial problems and their effect on other linked customers more accurately.

References

1. Bavelas, A.: A mathematical model for group structure. Appl. Anthropol. **7**, 16–30 (1948)
2. Crucitti, P., Latora, V., Porta, S.: Centrality measures in spatial networks of urban streets. Phys. Rev. E **73**, 036125 (2006)
3. Fletcher, J.M., Wennekers, T.: From structure to activity: Using centrality measures to predict neuronal activity. Int. J. Neural Syst. **28**(2), 1750013 (2018)
4. Bright, D.A., Greenhill, C., Reynolds, M., Rittler, A., Morselli, C.: The use of actor-level attributes and centrality measures to identify key actors: A case study of an Australian drug trafficking network. J. Contemp. Crim. Justice **31**(3), 262–278 (2015)
5. Inekwe, J.N., Jin, Y., Valenzuela, M.R.: Global financial network and liquidity risk. Aust. J. Manage. **43**(4), 593–613 (2018)
6. Miller, P.R., Bobkowski, P.S., Maliniak, D., Rapoport, R.B.: Talking politics on Facebook: Network centrality and political discussion practices in social media. Polit. Res. Q. **68**(2), 377–391 (2015)
7. Adosoglou, G., Park, S., Lombardo, G., Cagnoni, S., Pardalos, P.M.: Lazy network: a word embedding-based temporal financial network to avoid economic shocks in asset pricing models. Complexity **2022**, 9430919 (2022)
8. Adosoglou, G., Lombardo, G., Pardalos, P.M.: Neural network embeddings on corporate annual filings for portfolio selection. Expert Syst. Appl. **164**, 114053 (2021)
9. Hoberg, G., Phillips, G.: Text-based network industries and endogenous product differentiation. Expert Syst. Appl. **124**(5), 1423–1465 (2016)
10. Houston, J.F., Phillips, G.: Social networks in the global banking sector. J. Account. Econ. **65**(2–3), 237–269 (2018)
11. Newman, M.E.J: Networks: an Introduction. OUP Oxford (2010)
12. Schechtman, E., Schechtman, G.: The relationship between Gini methodology and the ROC curve. Available at SSRN: (2016). https://ssrn.com/abstract=2739245

Source Attribution and Emissions Quantification for Methane Leak Detection: A Non-linear Bayesian Regression Approach

Mirco Milletarì(✉), Sara Malvar, Yagna D. Oruganti, Leonardo O. Nunes, Yazeed Alaudah, and Anirudh Badam

Microsoft, Redmond, USA
mirco.milletari@microsoft.com

Abstract. Methane leak detection and remediation efforts are critical for combating climate change due to methane's role as a potent greenhouse gas. In this work, we consider the problem of source attribution and leak quantification: given a set of methane ground sensor readings, our goal is to determine the sources of the leaks and quantify their size in order to enable prompt remediation efforts and to assess the environmental impact of such emissions. Previous works considering a Bayesian inversion framework have focused on the over-determined (more sensors than sources) regime and a linear dependence of methane concentration on the leak rates. In this paper, we focus on the opposite, industry-relevant regime of few sources per sensor (under-determined regime) and consider a non-linear dependence on the leak rates. We find the model to be robust in determining the location of the major emission sources, and their leak rate quantification, especially when the signal strength from the source at a sensor location is high.

Keywords: Bayesian framework · Source attribution · Inverse problem · Leak quantification

1 Introduction

Methane (CH_4), the primary component of natural gas, is a potent greenhouse gas (GHG) with a Global Warming Potential (GWP) of 84–87 over a 20-year timescale [6]. The Intergovernmental Panel on Climate Change (IPCC) affirms that reduction of anthropogenic methane emissions is the most efficient way to curb a global temperature rise of 1.5 °C above pre-industrial levels by 2030 [17].

The global oil and gas industry is one of the primary sources of anthropogenic methane emissions, with significant leaks occurring across the entire oil and gas value chain, from production and processing to transmission, storage, and distribution. Examples of sources of methane leaks are malfunctioning clamps, flares, flow lines, tanks, pressure regulators, thief hatches, and valves. Capacity

G. Nicosia et al. (Eds.): LOD 2022, LNCS 13810, pp. 279–293, 2023.
https://doi.org/10.1007/978-3-031-25599-1_21

limitations in gathering, processing, and transportation infrastructure can also lead to the venting of excess methane. The International Energy Agency (IEA) estimates [10] that it is technically possible to avoid around 70% of today's methane emissions from global oil and gas operations. These statistics highlight the importance of leveraging various methane detection technologies and source attribution techniques to address this critical issue.

Most of these technologies rely on complex models of particulate transport in the atmosphere. Complexity is due to the interplay of multiple spatial scales (from the particle scale to near-source and long-range effects), multi-physics (coupling mass transport, turbulence, chemistry, and wet/dry deposition), and complex geometry (e.g., flow over topography or man-made structures). Atmospheric dispersion models have a long history, reaching back to Richardson's [16] and Taylor's [21] pioneering investigations of turbulent diffusion. However, maintaining accuracy is a prevalent challenge in dispersion modeling since many models have large uncertainties in effective parameters, such as the Monin-Obukhov length [15], atmospheric stability classes, or terrain roughness length.

Past research has mostly focused on improving forward transport models to evaluate downstream pollutant concentrations given source leak rates and meteorological variables. However, few works have focused on the source attribution problem, which belongs to the class of inverse problems. Methods for estimating source strength and/or location from measurements of concentration can be divided into two major categories depending on the physical scale of the problem. Researchers employed ground-based measurements and a high-resolution mesoscale air transport model to quantify GHG emissions at the urban, regional, and continental scales. They use a Bayesian statistical technique to predict emissions and the associated uncertainty by combining previous emission inventories with atmospheric measurements [14]. When the physical distance between the sources and sensor observations is minimal, using mesoscale air transport models for inversion becomes challenging. Typically, at such scales, atmospheric inversions are performed using plume dispersion and surface layer models. In this paper, we are primarily interested in observations taken relatively close to the source, and we limit ourselves to analyzing the uncertainty in inverse modeling linked to plume dispersion models.

A considerable number of inversion studies based on plume inversion models have been published in peer-reviewed journals. Several of these papers deal with uncertainty estimations [11,18]. Garcia et al. [8], for instance, considered a Bayesian regression model using a non-stationary forward operator while Lushi and Stockie [13] considered a positively constrained, linear least squared method together with the Gaussian Plume model to determine the leak rates of the sources. However, both studies considered a linear dependence on the leak rates and a design where the number of sensors (9) is much greater than the number of sources (4). Mathematically, the latter scenario results in an over-determined system, for which Linear Programming solvers work well.

In this paper, we propose a solution based on Bayesian optimization to identify the source of a methane leak and to quantify the size of the leak, using readings from a spatially sparse array of sensors, which corresponds to a mathematically under-determined system. This is a particularly relevant scenario for many industrial applications that require monitoring of large areas with costly ground sensors. The scenario that we consider in this study is for continuous monitoring of an Area of Interest (AoI), where an operator would be interested in detecting anomalous methane leaks, identifying their likely sources, and estimating leak size in near-real-time, to allow for prompt inspection and remediation. As a result, the proposed methodology focuses on achieving a reasonable trade-off between accuracy and a relatively low computational time.

2 Methods

Source attribution belongs to the class of inverse problems; it aims at finding the sources that generated a certain field configuration given readings of field values at some restricted number of points $\{\boldsymbol{x}_1, \boldsymbol{x}_2, \cdots, \boldsymbol{x}_M\} \in \mathbb{R}^d$, where d is the number of space dimensions. In this work, we consider the following scenario: during some observation time δt, some or all the sensors deployed in the field record methane concentration signals, exceeding a determined threshold. As there are multiple sources being monitored in the field, the sensors only record a compound signal that is assumed to be given by the linear combination of concentrations generated by a subset of sources at the each sensor location. The objective is therefore to find the decomposition of the compound signal to determine the contribution of each source. We are interested in determining the k sources that contribute the most to the signal and estimate their strength.

2.1 Bayesian Approach

The Bayesian approach relies on inverting the forward model using Bayes' principle and sampling algorithms, based on some form of Markov Chain Monte Carlo (MCMC) or Stochastic Variational Inference (SVI). In a physical model, all empirical parameters are subject to systematic and statistical errors; the former considers the measurement error associated with the instrument(s), while the latter encompasses statistical uncertainty in a set of measurements. In a Bayesian approach, this input uncertainty naturally propagates through the model in a non-parametric fashion. As a result, inferred parameters come with confidence levels that better reflect the physical reality of the model. This means that all quantities are expressed by probability distributions rather than single numbers. In general, given a parameter set $\boldsymbol{\theta}$, a variable set $\boldsymbol{q}$, and sensor readings $\boldsymbol{w}$, Bayes' principle reads: $P(\boldsymbol{q}, \boldsymbol{\theta}|\boldsymbol{w}) = P(\boldsymbol{w}|\boldsymbol{q}, \boldsymbol{\theta})P(\boldsymbol{q}, \boldsymbol{\theta})/Z(\boldsymbol{w})$, where $P(\boldsymbol{q}, \boldsymbol{\theta})$ is the prior, based on our current knowledge or assumptions on the form of the distribution, $P(\boldsymbol{w}|\boldsymbol{q}, \boldsymbol{\theta})$ is the likelihood, and $P(\boldsymbol{q}, \boldsymbol{\theta}|\boldsymbol{w})$ the posterior. Finally, $Z(\boldsymbol{w})$ is a normalization. In this work we restrict our analysis to a scenario where all the source locations are known. In this case, the unknowns are the leak rates of the N sources $\boldsymbol{q} = [q_1, q_2, \cdots, q_N]^T$ measured in kg/h. The methods presented here can be extended to scenarios with known and unknown sources, where the latter was considered in Wade and Senocak [24].

Let us call $A_{mn}(\mathbf{q}, \boldsymbol{\theta})$ the $M \times N$ (M sensors and N sources, $M \ll N$) forward operator mapping the concentration field from source to sensor location, parametrized by $\boldsymbol{\theta} = \{u, \phi, p\}$, where u is the modulus of the wind velocity $[m\,s^{-1}]$; ϕ its in-plane direction $[rad]$; and possibly other p parameters that depend on the details of the model. While $\boldsymbol{\theta}$ are measured quantities (with uncertainties), $\boldsymbol{q}$ are unknown and constitutes the fitting parameters of the model. Given an array of M sensors, let us call W_m the compound signal recorded at sensor m at time intervals δt. Then we have the relation:

$$w_m = \sum_{n=1}^{N} A_{mn}(q_n, \boldsymbol{\theta}) \equiv \mathcal{A}(\boldsymbol{q}, \boldsymbol{\theta}). \tag{1}$$

In general, this is a non-linear, time-dependent mapping, solution of the Diffusion-Advection partial differential equation (PDE). Following [8], we write it as:

$$\big(\partial_t + \boldsymbol{L}(\theta)\big)C(\boldsymbol{x}, t) = \sum_{n=1}^{N} q_n(t)\delta(\boldsymbol{x} - \boldsymbol{x}_n) \tag{2}$$

$$\boldsymbol{L}(\theta) = \boldsymbol{\nabla} \cdot \big(\boldsymbol{u}(\boldsymbol{x}, t) - \boldsymbol{D}(\boldsymbol{x})\boldsymbol{\nabla}\big), \tag{3}$$

where $\boldsymbol{L}(\boldsymbol{\theta})$ is a linear operator, possibly depending non-linearly on the parameters $\boldsymbol{\theta}$, comprising an advection and a diffusion term controlled by the diffusion matrix, $\boldsymbol{D}(\boldsymbol{x})$. The term $C(\boldsymbol{x}, t)$ is the concentration field at location $\boldsymbol{x} = (x, y, z)$ and time t. Finally, note that we are considering point-emission sources specified by the $\boldsymbol{x}_n$ coordinates in the Dirac delta function on the right hand side of Eq. (2). The solution implemented in the next section imposes a series of assumptions on the form of $C(\boldsymbol{x}, t)$ and, therefore, of $\mathcal{A}$ that makes the problem numerically manageable at different levels of complexity.

2.2 Non-linear Bayesian Regression: Stationary Model

To simulate the contribution of each source, we consider a forward operator based on the Gaussian plume model [23], which is a special solution of Eqs. (2) and (3) under the following simplifying assumptions:

1. The leak rates, $\boldsymbol{q}(t)$, vary slowly in time such that it can be considered constant over the measurement time scale, i.e. $\boldsymbol{q}(t) = \boldsymbol{q}$.
2. The wind velocity and direction are stationary and aligned along the x direction for $x \geq 0$, i.e. $\boldsymbol{u} = (u, 0, 0)$.
3. The diffusion matrix, $\boldsymbol{D}(\boldsymbol{x})$, is replaced by effective parameters based on the Pasquill stability class.

Boundary conditions include finiteness of the concentration field at the origin and infinity, together with the condition that the contaminant does not penetrate the ground, see [20] for details. Under these conditions, the PDE admits an analytical solution in the form of a Gaussian kernel:

$$C_n(\boldsymbol{x}) = \frac{q_n}{2\pi\, u\, \sigma_y\, \sigma_z} \exp\left\{ -\frac{(z-h)^2}{2\sigma_z^2} - \frac{(z+h)^2}{2\sigma_z^2} - \frac{y^2}{2\sigma_y^2} \right\}, \tag{4}$$

with the (scalar) concentration field measured in [kg m^{-3}], although we will often convert this to parts per million per volume (ppmv) in the rest of the paper. The σ_i are standard deviations, and h is the height of the source. Our implementation of the Gaussian plume model follows the one implemented in the `Chama`[1] open-source library [12], where the value of the standard deviations is re-defined to include heuristic information on the stability of the plume: $\sigma_i(x) = a_i\, x\,(1 + x\, b_i^{-1})^{-c_i}$, where the values of the parameters a_i, b_i, c_i depend on the atmospheric stability class (indexed from A to F) and are different for the y and z components. Weather stability classes are evaluated given the surface wind, cloud coverage, and the amount of solar radiation in the AoI on a specific date and time. Wind direction and source location are re-introduced respectively by rotating the simulation grid in-plane and by re-centering it on the source position. This expression, linear in the leak rate $\boldsymbol{q}$, was used in [13] as the diffusion/advection operator of a linear regression model. Following `Chama` [12], we consider buoyancy corrections to dispersion along the z-axis, introduced heuristically in Eq. (4) as:

$$z'_n = z + 1.6\,\frac{B_n^{1/3}\, x^{2/3}}{u}, \qquad B_n = \frac{g\, q_n}{\pi}\left(\frac{1}{\rho_{CH_4}} - \frac{1}{\rho_{air}}\right). \tag{5}$$

where g is the gravitational constant while ρ_{CH_4} and ρ_{air} are the density of methane and air measured in standard conditions. As such, we measure buoyancy in units of $m^4\, s^{-3}$. Note that this transformation introduces a non-linear dependence on $\boldsymbol{q}$, making our source attribution model non-linear.

To accelerate the Gaussian plume model for large-scale simulations, we leverage `PyTorch`[2] to parallelize evaluation over both sensors and sources. This allows for a massive speedup of over 50 times (on CPU) compared to the `Chama` implementation, with further speed-up possible by leveraging GPUs. Moreover, this enables gradient evaluation of training parameters in the model (in our case the leak rates and possibly the atmospheric data) necessary for the Bayesian optimization process using the Hamiltonian Montecarlo algorithm provided by the open-source Bayesian optimization library, `Pyro`[3] [4].

Assuming a normal distribution of the noise with covariance matrix Σ, the likelihood of the model reads:

$$P(\boldsymbol{w}|\boldsymbol{q},\boldsymbol{\theta}) = \frac{1}{(2\pi\ \det\Sigma)^{1/2}} e^{-\frac{1}{2}||\Sigma^{-1/2}(\boldsymbol{w} - \mathcal{A}(\boldsymbol{q},\boldsymbol{\theta}))||^2}. \tag{6}$$

Due to the $M \ll N$ regime we are interested in, corresponding to a low-density sensor placement, the problem is under-determined. While the parameters $\boldsymbol{\theta}$ depends on atmospheric conditions, the leak rates depend on the specifics of the physical process that led to the emission. Therefore, following García et al. [8], we reasonably assume $\boldsymbol{\theta}$ and $\boldsymbol{q}$ to be statistically independent; as a consequence, the prior distribution factorizes as $P(\boldsymbol{q},\boldsymbol{\theta}) = P(\boldsymbol{q})P(\boldsymbol{\theta})$, where the distribution on $\boldsymbol{\theta}$

[1] https://github.com/sandialabs/chama.
[2] https://github.com/pytorch/pytorch.
[3] https://github.com/pyro-ppl/pyro.

is obtained via direct measurement of the weather data at a particular location, together with their experimental (and possibly statistical) uncertainties due to temporal or spatial averaging. However, in the application considered in the next sections, we will make the simplifying assumption of $\boldsymbol{\theta}$ being deterministic; the reason for this choice is related to the considered industrial scenario, see Sect. 3 for details. In comparison, the scenario considered in Garcia [8] and Lushi [13] dealt with the detection of lead-zinc emission; in this case, sensors need to collect enough samples from direct deposition of the pollutants in a collection device, which may require hours, depending on the deposition velocity of the pollutant. In this case, ten minute averages of wind data were used over the measurement time. In the case of methane, sensors operate in near real-time using direct measurements based on a variety of techniques, such as mid/near infra-red lasers or metal oxide semiconductors to name a few. As we detail in Sect. 3 below, we are interested in a near real-time source attribution scenario; in this case, weather data will be taken at the time of measurement from the sensor's weather stations. By taking $\boldsymbol{\theta}$ as deterministic, we are therefore assuming that weather data are homogeneous across the AoI, in agreement with the same assumption used to obtain the Gaussian plume solution, and neglect systematic errors.

2.3 Ranking Model

To rank the source contributions, we use multi-point estimates of each leak rate's posterior distribution. As estimators, we take percentiles from 0 to 100 at steps of 2 lying in the 68% HPDI confidence interval, plus the sample average. Sampled point estimates are then used to reconstruct the source contribution to the signal measured at each sensor using again the forward model. For each prediction, we evaluate again the error with the observed value at each sensor location and evaluate the posterior predictive likelihood $\mathcal{P}_s \equiv P_s(\boldsymbol{w}|\boldsymbol{q}_s^\star, \boldsymbol{\theta})$, $q_{sn}^\star$ being the s-th point estimates of the n-th leak rate from the marginal posterior distribution; this will be used in the final ranking step to weight the goodness of the ranking solution. Note that we denote with $\star$ a variable or parameter fixed by a particular operation, e.g. optimization, sorting, or max.

Each one of the s samples from point estimates (also referred to as point samples) propose a different source reconstruction within the 68% HPDI of the marginal posteriors. By ranking emission sources by their contribution at each sensor, we obtain an ensemble of possible ranking:

$$R^s_{mn^\star} = \arg\operatorname*{sort}_n A^s_{mn}(q^\star_{sn}, \boldsymbol{\theta}), \tag{7}$$

where s is the point sample index and $A^s_{mn}(q^\star_{sn}, \boldsymbol{\theta})$ is the methane concentration value of source n measured from sensor m, obtained from the point estimate s of the leak rate. Each member of the ranking ensemble is weighted by the related predictive likelihood. The final ranking is obtained as a composite estimator. For each sensor, we take the proposed ranking with the highest likelihood:

$$R_{mn^\star} = \arg\max_{\mathcal{P}^s} \mathcal{P}^s_m \, R^s_{mn^\star}. \tag{8}$$

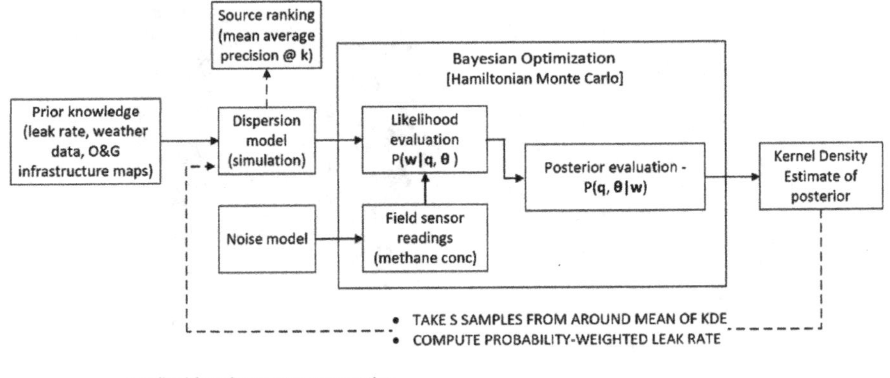

Fig. 1. Flow chart of the source attribution methodology

Finally, to each predicted ranking we can assign a probability obtained by multiplying the (selected) marginal posterior point estimate of the source and the predictive likelihood: $P(\boldsymbol{q}^{\star}_{s^\star}, \boldsymbol{\theta}|\boldsymbol{w}) \simeq P(\boldsymbol{w}|\boldsymbol{q}^{\star}_{s^\star}, \boldsymbol{\theta})\, P(\boldsymbol{q}^{\star}_{s^\star})$.

The end-to-end source attribution process is re-assumed in the flow chart of Fig. 1. In the next section, we apply the methods described here to a scenario of practical relevance.

3 Case Study

The scenario we considered is of direct practical relevance as it can be prohibitively expensive to monitor large areas of interest with a 1:1 or higher sensor-to-source ratio. IoT sensors transmit real-time data on methane concentration and weather readings. An anomaly detection algorithm is employed to detect abnormal methane emissions; if anomalous readings are detected, these are flagged to the source attribution system that returns the most likely location(s) of the leak. The setup of the experimental AoI is shown in Fig. 2. We consider an AoI of approximately 9 km^2, in the Permian Basin in West Texas and Southeastern New Mexico, for our study. The Permian Basin is one of the most prolific oil and gas basins in the US, and contains numerous oil and gas infrastructure assets, many of which are likely emitters of methane. The scenario that we consider for our study is one where 100 possible sources are monitored by 15 high resolution methane sensors in the AoI. Sensor locations have been determined using the sensor placement optimization procedure detailed in Wang [25]; the optimization output is shown in Fig. 2, where sensors are placed either close to ground level or at heights of 5 and 10 m. In the next sections, we discuss data collection and processing. We will also describe how the test scenario and sensor readings were simulated in the absence of field sensor data.

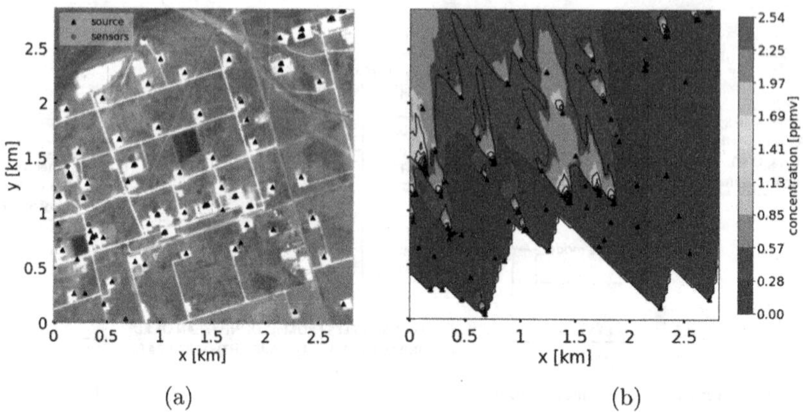

Fig. 2. a) Aerial view of the Area of Interest, showing locations of emission sources (black triangles) and sensors (red dots). b) Methane concentration (above background level) map in ppmv, on 20-07, at 3 pm. Level curve represent z direction. (Color figure online)

3.1 Data Collection and Processing

We gather various inputs required for the Bayesian analysis in the AoI, such as weather variables, historical methane leak rate data, and oil and gas facility maps. We obtain hourly weather data (wind speed, wind direction, temperature, pressure, cloud coverage) from the weather station closest to the study area, from the National Oceanic and Atmospheric Administration (NOAA) Integrated Surface Dataset (ISD) [1]. The wind rose diagram for our AoI is shown in Fig. 3a for a given test date and time. Methane emissions data can be obtained from aerial surveys or IoT sensor measurements or from historical knowledge of leaks from specific oil and gas assets. For our study, we leverage data from an extensive airborne campaign across the Permian Basin from September to November of 2019 [7] that quantified strong methane point source emissions (super emitters) at facility-scales. Since this data corresponds to leak rates from super emitters, we have a tunable parameter to scale the leak rates down. For our analysis, we scale it down by a factor of 3 to better represent the order of magnitude of leaks from normal methane emitters, while maintaining the heavy-tailed distribution shape from the original Permian Basin airborne campaign data set. We find the data to be in good agreement with an exponential distribution.

Oil and gas facilities locations data, including wells, natural gas pipelines and processing plants, available in the public domain, are ingested for the area of interest [3,22]. Satellite map of the AoI is obtained from Sentinel-2 data [2].

3.2 Scenario Simulation

Given the input data defined in the previous section, we use the forward model to simulate the methane concentration at each sensor location. The simulation is

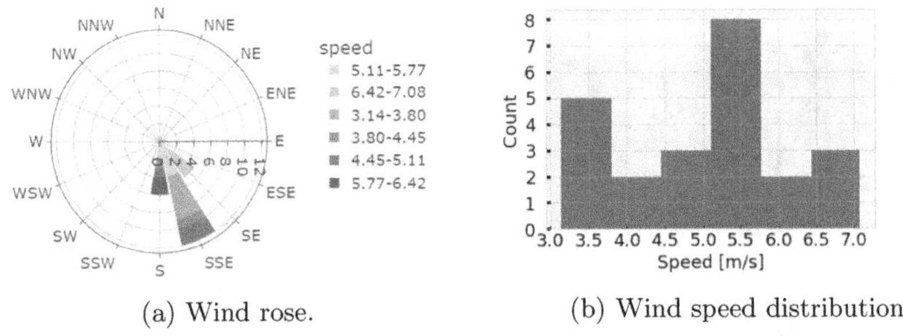

(a) Wind rose.

(b) Wind speed distribution.

Fig. 3. Wind speed and angle distribution for a specific day (07-20-2020).

performed on an area of approximately 9 km^2, and up to 200 m in the z-direction; the grid size (dx, dy, dz) is (25, 25, 5) m. The number of sources is 100, and the number of sensors is 15 (see Fig. 2). We sample the leak rate of each source from the fitted exponential leak rate distribution and use weather data at the time of detection as an input to the plume model defined in Sect. 2.2. As there are no interaction terms in Eq. 2, the concentration field at each point is assumed to be additive. As a consequence, the compound signal at a sensor location is evaluated via summation of individual source contributions. Finally, Gaussian noise is applied to the readings, with a standard deviation corresponding to the sensor's systematic error, together with a detection threshold; for both parameters, we have used values reported by the sensor's vendor of 0.002 ± 0.0001 ppmv over background level (estimated at 1.8 ppmv in the AoI). Throughout this paper, we always report concentration values over background. In Fig. 4a we show an example input data for the leak rates; this shows a typical pattern where most sources have low emissions with few of them being anomalous, i.e. outliers. This is one of the most challenging scenarios we encountered, and we present it here in detail; in practice, there are many possible scenarios, the most favourable being when all the sensors can capture a strong signal. We comment on these other results at the end of Sect. 3.3. In Fig. 4a we use Tukey's fence criteria to separate the bulk of the sample from the outliers; the shaded area in the plot is determined by the interval $[Q_1 - \alpha\,\mathrm{IQR}, Q_3 + \alpha\,\mathrm{IQR}]$, where $\mathrm{IQR} = Q_3 - Q_1$ is the inter quantile range, and Q_1, Q_3 the quantiles. A value of $\alpha = 1.5$ is used to determine the outliers, while $\alpha = 3$ determines extreme values. There are three outliers, corresponding to sources $[10, 80, 92]$, with source 10 being the highest emitter. The median separates low (50%) from average (47%) emitters, with high leak rate outliers constituting only the remaining 3%. However, not all emissions are measured by the sensors, as the concentration value depends on both weather conditions (determining dispersion) and sensor positioning. For the example considered here, one of the high leak rate outliers (80) is not captured at all by the sensors. The compound sensor readings are then used as an input to the source attribution algorithm as detailed in the next section.

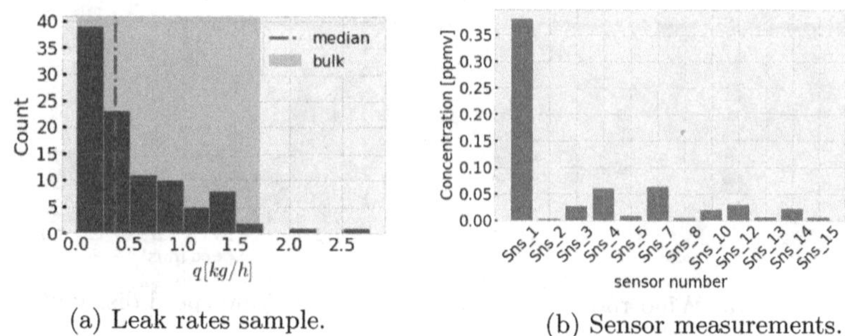

(a) Leak rates sample.

(b) Sensor measurements.

Fig. 4. a) Sample from the leak rate distribution, together with median and bulk; see text for details. b) Simulated measurement at sensor location on 07-20-2020 at 3 pm. This sample shows a typical pattern where most sources have low emissions, with few of them being super-emitters. In this scenario, only 12 of the deployed 15 sensors report above threshold readings.

3.3 Results

We leverage the `Pyro` [4] implementation of the No-U-Turn, Hamiltonian Monte Carlo [9] to sample the marginal posterior; the entire process is summarized in Fig. 1. We found that a relatively small collection of 1000 samples provides a good compromise between accuracy and computational time; the sampler returns the leak rate distribution for each of the 100 sources, at different degrees of convergence. As we are using priors obtained from empirical data, these are not necessarily conjugated, hence the marginal posterior distribution is unknown and needs to be fitted. Although it is possible to look for a continuous parametric fit, here we opt to use the Kernel Density Estimation (KDE) implementation in `scikit-learn`[4] using grid search with cross validation to fix the kernel and bandwidth of each leak rate distribution. In Fig. 5 we show two examples of distributions where convergence is achieved and where it is not. Each figure shows the histogram of the samples, the KDE fit, the 68% Highest Posterior Density Interval (HPDI), together with two vertical lines showing true value and sample average. Following the discussion of Sect. 2.3, we use 51 sample point estimates within the HPDI, for each leak rates marginal posterior distributions. In general, we found the posterior sample average to be a robust central estimator; in addition we use 50 percentiles points estimate (from 0 to 100 at steps of two). The point estimates are used in the forward model to estimate the predictive likelihood and the source contribution at each sensor. The latter are used in the ranking model to extract the top three sources, per sensor, contributing the most to the measured methane concentration, together with their ranking confidence. After this process, we are left with 51 ranking and concentration values for each sensor. In the final step, for each sensor, we select the maximum (predictive) likelihood value out of the 51 evaluated and use this as our best estimate for likely

[4] https://scikit-learn.org/.

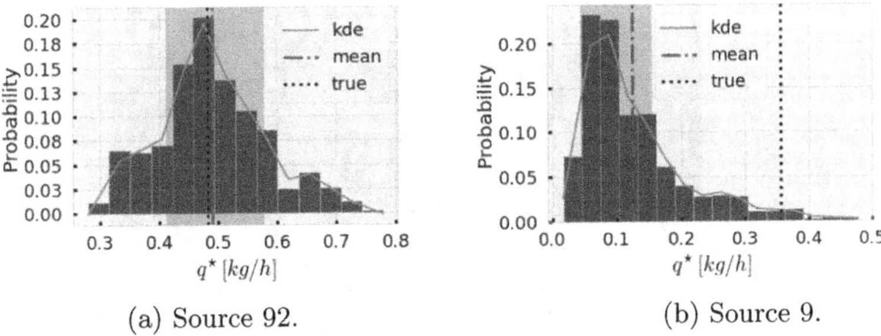

(a) Source 92.

(b) Source 9.

Fig. 5. Marginal posterior distributions: samples histogram and KDE fit are shown together with a 68% HPDI interval (shaded area), the sample mean and the true value. In Fig. 6a, the sample mean provides a good estimation of the true value, while this is not the case in Fig. 6b, where it lies in the tail of the distribution.

sources. We can visualize this result via the network map in Fig. 6, showing sensor to source connectivity for the three selected sources, weighted by source leak rate. Ultimately, this constitutes the model recommendation presented to the monitoring operator, to help them plan further field investigation and plan leak remediation by prioritizing the most likely source of leakage. For testing, we evaluate the mean average precision [19] at $k = 3$ (mAP@3); as our intent is to detect the highest emitting sources at each sensor, $k = 3$ represents a good compromise between keeping this focus while looking at mid-level emissions as well. As we explained later, the performance of the model decreases when including more sources, as optimization samples are dominated by the higher emitters. mAP@3 evaluates how many of the three proposed sources have been correctly ranked, and average the result over all available sensors. For the example above, we find mAP@3 = 0.86. Crucially, the ranking error depends on the relative magnitude of the source's leak rates, this being true also for the regression metrics presented in the next section. Figure 7 shows the true and predicted source contributions to the methane concentration signal and leak rates detected at a sensor.

Leak Rate Quantification. We have repeated the same analysis for 10 more days randomly sampled through the year at different times of the day. Depending on factors such as the weather, leak rate sample and crucially the number of sensors recording the signal (as low as 1), the mAP@3 may vary, although on average is still $\sim$0.83, showing the robustness of the model. Leak rate quantification and source attribution are both outputs from the Bayesian learning algorithm. Accurate leak size estimation is critical in quantifying the environmental footprint of methane leaks, and is also crucial from a regulatory and governance perspective, helping companies build trust with stakeholders and the public. Figure 7b shows the true and predicted leak rate estimates for the highest 3 emitters (sources) whose signal is detected at a sensor. We use Mean Absolute Percentage Error

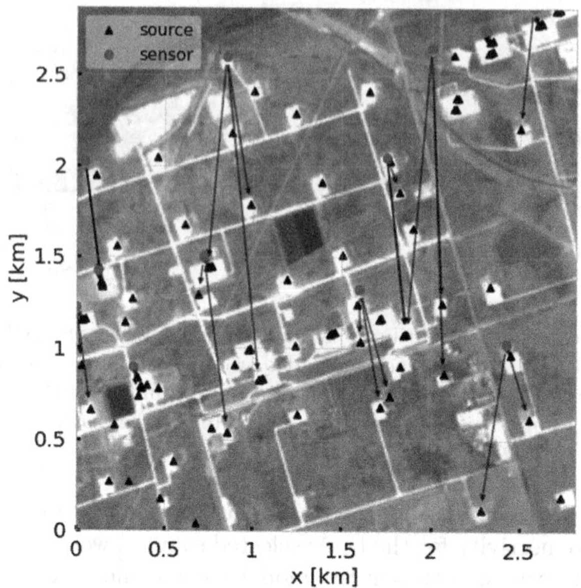

Fig. 6. Network Map: Connections between sensors and sources are used to visualize attribution

(MAPE) [5] as the metric to evaluate the performance of the leak rate quantification algorithm. When evaluated for sources that have been correctly classified as contributing to the signal at a sensor (as determined by the mAP@3 metric), we obtain a total MAPE $\simeq$ 29%. As we mentioned in the previous section, routine sources are more difficult to estimate due to their lower leak rates and the skew nature of the distribution; following Fig. 4a and the discussion of Sect. 3.2, when breaking down the error into low, medium and high leak rates (outliers) we find the corresponding MAPE to be: 50%, 24% and 1.7%, showing how medium and high leak rates can be reliably estimated. We found this behaviour to be consistent across different scenarios, see Sect. 3.3. This leak rate estimate can also be used to update the prior leak rate distribution, which can then be used for future analyses. This can be thought of as a Bayesian learning process that iteratively improves the source attribution and estimation process.

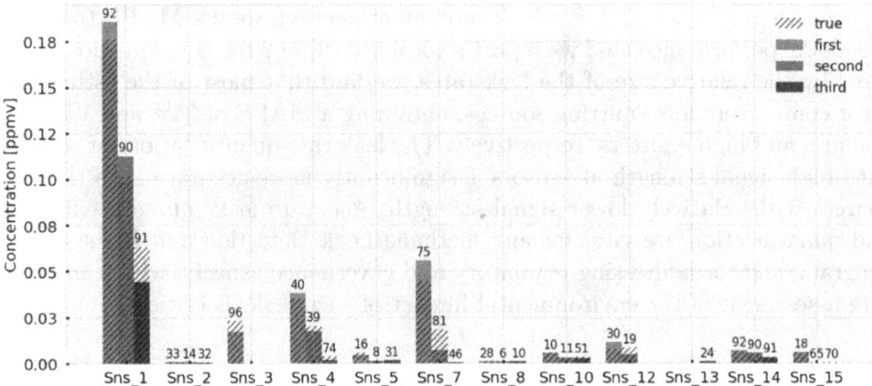

(a) True and predicted source contributions to the methane concentration signal (above background level) detected at a sensor.

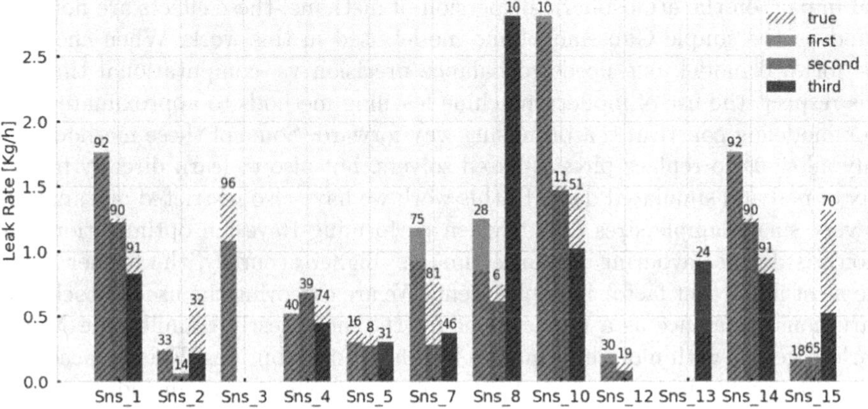

(b) True and predicted leak rate estimates for the highest 3 emitters (sources) whose signal is detected at a sensor.

Fig. 7. The number on top of each bar represents the source ID. Only correctly classified sources are shown.

4 Conclusions

We have presented a Bayesian source attribution and quantification model applied to the realistic situation of non linear dependence between concentration and leak rates, and a regime where the number of sources to monitor greatly exceeds the number of ground field sensors, mathematically corresponding to an under-determined system. We use the mean average precision at $k = 3$ (mAP@3) for evaluating the performance of the source attribution algorithm, and observe a mAP@3 = 0.86 for the experiments performed, which signifies that 86% of leaks detected at sensors were correctly attributed to the true sources; we found this result to be robust across different weather and leak rates sample scenarios, with an average mAP@3 ~0.83. For leak rate quantification, we use MAPE to evaluate model performance, and we report a total MAPE of 29%. Breaking down this error by the relative size of the leak rates, we find that most of the estimation error comes from low emitting sources, obtaining a MAPE of 24% and 1.7% for medium and high emitters, respectively. The leak rate quantification for sources with high signal strength at sensors is significantly more accurate than that for sources with relatively lower signal strength. Accurate leak source attribution and quantification are vital for any methane Leak Detection and Remediation program, and for addressing regulatory and governance aspects, where an accurate assessment of the environmental impact of such leaks is critical.

5 Future Work

Spatial heterogeneity in weather data and transient plume behavior have a crucial impact on the atmospheric dispersion of methane; these effects are not captured by the simple Gaussian plume model used in this work. When choosing the forward model, one needs to balance precision vs. computational time. In this respect, the use of modern machine learning methods to approximate complex modeling constitutes a promising way forward. Some of these methods not only allow us to replace physics-based solvers, but also to learn directly from a mix of real and simulated data. In this work we have also restricted our analysis to very small sample sizes (1000) when performing Bayesian optimization; the choice is due to favouring response time vs. higher accuracy, the former being the most important factor in deployment. We are exploring the use of Stochastic Variational Inference as a replacement for the more costly Hamiltonian Monte Carlo together with more informative likelihood distributions. Finally, access to sensor data will allow us to better estimate model parameters, including a more realistic account of total noise, beyond the systematic error currently modeled.

References

1. NOAA ISD datasets. https://www.ncei.noaa.gov/products/land-based-station/automated-surface-weather-observing-systems/
2. Sentinel-2 imagery. https://sentinel.esa.int/web/sentinel/missions/sentinel-2?msclkid=7c80fe6cc7ba11ec9c5499250f796bd7
3. Texas Railroad Commission datasets. https://www.rrc.texas.gov/resource-center/research/data-sets-available-for-download/. Accessed 13 Sept 2021
4. Bingham, E., et al.: Pyro: deep universal probabilistic programming. J. Mach. Learn. Res. **20**(1), 973–978 (2019)
5. Bowerman, B.L., O'Connell, R.T., Koehler, A.B.: Forecasting, Time Series and Regression: An Applied Approach. South-Western Pub, Southampton (2005)
6. de Coninck, H., et al.: Strengthening and implementing the global response. In: Global Warming of 1.5 C: Summary for Policy Makers, pp. 313–443. IPCC-The Intergovernmental Panel on Climate Change (2018)
7. Cusworth, D.H., et al.: Intermittency of large methane emitters in the Permian Basin. Environ. Sci. Technol. Lett. **8**(7), 567–573 (2021)

8. García, J.G., Hosseini, B., Stockie, J.M.: Simultaneous model calibration and source inversion in atmospheric dispersion models. Pure Appl. Geophys. **178**(3), 757–776 (2021)
9. Hoffman, M.D., Gelman, A., et al.: The no-U-turn sampler: adaptively setting path lengths in Hamiltonian Monte Carlo. J. Mach. Learn. Res. **15**(1), 1593–1623 (2014)
10. IEA: Driving down methane leaks from the oil and gas industry - technology report, January 2021. https://www.iea.org/reports/driving-down-methane-leaks-from-the-oil-and-gas-industry
11. Jeong, H.J., Kim, E.H., Suh, K.S., Hwang, W.T., Han, M.H., Lee, H.K.: Determination of the source rate released into the environment from a nuclear power plant. Radiat. Prot. Dosimetry. **113**(3), 308–313 (2005)
12. Klise, K.A., Nicholson, B.L., Laird, C.D.: Sensor placement optimization using Chama. Technical report, Sandia National Lab. (SNL-NM), Albuquerque, NM, USA (2017)
13. Lushi, E., Stockie, J.M.: An inverse gaussian plume approach for estimating atmospheric pollutant emissions from multiple point sources. Atmos. Environ. **44**(8), 1097–1107 (2010)
14. McKain, K., Wofsy, S.C., Nehrkorn, T., Stephens, B.B.: Assessment of ground-based atmospheric observations for verification of greenhouse gas emissions from an urban region. PNAS Earth Atmospheric Planetary Sci. **109**(22), 8423–8428 (1912)
15. Panofsky, H.A., Prasad, B.: Similarity theories and diffusion. Int. J. Air Wat. Poll. **9**, 419–430 (1965)
16. Richardson, L.F.: Atmospheric diffusion shown on a distance-neighbour graph. Proc. Roy. Soc. Lond. Ser. A Containing Papers of a Math. Phys. Character **110**(756), 709–737 (1926)
17. Rogelj, J., et al.: Mitigation pathways compatible with 1.5 c in the context of sustainable development. In: Global Warming of 1.5 C, pp. 93–174. Intergovernmental Panel on Climate Change (2018)
18. Rudd, A., Robins, A.G., Lepley, J.J., Belcher, S.E.: An inverse method for determining source characteristics for emergency response applications. Bound.-Layer Meteorol. **144**(1), 1–20 (2012)
19. Salton, G., McGill, M.J.: Introduction to Modern Information Retrieval. McGraw-Hill, New York (1983)
20. Stockie, J.M.: The mathematics of atmospheric dispersion modeling. SIAM Rev. **53**(2), 349–372 (2011)
21. Taylor, G.I.: Diffusion by continuous movements. Proc. Lond. Math. Soc. **2**(1), 196–212 (1922)
22. U.S. Energy Information Administration: Layer information for interactive state maps (2020). https://www.eia.gov/maps/layer_info-m.php
23. Veigele, V.J., Head, J.H.: Derivation of the Gaussian plume model. J. Air Pollut. Control Assoc. **28**(11), 1139–1140 (1978)
24. Wade, D., Senocak, I.: Stochastic reconstruction of multiple source atmospheric contaminant dispersion events. Atmos. Environ. **74**, 45–51 (2013)
25. Wang, S., et al.: Unsupervised machine learning framework for sensor placement optimization: analyzing methane leaks. In: NeurIPS 2021 Workshop on Tackling Climate Change with Machine Learning (2021). https://www.climatechange.ai/papers/neurips2021/70

Analysis of Heavy Vehicles Rollover with Artificial Intelligence Techniques

Filippo Velardocchia[1](✉), Guido Perboli[1], and Alessandro Vigliani[2]

[1] DIGEP - Politecnico di Torino, Turin, Italy
{filippo.velardocchia,guido.perboli}@polito.it
[2] DIMEAS - Politecnico di Torino, Turin, Italy
alessandro.vigliani@polito.it

Abstract. The issue of heavy vehicles rollover appears to be central in various sectors. This is due to the consequences entailed in terms of driver and passenger safety, other than considering aspects as environmental damaging and pollution. Therefore, several studies proposed estimative and predictive techniques to avoid this critical condition, with especially good results obtained by the using of Artificial Intelligence (AI) systems based on neural networks. Unfortunately, to conduct these kind of analyses a great quantity of data is required, with the same often difficult to be retrieved in sufficient numbers without incurring in unsustainable costs. To answer the problem, in this paper is presented a methodology based on synthetic data, generated in a specifically designed Matlab environment. This has been done by defining the characteristics of an heavy vehicle, a three axles truck, and making it complete maneuvers on surfaces and circuits purposely created to highlight rollover issues. After this phase, represented by the generation and processing of the data, follows the analysis of the same. This is the second major phase of the methodology, and contains the definition of a neural networks based algorithm. Referring to the nets, these are designed to obtain both the estimate and the prediction of four common rollover indexes, the roll angle and the Load Transfer Ratios (LTR, one for each axle). Very promising results were achieved in particular for the estimative part, offering new possibilities for the analysis of rollover issues both for the generation and the analysis of the data.

Keywords: Heavy vehicles · Rollover risk indicators · Neural networks

1 Introduction

The work presented will focus on the analysis of heavy vehicles rollover, in the specific case of an articulated truck (three axles), and the possibility of estimating and preventing the same through Artificial Intelligence (AI) techniques. The study of this issue is of particular interest since this type of vehicle, in Europe alone and in Canada and the United States, is responsible for the transport of more than 80% of goods [18]. In particular, referring to the U.S, rollover accounts

G. Nicosia et al. (Eds.): LOD 2022, LNCS 13810, pp. 294–308, 2023.
https://doi.org/10.1007/978-3-031-25599-1_22

for 13.9% of large truck fatal crashes [16]. As a matter of fact, statistics on road safety show that accidents involving at least one heavy vehicle are often more dangerous than those involving other types of vehicles [5,14,15], as can be noted specifically in [14], since they represent only 4.7% of accidents (in France in 2014) but cause more than 14% of fatalities. It has also to be highlighted that rollover accidents tend to have several implications, such as the damaging of roads or even environmental pollution [6]. This is a direct consequence of the fact that they are often correlated to heavy vehicles, characterized by a high center of gravity and articulated steering mechanism [1]. To prevent rollover accidents, it is therefore necessary to design a successful safety warning system to notice potential rollover danger [4,19]. This led to the development of different models and techniques, both from the point of active [3,7,11] and passive [10,17,20] rollover protection system, over than, specifically, studies focused on heavy vehicles characterized by articulated steering. In this last case, in particular, it is important to distinguish between simplified [8] and complex [12] models, with the latter characterized by a non suitable real-time application and the first ones with a limited range of operations. To solve these problems, recently, in the vehicle research, different models based on machine learning and empirical data were built [4,14], allowing the creation of data-driven models characterized by rapid and precise responses. The main issue in these cases is the fact that is not always possible to have a sufficient number of empirical data to generate an efficient algorithm, both for time and costs reasons [2]. In this paper, therefore, we developed a methodology based on Recurrent Neural Networks (RNN) receiving and analyzing synthetic data, provided to them by a realistic heavy vehicle model. This was done connecting these two aspects in a Matlab environment, using the Simscape and Deep Learning software to define, respectively, the three axles truck generating the data and the networks estimating and predicting rollover indicators from the same.

The main contributions are:

- The generation of data-driven models based on two different typologies of recurrent neural networks, one with estimation and the other with prediction purposes of four rollover risk indicators. These are represented by the roll angle and the three load transfer ratio (one for each axle), common parameters in this kind of analysis [9]. Recurring to this form of artificial intelligence allows having very flexible, adaptable and updatable models, other than the possibility of obtaining immediate response with good accuracy and reducing significantly time and economic costs.
- The creation of a realistic environment to generate a sufficient quantity of data for the neural networks. This kind of algorithm needs a high amount of data to work effectively, and it is often difficult to obtain the same from empirical operations, both for on-field problems and costs reasons. As a solution, in this paper is proposed a three axles articulated truck model, obtained using Simscape and, more precisely, a version originally developed by Steve Miller, Simscape Vehicle Templates. This allowed us to design maneuvers, surfaces, Pre and Post Processor codes, specifically developed to generate realistic data

regarding the movement and the conduct of the considered vehicle, as well as the implementation of the main modifications in a practical User Interface (UI) through the Matlab livescript.

The article is organized in three main sections (besides the introduction), inevitably linked to the two great phases of the work, the generation and elaboration of the data and the analysis of the same. In particular, after a brief overall overview in Sect. 2, Sect. 3 provides a more specific presentation of the methodology developed, in particular analyzing first the generation of the data (Subsect. 3.1) and, then, focusing on what procedure has been followed at an AI level (Subsect. 3.2). Finally, Sect. 4 concludes the paper, adding some considerations on possible future developments.

2 System Overview

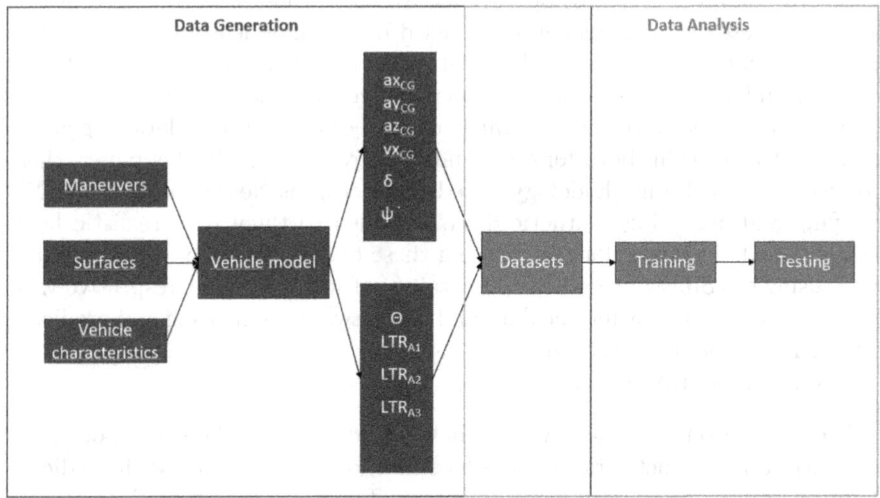

Fig. 1. System overview

As is possible to see, in Fig. 1 a brief overview of the overall system is provided. This is organized into two main parts, the data generation and the analysis of the same. Referring to the first phase, this is characterized by the definition of the simulation environment (Simscape), particularly in terms of maneuvers, surfaces and vehicle characteristics. Once designed and introduced them in a specific User Interface (UI), it is possible to utilize the latter to run the vehicle model previously composed, an articulated truck with three axles, obtaining different parameters of interest. These, retrieved by several Post Processor codes, detail the inputs and the outputs characterizing the AI algorithm. In particular, the first ones are represented by the steering angle (δ), the acceleration of the Center of Gravity (CG) in its three components, the longitudinal velocity and the yaw rate($\dot{\psi}$), while the second ones by the roll angle (θ) and the Load Transfer Ratio (LTR). Knowing the elements defining our data-driven models, after the generation of datasets containing the same, one for each specific maneuver requested to the truck (23 in total), it is possible to elaborate the RNN based on them. This process is characterized by different steps, for the sake of brevity reassumed in the training and testing blocks reported in Fig. 1. In the first one, in particular, is provided an architecture definition of the two typologies of neural networks (as anticipated, one for estimation, the other for predictive purposes), followed by the standardizing of the parameters and the effective training of the algorithms, based on 19 of the 23 datasets generated before. After this part, it is possible to continue with the testing phase of the algorithm, done with 4 datasets excluded by the training of the networks and, therefore, completely unknown to the same.

This system led to the obtaining of promising results especially for what regards the estimation algorithm, as it will be possible to see in the next sections more in detail.

3 Methodology

3.1 Simscape: Data Generation and Elaboration

In order to generate and obtain realistic data for the definition of the AI algorithms and their effectiveness, it is, as anticipated, proposed a methodology based on the Simscape software in the Matlab/Simulink environment. This instrument allows building models of physical components (such as electric motors, suspensions, etc.) realistically interacting with each other, thus resulting in a suitable program for this part of the discussion. It has to be noted that, to be able to affirm it, a sensitivity analysis was conducted on different types of vehicles and standard maneuvers (e.g., step steer, ramp steer etc.), obtaining results extremely similar to the ones that can be found in literature, thus confirming the reliability of the software. Regarding the same, a particular template was used, "Simscape Vehicle Templates", originally developed by Steve Miller. This contains several predefined components, presenting the possibility to design the model of interest without having the need of starting completely anew. In our specific case, this allowed us to rapidly design our vehicle (2), a three axles truck. This was built keeping in mind the goal of the research, highlighting rollover issues for heavy vehicles and proposing both a simple way to generate reliable data and an AI system to analyze the same.

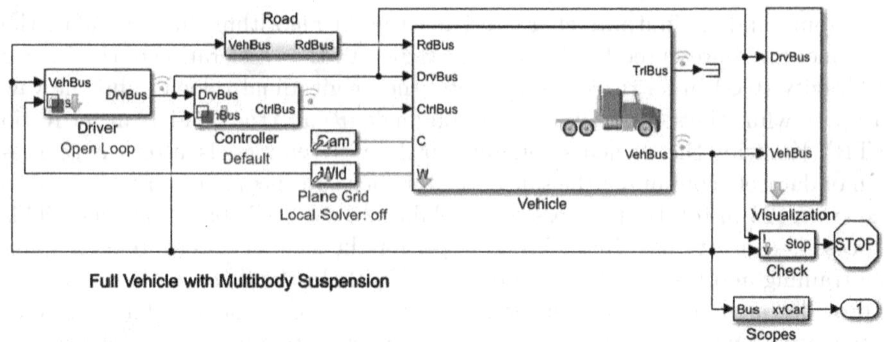

Fig. 2. Simulink model used for the data generation

As can be seen in Fig. 2, the model is characterized by the presence of numerous blocks interconnected with each other (e.g.: Road, Vehicle, Driver) and each having several subsystems composed of different elements. To interact effectively with them, other than recurring to the use of specific scripts, it is advisable to resort to a User Interface (UI). This is a simple yet powerful instrument that allows to rapidly define the key elements of the vehicle and its simulation environment (maneuvers, surfaces etc.), and it is here introduced in order to help the reader in the eventual replica of the methodology proposed. In particular, considering the original UI that allowed to define and recall the starting elements of our model, in Fig. 3 are reported its main characteristics for this research.

In order to understand the actual functioning of the UI, and how much the elements present within it are strongly correlated with the programming environment, it is highlighted that, by selecting a generic maneuver (e.g. WOT Braking, left tab of Fig. 3), there is an automatic interaction with a code (which can be accessed through the Code button) that takes care of setting components such as path, trajectory and surface on which the vehicle will have to perform its motion. To update any characteristic of the maneuver, like the steering angle, it is necessary to interact with the databases of the same (Fig. 3, central part) while, to observe the results, with specific codes in the Results section (right tab of Fig. 3).

Defined the model used and its correlated UI, it is now possible to present what effectively has been introduced to generate new data. The first issue was the creation of maneuvers and surfaces specifically designed for a three axles truck. As a matter of fact, Simscape Vehicle Templates provides already several maneuvers and surfaces in its predefined environment, but none of them are designed for the vehicle of interest, even more for the study of rollover eventualities. Therefore, we defined and introduced in the Simscape environment 23 new maneuvers purposely thought for our vehicle and the casuistry of this research, in order to be able to generate a sufficient quantity of quality data for the neural networks based algorithm. Each new maneuver required the definition of appropriate trajectory parameters, such as the steering angle or the longitudinal

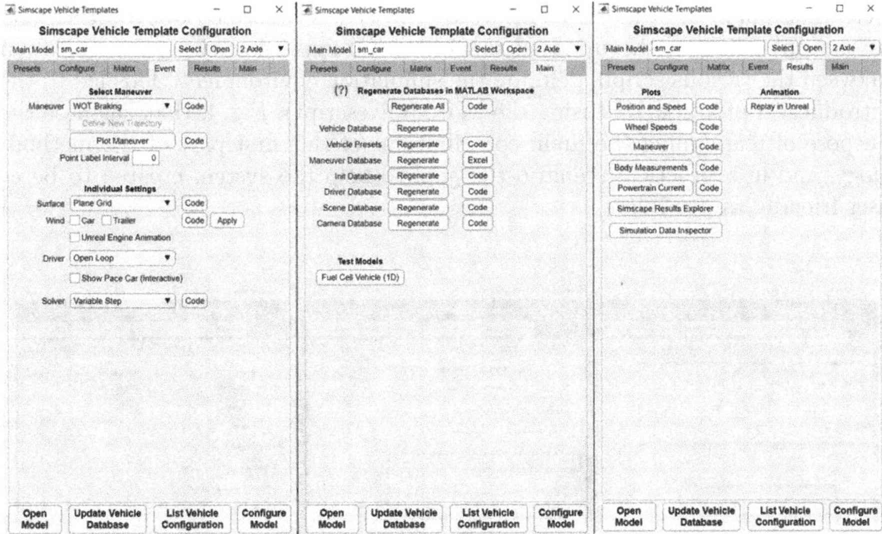

Fig. 3. User interface without addons

distance, depending on the characteristics of the aforementioned vehicle and of the circuits considered. The number of new maneuvers, 23, has been chosen in order to be sure to have enough data to be able to make effective estimates and predictions with the neural networks. As a matter of fact, each maneuver represents a database for the AI algorithm. The dimensions of these databases are very similar (approximately four thousands data each), to avoid changes in the weights of neural networks simply due to differences in the dimensions of the datasets. It has to be highlighted, moreover, that the maneuvers are very different one from another, leading to a very general composition of the data, that are composed by typical parameters of vehicle dynamics (e.g., acceleration, velocity, etc.). These, clearly, are time dependent ones, leading to the use, in Subsect. 3.2, of LSTM layers for defining the neural networks.

Once that the new maneuvers, and therefore our future datasets, were designed, there was the necessity to save the newly generated data. For this purpose, after the running of the simulation, we implemented new post processing codes that aimed both to save the data, and also to generate the same through analytical correlations starting from the ones provided by the final bus of the model (Fig. 2, in case the obtaining could not be done directly). Moreover, through the implementation of post processor codes, it was also possible to verify again, graphically, after the general sensitivity analysis, the correctness of the results obtained by the simulation in comparison with what proposed in literature. This confirmed that the vehicle dynamics parameters (36 in total for each dataset) attained were reliable and ready to be used. Several other additions were then made in order to facilitate the processes described above in aspects as the uploading of the updated maneuver database and the communication between the various scripts defining the simulation environment. This led to the introduction of a new UI (using the Matlab livescript), Fig. 4, that has here the purpose of reassuming the main contributions of this first part of the methodology, and in a practical scenario the goal to help the system created to be as user friendly as possible.

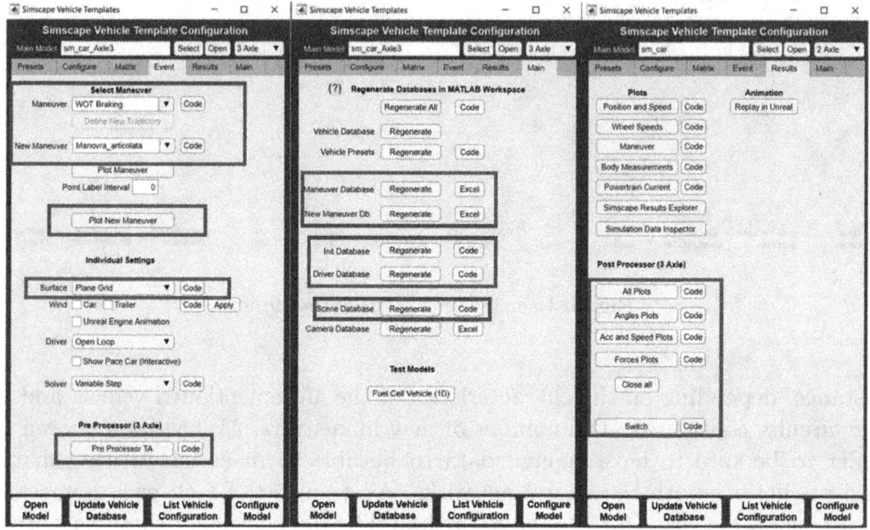

Fig. 4. Modified UI, Simscape

After completing the above steps, it was then possible to continue, knowing the datasets of interest and the tipping indicators to be obtained (Load Transfer Ratio of the three axles and roll angle), with the analysis of the same through AI using the generated synthetic data.

3.2 AI: Data Analysis and Results

After the generation and selection of the data of interest in the first part of the discussion, it is now necessary to understand how to analyze them to obtain the rollover index indicators. With this in mind, having seen the state of the art in literature [4], the second part of the methodology is presented. This consists, as previously anticipated, in the definition of an AI algorithm based on neural networks. This has been done in Matlab/Simulink environment using the Deep Learning Toolbox, allowing a very efficient communication between the nets and the Simscape environment by adequately coding the UI. Purpose of the algorithm developed is to achieve good estimates and predictions of the rollover indicators object of our research: the roll angle and the three LTR of each axles. This has to be done by analyzing the datasets whose generation has been described in the Subsect. 3.1. In particular, the analysis requires a training and a testing phase, considering that we are using neural networks.

Entering more in details, the process has been structured as follows. First, clearly, the training phase of the algorithm. This includes, at the beginning, the retrieving of the datasets containing the data of interest by using a dedicated script. After doing that, it is necessary to operate a selection on the 36 variables associated to each different maneuvers. Trying to simulate a realistic situation, we chose to train our networks on 6 main parameters, typical of trajectory studies and easily attainable by the ECU of the considered typology of heavy vehicle. This parameters, as mentioned in Sect. 2, are: the steering angle, the acceleration of the Center of Gravity (CG) in its three components, the longitudinal velocity and the yaw rate. Selected the parameters, before the definition and the training of the neural networks, it was necessary to standardize the same. For doing that, we used the mean and the standard deviation of each variable, calculable since, in the training phase, the entire dataset is known. Defined the parameters and their standardization, it is now possible to design the core of our algorithm: the architecture of the neural networks. First of all, we defined two different typologies of nets, one for estimative purposes, net_1, and the other for predictive ones, net_1_1. These networks share the same architecture, but are characterized by being one (net_1_1) subsequent to the other (net_1). As a matter of fact, hypothesising a real time application, purpose of net_1 is to estimate the four rollover indicators by using the six parameters aforementioned, therefore giving results at the same time of what are, effectively, its inputs. We then have a net that has 6 inputs (features) and 4 outputs (responses). This phase of estimate is strongly necessary because, otherwise, it would be impossible to attain rollover indicators with immediate response times (this is one of the reasons to use neural networks and not a classic dynamic model). Obtained the four rollover indexes, the same are then put out of phase forward in time compared to the input ones, in a predictive perspective. Synthesising, net_1 is characterized by having 6 different inputs and 4 different outputs, while net_1_4 inputs coinciding with the four outputs obtained by net_1, and 4 outputs that are simply the inputs shifted of time unities (in general corresponding to 3–4 s, but it is at discretion of the user). These two networks, as anticipated, share the same structure, and are characterized by being RNN, typical for time series analysis [13].

In particular, the architecture is composed of 4 different layers, with the first defined as a sequence input one (where the number of features is exploited), the second as an LSTM, the third as a fully connected (number of responses) and the fourth as a regression layer. It has to be noted that the LSTM layer is the one that learns long-term dependencies between time steps in time series and sequence data, and is therefore the one directly related to the number of neurons (hidden units) selected. In our case, we present neural networks with

100 neurons, number defined after having tried empirical rules to identify the same, and then after several tries and having seen that, with more than 100 of them, the results did not improve in accuracy and the process was becoming too time consuming (while with a minor number of neurons, the results start to present a decline in accuracy). The number and the typology of layers have been chosen following similar criteria. After having designed the main parts of the architecture, several training options were then defined. More specifically, we implemented an adaptive moment estimation with gradient threshold equal to 1, an initial learn rate of 0.01 with a drop factor of 0.2 after 120 epochs (with a maximum of 200), and an option that does not allow the data to be shuffled (since we want to maintain the original order). Other minor settings were than defined, but always after comparing the same with their alternative (for example, the stochastic gradient descent with momentum gave us results not suitable for any application, at least with our data) and evaluating the following results.

Designed the architecture, the training could take place. We used 19 of the 23 databases for this purpose, and the entirety of each dataset has been used (for what regards the 6 features) to train the networks and update their weights progressively. It is important to highlight again that, in a real case simulation, the first has the task of obtaining the overturning indicators (otherwise difficult to calculate quickly and precisely) and, by providing them in output, give way to the second type of network to predict an advanced trend over time. The training of the networks is done by trying to simulate this kind of situation, subjecting the two types of networks to a continuous update with the progressive addition of the data of each maneuver, in order to obtain a system as adaptable as possible. It has to be noted that we defined 2 different typologies of neural networks but, in reality, we obtained 19 of them. As a matter of fact, for each update of the networks, the script is programmed to save the new version in a specific folder, allowing us to evaluate, by seeing the graphs, if introducing a certain type of maneuver improves or damages the performance (in our case, there is a general improvement, obtained also by the elimination of seven other maneuvers that did not contribute to the improvement of the net). Clearly, during the training phase, it is exploited the fact of knowing in advance the target values of roll and LTR on which the networks must be trained. In our case, these are known from virtual simulation on Simscape, in case of not synthetic data a specific script should be prepared to calculate the same values from those available.

Completed the training phase, it is possible to describe the final part of the methodology developed, the testing of the AI system. This has been done by using 4 of the 23 maneuvers defined in the Subsect. 3.1, unknown to the networks and characterized by being very different one from each other, other than being of particular interest from the point of rollover analysis (strong stresses for the vehicle). The datasets tested are characterized by passing their 6 characteristics parameters to net_1, that obtains the 4 rollover indexes that will be the input for net_1_1. The testing phase is simulated in the perspective of highlighting the general behavior of the neural networks and the efficiency of the training. Therefore, as an example of the results obtained, in Figs. 5 and 6, it is possible to see the ones achieved with the network corresponding to the end of the training session (the nineteenth). It is important to highlight that the maneuver in exam was created simulating a movement of the truck on the F1 circuit of Suzuka (in a simplified version). It should be noted that, contrary to the training phase (where the times vary from minutes to hours, depending on the machine used and the possibility of using parallel computing), in this case, the time of response are immediate (order of tenths of a second), justifying the using of neural networks to obtain the indexes researched.

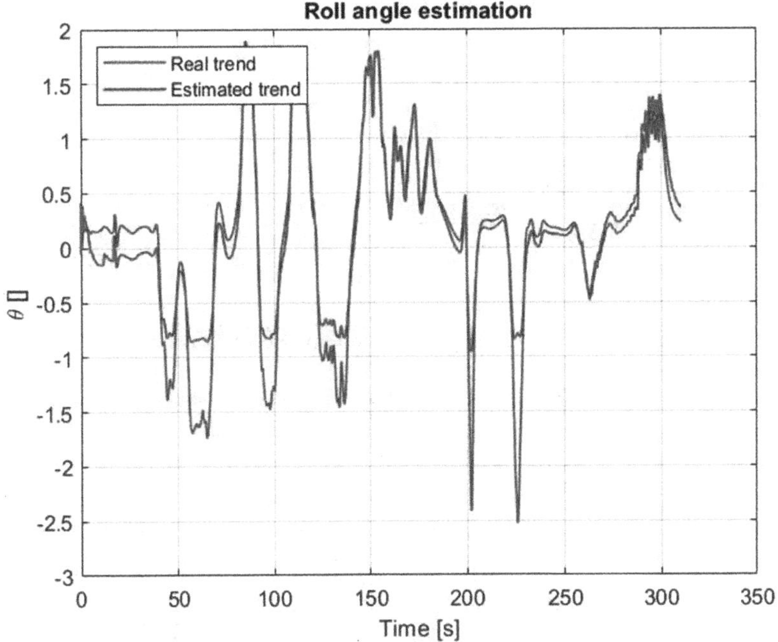

Fig. 5. Roll trend, net_1, maneuver on Suzuka circuit

Referring to Figs. 5 and 6, it is easy to observe really promising results from net_1. As a matter of fact, this neural network seems able, despite being trained in a general way (considering the variety in maneuvers of the datasets composing the training), to estimate with a very short time (tenth of seconds) and with good accuracy the requested indexes. This is confirmed even in a heavily conditioned noise situation, as can be seen in Fig. 7.

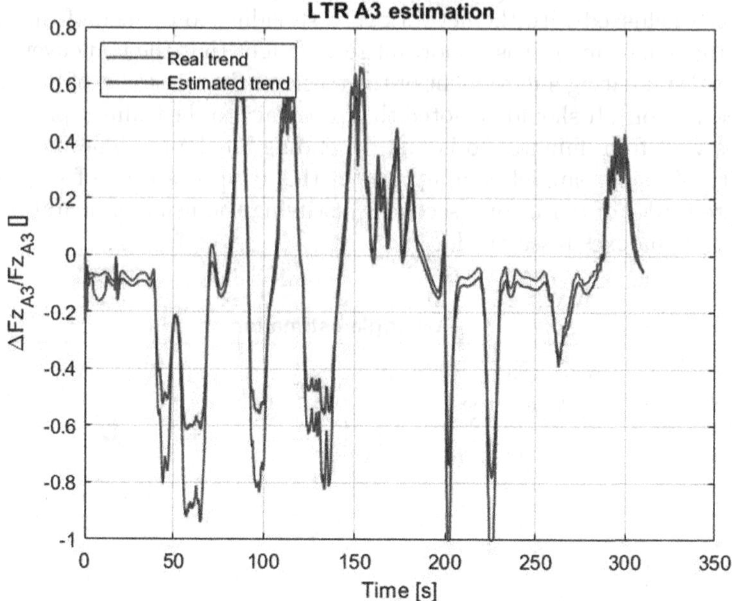

Fig. 6. LTR A3 trend, net_1, maneuver on Suzuka circuit

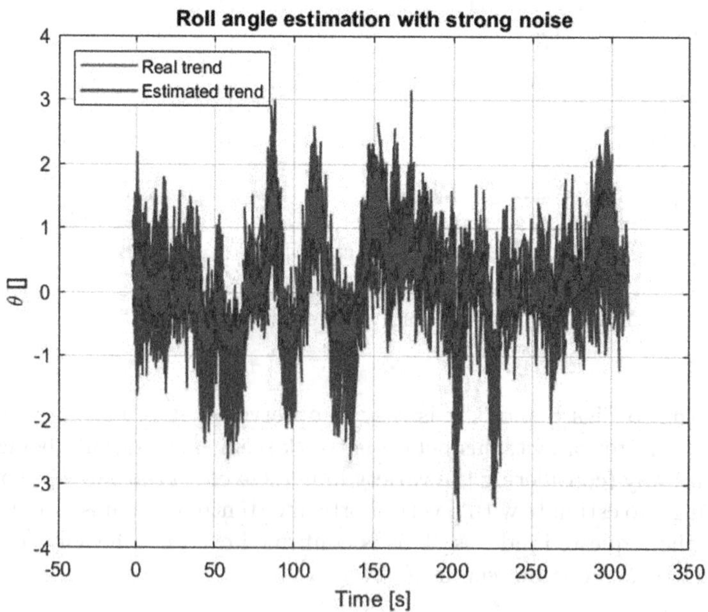

Fig. 7. Roll trend, net_1, Suzuka circuit, noise presence

It has to be noted that, since the datasets are composed almost in their entirety by maneuvers carried out on flat surfaces, net_1 tends to be able to correctly estimate the parameters of unknown ones but, if the same take place in terrain characterized by bumps or hills (respectively, Rough Road and Plateau, Fig. 9), as it is easily understandable, the estimation of the indicators loses in precision and accuracy.

Regarding what it has been obtained with net_1_1, as it is possible to observe in Fig. 8, the results still show the needing for an improvement of the same, probably both from code writing and, in particular, datasets training aspects.

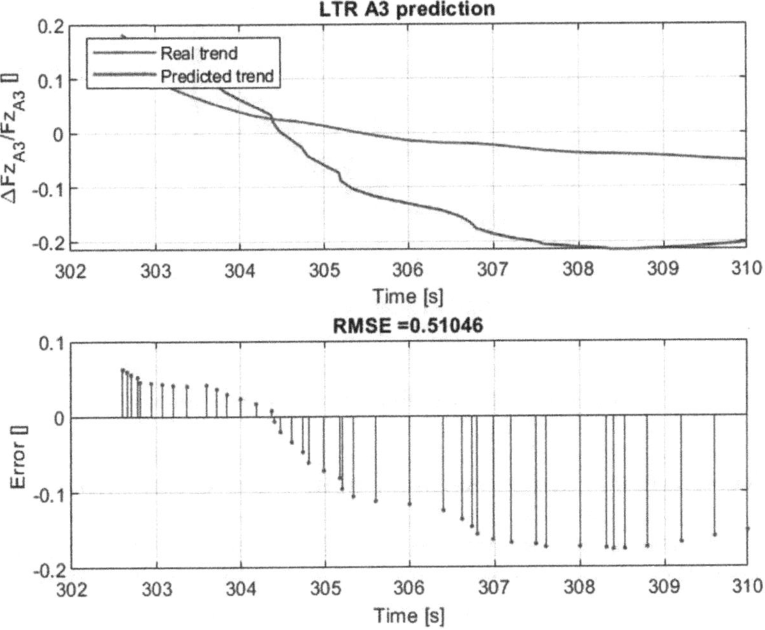

Fig. 8. LTRA3 trend, net_1_1, maneuver on Suzuka circuit

As anticipated, even in this case the main changes were introduced in the UI. This includes the possibility to interact immediately with the training and testing codes, modifying or adjusting them varying on necessities.

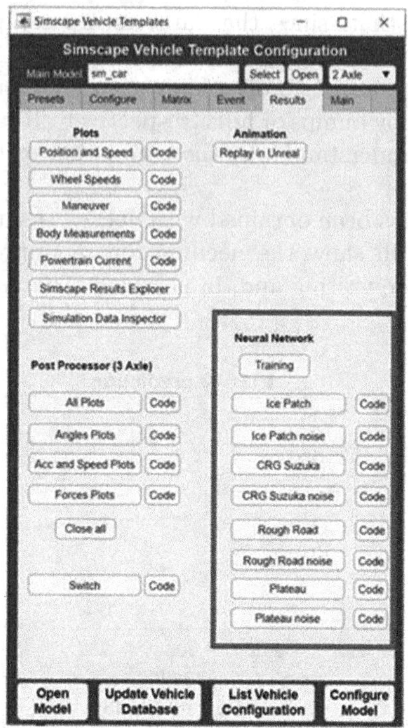

Fig. 9. UI with neural networks implementation

4 Conclusions and Possible Future Developments

The methodology presented revealed to be particularly effective in generating realistic synthetic data, representing a low cost alternative to data otherwise difficult to attain in acceptable time, quantity and costs. These data were then used to design an AI system based on two different typologies of neural networks, one for estimative, the other for predictive purposes. Even if the latter did not perform efficiently enough in terms of accuracy, the first one has already shown encouraging results as an estimator of rollover risk indicators, thanks to immediate responses and good accuracy. This suggests that the algorithm developed could be a valid opportunity for rollover risk estimation, representing a system that is easily adaptable, updatable and versatile. In order to improve the same we are expanding the databases and implementing new training techniques to obtain useful results also for the predictive part.

References

1. Azad, N., Khajepour, A., McPhee, J.: A survey of stability enhancement strategies for articulated steer vehicles. Int. J. Heavy Veh. Syst. **16**(1–2), 26–48 (2009). https://doi.org/10.1504/IJHVS.2009.023853

2. Baldi, M.M., Perboli, G., Tadei, R.: Driver maneuvers inference through machine learning. In: Pardalos, P.M., Conca, P., Giuffrida, G., Nicosia, G. (eds.) MOD 2016. LNCS, vol. 10122, pp. 182–192. Springer, Cham (2016). https://doi.org/10.1007/978-3-319-51469-7_15
3. Braghin, F., Cheli, F., Corradi, R., Tomasini, G., Sabbioni, E.: Active anti-rollover system for heavy-duty road vehicles. Veh. Syst. Dyn. **46**(SUPPL.1), 653–668 (2008). https://doi.org/10.1080/00423110802033064
4. Chen, X., Chen, W., Hou, L., Hu, H., Bu, X., Zhu, Q.: A novel data-driven rollover risk assessment for articulated steering vehicles using RNN. J. Mech. Sci. Technol. **34**(5), 2161–2170 (2020). https://doi.org/10.1007/s12206-020-0437-4
5. Evans, J.L., Batzer, S.A., Andrews, S.B.: Evaluation of heavy truck rollover accidents (2005)
6. Evans, J.L., Batzer, S.A., Andrews, S.B., Hooker, R.M.: Evaluation of heavy truck rollover crashworthiness. **2006**, 91–96 (2006)
7. Ghazali, M., Durali, M., Salarieh, H.: Path-following in model predictive rollover prevention using front steering and braking. Veh. Syst. Dyn. **55**(1), 121–148 (2017). https://doi.org/10.1080/00423114.2016.1246741
8. Iida, M., Nakashima, H., Tomiyama, H., Oh, T., Nakamura, T.: Small-radius turning performance of an articulated vehicle by direct yaw moment control. Comput. Electron. Agric. **76**(2), 277–283 (2011). https://doi.org/10.1016/j.compag.2011.02.006
9. Imine, H., Benallegue, A., Madani, T., Srairi, S.: Rollover risk prediction of heavy vehicle using high-order sliding-mode observer: Experimental results. IEEE Trans. Veh. Technol. **63**(6), 2533–2543 (2014). https://doi.org/10.1109/TVT.2013.2292998
10. Li, H., Zhao, Y., Wang, H., Lin, F.: Design of an improved predictive LTR for rollover warning systems. J. Braz. Soc. Mech. Sci. Eng. **39**(10), 3779–3791 (2017). https://doi.org/10.1007/s40430-017-0796-7
11. Li, L., Lu, Y., Wang, R., Chen, J.: A three-dimensional dynamics control framework of vehicle lateral stability and rollover prevention via active braking with MPC. IEEE Trans. Industr. Electron. **64**(4), 3389–3401 (2017). https://doi.org/10.1109/TIE.2016.2583400
12. Li, X., Wang, G., Yao, Z., Qu, J.: Dynamic model and validation of an articulated steering wheel loader on slopes and over obstacles. Veh. Syst. Dyn. **51**(9), 1305–1323 (2013). https://doi.org/10.1080/00423114.2013.800893
13. Rahman, A., Srikumar, V., Smith, A.D.: Predicting electricity consumption for commercial and residential buildings using deep recurrent neural networks. Appl. Energy **212**, 372–385 (2018). https://doi.org/10.1016/j.apenergy.2017.12.051
14. Sellami, Y., Imine, H., Boubezoul, A., Cadiou, J.C.: Rollover risk prediction of heavy vehicles by reliability index and empirical modelling. Veh. Syst. Dyn. **56**(3), 385–405 (2018). https://doi.org/10.1080/00423114.2017.1381980
15. Service de l'Observation et des Statistiques: Bilan social annuel du transport routier de marchandises. Technical report, SOeS (2015)
16. US Department of Transportation: Traffic safety facts 2016: a compilation of motor vehicle crash data from the fatality analysis reporting system and the general estimates system. Technical report, NHTSA (2017)
17. Zhang, X., Yang, Y., Guo, K., Lv, J., Peng, T.: Contour line of load transfer ratio for vehicle rollover prediction. Veh. Syst. Dyn. **55**(11), 1748–1763 (2017). https://doi.org/10.1080/00423114.2017.1321773

18. Zhu, T., Yin, X., Li, B., Ma, W.: A reliability approach to development of rollover prediction for heavy vehicles based on SVM empirical model with multiple observed variables. IEEE Access **8**, 89367–89380 (2020). https://doi.org/10.1109/ACCESS.2020.2994026
19. Zhu, T., Yin, X., Na, X., Li, B.: Research on a novel vehicle rollover risk warning algorithm based on support vector machine model. IEEE Access **8**, 108324–108334 (2020). https://doi.org/10.1109/ACCESS.2020.3001306
20. Zhu, T., Zong, C., Wu, B., Sun, Z.: Rollover warning system of heavy duty vehicle based on improved TTR algorithm. Jixie Gongcheng Xuebao/J. Mech. Eng. **47**(10), 88–94 (2011). https://doi.org/10.3901/JME.2011.10.088

Hyperparameter Tuning of Random Forests Using Radial Basis Function Models

Rommel G. Regis(✉)

Saint Joseph's University, Philadelphia, PA 19131, USA
rregis@sju.edu

Abstract. This paper considers the problem of tuning the hyperparameters of a random forest (RF) algorithm, which can be formulated as a discrete black-box optimization problem. Although default settings of RF hyperparameters in software packages work well in many cases, tuning these hyperparameters can improve the predictive performance of the RF. When dealing with large data sets, the tuning of RF hyperparameters becomes a computationally expensive black-box optimization problem. A suitable approach is to use a surrogate-based method where surrogates are used to approximate the functional relationship between the hyperparameters and the overall out-of-bag (OOB) prediction error of the RF. This paper develops a surrogate-based method for discrete black-box optimization that can be used to tune RF hyperparameters. Global and local variants of the proposed method that use radial basis function (RBF) surrogates are applied to tune the RF hyperparameters for seven regression data sets that involve up to 81 predictors and up to over 21K data points. The RBF algorithms obtained better overall OOB RMSE than discrete global random search, a discrete local random search algorithm and a Bayesian optimization approach given a limited budget on the number of hyperparameter settings to consider.

Keywords: Random forest · Hyperparameter tuning · Discrete optimization · Black-box optimization · Surrogate models · Radial basis functions

1 Introduction

Random forests are among the most popular and widely used machine learning tools that have shown good performance on many classification and regression problems. A *random forest (RF)* [5] is an ensemble of decision trees built from bootstrap samples of the set of training data. It is more general than bagged decision trees, which are also built from boostrap samples. The main difference between these decision tree ensembles is that, in an RF, not all predictors are considered at every split of each tree. This results in trees that are less correlated than those in bagged decision trees and yields improved generalization ability for the RF. For an introduction to RFs, see [12] or [16].

G. Nicosia et al. (Eds.): LOD 2022, LNCS 13810, pp. 309–324, 2023.
https://doi.org/10.1007/978-3-031-25599-1_23

The predictive performance of a random forest depends in part on the setting of the hyperparameters that control the learning process. In this paper, we focus on four RF hyperparameters: number of trees in the ensemble (ntree), number of predictor variables drawn randomly for each split (mtry), minimum number of samples in a node (min.node.size), and the fraction of the training set drawn randomly (with replacement) for building each tree (sample.frac). The goal is to find the settings of these hyperparameters that minimize the overall out-of-bag (OOB) prediction error of the RF. In general, a lower OOB prediction error leads to better predictive ability of the RF on new or unseen cases.

The general problem of hyperparameter tuning of machine learning models, including RFs, can be formulated as a black-box optimization problem where the decision variables are the hyperparameters and the objective function is an estimate of the generalization or prediction error of the model. In this paper, only discrete ordinal settings for the hyperparameters are considered. However, the proposed method can be modified to deal with hyperparameters that vary on a continuous scale. Formally, the hyperparameter tuning problem is to find an approximate optimal solution to the following discrete optimization problem:

$$\begin{aligned} & \min_{\Theta \in \mathbb{R}^d} \ \mathcal{E}(\Theta) \\ \text{s.t.} \quad & \\ & \theta^{(i)} \in D_i \subset \mathbb{R} \quad \text{for } i = 1, \ldots, d \end{aligned} \tag{1}$$

where $\Theta = (\theta^{(1)}, \ldots, \theta^{(d)})$ so that $\theta^{(i)}$ is the ith hyperparameter, and D_i is the finite set of the possible discrete ordinal settings of $\theta^{(i)}$. The elements in D_i do not have to be integers. For example, in the case of RFs, the hyperparameter sample.frac can take fractional values such as 0.7. Moreover, the objective function $\mathcal{E}(\Theta)$ is a black-box since an explicit analytical formula for this function in terms of the hyperparameters is not available. Moreover, $\mathcal{E}(\Theta)$ could be a random variable, resulting in a stochastic optimization problem. For example, in the case of RFs, the overall out-of-bag (OOB) prediction error will vary depending on the bootstrap samples used and the random sample of predictors considered at each of the splits. However, the above optimization problem is treated as deterministic for now by fixing the random seed when calculating $\mathcal{E}(\Theta)$. Future work will consider the case where $\mathcal{E}(\Theta)$ is a random variable and so we minimize the expected value of $\mathcal{E}(\Theta)$ over the same search space.

Among the most popular hyperparameter tuning strategies include grid search and random search [4]. However, when dealing with large datasets and machine learning models with many hyperparameters, the calculation of the objective function $\mathcal{E}(\Theta)$ can be computationally expensive. Hence, a complete grid search might not be feasible and simple uniform random search is not expected to yield good results. In the computationally expensive setting, a popular approach is to use surrogate-based optimization methods [27], including Bayesian optimization techniques [1]. For example, the HORD algorithm [11], which used RBF models, was used to optimize the hyperparameters for a convolutional neural network. Moreover, Bayesian optimization was used to tune the hyperparameters of a Deep Neural Network (DNN) and a Support Vector Machine (SVM) with RBF kernel [14]. Also, Bayesian optimization was used for hyperparameter tuning of a few machine learning models and was shown to be empirically superior to random search [26]. As for random forests, a recent paper on hyperparameter tuning strategies, including model-based methods, is given by [20].

This paper develops a surrogate-based algorithm for black-box optimization that can be used to tune hyperparameters of machine learning models when the hyperparameters take values on a discrete set. The proposed *B-CONDOR* algorithm is meant for the bound constrained discrete black-box optimization problem (1). According to [3], relatively few surrogate-based methods can handle discrete variables, so the proposed method is also a contribution in this area. In the numerical experiments, a global and a local variant of B-CONDOR are implemented using RBF surrogates and applied to tune the RF hyperparameters for seven regression data sets that involve up to 81 predictors and up to over 21K data points. Unlike Kriging or Gaussian process models, which are used in Bayesian optimization, RBF surrogates have not been widely used for the hyperparameter tuning of machine learning models, particularly with RFs. The B-CONDOR-RBF algorithms obtained better OOB RMSE for RF on the regression datasets compared with discrete global random search, a discrete local random search algorithm, and a Bayesian optimization approach based on the Expected Improvement (EI) criterion when given a limited computational budget. These results suggest that RBF models are promising for hyperparameter tuning of RFs and other machine learning models.

2 The B-CONDOR Algorithm for Hyperparameter Tuning of Random Forests

2.1 Algorithm Description

This paper proposes the surrogate-based *B-CONDOR* algorithm for the hyperparameter tuning of random forests and other machine learning models when the hyperparameters take values from a discrete set. The B-CONDOR algorithm attempts to solve the discrete black-box optimization problem in (1) and it is a modification of the *CONDOR (CONstrained Discrete Optimization using Response surfaces)* algorithm [23]. The original CONDOR algorithm was designed for computationally expensive discrete black-box optimization problems subject to black-box constraints and was successfully applied to a large-scale car structure design problem involving 222 discrete ordinal decision variables and 54 black-box inequality constraints. B-CONDOR is a modification of CONDOR that works for discrete black-box optimization problems where the decision variables are subject only to bound constraints.

As with other surrogate-based methods, B-CONDOR begins by selecting an initial set of points in the (discrete) search space where the function evaluations will take place. Here, a point represents one set of values for all the hyperparameters while one function evaluation corresponds to an evaluation of $\mathcal{E}(\Theta)$, which is an estimate of overall prediction error. For example, in the numerical experiments below on RFs for regression, $\mathcal{E}(\Theta)$ represents the overall out-of-bag (OOB) root mean square error (RMSE) of the RF for a given setting of the hyperparameters $\Theta = (\theta^{(1)}, \ldots, \theta^{(d)})$. The initial design points may be randomly selected uniformly from the discrete search space or they may be generated by using a space-filling design that is suitable for discrete search spaces. In the numerical experiments, the point corresponding to the recommended hyperparameters for a random forest are included among the initial design points. Then, the objective function $\mathcal{E}(\Theta)$ is evaluated at the initial points. The resulting objective function values are then used to fit the initial surrogates for $\mathcal{E}(\Theta)$.

In every iteration, B-CONDOR identifies the sample point with the best objective function value so far and fits a surrogate for the objective function. Next, B-CONDOR generates many random trial points within the search space $\prod_{i=1}^{d} D_i$ according to some probability distribution that may vary as the iterations progress. Two examples of how to generate trial points are described below. Then, among the trial points, the next sample point is selected to be the best trial point according to a weighted ranking between 0 and 1 of the scaled values of two criteria as in [22]: predicted value of the objective function (*surrogate criterion*), and minimum distance between the trial point and the previous sample points (*distance criterion*). The objective function is then evaluated at the best trial point, yielding a new data point that is used to build an updated surrogate for the objective function in the next iteration. The algorithm goes through the iterations until the maximum number of function evaluations has been reached.

Below is a pseudo-code that outlines the main steps of the B-CONDOR algorithm for the hyperparameter tuning problem of the form (1).

The B-CONDOR Algorithm

Inputs:

- Black-box function $\mathcal{E}(\Theta)$ defined for all $\Theta = (\theta^{(1)}, \ldots, \theta^{(d)})$ where $\theta^{(i)} \in D_i \subset \mathbb{R}$ for all $i = 1, \ldots, d$.
- Initial space-filling design $\mathcal{I}_0 = \{\Theta_1, \ldots, \Theta_{n_0}\} \subset \prod_{i=1}^{d} D_i$.
- A surrogate model to approximate the black box function $\mathcal{E}(\Theta)$ (e.g., RBFs).
- Number of random trial points to generate in each iteration: t_{rand}.
- Weight for the surrogate criterion: $0 < w_S < 1$.

(1) *(Evaluate Initial Points).* Evaluate $\mathcal{E}(\Theta)$ at an initial set of points $\mathcal{I}_0 = \{\Theta_1, \ldots, \Theta_{n_0}\}$ in the search space $\prod_{i=1}^{d} D_i$. Set the function evaluation counter $n \leftarrow |\mathcal{I}_0| = n_0$ and the sample points $\mathcal{P}_{n_0} = \mathcal{I}_0$.

(2) *(Iterate).* While $n < n_{\max}$ do:

- **(a)** *(Identify Current Best Point).* Among the sample points, let Θ_n^* be the one with the best objective function value.
- **(b)** *(Build Surrogate).* Build or update the surrogate for the objective $s_n^{\mathcal{E}}(\Theta)$ using info from all previous sample points $\mathcal{P}_n$.
- **(c)** *(Generate Trial Points).* Generate t_{rand} random trial points within the search space $\prod_{i=1}^{d} D_i$ according to some probability distribution.
- **(d)** *(Evaluate Surrogate at Trial Points).* Evaluate the surrogate for the objective $s_n^{\mathcal{E}}(\Theta)$ at the trial points in $\mathcal{T}_n$.

(e) *(Compute Minimum and Maximum of Surrogate at Trial Points).* Compute $s_n^{\mathcal{E},\min} = \min\{s_n^{\mathcal{E}}(\Theta) \mid \Theta \in \mathcal{T}_n\}$ and $s_n^{\mathcal{E},\max} = \max\{s_n^{\mathcal{E}}(\Theta) \mid \Theta \in \mathcal{T}_n\}$.
(f) *(Compute Minimum and Maximum of the Minimum Distance from Previous Sample Points).* Compute $\Delta_n^{\min} = \min\{\Delta_n(\Theta) \mid \Theta \in \mathcal{T}_n\}$ and $\Delta_n^{\max} = \max\{\Delta_n(\Theta) \mid \Theta \in \mathcal{T}_n\}$.
(g) *(Compute Surrogate Rank of Trial Points).* For each $\Theta \in \mathcal{T}_n$, compute

$$\mathcal{R}_n^S(\Theta) = \frac{s_n^{\mathcal{E}}(\Theta) - s_n^{\mathcal{E},\min}}{s_n^{\mathcal{E},\max} - s_n^{\mathcal{E},\min}}$$

provided $s_n^{\mathcal{E},\min} \neq s_n^{\mathcal{E},\max}$; otherwise, set $\mathcal{R}_n^S(\Theta) = 1$.
(h) *(Compute Minimum Distance Rank of Trial Points).* For each $\Theta \in \mathcal{T}_n$, compute

$$\mathcal{R}_n^{\Delta}(\Theta) = \frac{\Delta_n^{\max} - \Delta_n(\Theta)}{\Delta_n^{\max} - \Delta_n^{\min}}$$

provided $\Delta_n^{\min} \neq \Delta_n^{\max}$; otherwise, set $\mathcal{R}_n^{\Delta}(\Theta) = 1$.
(i) *(Compute Weighted Rank of Trial Points).* For each $\Theta \in \mathcal{T}_n$, compute

$$\mathcal{R}_n(\Theta) = w_S \mathcal{R}_n^S(\Theta) + (1 - w_S)\mathcal{R}_n^{\Delta}(\Theta)$$

(j) *(Select Next Evaluation Point).* Choose the next evaluation point Θ_{n+1} to be the trial point with the best weighted combination of surrogate rank and minimum distance rank:

$$\Theta_{n+1} = \text{argmin}_{\Theta \in \mathcal{T}_n} \mathcal{R}_n(\Theta)$$

(k) *(Evaluate Selected Point).* Evaluate $\mathcal{E}(\Theta_{n+1})$. Set $\mathcal{P}_{n+1} = \mathcal{P}_n \cup \{\Theta_{n+1}\}$ and reset $n \leftarrow n + 1$.

end.

(3) *(Return Best Point Obtained)* Return the best solution obtained Θ_n^* and the value $\mathcal{E}(\Theta_n^*)$.

In every iteration of B-CONDOR, multiple trial points are generated according to some probability distribution over the search space $\prod_{i=1}^{d} D_i$ that may adapt as the search progresses. In the numerical experiments, two variants are tested. In the *Global* variant, the trial points are generated uniformly at random throughout the entire discrete search space. In the *Local* variant, the trial points are generated in some neighborhood of the current best point Θ_n^* by modifying some of the variable settings of this current best point. As with the CONDOR algorithm [23], each component of the current best point is modified or perturbed with probability p_{pert}. Here, a perturbation of one component of the current best solution Θ_n^* consists of changing the current value of the corresponding discrete variable by either increasing or decreasing its value by a few steps. The neighborhood depth parameter, denoted by $depth_{\text{nbhd}}$, is the percentage of the number of settings that the variable is allowed to increase or decrease. For example, in the tuning of RFs below, the hyperparameter $\theta^{(2)}$ represents mtry, which is the number of predictor variables drawn randomly for each split. Suppose there are 80 possible settings of $\theta^{(2)}$ = mtry given by $D_2 = \{1, 2, \ldots, 80\}$ and suppose the current setting of $\theta^{(2)}$ at the current best point is 26. If $depth_{\text{nbhd}} = 10\%$, then $\theta^{(2)}$ is allowed to take 10% of the possible discrete settings for that variable above or below the current value. Hence, $\theta^{(2)}$ may be increased or decreased from 26 up to $\lceil 0.1(80) \rceil = 8$ discrete steps. This means that $\theta^{(2)}$ may take on the possible settings $\{18, 19, \ldots, 25, 27, 28, \ldots, 34\}$. Here, the current setting of 26 is excluded from the possible values to force a change in the value of $\theta^{(2)}$.

2.2 Radial Basis Function Interpolation

We propose using the interpolating radial basis function (RBF) model in Powell [19] as the surrogate for the B-CONDOR algorithm and refer to resulting algorithm as *B-CONDOR-RBF*. Similar to Kriging or Gaussian Process models, RBFs are ideal for the interpolation of scattered data. In this RBF model, each data point is a center and the basis functions can take various forms such as the cubic, thin plate spline, and Gaussian forms. Given n distinct points $\Theta_1, \ldots, \Theta_n \in \mathbb{R}^d$ with their function values $\mathcal{E}(\Theta_1), \ldots, \mathcal{E}(\Theta_n)$, we use the RBF model of the form:

$$s_n(\Theta) = \sum_{i=1}^{n} \lambda_i \phi(\|\Theta - \Theta_i\|) + c^T \Theta + c_0, \ \Theta \in \mathbb{R}^d,$$

where $\|\cdot\|$ is the 2-norm, $\lambda_i \in \mathbb{R}$ for $i = 1, \ldots, n$, $c \in \mathbb{R}^d$ and $c_0 \in \mathbb{R}$. Here, $c^T\Theta + c_0$ is simply a linear polynomial in d variables. Possible choices for the radial function $\phi(r)$ include: $\phi(r) = r^3$ (cubic), $\phi(r) = r^2 \log r$ (thin plate spline) and $\phi(r) = \exp(-\gamma r^2$ (Gaussian), where γ is a hyperparameter. Training this model is relatively straightforward since it simply involves solving a linear system with good numerical properties. See [19] for details. In the experiments, we choose the cubic RBF model due to its simplicity and previous success in expensive black-box optimization (e.g., see [2,6,22]).

3 Computational Experiments

3.1 Random Forest Hyperparameter Tuning Problems

The numerical experiments are conducted in R [21] via the RStudio platform [25] on an Intel(R) Core(TM) i7-7700T CPU @ 2.90 GHz, 2904 Mhz, 4 Core(s), 8 Logical Processor(s) Windows-based machine. The *ranger* R package [29] is used to train and evaluate the RFs on the data sets. This package provides a fast implementation of the RF algorithm that is well-suited for high-dimensional data. In particular, the codes for the objective functions used the ranger() function to calculate the OOB RMSE of the RF on a dataset.

The B-CONDOR-RBF algorithm is applied to RF hyperparameter tuning problems on seven regression data sets. The characteristics of these problems are summarized in Table 1. The *Concrete*, *Fuel Economy*, and *Chemical* data

Table 1. Characteristics of the RF hyperparameter tuning problems. The RFs are tuned on a fixed training set consisting of 70% of the entire data set.

Tuning problem	Number of predictors	Number of data points with no missing values	Size of the search space
Concrete	8	1030	7040
Fuel economy	13	95	11440
Boston	13	506	11440
Chemical	57	152	50160
Mashable	58	39644	51040
Ames	80	2930	70400
Superconductivity	81	21263	71280

sets are available through the AppliedPredictiveModeling R package [17]. The *Concrete* data set [30] was used to model the compressive strength of concrete formulations in terms of their ingredients and age. The *Fuel Economy* data set was used to model the fuel economy for 2012 passenger cars and trucks using various predictors such as engine displacement or number of cylinders. The *Chemical* data set was used to model the relationship between a chemical manufacturing process and the resulting final product yield. The *Boston* data set from the MASS R package [28] was used to model the median values of houses in the suburbs of Boston. The *Ames* data set [7] from the AmesHousing R package [15] was used to model the sale price of houses in Ames, Iowa. In addition, the *Mashable* and *Superconductivity* data sets come from the UC Irvine Machine Learning Repository [8]. The *Mashable* data set [9] was used to predict the popularity of news articles published by Mashable (www.mashable.com) in a period of two years. The *Superconductivity* data set [10] contains information on 21263 superconductors and was used to predict the critical temperature of a superconductor from 81 relevant features. The Fuel Economy, Mashable and Ames data sets all have both quantitative and qualitative predictor variables while the rest of the data sets only have quantitative predictors.

Recall that we wish to tune four RF hyperparameters: number of trees (ntree), number of predictor variables drawn randomly for each split (mtry), the minimum number of samples in a node (min.node.size), and the fraction of the training set drawn randomly (with replacement) for building each tree (sample.frac). This gives rise to a 4-dimensional discrete black-box optimization problem. The typical recommended hyperparameter settings for RF for regression [20] are: ntree $= 500$, mtry $= \lceil p/3 \rceil$ (where p is the number of predictors), min.node.size $= 5$, and sample.frac $= 1$ (since we are sampling with replacement). The possible values for these hyperparameters in the optimization problem are defined to include the recommended settings: ntree $\in \{100, 200, \ldots, 1000\}$, mtry $\in \{1, 2, \ldots, p\}$, min.node.size $\in \{3, 4, \ldots, 10\}$, and sample.frac $\in \{0.50, 0.55, 0.60, \ldots, 0.95, 1.0\}$. Hence, the search space has size $880p$. In the experiments, these variables are normalized to range from 0 to 1.

Before any experiments are performed, the rows with missing values are removed from the data sets. Table 1 shows the number of remaining rows in each data set. Then, each data set is split into a training set (70%) and a test set (30%) only once and the *objective function* is defined as the *overall out-of-bag (OOB) RMSE of the RF on the training set.* By using the same training set and by fixing the random seed in the calculation of the OOB RMSE for any setting of the RF hyperparameters, each RF tuning problem becomes deterministic. Later, after the optimization algorithms are used to minimize the objective function, the best RFs (in terms of OOB RMSE on the training set) obtained by the various algorithms are evaluated on the test set.

3.2 Experimental Setup

Two variants of the B-CONDOR-RBF algorithm with the cubic RBF model are implemented in R. One uses the global scheme (B-CONDOR-RBF-Global) while the other uses the local scheme (B-CONDOR-RBF-Local) to generate $100d$ random trial points in each iteration. For B-CONDOR-RBF-Local, the values of the perturbation probability p_{pert} and the neighborhood depth parameter $depth_{\text{nbhd}}$ used to generate trial points in the neighborhood of the current best solution are set to vary in a cycle that begins with a more global search and ends with a more local search as in CONDOR [23]. In particular, $(p_{\text{pert}}, depth_{\text{nbhd}})$ vary according to the following cycle: $\langle(1, 50\%), (0.5, 50\%), (0.5, 25\%), (0.5, 10\%), (0.5, 5\%)\rangle$. Also, for the two B-CONDOR-RBF algorithms, the weights used for the surrogate and distance criteria are 0.95 and 0.05, respectively, as in [22].

Since random search is a popular approach for hyperparameter tuning of RF, we compare B-CONDOR-RBF with a *Discrete Global Random Search* method and a *Discrete Local Random Search* method on the RF tuning problems. The former selects the next evaluation point uniformly at random throughout the entire discrete search space while the latter generates its evaluation point using the perturbation probability p_{pert} and the neighborhood depth parameter $depth_{\text{nbhd}}$ from B-CONDOR-RBF-Local and varying using the same cycle as above. The main difference between B-CONDOR-RBF-Local and the Discrete Local Random Search method is that the former uses an RBF to identify the most promising from multiple trial points while the latter generates only one trial point, which becomes the evaluation point.

We also compare the B-CONDOR-RBF algorithms with another surrogate-based method for RF hyperparameter tuning. Unfortunately, there does not seem to be any publicly available codes for surrogate-based optimization or Bayesian optimization that handles discrete variables. As mentioned earlier, there are relatively few surrogate approaches that can handle discrete variables in the literature. Hence, we implement a Bayesian optimization approach that handles ordinal discrete variables that is similar to the EGO algorithm [13]. It uses a Gaussian kernel for the covariance function and its next sample point is an approximate maximizer to the expected improvement (EI) function over the discrete search space. Moreover, the approximate maximizer of the EI is obtained by discrete uniform random search for $100d$ iterations. We use the km() and EI()

functions from the DiceKriging [24] and DiceOptim [18] R packages, respectively, to implement this alternative method, which we refer to as *Discrete-EI*.

The B-CONDOR-RBF algorithms and alternative methods are run for 30 trials on five of the problems and for 5 trials only on the more computationally expensive Mashable and Superconductivity problems. Each trial (for all algorithms) starts with the same $5(d+1)$ initial points (hyperparameter settings) where the function (overall OOB RMSE) is evaluated. The first initial point corresponds to the typical recommended values of the hyperparameters (given above) and the remaining $5d+4$ initial points are chosen uniformly at random throughout the search space. Moreover, in each trial, the algorithm is run up to a computational budget of $20(d+1)$ function evaluations. We use factors of $d+1$ since $d+1$ is the minimum number of points needed to fit a linear regression model, which is the simplest surrogate that can be used. The recommended hyperparameter settings for RF work reasonably well in many settings [20], and by including the point that corresponds to these settings, we can determine how much improvement is possible with tuning the hyperparameters given a budget on the number of function evaluations.

3.3 Results and Discussion

Table 2 shows the mean and standard error of the best objective function value (OOB RMSE on the fixed training set) obtained by the B-CONDOR-RBF algorithms and alternative methods at different computational budgets on the RF hyperparameter tuning problems. The means and standard errors are taken over 30 trials except on the Mashable and Superconductivity problems where the algorithms are run for only 5 trials. The best in each row of the table is indicated by a solid blue box while the second best is indicated by a dashed magenta box. Moreover, Fig. 1 shows the plots of the mean of the best overall OOB RMSE obtained by the five algorithms on four of the RF tuning problems for various computational budgets (function evaluations) beginning with the initial design points of size $5(d+1)$ up to a maximum of $20(d+1)$ function evaluations. The error bars represent 95% t confidence intervals for the mean. The other plots are not included due to space limitations.

The results in Table 2 and Fig. 1 indicate that the B-CONDOR-RBF algorithms are promising methods for the hyperparameter tuning of RFs. In particular, B-CONDOR-RBF-Global and B-CONDOR-RBF-Local have the best and second best results on most of the problems and computational budgets considered. Moreover, the two B-CONDOR-RBF algorithms consistently outperformed Discrete Global Random Search on all RF tuning problems after $10(d+1)$, $15(d+1)$ and $20(d+1)$ function evaluations. In fact, Fig. 1 shows that the two B-CONDOR-RBF algorithms are better than Discrete Global Random Search by a wide margin. Also, B-CONDOR-RBF-Global is better than Discrete Local Random Search on five of the seven RF tuning problems and is competitive with the latter on the Mashable and Superconductivity problems. In addition, B-CONDOR-RBF-Local is better than Discrete Local Random Search on five of the seven RF tuning problems and is competitive on the Boston and Mashable

Table 2. Mean and standard error of the best overall out-of-bag (OOB) RMSE found by B-CONDOR-RBF and alternative methods at different computational budgets on the RF tuning problems. The mean is taken over 30 trials except on the Mashable and Superconductivity problems where only 5 trials are available. The best in each row is indicated by a solid blue box while the second best is indicated by a dashed magenta box.

Tuning problem	Number of evaluations	B-CONDOR-RBF-local (cubic)	B-CONDOR-RBF-global (cubic)	Discrete-EI (Gaussian)	Discrete local random search	Discrete global random search
Concrete	10(d+1)	5.2096 (2.35e−03)	5.2075 (1.89e−03)	5.2058 (3.52e−03)	5.2266 (3.77e−03)	5.2576 (4.86e−03)
	15(d+1)	5.2043 (1.72e−03)	5.2029 (1.04e−03)	5.2007 (9.93e−04)	5.2099 (2.33e−03)	5.2472 (4.52e−03)
	20(d+1)	5.2014 (9.89e−04)	5.1996 (5.68e−04)	5.1997 (9.35e−04)	5.2050 (1.82e−03)	5.2433 (4.33e−03)
Fuel Economy	10(d+1)	3.8017 (3.69e−03)	3.7957 (1.51e−03)	3.8078 (6.11e−03)	3.8488 (9.03e−03)	3.9016 (8.06e−03)
	15(d+1)	3.7979 (3.02e−03)	3.7926 (1.04e−03)	3.8015 (5.82e−03)	3.8232 (6.10e−03)	3.8928 (7.80e−03)
	20(d+1)	3.7979 (3.02e−03)	3.7898 (4.08e−04)	3.8003 (5.86e−03)	3.8052 (3.08e−03)	3.8786 (6.96e−03)
Boston	10(d+1)	3.1327 (3.71e−03)	3.1288 (2.71e−03)	3.1362 (4.54e−03)	3.1459 (4.67e−03)	3.1683 (5.27e−03)
	15(d+1)	3.1274 (3.08e−03)	3.1173 (2.12e−03)	3.1181 (4.89e−03)	3.1356 (3.84e−03)	3.1601 (5.06e−03)
	20(d+1)	3.1251 (2.80e−03)	3.1123 (2.20e−03)	3.1064 (4.71e−03)	3.1217 (4.28e−03)	3.1586 (4.82e−03)
Chemical	10(d+1)	1.0324 (9.41e−04)	1.0305 (8.47e−04)	1.0339 (8.95e−04)	1.0371 (1.24e−03)	1.0449 (1.09e−03)
	15(d+1)	1.0284 (7.18e−04)	1.0283 (7.30e−04)	1.0329 (8.16e−04)	1.0321 (9.87e−04)	1.0429 (1.01e−03)
	20(d+1)	1.0272 (5.75e−04)	1.0265 (5.08e−04)	1.0314 (7.47e−04)	1.0303 (9.66e−04)	1.0409 (9.83e−04)
Mashable	10(d+1)	10684.2816 (5.51e−01)	10684.7162 (4.34e−01)	10685.9507 (1.43e+00)	10688.2028 (2.20e+00)	10697.2647 (2.70e+00)
	15(d+1)	10683.3741 (4.53e−01)	10684.4965 (6.03e−01)	10683.4231 (3.63e−01)	10683.2773 (9.14e−01)	10695.6533 (1.47e+00)
	20(d+1)	10683.3741 (4.53e−01)	10682.8340 (4.06e−01)	10683.0254 (1.61e−01)	10681.7804 (9.46e−01)	10693.2492 (1.57e+00)
Ames	10(d+1)	25600.86 (1.15e+01)	25594.66 (1.21e+01)	25616.18 (1.27e+01)	25616.30 (8.97e+00)	25687.90 (1.10e+01)
	15(d+1)	25578.04 (9.83e+00)	25563.83 (8.17e+00)	25601.22 (1.15e+01)	25599.28 (9.05e+00)	25684.10 (1.08e+01)
	20(d+1)	25564.09 (8.07e+00)	25550.18 (7.06e+00)	25595.47 (1.16e+01)	25581.75 (8.84e+00)	25671.08 (1.20e+01)
Superconductivity	10(d+1)	9.4741 (2.16e−03)	9.4833 (3.90e−03)	9.4822 (2.35e−03)	9.4959 (5.15e−03)	9.5048 (5.69e−03)
	15(d+1)	9.4718 (3.34e−04)	9.4794 (4.58e−03)	9.4789 (2.20e−03)	9.4830 (5.44e−03)	9.5000 (6.13e−03)
	20(d+1)	9.4716 (3.65e−04)	9.4779 (3.31e−03)	9.4768 (2.49e−03)	9.4762 (2.05e−03)	9.5000 (6.13e−03)

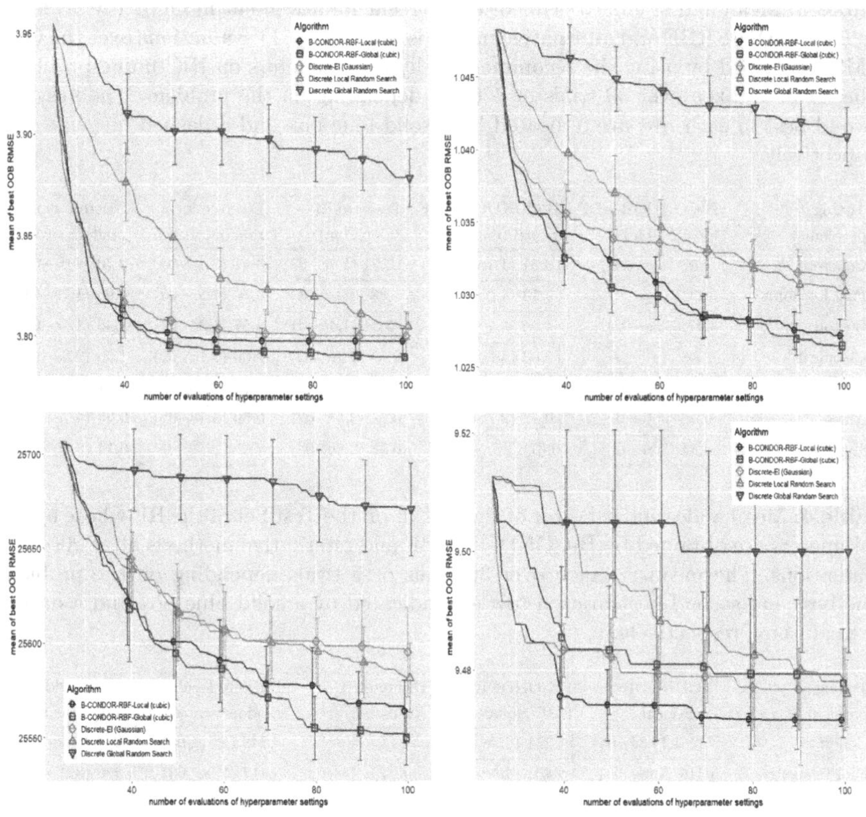

Fig. 1. Mean of the OOB RMSE obtained by the B-CONDOR-RBF algorithms and alternative methods at various computational budgets on four of the RF hyperparameter tuning problems: Fuel Economy (top left), Chemical Manufacturing Process (top right), Ames (bottom left) and Superconductivity (bottom right)

problems. In all cases where one of the B-CONDOR-RBF algorithms is overtaken by Discrete Local Random Search after a certain number of function evaluations, note that the results are not far apart.

Next, the B-CONDOR-RBF algorithms are better or at least as good as the Discrete-EI algorithm on the given problems. In particular, B-CONDOR-RBF-Global is better than or at least competitive with Discrete-EI on six of the seven tuning problems (all except on the Superconductivity problem). B-CONDOR-RBF-Local is also better than or at least competitive with Discrete-EI on six of the tuning problems (all except on the Concrete problem). These results are consistent with the plots in Fig. 1 for the Fuel Economy, Chemical, Ames and Superconductivity problems. Bayesian optimization approaches are among the state-of-the-art in surrogate-based optimization, so this makes B-CONDOR-RBF a promising alternative for expensive RF hyperparameter tuning problems.

Table 3. Mean and standard error of the percent improvement in OOB RMSE found by B-CONDOR-RBF and alternative methods after $20(d+1)$ evaluations over the OOB RMSE obtained by using the recommended hyperparameters on RF tuning problems. The mean is taken over 30 trials or 5 trials depending on the problem. The best and second best in each row are indicated by a solid blue box and a dashed magenta box, respectively.

Tuning problem	B-CONDOR-RBF-local (cubic)	B-CONDOR-RBF-global (cubic)	Discrete-EI (Gaussian)	Discrete local random search	Discrete global random search
Concrete	3.47 (1.8e−02)	3.51 (1.1e−02)	3.51 (1.7e−02)	3.41 (3.4e−02)	2.70 (8.0e−02)
Fuel Economy	6.94 (7.4e−02)	7.13 (1.0e−02)	6.88 (1.4e−01)	6.76 (7.5e−02)	4.96 (1.7e−01)
Boston	4.47 (8.6e−02)	4.86 (6.7e−02)	5.04 (1.4e−01)	4.57 (1.3e−01)	3.45 (1.5e−01)
Chemical	2.96 (5.4e−02)	3.02 (4.8e−02)	2.56 (7.1e−02)	2.66 (9.1e−02)	1.66 (9.3e−02)
Mashable	1.40 (4.2e−03)	1.41 (3.7e−03)	1.40 (1.5e−03)	1.41 (8.7e−03)	1.31 (1.5e−02)
Ames	0.66 (3.1e−02)	0.72 (2.7e−02)	0.54 (4.5e−02)	0.59 (3.4e−02)	0.25 (4.7e−02)
Superconductivity	0.44 (3.8e−03)	0.37 (3.5e−02)	0.38 (2.6e−02)	0.39 (2.2e−02)	0.14 (6.4e−02)

Table 4. Mean and standard error of the RMSE on the Test Set of the RF whose hyperparameters are obtained by B-CONDOR-RBF and alternative methods after $20(d+1)$ evaluations. The mean is taken over 30 trials or 5 trials depending on the problem. The best and second best in each row are indicated by a solid blue box and a dashed magenta box, respectively.

Tuning problem	B-CONDOR-RBF-local	B-CONDOR-RBF-global	Discrete-EI (Gaussian)	Discrete local random search	Discrete global random search
Concrete	5.241 (2.3e−03)	5.241 (2.2e−03)	5.243 (2.8e−03)	5.245 (2.5e−03)	5.257 (7.0e−03)
Fuel Economy	5.416 (7.6e−03)	5.401 (4.5e−04)	5.404 (2.5e−03)	5.417 (5.0e−03)	5.430 (8.9e−03)
Boston	3.584 (1.5e−02)	3.624 (1.3e−02)	3.659 (1.3e−02)	3.587 (1.7e−02)	3.590 (1.7e−02)
Chemical	1.300 (5.1e−04)	1.300 (5.6e−04)	1.304 (7.1e−04)	1.305 (1.5e−03)	1.312 (1.4e−03)
Mashable	2203.1 (2241)	2202.0 (2240)	2202.8 (2240)	2202.9 (2241)	2204.7 (2242)
Ames	24749.6 (32.1)	24620.2 (34.3)	24756.7 (29.4)	24668.6 (32.0)	24845.0 (41.7)
Superconductivity	1.568 (1.6)	1.567 (1.6)	1.570 (1.6)	1.567 (1.6)	1.569 (1.6)

Between the two B-CONDOR-RBF algorithms, the Global variant obtained better results than the Local variant on five of the problems and the results are competitive on the Mashable problem. However, the Local variant obtained the better result on the more difficult Superconductivity problem that involves 81 predictors. It would be interesting to investigate for which type of problems the Local variant is expected to yield better results compared to the global variant.

It is worth noting that the Discrete Local Random Search method is consistently better than Discrete Global Random Search on all problems. The former uses the same scheme to generate sample points as B-CONDOR-RBF-Local except there is no surrogate. This suggests that this local scheme is also promising for the RF hyperparameter tuning problem, especially when the problem is not computationally expensive. Moreover, on the Boston and Mashable problems, Discrete Local Random Search obtained better results than B-CONDOR-RBF-Local for larger numbers of function evaluations. This suggests that the RBF was not helpful at some point on these problems. However, within the computational budgets considered, the results for B-CONDOR-RBF-Local and Discrete Local Random Search are not too far apart, indicating that the use of the RBF does not result in significant deterioration in performance.

Recall that the typical recommended RF hyperparameter settings are included as the first point in the initial design of $5(d + 1)$ points used by all algorithms in a given trial. Table 3 shows the mean percentage improvement in the best overall OOB RMSE over the recommended hyperparameter settings obtained by the various algorithms after $20(d + 1)$ function evaluations on the RF tuning problems. As before, the best and second best for each tuning problem are indicated by a solid blue box and a dashed magenta box, respectively. When the mean values obtained are the same up to the second decimal place, the results are considered a tie. This table also shows that the B-CONDOR-RBF algorithms achieved the best and second best improvements over the recommended hyperparameters on the given problems. Note that these results are consistent with those obtained in Table 2. Moreover, the improvements are substantial with the highest mean percentage improvement at 7.13% obtained by B-CONDOR-RBF-Global on the Fuel Economy tuning problem.

Finally, Table 4 shows the mean and standard error of the RMSE on the Test Set of the RF whose hyperparameters are obtained by the B-CONDOR-RBF and alternative methods after $20(d + 1)$ evaluations. As before, the mean is taken over 30 trials on all problems except on the Mashable and Superconductivity problems where the mean is taken over only 5 trials. This table shows that the best hyperparameters found by B-CONDOR-RBF algorithms resulted in better Test Set RMSEs compared to the hyperparameters found by Discrete-EI and the random search algorithms. These results are expected since the B-CONDOR-RBF algorithms obtained the best overall OOB RMSEs on the training sets compared to the other methods.

4 Summary and Future Work

This paper considers the hyperparameter tuning of random forests (RFs) and presents the surrogate-based B-CONDOR algorithm as an alternative method to accomplish this task. We consider the case where the hyperparameters only take values on a discrete set. The hyperparameter tuning problems for RFs and other machine learning models can be formulated as a bound-constrained discrete black-box optimization problem, which is the general type of problem that B-CONDOR is designed to solve. However, B-CONDOR can be used for other discrete optimization problems beyond hyperparameter tuning.

We focused on four RF hyperparameters, and so, the resulting RF tuning problem has four decision variables and the objective is to minimize the overall out-of-bag (OOB) RMSE of the RF on the training set. We implemented a global and a local variant of the B-CONDOR algorithm using cubic RBF surrogates and the resulting B-CONDOR-RBF algorithms were applied to seven RF tuning problems on regression data sets involving up to 81 predictor variables and up to over 21K data points. The results showed that the B-CONDOR-RBF algorithms obtained consistently much better overall OOB RMSE than the Discrete Global Random Search method on all RF tuning problems. They also obtained better or at least competitive overall OOB RMSE than those obtained by the Discrete Local Random Search method and a Bayesian optimization approach on most of the RF tuning problems. RBFs are not as widely used as random search and Bayesian methods for hyperparameter tuning of machine learning models. In fact, this is one of the few, if not the only one, that used RBFs for the hyperparameter tuning of RFs.

In this study, only regression data sets were used. Hence, future work will also consider classification data sets. Moreover, other surrogate-based alternatives will be explored along with various types of radial functions for B-CONDOR-RBF. Also, it would be interesting to see if it is still possible improve the performance of the B-CONDOR-RBF algorithms on the more difficult Mashable, Ames and Superconductivity problems since the improvements in OOB RMSE obtained over the recommended hyperparameters are quite small. The method might need to be modified to work better with larger data sets involving many predictor variables. In addition, the performance of B-CONDOR on hyperparameter tuning problems involving other machine learning models, particularly deep neural networks and support vector machines, will also be explored.

References

1. Archetti, F., Candelieri, A.: Bayesian Optimization and Data Science. SO, Springer, Cham (2019). https://doi.org/10.1007/978-3-030-24494-1
2. Bagheri, S., Konen, W., Emmerich, M., Bäck, T.: Self-adjusting parameter control for surrogate-assisted constrained optimization under limited budgets. Appl. Soft Comput. **61**, 377–393 (2017)
3. Bartz-Beielstein, T., Zaefferer, M.: Model-based methods for continuous and discrete global optimization. Appl. Soft Comput. **55**, 154–167 (2017)
4. Bergstra, J., Bengio, Y.: Random search for hyper-parameter optimization. J. Mach. Learn. Res. **13**, 281–305 (2012)
5. Breiman, L.: Random forests. Mach. Learn. **45**, 5–32 (2001)
6. Costa, A., Nannicini, G.: RBFOpt: an open-source library for black-box optimization with costly function evaluations. Math. Program. Comput. **10**(4), 597–629 (2018). https://doi.org/10.1007/s12532-018-0144-7
7. De Cock, D.: Ames, Iowa: alternative to the Boston housing data as an end of semester regression project. J. Stat. Educ. **19**(3) (2011) https://doi.org/10.1080/10691898.2011.11889627
8. Dua, D., Graff, C.: UCI Machine Learning Repository. School of Information and Computer Science, University of California, Irvine (2019). https://archive.ics.uci.edu/ml
9. Fernandes, K., Vinagre, P., Cortez, P.: A proactive intelligent decision support system for predicting the popularity of online news. In: Proceedings of the 17th EPIA 2015 - Portuguese Conference on Artificial Intelligence, Coimbra, Portugal (2015)
10. Hamidieh, K.: A data-driven statistical model for predicting the critical temperature of a superconductor. Comput. Mater. Sci. **154**, 346–354 (2018)

11. Ilievski, I., Akhtar, T., Feng, J., Shoemaker, C.: Efficient hyperparameter optimization for deep learning algorithms using deterministic RBF surrogates. In: Proceedings of the AAAI Conference on Artificial Intelligence, vol. 31, no. 1 (2017)
12. James, G., Witten, D., Hastie, T., Tibshirani, R.: An Introduction to Statistical Learning with Applications in R. Springer, New York (2013). https://doi.org/10.1007/978-1-4614-7138-7
13. Jones, D.R., Schonlau, M., Welch, W.J.: Efficient global optimization of expensive black-box functions. J. Global Optim. **13**, 455–492 (1998)
14. Joy, T.T., Rana, S., Gupta, S., Venkatesh, S.: Hyperparameter tuning for big data using Bayesian optimisation. In: 2016 23rd International Conference on Pattern Recognition (ICPR), pp. 2574–2579 (2016). https://doi.org/10.1109/ICPR.2016.7900023
15. Kuhn, M.: AmesHousing: the Ames Iowa housing data. R package version 0.0.4 (2020). https://CRAN.R-project.org/package=AmesHousing
16. Kuhn, M., Johnson, K.: Applied Predictive Modeling. Springer, New York (2013). https://doi.org/10.1007/978-1-4614-6849-3
17. Kuhn, M., Johnson, K.: AppliedPredictiveModeling: functions and data sets for 'applied predictive modeling'. R package version 1.1-7 (2018). https://CRAN.R-project.org/package=AppliedPredictiveModeling
18. Picheny, V., Ginsbourger, D., Roustant, O.: DiceOptim: Kriging-based optimization for computer experiments. R package version 2.1.1 (2021). https://CRAN.R-project.org/package=DiceOptim
19. Powell, M.J.D.: The theory of radial basis function approximation in 1990. In: Light, W. (ed.) Advances in Numerical Analysis, Volume 2: Wavelets, Subdivision Algorithms and Radial Basis Functions, pp. 105–210. Oxford University Press, Oxford (1992)
20. Probst, P., Wright, M.N., Boulesteix, A.-L.: Hyperparameters and tuning strategies for random forest. WIREs Data Min. Knowl. Discov. **9**(3), e1301 (2019)
21. R Core Team: R: A language and environment for statistical computing. R Foundation for Statistical Computing, Vienna, Austria (2021). https://www.R-project.org/
22. Regis, R.G.: Stochastic radial basis function algorithms for large-scale optimization involving expensive black-box objective and constraint functions. Comput. Oper. Res. **38**(5), 837–853 (2011)
23. Regis, R.G.: Large-scale discrete constrained black-box optimization using radial basis functions. In: 2020 IEEE Symposium Series on Computational Intelligence (SSCI), pp. 2924–2931 (2020). https://doi.org/10.1109/SSCI47803.2020.9308581
24. Roustant, O., Ginsbourger, D., Deville, Y.: DiceKriging, DiceOptim: two R packages for the analysis of computer experiments by Kriging-based metamodeling and optimization. J. Stat. Softw. **51**(1), 1–55 (2012)
25. RStudio Team: RStudio: Integrated Development for R. RStudio, PBC, Boston (2020). https://www.rstudio.com/
26. Turner, R., et al.: Bayesian optimization is superior to random search for machine learning hyperparameter tuning: analysis of the black-box optimization challenge 2020. In: Proceedings of the NeurIPS 2020 Competition and Demonstration Track, in Proceedings of Machine Learning Research, vol. 133, pp. 3–26 (2021)
27. Vu, K.K., D'Ambrosio, C., Hamadi, Y., Liberti, L.: Surrogate-based methods for black-box optimization. Int. Trans. Oper. Res. **24**, 393–424 (2017)

28. Venables, W.N., Ripley, B.D.: Modern Applied Statistics with S, 4th edn. Springer, New York (2002). https://doi.org/10.1007/978-0-387-21706-2
29. Wright, M.N., Ziegler, A.: Ranger: a fast implementation of random forests for high dimensional data in C++ and R. J. Stat. Softw. **77**(1), 1–17 (2017)
30. Yeh, I.C.: Modeling of strength of high-performance concrete using artificial neural networks. Cem. Concr. Res. **28**(12), 1797–1808 (1998)

TREAT: Automated Construction and Maintenance of Probabilistic Knowledge Bases from Logs (Extended Abstract)

Ruiqi Zhu[1(✉)], Xue Li[1], Fangrong Wang[1], Alan Bundy[1], Jeff Z. Pan[1], Kwabena Nuamah[1], Stefano Mauceri[2], and Lei Xu[2]

[1] School of Informatics, University of Edinburgh, Edinburgh, UK
ruiqi.zhu@ed.ac.uk

[2] Huawei Ireland Research Centre, Dublin, Ireland

Abstract. Knowledge bases (KBs) are ideal vehicles for tackling many challenges, such as Query Answering, Root Cause Analysis. Given that the world is changing over time, previously acquired knowledge can become outdated. Thus, we need methods to update the knowledge when new information comes and repair any identified faults in the constructed KBs. However, to the best of our knowledge, there are few research works in this area. In this paper, we propose a system called TREAT (Tacit Relation Extraction and Transformation) to automatically construct a probabilistic KB which is continuously self-updating such that the knowledge remains consistent and up to date.

1 Motivation

Currently, it is common that data-driven machine learning models present one or more of the following limitations: (1) they can demand a large amount of high-quality training data [4]; (2) they require a significant amount of computational resources both at training and prediction time [4]; (3) it can be difficult for humans to interpret the rationale behind their output [1]. Our idea is to overcome these limitations by involving knowledge-driven methods. In line with the ambition of the so-called "Third wave of AI" [7], we envision a solution to use data-driven methods for building and continually maintaining a KB that can inform a variety of knowledge-driven methods.

A more concrete motivating problem is *automated KB construction from a 5G network system's logs and relevant documentations*. Logs contain information on system architecture and running status, which are good sources for obtaining knowledge. Yet, considering the logs are weakly structured, vary in different formats, and do not follow a natural language grammar, we need some novel techniques to acquire knowledge from them. There have been automated knowledge acquisition techniques by data-driven methods, e.g. LeKG [11]. However, even state-of-the-art techniques for open knowledge acquisition are error-prone, due to noise in the data. Thus, we also need to quantify the uncertainty of the acquired knowledge, and it is necessary to have some mechanisms for updating the knowledge given newly generated logs, and detecting and repairing the potential faults within the KB.

G. Nicosia et al. (Eds.): LOD 2022, LNCS 13810, pp. 325–329, 2023.
https://doi.org/10.1007/978-3-031-25599-1_24

To solve the motivating problem above, we proposed an integrated architecture named **TREAT (Tacit Relation Extraction and Transformation)** that is presented in Sect. 2. We claim that our research novelty lies in combining the existing technologies to tackle the knowledge acquisition and updating problem, as well as quantifying and mitigating the uncertainty within the knowledge.

2 The TREAT System

The TREAT system is designed to use the system logs to automatically construct a self-updating probabilistic KB for a running network instance. The overall architecture of the TREAT system is illustrated in Fig. 1. The "target network instance" continuously produces a stream of logs. The TREAT workflow *acquires* knowledge from the documentation that forms the initial KB, and from the logs that continuously update the KB. It also *assigns* probabilities to the newly obtained knowledge, and *updates* and *revises* the existing knowledge. The knowledge is put in the "PKB" in the diagram, which is a set of probabilistic triples $\{p_1 :: \phi_1, ..., p_n :: \phi_n\}$ where each piece of knowledge is a triple $\phi = (subj, pred, obj)$ associated with a probability value p indicating the degree of belief for this piece of knowledge. TREAT's four phases: *acquire, assign, update, revise* together form a never-ending loop so that the "PKB" can keep up-to-date along with the evolving target network instance. We define the four phases in the following subsections.

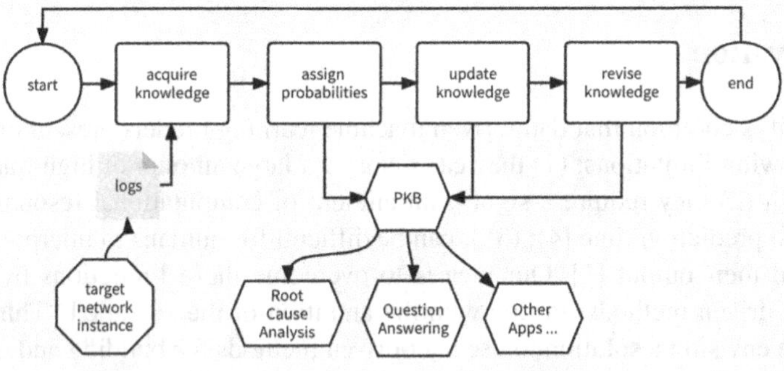

Fig. 1. Architecture and Workflow of the TREAT system

2.1 Knowledge Acquisition

This phase has two main tasks: *knowledge extraction*, which extracts useful knowledge from the system logs (and system manual), and *knowledge completion*, which enriches the KB based on the existing knowledge.

In the task of knowledge extraction, we developed a combination of text processing techniques called *LeKG* [11], which handles knowledge extraction from the documentation of natural language grammar, and logs generated from finite but unknown templates. The combined techniques involves template matching, Entity Recognition [9], Open Relation Extraction [2], and deductive reasoning with logical constraints. Finally, the output of *LeKG* is a collection of triples containing knowledge from the documentation and logs. Afterwards, we performed knowledge completion to expand the extracted KB.

In the task of knowledge completion, we mainly used a schema-aware iterative completion (SIC) pipeline recently developed by Wiharja [13]. Within this pipeline, we devised a set of functions that score the correctness of knowledge. During the completion, we generated candidate knowledge, and used the scoring functions (essentially knowledge graph embedding models [12]) to judge whether a given triple is correct. The higher the score, the more likely we can believe that the knowledge is correct, and thus could be put into the KB. However, these scores cannot be interpreted as probabilities, and they range outside the unit interval $[0, 1]$, but we subsequently leverage these scores to assign probability values in the next phase.

2.2 Initial Probability Assignment

How could we transform the scores into probabilities? In principle, we can view the scoring functions $S(\phi)$ of WP2 as feature extractors [10]. They take raw input data (i.e., the candidate triples) and produce scores as features of the input data.

Suppose that we have K scoring functions, $S_1, ..., S_K$, and set of ground truth triples $(\phi_1, y_1), ..., (\phi_m, y_m)$, where ϕ_i is a triple and $y_i \in \{0, 1\}$ is the correctness label of the ith triple, we can make a new dataset $[S_1(\phi_1), .., S_K(\phi_1), y_1], ...,$ $[S_1(\phi_n), ..., S_K(\phi_m), y_m]$ where $S_k(\phi_i)$ is the score of the ith triple given by the kth scoring function. In this new dataset, the triples with larger scores (cosine similarity) are more likely to be labelled 1, and vice versa.

Then, a simple probabilistic classifier $P : \mathbb{R}^K \rightarrow \{0, 1\}$, such as logistic regression, can be applied to learn from these extracted features, and output probabilities. We call this method post-calibration [10], meaning post-processing the output of the original model to produce well-calibrated probabilities.

2.3 Probability Updating

As explained in Sect. 2.1 and Sect. 2.2, the "target network instance" continuously produces a stream of logs, and by the procedures of Knowledge Acquisition and Initial Probability Assignment, we transform the continuous stream of logs to be a continuous stream of probabilistic triples. Collecting these probabilistic triples, we obtain a probabilistic KB $\mathcal{PKB} = \{p_1 :: \phi_1, ..., p_n :: \phi_n\}$. Yet a running system changes and evolves. To keep our probabilistic KB update to date, we need to equip it with continuous learning capability that integrates new information into the existing KB. The task of KB updating is to add new probabilistic triples $p :: \phi$ into the KB, or update the probability part p of existing probabilistic triples.

Before calling the updating procedure, we have an existing $\mathcal{PKB}$, and still, a continuous stream of probabilistic triples $\langle ..., q_j :: \phi_j, ... \rangle$ generated from the continuous log stream. For those newly arrived triples whose propositional part ϕ is not shown in $\mathcal{PKB}$, we can directly add all these new probabilistic triples into the $\mathcal{PKB}$. Otherwise, we can leverage the newly arrived probabilistic triples as new evidence to update the probabilities of the existing triples.

Updating probabilities is not so straightforward as directly adding a new probabilistic triple, but we have developed a preliminary method based on Jeffery's Conditionalisation [6]. The main idea is to regard every probability p_i as a random variable ρ_i and assume a distribution $P(\rho_i)$ (conventionally a Beta distribution $\rho_i \sim Beta(a_i, b_i)$ [5]). Then the updated distribution of the probability value is calculated based on the following formula

$$P_{new}(\rho) = q * P(\rho|\phi) + (1 - q) * P(\rho|\neg\phi) \tag{1}$$

where $P(\rho|\phi)$ and $P(\rho|\neg\phi)$ are the posterior probability calculated by Bayesian Conditionalisation and q is the probability of an observation ϕ being correct. The updated probability value is the expectation of the updated distribution $p_{new} = \mathbb{E}[P_{new}(\rho)]$.

2.4 Knowledge Revision

Apart from updating the probabilities with newly arrived information, an evolving $\mathcal{PKB}$ should also consider potential conflicts introduced to the knowledge, and fix them if any. In our data model, a piece of probabilistic knowledge $(p :: \phi)$ consists of the probability part p and the proposition part ϕ. Most of the literature on probabilistic knowledge (belief) revision focuses on how to revise the probability part p [3, 14], as we did in Sect.2.3. In the TREAT project, we not only developed the methods for updating probability p, but also attempted to figure out methods for modifying the proposition part ϕ. The TREAT system adopts Li's ABC [8] to perform revision, because the ABC provides more operations than just adding/deleting a proposition, but also changing the language (e.g., modifying predicates) or logical rules.

The ABC uses Datalog as the basic data model. It is convenient to convert triples into Datalog theories, so ABC perfectly fits our knowledge representation. It detects faulty knowledge given a Preferred Structure constituted by a set of positive triples and a set of negative triples such that all positive triples should be entailed by the KB while all the negative triples should not be entailed. We can place the knowledge extracted from documentation via templates in the Preferred Structure, or we can ask human experts to add their knowledge into the preferred structure, which also allows leveraging the valuable experts' knowledge (if any). To integrate the ABC in the TREAT system, a straightforward way is to convert the probabilistic triples into propositional ones by assigning *True* to the triples whose probabilities exceed 0.5, and *False* to the rest.

3 Conclusion

In this article, we investigated how to automatically construct an self-updating probabilistic KB from a continuous log stream. To tackle this problem, we proposed and implemented a system called TREAT, of which the components include novel methods like LeKG, ABC, for acquiring knowledge from logs, assigning probabilities to knowledge, updating and revising knowledge in the probabilistic context. We envision that the

$\mathcal{PKB}$ produced and maintained by the TREAT system can automatically update itself with newly acquired information and revise itself by detecting and repairing inconsistency and other faults, and thus the knowledge in the $\mathcal{PKB}$ can remain high-quality.

Acknowledgement. The authors would like to thank Huawei for supporting the research and providing data on which this paper was based under grant CIENG4721/LSC. The authors would also like to thank the anonymous reviewers for their helpful comments on improving the writing of this paper. For the purpose of open access, the author has applied a Creative Commons Attribution (CC BY) licence to any Author Accepted Manuscript version arising from this submission.

References

1. Arrieta, A.B., et al.: Explainable artificial intelligence (XAI): concepts, taxonomies, opportunities and challenges toward responsible AI. Inf. Fusion **58**, 82–115 (2020)
2. Bach, N., Badaskar, S.: A review of relation extraction. Literat. Rev. Lang. Stat. **II**(2), 1–15 (2007)
3. Chhogyal, K., Nayak, A., Schwitter, R., Sattar, A.: Probabilistic belief revision via imaging. In: Pham, D.-N., Park, S.-B. (eds.) PRICAI 2014. LNCS (LNAI), vol. 8862, pp. 694–707. Springer, Cham (2014). https://doi.org/10.1007/978-3-319-13560-1_55
4. Devlin, J., Chang, M.W., Lee, K., Toutanova, K.: Bert: pre-training of deep bidirectional transformers for language understanding (2019)
5. Gupta, A.K.: Beta Distribution. In: Lovric, M. (eds) International Encyclopedia of Statistical Science, pp. 144–145. Springer, Heidelberg (2011). https://doi.org/10.1007/978-3-642-04898-2_144
6. Jeffrey, R.C.: The Logic of Decision. University of Chicago Press, New York (1965)
7. Launchbury, J.: A DARPA perspective on artificial intelligence. Retrieved November 11, 2019 (2017)
8. Li, X., Bundy, A., Smaill, A.: ABC repair system for datalog-like theories. In: KEOD, pp. 333–340 (2018)
9. Mohit, B.: Named entity recognition. In: Zitouni, I. (ed.) Natural Language Processing of Semitic Languages. TANLP, pp. 221–245. Springer, Heidelberg (2014). https://doi.org/10.1007/978-3-642-45358-8_7
10. Rahimi, A., Gupta, K., Ajanthan, T., Mensink, T., Sminchisescu, C., Hartley, R.: Post-hoc calibration of neural networks. arXiv preprint arXiv:2006.12807 (2020)
11. Wang, F., et al.: Le Kg: a system for constructing knowledge graphs from log extraction. In: The 10th International Joint Conference on Knowledge Graphs. IJCKG 2021, New York, pp. 181–185 (2021)
12. Wang, Q., Mao, Z., Wang, B., Guo, L.: Knowledge graph embedding: a survey of approaches and applications. IEEE Trans. Knowl. Data Eng. **29**(12), 2724–2743 (2017)
13. Wiharja, K., Pan, J.Z., Kollingbaum, M.J., Deng, Y.: Schema aware iterative knowledge graph completion. J. Web Semant. **65**, 100616 (2020)
14. Zhuang, Z., Delgrande, J.P., Nayak, A.C., Sattar, A.: A unifying framework for probabilistic belief revision. In: IJCAI, pp. 1370–1376 (2017)

Sample-Based Rule Extraction for Explainable Reinforcement Learning

Raphael C. Engelhardt[1(✉)], Moritz Lange[2], Laurenz Wiskott[2], and Wolfgang Konen[1]

[1] Cologne Institute of Computer Science, Faculty of Computer Science and Engineering Science, TH Köln, Gummersbach, Germany
{Raphael.Engelhardt,Wolfgang.Konen}@th-koeln.de

[2] Faculty of Computer Science, Institute for Neural Computation, Ruhr-University Bochum, Bochum, Germany
{Moritz.Lange,Laurenz.Wiskott}@ini.rub.de

Abstract. In this paper we propose a novel, phenomenological approach to explainable Reinforcement Learning (RL). While the ever-increasing performance of RL agents surpasses human capabilities on many problems, it falls short concerning explainability, which might be of minor importance when solving toy problems but is certainly a major obstacle for the application of RL in industrial and safety-critical processes. The literature contains different approaches to increase explainability of deep artificial networks. However, to our knowledge there is no simple, agent-agnostic method to extract human-readable rules from trained RL agents. Our approach is based on the idea of observing the agent and its environment during evaluation episodes and inducing a decision tree from the collected samples, obtaining an explainable mapping of the environment's state to the agent's corresponding action. We tested our idea on classical control problems provided by OpenAI Gym using handcrafted rules as a benchmark as well as trained deep RL agents with two different algorithms for decision tree induction. The extracted rules demonstrate how this new approach might be a valuable step towards the goal of explainable RL.

Keywords: Reinforcement learning · Explainable RL · Rule learning

1 Introduction

One of the biggest downsides of powerful Deep Reinforcement Learning (DRL) algorithms is their opacity. The well-performing decision making process is buried in the depth of artificial neural networks, which might constitute a major barrier to the application of Reinforcement Learning (RL) in various areas. While

This research was supported by the research training group "Dataninja" (Trustworthy AI for Seamless Problem Solving: Next Generation Intelligence Joins Robust Data Analysis) funded by the German federal state of North Rhine-Westphalia.

G. Nicosia et al. (Eds.): LOD 2022, LNCS 13810, pp. 330–345, 2023.
https://doi.org/10.1007/978-3-031-25599-1_25

methods exist especially in the field of computer vision to increase explainability of deep networks, a general, simple and agent-agnostic method of extracting human-understandable rules from RL agents remains to be found. With this paper we propose a possible candidate for such a method and present results for classic control problems.

Instead of trying to specifically explain the inner mechanism of a trained RL agent, we approach the problem from a phenomenological perspective by constructing a set of simple, explainable rules, which imitate the behavior of the black-box RL agent. Our method consists of three steps:

1. A DRL agent is trained, and – for simpler problems – handcrafted (HC) policies are developed to solve the problem posed by the studied environment.
2. The agent (henceforth called "oracle") acting according to either HC rules or the trained policy is evaluated for a set number of episodes. At each timestep the state of the environment and action of the agent are logged.
3. A Decision Tree (DT) is induced from samples collected in the previous step.

The resulting tree is evaluated by applying its policy in a number of episodes and subsequently comparing average and standard deviation of the episodes' returns with the ones achieved by the oracle.

This approach has notable advantages:

- It is conceptually simple, as it translates the problem of explainable RL into a supervised learning setting.
- DTs are fully transparent and (at least for limited depth) offer a set of easily understandable rules.
- The approach is oracle-agnostic: it does not rely on the agent being trained by a specific RL algorithm, or any algorithm at all. A large enough training set of state-action pairs is sufficient.

This paper is structured in the following manner: In Sect. 2 we briefly discuss existing methods for improving explainability of RL. Section 3 explains in detail how our new approach works. Our experimental results are presented in Sect. 4.

2 Related Work

Over the years, several methods for rule deduction from other resources have been developed. Predicate invention [9,10], a subfield of inductive logic programming, is a well-known and established method for finding new predicates or rules from given examples and background knowledge in the symbolic domain. It is, however, difficult to apply to the non-symbolic domain (e.g. continuous observation and action spaces, continuous inputs and outputs of deep learning neural networks).

Rule extraction from opaque AI models has been widely studied in the context of explainable AI. There are many approaches, but none of the existing methods has proven conclusive for all applications. One method, called DeepRED [19], aims to translate feedforward networks layer-by-layer into rules and to simplify these rules. But the feedforward networks used in RL models do not always directly output policies (actions). Instead they output values or Q-functions, which are transformed by the surrounding RL mechanism into policies. Therefore, surrogate networks or policy networks, which transform the inputs directly into policies would have to be trained based on existing RL oracles in order to apply DeepRED.

Another paradigm for generating explainable models is imitation learning [13, 14] where a model with reduced complexity is learnt by guidance of a DRL oracle. The oracle serves as a form of expert demonstration for the desired behavior. Our approach is a form of imitation learning, as is the recent proposal by Verma et al. [17].

Verma et al. [17] propose a generative rule induction scheme, called Programmatically Interpretable RL (PIRL), through Neurally Directed Program Synthesis (NDPS). The key challenge is that the space of possible policies is vast and nonsmooth. This is addressed by using the DRL network to guide a local search in the space of programmatically synthesized policies. These policies are formed by a functional language that combines domain-specific operators and input elements. In the domain of classic control problems, these operators can for example mimic the well-known operators from PID control. As a result, this interesting method is able to generate complex yet interpretable rules. We will compare our results with those from [17]. Our approach differs from [17] insofar as it generates simpler rules (only one input per predicate) if we use traditional DTs like CART (of course at the price of a simpler policy space).

The method we describe in this paper is arguably closest to the mimic learning approach by Liu et al. [6]. Our approach differs to theirs in that it learns the policy directly rather than a Q-function. It does not require Q-values but merely recorded states and actions. Finally we use well-known simple DTs instead of the more complex Linear Model U-trees.

Coppens et al. [4] also use trees for policy learning, but train hierarchical filters to obtain trees with stochastic elements. While we provide explainability through interpretable rules, Coppens et al. visualize the filters and perform statistical analysis of action distributions.

3 Methods

Our method makes use of different RL environments, DRL training algorithms and DT induction techniques, which we briefly describe in the following.

3.1 Environments

We test our approach on four classic control problems made available as RL environments by OpenAI Gym [3]:

MountainCar. In this environment, first described by Moore [8, Chapter 4.3], a car initially positioned in a valley is supposed to reach a flag positioned on top of the mountain to the right as fast as possible. As the force of the car's motor is insufficient to simply drive up the slope, it needs to build up momentum by swinging back and forth in the valley (Fig. 1).

- Observation space: continuous, two-dimensional: car's position $x \in [-1.2, 0.6]$, car's velocity $v \in [-0.07, 0.07]$
- Action space: discrete, acceleration to the left, no acceleration, acceleration to the right, encoded as $0, 1, 2$
- Reward: -1 for each timestep as long as flag is not reached (flag is positioned at $x = 0.5$)
- Start state: x_0 random uniform in $[-0.6, -0.4]$, $v_0 = 0$
- Time limit: $t = 200$
- Solved if $\overline{R} \geq -110$ over 100 evaluation episodes

Fig. 1. MountainCar environment. Taken from [3]

MountainCarContinuous has the identical setup and goal as the discrete version described above. However, the action space is continuous and the reward structure is changed to favor less energy consumption. The differences to the discrete *MountainCar* are:

- Action space: continuous acceleration $a \in [-1, 1]$
- Reward: $-0.1 \cdot a^2$ for each timestep as long as the flag ($x = 0.45$) is not reached, additional $+100$ when the goal is reached
- Time limit: $t = 1000$
- Solved if $\overline{R} \geq 90$ over 100 evaluation episodes

MountainCarContinuous has the extra challenge that the cumulative reward of 0 (car stands still) is a local maximum.

CartPole described by Barto et al. [1, Chapter 5], consists of a pole balancing upright on top of a cart. By moving left or right, the cart should keep balancing the pole in an upright position for as long as possible, while maintaining the limits of the one-dimensional track the cart moves on. For better comparability with results of Verma et al. [17] mentioned in Sect. 2 we mostly use `CartPole-v0` (Fig. 2).

- Observation space: continuous, four-dimensional: car's position $x \in [-4.8, 4.8]$, car's velocity $v \in (-\infty, \infty)$, pole's angle $\theta \in [-0.42, 0.42]$ (in radians) and pole's angular velocity $\omega \in (-\infty, \infty)$
- Action space: discrete, accelerate to the left (0) or to the right (1)
- Reward: $+1$ for each timestep as long as $\theta \in [-0.21, 0.21]$ and $x \in [-2.4, 2.4]$
- Start state: $x_0, v_0, \theta_0, \omega_0$ random uniform in $[-0.05, 0.05]$
- Time limit: t = 200 (`CartPole-v0`) or t = 500 (`CartPole-v1`)
- Solved if $\overline{R} \geq 195$ (`CartPole-v0`) or ≥ 475 (`CartPole-v1`) over 100 evaluation episodes

Fig. 2. CartPole environment. Taken from [3]

Pendulum. The goal of this environment is to swing up an inverted pendulum, attached to an actuated hinge and keep it in this unstable equilibrium. The goal is the same as in CartPole, but Pendulum is more challenging, because the initial position can be at any angle and several swing-ups may be needed. (CartPole initializes at a near-upright position.) We use `Pendulum-v0` (Fig. 3).

- Observation space: continuous, three-dimensional: $(\cos\theta, \sin\theta, \omega)$ with pendulum angle θ (from positive y-axis) and its angular velocity ω.
- Action space: continuous, torque $a \in [-2, 2]$ applied to pendulum
- Reward: $-(\theta^2 + 0.1 \cdot \omega^2 + 0.001 \cdot a^2)$ at each timestep
- Start state: $\theta_0 \in [-\pi, \pi]$, $\omega_0 \in [-1, 1]$ (random uniform)
- Time limit: $t = 200$
- No official solved-condition is given for this environment

Fig. 3. Pendulum environment. Taken from [3]

3.2 Oracles

The samples on which we train the DTs are produced by two "families" of oracles, which we explain in the following.

Deep Reinforcement Learning Agents. For the training of agents we use DRL algorithms as implemented in the Python RL framework Stable Baselines3 [12]. Specifically, we work with DQN [7], PPO [15] and TD3 [5].

We apply all three algorithms to all environments, but show later on in Sect. 4 only the most successful one out of three. It should be noted that the environments are not necessarily easy for DRL: For example, on MountainCar, DQN succeeds but PPO does not. Vice versa, on CartPole, PPO succeeds but DQN does not.

Handcrafted Policies. For three problems, we present simple deterministic policies (HC), which serve a triple purpose: they constitute a proof of existence for simple, well-performing rules, they may be used as a surrogate for an oracle, and they serve as a benchmark for our core idea. They enable us to validate our approach. The fourth problem, Pendulum, currently has no HC policy. In the following, we briefly explain the HC rules for the different environments:

- MountainCar is solved by the rule set presented in Fig. 4a. The intuition behind it is as follows: The velocity is zero if and only if we are in the starting step of an episode. If $v = 0$, we push the car left or right, depending on the car's position relative to -0.4887. This constant was found empirically, it minimizes the number of swing-ups needed. In all other cases with $v \neq 0$ we push the car further in the direction of its current velocity to build up momentum. We will later see that this HC rule can be applied to MountainCarContinuous as well.

- CartPole: The following intuition motivates the rule set defined in Fig. 4b. Once the pole is in the upright position (both absolute angle $|\theta|$ and absolute angular velocity $|\omega|$ are small) it merely needs to maintain its state. In these cases the CartPole environment imitates 'no action' (which is not available) by executing the opposite of the previous action. The previous action is made available as an additional observable. If the absolute angle $|\theta|$ is small but absolute angular velocity $|\omega|$ large, the action reduces $|\omega|$. In all other cases the cart is pushed in the direction in which the pole leans, bringing it closer to the upright position.

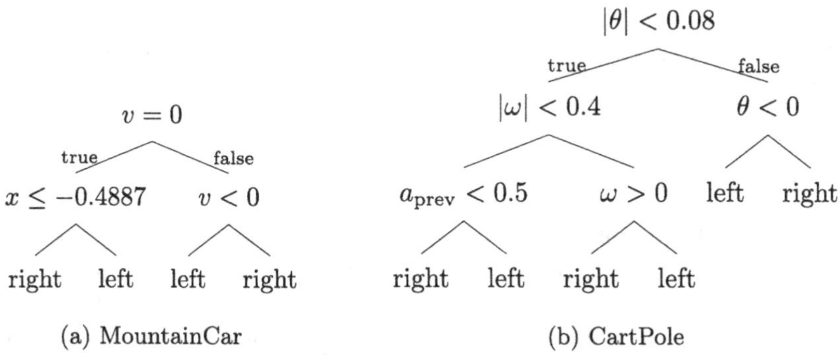

(a) MountainCar (b) CartPole

Fig. 4. HC policies for MountainCar, MountainCarContinuous, and CartPole

3.3 Decision Trees

For the induction of DTs we rely on two different algorithms:

- Classification and Regression Trees (CART) as described by Breiman et al. [2] and implemented in [11]. We use the classification or the regression form depending on whether the environment's action space is discrete or continuous, respectively.
- Oblique Predictive Clustering Trees (OPCT) as described in [16].

We chose OPCT as one representative of the whole class of oblique DTs (trees with slanted lines, such as CART-LC, OC1 and HHCART). We tested also HHCART [18] and found it to give similar results to OPCT.

4 Results

We present in this section first the results on the four selected environments individually. Section 4.5 gives a summarizing table and discusses common findings.[1]

[1] For details on the implementation and reproducibility the code can be found on https://github.com/RaphaelEngelhardt/sbreferl.

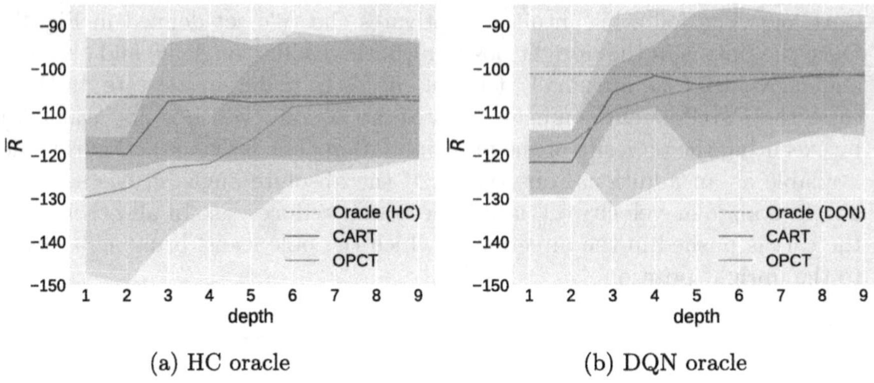

(a) HC oracle

(b) DQN oracle

Fig. 5. MountainCar. Performance comparison with different depths of trained DTs on samples produced by HC and DQN oracles.

4.1 MountainCar-v0

Our results for the MountainCar environment are as follows. The HC rule reaches an average return of $\overline{R} = -106 \pm 13$.[2] As shown in Fig. 5a, CART matches the oracle's performance at depth 3. OPCT only yields oracle-like performance at depths of at least 6, no longer considered to be easily interpretable. Trained with the DQN algorithm, the oracle yields a return of $\overline{R} = -101 \pm 10$. Again, CART has similar returns for depth ≥ 3 (Fig. 5b).

Given the two-dimensional state space of this environment, the characteristics of oracle and DT can be compared visually (see Fig. 6). Since the HC rule is formulated as a DT, it can be easily compared with the DT induced by CART from samples. Even if the trees in Fig. 4a and 7a are structured differently, the underlying rules are extremely similar. The latter shows how CART circumvents its inability to represent a predicate such as $v = 0$.

If we induce DTs from samples of a deep learning oracle, it is not a priori clear that similar rules emerge. As an example Fig. 7b shows the 'CART from DQN' rules at depth 3. It is interesting to note that the left sub-tree mainly contains leaves with action 'left'. If we replaced the 'no acceleration' leaf with 'left', a much simpler tree, structurally similar to Fig. 7a, would result with virtually same return $\overline{R} = -105 \pm 6$. The tree has just inserted the 'no acceleration' leaf to mimic the DQN samples, but it turns out not to be important for the overall return. It is noteworthy that the simplified tree is again structurally similar to both trees in Figs. 7a and 4a related to the HC policy.

[2] We report the average mean return $\pm$ average standard deviation of returns for the 5 repetitions with different seeds.

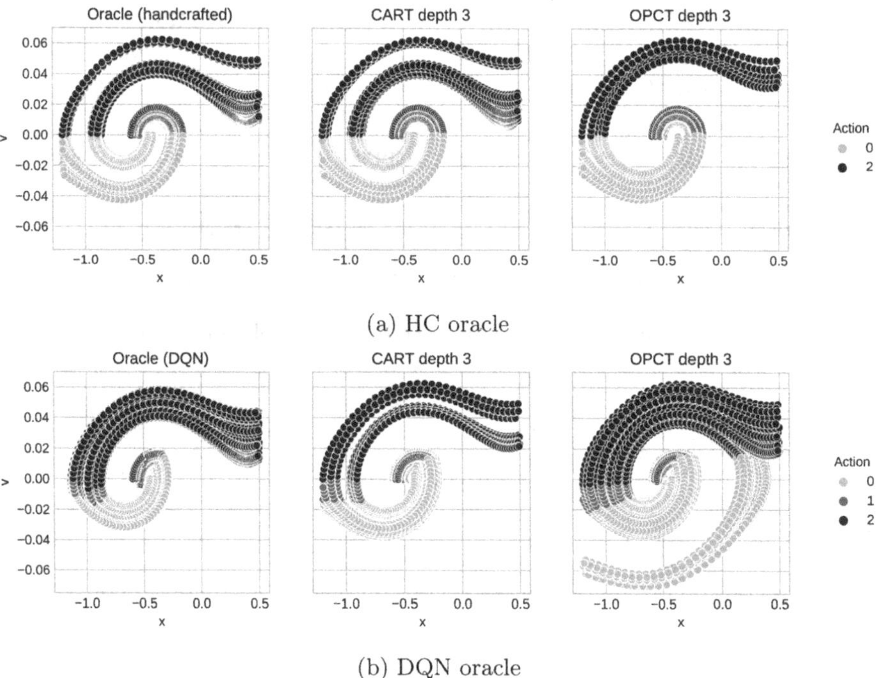

Fig. 6. MountainCar. Comparison between oracle and DT samples. Each point in the coordinate system represents the state of the environment at a timestep. The corresponding decision of the oracle or the induced DT is represented by the marker's color.

4.2 MountainCarContinous-v0

The continuous version of MountainCar is successfully solved by a TD3-trained DRL agent reaching a score of $\overline{R} = 91 \pm 2$. CART at depth 3 is able to reproduce this performance (see Fig. 8b). The performance of OPCT is somewhat weaker, but also almost satisfactory. The behaviours of oracle and trees can be compared in Fig. 9.

Given the structural similarity between the discrete and continuous version of MountainCar, it seems reasonable to apply the same HC rule, that solved the discrete version. With a performance of $\overline{R} = 91 \pm 1$ in fact, the HC rule is able to solve the continuous version as well. Our approach proves successful in that CART is able to reproduce the performance and to extract the known HC rule (neglecting minor numerical differences, the extracted DT is the same as the one shown in Fig. 7a). OPCT could not reach the oracle's performance at depth 3 (see Fig. 8a). Even at depth 1 both DTs show fairly good performance: CART finds a simplified version of the HC rule by splitting the state space at $v \approx 0$ and pushing the car in the direction of v.

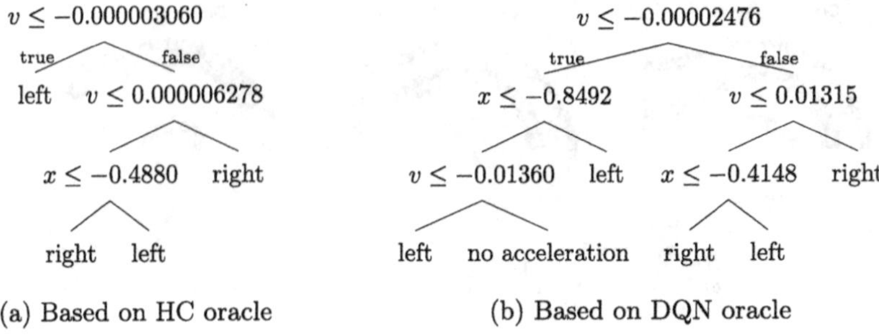

(a) Based on HC oracle (b) Based on DQN oracle

Fig. 7. MountainCar. DTs induced by CART from samples of different oracles

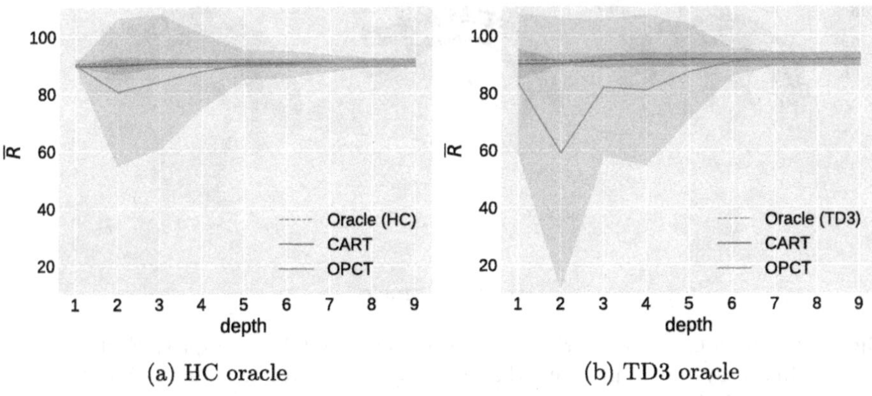

(a) HC oracle (b) TD3 oracle

Fig. 8. MountainCarContinuous. Performance comparison with different depths of trained DTs on samples produced by HC and TD3 oracles.

4.3 CartPole-v0

The CartPole environment is in a sense more challenging than the previous ones since it has four input dimensions. Interpretable models are less easy to visualize. On CartPole we test our approach with a HC oracle and with a PPO oracle.

Our HC oracle, taking into consideration the action of the previous timestep, reaches a perfect score of $\overline{R} = 200 \pm 0$ (Fig. 12). At depth 3, both CART and OPCT solve or nearly solve this environment with returns of $\overline{R} = 200 \pm 0$ and $\overline{R} = 192 \pm 15$, respectively, when trained on samples generated with said HC agent. The DRL oracle based on PPO performs as well as the HC oracle. In this case, the inputs are just the original CartPole variables x, v, θ, ω (no previous action a_{prev}). Both DT algorithms replicate the perfect score from depth 3 on.

Example rule sets found by depth-3 CART from HC and PPO agents are shown in Fig. 10. The tree in Fig. 10a is structurally different from the HC rule it is derived from (Fig. 4b). But upon careful inspection it becomes clear how it actually represents the same decision process, which explains the equal performance.

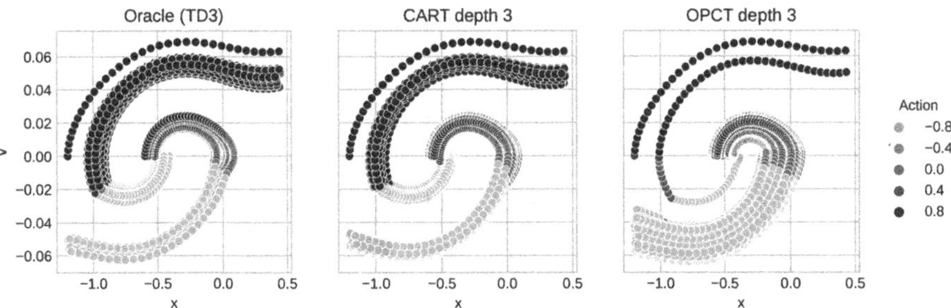

Fig. 9. MountainCarContinous. Comparison between TD3, CART, and OPCT.

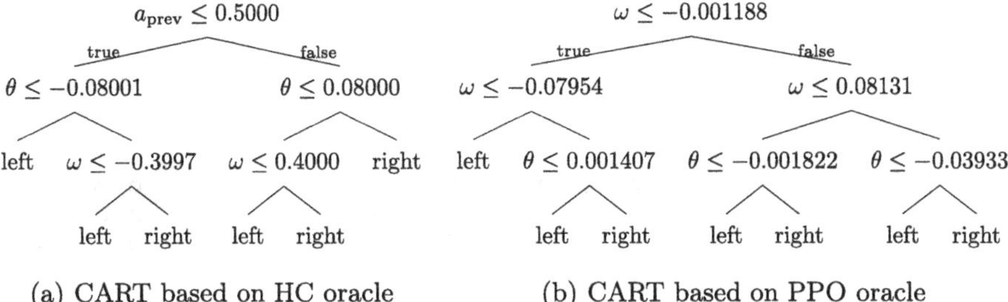

Fig. 10. CartPole. DTs induced by CART algorithm from samples of HC and PPO oracles. Note how, given the encoding of action 'left' and 'right' as 0 and 1, respectively, the condition of the root node in (a) effectively checks whether or not the previous action was 'left'.

Figure 10b is interesting in its own right: By careful inspection we see that the middle branches can be combined into the rule 'If $|\omega| < 0.08$ then apply action *left* or *right* depending on $\theta < 0$ or ≥ 0, respectively'. In other words, for small angular velocities push just in the direction where the pole is leaning. The outer branches mean 'For larger $|\omega|$ push into the direction that reduces $|\omega|$'.[3] We can summarize these findings as a new rule set (Fig. 11), which is even simpler than the one in Fig. 4b.

This new HC rule derived from *'CART based on PPO oracle'* has perfect performance $\overline{R} = 200$ when applied to the environment `CartPole-v0` and it has a higher performance $\overline{R} = 499$ on `CartPole-v1`, where the algorithm in Fig. 4b reaches only $\overline{R} = 488$.

As shown in Fig. 12b, OPCT trained on samples from PPO oracle reaches perfect performance at a depth of 1. This means it solves the CartPole problem with *one* single rule, shown in Fig. 13. While oblique DTs are generally more difficult to interpret, sign and magnitude of the coefficients might provide some helpful insights: In Fig. 13, θ has by far the largest coefficient, indicating that

[3] The condition $\omega > 0.08 \wedge \theta < -0.04$ is neglected here, since it can be assumed to appear very seldom: Positive ω will normally lead to positive θ.

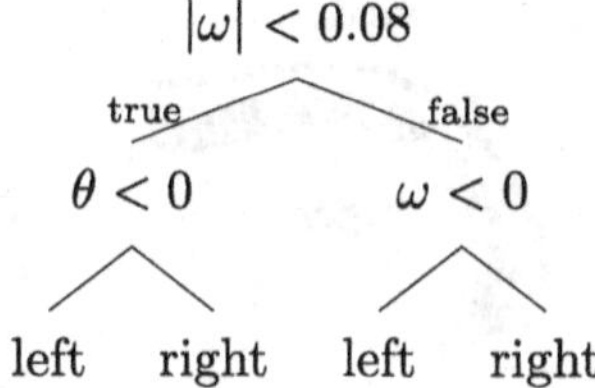

Fig. 11. CartPole. Simplified rule set derived from the *'CART based on PPO oracle'* tree shown in Fig. 10b.

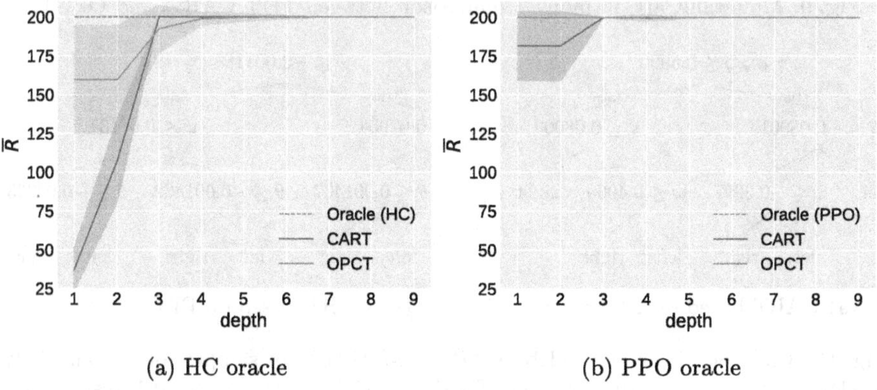

Fig. 12. CartPole. Performance comparison with different depths of trained DTs on samples produced by HC and PPO.

the model is most sensitive to this variable. The sign of θ will decide in most cases whether action *left* or *right* is chosen.

Interestingly, also the CART-induced tree performs quite well at depth 1 and 2 (better than its counterpart trained on HC samples at this depth). The respective tree decides solely on the basis of the sign of ω. The good performance shows that the PPO samples are easier to segment by linear cuts than the HC samples. This may be due to the fact that PPO avoids the non-linear function $|\cdot|$ (needs two cuts in a DT) and the extra input a_{prev} (needs another cut).

4.4 Pendulum-v0

Now we turn to the environment where our method so far is *not* able to produce good DTs at low depth. On the Pendulum environment a DRL oracle using TD3 reaches a good performance of $\overline{R} = -154 \pm 85$. The pole is stabilized in its unstable upright position in nearly every episode. Both tested DTs fail to reach TD3 oracle's performance at reasonable depths (see Fig. 14). While the performance clearly increases with the depth of the DTs, such trees no longer satisfy the criterion of explainability.

$$2.788x - 1.843v + 41.16\theta + 1.870\omega < -0.0080$$

true / false

left right

Fig. 13. Single rule solving the CartPole problem. This rule was found by OPCT based on PPO samples.

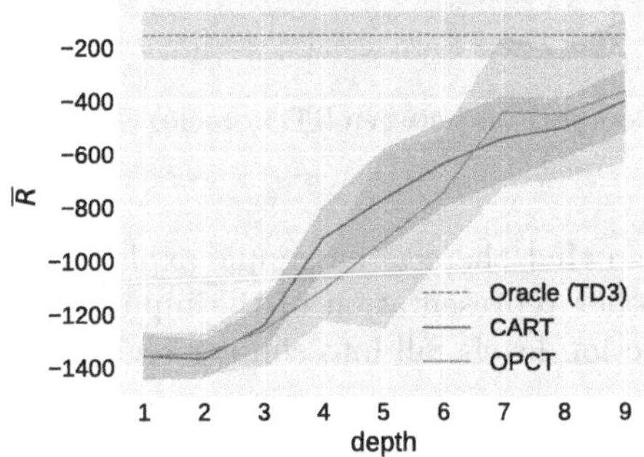

Fig. 14. Pendulum. Performance comparison with different depths of trained DTs on samples produced by TD3 oracle.

The original `Pendulum-v0` environment represents the angle as $\cos(\theta)$ and $\sin(\theta)$. We keep this representation throughout the experiments. For better visualization and without loss of information, we map $(\cos(\theta), \sin(\theta))$ back to the angle θ itself. In Fig. 15, the difference between the performances of oracle and DTs is reflected in the samples they produce: The TD3 oracle has learned to stabilize the pole in an upright position ($\theta \approx 0$) as can be seen from the white areas (absence of samples) for ($\theta \approx 0, |\omega| > 0$) in the left part of Fig. 15. The DTs fail to do so: there are many samples with $|\omega| > 0$ in region $\theta \approx 0$, meaning that the pole will rotate fast through the upright position.

It is an open question *why* our method is not able to produce good DTs at low depth for this environment. We speculate that the complex functional relationship between input space and continuous action space might be responsible for this. If we compare the TD3 picture in the left part of Fig. 15 with the other continuous-action environment MountainCarContinuous in the left part of Fig. 9, we see in the former intricate overlaps of many different action levels while in the latter mainly two action levels appear, which can be separated by a few cuts. This argument is supported by the fact that DTs with higher depths (more complex functions) come closer to the oracle's performance in Fig. 14.

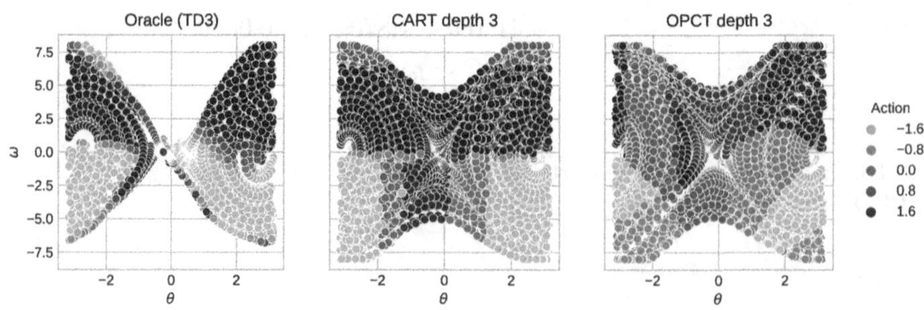

Fig. 15. Pendulum. Comparison between TD3 oracle, CART, and OPCT samples.

This does not mean that there *is* no simple explainable model: It might be that there exists another representation of the input space (yet to be found!) where the desirable action levels fall into simpler regions.

4.5 Comparison of Approaches

We summarize all obtained results in Table 1. All oracles meet the *solved* condition of the environments and all DTs of only depth 3 meet it as well or come at least close (with the exception of the Pendulum environment).

With two slightly different optimization schemes, Verma et al. [17, App. A, Tab. 6] report two numbers, which we report in row *[Verma]* of Table 1. The rules extracted by means of our method exhibit better performance than the rules obtained by the PIRL approach of Verma et al. [17]. Furthermore, we could not reproduce these results: When we apply the rules explicitly stated in [17, App. B, Figs. 9 and 10] to the respective environments, we obtain only the lower results shown in rows *[Verma_R]* of Table 1. Finally, the rules given in [17] are not easy to interpret.

Given their capacity to find optima in a broader class of splits, OPCT models were expected to map DRL oracles' behaviours better (at lower depth) than CART, which is limited to axis-parallel cuts. These expectations are *not* fulfilled by our results. OPCT often only had a similar or lower performance than CART at the same depth level. In the rare cases where OPCT is better than CART (e.g. CartPole at depth 1, Fig. 12), the single OPCT rule obtained (Fig. 13) is compact, but still difficult to interpret.

Table 1. Results for combinations of environment, oracle and DT algorithm. The DTs are of depth 3 in all cases. Reported are the average mean return $\overline{R}$ ± average standard deviation of returns $\overline{\sigma}$ for 5 repetitions with different seeds, each evaluated for 100 episodes. Agents with $\overline{R} \geq R_{solved}$ are said to *solve* an environment. See main text for the meaning of rows *[Verma]* and *[Verma_R]*.

Environment (R_{solved})	Oracle	Oracle's $\overline{R} \pm \overline{\sigma}$	DT	DT's $\overline{R} \pm \overline{\sigma}$
MountainCar-v0 (−110)	HC	-106.42 ± 13.57	CART	-107.46 ± 13.47
			OPCT	-122.80 ± 15.55
	DQN	-101.13 ± 10.45	CART	-105.29 ± 6.11
			OPCT	-109.63 ± 22.06
	[Verma]	-143.9, -108.1		
	[Verma_R]	-162.6 ± 3.8		
MountainCarCont.-v0 (90)	TD3	91.40 ± 2.19	CART	91.08 ± 2.16
			OPCT	81.89 ± 23.93
	HC	90.51 ± 1.33	CART	90.60 ± 1.33
			OPCT	84.20 ± 23.38
CartPole-v0 (195)	HC	200.00 ± 0.00	CART	200.00 ± 0.00
			OPCT	192.30 ± 15.20
	PPO	200.00 ± 0.00	CART	200.00 ± 0.00
			OPCT	200.00 ± 0.00
	[Verma]	143.2, 183.2		
	[Verma_R]	106.0 ± 16.9		
Pendulum-v0	TD3	-154.17 ± 85.02	CART	-1241.24 ± 80.38
			OPCT	-1258.98 ± 136.37

5 Conclusion

In this paper we provide a new method of extracting plain, human-readable rules from DRL agents. The method is simple, provides transparent rules and is not specific to any particular kind of oracle. We show that for the problems MountainCar, MountainCarContinuous and CartPole our approach is able to extract simple rule sets, which perform as well as the oracles they are based on. Our method currently still has limitations in the Pendulum environment, where it is not yet possible to automatically extract simple and well-performing rules.

We show that in several cases the automatically induced trees are 'understandable' in the sense that they are similar or equivalent to handcrafted rule sets that we provided as benchmarks.

Another aspect is that given the transparency of the extracted rule sets, these can later be inspected and possibly optimized or simplified (MountainCar, discussion of Fig. 7b). In one case (CartPole, CART based on DRL oracle PPO, Fig. 10b), we could simplify the tree and obtain an even simpler and better-performing handcrafted rule than the previously known one.

Future work will consist in determining the reason for the failure of the method when applied to the Pendulum problem. The method should then be improved and applied to more complex environments. We also want to study in all these environments the unsupervised learning of representations, which are then fed into DTs as additional features.

We hope that this work and our future research will help to establish better explainable models in the field of deep reinforcement learning.

References

1. Barto, A.G., Sutton, R.S., Anderson, C.W.: Neuronlike adaptive elements that can solve difficult learning control problems. IEEE Trans. Syst. Man Cybern. **13**(5), 834–846 (1983). https://doi.org/10.1109/TSMC.1983.6313077
2. Breiman, L., Friedman, J.H., Olshen, R.A., Stone, C.J.: Classification and Regression Trees. Routledge, London (1984)
3. Brockman, G., et al.: OpenAI Gym (2016). https://arxiv.org/abs/1606.01540
4. Coppens, Y., Efthymiadis, K., et al.: Distilling deep reinforcement learning policies in soft decision trees. In: Proceedings of IJCAI 2019 Workshop on Explainable Artificial Intelligence, pp. 1–6 (2019)
5. Fujimoto, S., van Hoof, H., Meger, D.: Addressing function approximation error in actor-critic methods. In: Dy, J., Krause, A. (eds.) Proceedings of 35th ICML, pp. 1587–1596. PMLR (2018)
6. Liu, G., Schulte, O., Zhu, W., Li, Q.: Toward interpretable deep reinforcement learning with linear model U-trees. In: Berlingerio, M., Bonchi, F., Gärtner, T., Hurley, N., Ifrim, G. (eds.) ECML PKDD 2018. LNCS (LNAI), vol. 11052, pp. 414–429. Springer, Cham (2019). https://doi.org/10.1007/978-3-030-10928-8_25
7. Mnih, V., Kavukcuoglu, K., et al.: Playing Atari with deep reinforcement learning (2013). https://arxiv.org/abs/1312.5602
8. Moore, A.W.: Efficient memory-based learning for robot control. Technical report, University of Cambridge (1990)
9. Muggleton, S.: Predicate invention and utilization. J. Exp. Theor. Artif. Intell. **6**(1), 121–130 (1994). https://doi.org/10.1080/09528139408953784
10. Muggleton, S.H., Lin, D., Tamaddoni-Nezhad, A.: Meta-interpretive learning of higher-order dyadic datalog: predicate invention revisited. Mach. Learn. **100**(1), 49–73 (2015). https://doi.org/10.1007/s10994-014-5471-y
11. Pedregosa, F., Varoquaux, G., et al.: Scikit-learn: machine learning in Python. J. Mach. Learn. Res. **12**(85), 2825–2830 (2011)
12. Raffin, A., Hill, A., et al.: Stable-baselines3: reliable reinforcement learning implementations. J. Mach. Learn. Res. **22**(268), 1–8 (2021)
13. Ross, S., Gordon, G., Bagnell, D.: A reduction of imitation learning and structured prediction to no-regret online learning. In: Proceedings of 14th International Conference on Artificial Intelligence and Statistics, vol. 15, pp. 627–635 (2011)
14. Schaal, S.: Is imitation learning the route to humanoid robots? Trends Cogn. Sci. **3**(6), 233–242 (1999). https://doi.org/10.1016/S1364-6613(99)01327-3
15. Schulman, J., Wolski, F., Dhariwal, P., Radford, A., Klimov, O.: Proximal policy optimization algorithms (2017). https://arxiv.org/abs/1707.06347
16. Stepišnik, T., Kocev, D.: Oblique predictive clustering trees. Knowl.-Based Syst. **227**, 107228 (2021). https://doi.org/10.1016/j.knosys.2021.107228

17. Verma, A., Murali, V., et al.: Programmatically interpretable reinforcement learning. In: Dy, J., Krause, A. (eds.) Proceedings of 35th ICML, pp. 5045–5054. PMLR (2018)
18. Wickramarachchi, D., Robertson, B., Reale, M., Price, C., Brown, J.: HHCART: an oblique decision tree. Comput. Stat. Data Anal. **96**, 12–23 (2016). https://doi.org/10.1016/j.csda.2015.11.006
19. Zilke, J.R., Loza Mencía, E., Janssen, F.: DeepRED – rule extraction from deep neural networks. In: Calders, T., Ceci, M., Malerba, D. (eds.) DS 2016. LNCS (LNAI), vol. 9956, pp. 457–473. Springer, Cham (2016). https://doi.org/10.1007/978-3-319-46307-0_29

ABC in Root Cause Analysis: Discovering Missing Information and Repairing System Failures

Xue Li[1(✉)], Alan Bundy[1], Ruiqi Zhu[1], Fangrong Wang[1], Stefano Mauceri[2], Lei Xu[2], and Jeff Z. Pan[1]

[1] School of Informatics, University of Edinburgh, Edinburgh, UK
{Xue.Shirley.Li,A.Bundy,Ruiqi.Zhu,Sylvia.Wang,J.Z.Pan}@ed.ac.uk
[2] Huawei Ireland Research Centre, Dublin, Ireland
{Stefano.Mauceri1,xulei139}@huawei.com

Abstract. Root-cause analysis (RCA) is a crucial task in software system maintenance, where system logs play an essential role in capturing system behaviours and describing failures. Automatic RCA approaches are desired, which face the challenge that the knowledge model (KM) extracted from system logs can be faulty when logs are not correctly representing some information. When unrepresented information is required for successful RCA, it is called missing information (MI). Although much work has focused on automatically finding root causes of system failures based on the given logs, automated RCA with MI remains under-explored. This paper proposes using the Abduction, Belief Revision and Conceptual Change (ABC) system to automate RCA after repairing the system's KM to contain MI. First, we show how ABC can be used to discover MI and repair the KM. Then we demonstrate how ABC automatically finds and repairs root causes. Based on automated reasoning, ABC considers the effect of changing a cause when repairing a system failure: the root cause is the one whose change leaves the fewest failures. Although ABC outputs multiple possible solutions for experts to choose from, it hugely reduces manual work in discovering MI and analysing root causes, especially in large-scale system management, where any reduction in manual work is very beneficial. This is the first application of an automatic theory repair system to RCA tasks: KM is not only used, it will be improved because our approach can guide engineers to produce KM/higher-quality logs that contain the spotted MI, thus improving the maintenance of complex software systems.

Keywords: Root cause analysis · Missing information · System management · Automatic theory repair

1 Introduction

Software system failures are unavoidable, so fast diagnosis and repair are crucial in system maintenance, where root cause analysis (RCA) is required. Traditionally, experts manually analyse raw system logs for RCA, where the challenges

G. Nicosia et al. (Eds.): LOD 2022, LNCS 13810, pp. 346–359, 2023.
https://doi.org/10.1007/978-3-031-25599-1_26

include that 1) the first arisen failure is usually not the root cause; 2) single causes can trigger multiple failures; 3) the system is enormous. Thus, manual RCA is difficult; 4) the model of the system may be inaccurate so that automatic methods become unreliable.

Much work has focused on automatically mining logs and assisting experts in discovering the root cause of system failures, including log filtering that collects the most relevant logs [20], log extraction as knowledge graphs (KG) [19], clustering logs [11,15], mining and representing information from logs [6,8]; automating network management in software based on KG [21] and analysing causality patterns among components in a software system [3,12,14], where the last takes extra tests to learn and validate dependencies by experts manually. It can be seen that log-based RCA is popular because logs are arguably the most straightforward source of information about the system. However, these log-based RCA methods' performance is restricted by the quality of logs, e.g., whether they cover all essential information for diagnosing root causes and whether they are written in ways that data-driven RCA pipelines can effectively consume.

The higher quality the log is, the more accurate the corresponding model of the system built from that log is. If the symptoms of the root cause can be inferred from the current model of the system, then this model is seen accurate and we can analyse that inference to discover the root cause. Otherwise, the current model is inaccurate, which is missing some crucial information, i.e. MI, about the current state of the system.

We argue that in time, continuous improvement of system logs, in the sense of incorporating MI[1] as necessary, could significantly enhance the performance of automated RCA pipelines or any other task related to log-mining. In manual RCA, experts sometimes need extra system tests to identify MI. Thus, an automated RCA not only needs to be aware of the MI in the system logs, but also should account for the domain experts' knowledge and experience [17].

Accounting for these considerations, the Abduction, Belief Revision and Conceptual Change system (ABC) [10,16], which repairs faulty logical theories based on a given benchmark, is here employed to automate RCA[2]. The main input is the system's KM that comprises of: 1) an automatically extracted KG from both system logs and the manual, and 2) rules manually formalised to introduce domain knowledge. Based on the KM and the given observed system failures, ABC discovers MI first and repairs the system model so that it predicts the failures, and then ABC analyses the repaired model to find the root cause. When there are multiple possible MI, ABC provides all possibilities to experts so that the correct MI can be found interactively.

This paper demonstrates the first application of a theory repair system to RCA tasks. The main contribution includes the follows:

1. An approach to discover log MI that is instrumental to RCA and the fault recovery process.

[1] The damage caused by MI in RCA is described in Fig. 2 and further discussed in the next section.

[2] ABC's code is available on GitHub https://github.com/XuerLi/ABC_Datalog.

2. An approach to automatically identify the root cause and suggest data-driven repairs.
3. An approach to guide experts to enrich the KM or to extend the system logs as to continuously improve the performance of the RCA pipeline.

Essential definitions of the ABC repair mechanism are given in Sect. 2. Root-cause analysis is introduced in Sect. 3, where MI is repaired by ABC in Sect. 3.1 first, and then the discovering and repairing of root causes are discussed in Sect. 3.2. An initial evaluation is given in Sect. 4, followed by the conclusion in Sect. 5.

2 ABC Repair Mechanism

ABC represents environments using logical theories based on the DataLog logic programming language [1], where axioms are Horn clauses. In *Kowalski Form*, these clauses take one of the following forms, i.e.,

$$Q_1 \wedge \ldots \wedge Q_m \implies P \tag{1}$$

$$Q_1 \wedge \ldots \wedge Q_m \implies \tag{2}$$

$$\implies P \tag{3}$$

$$\implies \tag{4}$$

where m is a natural number; Q_j, $1 \leq j \leq m$ and P are propositions. Then the above clauses represent a rule, a goal of m sub-goals, an assertion, and the empty clause, respectively[3].

In DataLog, the arguments of propositions are either constants or variables, i.e., there are no non-nullary functions. This makes Selected Literal Resolution (SL) [9] with a fair search strategy a decision procedure for DataLog theories. Decidability is important for establishing certainty in the detection of faults. Example 1 in Sect. 4 illustrates a DataLog theory. Note that the $\implies$ arrow is retained even when $m = 0$.

Figure 1 shows ABC's workflow. The inputs to ABC are a Datalog theory $\mathbb{T}$ and the preferred structure $\mathbb{S}$ which consists of a pair of sets of ground propositions: those propositions that are observed to be true $\mathcal{T}(\mathbb{S})$ and those observed to be false $\mathcal{F}(\mathbb{S})$. The pre-process in C1 reads and rewrites inputs into the internal format for later use. Then in C2, ABC applies SL to $\mathbb{T}$ to detect incompatibility and insufficiency faults based on $\mathcal{F}(\mathbb{S})$ and $\mathcal{T}(\mathbb{S})$, defined below. Incompatibilities are conflicts between the theory and the observations and insufficiencies are the failure of the theory to predict observations.

Definition 1 (Types of Fault). *Let* $\mathbb{T}$ *be a DataLog theory.*

Incompatibility: $\exists \phi.\ \mathbb{T} \vdash \phi \wedge \phi \in \mathcal{F}(\mathbb{S})$;
Insufficiency: $\exists \phi.\ \mathbb{T} \nvdash \phi \wedge \phi \in \mathcal{T}(\mathbb{S})$

[3] Keeping $\implies$ is required by the inference of refutation.

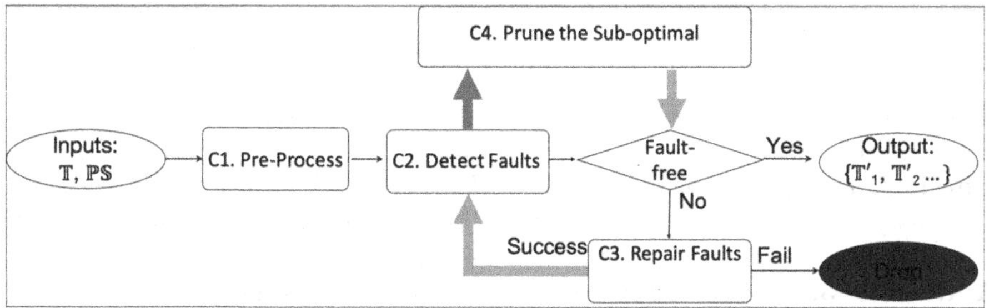

Fig. 1. Flowchart of the ABC: green arrows deliver a set of theories one by one to the next process; the blue arrow collects and delivers theories as a set; When a faulty-theory is not repairable, it will be dropped from the repair process. (Color figure online)

As ABC uses SL-resolution [9] which is not only sound and complete [5], but also decidable [13] for Datalog theories so that proofs can always be detected if there are any.

In C3, repairs are generated to fix detected faults. An insufficiency is repaired by unblocking a proof with additional necessary SL steps, while an incompatibility is repaired by blocking all its proofs, which can be done by breaking one SL step in each of them [16]. ABC repairs faulty theories using eleven *repair operations*. There are five for repairing incompatibilities and six for repairing insufficiencies, defined below.

Definition 2 (Repair Operations for Incompatibility). *In the case of incompatibility, the unwanted proof can be blocked by causing any of the SL steps to fail. Suppose the targeted SL step is between a goal, $P(s_1, \ldots, s_n)$, and an axiom, Body $\implies P(t_1, \ldots, t_n)$, where each s_i and t_i pair can be unified. Possible repair operations are as follows:*

Belief Revision 1: *Delete the targeted axiom: Body $\implies P(t_1, \ldots, t_n)$.*

Belief Revision 2: *Add an additional precondition to the body of an earlier rule axiom which will become an unprovable subgoal in the unwanted proof.*

Reformation 3: *Rename P in the targeted axiom to either a new predicate or a different existing predicate P'.*

Reformation 4: *Increase the arity of all occurrences P in the axioms by adding a new argument. Ensure that the new arguments in the targeted occurrence of P, are not unifiable. In Datalog, this can only be ensured if they are unequal constants at the point of unification.*

Reformation 5: *For some i, suppose s_i is C. Since s_i and t_i unify, t_i is either C or a variable. Change t_i to either a new constant or a different existing constant C'.*

Definition 3 Repair Operations for Insufficiency). *In the case of insufficiency, the wanted but failed proof can be unblocked by causing a currently failing SL step to succeed. Suppose the chosen SL step is between a goal $P(s_1, \ldots, s_m)$*

and an axiom Body $\Longrightarrow P'(t_1, \ldots, t_n)$*, where either* $P \neq P'$ *or for some* i*,* s_i *and* t_i *cannot be unified. Possible repair operations are:*

Abduction 1: *Add the goal* $P(s_1, \ldots, s_m)$ *as a new assertion and replace variables with constants.*

Abduction 2: *Add a new rule whose head unifies with the goal* $P(s_1, \ldots, s_m)$ *by analogising an existing rule or formalising a precondition based on a theorem whose arguments overlap with the ones of that goal.*

Abduction 3: *Locate the rule axiom whose precondition created this goal and delete this precondition from the rule.*

Reformation 4: *Replace* $P'(t_1, \ldots, t_n)$ *in the axiom with* $P(s_1, \ldots, s_m)$*.*

Reformation 5: *Suppose* s_i *and* t_i *are not unifiable. Decrease the arity of all occurrences* P' *by 1 by deleting its* i^{th} *argument.*

Reformation 6: *If* s_i *and* t_i *are not unifiable, then they are unequal constants, say,* C *and* C'*. Either (a) rename all occurrences of* C' *in the axioms to* C *or (b) replace the offending occurrence of* C' *in the targeted axiom by a new variable.*

ABC has a protection heuristic that allows the user to specify any term that should be protected from being changed. Usually a faulty theory requires multiple repairs to be fully repaired. Due to the diverse repairs, ABC tends to be over-productive [16]. Thus, only those with the fewest faults are selected as the optimal among alternatives [10,18] in C4. ABC repeats its repair process until there is no fault left.

As aforementioned, the KM in this paper contains a KG, which needs to be translated to Datalog first. The translation is straightforward as a triple is an assertion of a binary predicate and the TBox contains logical rules. To be succinct, we omit this format translation and directly represent the KM as a Datalog theory in this paper.

3 ABC in Root Cause Analysis

Given the assertion representing a failure as the goal[4], a cause of that failure is an axiom involved in the subset of KM which entails the goal. Thus, the root causes of a set of failures are defined as follows.

Definition 4 (Root Causes $\mathbb{R}$). *Given a set of system failures* $\mathbb{E}$*, and KM of the system* $\mathbb{T}$*, the root causes of* $\mathbb{E}$ *are a minimal set of axioms* $\mathbb{R}$ *involved in the proofs of* $\mathbb{E}$*.*

$$\forall \beta \in \mathbb{E}, \mathbb{T} \setminus \mathbb{R} \nvdash \beta \bigwedge \forall \alpha \in \mathbb{R}, \exists \beta \in \mathbb{E}, \mathbb{T} \setminus \{\alpha\} \nvdash \beta \tag{5}$$

The first half of Eq. (5) says that a system failure β won't exist if the system does not contain root causes $\mathbb{E}$. The second half represents that any axiom α

[4] For example, a triple represents an alarm about a system failure.

which representing root causes are necessary in terms of resulting at least one system failure.

ABC starts its RCA by finding MI and then repairing the KM to include it. Figure 2 shows the damage caused by MI in RCA. All nodes are explicitly in KM except the dashed node[5]. Due to MI, the green node will not be diagnosed as the root cause of all four failures, as it should be.

Figure 3 depicts ABC's workflow in RCA, where $\mathbb{T}$ is the faulty model of the network which lacks of MI, $\mathbb{T}_1$ is correct model of the faulty network and $\mathbb{T}_2$ is how the repaired network will look.

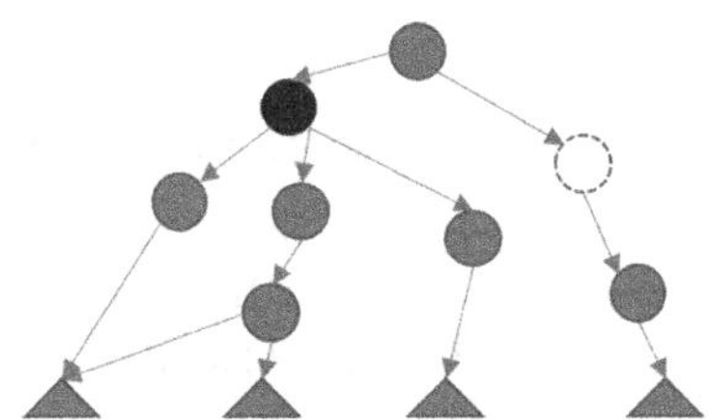

Fig. 2. Failed RCA due to MI: triangles are propositions describing system failures; circle nodes are axioms or theorems representing system behaviours; an arrow starts from a behaviour's representation to its logical consequence's; the dashed node corresponds to the axiom that should be added to represent MI, which is not in the original KM.

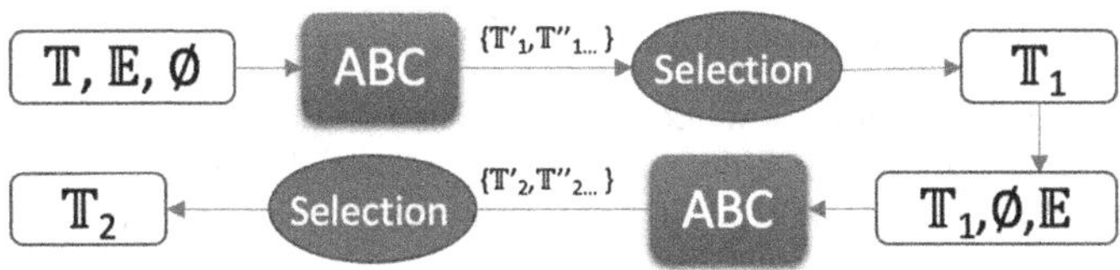

Fig. 3. RCA's flowchart: RCA's input are 1) KM $\mathbb{T}$; 2) the observed system failures as a set of assertions $\mathbb{E}$. RCA's output is the repaired KM $\mathbb{T}_2$ where the root cause is addressed. Here ABC's inputs are a KM, $\mathcal{T}(\mathbb{S})$ and $\mathcal{F}(\mathbb{S})$ in turn: in the first step $\mathcal{T}(\mathbb{S}) = \mathbb{E}$, $\mathcal{F}(\mathbb{S}) = \emptyset$; ABC outputs potential repairs $\{\mathbb{T}_1', \mathbb{T}_1''...\}$, from which the selected $\mathbb{T}_1$ is the input KM of the second step, where $\mathcal{T}(\mathbb{S}) = \emptyset$, $\mathcal{F}(\mathbb{S}) = \mathbb{E}$.

Before the detailed discussion in Sect. 3.1 and Sect. 3.2, a general introduction of the workflow is given here. The main input KM, $\mathbb{T}$ written in Datalog, contains two parts: 1) knowledge extracted from system logs and user manual; and 2) rules representing experts' domain knowledge. Firstly, ABC repairs $\mathbb{T}$ w.r.t. the MI. Based on the enriched KM $\mathbb{T}_1$, which infers all failures, ABC repairs $\mathbb{T}_1$ and then system failures can be fixed in $\mathbb{T}_2$. Consequently, the knowledge changed in

[5] A cause may be missing while its logical consequence exists in a KM, e.g., only the latter is recorded in the log.

this repair constitutes the root cause. As ABC outputs multiple repaired KMs, which are correct representations for the system in different scenarios. So the 'Selection' in Fig. 3 allows domain experts to choose the one that represents the target system correctly. In future, this selection can be automated by employing probability, where the selected KM should have the most significant sum of axiom probability that represents how much an axiom is trusted in describing the system accurately.

The input KG and the assertions of observed system failures are extracted from system logs and the user manual by [19] and Datalog rules representing domain knowledge are formalised by domain experts or the results of rule mining tools after being validated by experts.

3.1 Repairing Knowledge Model to Include MI

A KM containing all essential information about a system failure is the base of RCA. Thus, it is important to find MI first before analysing root causes. This section introduces why a KM has MI; how ABC detects MI and repairs KM to cover MI.

A system failure is a logical consequence of that system's setup and behaviour. Therefore, a correct representation for modelling the system ($\mathbb{M}$) should entail all of the theorems ($\mathbb{E}$) which represent the system's failures, shown by Eq. (6). Otherwise, that model is incorrect. The information that is well represented in a correct model but not in an incorrect model is called MI.

$$\forall \beta \in \mathbb{E},\ \mathbb{M} \implies \beta \tag{6}$$

Typical causes of MI are summarised as the following.

- KM is incomplete where new axioms need to be added.
- KM is inconsistent where old axioms need to be deleted
- KM is poorly written so that its representation needs to be adjusted, e.g., rewriting a misspelled constant.

ABC system is chosen to address MI because it has rich operations of axiom deletion/addition and representation changes. By giving assertions of observed system failures as ABC's $\mathcal{T}(\mathbb{S})$, ABC checks whether each of them is a logical consequence of the system's KM. If a failure $\alpha \in \mathcal{T}(\mathbb{S})$ is not a theorem, then ABC repairs KM to build a proof of α. These repairs represent MI about system failures so that all essential information about the observed system failures are well represented in the repaired model.

Definition 5 (Repair KM for MI). *Given the input of a KM $\mathbb{T}$ to repair, all assertions of the observed system failures $\mathbb{E}$ as the true set of preferred structure ($\mathcal{T}(\mathbb{S})$) and empty set as the false set of preferred structure ($\mathcal{F}(\mathbb{S})$), ABC generates repaired KM ($\mathbb{T}_1$) that logically entails system failures.*

$$\mathbb{T}_1 = \begin{cases} \mathbb{T}: \ \forall \alpha \in \mathbb{E}, \mathbb{T} \vdash \alpha \\ \mathbb{T}': \mathbb{T}' \in \nu(\mathbb{T}, \mathbb{E}, \emptyset),\ \exists \alpha \in \mathbb{E}, \mathbb{T} \not\vdash \alpha \end{cases} \tag{7}$$

where ν is ABC's repair function whose inputs are the knowledge model, preferred structure: $\mathcal{T}(\mathbb{S})$ and $\mathcal{F}(\mathbb{S})$ in turn.

If the current $\mathbb{T}$ has all essential information for observed system failures, then $\mathbb{T}$ entails all these failures so it does not need to be repaired, which is the first case in Eq. (7). Otherwise, ABC will repair $\mathbb{T}$ and generate multiple possible KMs, among which, experts can select the accurate one that describes the target system correctly. Traditionally, experts need to brainstorm all system settings or behaviours which maybe relevant to the system failures [4]. Thus, ABC's multiple solutions are not trivial: they provide directions for experts to further discover the relevant information about system failures.

However, this step examines each system failure individually, but not combining all of them to find the root cause: it is an important preparation of the RCA discussed in Sect. 3.2.

3.2 Root-Cause Discovering and Repairing

After the last step, the KM contains all essential information for system failures. This step aims to find the root cause of all failures by theory repair. The effect of repairing a cause reveals root causes: *the root cause is one whose repair leaves the fewest failures.* Based on the sub-optimal pruning mechanism [10,18], the optimal repairs that use the minimal number of operations but solve the maximal number of faults are the ones that fix the root causes.

Definition 6 (RCA by ABC Theory Repair). *Let $\mathbb{T}_1$ be a KM to repair, which entails all observed system failures: $\forall \beta \in \mathbb{E}, \mathbb{T}_1 \implies \beta$, and all assertions of observed system failures $\mathbb{E}$ as the false set of preferred structure ($\mathcal{F}(\mathbb{S})$) and empty set as the true set of preferred structure ($\mathcal{T}(\mathbb{S})$). ABC generates repaired KM $\mathbb{T}'$ that fixes system failures, where $\mathbb{T}' \in \nu(\mathbb{T}_1, \emptyset, \mathbb{E})$, from which experts can choose the best KM for the failure-free system $\mathbb{T}_2$. Then the knowledge changed in $\mathbb{T}_2$ is the root cause. Here ν is the same function defined in Eq. (7)*

$$\mathbb{T}_2 = \mathbb{T}', \ \textit{where}\ \mathbb{T}' \in \nu(\mathbb{T}_1, \emptyset, \mathbb{E}) \bigwedge \forall \beta \in \mathbb{E}, \mathbb{T}_1 \implies \beta \tag{8}$$

We claim that given the KM $\mathbb{T}_1$ as the input theory to repair, and assertions of system failures $\mathbb{E}$ as $\mathcal{F}(\mathbb{S})$, ABC blocks all proofs of $\mathbb{E}$ with the minimal repair operations, so root causes are identified by examining the repairs that need to be made to remove them in its repaired KM $\mathbb{T}_2$. Thus, the changed parts of these repairs are the root causes.

As the KM contains some fundamental information about a system that is always correct, e.g., IP address format, ABC's protection heuristic is useful to protect them from being repaired. This protection avoids incorrect repairs so prunes fake root causes.

Example 1 *Knowledge Model $\mathbb{T}$ with MI suffering from all failures.*

$$
\begin{aligned}
&\implies microservice(id1, s1) &(1)\\
&\implies microservice(id2, s1) &(2)\\
&\implies microservice(id3, s2) &(3)\\
&\implies full(d1) &(4)\\
&\implies createOn(id2, d1) &(5)\\
&\implies sameRoute(id2, id3) &(6)\\
full(X) \wedge createOn(Y, X) &\implies fail(Y) &(7)\\
ms(X, s1) \wedge ms(Y, s2) \wedge &&\\
sameRoute(X, Y) &\implies depend(Y, X) &(8)\\
depend(X, Y) \wedge fail(Y) &\implies fail(X) &(9)
\end{aligned}
$$

$$\mathcal{T}(\mathbb{S}) = \{fail(id1), fail(id2), fail(id3)\}$$
$$\mathcal{F}(\mathbb{S}) = \emptyset$$

As aforementioned, ABC is guaranteed to find fault proofs when there are any. Once a theory is detected as faulty, then ABC will recursively try to repair its faults. If this repair process terminates with success, we can guarantee that a root cause has been found. But there is no guarantee that this repair process will always terminate successfully. It could introduce new faults at least as fast as it removes them theoretically: no such a case occurs so far.

4 Evaluation

In this section, our solution is validated in the context of Huawei's 5G network where the failure of one microservice could have a waterfall effect. As our method of repairing system model when tackling RCA tasks is unique, there will be no comparison but a detailed example to illustrate our method in this evaluation, especially in terms of the model repairing. Finally, we will discuss how ABC finds and repairs the root cause. The comparison with other RCA systems that do not addressing MI is a future work, which will be given as a part of the evaluation of the TREAT project [22].

Example 1 describes the KM, $\mathbb{T}$, written in Datalog by following the convention given by Eq. (1): three microservice instances' IDs are *id*1, *id*2 and *id*3, respectively, and the first two are of type *s*1 and the last *s*2, represented by axiom (1–3). Axioms (4–5) say that *id*2 is created on *d*1 and that the device *d*1 is already full. Rule (7) tells that a microservice fails when it is created on a full device, $fail(id2)$ is a theorem of this $\mathbb{T}$. Note that there is an inconsistent representation between microservice in (1–3) and its acronym 'ms' in (8), which makes the reasoner fail in unifying $microservice(X, Y)$ and $ms(X, Y)$. Otherwise, based on (2, 3, 6, 8), we would know that $id2$ and $id3$ were for the same route and $depend(id3, id2)$. By combining (9), $fail(id3)$ would be concluded. Thus, among all system failures of $fail(id1)$, $fail(id2)$ and $fail(id3)$ given by $\mathcal{T}(\mathbb{S})$, only $fail(id2)$ can be predicted by $\mathbb{T}$.

Assume that the full device $d1$ causes two newly created microservice instances $id1$ and $id2$ to fail. In addition, another microservice $id3$ also fails because it depends on the failed $id2$. However, the incomplete log only contains the information about creating instance $id2$ and misses creating $id1$, e.g., it is deleted due to the log's limit being reached. Thus, RCA needs to discover the MI about $id1$'s creation and then diagnose that the full device is the root cause of these three failures. In addition, the inconsistent use of $microservice$ and ms need to be corrected as well.

Example 2 ***Enriched KM $\mathbb{T}_1$ suffering from all failures.***

$$\Longrightarrow microservice(id1, s1) \quad (1)$$
$$\Longrightarrow microservice(id2, s1) \quad (2)$$
$$\Longrightarrow microservice(id3, s2) \quad (3)$$
$$\Longrightarrow full(d1) \quad (4)$$
$$\Longrightarrow createOn(id2, d1) \quad (5)$$
$$\Longrightarrow createOn(id1, d1) \quad (5^*)$$
$$\Longrightarrow sameRoute(id2, id3) \quad (6)$$
$$full(X) \wedge createOn(Y, X) \Longrightarrow fail(Y) \quad (7)$$
$$microservice(X, s1) \wedge microservice(Y, s2) \wedge sameRoute(X, Y) \Longrightarrow depend(Y, X) \quad (8')$$
$$depend(X, Y) \wedge fail(Y) \Longrightarrow fail(X) \quad (9)$$

$$\mathcal{T}(\mathbb{S}) = \emptyset$$
$$\mathcal{F}(\mathbb{S}) = \{fail(id1), fail(id2), fail(id3)\}$$

In this first step of RCA, $\mathbb{T}$ needs to be repaired so that it can predict not only $fail(id2)$, but also $fail(id1)$ and $fail(id3)$. Among all ABC's repairs that build a proof for $fail(id1)$, adding $createOn(id1, d1)$ is the correct cause of $fail(id1)$ in this scenario, which is generated by Abduction 1 in Definition 3. Meanwhile, by renaming ms into $microservice$ in axiom (8), the inconsistency is repaired, which is generated by Reformation 6 in Definition 3. Then $\mathbb{T}_1$ in Example 2 is selected as the enriched model for the next step of RCA, where changes are highlighted in red.

In the second step of RCA, $\mathbb{T}_1$ is the input and failures are given as $\mathcal{F}(\mathbb{S})$. Then ABC generates repairs that block all proofs of three failures, among which deleting axiom (4) and rewriting (4) as a new axiom $\Longrightarrow dummy\_full(d1)$ fix the root cause of $full(d1)$. These two solutions are shown in Example 3 and 4 whose repair operations are from Belief Revision 1 and in Reformation 3 in Definition 2, respectively. The repaired axiom with the new predicate $dummy\_full$ represents that the root cause is board $d1$ being full, and the operations that can change $d1$'s status from full to not full can address these failures.

Example 3 ***Repaired KM $\mathbb{T}_2$ derives no failures.***

$$\begin{aligned}
&\Longrightarrow microservice(id1, s1) && (1)\\
&\Longrightarrow microservice(id2, s1) && (2)\\
&\Longrightarrow microservice(id3, s2) && (3)\\
&\Longrightarrow dummy_full(d1) && (4)\\
&\Longrightarrow createOn(id2, d1) && (5)\\
&\Longrightarrow createOn(id1, d1) && (5^*)\\
&\Longrightarrow sameRoute(id2, id3) && (6)\\
full(X) \wedge createOn(Y, X) &\Longrightarrow fail(Y) && (7)\\
microservice(X, s1) \wedge microservice(Y, s2) \wedge &\\
sameRoute(X, Y) &\Longrightarrow depend(Y, X) && (8')\\
depend(X, Y) \wedge fail(Y) &\Longrightarrow fail(X) && (9)
\end{aligned}$$

$\mathcal{T}(\mathbb{S}) = \emptyset$
$\mathcal{F}(\mathbb{S}) = \{fail(id1), fail(id2), fail(id3)\}$

This example shows how the ABC repair system detects root-causes when there is missing information: it extend the KM with MI about a failure and then diagnoses and repairs the observed failures' root cause. To our best knowledge, other RCA methods do not deal with MI so cannot find correct root causes when there is MI.

In addition, discovering the missing information about *createOn* is guidance for experts to optimise system logs: records about *createOn* are important in RCA so they should be protected to avoid being deleted in future. On the other hand, it spots and solves the representation inconsistency of using both *microservice* and *ms* by replacing all occurrences of the latter with the former.

Example 4 *Repaired KM $\mathbb{T}_3$ by Belief Revision derives no failures.*

$$\begin{aligned}
&\implies microservice(id1, s1) &(1)\\
&\implies microservice(id2, s1) &(2)\\
&\implies microservice(id3, s2) &(3)\\
&\implies createOn(id2, d1) &(5)\\
&\implies createOn(id1, d1) &(5^*)\\
&\implies sameRoute(id2, id3) &(6)\\
full(X) \wedge createOn(Y, X) &\implies fail(Y) &(7)\\
microservice(X, s1) \wedge microservice(Y, s2) \wedge &\\
sameRoute(X, Y) &\implies depend(Y, X) &(8)\\
depend(X, Y) \wedge fail(Y) &\implies fail(X) &(9)
\end{aligned}$$

$$\mathcal{T}(\mathbb{S}) = \emptyset$$
$$\mathcal{F}(\mathbb{S}) = \{fail(id1), fail(id2), fail(id3)\}$$

5 Conclusion

This paper introduces a novel RCA mechanism for system failures by applying ABC to the system's KM and system failures: firstly, ABC automatically discovers possible MI when failures cannot be deduced from the given KM. After adding the interactively selected MI to KM, ABC automatically repairs the enriched KM to fix root causes so that failures are not logical consequences.

Automatically discovering possible MI makes ABC less restricted by the completeness of the KM than other approaches. It also guides experts to enrich the KM by extending the system logs w.r.t. MI as to continuously improve the performance of the RCA pipeline.

Limitations reveal the future work: 1) ABC's scalability limit for massive KM. Possible solutions include optimising its computation flow and minimising its input by pruning axioms that are irrelevant to the failures, e.g. clustering logs [11]. 2) Experts are needed to select the best answer from ABC's output currently. To minimise manual work, ABC's output can be ranked by incorporating probability in future; 3) a more sophisticated evaluation is crucial, ideally, with open-source data [2,7].

Acknowledgment. The authors would like to thank Huawei for supporting the research and providing data on which this paper was based under grant CIENG4721/LSC. Also we gratefully acknowledge UKRI grant EP/V026607/1 and the support of ELIAI (The Edinburgh Laboratory for Integrated Artificial Intelligence)

EPSRC (grant no EP/W002876/1). Thanks are also due to Zhenhao Zhou for the valuable discussions around network software systems. In addition, anonymous reviewers also gave us very useful feedback that improved the quality of this paper.

References

1. Ceri, S., Gottlob, G., Tanca, L.: Logic Programming and Databases. Surveys in Computer Science, Springer, Berlin (1990). https://doi.org/10.1007/978-3-642-83952-8
2. Chapman, A., et al.: Dataset search: a survey. VLDB J. **29**(1), 251–272 (2020)
3. Cherrared, S., Imadali, S., Fabre, E., Gössler, G.: SFC self-modeling and active diagnosis. IEEE Trans. Network Serv. Manage. **18**, 2515–2530 (2021)
4. Dalal, S., Chhillar, R.S.: Empirical study of root cause analysis of software failure. ACM SIGSOFT Software Engineering Notes **38**(4), 1–7 (2013)
5. Gallier, J.: SLD-Resolution and Logic Programming. Chapter 9 of Logic for Computer Science: Foundations of Automatic Theorem Proving (2003). originally published by Wiley 1986
6. He, P., Zhu, J., He, S., Li, J., Lyu, M.R.: An evaluation study on log parsing and its use in log mining. In: 2016 46th Annual IEEE/IFIP International Conference on Dependable Systems and Networks (DSN), pp. 654–661. IEEE (2016)
7. He, S., Zhu, J., He, P., Lyu, M.R.: Loghub: a large collection of system log datasets towards automated log analytics. arXiv preprint arXiv:2008.06448 (2020)
8. Jia, T., Chen, P., Yang, L., Li, Y., Meng, F., Xu, J.: An approach for anomaly diagnosis based on hybrid graph model with logs for distributed services. In: 2017 IEEE International Conference on Web Services (ICWS), pp. 25–32. IEEE (2017)
9. Kowalski, R.A., Kuehner, D.: Linear resolution with selection function. Artif. Intell. **2**, 227–60 (1971)
10. Li, X.: Automating the Repair of Faulty Logical Theories. Ph.D. thesis, School of Informatics, University of Edinburgh (2021)
11. Lin, Q., Zhang, H., Lou, J.G., Zhang, Y., Chen, X.: Log clustering based problem identification for online service systems. In: 2016 IEEE/ACM 38th International Conference on Software Engineering Companion (ICSE-C), pp. 102–111. IEEE (2016)
12. Lu, J., Dousson, C., Krief, F.: A self-diagnosis algorithm based on causal graphs. In: The Seventh International Conference on Autonomic and Autonomous Systems, ICAS, vol. 2011 (2011)
13. Pfenning, F.: Datalog. Lecture 26, 15–819K: Logic Programming (2006). https://www.cs.cmu.edu/~fp/courses/lp/lectures/26-datalog.pdf
14. Qiu, J., Du, Q., Yin, K., Zhang, S.L., Qian, C.: A causality mining and knowledge graph based method of root cause diagnosis for performance anomaly in cloud applications. Appl. Sci. **10**(6), 2166 (2020)
15. Shima, K.: Length matters: clustering system log messages using length of words. arXiv preprint arXiv:1611.03213 (2016)
16. Smaill, A., Li, X., Bundy, A.: ABC repair system for Datalog-like theories. In: KEOD, pp. 333–340 (2018)
17. Solé, M., Muntés-Mulero, V., Rana, A.I., Estrada, G.: Survey on models and techniques for root-cause analysis. arXiv preprint arXiv:1701.08546 (2017)

18. Urbonas, M., Bundy, A., Casanova, J., Li, X.: The use of max-sat for optimal choice of automated theory repairs. In: Bramer, M., Ellis, R. (eds.) SGAI 2020. LNCS (LNAI), vol. 12498, pp. 49–63. Springer, Cham (2020). https://doi.org/10.1007/978-3-030-63799-6_4
19. Wang, F., et al.: LEKG: a system for constructing knowledge graphs from log extraction. In: The 10th International Joint Conference on Knowledge Graphs (2021)
20. Zawawy, H., Kontogiannis, K., Mylopoulos, J.: Log filtering and interpretation for root cause analysis. In: 2010 IEEE International Conference on Software Maintenance, pp. 1–5. IEEE (2010)
21. Zhou, Q., Gray, A.J., McLaughlin, S.: Seanet-towards a knowledge graph based autonomic management of software defined networks. arXiv preprint arXiv:2106.13367 (2021)
22. Zhu, R., et al.: TREAT: automated construction and maintenance of probabilistic knowledge bases from logs (extended abstract). In: The 8th Annual Conference on machine Learning, Optimization and Data Science (LOD) (2022)

Forecasting Daily Cash Flows in a Company - Shortcoming in the Research Field and Solution Exploration

Bartłomiej Małkus(✉) and Grzegorz J. Nalepa

Institute of Applied Computer Science, Jagiellonian University, 31-007 Kraków, Poland

bartlomiej.malkus@doctoral.uj.edu.pl, grzegorz.j.nalepa@uj.edu.pl

Abstract. Daily cash flow forecasting is important for maintaining company financial liquidity, improves resource allocation, and aids financial managers in decision making. On the one hand, it helps to avoid excessive amounts of cash outstanding on company's account, and on the other hand, it helps to avoid liquidity issues. This area isn't popular for research though - literature on the topic is limited and has a major shortcoming - in most cases publicly unavailable datasets are used. It can be attributed to generally smaller availability of financial data than in other fields, but there are two issues arising from such situation - not reproducible results of existing work and the area being less approachable by new potential researchers. The goal of this paper is two-fold. Firstly, it is reviewing existing literature, methods, and datasets used, together with details provided on those datasets. Secondly, it is exploring publicly available datasets to be used in further research, containing either cash flows directly or data from which cash flows can be derived.

Keywords: Daily cash flows · Forecasting · Time series · Cash flows datasets

1 Introduction

Predicting daily cash flows refers to estimating the amount of cash flowing into the company account on a given day. Day-to-day cash flow forecasting is important for maintaining financial liquidity, improves resource allocation, and aids financial managers in decision-making, both investment and payment ones. On one hand, it helps to avoid excessive amounts of cash outstanding on the company account, for which investment opportunities would have been missed, and allows for better investment planning. Depending on the horizon of predictions, it may also improve payment decisions, which may in turn lead to financial gains, like early payment bonuses or better conditions on future deals. On the other hand, cash flow forecasts help to avoid liquidity issues and lower interest costs by reducing the need for external funding, such as short-term loans or an overdraft facility.

G. Nicosia et al. (Eds.): LOD 2022, LNCS 13810, pp. 360–369, 2023.
https://doi.org/10.1007/978-3-031-25599-1_27

When researching this topic, it is noticeable that the literature related to the prediction of daily cash flows is very limited. Further observation is that most articles on this topic use publicly unavailable datasets and only briefly describe details of methods used (e.g. architecture of neural networks). We did not find any article which used publicly available data and at the same time utilized machine learning methods. One reason for that could be the fact that financial data is less available than other kinds of data, as companies have no interest in publishing more data than they are obligated to.

When it comes to research, the problem of using publicly unavailable datasets is the fact that the results are not reproducible and there is no easy way to compare existing methods with newly researched ones. The objective of this paper is from one side an analysis of the existing literature and methods used, and from the other side it is an analysis of publicly available datasets on cash flows or datasets which could be used to derive such data for the research purposes. The rest of the paper is organized as follows: In Sect. 2 we discuss different types of cashflows and their interpretation as cashflow. The in Sect. 3 we elaborate on the related work in tis area. Section 4 is devoted to the discussion of datasets discovered related to cash flows. We conclude this short paper in Sect. 5.

2 Cash Flows as Timeseries

2.1 Cash Flows

There are three main types of cash flows: cash flow from operations, cash flow from investing, and cash flow from financing. First type, cash flow from operations, is related to all of company's regular business activities, like manufacturing, sales or providing services to customers. It contains bills or invoices payment, prepayments, salaries etc. Cash flow from investing, as the name suggests, is related to various investment-related activities, it may come from speculative transactions on assets, securities, real estate, but also from research and development activities. Last type of cash flows, cash flow from financing is related to company's funding and capital, it comes from equity, debt, and dividend transactions. This paper focuses on the first type of cash flows – cash flow from operations.

Cash flows may be further divided into incoming and outgoing. The first type is harder to predict due to different types of deals, payments, varying customers' payment time and solvency. Outgoing cash flows are more manageable by a company or its financial managers and, in a way, are dependent on incoming cash flows as they depend on available cash. In this paper, the prediction of only the incoming part of the cash flow (cash inflow) is considered.

In addition to day-to-day cash flows, the topic of yearly cash flow predictions can be encountered in the literature. Companies listed on regulated markets most often are required to publish financial results periodically, and such values/predictions may be helpful in assessing company's condition. However, such predictions are beyond the scope of this article.

2.2 Incomes and Cash Flows

When a company operates in the field of manufacturing, trade, or services, a large part of its cash flows come from transactions with customers. Whenever a transaction is finalized, an invoice is issued, and the company records an income. However, such invoices may have various forms of payment and translate into cash flows in a different way.

When a good or service provided is paid for immediately, transaction and income result in immediate cash flow, so the problem of predicting cash flows may be reduced to the problem of predicting sales (which is a more common practice). Companies with such business model are rare, however. The much more common case is a flexible policy of accepting payments where e.g. delayed or installment payments are allowed, such approach generates risk of late payment and uncollectible debts, which result in no easy translation between sales and cash flows. Some companies, due to the nature of their product or service, require prepayments, which adds to the complexity of the problem. As a result, forecasting cash flows may be seen as a separate problem.

2.3 Time Series and Their Forecasting

Cash flow data, as a large part of financial data, is represented in the form of time series. Depending on the company, they may be not directly stored, but if company have enough other data, it may be transformed to cash flow data. Raw data may be in the form of list of invoices, or list of transactions with price, type, date and due date. To transform it into cash flow data for forecasting purposes (as considered in this paper), a simple accumulation by date is applied, resulting in daily cash flow data. From the forecasting point of view, it is important for the raw data to be complete (to contain all the transactions), otherwise, due to accumulation operation, daily cash flows and resulting prediction model will likely be incorrect.

The most common approach to predicting daily cash flows found in the literature is to predict future cash flows based on their past values, sometimes extended by additional data, such as an indicator of whether a particular day is weekend or holiday. Such an approach boils down to autoregressive time-series forecasting problem and can be solved as such.

Main concerns when analyzing and forecasting economic time series are: trend, and closely related topic of stationarity and non-stationarity of time series [16], seasonality, holidays effect. Others, less specific to time series, are outliers and missing data handling.

3 Related Work

As mentioned above, the literature related to cash flow prediction is limited. Data availability poses a challenge here, as it is with many researches operating on financial data. Companies do gather data, but they have no interest in publishing it, and in many cases it could even be undesirable from them, as it reveals internal operating mechanism of a company. There is an information obligation for companies listed on regulated markets, but it concerns periodic data (quaterly, yearly), which is of no use for day-to-day data analysis. Most of the works in literature base on publicly unavailable datasets, which are provided e.g. by an employee of a company, who is coauthoring the paper or by companies to researchers in exchange for the models created. In the case of such research, both the datasets and details of the methods used are not published together with the paper, so it is impossible to reproduce the results.

When it comes to the methods used, for the purpose of predicting time series (especially financial), autoregressive processes like ARIMA (autoregressive integrated moving average) and its derivatives are among the most widely used models. In the literature related to cash flow forecasts ARIMA (and its derivatives) are often used as a benchmark, other methods are neural networks in the form of vanilla feed-forward MLPs or recurrent neural networks like LSTM/GRU.

Moving on to a literature review, in [21], authors compare ARIMA model with Prophet framework, feedforward MLPs and LSTM networks. They introduce cost function alternative to most often used mean squared error - IOC, Interest Opportunity Cost, depending on the investment return rate and credit cost, which "expresses in monetary terms how much money is lost using a model's imperfect predictions compared to the hypothetical case of perfect predictions". Data used in the paper consists of transactions of a German company, spans to the period of 33 months, and contains around 5 millions records, not all of them are cash flows, as it contains e.g. credit nodes as well. In addition to basic transaction details, such as date and amount, it contains holiday data. Data is not provided by the authors and not publicly available. Payments are aggregated by days and synthetic features are introduced – weekends and holidays indicators. The best results was achieved using a LSTM network optimized for IOC, both under IOC and MSE measures. ARIMA gave the worst results, Prophet was comparable to networks optimized for MSE, but when networks were optimized for IOC, they were performing better. MSE and MAE values are provided, but due to the dataset not being available, it is not easy to compare them with other works.

Prophet [19] is time series forecasting framework introduced by Facebook (currently Meta) in 2017 and promises to be easy to use, even by users unfamiliar with time series forecasting, handle multiple seasonality, holidays effects, and to deal better with outliers, missing data, and trend changes. It uses an additive decomposition model of time series, which divides it into three main components – trend, seasonality, and holidays, and works on the basis of curve fitting. Therefore, unlike ARIMA, does not require data to be evenly spaced in time.

In [12], authors analyze the application of recurrent neural networks, GRU in particular, to the prediction of daily cash flows in the electric energy industry. Past payment data used in the paper consists of details (date and payment amount) of customer payments of a Chinese company for one of Chinese provinces. The dataset is not publicly available. Its time span is 74 months, there are around 37 millions of records. Dataset is processed to represent amount of cash inflows per day, some synthetic features are introduced to improve seasonality handling, like month, day of week or holiday indicators. The results of prediction with GRU units based neural network are compared with ARIMA model fitted to the same data. According to the numbers in the paper, the neural network gave significantly better predictions than ARIMA model, however, MAPE (mean absolute percentage error) was still around 25% for network based model. No cross-validation results are provided, results are provided only for the first month of the test set.

In [20], authors compare Grey Neural Network, combination of Grey Model [15] and MLP network with both of these components separately. The dataset is not publicly available, comes from a commercial bank and covers two years, no more details are provided. Each month is divided for the purpose of creating training and test sets – 24 days go to training set, remaining 6 are used as a test set. The proposed approach (combination of Grey Model with neural networks) gives the best results, though the numbers provided do not allow one to compare the results with other research, as they depend on values in the dataset used.

In contrast to predicting daily cash flows for a company, there is some literature on predicting cash flows for mid- or long-term projects. Specifics of such forecasting problem are a bit different, as they include prediction of both incoming and outgoing cash flows and granularity of such forecasts is much smaller, as they are often predicted monthly or per project stages. Such predictions are not the topic of this paper, but due to similarities of these two issues, the methodology of project cash flows prediction is worth mentioning. Besides the methods that were already mentioned, combination of fuzzy logic and SVM, fuzzy SVM is proposed in [14]. In [13], authors show usage of time-dependent LS-SVM – data points are weighted based on the time period they come from. Datasets for both of these papers are not publicly available.

4 Cash Flow Related Datasets

As mentioned above, public cash flows datasets are hardly available, and most researchers operate on publicly unavailable ones. Datasets which can be used for the purpose of predicting future cash flows can exist in different forms. The first is a direct one, i.e. cash flows data from different kind of transactions and operations, this form is unlikely to be encountered though, as other data, like transactions or payments would have to be transformed to cash flows and there is not much of a value for companies to store cash flows data in the direct form. The second is an indirect one – it can be a transactions or invoices list with payment details (amount, date of payment). Data in such form is more likely to be encountered, as details about transactions and invoices are stored in companies' accounting systems.

We made an effort to find publicly available cash flow datasets, or datasets which can be transformed into cash flows. Later, after unsatisfactory results of the search, it was broadened to sales datasets (more details on this further in the paper). The search was carried out on multiple sites and services offering machine learning datasets:

- Kaggle [4],
- data.world open datasets [3],
- UCI Machine Learning Repository [10],
- Nasdaq Data Link [6] (contains different kinds of financial datasets),
- Dataset search engine from Google [2],
- Registry of Open Data on AWS [9],
- regular Google search.

Table 1 gives a summary of keywords used in the search, together with a brief description of the kind of results that each keyword or group of keywords yielded. Initial idea for the search was to either find cash flows dataset in the direct form, or to find a list of company's transactions, invoicing or payment data and to transform it to cash flows. Such results did not yield satisfactory results though, so the search was extended to include sales data. Cash flows can be derived from sales data under certain assumption – that all payments, are immediate, (i.e. sale result in an immediate cash flow). As an example, such assumption holds quite well for retail stores.

Datasets found in the search process are described in Table 2. The initial search for cash flow or transactions data yielded only two results, of which one is of really low quality as a data source, as it is an unlabeled set found on GitHub. After finding the results unsatisfactory, the search was extended to cover sales data as well, as described above. This yielded more results, however, only few of them were matching criteria of using them as cash flow data – others did not contain price information of sales, were from a field where assumption regarding immediate payments did not hold, or were from a short period of time (e.g. couple of months), which does not allow to properly capture seasonality of data. The table presents only datasets covering more than one year.

Table 1. Keywords used in the dataset search together with a comment and a brief summary of the results each keyword or group of keywords yielded.

Keyword(s)	Comments and results summary
cash flows company cash flows incoming cash flows	Small amount of data, mostly accumulated cash flows from financial reports of companies
transactions customer transactions	The idea was to transform transaction dataset with payment date available to cash flow data by aggregating payments received on each day. The problem is that data can be incomplete (not containing all the transactions for particular day, such completeness is not needed for predicting e.g. payment time), which poses an issue on aggregation. Datasets found this way was mostly credit card transactions, bank customers' transactions, ATM withdrawals, capital market and real estate transactions
account receivables	Small amount of data. The idea and issue were similar to the above – to aggregate payments for each date to get cash flow data, the problem of potentially incomplete data holds as well
invoice payments customer payments	More data, though mostly on outgoing payments, not incoming ones. Large part of the datasets were from the field of government and public entities payments, as such unit sometimes publish detailed reports on their financial operations
sales sales forecasting retail sales	This was a part of an extended search after the initial search yielded unsatisfactory results. The idea was to aggregate sales data to cash flow data under certain assumptions (that all payments are immediate, such assumption holds e.g. for retails stores sales). More datasates were found this way, though not all were matching desired profile - some of them did not contain price data, just demand; some of them were from companies where immediate payment assumption does not hold; and some of them were from a short period of time (e.g. couple of month), so not all seasonal dependencies could be captured

Table 2. Datasets that can be considered to be transformed to cash flow data, found in the searched sources with a comment and details.

Dataset	Number of records	Time span	Comment
IBM Late Payment Histories [1]	2 466	727 days	No guarantee of dataset being complete, as it is not known what subset of all the invoices is present in the data (such incompletness is undesirable when aggregating data daily). Can be considered to be used though
Payment Date Prediction (unlabelled dataset on GitHub) [11]	50 000	504 days	Unknown origin of the dataset, likely incomplete due to a round number of records. Legal issues may arise when utilizing this, as there is no information on license. Low usability in general
Online Retail II [7]	1 067 371	738 days	Sales (invoice) data from online transactions of UK-based company selling all-occasion gift-ware. Some of company's customers are wholesalers, so immediate cash flow assumption is up for discussion here, as part of the sales may be credited. Dataset may be considered to be used though
Historical sales data (used for Coursera course) [8]	2 935 849	1073 days	Sales data from multiple stores of an unknown industry. Amount of the sales suggest that the stores are retail, so immediate cash flow assumption may hold and cash flows can be derived from it
M5 Forecasting competition dataset [5]	30 490	1941 days	Sales data provided by Walmart, does not cover all products sold, focuses on the ones with low or intermittent demand, and because of that, may not represent overall cash flow of the store(s) well

5 Summary

The objective of the paper is to develop a work oriented at providing more reproducible research results in the field of daily cash flow forecasts. The use of publicly available datasets we review in this paper is of great value to future researchers in the field. As it is expected with financial data, there are only a few publicly available datasets, though some of them have a potential to be used in the research focusing on forecasting cash flow data. They require some processing before being used, though deriving cash flows data from them is a straightforward operation.

Our future work will include the application of the mentioned prediction methods and comparison their effectiveness on public datasets. Furthermore, we will explore additional potential methods to be used, including the ones that are already used in the field of time-series prediction, like regression kNN [17], fuzzy methods [18] or already mentioned methods used for project cash flow forecasting, and eventually to improve, extend those methods or introduce a new one. Finally, another potential workaround to the problem of small amount of data is generation of such data. Multiple approaches could be taken here as well, but as there are some sales datasets available, generating more sales data based on the existing datasets and transforming them to cash flows seems like a viable approach.

References

1. Account Receivables Dataset on Kaggle (IBM Late Payment Histories). https://www.kaggle.com/hhenry/finance-factoring-ibm-late-payment-histories. Accessed 10 Mar 2022
2. Dataset Search from Google. https://datasetsearch.research.google.com/. Accessed 10 Mar 2022
3. data.world datasets. https://data.world/datasets/open-data. Accessed 10 Mar 2022
4. Kaggle datasets. https://www.kaggle.com/datasets. Accessed 10 Mar 2022
5. M5 Forecasting competition dataset. https://www.kaggle.com/c/m5-forecasting-accuracy/data. Accessed 10 Mar 2022
6. Nasdaq Data Link. https://data.nasdaq.com/. Accessed 10 Mar 2022
7. Online retail II dataset on UCI Machine Learning Repository. https://archive-beta.ics.uci.edu/ml/datasets/online+retail+ii. Accessed 10 Mar 2022
8. Predict Future Sales (final project dataset for Coursera course). https://www.kaggle.com/c/competitive-data-science-predict-future-sales/data. Accessed 10 Mar 2022
9. Registry of Open Data on AWS. https://registry.opendata.aws/. Accessed 10 Mar 2022
10. UCI Machine Learning Repository. https://archive.ics.uci.edu/. Accessed 10 Mar 2022
11. Unlabelled dataset on GitHub. https://github.com/SkywalkerHub/Payment-Date-Prediction/blob/main/Dataset.csv. Accessed 10 Mar 2022
12. Chen, S., et al.: Cash flow forecasting model for electricity sale based on deep recurrent neural network. In: 2019 IEEE International Conference on Power Data Science (ICPDS), pp. 67–70. IEEE (2019)
13. Cheng, M.Y., Hoang, N.D., Wu, Y.W.: Cash flow prediction for construction project using a novel adaptive time-dependent least squares support vector machine inference model. J. Civ. Eng. Manage. **21**(6), 679–688 (2015)
14. Cheng, M.Y., Roy, A.F.: Evolutionary fuzzy decision model for cash flow prediction using time-dependent support vector machines. Int. J. Project Manage. **29**(1), 56–65 (2011)
15. Kayacan, E., Ulutas, B., Kaynak, O.: Grey system theory-based models in time series prediction. Expert Syst. Appl. **37**(2), 1784–1789 (2010)
16. Manuca, R., Savit, R.: Stationarity and nonstationarity in time series analysis. Physica D **99**(2–3), 134–161 (1996)

17. Martínez, F., Frías, M.P., Pérez-Godoy, M.D., Rivera, A.J.: Dealing with seasonality by narrowing the training set in time series forecasting with knn. Expert Syst. Appl. **103**, 38–48 (2018)
18. Nguyen, L., Novák, V.: Forecasting seasonal time series based on fuzzy techniques. Fuzzy Sets Syst. **361**, 114–129 (2019)
19. Taylor, S.J., Letham, B.: Forecasting at scale. Am. Stat. **72**(1), 37–45 (2018)
20. Wang, J.S., Ning, C.X., Cui, W.H.: Time series prediction of bank cash flow based on grey neural network algorithm. In: 2015 International Conference on Estimation, Detection and Information Fusion (ICEDIF), pp. 272–277. IEEE (2015)
21. Weytjens, H., Lohmann, E., Kleinsteuber, M.: Cash flow prediction: MLP and LSTM compared to ARIMA and Prophet. Electron. Commer. Res. **21**(2), 371–391 (2021)

Neural Network Based Drift Detection

Christofer Fellicious[✉], Lorenz Wendlinger, and Michael Granitzer

University of Passau, 94032 Passau, Germany
{christofer.fellicious,lorenz.wendlinger,michael.granitzer}@uni-passau.de

Abstract. The unprecedented growth in machine learning has shed light on its unique set of challenges. One such challenge is apparent changes in the input data distribution over time known as Concept Drifts. In such cases, the model's performance degrades according to the changes in the data distribution. The remedy for concept drifts is retraining the model with the most recent data to improve the model's performance. The significant issue is identifying the precise point at which the model must be updated for maximum performance benefits with minimum retraining effort. This problem is challenging to address in unsupervised detection methods with no access to label data to identify the changing distributions for the targets of the input data. Here, we present our unsupervised method based on a Generative Adversarial Network and a feed forward neural network for detecting concept drifts without the need for target labels. We demonstrate that our method is better at identifying concept drifts and outperforms the baseline and other comparable methods.

Keywords: Concept drift detection · Unsupervised drift detection · Neural networks · Machine learning

1 Introduction

Many machine learning approaches assume that the underlying data distributions of the training and testing datasets are identical. However, this is not the case in application domains where data collection occurs over extended periods, such as shopping behaviour, network intrusion detection, credit card fraud detection, and so forth. In such scenarios, models trained only on a subset of the initial data might face performance degradation according to the changing data distributions. A solution to this problem is retraining the model at different points in the timeline with the change in the data distribution. Identifying concept drifts as soon as possible helps improve the model's performance while keeping the retraining effort low.

Realistically, we cannot predict how the input data distribution would change over time, and it is computationally infeasible to prepare for all possible scenarios. However, identifying a concept drift as early as possible and retraining the models can mitigate such problems. Another solution to this could be retraining the models as specific intervals without considering whether a concept drift has

G. Nicosia et al. (Eds.): LOD 2022, LNCS 13810, pp. 370–383, 2023.
https://doi.org/10.1007/978-3-031-25599-1_28

occurred or not. Nevertheless, this is not desirable as the data distribution could have been stationary, and retraining is wasteful. It could also be that a drift and retraining window does not align, degrading the model performance despite retraining.

Generative Adversarial Networks are proven to be powerful in several use cases such as image synthesis [25], image upscaling [24] to even audio synthesis [5]. Our method uses Generative Adversarial Networks and feed forward neural networks. We use the expressive power of neural networks to understand the subtle differences between different concepts and identify concept drifts. We use Generative Adversarial Networks to generate synthetic data and discriminate between data distributions. After identifying a drift using the discriminator, we switch to a neural network based method for better performance. Our method outperforms comparative supervised and unsupervised methods on the multiple datasets in outright performance. We also show that the method can easily generalize to synthetic and real-world datasets with minimal changes to the procedure.

2 Related Work

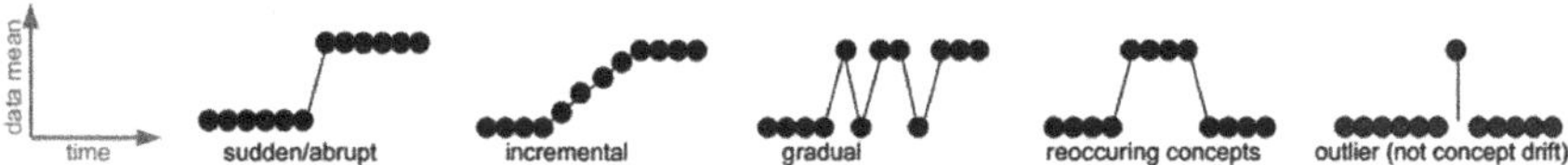

Fig. 1. Different kinds of concept drift as shown by [8]

Concept drifts in data streams can be classified into different types based on how the input distribution changes or based on how the target values change over time. We define a concept as a sequence of input feature vectors which has the same underlying probability distribution. A survey by Gama et al. [8] describes the different types of concept drifts based on the changes to targets with respect to time. Figure 1 shows the different types of drifts on a one dimensional toy dataset as explained by Gama et al. An incremental drift occurs when input features change slowly over time such as with wear and tear of components and an abrupt drift occurs when the input distribution changes suddenly. Gama et al. also explain about gradual drift where the input features oscillate between two distributions before settling into the drifted feature distribution. A survey by Bayram et al. [3] explain the different types of drift in detail along with the different methods and the current challenges in the field. Input distributions can also recur such as the buying behaviour of people where people tend to shop more at specific times of the year. One of the main challenges of drift detection algorithms is with outliers and to not mistake outliers as the initiation of concept drift. Outliers are short lived bursts of erratic data which are entirely different from the trained data distribution but they are not considered a concept drift as the input data distribution settles back. An outlier in a toy dataset can be seen in Fig. 1. For the reason stated above the model is not updated for outliers.

Another type of concept drift occurs when the higher dimensional space where the initial concept resided in, itself changes. In Fig. 2, we see the initial dataset that we trained on, but after a period of time the concept changed. While the data distribution remains more or less constant, the targets of the distribution has changed. Such changes mostly cannot be detected by unsupervised methods, as these methods look only at the input feature vectors and their distributions. This type of concept drift can be gradual, abrupt or incremental as explained earlier.

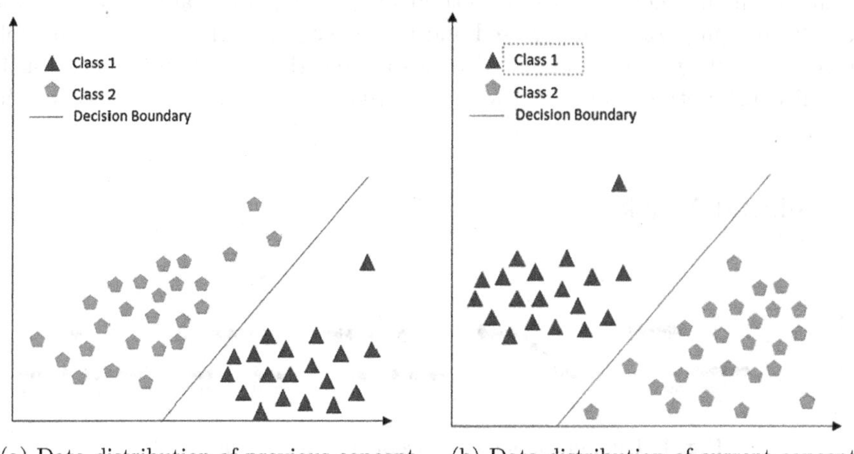

(a) Data distribution of previous concept (b) Data distribution of current concept

Fig. 2. Changing of concepts

A supervised method known as the Drift Detection Method (DDM) implemented by Gama et al. [7] uses the error rate of the trained model to detect when a drift occurs in the data stream. The hypothesis is that in a data stream with no change in the input distributions, the error rate will decrease with more input data. However, a change in the input distribution will adversely affect the model and generate more erroneous predictions. By monitoring this change in the error rate, and identifying the point at which the error rate starts to increase, we can find the initiation points of drifts. DDM uses a two tiered drift error threshold to identify a concept drift. When predicting on the data stream, if the error rate increases and reaches the lower threshold a drift warning is signalled. If the error rate continues to increase and reaches the second threshold, a concept drift is signalled. At this point, the model is retrained based on the window from the point where the drift warning was issued until the drift was detected. One of the key issues is defining a window size to retrain the model. If the window size is too large, the window might have instances from the previous "concept". A smaller window size means that the algorithm might not have enough instances to generalize and might adversely affect the performance of the model. To counter this, an idea of context is introduced. Contexts are groups of features that are contiguous in time where the underlying data distribution is assumed to be stationary. For the DDM method the context is defined as the input between when the drift warning is issued and until the drift is detected. The model is then retrained based on this context window. A drawback of this method is that if a drift warning is signalled earlier to a slow incremental drift, the retraining takes place over a very large window and the model does not generalize enough to the newest concept.

Early Drift Detection Method (EDDM) [2] is another popular method for supervised drift detection. The algorithm improves performance on gradual concept drift while also performing well on abrupt concept drifts. Identifying proper windows for detecting drifts is a challenging task. The Adaptive Windowing (ADWIN) [4] in this direction for drift detection. ADWIN uses statistical tests on different windows and detects concept drifts based on the error rates, similar to that of DDM. Fisher [6] developed a statistical method that can be used to analyze contingency tables. It is particularly useful as it can be used when the sample size is small and can analyze the deviation from the null hypothesis. A method using Fisher's Exact Test was proposed by [16]. This method analyzed two separate context windows and the assumption that if there was no drift, the errors will be equally distributed between the two windows. They make use of the generalized contingency table from Fisher's Exact test to test the null hypothesis. These above mentioned methods are all supervised and make use of the labels of the input data to identify the presence of a concept drift.

Hu et al. [13] explain why detecting concept drifts while maintaining high performance is not a trivial matter. The paper explains that high performance can be achieved with labelled data but labelling data streams could prove to be expensive. Unsupervised drift detection is difficult as we don't have access to the labels. Most unsupervised methods have to rely on the input data distributions and detect changes in them for identify concept drifts. Statistical tests like the Kolmogorov-Smirnov (K-S) test the possibility of the two different samples belonging to the same distribution. The K-S test is a nonparametric test that can be used to ascertain whether a pair of samples belong to the same data distribution. The problem with the K-S test is its time complexity that grows with the number of samples. The K-S test works with univariate data which means we have to apply the K-S test over all feature dimensions. A faster alternative was proposed by Reis et al. [21] using an incremental K-S test. This implementation makes use of a custom tree data structure called "*treap*". The algorithm uses two windows, where the first one is fixed and the second one is sliding. The first window contains the original concepts that the model is trained on, while the second one slides and checks whether a concept drift has occurred. The implementation also applies the test for each attribute independently instead of using a multivariate test method. While the K-S test remains popular for univariate data and an expansion of the K-S test for multivariate data was proposed by Justel et al. [14]. Babu and Feigelson [1] explain the pitfalls of using the K-S test. They also explain why using the K-S test for multivariate data could be erroneous. They demonstrate that the K-S test is most sensitive to large scale differences in location between the model and data. When considering a bivariate distribution F, "The distribution of the K-S statistic varies with the unknown F and hence is not distribution-free when two or more dimensions are present. The K-S statistic still is a measure of *distance* between the data and model, but probabilities can not be assigned to a given value of the statistic without detailed calculation for each case under consideration".

Suprem et al. [23] explain about two different kinds of drift, namely task drift and domain drift. "Task drift reflects real changes in the world. Formally, this corresponds to the drift in the conditional distribution of labels given the input data (i.e., P(Y | X)), often resulting from an updated definition of the task necessitating a change in the predictive function from the input space to label space. Domain drift does not occur in reality but rather occurs in the ML model reflecting this reality. In practice, this type of drift arises when the model does not identify all the relevant features or cannot cope with class imbalance. Formally, this corresponds to the drift in the marginal distribution of the input data (i.e., P(X)), with an additional assumption that P(Y | X) remains the same". Their method named ODIN uses auto encoders for identifying drifts and to select trained classifiers based on the drift. Their method uses autoencoders to generate latent representations and a decoder to generate the decoder coupled to a discriminator. The loss function is computed as the weighted sum of the loss of the latent discriminator loss, the image discriminator and the standard reconstruction loss.

Gözüaçık et al. [10] introduced an unsupervised method for identifying concept drifts using a base classifier and its ability to discriminate between an older context window and newer window with a certain accuracy. If the classifier is able to distinguish between the older concept and the current data, then the input data distribution has changed and this signals the beginning of a new concept and thus requires retraining. This method employs two different classifiers, one for discriminating between the old and new concepts and another one for predicting the labels on the data. This method makes use of unbalanced concept windows with a larger window for old concept (100 instances) and a smaller window of 10 instances for the newer concept. This could lead to overfitting by the drift detection model on the trained window and detect concept drifts where none is present if the Area Under Curve metric value is set to a low value.

3 Neural Network Based Drift Detection

Generative Adversarial Networks (GAN) introduced by Goodfellow et al. [9] significantly improved the previous generative models by including an adversarial component. GAN works by training two separate networks, a generator that produces synthetic data similar to the real data and a discriminator that tries to determine whether the data given to it is real input data or synthetic data from the generator. Usually, the generator is trained for several steps until the discriminator cannot distinguish between the generated data and real data. Then the discriminator is trained until the losses drop below a threshold value. GAN presents the advantage that the generator part could theoretically be trained to represent any data distribution. This property of the GAN is exploited to generate synthetic data similar to real data. On the other hand, a discriminator could be trained to identify subtle changes in distributions between the real data and the generated synthetic data. We take advantage of these properties in our approach of using GAN to detect concept drifts in data streams. From the works of Liu and Tuzel [17] for generating pairs of images, Gupta et al. [11] for trajectory prediction, Zhan et al. [25] in image synthesis, Wang et al. [24] in image upscaling, Engel et al. [5] for audio synthesis, we understand that GANs are capable of learning complex patterns very well. However, concept drifts are also complex, can appear in different forms, and are unpredictable. Since we cannot be sure when a concept drift might occur or what type of concept drift could occur, we need a model than can learn the underlying distribution of the current concept window. In our use case, GAN provides us with the advantage of having the ability to generate synthetic data using the generator. When trained for a small number of epochs, this generated data simulates data that is similar to the real data but slightly out of distribution. We use this data to simulate different drifts and the discriminator to distinguish between the current data distribution and drifted data which has a different underlying distribution. We use the generator part to simulate drifted data and the discriminator to distinguish between the probability distributions of the normal data and the drifted data. However, this approach consumes more time, so we switch to a neural network based approach after the first drift is identified. This switch speeds up the execution and provides better results on the Hoeffding Tree Classifier's secondary classifier.

Our approach uses a neural network for detecting drifts and a Hoeffding Tree classifier to predict the labels on the input data stream. The Hoeffding Tree Classifier was chosen as the algorithm for predicting the streaming data labels because it is one of the most used classifiers for concept drift detection. We want to compare our algorithm with others on the same level. The classifier is initialized with only the default hyperparameters and only the number of classes passed during the initialization.

3.1 Network Architectures

For our method, we require three different neural networks, one which we call the Generator ($G(x)$), another the Discriminator ($D(x)$) and the last one is the Drift Detector ($DD(x)$). The first two networks form the GAN, where $G(x)$ is used to generate synthetic data and $D(x)$ is used to identify actual data from synthetic data. The third network, $DD(x)$ is a network that we use to identify drifts after the first drift is detected. After identifying a drift, we have access to data distributions from the older data distribution (Old Concept) and the new data distribution (New Concept). We use these two different data distributions to create a training dataset with the **Old Concept** which is the data where the probability distribution has already changed, and the **New Concept** which is the current probability distribution of the data. $G(x)$ consists of an input layer, a hidden layer and an output layer, each having 128 neurons. The input layer uses the Rectified Linear Unit (ReLU) used by Krizhevsky et al. [15] and the Sigmoid function as the non-linearity for the network. The experiments show that this model has the expressive power to generate the required synthetic data.

$D(x)$ is also a fully connected feed forward network of two hidden layers compared to the single hidden layer of $G(x)$. From our experiments, we found out that two hidden layers were the minimum depth required to have the power to discriminate between multiple types of drifts for all the test datasets. Therefore, we concentrate more on the power of $D(x)$ to perform well rather than $G(x)$. The layers also double in size from the input layer starting at 128 neurons to the first hidden layer with 256 neurons and the second with 512 neurons. We cut off any negative output from the first layer using the ReLU non-linearity. $D(x)$ should not make decisions based only on a single neuron getting activated. Furthermore, to reduce such predictions linked to the activation of specific neurons or groups of neurons, a layer introduced by Srivastava et al. [22] called Dropout is used to prevent $D(x)$ from overfitting to the data. The model predicts the probability of the input data to be a drift or not, and we use a Sigmoid layer to compress the data between 0 and 1.

$DD(x)$ is identical to $D(x)$, except we apply batch normalization of the data before the input layer. The number of parameters for each network is shown in Table 1.

Table 1. Network parameters

Network	Parameters
Generator $G(x)$	18,438
Discriminator $D(x)$	166,017
Drift Detector $DD(x)$	166,017

3.2 Training

We compare our results directly with state-of-the-art in the unsupervised method, and therefore, we use the same datasets which the authors of the D3 algorithm provide. The datasets are available in a GitHub repository provided by the authors of the D3 algorithm[1]. We start the training with a single concept obtained from the initial training window (W_0). This initial concept consists of all the features contained in W_0. As no prior information about other concepts is present initially, the GAN is used to simulate different concept drifts. $D(x)$ learns to identify the artificial data, and this is required as initially, we do not possess any knowledge about how the distribution will change as time passes.

[1] https://github.com/ogozuacik/concept-drift-datasets-scikit-multiflow.

Hence the simulation of different data distributions outside the real data using $G(x)$ of the GAN architecture. Here, $G(x)$ learns to simulate probability distributions to that of W_0, while $D(x)$ learns to differentiate between distributions from W_0 and $G(x)$. Here the aim is to force $D(x)$ to distinguish any other distribution outside the real data. Synthetic data generated by G(x) represents the drifted data, and identifying the synthetic data by $D(x)$ signals a concept drift.

Once the training is complete, $D(x)$ can detect input data with a different data distribution than the train data and signal a concept drift. Detecting a concept drift gives us access to two concepts, the data that $D(x)$ trained on and the successive data after detecting the concept drift. We then switch over to the fully connected network for differentiating between concepts, and the training data is created using two different windows representing the two different concepts. For example, for the concept drift detected at index i, we create two different concept windows W_k and W_{k+1}.

$$W_k = data[i-\rho]...data[i-1] \tag{1}$$

$$W_{k+1} = data[i]...data[i+\rho-1] \tag{2}$$

where ρ is the window size

W_k is a series of contiguous inputs from $i-\rho$ until $i-1$, and this window of data represents the past concept. Similarly, W_{k+1} is a series of contiguous inputs from i until $i+\rho$, and this window represents the current data distribution or concept. We specify the old concept as the data stream of size ρ into the past Eq. 1. The new concept is the data obtained from the point of drift detection. It is of the same size ρ as the old concept given in Eq. 2. As long as the input data stream conforms to the same distribution of W_{k+1}, no concept drift will be detected. However, if the prediction of the network exceeds a certain threshold, it indicates a drift. This indication means that the tested input data deviates from the probability distribution trained with W_{k+1}, and the model must undergo retraining.

From this point on, where the first drift is signalled, we do not use $D(x)$ but only $DD(x)$ as we have data suitable for identifying concepts from W_k and W_{k+1}. We train $DD(x)$ using W_k and W_{k+1} assigning the features labels 1 and 0. A feature vector Z is assigned a label based on the following condition

$$x = \begin{cases} 1, if Z \in W_k \\ 0, if Z \in W_{k+1} \end{cases}$$

For the GAN, during the initial concept, we divide the input data by the mean (μ) of W_0. For all subsequent drifts, the data is standardized using the mean (μ) and standard deviation (σ) from W_{k+1}. This process repeats whenever a concept drift is detected. The training algorithm for drift detection is outlined in Algorithm 1, and the code is available in an online repository[2].

We define a specific window size for the input data to train the networks. We chose Stochastic Gradient Descent (SGD) as the optimizer with a learning rate

[2] https://github.com/cfellicious/NeuralNetworkbasedDriftDetection.

of 0.0001 and momentum of 0.9. The reason for this choice behind the optimizer is that SGD is stable and does not fluctuate compared to other optimizers such as Adadelta. We use the Binary Cross Entropy loss function to compute the loss for the networks.

Algorithm 1: Neural Network based Drift Classification

```
FindDrifts(ρ, τ, k, epochs) → Window size, Threshold, batch size and training
  epochs respectively ;
W = data[0:ρ]          → Obtain ρ samples for training the GAN;
μ = 0 ;
σ = standard_deviation(W) ;
model, Generated_Data = TrainGan(W, epochs, k)          → Train the GAN;
while isDataBatchAvailable do
    Data = (DataBatch − μ)/σ ;
    if model(Data) < τ then
        Batch prediction below theshold(τ), Drift Detected ;
        old_context = W            → Create the old context window ;
        W = ∅ ;
        while W is not full do
            W = W ∪ DataBatch            → Update the window with new
              context ;
        end
        ℓ = ∅                        → Initialize labels vector ;
        ℓ = ℓ ∪ 0 for [1, ρ]         → Labels for old context ;
        ℓ = ℓ ∪ 1 for [1, ρ]         → Add the labels for new context ;
        f = old_context ∪ W          → Combine old and new contexts ;
        μ = mean(W) ;
        σ = standard_deviation(W) ;
        f = (f - μ) / σ            → Standardize the data ;
        model = TrainNetwork(W, ℓ, f, epochs)          → Train the model ;
    else
        W = W ∪ DataBatch          → Add the current batch to the window ;
        Drop k number of old data elements from of W;
    end
end
```

4 Results and Discussion

The experiments were executed on a desktop computer with an Intel i7-7700 CPU and 32 GB of system memory. In addition, the system had an nVidia GTX1080 with 8 GB of VRAM. We evaluate our approach on different real-world and synthetic datasets. For our evaluation, we use a secondary classifier that predicts the actual labels on the incoming data. Concept drifts occur in many forms as shown in Fig. 1. Moreover, the datasets do not contain labels

Table 2. Hyperparameter values tested

Hyperparameter	Values tested	Selected value
Window size (ρ)	[25, 50, 60, 75, 100, 125, 150, 175, 200, 250, 500]	100
Batch size (k)	[1, 2, 4, 8, 16, 32, 64]	4
Epochs	[10, 20, 25, 50, 100, 125, 150, 200, 250, 300, 400, 500]	200
Threshold (τ)	[0.45, 0.46, 0.47, 0.48, 0.49, 0.50, 0.51, 0.52, 0.53, 0.54]	[0.47, 0.53]

identifying drifts. Therefore we assess the performance of the drift detection algorithm based on the performance of the secondary classifier, which is a supervised learning algorithm. The better the performance of the secondary classifier, the better the drift detection algorithm is also better. The assumption is that the retraining points are better identified and thus allow the secondary classifier to retrain better. Our approach used the default parameters for the secondary classifier without any hyperparameter tuning on that part.

Our comparison is based on the results by published by Gözüaçık et al. [10]. We use accuracy as the metric for evaluating all the datasets as it is the most prevalent metric used for them. We use the implementation of scikit-learn [20]. As for the Hoeffding Tree, the scikit-multiflow library is used [19]. This library caters to analysing streaming datasets. For the evaluation, we use the Interleaved-Test-Then-Train method outlined in the paper by Gözüaçık et al. [10]. This method tests every incoming feature vector and then adapts the classifier to the tested feature vector. The Hoeffding tree classifier is reset and retrained every time a drift is detected.

We evaluated our hyperparameters such as the window size, batch size, threshold and epochs at different values as shown in Table 2.

4.1 Baseline Methods

It is essential for any experiment is to introduce baselines and do a comparison of the experiment with the baselines and also the state of the art. We devised three different baseline measures for our tests. The first method we chose is to train the classifier on a small window (ρ) of input feature vectors and then use that model to test on the entirety of the data. We call this method the *Initial Train* method as we train it only on the initial window. This method assumes that the data is stationary and no concept drift occurs. For the second method, we consider the effect of updating the classifier at every window of size ρ. We test the model on a window of input feature vectors and update the model on that data window without fully retraining the model. We call this method the *Regular Update* method and the objective of this baseline is to identify the impact of updating the model at fixed intervals. With this method, we train on all the data as soon as we test it. This allows for an expanded training corpus even if the training is done on data belonging to different distributions. The assumption here is that the underlying input data distribution is stationary similar to the

Table 3. Comparison of accuracy values of different methods

Dataset	Initial train	Regular update	Regular retrain	D3	Our method
ELEC	56.44	77.69	75.4	86.69	**86.81**
Poker	50.12	74.08	63.05	75.59	**76.63**
Rialto	09.95	31.373	51.02	52.39	**54.84**
COVTYPE	47.69	82.49	55.48	**87.17**	86.00
Rotating hyperplane	74.00	84.09	82.16	85.29	**86.95**
Moving squares	30.01	33.63	81.74	66.28	**91.20**
Moving RBF	25.91	33.94	51.06	**51.59**	48.03
Interchanging RBF	15.75	25.78	**96.04**	82.81	84.74

Initial Train baseline method. The last method consists of using the model to predict of a window of input feature vectors and then retraining the model using the window that we just tested. This method involves retraining the model from scratch at every window once the window is tested and is called the *Regular Retrain* method.

4.2 Results and Discussion

We set the training window size (W) as 100 based on trial and error methods. A smaller window size does not adequately generalise features for the network to discriminate between the old and new contexts adequately. The training epochs, e is 200 and the batch size, b is 4 for the GAN and 16 for the Drift Detector network ($DD(x)$). We set the threshold (ρ) to 0.53. The threshold is slightly larger than 0.5 because, from our experiments, we noticed that the value for detecting older context is never almost zero but slightly above it. In contrast, the new context was very close to zero, proving effective for real-world datasets. However, for the artificial datasets, the threshold was lowered to 0.47 for the same reasons. We assume that the noise added to the datasets causes this, thus necessitating a lower threshold.

We compare our results with the Discriminative Drift Detector (D3) method as that is the current best unsupervised method as reported by Gözüaçık et al. [10]. We also compare the results against different baselines to investigate the different retraining methods without concept drift detection. Another aspect of consideration in our experiments is the effect of randomness. We executed the experiment for ten independent runs with different random seeds and tabulated the results. The best result from the ten independent runs is in Table 3. From that, we can see that our method outperforms the D3 unsupervised method in all but two datasets: the Moving RBF and the COVTYPE datasets. Our method also improves on the Moving Squares dataset, as published by Losing et al. [18], by a margin of more than 20%. The only outlier result is in the Interchanging RBF dataset, where the *Regular Retrain* method produces the best result by

Table 4. Mean accuracies over 10 runs with mean (μ) and standard deviation (σ)

Dataset	Our method	
	μ	σ
ELEC	86.53	0.11
Poker	76.50	0.11
Rialto	54.63	1.03
COVTYPE	83.53	1.66
Rotating hyperplane	86.90	0.13
Moving Squares	90.69	1.42
Moving RBF	47.81	0.18
Interchanging RBF	84.47	0.26

scoring 96.04% accuracy value. This result demonstrates that for this specific kind of dataset, having the model trained on the most recent data is the best. Looking at the average values shown in Table 4, we can see that our method outperforms the D3 method in most datasets. The only outlier for the method occurs in the ELEC dataset [12], where we report a lower accuracy than the D3 method.

The raw performance of our models on the dataset is undeniable. However, the question is whether it is worth the extra time to train a model to detect concept drifts, retrain models for the streaming dataset, and so on. If we look at the real-world datasets, the concept drift methods based classification performs much better than the baseline methods. When the model is trained only on a few instances (Initial Train method) at the beginning of the data stream, the results are the worst of all the methods. These results clearly show that updating the model for the data stream is vital to ensure good real-world performance. The Regular Update method of adding the incoming data instances to the model does improve results significantly compared to the first baseline method. Compared to the D3 method, they exhibit the same raw performance for a few datasets like Poker hand and Rotating hyperplane. The data distributions in these datasets did not vary wildly, and the variations could be mitigated with partial updates. Our third baseline method, the Regular Retrain method, which resets the model and retrains in every window, produced the best result in the Interchanging RBF dataset beating even the methods that detected concept drift. This dataset simulates abrupt data drift with multiple Gaussian distributions. The presence of abrupt drifts best explains why keeping the latest data in the memory is the best method for this scenario. This method performs better than the D3 method for the Moving Squares and is the second best for the Moving RBF dataset. An issue with this training mode is the cost incurred when retraining the models at fixed intervals. In real-world datasets, drift boundaries are often not clearly defined; thus, the drift will not occur at a predefined boundary. This causes overlaps in the training windows where there is more data from the older context than the newer context, thus adversely affecting the model's performance. This data inclusion from older data distributions when retraining is one of the main reasons why this baseline method performs considerably worse in real-world datasets.

5 Conclusion

This paper introduces a method for identifying concept drifts in streaming data using Generative Adversarial Networks and feed forward neural networks. We show that our method combining a GAN and a feed forward neural network helps in identifying drifts. Furthermore, our experimental results show that the proposed method outperforms comparable unsupervised and many supervised methods on identical real-world and artificial datasets.

In the future, we look to identify if any recurring patterns are present in the data stream and to use older classifiers as an ensemble method to improve the classifier's performance on the streamed data. We also explore methods of incorporating synthetic data from the generator to improve the method's performance.

Acknowledgement. This work was partially funded by the Bundesministerium für Bildung und Forschung (BMBF, German Federal Ministry of Education and Research) – project 01IS21063A-C (SmartVMI).

References

1. Babu, G., Feigelson, E.: Astrostatistics: goodness-of-fit and all that! In: Astronomical Data Analysis Software and Systems XV, vol. 351, p. 127 (2006)
2. Baena-Garcıa, M., del Campo-Ávila, J., Fidalgo, R., Bifet, A., Gavalda, R., Morales-Bueno, R.: Early drift detection method. In: Fourth International Workshop on Knowledge Discovery from Data Streams, vol. 6, pp. 77–86 (2006)
3. Bayram, F., Ahmed, B.S., Kassler, A.: From concept drift to model degradation: an overview on performance-aware drift detectors. Knowl.-Based Syst. 108632 (2022)
4. Bifet, A., Gavalda, R.: Learning from time-changing data with adaptive windowing. In: Proceedings of the 2007 SIAM International Conference on Data Mining, pp. 443–448. SIAM (2007)
5. Engel, J., Agrawal, K.K., Chen, S., Gulrajani, I., Donahue, C., Roberts, A.: Gansynth: adversarial neural audio synthesis. arXiv preprint arXiv:1902.08710 (2019)
6. Fisher, R.A.: Statistical methods for research workers. In: Kotz, S., Johnson, N.L. (eds.) Breakthroughs in Statistics. Springer Series in Statistics, pp. 66–70. Springer, New York (1992). https://doi.org/10.1007/978-1-4612-4380-9_6
7. Gama, J., Medas, P., Castillo, G., Rodrigues, P.: Learning with drift detection. In: Bazzan, A.L.C., Labidi, S. (eds.) SBIA 2004. LNCS (LNAI), vol. 3171, pp. 286–295. Springer, Heidelberg (2004). https://doi.org/10.1007/978-3-540-28645-5_29
8. Gama, J., Žliobaitė, I., Bifet, A., Pechenizkiy, M., Bouchachia, A.: A survey on concept drift adaptation. ACM Computi. Surv. (CSUR) **46**(4), 1–37 (2014)
9. Goodfellow, I.J., et al.: Generative adversarial networks. arXiv preprint arXiv:1406.2661 (2014)

10. Gözüaçık, Ö., Büyükçakır, A., Bonab, H., Can, F.: Unsupervised concept drift detection with a discriminative classifier. In: Proceedings of the 28th ACM International Conference on Information and Knowledge Management, pp. 2365–2368 (2019)
11. Gupta, A., Johnson, J., Fei-Fei, L., Savarese, S., Alahi, A.: Social GAN: socially acceptable trajectories with generative adversarial networks. In: Proceedings of the IEEE Conference on Computer Vision and Pattern Recognition, pp. 2255–2264 (2018)
12. Harries, M., Wales, N.S.: Splice-2 comparative evaluation: electricity pricing (1999)
13. Hu, H., Kantardzic, M., Sethi, T.S.: No free lunch theorem for concept drift detection in streaming data classification: a review. Wiley Interdisc. Rev. Data Min. Knowl. Discov. **10**(2), e1327 (2020)
14. Justel, A., Peña, D., Zamar, R.: A multivariate Kolmogorov-Smirnov test of goodness of fit. Stat. Probab. Lett. **35**(3), 251–259 (1997)
15. Krizhevsky, A., Sutskever, I., Hinton, G.E.: Imagenet classification with deep convolutional neural networks. Adv. Neural. Inf. Process. Syst. **25**, 1097–1105 (2012)
16. de Lima Cabral, D.R., de Barros, R.S.M.: Concept drift detection based on fisher's exact test. Inf. Sci. **442**, 220–234 (2018)
17. Liu, M.Y., Tuzel, O.: Coupled generative adversarial networks. Adv. Neural. Inf. Process. Syst. **29**, 469–477 (2016)
18. Losing, V., Hammer, B., Wersing, H.: KNN classifier with self adjusting memory for heterogeneous concept drift. In: 2016 IEEE 16th International Conference on Data Mining (ICDM), pp. 291–300. IEEE (2016)
19. Montiel, J., Read, J., Bifet, A., Abdessalem, T.: Scikit-multiflow: a multi-output streaming framework. J. Mach. Learn. Res. **19**(1), 2914–2915 (2018)
20. Pedregosa, F., et al.: Scikit-learn: machine learning in Python. J. Mach. Learn. Res. **12**, 2825–2830 (2011)
21. dos Reis, D.M., Flach, P., Matwin, S., Batista, G.: Fast unsupervised online drift detection using incremental Kolmogorov-Smirnov test. In: Proceedings of the 22nd ACM SIGKDD International Conference on Knowledge Discovery and Data Mining, pp. 1545–1554 (2016)
22. Srivastava, N., Hinton, G., Krizhevsky, A., Sutskever, I., Salakhutdinov, R.: Dropout: a simple way to prevent neural networks from overfitting. J. Mach. Learn. Res. **15**(1), 1929–1958 (2014)
23. Suprem, A., Arulraj, J., Pu, C., Ferreira, J.: Odin: automated drift detection and recovery in video analytics. arXiv preprint arXiv:2009.05440 (2020)
24. Wang, X., et al.: ESRGAN: enhanced super-resolution generative adversarial networks. In: Leal-Taixé, L., Roth, S. (eds.) ECCV 2018. LNCS, vol. 11133, pp. 63–79. Springer, Cham (2019). https://doi.org/10.1007/978-3-030-11021-5_5
25. Zhan, F., Zhu, H., Lu, S.: Spatial fusion GAN for image synthesis. In: Proceedings of the IEEE/CVF Conference on Computer Vision and Pattern Recognition, pp. 3653–3662 (2019)

Aspects and Views on Responsible Artificial Intelligence

Boštjan Brumen[1(✉)], Sabrina Göllner[2], and Marina Tropmann-Frick[2]

[1] Faculty of Electrical Engineering and Computer Science, University of Maribor, Maribor, Slovenia
bostjan.brumen@um.si

[2] Department of Computer Science, Hamburg University of Applied Sciences, Hamburg, Germany
{sabrina.goellner,marina.tropmann-frick}@haw-hamburg.de

Abstract. Background: There is a lot of discussion in EU politics about trust in artificial intelligence (AI). Because it can be used as a lethal weapon we need (EU) regulations that take care of setting up a framework. Companies need guidance to develop their AI-based products and services in an acceptable manner. The research should help AI and machine learning practitioners to prepare for what is coming next, and which aspects they should focus on. **Objective:** In the present research, we aim to understand the role of "Responsible AI" from different perspectives, what constitutes the umbrella term "Responsible AI" and what terms define it. The research question is: "What are the aspects defining the 'Responsible AI'?" **Method:** A structured literature review (SLR) was used as the research method. We searched four databases for relevant research results on "Responsible AI" in the last two years. 118 research papers were finally included in our study. **Results:** We found only three papers that try to define "Responsible AI". They use concepts such as Fairness, Privacy, Accountability, Transparency, Ethics, Security & Safety, Soundness, and Explainability to define the "Responsible AI". After studying all the 118 analyzed papers we strongly believe that the terms that are included in those 3 definitions are not enough; some are ambiguous, missing, or used as synonyms. We developed a four-dimensional representation of similarities and differences for defining "Responsible AI". "Responsible AI" must be a human-centered approach and the concept must include the implementation of AI methods that focus on ethics, explainability of models as well as privacy, security, and trust.

Keywords: Structured literature review · Artificial intelligence · Responsible AI · Privacy-preserving AI · Explainable AI · Ethical AI · Trustworthy AI

1 Introduction

In the past years, a considerable amount of research has been conducted in order to improve Artificial Intelligence (AI) even further, as it is already being used

G. Nicosia et al. (Eds.): LOD 2022, LNCS 13810, pp. 384–398, 2023.
https://doi.org/10.1007/978-3-031-25599-1_29

in many aspects of social life and industry. The European Commission supports the initiatives and is itself proactive in the field. The EU Commission published a series of papers [10–12] in 2020 and 2021 in which they address their strategy for AI. Numerous efforts in the field of AI are presented therein. It becomes clear that the EU Commission is seeking for a framework that reflects the EU values, is sustainable, secure, and with a special focus on trustworthiness. However, despite their detailed nature, the political documents still lack of details and definitions, and agreed-upon terms that are equally understood by relevant stakeholders. The Commission's goal is to *"strengthen competitiveness, the capacity for innovation and the responsible use of AI in the EU."* As it is a very important political aspect, we urgently need a unified AI framework for "Responsible AI". Since trustworthy AI is only a part of the whole picture, it is crucial to include also other aspects of responsibility in AI. To support this, we will take a step towards a clarification of the term "Responsible AI".

For this contribution, we conducted a structured review that aims to provide a clear answer to the questions how "Responsible AI" is defined and what is the state of the art in the literature. In our analysis, we found that there is a high level of inconsistency in the terminology, definitions, and principles for responsible AI. The problem also exists with the content-wise similar expressions to "Responsible AI". This complicates the understanding of the term. There are already many approaches in the analyzed areas of trustworthy, ethical, explainable, privacy-preserving, and secure AI, but there are still many open problems to be solved in the future. To the best of our knowledge, this is the first detailed and structured review dealing with the term and the aspects of "Responsible AI".

The rest of the paper is structured as follows: The next section deals with our research methodology. This includes defining our research aims, objectives and question, as well as specifying the databases and research query we used. The third section is the analysis part, where we first find out what definitions exist for "Responsible AI" as well as for content-wise similar expressions. These findings are collected and then compared with each other. The subsequent section then summarizes the key findings in the previously defined scopes namely "Human-centered, Trustworthy, Ethics, Explainability, Privacy and Security".

In addition, in the discussion section, we have summarized the key points that form the pillars for the development of "Responsible AI". This forms the basic idea of a framework for "Responsible AI". Finally, we describe the limitations of our work and conclude with an outlook on future work.

2 Research Methodology

To answer the research questions, a structured literature review (SLR) was performed based on the guidelines developed in [22]. The process of doing the structured literature review in our research is described in detail in the following subsections.

2.1 Research Aims and Objectives

In the presented research, we aim 1) to understand the role of "Responsible AI" from different perspectives and 2) what constitutes the umbrella term "Responsible AI". Based on the aims of the research, we state the following research questions: What are the aspects defining Responsible AI?

In order to get best results when searching for the relevant studies, we used the indexing data sources. These sources enabled a wide search of publications that would otherwise be overlooked. The following databases were searched: ACM Digital Library, IEEE Explore, SpringerLink, and Elsevier ScienceDirect.

2.2 Studies Selection

To search for documents, the following search query was used in the different databases: `("Artificial Intelligence" OR "Machine Learning" OR "Deep Learning" OR "Neural Network" OR "AI" OR "ML") AND (Ethic* OR Explain* OR Trust*) AND (Privacy*)`. Considering that inconsistent terminology is used for "Artificial Intelligence", the terms "Machine Learning", "Deep Learning" and "Neural Network" were added, which should be considered as synonyms. Because there are already many papers using the abbreviations "AI" and "ML", these terms were included to the set of synonyms. The phrases "Ethic", "Trust" and "Explain" as well as "Privacy" were included with an asterisk (*), for all combinations of the terms following the asterisk are included in the results (e.g. explain*ability).

The search strings were combined using the boolean operator "OR" for inclusiveness and the operator "AND" for the intersection of all sets of search strings. These sets of search strings are put within parentheses. The selection of the period of publication was set to the last two years: 2020 and 2021 to get all of the state-of-the-art papers. The search was performed in December 2021. The results were sorted by relevancy prior to the inspection, which was important because the lack of advanced options in some search engines returned many non-relevant results. To exclude irrelevant papers, the authors followed a set of guidelines during the screening stage. Papers did not pass the screening if they:

- mention AI in the context of cyber-security, embedded systems, robotics, autonomous driving or internet of things,
- have content that, by the definition of our terms, does not belong to Responsible AI, but to other AI studies,
- only consist of an abstract or a posters.

These defined guidelines were used to greatly decrease the number of full-text papers to be evaluated in subsequent stages, allowing the examiners to focus only on potentially relevant papers. The initial search produced 6.711 papers of which 2.465 were retrieved from ACM, 629 from IEEE, 871 from Elsevier Science Direct and 2.746 from Springer Link. The screening using the title, abstract and keywords removed 6.507 papers. During the check of the remaining papers for eligibility we excluded 77 irrelevant studies and 9 inaccessible papers. We ended up with 118 papers.

3 Analysis

This section includes the analysis part in which we first find out which definitions for "Responsible AI" exist in the literature so far. Afterwards, we explore the content-wise similar expressions of "Responsible AI" and look also for their definitions in the literature. These definitions are then compared with each other and overlaps are being highlighted.

3.1 Responsible AI

Out of all 118 analyzed papers we only found 3 scientific papers which explicitly define "Responsible" AI. The three papers use the following terminology in the context of "Responsible AI":

Terms Defining Responsible AI

- Fairness, Privacy, Accountability, Transparency and Soundness [24]
- Fairness, Privacy, Accountability, Transparency, Ethics, Security& Safety [3]
- Fairness, Privacy, Accountability, Transparency, Explainability [9]

However, after carefully studying all the 118 papers we found out that the terms that are included in those 3 definitions are not enough. The following terms are missing, ambiguous or used as synonyms:

- We found that "Responsible AI" should also include concepts, such as 'trustworthiness' and 'Human-centered', which are not included in any of the above definitions,
- 'Fairness' and 'Accountability' should have been included within ethics, and
- 'Soundness', interpreted as 'technical robustness', is a part of the security and safety.
- 'Transparency' can be treated as a synonym for explainability.

However, only defining the terms on their own is also not enough. In addition to the clarification of the constituent terms, their meaning and dependencies between each other must be analyzed, too. Therefore, we analyze and discuss the aforementioned terms in the context of AI in the literature in Sect. 3.

Content-Wise Similar Expressions for Responsible AI. During the analysis we also found content-wise similar expressions to "Responsible AI", which need to be included into the study.

The term "Responsible AI" is often used interchangeably with the terms "Ethical AI" or "Trustworthy" AI, for example in [9]. Another example of a content-wise similar expression is "Human-Centered AI". Therefore, we treat the terms

- "Trustworthy AI", e.g., in [21]
- "Ethical AI", e.g., in [19]
- "Human-Centered AI", e.g. in [29]

as the content-wise similar expressions for "Responsible AI" hereinafter.

Analysis of Definitions. In this subsection we analyze the definitions for "Responsible AI" and also the definitions of the content-wise similar expressions to "Responsible AI", which we have defined in the previous section.

Methodology: We compare a total of 15 different definitions for this topic. We found 3 definitions of "Responsible AI", 6 definitions of "Trustworthy AI", 5 of "Ethical AI" and 1 of "Human Centered AI" in the analyzed literature of the 118 papers. Now we have visualized these definitions in a venn diagram to investigate the overlaps and differences (Fig. 1):

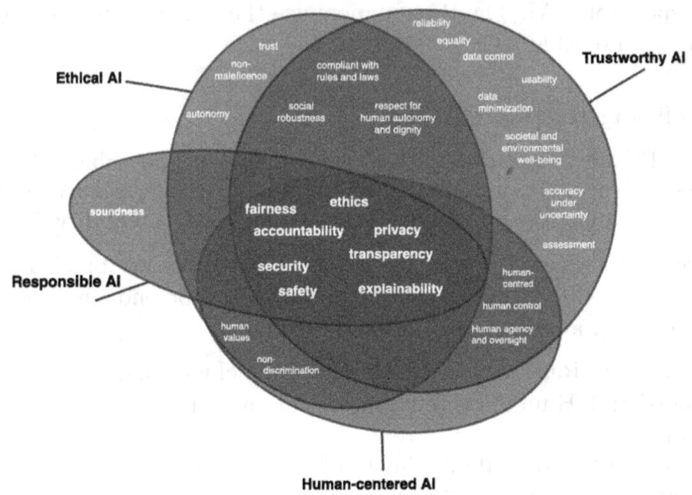

Fig. 1. Four dimensional Venn diagram to visually represent the similarities and differences found for defining "responsible," "ethical," "trustworthy," and "human-centered" AI.

Analysis: There is a high overlap but no consensus in all of the definitions. Similar terms or even synonyms are used in each of them. Also the intersections of the four sets often contain similar terms. One set often contains an element of another set, for example the term "trust" occurs within the definition of "Ethical AI". And the other way around definitions about trust also contain ethical terms.

The common terms in all definitions are: "Fairness", "Ethics", "Accountability", "Privacy", "Security", "Safety", "Transparency", and "Explainability".

Conclusions: The fact that there are so many different views on the topic of interest shows on the one hand that the concept of "Responsible AI" is difficult to grasp and secondly, that there is no consensus on the definition. This clearly shows the importance and relevance of the present work to clarify the definition.

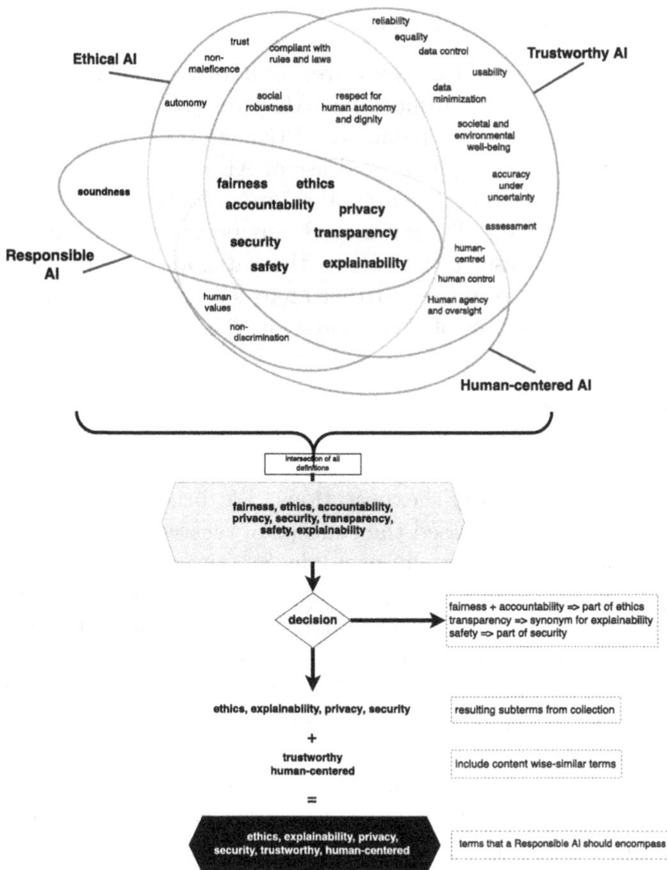

Fig. 2. Flowchart for extracting and combining the different terms.

The Fig. 2 is the process for our definition of "Responsible AI". To define the "Responsible AI" we take into the account the common terms of all the definitions we found in the 118 papers, and also the terms that are content-wise-similar to "Responsible AI", since these have the highest significance. After clarifying synonyms and combining terms into generic terms, we came up with the following six terms, which are the terms that a Responsible AI should encompass:

- Human-centered,
- Ethics,
- Trustworthiness,
- Security,
- Privacy, and
- Explainability.

3.2 Aspects of Responsible AI

According to our analysis of the literature, we have identified several categories in Sect. 3 in connection to Responsible AI, namely "Human-centered, Ethical, Trustworthy, Secure, Explainable, and Privacy-preserving AI" which should ensure the responsible development and use of AI.

In this subsection we put the categories into a context and further describe their meaning. We studied all relevant 118 papers we found using our methodology (see Sect. 2) and assigned them to the categories listed in the previous section.[1]. These papers were then further analyzed based on their content, and we highlight the key features of these papers in this section.

Trustworthy AI. The use of AI can have vital consequences for people and the society. Trust is a goal that should be achieved and perceived through appropriate implementation. If an AI is trusted by the user, then it delivers on its promises and fulfills the user's expectations. We briefly summarize in which ways the literature has addressed the concept of trustworthiness.

Some of the analyzed papers, for example [38], survey trustworthy AI in their study and summarize some of the best practices. The work of [35] highlights some of the difficulties in this area as well as how to increase the trust of users.

Other publications, such as [2] or [23], address how people acquire trust in AI and give an example of automated decision-making in the context of healthcare.

Frameworks for developing Trustworthy AI were proposed, for example in the work of [33].

There are several papers that handle very special topics related to "Trustworthy AI". For example [20] examines how trust can be formalized and what prerequisites are needed for trusting the AI.

Ethical AI. We summarize the most important key points that came up while analyzing the literature concerning the ethics in AI. In our opinion, the following definition best describes the ethics in conjunction with AI:

> *"AI ethics is the attempt to guide human conduct in the design and use of artificial automata or artificial machines, aka computers, in particular, by rationally formulating and following principles or rules that reflect our basic individual and social commitments and our leading ideals and values* [16]*."*

Implementing Ethical AI is often discussed and structured in frameworks because the difficulty in moving from principles to practice presents a significant challenge to implementing ethical guidelines. In their study, papers like the one of [26] or [4] have proposed ethically-aligned frameworks for AI.

AI brings up many dilemmas and complex socio-technical challenges, e.g., the lack of transparency, human alienation, privacy disclosure, and developers'

[1] In this paper will not cover the topic of 'Human-Centered AI', this will be analyzed in a future work.

responsibility issues. Some of these topics are discussed by [36] and [27] and also in other works.

Ethical AI is a very diverse field of study. Therefore, we also found papers like [32] handling very special approaches to ethics that could not be categorized.

In addition, fairness, robust laws and regulations, and transparency were also frequently seen as key aspects of ethical AI. The latter aspect is often achieved using explainability (XAI) methods. Therefore, it is discussed separately in the next subsection.

Explainable AI. Decisions made by AI systems or by humans using AI can have a direct impact on the well-being, rights, and opportunities of those affected by the decisions. This is what makes the problem of explainability of AI such a significant ethical, practical, and technological problem.

A concise definition, which in the best way in our opinion describes the explainable AI, is the following:

> *"Given a certain audience, explainability refers to the details and reasons a model gives to make its functioning clear or easy to understand* [3] ."

It became clear that the target groups receiving the explanations need to be analyzed and their individual requirements are of great importance, as well as the usability of the software presenting these explanations. This topic of user-centric explanations and stakeholder needs is discussed for example by the authors of [30] and [31].

Despite the many different XAI techniques discussed in the literature, the papers propose some new approaches in the field of XAI, for example, new ideas for the Reinforcement Learning by the authors of [18].

Some of the relevant papers presented frameworks for analyzing explainable AI systems based on different approaches. [25] presents a framework for analyzing the robustness, fairness, and explainability of a model.

Certain publications deal with other topics in this field of research, like for example the evaluation of explainability tools and methods by [15].

The "right to explanation" in the context of AI systems that directly affect individuals through their decisions, especially in legal and financial terms, is one of the themes of the General Data Protection Regulation (GDPR). Therefore, we need to protect the data through secure and privacy-preserving AI-methods. We present the state of the art in the area of privacy-preserving and secure AI found in the literature in the next section.

Privacy-Preserving and Secure AI. As was noted before, privacy and security are seen as central aspects of building trust in AI. However, the fuel for the good performance of the AI models is data, especially sensitive data. This has led to growing privacy concerns, such as unlawful use of private data and disclosure of sensitive data. [8]. We, therefore, need comprehensive privacy protection through holistic approaches to privacy protection that can also take into account the specific use of data and the transactions and activities of users [5].

Some papers give an overview of privacy-preserving machine learning (PPML)-techniques through a review like for example in [6]. We found several papers dealing with approaches using and combining different PPML-techniques. For Example Differential Privacy, Secure Multiparty Computation, and especially Federated Learning.

The concept of Differential Privacy (DP) is used in many approaches, mostly in combination with other techniques, to preserve privacy. For example, [14] discusses this topic through a case study.

Homomorphic Encryption (HE) allows performing computations directly with encrypted data (ciphertext) without the need to decrypt them. The technique is used in the study of [37].

Secure Multiparty Computation (MPC/SMPC) is a cryptographic protocol that distributes computation among multiple parties, where no single party can see the other parties' data. The approach was used by [34].

Federated Learning (FL) is a popular framework for decentralized learning and FL is the most common method for preserving privacy found in this analysis. The paper of [7,13,17] along with many others discuss this in their approaches and mostly extend this concept with some of the aforementioned techniques and individual extensions.

Many new papers are looking at hybridizing approaches as this could be promising solutions for the future. A hybrid PPML approach can take advantage of each component, providing an optimal trade-off between ML task performance and privacy overhead. For example, [21,28] use a hybrid approach in their works. Several other approaches are presenting their research in the area of computer vision, see for example [1] and deal with privacy-preserving.

We conclude that there is a great amount of research related to privacy and security in the field of AI. However, no single approach can result in a perfect privacy-preserving and secure AI. Many open challenges are left to be explored and solved.

4 Discussion

After the analysis part, we continue with the key points that emerged during the analysis and build upon these findings to present our idea of a Responsible AI framework.

4.1 Pillars of Responsible AI

Human-centered, trustworthy, ethical, explainable, privacy-preserving and secure AI form the pillars of a Responsible AI framework. These are also the criteria that should be considered to develop Responsible AI

Key requirements for a trustworthy AI should meet are providing an understandable reasoning process to the user, giving a probabilistic accuracy under uncertainty, as well as understandable usability (interface). The AI should act "as intended" when facing a given problem and be perceived as fair, useful, and reliable.

For Ethical AI it is required that the results are fair, which means non-biased and non-discriminating in every way, accountability is included in the sense of justifying the decisions and actions, the AI solutions are following the principles of sustainable development goals, as well as compliant with robust laws and regulations.

XAI requires a technically correct implementation of AI, which includes explanations that must be tailored to the needs of the users and the target group with the help of intuitive user interfaces. There is a high impact of explanations on the decision-making process. Explainable AI is also a feature describing how well the system does its work (so-called non-functional requirement).

Important requirements for privacy and security techniques are the need to comply with data protection regulations (e.g. GDPR), and the need to be complemented by proper organizational processes. They must be used depending on tasks to be executed on the data and on specific transactions a user is executing.

A promising way is using hybrid PPML-approaches because they can take an advantage of each component, providing an optimal trade-off between AI task performance and privacy overhead. Solutions should focus on the reduction of communication and computational cost (especially in distributed approaches).

We believe, that "Responsible AI" as a whole, is an interdisciplinary and dynamic process because it goes beyond technology and includes laws (compliance and regulations) and society standards such as ethics guidelines as well as sustainability.

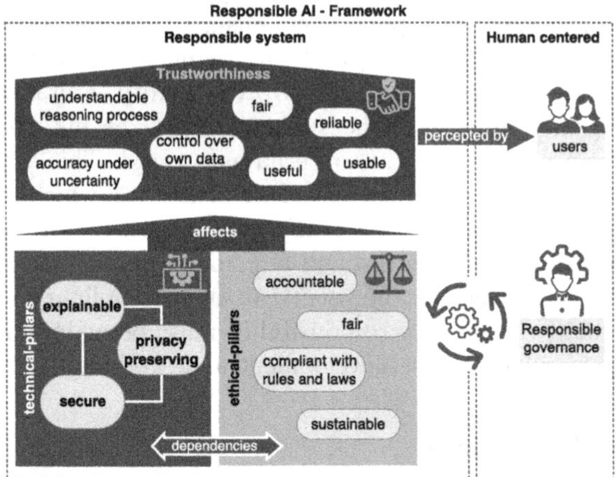

Fig. 3. Pillars for a responsible AI-framework

In general, the framework presented Fig. 3 consists of two main facets: The left side of the diagram shows the so-called "responsible system". This contains all the conditions that such an AI system must contain, including technical aspects, ethical aspects, and trust aspects.

The right side shows the human-centered aspects of the framework consisting of the target groups (users) and also the developers and maintainers of the system (Human in the Loop).

Inside the responsible system, one can find firstly, the technical pillars on the bottom left side which enable the technical aspects of AI in the first place. Secondly, there are ethical pillars on the bottom right side. These pillars both have a high impact on trustworthiness and are dependent on each other. This can be understood as if the conditions for the technical as well as the ethical side are fulfilled, then users' trust can be expected. Trust can be seen as a constantly required perception of users.

In each of the pillars terms are acting as the "supporting bars": On the technical side, we have "explainability, privacy-preserving, and security". These are also dependent on each other: Explainability methods must value privacy and security, which means they must not have excess access to a model or its underlying data so that it results in a privacy breach. Privacy-preserving is dependent on the secure implementation of the system. Nevertheless, the methods of explainability must be reliable.

In the case of the ethical pillar, these are "accountability, fairness, sustainability, and compliance with robust laws and regulations". They are crucial for proving that the present AI meets ethical and legal requirements.

On the human-centered side, there are two important main groups of persons. On the one hand, there are the developers and on the other hand the users/stakeholders. The developers must ensure that the system meets the requirements by constantly testing and improving it. We call this "responsible development and governance". This involves quality assurance through continuous checking and improving the system during the life cycle. The whole responsible system needs therefore to be human-tested. This 'human-centered' approach affects the perception of trust of the users, too which is mandatory.

In summary, the Responsible AI-Framework consists of the responsible system, the responsible governance through humans (Human in the loop), and the perception of the end-users. It represents an interdisciplinary and continuous process. As a result, the state of responsibility must be maintained throughout the whole lifecycle, which means development and production.

4.2 Research Limitations

In the current study, we have included the literature available through various databases and tried to provide a comprehensive and detailed survey of the literature in the field of responsible AI. In conducting the study, we, unfortunately, had the limitation that some publications were not freely accessible. Although we made a good effort to obtain the information needed for the study on responsible AI from various international journals, accessibility was still a problem. It is also possible that some of the relevant research publications are not listed in the databases we used for searching. Another limitation is the period of publication which was limited to the last two years. However, to ensure the currency and state of the art, and to include the latest research, we traded off the wider scope for the contemporaneity of the results.

5 Conclusion

This research made important contributions to the concept of responsible AI. The field of AI is a fast-changing area and a comprehensive legal framework for responsible AI is urgently needed. Starting from reading the series of EU-Papers on Artificial Intelligence and those of the last two years we found that there is a lot of confusion and misinterpretation of the field. There is no common terminology, and the resulting legal framework could easily have intentions to regulate one aspect, but end up regulating the wrong one.

At the moment, it is still not clear how a regulatory framework is going to be defined neither in general nor in detail. We also noticed that solely the trust as a goal to define a framework and regulation for AI is not sufficient. Regulations for "Responsible AI" need to be defined instead. As the EU is a leading authority when it comes to setting standards (as was the case with the GDPR) we find it necessary to give the politicians professional guidance in what is important to consider and regulate.

Also important is the need to help the practitioners and companies to prepare for what is coming next in both legal frameworks and the research. Due to the current phase of building necessary information architectures, the organizations have an opportunity to make the change, which should be characterized by Responsible AI.

Our research question, "What are the terms defining Responsible AI?" is answered in this paper. They are "Human-centered," "Ethics," "Trustworthiness," "Security," "Privacy-preserving," and "Explainability".

By analyzing these terms and the state of the art of aspects connected to those terms in the recent research papers we have shown which are the most important parts to consider when developing AI products. Thus, conducting the first structured literature review of this field and identifying 118 relevant works is an important contribution to the field. In the last section, we also have elaborated on and discussed the pillars for developing a framework in the context of Responsible AI based on the knowledge of the analysis part. The pillars should be used in the future context of both, the legal frameworks and the practical implementations of AI methods in products and services.

In future research, we will deal with specifying a concise definition of responsible AI and examining how different aspects reflected in the terms found in this research influence Responsible AI. We will provide a more detailed review of what is the state of the art by including a qualitative and quantitative analysis of the identified 118 papers.

The topics that still need to be covered in the context of responsible AI are the concepts of "Human-Centered AI" and "Human-in-the-loop" to further complete the present research. Other important topics to be worked on are the measurement approaches for responsible AI.

References

1. Agarwal, A.: Privacy preservation through facial de-identification with simultaneous emotion preservation. Signal Image Video Process. (2020)
2. Araujo, T., Helberger, N., Kruikemeier, S., de Vreese, C.H.: In AI we trust? Perceptions about automated decision-making by artificial intelligence. AI Soc. **35**(3), 611–623 (2020)
3. Arrieta, A.B., et al.: Explainable artificial intelligence (XAI): concepts, taxonomies, opportunities and challenges toward responsible AI. Inf. Fusion **58**, 82–115 (2020)
4. Benjamins, R.: A choices framework for the responsible use of AI. AI Ethics **1**(1), 49–53 (2021)
5. Bertino, E.: Privacy in the era of 5G, IoT, big data and machine learning. In: 2020 Second IEEE International Conference on Trust, Privacy and Security in Intelligent Systems and Applications (TPS-ISA), pp. 134–137 (2020)
6. Boulemtafes, A., Derhab, A., Challal, Y.: A review of privacy-preserving techniques for deep learning. Neurocomputing **384**, 21–45 (2020)
7. Chai, Z., Chen, Y., Anwar, A., Zhao, L., Cheng, Y., Rangwala, H.: FedAT: a high-performance and communication-efficient federated learning system with asynchronous tiers. In: Proceedings of the International Conference for High Performance Computing, Networking, Storage and Analysis. SC '21, pp. 1–16. ACM, NY, USA, November 2021
8. Cheng, L., Varshney, K.R., Liu, H.: Socially responsible AI algorithms: issues, purposes, and challenges. J. Artif. Int. Res. **71**, 1137–1181 (2021)
9. Eitel-Porter, R.: Beyond the promise: implementing ethical AI. AI Ethics **1**(1), 73–80 (2021)
10. EU-Commission: White paper on artificial intelligence a European approach to excellence and trust (2020)
11. EU-Commission: Coordinated plan on artificial intelligence 2021 review (2021)
12. EU-Commission: Proposal for a regulation of the European parliament and of the council laying down harmonised rules on artificial intelligence (artificial intelligence act) and amending certain union legislative acts (2021)
13. Fereidooni, H., et al.: SafeLearn: secure aggregation for private federated learning. In: 2021 IEEE Security and Privacy Workshops (SPW), pp. 56–62 (2021)
14. Guevara, M., Desfontaines, D., Waldo, J., Coatta, T.: Differential privacy: the pursuit of protections by default. Commun. ACM **64**(2), 36–43 (2021)
15. Hailemariam, Y., Yazdinejad, A., Parizi, R.M., Srivastava, G., Dehghantanha, A.: An empirical evaluation of AI deep explainable tools. In: 2020 IEEE Globecom Workshops (GC Wkshps), pp. 1–6 (2020)
16. Hanna, R., Kazim, E.: Philosophical foundations for digital ethics and AI ethics: a dignitarian approach. AI Ethics (2021)
17. Hao, M., Li, H., Xu, G., Chen, H., Zhang, T.: Efficient, private and robust federated learning. In: Annual Computer Security Applications Conference. ACSAC, pp. 45–60. ACM, New York, NY, USA, December 2021

18. Heuillet, A., Couthouis, F., Díaz-Rodríguez, N.: Explainability in deep reinforcement learning. Knowl.-Based Syst. **214**, 106685 (2021)
19. Hickok, M.: Lessons learned from AI ethics principles for future actions. AI Ethics **1**(1), 41–47 (2021)
20. Jacovi, A., Marasović, A., Miller, T., Goldberg, Y.: Formalizing trust in artificial intelligence: prerequisites, causes and goals of human trust in AI. In: Proceedings of the 2021 ACM Conference on Fairness, Accountability, and Transparency. FAccT '21, pp. 624–635. ACM, New York, NY, USA (2021)
21. Jain, S., Luthra, M., Sharma, S., Fatima, M.: Trustworthiness of artificial intelligence. In: 2020 6th International Conference on Advanced Computing and Communication Systems (ICACCS), pp. 907–912 (2020)
22. Kitchenham, B., Brereton, O.P., Budgen, D., Turne, M., Bailey, J., Linkman, S.: Systematic literature reviews in software engineering - a systematic literature review. Inf. Softw. Technol. **51**, 7–15 (2009)
23. Lee, M.K., Rich, K.: Who is included in human perceptions of AI?: trust and perceived fairness around healthcare AI and cultural mistrust. In: Proceedings of the 2021 CHI Conference on Human Factors in Computing Systems. ACM, New York, NY, USA (2021)
24. Maree, C., Modal, J.E., Omlin, C.W.: Towards responsible AI for financial transactions. In: 2020 IEEE Symposium Series on Computational Intelligence, pp. 16–21 (2020)
25. Mohseni, S., Zarei, N., Ragan, E.D.: A multidisciplinary survey and framework for design and evaluation of explainable AI systems. ACM Trans. Interact. Intell. Syst. **11**(3–4) (2021)
26. Morley, J., Elhalal, A., Garcia, F., Kinsey, L., Mökander, J., Floridi, L.: Ethics as a service: a pragmatic operationalisation of AI ethics. Minds Mach. (2021)
27. Rochel, J., Evéquoz, F.: Getting into the engine room: a blueprint to investigate the shadowy steps of AI ethics. AI Soc. (2020)
28. Rodríguez-Barroso, N., et al.: Federated learning and differential privacy: software tools analysis, the sherpa.ai fl framework and methodological guidelines for preserving data privacy. Inf. Fusion **64**, 270–292 (2020)
29. Shneiderman, B.: Bridging the gap between ethics and practice: guidelines for reliable, safe, and trustworthy human-centered AI systems. ACM Trans. Interact. Intell. Syst. **10**(4) (2020)
30. Sun, L., Li, Z., Zhang, Y., Liu, Y., Lou, S., Zhou, Z.: Capturing the trends, applications, issues, and potential strategies of designing transparent AI agents. In: Extended Abstracts of the 2021 CHI Conference on Human Factors in Computing Systems. CHI EA '21. ACM, New York, NY, USA (2021)
31. Suresh, H., Gomez, S.R., Nam, K.K., Satyanarayan, A.: Beyond expertise and roles: a framework to characterize the stakeholders of interpretable machine learning and their needs. In: Proceedings of the 2021 CHI Conference on Human Factors in Computing Systems. ACM, New York, NY, USA (2021)
32. Tartaglione, E., Grangetto, M.: A non-discriminatory approach to ethical deep learning. In: 2020 IEEE 19th International Conference on Trust, Security and Privacy in Computing and Communications (TrustCom), pp. 943–950 (2020)
33. Toreini, E., Aitken, M., Coopamootoo, K., Elliott, K., Zelaya, C.G., van Moorsel, A.: The relationship between trust in AI and trustworthy machine learning technologies. In: Proceedings of the 2020 Conference on Fairness, Accountability, and Transparency. FAT* '20, pp. 272–283. ACM, New York, NY, USA (2020)

34. Tran, A.T., Luong, T.D., Karnjana, J., Huynh, V.N.: An efficient approach for privacy preserving decentralized deep learning models based on secure multi-party computation. Neurocomputing **422**, 245–262 (2021)
35. Wing, J.M.: Trustworthy AI. Commun. ACM **64**(10), 64–71 (2021)
36. Xiaoling, P.: Discussion on ethical dilemma caused by artificial intelligence and countermeasures. In: 2021 IEEE Asia-Pacific Conference on Image Processing, Electronics and Computers (IPEC), pp. 453–457 (2021)
37. Yuan, L., Shen, G.: A training scheme of deep neural networks on encrypted data. In: Proceedings of the 2020 International Conference on Cyberspace Innovation of Advanced Technologies. CIAT 2020, pp. 490–495, ACM, USA (2020)
38. Zhang, T., Qin, Y., Li, Q.: Trusted artificial intelligence: technique requirements and best practices. In: 2021 International Conference on Cyberworlds (CW), pp. 303–306, September 2021

A Real-Time Semantic Anomaly Labeler Capturing Local Data Stream Features to Distinguish Anomaly Types in Production

Philip Stahmann(✉), Maximilian Nebel, and Bodo Rieger

University of Osnabrueck, Osnabrueck, Germany
{pstahmann,maxnebel,brieger}@uni-osnabrueck.de

Abstract. The digitalization entails a significant increase of information that can be used for decision-making in many business areas. In production, the proliferation of smart sensor technology leads to the real-time availability of manifold information from entire production environments. Due to the digitalization, many decision processes are automated. However, humans are supposed to remain at the center of decision-making to steer production. One central area of decision-making in digitalized production is real-time anomaly detection. Current implementations mainly focus on finding anomalies in sensor data streams. This research goes a step further by presenting design, prototypical implementation and evaluation of a real-time semantic anomaly labeler. The core functionality is to provide semantic annotations for anomalies to enable humans to make more informed decisions in real-time. The resulting implementation is flexibly applicable as it uses local data features to distinguish kinds of anomalies that receive different labels. Demonstration and evaluation show that the resulting implementation is capable of reliably labelling anomalies of different kinds from production processes in real-time with high precision.

Keywords: Digitalization of production · Real-time labeling · Semantic labeling

1 Introduction

In many sectors, the digitalization increases the pressure to make complex decisions fast [5]. The production industry is one example, where the digitalization radically impacts value-adding processes as autonomy and flexibility of production machines increase [7]. Especially advances in sensor technology change timeliness and extent of available information [14]. Smart sensors are capable of perceiving and communicating data in real-time. The emitted data enable close production supervision and control. One aspect with rising interest in research and practice is real-time anomaly detection [15]. Algorithms continuously analyze data streams from production processes to detect indications for anomalies and potentially costly production failures. Decision-makers need to keep track to act on anomaly detections. To make informed decisions, humans require the right information on anomalies as soon as possible. Current implementations mainly emphasize anomaly detection, but not the communication of more descriptive information, e.g.

G. Nicosia et al. (Eds.): LOD 2022, LNCS 13810, pp. 399–413, 2023.
https://doi.org/10.1007/978-3-031-25599-1_30

on types of anomalies or reactions that prevented or mitigated detected anomalies (e.g. [6, 13, 22]). Having more information on anomalies might reduce production failure and support efficiency. This research targets design, prototypical implementation and evaluation of a real-time semantic anomaly labeler based on machine learning classification. We formulate the following research question: *How can a real-time semantic labeler of production anomalies be designed and implemented?*

After conducting a comprehensive structured literature review (cf. Sect. 4), we find that there is no implementation that is equivalent to our result. We use the design science research approach to answer the research question in the following seven sections [12]. Section 2 outlines related work on sematic labelling and anomaly detection in production. Section 3 details the steps of the design science research process. Section 4 contains problem characteristics and design principles as basis for the developed real-time sematic anomaly labeler, which is presented in Sect. 5. Subsequently, the result is comprehensively demonstrated and evaluated using common metrics and a study to compare automated labelling with human labelling. Lastly, Sects. 7 and 8 finish with a presentation of the research contribution as well as limitations and starting points for future work.

2 Related Work

The purpose of this section is to present the state of research on the core topics relevant to this paper. The sources for the research are the result of the structured literature research, which is detailed in Sect. 4.

2.1 Semantic Labelling in Data Streams

Labelling streaming data is a research topic of increasing interest. Labels shall contribute to better communication of real-time data. Without labels, potentials that reside in context information may remain untapped [17]. Efforts to automate semantic labelling target the reduction of manual annotations by experts, so that errors can be reduced and time can be saved [17]. Despite the topic's relevance in various domains, there is no common framework for useful real-time semantic labelling for complexity reasons.

In general, there are two interdependent approaches in literature for automated semantic labelling in data streams. Firstly, ontologies can be created that serve as basis for standardized labelling. E.g., the Stream Annotation Ontology has the purpose to provide general orientation and semantics for the researchers and practitioners in the field of real-time labelling [8]. Yet, ontology-based approaches require specific knowledge on the data to assign correct labels [9]. Otherwise, the high abstraction level of ontologies may make it difficult to obtain results that are useful in specific applications. Secondly, machine learning approaches capture stream-specific properties to annotate labels to data. The latter approach requires data to train machine learning models. Historical data or simulations may serve as sources for training data [15]. Compared to ontological approaches, machine learning models are capable of capturing properties that are very specific to input data streams. On the downside, models are hardly generalizable over different streams.

Our research focuses on the latter approach by providing a real-time semantic labeler with a machine learning model at its core. In line with propositions from literature, the model relies on local structural information from the input data stream [17]. To overcome downsides of low generalizability, the tool provides high flexibility towards machine learning model selection, parametrization and input data, so that it can be applied to various streams. As further outlined in the following, we decided for a design, prototypical implementation and evaluation in the domain of manufacturing.

2.2 Anomaly Detection in Digitalized Production

Real-time anomaly detection is used to find potential production failures and failure causes [15]. It may support decision-makers in production with the capability to react immediately on anomalies. To understand anomalies and decide on timely reactions, they need to be explainable to humans, as is evident in the research field of expert systems [2]. Semantic labeling may increase the explainability of anomalies. E.g., standardized labels can improve the understanding of anomalies in different production environments and therefore support consistency of decision making.

Anomalies characterize through their variation from expectation that is created by previous data analysis [3]. The most common types of anomalies in the found literature are collective anomalies, context anomalies and point anomalies (ibid.). Collective anomalies are groups of related data points that in combination are anomalous. I.e., they can only be detected in combination with adjacent data points. To detect a context anomaly, two data features must be defined. Firstly, there has to be a context feature, such as measured time. The context feature defines the position of data points in a stream. Secondly, a behavior feature is considered over time, e.g. the temperature of a production machine.

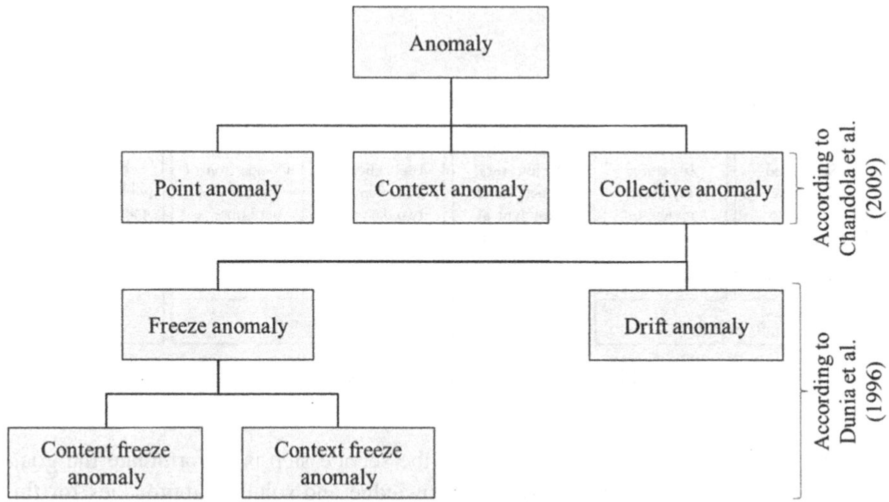

Fig. 1. Relation of types of anomalies.

In context anomalies, the behavior feature does not match expectations in relation to the given context feature [3]. Point anomalies refer to a single specific data point [16]. In particular, they differ from collective anomalies in that they are exclusively single data points. As possible extension to this classification, two more anomaly types appear in the found literature, namely drift and freeze anomalies. These two types specify kinds of collective anomalies [4]. Regarding drift anomalies, data shows a linearly deteriorating behavior [1]. Freeze anomalies are static, because the anomalous data do not change over time, contrary to expectations [1]. They can be divided into content and context freeze anomalies according to their cause. Figure 1 shows the relation of the named anomalies.

3 Applied Design Science Research Methodology

The design science research methodology (DSRM) is used to design and prototypically implement the real-time semantic anomaly labeler [12]. The methodological process supports the structuring of the procedure and its documentation. In line with the DSRM, the targeted prototypical result is called *artifact* in the following. The entire development process is based on six phases, which are shown in Fig. 2. In principle, each phase can be used as entry point. From the selected entry point, the project can then be developed further until it is completed [12]. In the case of this work, the nominal process sequence is used, starting from the left in Fig. 2.

First, the problem at hand is identified so that it can be solved in the subsequent process steps. The problem identification shows the value of an adequate solution. The highlighting of the problem is crucial for the DSRM to motivate the researcher to solve the found problem. In this phase, the division of the problem into several subproblems can be useful if the atomic view helps to better understand the problem and the solution [12]. The problem is defined in Sect. 4 on the basis of a structured literature review. This methodology is suitable for looking at the existing solution space and finding starting points for its expansion [20].

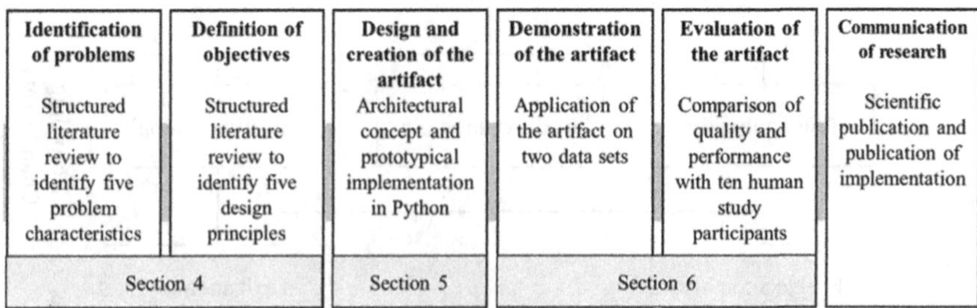

Identification of problems
Structured literature review to identify five problem characteristics
Definition of objectives
Structured literature review to identify five design principles
Design and creation of the artifact
Architectural concept and prototypical implementation in Python
Demonstration of the artifact
Application of the artifact on two data sets
Evaluation of the artifact
Comparison of quality and performance with ten human study participants
Communication of research
Scientific publication and publication of implementation
Section 4
Section 5
Section 6

Fig. 2. Applied DSRM (following [12]).

After the problem has been considered, the second step is to formulate the goals to realize the desired solutions. Existing knowledge and solution approaches for this problem are considered and evaluated to examine the implementation possibilities and the meaningfulness of the potential solutions [12]. In Sect. 4, the goals are identified in the form of design principles. The definition of the design principles is also based on the structured literature research from the previous phase.

The third step is the guided design and creation of the artifact. It includes both conceptual design of the artifact's architecture and a prototypical implementation. This third step in particular has an iterative character due to the large number of design options [12]. Section 5 presents the developed architecture and describes the prototypical implementation.

After the design and creation process of the artifact is completed, the fourth step is the demonstration of the artifact. In this step, the prototype is applied to solve the problem. In particular, this phase shall point out the problem-solving competence of the artifact in the targeted application scenario [12]. Section 6 includes the demonstration on two data sets simulated based on real production processes. The artifact is applied to the data sets to demonstrate its functionality [18].

The fifth step is the evaluation of the results. Here, the success of the artifact is evaluated by empirical measurement. The evaluation represents a target-performance comparison with regards to the design principles from the second step. It is compared to what extent the resulting artifact is able to solve the problem and whether the quality corresponds to the anticipation. Various metrics can be used here, such as precision and time requirements. If the results of the evaluation are not satisfactory, it is possible to return to the third phase [12]. The evaluation is performed in Sect. 7 and also focuses on the functionality of the prototype implementation. Additional to statistical metric, in the evaluation, quality and performance of the artifact are compared with those of ten human participants in an experimental study.

The last step is communication. The goal is to make all necessary information about the artifact and the problem available, as well as to inform about the importance and usefulness of this artifact [12]. The communication is done through this publication. In addition, the implementation is made openly available in the following repository for better comprehensibility and extensibility: https://github.com/anonymousPublisher1793/RTSAL.

4 Definition of Problem Characteristics and Design Principles

A structured literature review was conducted to characterize the problem characteristics (PCs) and design principles (DPs) [20]. The following search string was used:

("Real time analytics" OR "streaming analytics") AND ("semantic labeling" OR "semantic role labeling" OR "semantic segmentation" OR "anomaly detection" OR "outlier detection") AND ("industrie 4.0" OR "industrial internet of things" OR "industry 4.0" OR "intelligent manufacturing" OR "smart factory" OR "smart manufacturing" OR ") AND "machine learning".

The search string consists of four blocks that are connected by the keyword "AND". The first block refers to the requirement to analyze data in real-time. The second block involves anomalies and labeling. The third block uses equivalent terms to digitalized production. Finally, the term "machine learning" is included, as this is an essential requirement for the design of the artifact. The search string was used in the databases *AISel, Google Scholar, IEEE Xplore, Science Direct, Springer Link, Wiley Online Library* and *Web of Science*. These databases were selected, because they host major outlets of information systems including practically focused publications. A total of 122 search results were obtained. Subsequently, literature was systematically excluded as Fig. 3 shows [20]. The exclusion of literature is done sequentially by title, duplicates, abstract, and full text. This procedure results in 15 full texts. These are used as the basis for a backward search. This step increases the number of relevant publications to 19.

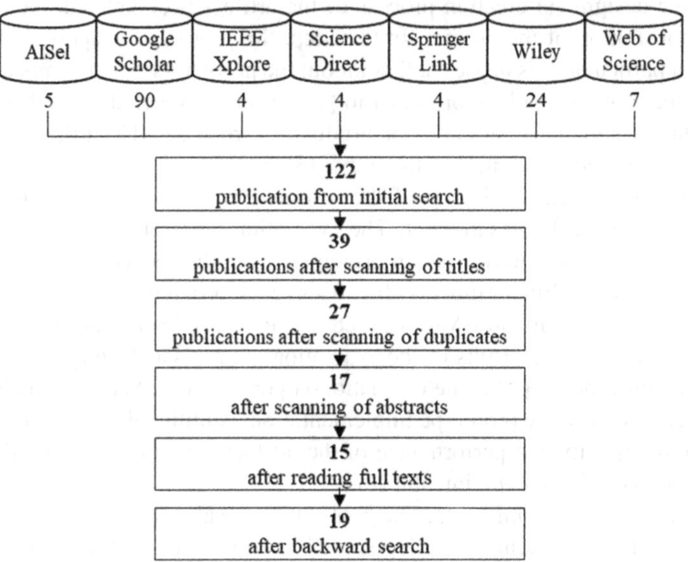

Fig. 3. Search and elimination steps from structured literature review.

The concepts covered by the resulting publications were grouped to have a basis for the derivation of PCs and DPs [21]. Table 1 contains the five main PCs identified in literature. PC1 refers to the lack of differentiation of types of anomalies implementations of machine learning based labelers [9]. Designing the artifact based on the types of anomalies from Sect. 2 can lead to more differentiated reactions. PC2 and PC3 refer to the recurrence of anomalies in one data set, but also in data sets of different production processes. Uniform semantics for labeling recurring anomalies may help to standardize the understanding of anomalies and reactions [2]. PC4 outlines that labelling cannot be implemented consistently by humans due to the real-time requirements and the effort. The artifact to be developed will enable the permanent application of labels. In addition, the artifact should provide the user with probabilities for the correctness of each assigned label, so that the user can decide whether to act in response to the label. Derived from literature and the PCs from Table 1, Table 2 shows the five design principles. DP1 focuses on the ability of the artifact to differentiate the five types of anomalies presented in Sect. 2 (cf. **PC1**). Since classification methods specific to local structures in datasets are to be used to determine the correct labels, DP2 elaborates on local features (cf. **PC2** and **PC3**). Label assignment shall depend on the values of relevant features, such as slope or expected values. DP3 draws the comparison to humans and their ability to classify anomalies (cf. **PC4** and **PC5**).

Table 1. Problem characteristics identified in literature.

	Description
PC1	Lacking differentiation between kinds of anomalies in production
PC2	Recurring anomalies are not recognized as familiar
PC3	The same anomalies in different data sets from various production machines are assigned different labels
PC4	Labelling of anomalies not possible by humans due to effort and time restrictions
PC5	Precision of labels that are relevant for decision-making in production is unclear, which may impact perceived reliability

Additionally, there is an existing variation and the possibility for individual extension of classification algorithms. DP4 therefore states that the artifact should be flexible with respect to the classification of data for label assignment (cf. **PC5**). DP5 addresses the unbalanced nature of data sets with anomalies (cf. **PC4** and **PC5**). It can be assumed that the essential part of the data set contains expectable values and anomalies are exceptions. Also, for the labels to be assigned, no equal distribution can be expected. The artifact is robust in terms of unbalanced label assignment.

Table 2. Design principles identified from structured literature review.

#DP	Definition
DP1	Differentiation of production anomaly kinds including collective, context, drift, freezing and point anomalies
DP2	Consideration of local features in production data, such as slope and expected value for labelling
DP3	Exhibition of better precision and performance than labelling by humans
DP4	Exhibition of flexibility in terms of classification algorithms
DP5	Dealing with unbalanced classes in realistic production data sets

5 Real-Time Anomaly Labeler

The third step of the DSRM is the creation of the artifact. To show the origins of our solution, we refer to the DPs from Table 2 in this section. The iterative creation process of the artifact resulted in the architecture shown in Fig. 4. It is divided into two halves and four subsections that interact with each other. The starting point is the production plant where sensors permanently measure and communicate production-relevant data. Real-time anomaly detection algorithms analyze the sensor-emitted data for unexpected values [10]. To provide data labels, which are later used for training, the data are first processed by experts or taken from historical data or simulations. The processed training data is used to create a machine learning model for classification. In parallel, local features are calculated from the sensor data in real-time. The calculated features are also included in the machine learning model. A connection between local data structures and labels is established. This enables the differentiability of types of anomalies specific to the input data (cf. **DP1** and **DP2**). After training, the semantic labeler is able to label algorithmically determined anomalies in real-time.

The semantic labeler is implemented in such a way that different algorithms can be used for classification (cf. **DP4**). When starting the semantic labeler, the user can select which algorithm to use.

After the labels and features for the machine learning model are available, various classification algorithms can be tested. The primary purpose of these tests is to investigate a basic suitability of the algorithms for the data before then comparing the precision of the different algorithms (cf. **DP 3**). Supervised machine learning algorithms can be divided into algorithms that natively support multi-class classification and algorithms that are normally designed for two classes. However, heuristics can also be used to apply classification algorithms designed for two classes to several classes.

To deal with the potentially imbalanced character of the anomalies, the semantic labeler applies stratified sampling when building the machine learning model (cf. **DP5**). Stratified sampling aims at balancing the distribution of classes. For this purpose, a sample is created from the data in which either the large classes are reduced in size or the small classes are artificially increased in size [19].

Apart from choice of classification algorithms, users can make three more configurations to appropriate the use of the semantic labeler according to their need. First, users can define the features to be included or excluded from the creation of the machine learning model. This setting allows the individualization of the semantic labeler to different data sets. Secondly, the user can set the number of local data used for feature calculation. A small number includes few data, so that features are calculated more frequently, which increases the required calculation capacities. Thirdly, users can set the number of applications of the classification model when applied to the data stream. The more often the model is applied to the same data, the more secure is the provided probability that is assigned to each label. Security of the probability may increase perceived reliability of the semantic labeler.

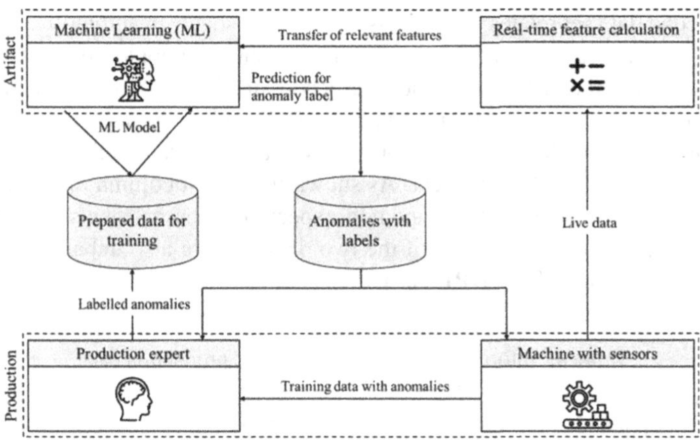

Fig. 4. Architecture of the semantic anomaly labeler.

6 Demonstration and Evaluation

In this section, the semantic anomaly labeler is applied to two datasets from the industrial production domain. They are the only openly available production data sets with semantically labelled anomalies.

The first data set includes sensor data from a production step in the manufacture of metal gears. In this step, gears are placed in a tank with a chemical water mixture to clean them and to harden the material. The data set contains temperature data of the liquid measured by sensors in the tank. The temperature is regulated by immersion heaters and by the continuous addition of cold water. The data set contains 191,672 measurements. Four anomalies are distinguishable. Table 3 contains the descriptions, the causes and the numbers of the anomalies.

Table 3. Information on anomalies of gear production data.

Anomaly description	Type	Cause description	Occurrences
Threshold violation	All	Immersion heaters do not work as expected	125
Unexpected jumps without threshold violation	Point	Dirt contaminates sensors	33
Deviating slope	Drift	Immersion heaters are not adjusted correctly	311
Invariant measurements	Freeze	Dirt contaminates sensors	1,503

The second data set relates to the pressing of metal sheets in automotive production. It contains sensor measurements of the pressure, which must not leave a certain range in order not to deform the metal parts. The dataset contains 200,000 measurements. Table 4 contains the descriptions, causes and numbers of anomalies. In particular, the data sets are found to be suitable for demonstration and evaluation, because all the different types of anomalies from Sect. 2 are present. As shown in the last column of Tables 3 and 4, anomalies are scarce, so expectable and non-expectable measurements are unbalanced. In addition, the different anomalies of the two data sets are also unbalanced. The data sets' characteristics thus fit to **DP1** and **DP5**.

Table 4. Information on anomalies of car production data.

Anomaly description	Type	Cause description	Occurrences
Press comes to an unexpected halt	Freeze	Metal components not adjusted properly	4,133
Threshold violation	All	Machine not adjusted correctly	4,086
Speed of press movement unexpected	Drift	Maladjustment of machine parameters	1,640
Jumps in the measurements	Point	Measurement errors	256

Four different algorithms are used to label the anomalies (cf. **DP4**). The algorithms were implemented in Python 3.9 using scikit-learn [11]. For the creation of the models, features were selected with a window length of 1,000 measurements. Furthermore, a second window covering five data points is added to capture local structures more closely. All features, which were calculated from the given data, are shown in Table 5. Choice of features and assignment of windows result from extensive testing (cf. **DP2**).

Table 6 shows the percentage of correctly assigned labels in both data sets for the four algorithms. The labels are assigned according to which label per anomaly has the highest probability of being correct. The probabilities of the assigned labels can be viewed in real-time. In both datasets, the decision tree classifier shows the best results. Therefore, the decision tree classifier is used to compute metrics for evaluation and comparison in the following.

For the evaluation of the semantic anomaly labeler, false positive rate (FPR), Matthew's correlation coefficient (MCC) and F1-score are used. These three metrics were chosen because they are robust against unbalanced data, e.g. compared to accuracy [10]. For the calculation of the metrics, four cases are distinguished in labelling the anomalies. True positives (TP) indicate whether a correct label was assigned, true negatives (TN) indicate whether a wrong label was correctly not assigned, false positives (FP) indicate whether a wrong label is used by the algorithm and false negatives (FN) indicate whether the correct label was incorrectly assigned. Based on these four cases, the key figures are calculated as follows:

Table 5. Considered attributes and their descriptions.

Window	Name	Description
1,000	Average	Arithmetic mean of non-anomalous values inside the window
	Standard deviation	Calculated using non-anomalous values inside the window
	Amplitude	Half of the range of the data inside the window
5	Average	Arithmetic mean of the last five values including anomalies
	Standard deviation	Calculated using the last five values including anomalies
	Skewness	Calculated using the last five values including anomalies
None	Slope	Incline or decline compared to the previous value
	Expected value	Calculated from the previous value and its slope
	Is collective	Binary information whether an anomaly is collective

Table 6. Results of four different algorithms applied to two data sets.

Algorithm	% of correct labels	
	Gear production	Car production
Decision tree classifier	89.44%	80.56%
K-nearest neighbors	84.46%	78.77%
Multi-layer perceptron	72.73%	79.49%
Naïve bayes	63.93%	66.39%

$$FPR = \frac{FP}{FP + TN} \quad (1) \qquad Recall = \frac{TP}{TP + FN} \quad (2)$$

$$Precision = \frac{TP}{TP + FP} \quad (3) \qquad F - score = \frac{Precision * Recall}{Precision + Recall} \quad (4)$$

$$MCC = \frac{TP * TN - FP * FN}{\sqrt{(TP + FP)(TP + FN)(TN + FP)(TN + FN)}} \quad (5)$$

Table 7 shows the metrics for the decision tree classifier for both data sets differentiated by types of anomalies. The model shows very good results overall for all three metrics in both data sets. The FPR is consistently very low. The MCC score shows the worst value of 0.8 for the context anomaly in the gear dataset. For the car dataset, point and again context anomalies are lowest for the MCC. The MCC of 0.17 for the context anomalies is especially low compared to the results for the other anomaly types. The relatively lower values for context anomalies are due to the fact that the semantic anomaly labeler initially has to learn the relevant context, which takes longer than the recognition for the other types of anomalies. Therefore, context anomalies are initially mislabeled more often. This effect is not reflected in context freeze anomalies, because freezes are most likely identified directly and then only the freeze types of anomalies have to be distinguished.

During calculations, the average runtime was recorded. To determine the real-time capability, the artifact was applied 40 times to the data sets. Different runtimes are caused by stochastic components of the algorithm. The calculations took place on a system with Windows 10, an i5-6300HQ CPU @2.30GHz and 16GB of RAM. On average, 622.89 data per second could be processed, which demonstrates the artifact's real-time capability.

Table 7. Metrics for the decision tree classifier shown by anomaly type for both data sets.

		FPR	MCC	F1-Score
Gear	Point	0	0.99	0.98
	Context	0.01	0.80	
	Context freeze	0	0.98	
	Content freeze	0	0.99	
	Drift	0	0.92	
Car	Point	0	0.78	0.94
	Context	0.01	0.17	
	Context freeze	0.03	0.93	
	Content freeze	0.03	0.90	
	Drift	0.01	0.93	

To verify the implementation of **DP3**, the artifact is compared to human labelling in an experimental study. The goal is to compare the quality of human labelling with the semantic anomaly labeler. The experiment requires participants with expertise in production or information processing. The participants were asked to manually label a selected extract of the gear data set. The size of the extract contains a total of 32 anomalies with 1,155 anomalous data points. Ten participants were acquired through private channels. Initially, a pretest was carried out to critically reflect on the planned implementation and to revise it if necessary. During the pretest, mainly minor adjustments were made to the information sheet.

In the experimental study, four participants come from the field of information systems, three from industrial engineering and three are mechanical engineers. The participants were provided with all necessary information before the start of their labelling. An information sheet has been handed out to all participants with explanations for all anomaly types. Any questions were clarified, after the participants have read the information. The test was started only when the participant feels capable of completing the task. In addition, the test was guided if desired. This does not mean that solutions were provided or that participants were guided to a particular solution.

Among the participants, two people managed to label the given data without errors. Averaged over the entire test group, the F1-score is 0.77. Most of the errors result from

Table 8. Metrics for both data sets using the decision tree classifier.

	Semantic anomaly labeler			Human labelers (avg)		
	FPR	MCC	F1-score	FPR	MCC	F1-score
Point	0	0.97	0.99	0.01	0.81	0.77
Context	0	0.67		0	0.54	
Context freeze	0	0.99		0.23	0.57	
Content freeze	0	0.99		0.16	0.6	
Drift	0	0.95		0	0	

swapping the contextual and content freezing anomalies. Table 8 shows the metrics for the semantic anomaly labeler and the average of the human labelers. The lowest MCC is again yielded for context anomalies. The semantic anomaly labeler performs overall better than the participants.

7 Results and Contribution

Complexity of decision-making increases in many domains such as in industrial production. Major reasons for this development are increasingly individualized customer requirements and the digitalization that makes real-time data available and enables production machine flexibility. Explainability of the production processes at any time may support keeping the human in the loop as decision makers. Research advancements on expert systems show that human readable, semantic explanations support comprehensibility [2]. To this end, we developed a real-time semantic labeler to explain anomalies in production. We use the design science research approach and make three contributions [12]. Firstly, we conducted a structured literature review to identify problem characteristics and design principles specific for real-time anomaly labelling in the production domain [20]. Secondly, we conceptualized and prototypically implemented the real-time semantic anomaly labeler and published its code open source. The prototype can be individualized according to users' need. The multi-class classification algorithms can be interchanged, so that users can compare results of different models. Users can also vary input features that are calculated to capture local data structures in data sets, which affects label assignment. There are no restrictions to feature engineering. The resulting implementation is capable of distinguishing different kinds of anomalies in production. Thirdly, this research covers extensive demonstration and evaluation of the semantic anomaly labelers capabilities guided by the design principles. Metrics suitable for unbalanced class distributions are calculated to demonstrate labeling capabilities. Additionally, the labeling quality is compared to human labeling that resulted from an experimental study.

8 Limitations and Future Work

The research is not free from limitations. There are three limitations as outlined in the following. Firstly, the structured literature review is subjective with respect to the elimination process of the publications. Likewise, the derivation of PCs and DPs is influenced by the interpretation of the researchers. Secondly, the presented results are hardly generalizable. There were only two data sets used for demonstration and evaluation. This is due to the scarcity of data sets with labelled anomalies. The comparison of the quality of the semantic labeler with human labelling is also not generalizable because the experimental group is small and only a small data sample was used. Future research can contribute with more labeled data sets to investigate the results of the semantic anomaly labeler more sophisticatedly. Thirdly, labeling all instances in a data set requires expertise in the production process. The data labelling process is tedious and time-consuming. However, the use of the semantic anomaly labeler can reduce this effort by automated real-time labelling once it is trained.

The code has been published for comprehensibility as well as for further testing and improvement. Practitioners and researchers can extend the semantic anomaly labeler and use it for their own purposes. One possible extension can use the implemented flexibility of using multiclass classification models and improve the classification results by an ensemble learning approach. Additionally, the semantic labeler may be integrated with new or existing ontologies for data stream annotation. Another possible extension is the creation of a graphical user interface. An interface can increase the usability of the implementation as well as the user's understanding of the labels in their production context.

References

1. Barbariol, T., Feltresi, E., Susto, G.A.: Self-diagnosis of multiphase flow meters through machine learning-based anomaly detection. Energies **13**(12), 3136 (2020)
2. Beierle, C. and Kern-Isberner, G. 2019. Methoden wissensbasierter Systeme. Springer Fachmedien Wiesbaden, Wiesbaden
3. Chandola, V., Banerjee, A., Kumar, V.: Anomaly detection. ACM Comput. Surv. **41**(3), 1–58 (2009)
4. Dunia, R., Qin, S.J., Edgar, T.F., McAvoy, T.J.: Identification of faulty sensors using principal component analysis. AIChE J. **42**(10), 2797–2812 (1996)
5. Hoßfeld, S.: Optimization on decision making driven by digitalization. JEW **5**, 2 (2017)
6. Jin, X., Chow, T.W.S., Tsui, K.-L.: Online anomaly detection of brushless DC motor using current monitoring technique. Int. J. Performab. Eng. **10**, 263 (2014)
7. Kagermann, H., Wahlster, W., Helbig, J.: Umsetzungsempfehlungen für das Zukunftsprojekt Industrie 4.0. Abschlussbericht des Arbeitskreises Industrie 4.0 (2013)
8. Kolozali, S., Bermudez-Edo, M., Barnaghi, P.: Stream Annotation Ontology (2016). http://iot.ee.surrey.ac.uk/citypulse/ontologies/sao/sao
9. Kolozali, S., Bermudez-Edo, M., Puschmann, D., Ganz, F., Barnaghi, P.: A knowledge-based approach for real-time IoT data stream annotation and processing. In: 2014 IEEE International Conference on Internet of Things(iThings), and IEEE Green Computing and Communications (GreenCom) and IEEE Cyber, Physical and Social Computing (CPSCom), pp. 215–222. IEEE (2014). https://doi.org/10.1109/iThings.2014.39

10. Lavin, A., Ahmad, S.: Evaluating Real-Time Anomaly Detection Algorithms - The Numenta Anomaly Benchmark, pp. 38–44 (2015)
11. Pedregosa, F., et al.: Scikit-learn: machine learning in python. J. Mach. Learn. Res. **12**, 2825–2830 (2011)
12. Peffers, K., Tuunanen, T., Rothenberger, M.A., Chatterjee, S.: A design science research methodology for information systems research. J. Manag. Inf. Syst. **24**(3), 45–77 (2007)
13. Rousopoulou, V., Nizamis, A., Vafeiadis, T., Ioannidis, D., Tzovaras, D.: Predictive maintenance for injection molding machines enabled by cognitive analytics for industry 4.0. Front. Artif. Intell. **3**, 578152 (2020)
14. Schütze, A., Helwig, N., Schneider, T.: Sensors 4.0 – smart sensors and measurement technology enable Industry 4.0. J. Sens. Sensor Syst. **7**(1), 359–371 (2018)
15. Stahmann, P., Rieger, B.: Requirements identification for real-time anomaly detection in industrie 4.0 machine groups: a structured literature review. In: Proceedings of the 54th Hawaii International Conference on System Sciences (2021)
16. Stojanovic, L., Dinic, M., Stojanovic, N., Stonjanovic, A.: Big-data- driven Anomaly Detection in Industry (4.0): an approach and a case study. In: IEEE (2016)
17. Sun, J., Kamiya, M., Takeuchi, S.: Introducing hierarchical clustering with real time stream reasoning into semantic-enabled IoT. In: 2018 IEEE 42nd Annual Computer Software and Applications Conference (COMPSAC), pp. 540–545. IEEE (2018). https://doi.org/10.1109/COMPSAC.2018.10291
18. Venable, J., PriesHeje, J., Baskerville, R.: FEDS: a framework for evaluation in design science research. Eur. J. Inf. Syst. **25**(1), 77–89 (2016)
19. Vilariño, F., Spyridonos, P., Vitrià, J., Radeva, P.: Experiments with SVM and stratified sampling with an imbalanced problem: detection of intestinal contractions. In: Singh, S., Singh, M., Apte, C., Perner, P. (eds.) Pattern Recognition and Image Analysis. LNCS, vol. 3687, pp. 783–791. Springer, Heidelberg (2005). https://doi.org/10.1007/11552499_86
20. vom Brocke, J., Simons, A., Riemer, K., Niehaves, B., Plattfaut, R., Cleven, A.: Standing on the shoulders of giants: challenges and recommendations of literature search in information systems research. Commun. Assoc. Inf. Syst. **37**, 1–9 (2015). https://doi.org/10.17705/1CAIS.03709
21. Webster, J., Watson, R.T.: Analyzing the past to prepare for the future: writing a literature review. MIS Q. **26**(2), xiii–xxiii (2002)
22. Xie, L., et al.: Soft sensors for online steam quality measurements of OTSGs. J. Process Control **23**(7), 990–1000 (2013)

Generalising via Meta-examples for Continual Learning in the Wild

Alessia Bertugli[1(✉)], Stefano Vincenzi[2], Simone Calderara[2], and Andrea Passerini[1]

[1] Università di Trento, Trento, Italy
{alessia.bertugli,andrea.passerini}@unitn.it
[2] Università di Modena e Reggio Emilia, Modena, Italy
{stefano.vincenzi,simone.calderara}@unimore.it

Abstract. Future deep learning systems call for techniques that can deal with the evolving nature of temporal data and scarcity of annotations when new problems occur. As a step towards this goal, we present FUSION (Few-shot UnSupervIsed cONtinual learning), a learning strategy that enables a neural network to learn quickly and continually on streams of unlabelled data and unbalanced tasks. The objective is to maximise the knowledge extracted from the unlabelled data stream (unsupervised), favor the forward transfer of previously learnt tasks and features (continual) and exploit as much as possible the supervised information when available (few-shot). The core of FUSION is MEML - Meta-Example Meta-Learning - that consolidates a meta-representation through the use of a self-attention mechanism during a single inner loop in the meta-optimisation stage. To further enhance the capability of MEML to generalise from few data, we extend it by creating various augmented surrogate tasks and by optimising over the hardest. An extensive experimental evaluation on public computer vision benchmarks shows that FUSION outperforms existing state-of-the-art solutions both in the few-shot and continual learning experimental settings. (The code is available at https://github.com/alessiabertugli/FUSION).

1 Introduction

Human-like learning has always been a challenge for deep learning algorithms. Neural networks work differently than the human brain, needing a large number of independent and identically distributed (iid) labelled data to face up the training process. Due to their weakness to directly deal with few, online, and unlabelled data, the majority of deep learning approaches are bounded to specific applications. Continual learning, meta-learning, and unsupervised learning try to overcome these limitations by proposing targeted solutions. In particular, continual learning has been largely investigated in the last few years to solve the

Supplementary Information The online version contains supplementary material available at https://doi.org/10.1007/978-3-031-25599-1_31.

G. Nicosia et al. (Eds.): LOD 2022, LNCS 13810, pp. 414–429, 2023.
https://doi.org/10.1007/978-3-031-25599-1_31

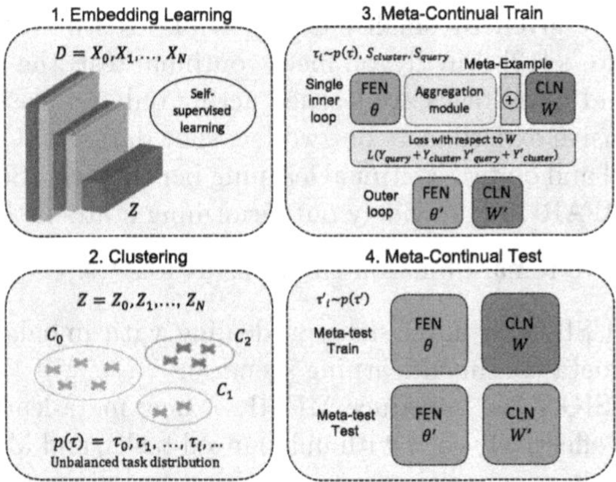

Fig. 1. Overview of FUSION learning strategy. Further details in Sect. 3.

catastrophic forgetting problem that affects neural networks trained on incremental data. When data are available as a stream of tasks, neural networks tend to focus on the most recent, overwriting their past knowledge and consequently causing forgetting. Several methods [8,15,19,21,22,27] have been proposed to solve this issue involving a memory buffer, network expansion, selective regularisation, and distillation. Some works take advantage of the meta-learning abilities of generalisation on different tasks and rapid learning on new ones to deal with continual learning problems, giving life to meta-continual learning [3,14] and continual-meta learning [5]. Due to the complex nature of the problem, the proposed approaches generally involve supervised or reinforcement learning settings. Moreover, the majority of continual learning solutions assume that data are perfectly balanced or equally distributed among classes. This problem is non-trivial for continual learning since specific solutions have to be found to preserve a balanced memory in presence of an imbalanced stream of data.

In this paper, we introduce FUSION (standing for Few-shot UnSupervIsed cONtinual learning), a new learning strategy for unsupervised meta-continual learning that can learn from small datasets and does not require the underlying tasks to be balanced. As reported in Fig. 1, FUSION is composed of four phases: embedding learning, clustering, meta-continual train and meta-continual test. In the embedding learning phase, a neural network is trained to generate embeddings that facilitate the subsequent separation. Then clustering is applied on these embeddings, and each cluster corresponds to a task (i.e., a class) for the following phase. As clustering is not constrained to produce balanced clusters, the resulting tasks are also unbalanced. For the meta-continual training phase, we introduce a novel meta-learning based algorithm that can effectively cope with unbalanced tasks. The algorithm, called MEML (for Meta-Example Meta-Learning), relies on a single inner loop update performed on an aggregated attentive representation, that we call meta-example. In so doing, MEML learns meta-representations that enrich the general features provided by large clusters with the variability given by small clusters, while existing approaches simply discard small clusters [13]. Finally, on meta-continual test, the learned representation is frozen and novel tasks are learned acting only on classifications layers. We perform extensive experiments on two few-shot datasets, Omniglot [17] and Mini-ImageNet [9] and on two continual learning benchmarks, Sequential MNIST and Sequential CIFAR-10 [16] widely outperforming state-of-the-art methods.

Contributions. We remark our contributions as follows:

- We propose FUSION, a novel strategy dealing with unbalanced tasks in an unsupervised meta-continual learning scenario;
- As part of FUSION, we introduce MEML, a new meta-learning based algorithm that can effectively cope with unbalanced tasks, and MEMLX, a variant of MEML exploiting an original augmentation technique to increase robustness, especially when dealing with undersized datasets;
- We test FUSION on an unsupervised meta-continual learning setting reaching superior performance compared to state-of-the-art approaches. Ablations studies empirically show that the imbalance in the task dimension does not negatively affect the performance, and no balancing technique is required;
- We additionally test MEML, our meta-continual learning method, in standard supervised continual learning, achieving better results with respect to specifically tailored solutions.

2 Related Work

2.1 Continual Learning

Background. Continual learning is one of the most challenging problems in deep learning research since neural networks are heavily affected by catastrophic forgetting when data are available as a stream of tasks. In more detail, neural networks tend to focus on the most recent tasks, overwriting their past knowledge and consequently causing forgetting. As theoretically exposed in [26], there are three main evaluation protocols for comparing methods' performance: Task-IL, Domain-IL and Class-IL. Task-IL is the easiest scenario since task-identity is always provided, even at test-time; Domain-IL only needs to solve the current task, no task-identity is necessary; Class-IL instead intends to solve all tasks seen so far with no task identity given. Much of the recent literature [1,4,14] is directed towards methods that do not require the detection of the task change. Our proposed approach follows this line of research, using a rehearsal technique to avoid forgetting without the need for task identity and targeted solutions to find them. Finally, continual learning methods can be divided into three main categories.

Architectural Strategies. They are based on specific architectures designed to mitigate catastrophic forgetting [24]. Progressive Neural Networks (PNN) [24] are based on parameter-freezing and network expansion, but suffers from a capacity problem because it implies adding a column to the neural network at each new task, so growing up the number of tasks training the neural network becomes more difficult due to exploding/vanishing gradient problems.

Regularisation Strategies. They rely on putting regularisation terms into the loss function, promoting selective consolidation of important past weights [15, 27]. Elastic Weights Consolidation (EWC) [15] uses a regularization term to control catastrophic forgetting by selectively constraining the model weights that are important for previous tasks through the computation of Fisher information matrix of weights importance. Synaptic Intelligence (SI) [27] can be considered as a variant of EWC, that computes weights importance online during SGD.

Rehearsal Strategies. Rehearsal strategies focus on retaining part of past information and periodically replaying it to the model to strengthen connections for memories, involving meta-learning [22], combination of rehearsal and regularisation strategies [8,19], knowledge distillation [18] and generative replay [25]. Experience Replay [23] stores a limited amount of information of the past and then adds a further term to the loss that takes into account loss minimization on the buffer data, besides the current data. Meta-Experience Replay (MER) [22] induces a meta-learning update in the process and integrate an experience replay buffer, updated with reservoir sampling, facilitating continual learning while maximizing transfer and minimizing interference. Gradient Episodic Memory (GEM) [19] and its more efficient version A-GEM [8] is a mix of regularization and rehearsal strategies.

2.2 Meta-learning

Background. Meta-learning, or learning to learn, aims to improve the neural networks ability to rapidly learn new tasks with few training samples. The tasks can comprise a variety of problems, such as classification, regression and reinforcement learning, but differently from Continual learning, training doesn't occur with incremental tasks and models are evaluated on new unseen tasks. The majority of meta-learning approaches proposed in the literature are based on Model-Agnostic Meta-Learning (MAML) [10].

MAML. By learning an effective parameter initialisation, with a double loop procedure, MAML limits the number of stochastic gradient descent steps required to learn new tasks, speeding up the adaptation process performed at meta-test time. The double loop procedure acts as follow: an inner loop that updates the parameters of the neural network to learn task-specific features and an outer loop generalizing to all tasks. The success of MAML is due to its model-agnostic nature and the limited number of parameters it requires. Nevertheless, it suffers from some limitations related to the amount of computational power it needs during the training phase. To solve this issue, the authors propose a further version, First Order MAML (FOMAML) that focus on removing the second derivative causing the need for large computational resources. ANIL [20] investigates the success of MAML finding that it mostly depends on feature reuse rather than rapid learning. This way, the authors propose a slim version of MAML, removing almost all inner loops except for task-specific heads.

Unsupervised Meta-learning. Although MAML is suitable for many learning settings, few works investigate the unsupervised meta-learning problem. CACTUs [13] proposes a new unsupervised meta-learning method relying on clustering feature embeddings through the k-means algorithm and then builds tasks upon the predicted classes. The authors employ representation learning strategies to learn compliant embeddings during a pre-training phase. From these learned embeddings, a k-means algorithm clusters the features and assigns pseudo-labels to all samples. Finally, the tasks are built on these pseudo-labels.

2.3 Meta-learning for Continual Learning

Meta-learning has been extensively merged with continual learning for different purposes. We highlight the existence of two strands of literature [5]: *meta-continual learning*, that aims to incremental task learning, and *continual-meta learning* that instead focuses on fast remembering. To clarify the difference between these two branches, we adopt the standard notation that denotes S as the support set and Q as the query set. The sets are generated from the data distribution of the context (task) C and respectively contain the samples employed in the inner and outer loops (e.g. in a classification scenario, both S and Q contain different samples of the same classes included in the current task). We define the meta-learning algorithm as ML_ϕ and the continual learning one with CL_ϕ.

Continual-Meta Learning. Continual-meta learning mainly focuses on making meta-learning algorithms online, to rapidly remember meta-test tasks. In detail, it considers a sequence of tasks $S_{1:T}$, $Q_{1:T}$, where the inner loop computation is performed through $f_{\theta t} = ML_\phi(S_{t-1})$, while the learning of ϕ (outer loop) is obtained using gradient descent over the $l_t = \mathcal{L}(f_{\theta t}, S_t)$. Since local stationarity is assumed, the model fails on its first prediction when the task switches. At the end of the sequence, ML_ϕ recomputes the inner loops over the previous supports and evaluate on the query set $Q_{1:T}$.

Meta-continual Learning. More relevant to our work are meta-continual learning algorithms [1,14], which use meta-learning rules to "learn how not to forget". Resembling the notation proposed in [5], given K sequences sampled i.i.d. from a distribution of contexts C, $S_{i,1:T}, Q_{i,1:T} \sim X_{i,1:T}|C_{i,1:T}$, CL_ϕ is learned with $\nabla_\phi \sum_t \mathcal{L}(CL_\phi(S_t), Q_t)$ with $i < N < K$ and evaluated on the left out sets $\sum_{i=N}^{K} \mathcal{L}(CL_\phi(S_t), Q_t)$. In particular, OML [14] and its variant ANML [1] favour sparse representations by employing a trajectory-input update in the inner loop and a random-input update in the outer one. The algorithm jointly trains a representation learning network (RLN) and a prediction learning network (PLN) during the meta-training phase. Then, at meta-test time, the RLN layers are frozen and only the PLN is updated. ANML replaces the RLN network with a neuro-modulatory network that acts as a gating mechanism on the PLN activations following the idea of conditional computation.

3 Few-Shot Unsupervised Continual Learning

Meta-continual learning [1,14] deals with the problem of allowing neural networks to learn from a stream of few, non i.i.d. examples and quickly adapt to new tasks. It can be considered as a few-shot learning problem, where tasks are incrementally seen, one class after the others. Formally, we define a distribution of training classification tasks $p(\mathcal{T}) = \mathcal{T}_0, \mathcal{T}_1, ..., \mathcal{T}_i, ...$. During meta-continual training, the neural network sees all samples belonging to $\mathcal{T}_0$ first, then all samples belonging to $\mathcal{T}_1$, and so on, without shuffling elements across tasks as in traditional deep learning settings. The network should be able to learn a general representation, capturing important features across tasks, without catastrophic forgetting, meaning to overfit on the last seen tasks. During the meta-test phase, a different distribution of unknown tasks $p(\mathcal{T}') = \mathcal{T}'_0, \mathcal{T}'_1, ..., \mathcal{T}'_i, ...$ is presented to the neural network again in an incremental way. The neural network, starting from the learned representation, should quickly learn to solve the novel tasks. In this paper, differently from standard meta-continual learning, we focus on the case where no training labels are available and tasks have to be constructed in an unsupervised way, using pseudo-labels instead of the real labels in the meta-continual problem. To investigate how neural networks learn when dealing with a real distribution and flow of unbalanced tasks, we propose FUSION, a novel learning strategy composed of four phases.

3.1 Embedding Learning

Rather than requiring the task construction phase to directly work on high dimensional raw data, an embedding learning network, which is different from the one employed in the following phases, is used to determine an embedding that facilitates the subsequent task construction. Through an unsupervised training [2,6], the embedding learning network produces an embedding vector set $Z = Z_0, Z_1, ..., Z_N$, starting from the N data points in the training set $D = X_0, X_1, ..., X_N$ (see Fig. 1.1). Embeddings can be learned in different ways, through generative models [2] or self-supervised learning [6]. In Fig. 1.1 an illustration of an unsupervised embedding learning based on self-supervised learning is shown.

3.2 Clustering

As done in [13], the task construction phase exploits the k-means algorithm over suitable embeddings obtained with the embedding learning phase previously described. This simple but effective method assigns the same pseudo-label to all data points belonging to the same cluster. This way, a distribution $p(\mathcal{T}) = \mathcal{T}_0, \mathcal{T}_1, ..., \mathcal{T}_i, ...$ of tasks is built from the generated clusters as reported in Fig. 1. Applying k-means over these embeddings leads to unbalanced clusters, which determine unbalanced tasks. This is in contrast with typical meta-learning and continual learning problems, where data are perfectly balanced. To recover a balanced setting, in [13], the authors set a threshold on the cluster dimension, discarding extra samples and smaller clusters. We believe that these approaches are sub-optimal as they alter the data distribution. In an unsupervised setting, where data points are grouped based on the similarity of their features, variability is an essential factor. In a task imbalanced setting, the obtained meta-representation is influenced by both small and large clusters.

3.3 Meta-continual Train

Motivation. The adopted training protocol is related to the way data are provided at meta-test train time. In that phase, the model receives as input a stream of new unseen tasks, each with correlated samples; we do not assume access to other classes (opposed to the training phase) and only the current one is available. In this respect, since the network's finetuning occur with this stream of data, during training we reproduce a comparable scenario. In particular, we need to design a training strategy that is sample efficient and directly optimize for a proper initial weights configuration. These suitable weights allow the network to work well on novel tasks after a few gradient steps using only a few samples. In the context of meta-learning, MAML relies on a two-loop training procedure performed on a batch of training tasks. The inner loop completes N step of gradient updates on a portion of samples of the training tasks, while the outer loop exploits the remaining ones to optimize for a quickly adaptable representation (meta-objective). Recent investigations on this algorithm explain that the real reason for MAML's success resides in feature reuse instead of rapid learning [20], proving that learning meaningful representations is a crucial factor.

Procedure. The created tasks are sampled one at a time $\mathcal{T}_i \sim p(\mathcal{T})$ for the unsupervised meta-continual training phase as shown in Fig. 1. The training process happens in a class-incremental way - where one task corresponds to one cluster - following a two-loop update procedure. The inner loop involves samples belonging to the ongoing task, while the outer loop contains elements sampled from both the current and other random clusters. In fact, during this stage, the network may suffer from the catastrophic forgetting effect on the learned representation if no technique is used to generalise or remember. To this end, the query set, used to update parameters in the outer loop, have to be designed to simulate an iid distribution, containing elements belonging to different tasks. The unbalanced case takes two-third of the current cluster data for the inner loop and adds one-third to a fixed number of random samples for the outer loop. The balanced case - usually adopted with supervised data - instead takes the same number of samples among tasks for both the inner and the outer loop. To deal with the meta-continual train in FUSION (Fig. 1.3), we propose MEML, a meta-learning procedure based on the construction of a meta-example, a prototype of the task obtained through self-attention. The whole architecture is composed of a Feature Extraction Network (FEN), an aggregation module and a CLassification Network (CLN). The FEN is updated only in the outer loop (highlighted in blue in the figure), while frozen during the inner (grey). Both the aggregation module and the CLN are renewed in the inner and outer loop.

MEML. We remove the need for several inner loops, maintaining a single inner loop update through a mechanism for aggregating examples based on self-attention. This way, we considerably reduce the training time and computational resources needed for training the model and increases global performance. The use of a meta-example instead of a trajectory of samples is particularly helpful in class-incremental continual learning to avoid catastrophic forgetting. In fact, instead of sequentially processing multiple examples of the same class and updating the parameters at each one (or at each batch), the network does it only once per class, reducing the forgetting effect. At each time-step, a task $\mathcal{T}_i = (\mathcal{S}_{cluster}, \mathcal{S}_{query})$ is randomly sampled from the task distribution $p(\mathcal{T})$. $\mathcal{S}_{cluster}$ contains elements of the same cluster as indicated in Eq. 1, where $Y_{cluster} = Y_0 = ... = Y_k$ is the cluster pseudo-label and K is the number of data points in the cluster. Instead, $\mathcal{S}_{query}$ (Eq. 2) contains a variable number of elements belonging to the current cluster and a fixed number of elements randomly sampled from all other clusters, where Q is the total number of elements in the query set.

$$\mathcal{S}_{cluster} = \{(X_k, Y_k)\}_{k=0}^{K}, \quad (1) \qquad \mathcal{S}_{query} = \{(X_q, Y_q)\}_{q=0}^{Q}. \quad (2)$$

$\mathcal{S}_{cluster}$ is used for the inner loop update, while $\mathcal{S}_{query}$ is used to optimise the meta-objective during the outer loop. All the elements belonging to $\mathcal{S}_{cluster}$ are processed by the FEN, parameterised by θ, computing the feature vectors $R_0, R_1, ..., R_K$ in parallel for all task elements (see Eq. 3). The obtained embeddings are refined in Eq. 4 with an attention function, parameterised by ρ, that computes the attention coefficients $\boldsymbol{a}$ from the features vectors. Then, the final aggregated representation vector R_{ME} (Eq. 5), for *meta-example* representation, captures the most salient features.

$$R_{0:K} = f_\theta(X_{0:K}), \quad (3) \quad \boldsymbol{a} = Softmax[f_\rho(R_{0:K})], \quad (4) \quad R_{ME} = \boldsymbol{a}^\intercal R_{0:K}. \quad (5)$$

The single inner loop is performed on this meta-example, which adds up the weighted-features contribution of each element of the current cluster. Then, the cross-entropy loss $\mathcal{L}$ between the predicted label and the pseudo-label is computed and both the classification network parameters W and the attention parameters ρ are updated with a gradient descent step (as indicated in Eq. 6), where $\psi = \{W, \rho\}$ and α is the inner loop learning rate. Finally, to update the whole network parameters $\phi = \{\theta, W, \rho\}$, and to ensure generalisation across tasks, the outer loop loss is computed from S_{query}. The outer loop parameters are thus updated as shown in Eq. 7 below, where β is the outer loop learning rate.

$$\psi \leftarrow \psi - \alpha \nabla_\psi \mathcal{L}(f_\psi(R_{ME}), Y_{cluster}), \quad (6) \qquad \phi \leftarrow \phi - \alpha \nabla_\phi \mathcal{L}(f_\phi(X_{0:Q}), Y_{0:Q}). \quad (7)$$

Note that with the aggregation mechanism introduced by MEML, a single inner loop is made regardless of the number of examples in the cluster, thus eliminating the problem of unbalancing at the inner loop level.

MEMLX. Since the aim is to learn a representation that generalises to unseen classes, we introduce an original augmentation technique inspired by [12]. The idea is to generate multiple sets of augmented input data and retain the set with maximal loss to be used as training data. Minimising the average risk of this worst-case augmented data set enforces robustness and acts as a regularisation against random perturbations, leading to a boost in the generalisation capability. Starting from the previously defined $\mathcal{S}_{cluster}$ and $\mathcal{S}_{query}$ we generate m sets of augmented data:

$$\{\mathcal{S}^i_{cluster}, \mathcal{S}^i_{query}\}^m_{i=1} \leftarrow A(\mathcal{S}_{cluster}), A(\mathcal{S}_{query}), \tag{8}$$

where A is an augmentation strategy that executes a combination of different data transformations for each $i \in m$. Hence, for each of these newly generated sets of data we perform an evaluation forward pass through the network and compute the loss, retaining the $\mathcal{S}^{i_c}_{cluster}$ and $\mathcal{S}^{i_q}_{query}$ sets giving the highest loss to be used as input to MEML for the training step:

$$\begin{aligned} i_c &= \text{argmax}_{i \in 1,..m} \mathcal{L}(f(\mathcal{S}^i_{cluster}), Y_{cluster}), \\ i_q &= \text{argmax}_{i \in 1,..m} \mathcal{L}(f(\mathcal{S}^i_{query}), Y_{0:Q}). \end{aligned} \tag{9}$$

Three different augmented batches are created starting from the input batch, each forwarded through the network producing logits. The Cross-Entropy losses between those latter and the targets are computed, keeping the augmented batch corresponding to the highest value. In detail, we adopt the following augmentation: vertical flip, horizontal flip for batch 1; colour jitter (brightness, contrast, saturation, hue) for batch 2; random affine, random crop for batch 3.

3.4 Meta-continual Test

At meta-continual test time, novel and unseen tasks $\mathcal{T}'_i \sim p(\mathcal{T}')$ from the test set are provided to the network, as illustrated in Fig. 1. Here $p(\mathcal{T}')$ represents the distribution of supervised test tasks and $\mathcal{T}'_i$ corresponds to a sampled test class. The representation learned during meta-train remains frozen, and only the prediction layers are fine-tuned. The test set is composed of novel tasks, that can be part of the same distribution (e.g. distinct classes within the same dataset) or even belong to a different distribution (e.g. training and testing performed on different datasets).

4 Experiments

4.1 Few-Shot Unsupervised Continual Learning

Datasets. We employ Omniglot and Mini-ImageNet, two datasets typically used for few-shot learning evaluation. The Omniglot dataset contains 1623 characters from 50 different alphabets with 20 greyscale image samples per class. We use the same splits as [13], using 1100 characters for meta-training, 100 for meta-validation, and 423 for meta-testing. The Mini-ImageNet dataset consists of 100 classes of realistic RGB images with 600 examples per class. We use 64 classes for meta-training, 16 for meta-validation and 20 for meta-test.

Architecture. Following [14], we use for the FEN a six-layer CNN interleaved by ReLU activations with 256 filters for Omniglot and 64 for Mini-ImageNet. All convolutional layers have a 3×3 kernel (for Omniglot, the last one is a 1×1 kernel) and are followed by two linear layers constituting the CLN. The attention mechanism is implemented with two additional linear layers interleaved by a Tanh function, followed by a Softmax and a sum to compute attention coefficients and aggregate features. We use the same architecture for competitive methods. We do not apply the Softmax activation and the final aggregation but we keep the added linear layers, obtaining the same number of parameters. The choice in using two simple linear layers as attention mechanism is made specifically since the aim of the paper is to highlight how this kind of mechanism can enhance performance and significantly decrease both training time and memory usage.

Table 1. Meta-test test accuracy on Omniglot.

Algorithm/Tasks	Omniglot					
	10	50	75	100	150	200
Oracle OML [14]	88.4	74.0	69.8	57.4	51.6	47.9
Oracle ANML [1]	86.9	63.0	60.3	56.5	45.4	37.1
Oracle MEML (Ours)	**94.2**	**81.3**	**80.0**	**76.5**	**68.8**	**66.6**
Oracle MEMLX (Ours)	**94.2**	75.2	75.0	67.2	58.9	55.4
OML	74.6	32.5	30.6	25.8	19.9	16.1
ANML	72.2	46.5	43.7	37.9	26.5	20.8
MEML (Ours)	**89.0**	48.9	46.6	37.0	29.3	25.9
MEMLX (Ours)	82.8	**50.6**	**49.8**	**42.0**	**34.9**	**31.0**

Table 2. Meta-test test accuracy on Mini-ImageNet.

Algorithm/Tasks	Mini-ImageNet				
	2	4	6	8	10
Oracle OML [14]	50.0	31.9	27.0	16.7	13.9
Oracle MEML (Ours)	66.0	33.0	28.0	29.1	21.1
Oracle MEMLX (Ours)	**74.0**	**60.0**	**36.7**	**51.3**	**40.1**
OML	49.3	41.0	19.2	18.2	12.0
MEML (Ours)	70.0	**48.4**	36.0	34.0	21.6
MEMLX (Ours)	**72.0**	45.0	**50.0**	**45.6**	**29.9**

Training. For Omniglot, we train the model for 60000 steps while for Mini-ImageNet for 200000, with meta-batch size equals to 1. The outer loop learning rate is set to $1e^{-4}$ while the inner loop learning rate is set to 0.1 for Omniglot and 0.01 for Mini-ImageNet, with Adam optimiser. As embedding learning networks, we employ Deep Cluster [6] for Mini-ImageNet and ACAI [2] for Omniglot. Since Mini-ImageNet contains 600 examples per class, after clustering, we sample examples between 10 and 30, proportionally to the cluster dimension to keep the imbalance between tasks. We report the test accuracy on a different number of unseen classes, which induces increasingly complex problems as the number increase. Following the protocol employed in [14], all results are obtained through the mean of 50 runs for Omniglot and 5 for Mini-ImageNet.

Performance Analysis. In Tables 1 and 2, we report results respectively on Omniglot and Mini-ImageNet, comparing our model with competing methods. To see how the performance of MEML within our FUSION is far from those achievable with the real labels, we also report for all datasets the accuracy reached in a supervised case (*oracles*). We define Oracle OML [14] and Oracle ANML [1] [2] as supervised competitors, and Oracle MEML the supervised version of our model. MEML outperforms OML on Omniglot and Mini-ImageNet and ANML on Omniglot, suggesting that the meta-examples strategy is beneficial on both FUSION and fully supervised cases. MEMLX, the advanced version exploiting a specific augmentation technique is able to improve the MEML results in almost all experiments. In particular, MEMLX outperforms MEML on both Omniglot and Mini-ImageNet in FUSION and even in the fully supervised case on Mini-ImageNet. The only experiment in which MEML outperforms MEMLX is on Omniglot, in the supervised case. In our opinion, the reason is to be found in the type of dataset. Omniglot is a dataset made up of 1100 classes and therefore characters are sometimes very similar to each other. Precisely for this reason, applying augmentation can lead the network to confuse augmented characters for a class with characters belonging to other classes. In the unsupervised case, the clusters are grouped by features, which should better separate the data from a visual point of view, thus favouring our augmentation technique.

Table 3. Training time and GPU usage of MEML and MEMLX compared to OML on Omniglot and Mini-ImageNet.

Algorithm	Omniglot		Mini-ImageNet	
	Time	GPU	Time	GPU
OML [14]	1 h 32 m	2.239 GB	7h 44 m	3.111 GB
MEML	47 m	0.743 GB	3 h 58 m	1.147 GB
MEMLX	1 h 1 m	0.737 GB	4 h 52 m	1.149 GB

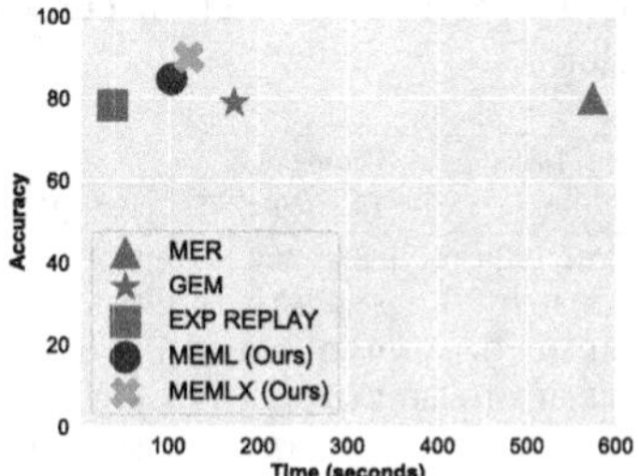

Fig. 2. Training time comparison with respect to the accuracy between the most important state-of-the-art continual learning methods.

Time and Computation Analysis. In Table 3, we compare training time and computational resources usage between OML, MEML and MEMLX on Omniglot and Mini-ImageNet. We measure the time to complete all training steps and the computational resources in gigabytes occupied on the GPU. Both datasets confirm that our methods, adopting a single inner update, train considerably faster and uses approximately one-third of the GPU resources with respect to OML. MEMLX undergoes minimal slowdown, despite the use of our augmentation

[2] Our results on Oracle ANML are different from the ones presented in the original paper due to a different use of data. To make a fair comparison we use 10 samples for the support set and 15 for the query set for all models, while in the original ANML paper the authors use 20 samples for the support set and 64 for the query set. We do not test ANML on Mini-ImageNet due to the high computational resources needed.

strategy. To a fair comparison, all tests are performed on the same hardware equipped with an NVIDIA Titan X GPU.

4.2 Supervised Continual Learning

To further prove the effectiveness of our meta-example strategy, we put MEML and MEMLX in standard supervised continual learning and show its performance compared to state-of-the-art continual learning approaches.

Datasets. We experiment on Sequential MNIST and Sequential CIFAR-10. In detail, the MNIST classification benchmark and the CIFAR-10 dataset [16] are split into 5 subsets of consecutive classes composed of 2 classes each.

Architecture. For tests on Sequential MNIST, we employ as architecture a fully-connected network with two hidden layers, interleaving with ReLU activation as proposed in [19,22]. For tests on CIFAR-10, we rely on ResNet18 [21].

Training. We train all models in a class-incremental way (Class-IL), the hardest scenario among the three described in [26], which does not provide task identities. We train for 1 epoch for Sequential MNIST and 50 epochs for Sequential CIFAR-10. SGD optimiser is used for all methods for a fair comparison. A grid search of hyperparameters is performed on all models taking the best ones for each. For rehearsal-based strategies, we report results on buffer size 200, 500 and 5120. The standard continual learning test protocol is used for all methods, where the accuracy is measured on test data composed of unseen samples of all training tasks at the end of the whole training process. We adapt our meta-example strategy to a double class per task making a meta-example for each class corresponding to two inner loops. The query set used within FUSION mirror the memory buffer in continual learning. The memory buffer contains elements from previously seen tasks, while the query set samples elements from all training tasks. For MEMLX, we apply our augmentation technique on both current task data and buffer data.

Performance Analysis. In Table 4 we show accuracy results on Sequential MNIST and Sequential CIFAR-10[3] respectively. MEML and MEMLX consistently overcome all state-of-the-art methods on both datasets. We denote that MEML is significantly different from MER, processing one sample at a time and making an inner loop on all samples. This greatly increases the training time, making this strategy ineffective for datasets such as CIFAR-10. On the contrary, MEML makes as many inner loops as there are classes per task and finally a single outer loop on both task data and buffer data. This way, MEML training time is comparable to the other rehearsal strategy, but with the generalisation benefit given from meta-learning. To further confirm the beneficial role of the meta-learning procedure, we observe that EXP REPLAY, using only one loop, reaches lower performance. In Table 5 we report results on additional continual

[3] Due to high training time we do not report MER results on Sequential CIFAR-10.

learning metrics: forward transfer, backward transfer and forgetting. In particular, **forward transfer** (FWT) measure the capability of the model to improve on unseen tasks with respect to a random-initialized network. It is computed making the difference between the accuracy before the training on each task and the accuracy of a random-initialized network, averaged on all tasks. **Backward transfer** [19] (BWT) is computed making the difference between the current accuracy and its best value for each task, making the assumption that the highest value of the accuracy on a task is the accuracy at the end of it. Finally, **forgetting** is similar to BTW, without the letter assumption. We compare the best performer algorithms on both Sequential MNIST and Sequential CIFAR. MEML and MEMLX outperform all the other methods on BWT and forgetting, while little lower performance are reached on FWT. Since results are consistent for all buffer dimensions, we report results on buffer 5120.

Time Analysis. We make a training time analysis (see Fig. 2) between the most relevant state-of-the-art continual learning strategy on Sequential MNIST. We measure the training time in seconds since the last task. We find that MEML and MEMLX are slower only compared to EXP REPLAY due to the meta-learning strategy, but they are faster with respect to both GEM and MER, reaching higher accuracy.

Meta-example Single Update *vs.* Multiple Updates. To prove the effectiveness of our method - MEML - based on meta-examples, we compare it with: OML [14] - performing multiple updates, one for each element of the cluster; OML with a single update - adopting a single update over a randomly sampled data point from each task; MEML with mean ME - a version exploiting the mean between the feature vector computed by the FEN. In Fig. 3, we show that MEML and MEMLX consistently outperform all the other baselines on Omniglot. OML with a single update gives analogous performance to the multiple updates one, confirming the idea that the strength of generalisation relies on the feature reuse.

Table 4. MEML and MEMLX compared to state-of-the-art continual learning methods on Sequential MNIST and Sequential CIFAR-10 in class-incremental learning.

Algorithm/Buffer	Sequential MNIST				Sequential CIFAR-10			
	None	200	500	5120	None	200	500	5120
LWF [18]	19.62	–	–	–	19.60	–	–	–
EWC [15]	20.07	–	–	–	19.52	–	–	–
SI [27]	20.28	–	–	–	19.49	–	–	–
SAM [11]	62.63	–	–	–	–	–	–	–
iCARL [21]	–	–	–	–	–	51.04	49.08	53.77
HAL [7]	–	79.80	86.80	88.68	–	32.72	46.24	66.26
GEM [19]	–	78.85	85.86	95.72	–	28.91	23.81	25.26
EXP REPLAY [23]	–	78.23	88.67	94.52	–	47.88	59.01	83.65
MER [22]	–	79.90	88.38	94.58	–	–	–	–
MEML (Ours)	–	84.63	90.85	**96.04**	–	**54.33**	**66.41**	83.91
MEMLX (Ours)	–	**89.94**	**92.11**	94.88	–	51.98	63.25	**83.95**

Table 5. Forward transfer, backward transfer and forgetting comparison on Sequential MNIST and Sequential CIFAR-10 in class-incremental learning.

Algorithm/Metric	Sequential MNIST			Sequential CIFAR-10		
	FWT	BWT	Forgetting	FWT	BWT	Forgetting
HAL [7]	–10.06	–6.55	6.55	–10.34	–27.19	27.19
GEM [19]	**–9.51**	–4.14	4.30	–9.18	–75.27	75.27
EXP REPLAY [23]	–10.97	–6.07	6.08	**–8.45**	–13.99	13.99
MER [22]	–10.50	–3.22	3.22	–	–	–
MEML (Ours)	–9.74	–3.12	3.12	–12.68	**–10.97**	**10.97**
MEMLX (Ours)	–9.74	**–1.72**	**1.92**	–12.74	–12.42	12.42

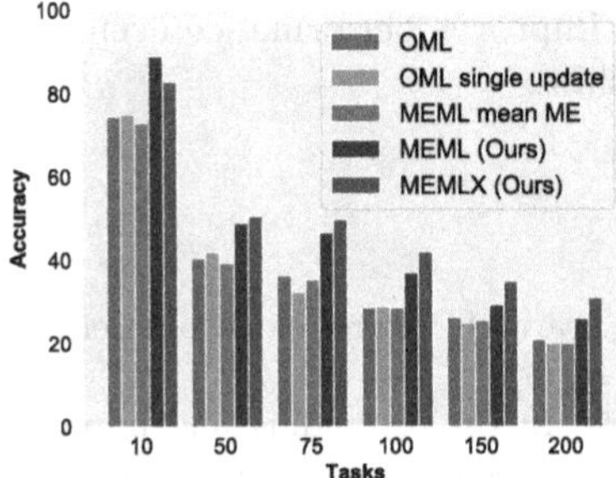

Fig. 3. The capability of meta-example on Omniglot.

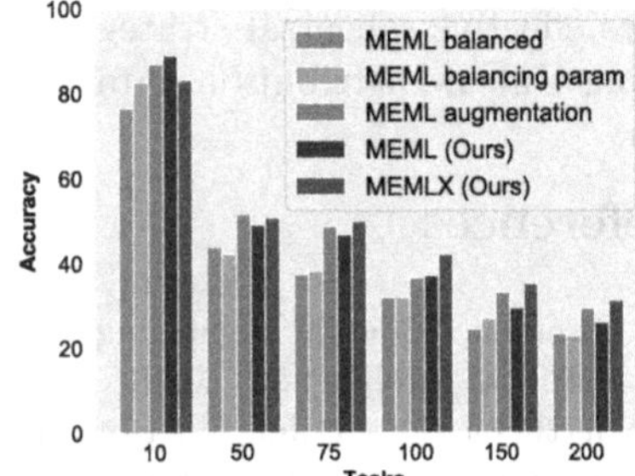

Fig. 4. Unbalanced *vs.* balanced settings on Omniglot.

Also, the MEML with mean ME has performance comparable with the multiple and single update ones, proving the effectiveness of our aggregation mechanism to determine a suitable and general embedding vector for the CLN.

Balanced *vs.* Unbalanced Tasks. To justify the use of unbalanced tasks and show that allowing unbalanced clusters is more beneficial than enforcing fewer balanced ones, we present some comparisons in Fig. 4. First of all, we introduce a baseline in which the number of clusters is set to the true number of classes, removing from the task distribution the ones containing less than N elements and sampling N elements from the bigger ones. We thus obtain a perfectly balanced training set at the cost of less variety within the clusters; however, this leads to poor performance as small clusters are never represented. To verify if maintaining variety and balancing data can lead to better performance, we try two balancing strategies: augmentation, at data-level, and balancing parameter, at model-level. For the first one, we keep all clusters, sampling N elements from the bigger and using data augmentation for the smaller to reach N elements. At model-level, we multiply the loss term by a balancing parameter to weigh the update for each task based on cluster length. These tests result in lower performance with respect to MEML and MEMLX, suggesting that the only thing that matters is cluster variety and unbalancing does not negatively affect the training.

5 Conclusion and Future Work

We tackle a novel problem concerning few-shot unsupervised continual learning, proposing an effective learning strategy based on the construction of unbalanced tasks and meta-examples. With an unconstrained clustering approach, we find that no balancing technique is necessary for an unsupervised scenario that needs to generalise to new tasks. Our model, exploiting a single inner update through meta-examples, increase performance as the most relevant features are selected. In addition, an original augmentation technique is applied to reinforce its strength. We show that MEML and MEMLX not only outperform the other baselines within FUSION but also exceed state-of-the-art approaches in class-incremental continual learning. Interesting future research is to investigate a more effective rehearsal strategy that further improves performance even when facing Out-of-Distribution data and domain shift.

References

1. Beaulieu, S., et al.: Learning to continually learn. In: European Conference on Artificial Intelligence (2020)
2. Berthelot, D., Raffel, C., Roy, A., Goodfellow, I.: Understanding and improving interpolation in autoencoders via an adversarial regularizer. In: International Conference on Learning Representations (2019)
3. Bertugli, A., Vincenzi, S., Calderara, S., Passerini, A.: Few-shot unsupervised continual learning through meta-examples. In: Neural Information Processing Systems Workshops (2020)
4. Buzzega, P., Boschini, M., Porrello, A., Abati, D., Calderara, S.: Dark experience for general continual learning: a strong, simple baseline. In: Neural Information Processing Systems (2020)
5. Caccia, M., et al.: Online fast adaptation and knowledge accumulation: a new approach to continual learning. In: Neural Information Processing Systems (2020)
6. Caron, M., Bojanowski, P., Joulin, A., Douze, M.: Deep clustering for unsupervised learning of visual features. In: Ferrari, V., Hebert, M., Sminchisescu, C., Weiss, Y. (eds.) Computer Vision – ECCV 2018. LNCS, vol. 11218, pp. 139–156. Springer, Cham (2018). https://doi.org/10.1007/978-3-030-01264-9_9
7. Chaudhry, A., Gordo, A., Dokania, P., Torr, P., Lopez-Paz, D.: Using hindsight to anchor past knowledge in continual learning. In: AAAI Conference on Artificial Intelligence (2021)
8. Chaudhry, A., Ranzato, M., Rohrbach, M., Elhoseiny, M.: Efficient lifelong learning with A-GEM. In: International Conference on Learning Representations (2019)
9. Deng, J., Dong, W., Socher, R., Li, L.J., Li, K., Fei-Fei, L.: ImageNet: a large-scale hierarchical image database. In: IEEE International Conference on Computer Vision and Pattern Recognition (2009)
10. Finn, C., Abbeel, P., Levine, S.: Model-agnostic meta-learning for fast adaptation of deep networks. In: International Conference on Machine Learning (2017)
11. Sokar, G., Mocanu, D.C., Pechenizkiy, M.: Self-attention meta-learner for continual learning. In: International Conference on Autonomous Agents and Multiagent Systems (2021)
12. Gong, C., Ren, T., Ye, M., Liu, Q.: MaxUp: a simple way to improve generalization of neural network training. arXiv preprint arXiv:2002.09024 (2020)
13. Hsu, K., Levine, S., Finn, C.: Unsupervised learning via meta-learning. In: International Conference on Learning Representations (2019)
14. Javed, K., White, M.: Meta-learning representations for continual learning. In: Neural Information Processing Systems (2019)
15. Kirkpatrick, J.N., et al.: Overcoming catastrophic forgetting in neural networks. Proceed. Nat. Academy Sci. United States Am. **114**(13), 3521–3526 (2016)
16. Krizhevsky, A.: Learning multiple layers of features from tiny images. Tech. rep, Canadian Institute for Advanced Research (2009)
17. Lake, B.M., Salakhutdinov, R., Tenenbaum, J.B.: Human-level concept learning through probabilistic program induction. Am. Assoc. Advan. Sci. **350**(6266), 1332–1338 (2015)

18. Li, Z., Hoiem, D.: Learning without forgetting. IEEE Trans. Pattern Anal. Mach. Intell. **40**, 2935–2947 (2018)
19. Lopez-Paz, D., Ranzato, M.A.: Gradient episodic memory for continual learning. In: Neural Information Processing Systems (2017)
20. Raghu, A., Raghu, M., Bengio, S., Vinyals, O.: Rapid learning or feature reuse? towards understanding the effectiveness of MAML. In: International Conference on Learning Representations (2020)
21. Rebuffi, S.A., Kolesnikov, A.I., Sperl, G., Lampert, C.H.: iCaRL: incremental classifier and representation learning. IEEE International Conference on Computer Vision and Pattern Recognition (2017)
22. Riemer, M., et al.: Learning to learn without forgetting by maximizing transfer and minimizing interference. In: International Conference on Learning Representations (2019)
23. Rolnick, D., Ahuja, A., Schwarz, J., Lillicrap, T., Wayne, G.: Experience replay for continual learning. In: Neural Information Processing Systems (2019)
24. Rusu, A.A., et al.: Progressive neural networks. ArXiv pre-print (2016)
25. Shin, H., Lee, J.K., Kim, J., Kim, J.: Continual learning with deep generative replay. In: Neural Information Processing Systems (2017)
26. van de Ven, G.M., Tolias, A.S.: Three scenarios for continual learning. In: Neural Information Processing Systems Workshops (2018)
27. Zenke, F., Poole, B., Ganguli, S.: Continual learning through synaptic intelligence. In: International Conference on Machine Learning (2017)

Explainable Deep-Learning Model Reveals Past Cardiovascular Disease in Patients with Diabetes Using Free-Form Visit Reports

Alessandro Guazzo[1(✉)], Enrico Longato[1], Gian Paolo Fadini[2], Mario Luca Morieri[2], Giovanni Sparacino[1], and Barbara Di Camillo[1,3]

[1] Department of Information Engineering, University of Padova, Padova, Italy
alessandro.guazzo@phd.unipd.it
[2] Department of Medicine, University of Padova, Padova, Italy
[3] Department of Comparative Biomedicine and Food Science, University of Padova, Legnaro, Italy

Abstract. Writing notes remains the most widespread method by which medical doctors record clinical events. As a result, a relevant portion of the information concerning a patient's health status is recorded in the form of text. This is particularly relevant in the case of people affected by chronic diseases, such as diabetes, who need frequent medical attention, and, thus, produce substantial amounts of textual data. Natural language processing methods are valuable in this context, as they can enable the automatic conversion of free-form text into structured data. However, in everyday clinical practice, these approaches, which often rely on artificial intelligence, are not widely used as they are perceived as black boxes. In this study, we used a typical electronic health record dataset, paired with a subset of a hospital discharge registry, to develop a tool to automatically identify past cardiovascular disease (CVD) hospitalisations using the free-form text of routine visits of diabetic patients. Thanks to its high performance (F1-score = 0.927), the proposed tool may be reliably used to help clinicians to fill medical records in a structured fashion by raising an alert when the model identifies a CVD hospitalisation, but no history of CVD is annotated within the patient's electronic health record. We also provided a robust interpretability framework following the trustworthy artificial intelligence principles of i) human agency, ii) human oversight, and iii) technical robustness and safety. Leveraging the provided interpretability framework, it is also possible to understand the most relevant factors that led to the alert, empowering the user to decide whether to follow up on the algorithm's suggestion or ignore it.

Keywords: Cardiovascular diseases · Deep learning · Diabetes mellitus · Natural language processing

G. Nicosia et al. (Eds.): LOD 2022, LNCS 13810, pp. 430–443, 2023.
https://doi.org/10.1007/978-3-031-25599-1_32

1 Introduction

Diabetes prevalence is estimated to be 9.8% worldwide and it is expected to increase to 11.2% before 2045 [1]. High blood glucose levels that characterize diabetes, if not properly controlled with personalised therapeutic approaches [2], may lead to various complications such as cardiovascular disease (CVD), retinopathy, and nephropathy, which contribute to make this disease a major cause of death globally [3]. To delay complications, patients with diabetes are typically followed up by general practitioners or endocrinologists and therefore undergo periodic routine visits [4]. From a data perspective, this longitudinal nature of diabetes leads to the need of describing the course of the disease over time, thus producing a very large stream of heterogeneous data, usually handled via digital electronic health records (EHR).

Datasets extracted from EHR systems have been largely used in the literature to develop predictive models for relevant endpoints, with the aim of stratifying the population of patients based on predicted risk [5], or conduct retrospective and prospective population studies on the epidemiology of diabetes and its complications [6]. It is thus of utmost importance for clinicians to record information about disease prognosis such as therapies, healthcare encounters, previous pathologies, and complications so as to build rich, and reliable, datasets that describe as accurately as possible the disease course for large cohorts of patients. However, most of the information is locked behind free-form text [7] since, for clinicians, writing down text notes remains the simplest and fastest method to record clinical events [8]. Unstructured clinical free-form text is thus dominant over structured data [9, 10], and datasets must be manually reviewed by experts to extract relevant information before they can be re-used and re-purposed, leading to possible scalability and cost issues [11].

Developing natural language processing (NLP) models has become key as these technologies provide a solution to transform free-form text into structured clinical data automatically [12]. However, these approaches, and especially those that rely on deep learning (DL), are perceived as black boxes by clinicians, who struggle to trust and use approaches that they do not fully understand. Model explainability and result interpretability have become cardinal points of trustworthy artificial intelligence (AI) [13]. These aspects must be taken into consideration when developing tools that aim at being used in everyday clinical practice as they allow the end user to be in control over the algorithm so that the latter ends up being correctly perceived as a helpful tool instead of an unknown entity.

Among all complications caused by diabetes, CVD is one of the most relevant from an automatic data analysis point of view, as it is a major cause of mortality and disability [14], but also relatively easy to identify via hospitalisations data. Hence, in this study we aimed at developing a DL NLP tool, based on the long short-term memory (LSTM) architecture, to automatically identify past CVD hospitalisations using free-form text of routine visits of diabetic patients. We developed this prototype according to the trustworthy AI principles of i) human agency, ii) human oversight, and iii) technical robustness and safety by providing a result interpretability framework that allows the user to fully understand the reasoning behind the algorithm's decisions, as well as to have full control over its suggestions.

2 Methods

2.1 Dataset

We used the database of the Diabetic Outpatient Clinic of the University Hospital of Padova (Italy) containing the text of 197,411 routine visits belonging to 16,876 patients with diabetes. We then augmented this database with CVD hospital admission and discharge data by considering the hospital discharge registry of the Veneto Region, which contains records of 16,292 patients with diabetes who were treated at the University Hospital of Padova. We harmonised the two databases by considering an observation period spanning from January 1st, 2011, to September 30th, 2018. Moreover, to avoid sporadic entries and false negative caused by patients from neighbouring regions who visited the Diabetes Outpatient Clinic of the University Hospital of Padova for routine visits, but whose other healthcare needs were met in their region of origin, we considered only patients that are Italian citizens registered as healthcare beneficiaries in the Veneto Region; only visits who were carried out during the patients' healthcare eligibility periods in the Veneto Region; and patients with less than a visit per year in three different years were excluded.

After the harmonisation step, we linked visits and CVD hospitalisations. We identified CVD hospitalisations using ICD-9-CM diagnosis codes [15] from 390 to 459, or ICD-9-CM intervention codes associated to operations on the cardiovascular system (00.61–66, 36.03, 36.06–07, 36.10–19, 00.55, 39.50, 39.52, 38.48, 39.71, 39.90). For each subject and each visit, we automatically searched for a CVD hospitalisation discharge within a time window going from January 1st, 2011 (the start of the observation period) to the visit's date. We labelled with "1" visits for which a previous hospitalisation was found and with "0" those for which there was no record of past hospitalisation.

When we evaluated model performance, we considered both a by-visit perspective, described above, and a by-patient one, that we produced by assigning a positive label (1) to patients with at least one of their visits labelled with a "1," and a negative label (0) to those with no record of past CVD hospitalisations. According to the by-visit perspective, we considered each visit as an independent entry, whereas, for the by-patient perspective, we only distinguished between patients with and without CVD hospitalisations before any visit.

From the whole dataset we obtained three subsets: a training set (~80% of the total sample size), a validation set (10%), and a test set (10%) ensuring that all the visits belonging to the same patient were part of the same subset.

Finally, we pre-processed the dataset according to the following typical steps [16]:

- Deletion of Italian stop words.
- Word stemming.
- Deletion of the 1% least frequent words.
- Exclusion of visits consisting of less than 3 words.

Some relevant characteristics of the final dataset and its subsets (training, validation, and test) are shown in Table 1. Note how this is a highly unbalanced dataset as patients undergo many visits but only one is performed after a hospitalisation, moreover, only a small fraction of patients is hospitalised for cardiovascular diseases.

Table 1. Dataset subsets characteristics considering number of subjects, total number of visits, and number of positive visits. Frequency of positive visits is reported within round brackets in the last column.

Subset	N. subjects	N. visits	N. positive visits
Training	5,056	55,765	1,940 (3.5%)
Validation	632	7,346	252 (3.4%)
Test	632	6,760	231 (3.4%)
Total	**6,320**	**69,871**	**2,423 (3.5%)**

2.2 Models and Hyperparameters Optimisation

We developed a long short-term memory (LSTM) neural network [17]. We organised the network architecture as a sequence starting with an embedding layer [18], followed by a bidirectional LSTM layer with tanh (output) and sigmoid (recurrent) activation functions [19], and a set of dense layers ending on a single output node. We considered as hyper-parameters: the dimension of the embedding layer (32, 64, 128, or 256), the dimension of the LSTM layer (16, 32, 64, or 128), the dropout of the LSTM layer (0, 0.15, or 0.3 non-recurrent), the number of dense layers (2–6) and dimension of the first one (16, 32, 64, or 128; with layers progressively halving in size towards the output), and the dropout of dense layers (no dropout or 0.1; the same for all dense layers).

As a benchmark model, we used a simple logistic regression (LR) trained on a bag-of-words (BOW) [20] version of the dataset. For this model, we considered a L2 regularisation loss-function with a single hyperparameter that needed optimisation, the inverse of the regularisation strength C.

We performed hyperparameter optimisation considering only the training set using a 5-fold cross validation [21] and a random search approach [22] accounting for 200 combinations of hyperparameters for the LSTM and, for the LR, 10,000 values of C, randomly sampled from a log uniform distribution ranging from 10^{-4} to 10^{2}. We selected the best hyperparameters as those that led to the minimum average loss across the 5 folds (binary cross-entropy for LSTM and L2 loss for LR). For the network training we used Adam as an optimisation algorithm [23], with the initial learning rate set to 5×10^{-5} and a decay rate equal to the initial learning rate divided by the maximum number of epochs (200). We also set an early stopping criterion (20 consecutive epochs with no improvement) to avoid overfitting while reducing computational time [24]. After this first optimisation step, we re-trained both models (LSTM and LR) with optimal hyperparameters on the whole training set. For the LSTM, we repeated the re-training process 100 times with different randomised initialisations, and we choose the best performing network according to the loss computed on the validation set.

To transform models from rankers into classifiers, useable in practice for CVD identification, we implemented two thresholding approaches. In the first one we used one probability threshold (th) to discriminate between the positive (1, if predicted probability $p \geq th$) and negative (0, if $p < th$) predictions. For the second approach we considered two thresholds, a low (th_{low}) and a high (th_{high}) one, to discriminate between positive

(1, if $p \geq th_{high}$), negative (0, if $p \leq th_{low}$), and uncertain (−1, if $th_{low} < p < th_{high}$) predictions. For the single-threshold approach, we selected the optimal threshold by using each probability value predicted for visits in the validation set as a threshold and choosing the one associated to the maximum F1-score. For the two-threshold approach, we considered 4 different target levels of uncertainty: 5%, 10%, 15%, and 20%. For each level, we considered 500,000 different combinations of thresholds in the validation set and chose the best two thresholds as those that allow to reach the highest F1-score while excluding a fraction of patients close to the target uncertainty level, thus corresponding to the minimum of the following cost function J_{th}:

$$J_{th} = |F1_{th} - 1| + |U_{th} - U| \tag{1}$$

where: $F1_{th}$ is the F1-score, U_{th} is the actual uncertainty level, and U ∈ (0.05, 0.1, 0.15, 0.2) is the target uncertainty level. By setting a level of uncertainty we ignore some model predictions and allow for a subsequent manual evaluation to be performed by clinicians on a small subset of uncertain visits.

2.3 Performance Evaluation

In the by-visit perspective we evaluated the discrimination performance of the models via four metrics: the area under the precision-recall curve (AUPRC) [25] for the continuous network output; and precision, recall, and F1-score after applying the thresholding approaches. In the by-patient perspective, we did not consider the AUPRC as predicted labels were assigned by conglomerating by-visit outputs after thresholding.

Despite not being as complete and reliable as the hospital discharge registry, some information on CVD hospitalisations is present also in the Diabetes Outpatient Clinic Database. Hence, positive patients identified by our model may have previous history of CVD already annotated in their structured electronic healthcare records. To better evaluate the practical impact of our models as tools usable in clinical practice, i.e., to understand how many patients without CVD history reported in the clinic's database would be correctly identified and brought to the clinician's attention with an alert, we considered two additional metrics in the patent-oriented perspective, namely the true positive (TP) alert rate (i.e., the fraction of true positives identified by the models but not already reported in the clinic's database), and alerts raised (i.e., the fraction of alerts that are raised with respect to all those that are needed).

When we considered uncertainty, in the by-visit perspective we independently excluded uncertain visits from metrics computation, whereas, in the by-patient one, we excluded only patients whose visits were all classified as uncertain.

2.4 Model Explainability

To provide a robust interpretability framework and understand which words drive prediction towards a positive or negative outcome we computed word-specific weights for each word in the vocabulary. In the BOW + LR approach, we weighted each word's contribution by its regression coefficient. In the LSTM approach, which lacks coefficients directly associated to the importance of each word, we used shapely values instead [26].

For each element of a given input sequence, its shapely value is computed by evaluating how prediction is affected when its value is permutated with those of other elements randomly sampled from other sequences constituting the background set. As a result, we can obtain a shapely value for each word of a given sequence, however, this value is not word specific, as the same word may be associated to different shapely values within the different sequences where it appears. In order to compute word-specific shapely values, we independently interpreted the text of all visits belonging to the test set (N = 6,760) using, for each one, 600 elements randomly sampled from the training set as a background set [27]. After this first step, each word had several shapely values associated to it, one for each time the word appeared in the visits' text. We computed each word's final weight as the average of the shapely values thus obtained.

To represent the relative impact of individual words for overall prediction, we produced two word-clouds for each developed model (Sect. 3.3, Fig. 1), to visualise words that, on average, drive predictions towards positive and negative labels, respectively. The size of each word was proportional to the absolute value of the word's weight.

Finally, to allow the end user to understand how each word of a single visit's free-form text influenced the associated prediction, we used the computed word-specific weights to obtain force plots (Sect. 3.3, Fig. 2) specific for each visit [28]. In the force plots, the predicted score for a given visit is displayed in bold, each word is displayed next to a bar of size proportional to the word's weight: the wider the bar the more important the word. Bars are also colour-coded, with red and blue representing a push towards, respectively, positive and negative predictions. Words that had more of an impact on the score are located closer to the dividing boundary between red and blue.

3 Results

3.1 Classification Results: By-Visit Perspective

The performance metrics obtained in the by-visit perspective, where we considered each visit independently from the others, are shown in Table 2. The table is divided in two sections, one for the benchmark model (BOW + LR) and the other for the LSTM network. Within each section, we reported F1-score, precision, and recall according to the considered levels of uncertainty: 0% for the single threshold approach (first row) and 5%, 10%, 15%, 20% for the two-threshold approach (rows 2 to 5).

Without considering uncertainty, the LR and LSTM discrimination performance were comparable (F1-score ~ 0.8), despite AUPRC being higher for the LSTM (0.842 vs. 0.822). On the other hand, when we used a two-threshold approach and a low uncertainty level (5% target, corresponding to ~ 340 visits to be manually classified by clinicians), the LR led to slightly higher precision (0.970 vs. 0.957) at the cost of lower recall (0.791 vs. 0.828) and a higher uncertainty (6.1% vs 5.5%). We could achieve better results by further increasing the uncertainty level (max F1-score = 0.938, LSTM, 21.4% uncertainty ~ 1,300 excluded visits).

Overall, the LSTM network performed better than the LR benchmark, especially according to recall, the most challenging metric in this context. Indeed, i) our models needed to correctly identify few positive visits among many negatives (only 3.4% of visits are positive as per Table 1); and ii) several visits labelled as positive do not contain any information related to a previous CVD hospitalisation in the free-form text. This may happen, e.g., when the specialist dedicates a visit to discussing a patient's general health status, glycaemic control, or other possible issues, but not previous hospitalisations.

Table 2. By-visit performance evaluation metrics computed on the test set (N = 6,760). The percentage of visits in the test set classified as uncertain when using two thresholds is reported in the U column. The table is divided in two sections, one for the BOW + LR and the other for the LSTM. Within each section, F1-score, precision, and recall are reported according to the considered levels of uncertainty.

Model	AUPRC	U (%)	F1-score	Precision	Recall
BOW + LR	0.822	0	0.805	0.858	0.758
		6.1	0.872	0.970	0.791
		11.4	0.886	0.971	0.815
		16.5	0.896	0.970	0.832
		22.5	0.901	0.970	0.842
LSTM	0.842	0	0.803	0.854	0.758
		5.5	0.888	0.957	0.828
		11.0	0.911	0.957	0.870
		16.3	0.925	0.972	0.881
		21.4	0.938	0.973	0.905

3.2 Classification Results: By-Patient Perspective

The performance metrics obtained in the by-patient perspective, where we conglomerated visits according to the patients they belonged to, are shown in Table 3. The table is divided, again, in two sections, one for the benchmark model (BOW + LR) and the other for the LSTM network and stratified by uncertainty level. Within each section, in addition to the same metrics presented in Table 2 (except the AUPRC, as per Sect. 2.3), we also reported the TP alert rate, and the number of alerts that would be raised.

With no uncertainty, the LR slightly outperformed the LSTM according to all considered metrics. However, when we used a double thresholding approach and a low uncertainty level (5% target, 3 patients excluded), the best performing model was the LSTM, especially in terms of recall and alerts raised (0.896 vs. 0.874 and 87.1% vs. 83.9% respectively). Further increasing the uncertainty level yielded even better results (max F1-score = 0.958, LSTM, 6 patients excluded).

TP alert rate was > 67% for both models, meaning that of all patients for which we were able to correctly identify their CVD status with our models, 2 out of 3 had no previous history of CVD correctly reported in the clinic's database and would have ended up generating an alert. The number of alerts raised was also high (>85%), meaning that of all alerts that should have been raised for patients with previous CVD hospitalisations not reported in the clinic's database, the vast majority would have been correctly brought to the clinician's attention.

Even if we used a high level of uncertainty at the visit level (~20%), at the patient level few subjects had to be excluded due to having all their visits classified as uncertain (5–6). These results justify the use of a two-threshold scheme and a relatively low uncertainty level (5–10%), especially with the LSTM network as it is the best performing methodological approach in this scenario.

Table 3. By-patient performance evaluation metrics computed on the test set (N = 6,760). The percentage of visits in the test set classified as uncertain when using two thresholds is reported in the U column. The number of patients that had to be excluded from the performance metric computation since all their visits were classified as uncertain in the two-threshold scheme is reported in the Excluded patients column instead.TP alert rate (i.e., the fraction of true positives identified by the models but not already reported in the clinic's database) and Alerts raised (i.e., the fraction of alerts that are raised with respect to all those that are needed) are considered instead of AUPRC in this setting. The table is divided in two sections, one for the BOW + LR and the other for the LSTM. Within each section, F1-score, precision, and recall are reported according to the considered levels of uncertainty.

Model	U (%)	Excluded patients	F1-score	Precision	Recall	TP alert rate (%)	Alerts raised (%)
BOW + LR	0	0	0.888	0.903	0.874	67.6	85.5
	6.1	3	0.911	0.964	0.864	67.6	83.9
	11.4	3	0.921	0.965	0.880	69.1	88.1
	16.5	4	0.931	0.964	0.900	68.5	88.1
	22.5	5	0.935	0.964	0.908	68.5	88.1
LSTM	0	0	0.879	0.890	0.868	67.4	84.5
	5.5	3	0.927	0.960	0.896	66.9	87.1
	11.0	3	0.945	0.960	0.931	66.9	91.0
	16.3	6	0.950	0.975	0.943	68.1	91.9
	21.4	6	0.958	0.966	0.950	67.8	92.9

3.3 Model Explainability Results

The word clouds showing words that drive prediction toward the positive or negative label are shown in Fig. 1 for both BOW + LR and LSTM models. The figure shows four word clouds: positive words used by the BOW + LR (top left), positive words used by the LSTM (top right), negative words used by the BOW + LR (bottom left), and negative words used by the LSTM (bottom right).

Positive words were similar among the two models, in particular the most important words that, if present in the text of a visit, contributed towards a positive classification (previous CVD), were related to "hospital admission" (Italian stem: *"ricover"*) or

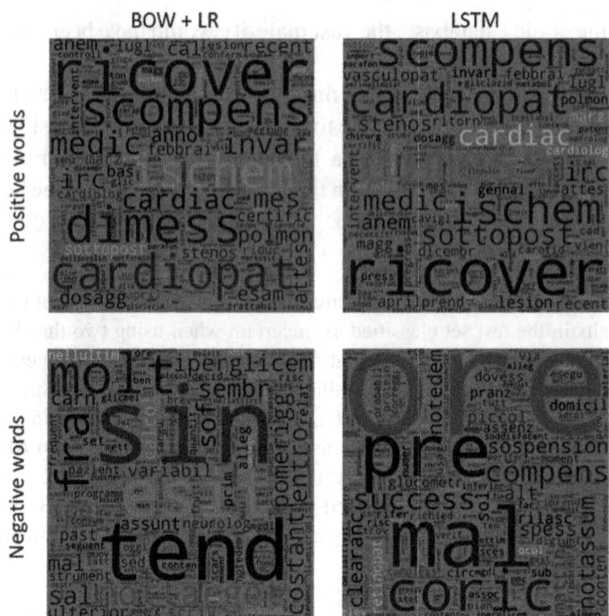

Fig. 1. Word clouds for model explainability. Each word dimension depends on its corresponding weight computed as per Sect. 2.4. Important words for the BOW + LR approach are displayed on the left while those important for the LSTM network are displayed on the right. Positive words, associated to a positive weight, are reported in the top part of the figure while negative words, those associated to a negative weight, are reported at the bottom. The reported words are all stemmed Italian words.

"discharge" (Italian stem: *"dimess"*) and CVD pathologies that often lead to hospitalisations such as "heart failure," "ischemia," or "cardiopathy" (Italian stems: *"scompens," "cardiac," "ischem,"* and *cardiopat"* respectively). The most relevant words, driving prediction towards the negative class (no previous CVD hospitalisation), were instead different among the two models. In fact, the BOW + LR approach considered more words related to glycaemic control such as "hyperglycaemia," "not in target," and "tendency" (Italian stems: *"iperglicemic," "nottarget,"* and *"tend"* respectively) while the LSTM considered more words related to the general health status of the patient and their habits such as "time of meal," "pain," and "compensation," as well as information about routine visits with "previous" and "next" (Italian stems: *"ore," "mal," "compens," "pre,"* and *"success"* respectively). From these results, it appears that the LSTM better captures important relationships among the words within the phrase, meanwhile the BOW + LR focuses more on the presence/absence of relevant words. This is to be expected as the BOW approach does not use information about words position in the sentence.

Once word-specific weights are available, it is possible to independently interpret how the text of each visit influenced the algorithm prediction by using force plots as shown in Fig. 2 using the LSTM model as an example. Starting from the top, the first force plot shown in Fig. 2 is obtained from a visit that was correctly classified as positive (TP) but belonging to a patient with no record of CVD hospitalisation in the clinic's

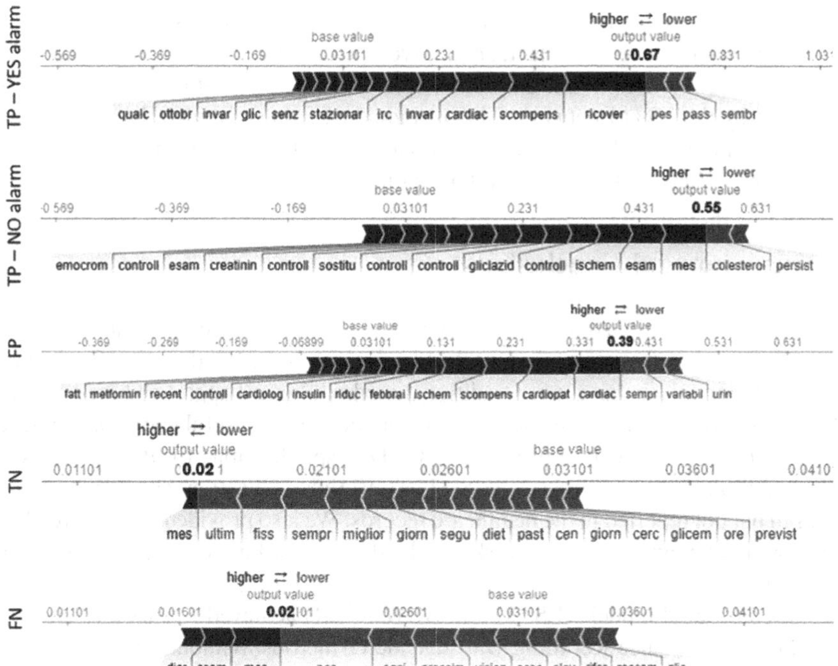

Fig. 2. Force plots for LSTM model explainability. Each word of a visit is depicted with its weight represented by a bar, the wider the bar the higher the corresponding weight. Positive words are displayed with a red bar while negative words are displayed with a blue bar. From the top we show a visit correctly classified as positive that would generate an alert as no previous history of CVD was recorded within the patient's healthcare record (TP-YES alarm), a visit correctly classified as positive that would not generate an alert as previous history of CVD was already reported within the patient's healthcare record (TP-NO alarm), a negative visit classified as positive (false positive FP), a visit correctly classified as negative (true negative TN), and a positive visit classified as negative (false negative FN) (Color figure online).

database. This visit would have generated an alert signalling an information gap to the clinician, possibly prompting corrective action. The second force plot shown in Fig. 2, instead, was obtained from a visit correctly classified as positive (TP) that belonged to a patient with a record of CVD hospitalisation in the clinic's database. This visit, despite being positive, would not have generated an alert. It is interesting to note the differences between these two visits: while the former includes mentions of both CVD and hospitalisations, the latter mentions CVD, but no words related to hospitalisation are present. This result suggest that clinicians may record hospitalisations in free-form when they do not record it in other sections of the healthcare record. The third force plot shown in Fig. 2 shows the text of a visit that did not follow a CVD hospitalisation but was classified as positive anyway. As we can note from the visit's text there are several mentions of CVD that led the LSTM towards a positive prediction, this visit may be of a patient followed at the clinic in Padova who had a CVD hospitalisation in a different region and was, thus, not in the discharge registry of the Veneto Region, the only one available for this study. The fourth force plot was obtained from a visit that was correctly classified as negative, being, in fact, about glycaemic control. Finally, the last force plot shown in Fig. 2 refers to a visit that followed a CVD hospitalisation but was classified as negative by the LSTM network. This is a good example of a visit that, as discussed in Sect. 3.1, might have been difficult to classify correctly even by an expert due to the lack of words related to CVD or hospitalisations.

4 Discussion and Conclusions

In this study, we developed a LSTM network to associate the free-form text of routine visits of diabetic patients to previous CVD hospitalisations. We then benchmarked the proposed AI model against a more traditional methodological approach based on BOW and LR. Finally, to transform the proposed model into a useful tool that could aid clinicians in the identification of previous CVD hospitalisations for their patients, we implemented an alert system that triggers when CVD hospitalisations are identified by the algorithm but not reported in the healthcare records. We also provided an interpretability framework to explain how the network weighted each word in the context of individual predictions.

Both considered models (LSTM and BOW + LR) achieved remarkably good discrimination performance according to all considered metrics. Without considering uncertainty, both models achieved an F1-score greater than 0.8. Moreover, with an uncertainty level as low as 5.5% and only 3 patients excluded, performance sharply increased with the F1-score reaching values above 0.9. The BOW + LR model, used as benchmark, led to surprisingly good results, however, despite having similar performance to the LSTM in terms of F1-score, the latter was preferable in terms of recall, uncertainty, TP alert rate, and alerts raised. The higher performance of the AI approach is mostly due to its ability to learn how words are used as part of a text, the BOW approach, instead, does not retain information on word positioning within the original text. In fact, in Italian as much as in English, a word's position is often informative in and of itself, as it pertains to the word's meaning in context. Thus, it is reasonable that the LSTM model yielded better predictions by combining word meaning and word order with respect to the BOW + LR approach, which only considered the former aspect.

The proposed LSTM-based tool may be useful for clinicians as it addresses the known problem that, when faced with a choice between a structured and an unstructured field, physicians tend to prefer the latter, which is more in line with their attitude and training. Case in point, in the database of the Diabetes Outpatient Clinic of the University Hospital of Padova used in this study, among patients who had a CVD hospital discharge, only one in three had previous history of CVD correctly reported in the structured section of their healthcare record, despite almost all having at least one visit with mentions of hospitalisations or CVD. Payers and administrators may also benefit from the use of such a tool as they could better investigate whether relevant information, such as previous pathologies, are coherently reported in healthcare records, and possibly implement mitigation strategies, e.g., if they are not satisfied with the reporting rate. A tool that automatically reads all of a patient's visits and raises an alert if some are related to CVD hospitalisations and no history of previous CVD is recorded may be key to help clinicians easily fill their patients' medical records in a more structured way, thus leading to more informative and re-usable datasets. In addition, it will overcome the fatigue clinicians may encounter in re-reading bulks of text buried in different records of longitudinal patient's EHR to recall the history of a prior CVD hospitalisation to which the patient may not properly refer.

According to the principle of trustworthy AI, humans must always have full control over algorithms [13]. Thus, the interpretability of the results is another key feature that the tool must have in order to be reliably used. For instance, when an alert is raised, the most relevant factors that influenced the prediction ought to be displayed. In this way, clinicians can understand the model's reasoning and decide whether to accept its suggestion and act or ignore it. Finally, uncertainty is also relevant, especially when the number of visits to be manually classified is low enough, as it is arguably better to allow the algorithm to highlight a few uncertain visits where the clinician should make the final choice than providing a wrong classification.

The proposed LSTM-based tool follows the trustworthy AI principles of i) human agency, ii) human oversight, and iii) technical robustness and safety [13] as i) users are given the knowledge and tools to comprehend the system thanks to a powerful (if not, possibly, arbitrarily scalable) interpretability framework; ii) the human is always in command since clinicians can overrule AI suggestions after seeing the model's interpretability reports (e.g., force plots); and iii) the model achieves excellent discrimination performance, is reliable, and reproducible. A possible limitation of the proposed model lies in the fact that, in the Diabetes Outpatient Clinic of the University Hospital of Padova, visits are typically carried out according to similar approaches by different clinicians. There is a definite possibility that the proposed LSTM may struggle with visits coming from other clinics, where clinicians may follow slightly different protocols and approaches to record relevant details in free-form text. At the time of writing, we have not been able to obtain access to data collected from different clinics, however, as validating the proposed model on a different dataset is an extremely interesting and important task for future work, there are ongoing talks with other partners with the aim of obtaining access to external datasets. Note that this is a preliminary work, and LSTM neural networks were preferred to other DL architectures as their performance proved to be superior in similar NLP applications [29]. Further extensions are planned where we will consider other, more advanced, NLP models, explore a more exhaustive set of hyperparameters, perform significance tests when comparing results, and investigate other trustworthy AI principles.

In conclusion, following the main principles of trustworthy AI, we proposed an explainable DL model, based on the LSTM neural network architecture, to identify previous CVD hospitalisations of diabetic patients by mining the free-form text of their routine visits. The prototype may be useful in everyday clinical practice to fill medical records in a structured fashion by raising an alert when the algorithm identifies a CVD hospitalisation, but no history of CVD is annotated within the patient's electronic health record. Thanks to the interpretability framework, it is possible to observe the most relevant words that led to the alarm, the clinician is thus given all the information needed to decide whether to act or ignore the algorithm suggestion.

Acknowledgements. This work was supported in part by MIUR (Italian Ministry for Education) under the initiative "Department of Excellence" (Law 232/2016).

References

1. Global diabetes data report 2000–2045. https://diabetesatlas.org/data/. Accessed 04 Apr 2022
2. Papatheodorou, K., Banach, M., Bekiari, E., Rizzo, M., Edmonds, M.: Complications of diabetes 2017. J. Diabetes Res. **2018**, 3086167 (2018). https://doi.org/10.1155/2018/3086167
3. Saeedi, P., et al.: Mortality attributable to diabetes in 20–79 years old adults, 2019 estimates: results from the International Diabetes Federation Diabetes Atlas, 9th edition. Diabetes Res. Clin. Pract. **162**, 108086 (2020). https://doi.org/10.1016/j.diabres.2020.108086
4. Powell, P.W., Corathers, S.D., Raymond, J., Streisand, R.: New approaches to providing individualized diabetes care in the 21st century. Curr. Diabetes Rev. **11**(4), 222–230 (2015)
5. Ravaut, M., et al.: Predicting adverse outcomes due to diabetes complications with machine learning using administrative health data. NPJ Digit. Med. **4**(1), 1 (2021). https://doi.org/10.1038/s41746-021-00394-8
6. Aune, D., et al.: Diabetes mellitus, blood glucose and the risk of heart failure: a systematic review and meta-analysis of prospective studies. Nutr. Metab. Cardiovasc. Dis. **28**(11), 1081–1091 (2018). https://doi.org/10.1016/j.numecd.2018.07.005
7. Jensen, K., et al.: Analysis of free text in electronic health records for identification of cancer patient trajectories. Sci. Rep. **7**(1), 1 (2017). https://doi.org/10.1038/srep46226
8. Sheikhalishahi, S., Miotto, R., Dudley, J.T., Lavelli, A., Rinaldi, F., Osmani, V.: Natural language processing of clinical notes on chronic diseases: systematic review. JMIR Med. Inform. **7**(2), e12239 (2019). https://doi.org/10.2196/12239
9. Wei, W.-Q., Teixeira, P.L., Mo, H., Cronin, R.M., Warner, J.L., Denny, J.C.: Combining billing codes, clinical notes, and medications from electronic health records provides superior phenotyping performance. J. Am. Med. Inform. Assoc. JAMIA **23**(e1), e20-27 (2016). https://doi.org/10.1093/jamia/ocv130
10. Ohno-Machado, L., Nadkarni, P., Johnson, K.: Natural language processing: algorithms and tools to extract computable information from EHRs and from the biomedical literature. J. Am. Med. Inform. Assoc. JAMIA **20**(5), 805 (2013). https://doi.org/10.1136/amiajnl-2013-002214
11. Jonnagaddala, J., Liaw, S.-T., Ray, P., Kumar, M., Dai, H.-J., Hsu, C.-Y.: Identification and progression of heart disease risk factors in diabetic patients from longitudinal electronic health records. BioMed Res. Int. **2015**, 636371 (2015). https://doi.org/10.1155/2015/636371
12. 'Overcoming barriers to NLP for clinical text: the role of shared tasks and the need for additional creative solutions - PubMed'. https://pubmed.ncbi.nlm.nih.gov/21846785/. Accessed 31 Jan 2022
13. Markus, A.F., Kors, J.A., Rijnbeek, P.R.: The role of explainability in creating trustworthy artificial intelligence for health care: a comprehensive survey of the terminology, design choices, and evaluation strategies. J. Biomed. Inform. **113**, 103655 (2021). https://doi.org/10.1016/j.jbi.2020.103655
14. Shah, A.D., et al.: Type 2 diabetes and incidence of cardiovascular diseases: a cohort study in 1·9 million people. Lancet Diabetes Endocrinol. **3**(2), 105–113 (2015). https://doi.org/10.1016/S2213-8587(14)70219-0
15. ICD - ICD-9-CM - International Classification of Diseases, Ninth Revision, Clinical Modification, 03 November 2021.https://www.cdc.gov/nchs/icd/icd9cm.htm. Accessed 15 Feb 2022

16. Kathuria, A., Gupta, A., Singla, R.K.: A review of tools and techniques for preprocessing of textual data. In: Singh, V., Asari, V.K., Kumar, S., Patel, R.B. (eds.) Computational Methods and Data Engineering. AISC, vol. 1227, pp. 407–422. Springer, Singapore (2021). https://doi.org/10.1007/978-981-15-6876-3_31
17. Staudemeyer, R.C., Morris, E.R.: Understanding LSTM -- a tutorial into long short-term memory recurrent neural networks. arXiv:190909586 Cs, September 2019. http://arxiv.org/abs/1909.09586. Accessed 01 Feb 2022
18. Mandelbaum, A., Shalev, A.: Word embeddings and their use in sentence classification tasks. arXiv:161008229 Cs, October 2016. http://arxiv.org/abs/1610.08229. Accessed 01 Feb 2022
19. Ding, B., Qian, H., Zhou, J.: Activation functions and their characteristics in deep neural networks. In: 2018 Chinese Control And Decision Conference (CCDC), June 2018, pp. 1836–1841 (2018). https://doi.org/10.1109/CCDC.2018.8407425
20. Qader, W.A., Ameen, M.M., Ahmed, B.I.: An overview of bag of words; importance, implementation, applications, and challenges. In: 2019 International Engineering Conference (IEC), June 2019, pp. 200–204 (2019). https://doi.org/10.1109/IEC47844.2019.8950616
21. Berrar, D.: Cross-validation. In: Ranganathan, S., Gribskov, M., Nakai, K., Schönbach, C. (eds.) Encyclopedia of Bioinformatics and Computational Biology, pp. 542–545. Academic Press, Oxford (2019). https://doi.org/10.1016/B978-0-12-809633-8.20349-X
22. Bergstra, J., Bengio, Y.: Random search for hyper-parameter optimization. J. Mach. Learn. Res. **13**(10), 281–305 (2012)
23. Kingma, D.P., Ba, J.: Adam: a method for stochastic optimization. arXiv:14126980 Cs, January 2017. http://arxiv.org/abs/1412.6980. Accessed 01 Feb 2022
24. Prechelt, L.: Early stopping — but when? In: Montavon, G., Orr, G.B., Müller, K.-R. (eds.) Neural Networks: Tricks of the Trade. LNCS, vol. 7700, pp. 53–67. Springer, Heidelberg (2012). https://doi.org/10.1007/978-3-642-35289-8_5
25. Boyd, K., Eng, K.H., Page, C.D.: Area under the precision-recall curve: point estimates and confidence intervals. In: Blockeel, H., Kersting, K., Nijssen, S., Železný, F. (eds.) ECML PKDD 2013. LNCS (LNAI), vol. 8190, pp. 451–466. Springer, Heidelberg (2013). https://doi.org/10.1007/978-3-642-40994-3_29
26. Lundberg, S.M., Lee, S.-I.: A unified approach to interpreting model predictions. In: Proceedings of the 31st International Conference on Neural Information Processing Systems, Red Hook, NY, USA, December 2017, pp. 4768–4777 (2017)
27. shap.DeepExplainer—SHAP latest documentation. https://shap-lrjball.readthedocs.io/en/latest/generated/shap.DeepExplainer.html. Accessed 07 Mar 2022
28. shap.plots.force—SHAP latest documentation. https://shap.readthedocs.io/en/latest/generated/shap.plots.force.html. Accessed 09 Mar 2022)
29. Guan, M., Cho, S., Petro, R., Zhang, W., Pasche, B., Topaloglu, U.: Natural language processing and recurrent network models for identifying genomic mutation-associated cancer treatment change from patient progress notes. JAMIA Open **2**(1), 139–149 (2019). https://doi.org/10.1093/jamiaopen/ooy061

Helping the Oracle: Vector Sign Constraints for Model Shrinkage Methodologies

Ana Boskovic[1,2(✉)] and Marco Gross[3]

[1] ETH Zurich, Ramistrasse 92, 8092 Zurich, Switzerland
ana.boskovic@ec.europa.eu

[2] Joint Research Center, European Commission, Via Enrico Fermi, 2749 21027 Ispra, VA, Italy

[3] International Monetary Fund, 700 19th Street, N.W., Washington, D.C. 20431, USA
mgross@imf.org

Abstract. Motivated by the need for obtaining econometric models with theory-conform signs of long-run multipliers or other groups of predictor variables for numerous purposes in economics and other scientific disciplines, we develop a vector sign constrained variant of existing model shrinkage methodologies, such as (Adaptive) Lasso and (Adaptive) Elastic Net. A battery of Monte Carlo experiments is used to illustrate that the addition of such constraints "helps the Oracle property" (the ability to identify the true model) for those methods that do not initially carry it, such as Lasso. For methods that possess it already (the adaptive variants), the constraints help increase efficiency in finite data samples. An application of the methods to empirical default rate data and their macro-financial drivers for 16 countries from Europe and the U.S. corroborates the avail of the sign constrained methodology in empirical applications for obtaining more robust economic models.

Keywords: Machine learning · Model selection · Oracle property

1 Introduction

Model selection methodologies—as one element in the field of machine learning—are widely used in many scientific disciplines, including geophysics, economics, finance, network analysis, image recognition, and others, and keep gaining in relevance amid the accumulation of "big data". They serve to help identify a subset of relevant predictors for some target variable, to corroborate (accept, reject, refine) theories as well as for designing econometric models to be of avail for out-of-sample forecasting. Well known model selection methods include the least absolute shrinkage and selector operator (LASSO) [22], Adaptive Lasso [29], Elastic Net [30], Adaptive Elastic Net [31] and other variants which we will reference later. They co-exist with stepwise selection algorithms [3, 14, 21]. Useful entry points to the literature on the variety of model building techniques include textbooks such as [7, 16, 20], among numerous others.

All such methodologies are generally employed in a reduced-form manner, following a "let the data speak" philosophy. There is conventionally no role for theory to inform

G. Nicosia et al. (Eds.): LOD 2022, LNCS 13810, pp. 444–458, 2023.
https://doi.org/10.1007/978-3-031-25599-1_33

neither the relative importance of predictors nor the expected signs of causal relationships when such methods are used, which would be instrumental, however, to obtain economically meaningful models. If data were abundant and not inflicted with noise, this should be unproblematic. However, empirical analyses often operate with data of insufficient quality along some dimensions, including limited time series and/or cross-section observation coverage and noise of different kinds. Against this background, the aim of the paper is to promote the incorporation of theory-implied constraints on the signs of coefficients or linear combinations of them to thereby render existing model selection methods even more valuable. The value of doing so lies in (1) helping the methodologies find the true model, conditional on the assumptions implied by theory being correct, and (2) possibly helping to enhance econometric efficiency. Both are relevant considerations especially when dealing with data of the kind as hinted to above (noise, limited observations, etc.).

The empirical application that we will present relates to financial system stress testing, in particular bank stress testing, whose importance has increased since the global financial crisis of 2007–09. Stress testing entails establishing a link between banks' risk parameters (and hence their profit and loss and balance sheet evolution) with their macro-financial environment. This link is often established in an econometric fashion, to then be able to translate macro-financial scenarios to the banks. Common practice for designing such models is still to develop them in a "manual manner" (i.e., by "handpicking" the equations), through choosing predictor sets and lag structures in a way that will let the models look acceptable from an economic and econometric point of view, with the latter including the conformity of the signs of estimated coefficients with theory. Here, the sign constrained shrinkage methods that we explore can contribute with their appeal for imposing theory-implied constraints in the first instance, to support the model building process.[1]

The paper is structured as follows: We outline the econometric setting and the sign constraints methodology in Sect. 2 and flank them with various relevant references to the literature. A series of Monte Carlo experiments is presented in Sect. 3 to show that sign constraints can enhance model identification and econometric efficiency. The empirical application in Sect. 4, involving data for nonfinancial firm default rates and a pool of macro-financial predictor variables from 16 countries, is used to show that the unconstrained model selection methods often result in models whose estimates are not conform with theory, and that the imposition of the sign constraints well serves its intended purpose.

2 Econometric Setting

2.1 Methods for Variable Selection and Regularization

Model selection methodologies involve a form of regularization, trading off variance and bias. The econometric structures of various model shrinkage methods are summarized in Table 1. The methods can be upfront categorized regarding (1) their adherence to the

[1] See [10] for a discussion of model uncertainty and the use of a sign constrained Bayesian Model Averaging (BMA) methodology.

so-called Oracle property; and (2) whether they can handle the presence of correlated predictors that otherwise pose a difficulty in identifying the correct regressors. The schematic in Fig. 1 depicts where the different methodologies are positioned along these two dimensions.

For a method to hold the Oracle property, two conditions should be satisfied [5, 6]: (1) It should be able to identify the correct model predictors, and (2) it should have the optimal estimation error rate. To be deemed optimal, a procedure should possess these properties along with continuous shrinkage with respect to the regularization parameter.

Table 1. General forms of various shrinkage and model selection methodologies

Method	Form	
Ridge	$\beta = argmin_{\beta}\left\{\|\|y - X\beta\|\|_2^2 + \lambda\sum_{j=1}^{p}\beta_j^2\right\}$	(1)
Non-negative garrote	$\beta = argmin_{\beta}\frac{1}{2}\|\|Y - Zd\|\|^2 + n\lambda\sum_{j=1}^{p}d_j$ subject to $d_j > 0\forall j$, where $Z = (Z_1, \ldots, Z_p)$ and $Z_j = X_j\beta_j^{LS}$	(2)
Lasso	$\beta = argmin_{\beta}\left\{\|\|y - X\beta\|\|_2^2 + \lambda\|\|\beta\|\|_1\right\}, \lambda > 0$	(3)
Adaptive Lasso	$\beta = argmin_{\beta}\left\{\|\|y - X\beta\|\|_2^2 + \lambda\sum_{j=1}^{p}\widehat{w}_j\|\beta_j\|\right\}$	(4)
Elastic Net	$\beta = argmin_{\beta}\left\{\|\|y - X\beta\|\|_2^2 + \lambda_2\|\|\beta\|\|_2^2 + \lambda_1\|\|\beta\|\|_1\right\}$	(5)
Adaptive ElNet	$\beta = argmin_{\beta}\left\{\|\|y - X\beta\|\|_2^2 + \lambda_2\|\|\beta\|\|_2^2 + \lambda_1^*\widehat{w}_j\|\beta_j\|\right\}$	(6)

Lasso shrinks some of the model coefficients to zero and thereby performs variable selection. It minimizes the sum of squared errors, with a bound on the sum of the absolute values of the coefficients. While Lasso uses the L1 regularization term ($\lambda||\beta||_1$, where λ is a shrinkage parameter), Ridge regression [11, 12] uses the L2 regularization ($\lambda\sum_{j=1}^{p}\beta_j^2$), adding the "squares" of the coefficient as the penalty term to the loss function, thereby avoiding shrinking any of the regressors to zero. Elastic net (ElNet) is a methodological combination of the two, that at the same time produces a parsimonious model and, in addition to Lasso, accounts for correlated regressors.

Since Lasso penalizes all coefficients in the same manner, resulting in over-penalization of large coefficients, it is proven to not always be consistent, unless certain conditions are satisfied [19]. To address this issue, [5] developed a concave penalty function for a method they call Smoothly Clipped Absolute Deviation (SCAD). They proved that this method has the Oracle property.

The Adaptive Lasso is a convex optimization problem with the L1 constraint. Therefore, it can be solved by the same efficient algorithm that is used for solving the Lasso (that is, the LARS algorithm, see [4]). [29] proves that Adaptive Lasso is at least as competitive as methods with other concave penalties, while being computationally more attractive. Following the same principles, [9] proposes an adaptive version of Elastic Net, combining the advantages of Elastic Net over its predecessors in dealing with correlated

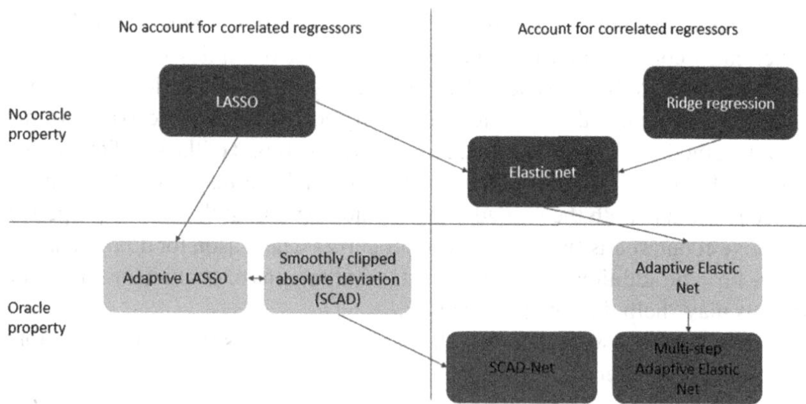

Fig. 1. Categorization of model shrinkage and selection methodologies

regressors, then with an Oracle property. [26] propose a multi-step Adaptive Elastic Net, which, in addition to previously established properties of the method, reduces false positives in high-dimensional variable selection problems.

Several other extensions of the Lasso methodology were developed. For instance, [27] proposed the group Lasso, to address the model selection when groups of regressors are jointly relevant and need to be selected into the models as a group. [24] propose an Adaptive Group Lasso that addresses the same issue, while overcoming the inconsistency of Group Lasso. [23] propose an order-constrained version of L1-regularized regression to address prediction problems involving time-lagged variables, to therefore be suitable for time series model selection.

There are some papers which show that the imposition of sign-constraints can as such be an efficient means of regularization. [18] shows that a simple sign-constrained least squares estimation is an effective regularization technique for a certain class of high-dimensional regression problems, while estimation rates converge with the LASSO procedure and not necessitating any tuning parameters. However, for the purpose of our application and the empirical needs we aim to address, we need both the regularization and model selection, rendering our extension of sign constraints to penalized estimation methods necessary.[2]

We consider an extension for shrinkage-based methods by adding linear inequality constraints to the coefficient sets of a model, to thereby weave prior information into the model selection procedure. The following section summarizes how to go about this.

2.2 Vector Sign Constraints

The idea to impose constraints on the signs of sums of coefficients may arise in time series econometric applications, where different groups of predictor variables may reflect the

[2] Similarly, although an interesting and relevant strand of work, the machine learning methods that use data to learn the constraints independently and solve the optimization problem [17] would not be useful for the empirical application we consider, which involves short and noisy data.

contemporaneous and lagged inclusion of a given predictor. Considering such inclusion of predictors in contemporaneous and lagged form also implies that a methodology that performs well in the presence of correlated regressors is particularly warranted, since many macro-financial time series are serially correlated to some extent. Hence, an Elastic Net methodology involving an L2 regularization is part of the set of methods upon which we build the vector-sign constraint mechanism. In terms of the Oracle property, we choose to include both the non-adaptive versions (without the Oracle property) and their adaptive counterparts (with the Oracle property). The reason for doing so is that we wish to see how the addition of prior knowledge on the signs of coefficients or coefficient vector sums may "help the Oracle property".

The constraints that we impose on all four focus methods (the last four in Table 1) can generically be written as:

$$C\beta \leq d \tag{7}$$

where β is, as above, a $K \times 1$ vector of regression coefficients, C is a $G \times K$ constraints matrix which ought to have full rank (G counts the number of constraints imposed), and d is a $G \times 1$ vector that completes the constraints.

We have identified five papers that relate to the imposition of linear equality or inequality constraints in the literature: [15] develop what they call the Penalized and Constrained (PaC) algorithm which can be used to estimate Lasso estimation problems with linear equality and inequality constraints. The inequality constraints are placed in an annex and with the emphasis being on the algorithm for estimating such model structures. They consider, as an application, the case of high-dimensional website advertising data. [26] propose a path-following algorithm for quadratic programming that replaces hard constraints by what are called exact penalties. [13] derive the dual of the linear constrained generalized Lasso and propose a coordinate descent algorithm for calculating primal and dual solutions. They suggest that coordinate descent can be replaced with quadratic programing when its efficient and stable implementation is possible. Their simulation exercises pertain to the performance regarding the degree of freedom estimation and tuning parameter selection (via AIC vs. BIC). [8] employ three algorithms for estimating constrained Lasso: quadratic programming, the alternating direction method of multipliers, and their own solution path algorithm. They focus on the comparative computational efficiency of the three estimation methods, including with a view to runtime when larger scale applications are considered. Finally, [25] propose an L1 penalized constrained least absolute deviation (LAD) method. The motivation for doing so is seen in cases where heavy-tailed errors or outliers are present, for the variance of the errors to become unbound, in which case constrained Lasso is no longer applicable.

This array of research has overall dealt with estimation algorithms and computational efficiency of constrained Lasso-type problem. The ability of the imposition of constraints to boost their predictive ability, including with a view to econometric efficiency, has not been addressed as far as we see. This is where we aim to contribute: By exploring the avail of sign constraints systematically for a range of methods, with a focus on the gain they may imply for enhancing the Oracle property for methods that do not initially carry it, for improving efficiency for those that do possess it, as well as by flanking our analytical work with an empirical application that is relevant in the field of bank stress testing.

3 Monte Carlo Simulation

To explore the performance of the regular and the vector sign constrained variants of the four model selection methods, we run an array of simulations. We consider the following data generating process (DGP):

$$Y = \beta_1 X_1 + \beta_2 X_2 + \beta_3 X_3 + \beta_4 X_4 + \beta_5 X_5 + \beta_6 X_6 + \beta_7 X_7 + \beta_8 X_8 + \varepsilon \quad (8)$$

where ε is is i.i.d. Gaussian disturbance with mean zero and variance Σ. To explore the performance of each method in presence of different levels of the cross-correlation among the regressors, we generate the coefficient correlation matrix using the following equation:

$$corr(i, j) = \sigma_{i-j} \quad (9)$$

where i and j count the predictor variables contained in X. Table 2 summarizes the parameterization of the DGP that we take as a starting point for all simulations. Two cases are considered for the covariance structure of the predictors: zero covariance vs. strong covariance. Three cases are considered in terms of error variance and implied signal to noise ratios (high, medium and low): the error variances at 0.5, 1, and 2 let the implied signal to noise ratios amount to 29.7, 8.2 and 2.8, respectively.

Table 2. Data generating process

Coefficients: $(\beta_1, \ldots, \beta_8) = (2, -0.7, 0, 0, -1.5, 0.5, 0, 0)$
Noise levels: $\rho = (0.5, 1, 2)$
Sample size: $n = (40, 120, 500)$
Parameter for calibrating the data correlation matrix: $\sigma = (0, 0.75)$
Replication datasets: $D = 20{,}000$

For the sign constrained versions of the four methods, we impose the constraints on two sets of coefficients, defining C and d from Eq. (7) for individual constraints as in Eq. (10) and for joint coefficient constraints as under Eq. (11):

$$C^{\text{indiv}} = \begin{bmatrix} 1 & 0 & 0 & 0 & 0 & 0 & 0 & 0 \\ 0 & 0 & 0 & 0 & 0 & 1 & 0 & 0 \\ 0 & -1 & 0 & 0 & 0 & 0 & 0 & 0 \\ 0 & 0 & 0 & 0 & -1 & 0 & 0 & 0 \end{bmatrix}, \quad d^{\text{indiv}} = \begin{bmatrix} 0 \\ 0 \\ 0 \\ 0 \end{bmatrix} \quad (10)$$

$$C^{\text{joint}} = \begin{bmatrix} 1 & 1 & 0 & 0 & 0 & 0 & 0 & 0 \\ 0 & 0 & 0 & 0 & -1 & -1 & 0 & 0 \end{bmatrix}, \quad d^{\text{joint}} = \begin{bmatrix} 0 \\ 0 \end{bmatrix} \quad (11)$$

From the set of six true models (one mean coefficient vector, two predictor correlation structures, three error variance settings), we simulate 20,000 artificial data sets for all predictors and the implied dependent variable, of sample size $n = (40, 120, 500)$ for each of them. For all 20,000 data sets under all six settings, the first 75% of the sample are used for variable selection and estimation, while the remaining 25% of the observations per data set are considered as an out-of-sample portion based on which the predictive accuracy is assessed. For the in-sample variable selection and estimation, we employ a five-fold cross-validation methodology for obtaining the optimal shrinkage parameter λ. For the Elastic Net methods, we set the tuning parameter (alpha) to 0.5, which implies an equal weight for the L1 and L2 penalty terms. In that way, we combine the feature selection capability of Lasso and the ability of ridge regressions to handle multicollinearity in the data. Predictive accuracy is judged with a view to (1) the percentage of correctly identified predictor sets, and (2) their root mean square error (RMSE).

Tables 3 and 4. show the percentage of correctly identified predictor sets over the 20,000 simulation rounds (median) for the samples with no and high correlation between regressors, respectively. C* represent constraints applied to individual coefficients. C** are constraints applied to groups of coefficients.

Imposing individual constraints on coefficients is found to perform better in terms of model identification than their unconstrained counterparts, showing that the imposition of constraints improves the Oracle property in the domain of correct identification of the model. This is especially true for Adaptive Lasso, which improves the model identification by up to 33% in small samples with no correlation between regressors. This is not surprising, as it has been proven that Adaptive Lasso has the Oracle property which the simple Lasso does not, and thus is superior in terms of its predictive performance and model identification. Despite the adaptive variants having the Oracle property, the addition of sign constraints still helps in small samples, and/or with notable noise.

Applying individual constraints enhances the ability of the selected methods in identifying the correct model even when the correlation between regressors is high (Table 4). In this setting, only in small sample settings with high noise there is no visible improvement. In line with its theoretical properties, Elastic Net methods perform best in model selection when the correlation between regressors is high. The adaptive versions outperform the non-adaptive methods, in particular in small samples with correlated regressors. Therefore, it is not surprising that Adaptive Elastic Net performs the best in model identification, as it combines the features of Elastic Net and Adaptive Lasso [9].

In both settings, applying joint constraints does not improve the success rate in correct variable identification. However, although joint constraints methods do not outperform their unconstrained counterparts, they perform no worse. Their value-added lies in their importance in empirical applications, when econometric models require such grouping of coefficients, like in the case of financial stress testing, as will be exemplified in the section dedicated to the empirical application.

Table 3. Percentage of correctly selected predictor sets, no correlation between regressors

	n = 40 $\delta = 0.5$	n = 40 $\delta = 2$	n = 120 $\delta = 0.5$	n = 120 $\delta = 1$	n = 500 $\delta = 0.5$	n = 500 $\delta = 2$
LASSO	0.94	0.29	1.00	0.95	1.00	0.97
LASSO-C*	0.96	0.41	1.00	0.98	1.00	0.99
LASSO-C**	0.94	0.29	1.00	0.95	1.00	0.97
aLASSO	0.89	0.37	0.96	0.89	0.96	0.92
aLASSO-C*	0.97	0.47	1.00	0.98	1.00	0.99
aLASSO-C**	0.89	0.37	0.96	0.89	0.96	0.92
ElNet	0.95	0.44	1.00	0.97	1.00	0.98
ElNet-C*	0.97	0.55	1.00	0.98	0.99	0.99
ElNet-C**	0.95	0.44	1.00	0.97	1.00	0.98
aElNet	0.93	0.47	0.99	0.95	1.00	0.96
aElNet-C*	0.97	0.53	1.00	0.98	1.00	0.99
aElNet-C**	0.93	0.47	0.99	0.95	1.00	0.96

Table 4. Percentage of correctly selected predictor sets, high correlation between regressors

	n = 40 $\delta = 0.5$	n = 40 $\delta = 2$	n = 120 $\delta = 0.5$	n = 120 $\delta = 2$	n = 500 $\delta = 0.5$	n = 500 $\delta = 2$
LASSO	0.70	0.18	0.99	0.32	1.00	0.79
LASSO-C*	0.76	0.18	0.99	0.40	1.00	0.84
LASSO-C**	0.70	0.17	0.99	0.32	1.00	0.80
aLASSO	0.70	0.27	0.96	0.43	0.97	0.76
aLASSO-C*	0.80	0.23	0.99	0.48	1.00	0.84
aLASSO-C**	0.70	0.25	0.96	0.42	0.97	0.76
ElNet	0.74	0.28	0.99	0.44	1.00	0.83
ElNet-C*	0.80	0.22	0.99	0.46	1.00	0.89
ElNet-C**	0.74	0.27	0.99	0.46	1.00	0.83
aElNet	0.81	0.31	0.98	0.50	0.99	0.85
aElNet-C*	0.86	0.25	1.00	0.52	1.00	0.90
aElNet-C**	0.82	0.29	0.98	0.50	0.99	0.86

Table 5 shows the out-of-sample RMSE estimates resulting from all methodologies in a setting when there is no correlation between regressors. An asterisk indicates those cases where the distribution of RMSEs significantly differs from its unconstrained counterpart (at 1% significance level). OLS model RMSEs are included as a benchmark here, where irrelevant regressors are included by design, which impinges on econometric efficiency, while being free of any source of potential bias.

Table 5. Median RMSEs, all methods, different noise levels and sample size when there is no correlation between regressors

	n = 40 δ = 0.5	n = 40 δ = 2	n = 120 δ = 0.5	n = 120 δ = 2	n = 500 δ = 0.5	n = 500 δ = 2
OLS	0.57	2.28	0.52	2.08	0.50	2.02
LASSO	0.76	2.54	0.75	2.28	0.75	2.09
LASSO-C*	0.73*	2.37*	0.74*	2.16*	0.75*	2.07*
LASSO-C**	0.76	2.53	0.75	2.26*	0.75	2.09
aLASSO	0.85	2.47	0.90	2.29	0.78	2.24
aLASSO-C*	0.57*	2.28*	0.52*	2.09*	0.50*	2.02*
aLASSO-C**	0.79	2.44*	0.88	2.25*	0.79	2.23
ElNet	1.00	2.68	1.01	2.41	1.24	2.29
ElNet-C*	0.97*	2.49*	0.98*	2.30*	1.18*	2.26*
ElNet-C**	1.00	2.68	1.01	2.39*	1.24	2.29
aElNet	0.99	2.65	0.94	2.38	0.92	2.22
aElNet-C*	0.86*	2.47*	0.81*	2.22*	0.79*	2.13*
aElNet-C**	0.97*	2.62*	0.94	2.36*	0.92	2.22

For all methods under scrutiny, adding the individual sign constraints decreases the RMSE and therefore enhances prediction accuracy, that is, they help the Oracle property in its second dimension. This is most notable in case of aLasso whose RMSEs perform as well as or better than from OLS. The addition of joint constraints to selected methods only occasionally improves their performance in terms of their out-of-sample RMSEs. When considering no correlation of regressors, this is the case in small samples, combined with high noise. However, the improvement is not as substantial as it is with the individual constraints. In the sample with correlated regressors (Table 6), the evidence is mixed: joint constraints lead to slight improvements in samples with moderate noise.

Although some methods come close to the OLS performance when we observe their RMSE performance, most methods will expectedly perform worse than the OLS and still imply some bias by their design. To empirically make their best use, OLS estimates can be refitted based on the subset of variables chosen by each method—as is common practice by many practitioners in the field—to reduce the bias of those methods.

In conclusion, we show that imposing individual constraints on a set of model selection methods improves their Oracle property, meaning enhanced predictive accuracy and correct model identification in small data samples. The imposition of joint constraints does not necessarily do so, but even when they do not imply an improvement, they do not cause any deterioration in predictive performance either. Their most notable value

Table 6. Median RMSEs, all methods, different noise levels and sample sizes when there is high correlation between regressors

	n = 40 δ = 0.5	n = 40 δ = 2	n = 120 δ = 0.5	n = 120 δ = 2	n = 500 δ = 0.5	n = 500 δ = 2
OLS	0.57	2.28	0.52	2.08	0.50	2.02
LASSO	0.68	2.32	0.57	2.18	0.57	2.06
LASSO-C*	0.64*	2.25*	0.56	2.13*	0.57	2.03*
LASSO-C**	0.67	2.31	0.57	2.17*	0.57	2.05
aLASSO	0.74	2.33	0.70	2.21	0.69	2.12
aLASSO-C*	0.60*	2.25*	0.54*	2.11*	0.52*	2.03*
aLASSO-C**	0.69*	2.32	0.69	2.18*	0.70	2.11*
ElNet	0.83	2.42	0.73	2.27	0.69	2.12
ElNet-C*	0.83	2.35*	0.74	2.21*	0.69	2.11*
ElNet-C**	0.83	2.42	0.73	2.25*	0.69	2.12
aElNet	0.84	2.43	0.73	2.30	0.70	2.15
aElNet-C*	0.73*	2.37*	0.65*	2.20*	0.63*	2.07*
aElNet-C**	0.81*	2.42	0.73	2.27*	0.70	2.14*

remains to be seen in empirical applications, as incorporating prior knowledge in model identification should help align the model's structure with theory.

4 Empirical Application

To examine the performance of the sign constrained (a) Lasso and (b) Elastic Net, we employ them for an empirical application in the field of financial stress testing. Financial sector stress testing is an important tool for assessing the resilience of the financial system and for gauging risks arising at system-wide level from a macroprudential perspective. The financial crisis and its aftermath led to a greater use of financial stress tests, including with a view to informing the timing and calibration of macroprudential policies. Stress testing involves selecting the models that are used to establish a link between bank risk parameters with the macro and financial factors defined in a scenario, to use such econometric models to project the risk parameters' evolution conditional on the scenario into the future. Useful entry point to stress test methodologies used at major central banks includes [2].

The dependent variable that we consider is a probability of default (PD) for non-financial listed corporates, sourced from Moody's KMV, with a quarterly frequency spanning the 2002Q1-2019Q4 period and comprising 16 countries for which the PDs are aggregated using firm assets as weights: Austria, Belgium, Czech Republic, Finland, France, Germany, Greece, Ireland, Italy, Luxembourg, the Netherlands, Portugal, Spain, Sweden, Great Britain, and the United States. Models for PDs, next to a loss given default (LGD) component (see [2], Chapter 4) as the second element to complete a credit risk assessment, play an important role in the overall stress test model suites, because loan losses constitute a major component of banks' profit and loss flows, next to interest income and expenses. It is, therefore, important that the credit risk models are developed in a robust manner, to ensure that they provide precise estimates for scenario-conditional PD paths.

Since for the application that follows we operate in a time series context, we define the notion of a long-run multiplier (LRM) for predictor X^k explicitly, on which we will impose the sign constraints:

$$\sum_{l=0}^{\infty} \frac{\delta E(Y_{t+1})}{\delta X_t^k} = \frac{\widehat{\beta_0^k} + \cdots + \widehat{\beta_q^k}}{1 - \rho_1 - \cdots - \rho_p} \equiv \theta^k \tag{12}$$

Table 7 lists the ten potential predictor variables that we consider, along with the LRM sign constraints that will be imposed.

Table 7. Predictor variables and LRM sign constraints

Acronym	Variable name	LRM sign constraint
RGRP	Real GDP growth	−
ITR	Real investment growth	−
CAPUTIL	Capacity utilization rate	−
URX	Unemployment rate	+
CREDIT	Nonfinancial private sector credit growth	−
TS	Term spread	+
RS	Risk spread	+
CPI	Consumer price inflation	+
EER	Effective exchange rate growth	+
OIL	Oil price	+

We consider the time contemporaneous and lagged inclusion of these ten variables. The empirical environment represents a case of non-negligible correlation among the potential predictors, including through the allowance for lagged terms for variables that are relatively persistent, such as price inflation and capacity utilization. To address the presence of such correlations, we set alpha, the Elastic Net parameter, to 0.5. Two autoregressive lags of the dependent variables are considered as well. They are included in the set of potential predictors, resulting in a total of $2 \times 10 + 2 = 22$ effective right hand-side variables, not counting the intercept. We use 5-fold cross validation to determine the optimal shrinkage parameter for each method.

Regarding the LRM sign constraints (Table 7): Stronger GDP growth, real investment, capacity utilization, inflation, and credit growth shall all come along with lower PDs. The opposite is true for the unemployment rate. Term and risk spreads gradually fall during booms and widen during ensuing recessions, hence implying the positive sign we impose. As all the countries included in the sample are net oil importers, they will all be negatively affected by an increase in oil price, thus raising firms' PDs. The sign of the effective exchange rate growth variable will depend on whether the country is net importer or net exporter, which can be time varying. When a country is a net importer, exchange rate appreciation will cause their exports to be more expensive, dampening a part of external demand and leading to higher PDs, while the net importers benefit from appreciation through stronger purchasing power abroad which implies lower PDs. We assign the positive constraint on exchange rate growth, but we prepare the data in such way to flip their sign in periods when the countries' net exports were negative, making a de facto negative constraint for these cases.

Imposing the constraints on LRMs means that we are imposing joint sign constraints on contemporaneous and lagged coefficients of a given predictor variable. The extensive simulations, reported on in the previous section, have shown that individual constraints improve the Oracle property of our methods in its two dimensions, while the imposition of joint constraints does not always lead to such improvements. Keeping in mind that the double constrained methods still perform at least as well as unconstrained, in our empirical exercise we focus on their value in choosing meaningful economic models.

To judge the materiality of obtaining wrongly signed coefficients, LRMs respectively, we examine the coefficients' signs in models selected by the unconstrained (a)Lasso and (a)Elastic Net variants. We report the portion of variables with incorrect signs in the cross-section of countries in Table 8. The percentages reflect the share of LRMs with incorrect signs (sign constraints in Table 7) on average across countries. The OLS column is included as a benchmark. The average across all variables and the four methods (excluding OLS) amounts to 39.8%. Except for the risk spread, this percentage is notable for all variables and all methods. Some variables, such as real GDP growth, would enter the models with an incorrect sign up to 75% of the times (with aLasso).

The sizable occurrence of theory non-conform signs is problematic and limiting the use of unconstrained model selection methods for economic forecasting, including stress testing (conditional forecasting). The imposition of sign constraints addresses this challenge, by allowing for incorporating prior theoretical and expert-based economic reasoning. Importantly, model and estimation uncertainty are often aggravated for stress testers, when operating in weak data environments (short time series, noisy/imperfect data, etc.). Hence, the imposition of prior assumptions on the sign of relationships should be beneficial from that perspective.

Table 8. Percentage of incorrect signs on LRMs when employing unconstrained model selection methods

	OLS	Lasso	aLasso	ElNet	aElNet
RGRP	68.8%	25.0%	68.8%	62.5%	56.3%
ITR	18.8%	37.5%	18.8%	43.8%	31.3%
URX	43.8%	0.0%	25.0%	6.3%	25.0%
CAPUTIL	31.3%	50.0%	31.3%	56.3%	31.3%
CREDIT	56.3%	56.3%	87.5%	43.8%	68.8%
TS	56.3%	68.8%	62.5%	56.3%	68.8%
RS	6.3%	6.3%	12.5%	6.3%	0.0%
CPI	31.3%	25.0%	25.0%	37.5%	25.0%
EER	62.5%	43.8%	56.3%	50.0%	75.0%
OIL	12.5%	12.5%	31.3%	31.3%	37.5%

5 Conclusions

The objective of this paper was to promote the idea of imposing sign constraints on individual model coefficients or linear combinations thereof when employing otherwise conventional (adaptive) Lasso and (adaptive) Elastic-Net model selection methodologies. The purpose of doing so lies in enhancing the model identification and predictive accuracy in small data samples, which are possibly inflicted with noise.

Our Monte Carlo simulations suggest that this a valuable strategy: the addition of sign constraints "helps the Oracle property," that is, the ability to identify the true model, for those methods that do not initially carry it, such as Lasso, in finite data samples. For methods that possess it already (the adaptive variants), the constraints help increase efficiency in small samples, conditional in all cases on the constraints being correct.

We examine the use of inequality constraints on the signs of both individual coefficients and joint coefficients. The latter can be useful in empirical applications, which we illustrate with a time series application where the joint set pertains to the long-run multipliers of the respective predictor variables. The empirical analysis entailed the use of probability of default metrics, which are one central element in large-scale bank stress test model suites, and which in practical applications is often challenged by short, noisy data. Having model selection methods that allow pre-informing the structure and estimates of the equations shall therefore be instrumental to obtain as robust models as feasible.

Acknowledgements. This paper has benefited from valuable comments received by the participants of an IMF-internal seminar in summer 2019.

References

1. Breiman, L.: Better subset regression using the nonnegative garrote. Technometrics **37**(4), 373–384 (1995)
2. Dees, S., Henry, J., Martin, R. (eds.): STAMP€: Stress-Test Analytics for Macroprudential Purposes in the Euro Area. European Central Bank, e-book (2017)
3. Derksen, S., Keselman, H.J.: Backward, forward and stepwise automated subset selection algorithms: frequency of obtaining authentic and noise variables. Br. J. Math. Stat. Psychol. **45**(2), 265–282 (1992)
4. Efron, B., Hastie, T., Johnstone, I., Tibshirani, R.: Least angle regression. Ann. Stat. **32**(2), 407–99 (2004)
5. Fan, J., Li, R.: Variable selection via nonconcave penalized likelihood and its oracle properties. J. Am. Stat. Assoc. **96**(456), 1348–1360 (2001)
6. Fan, J., Peng, H.: Nonconcave penalized likelihood with a diverging number of parameters. Ann. Stat. **32**(3), 928–961 (2004)
7. Forsyth, D.: Applied Machine Learning. Springer, Cham (2019). https://doi.org/10.1007/978-3-030-18114-7
8. Gains, B.R., Kim, J., Zhou, H.: Algorithms for fitting the constrained Lasso. J. Comput. Graph. Stat. **27**(4), 861–71 (2016)
9. Ghosh, S.: Adaptive elastic net: an improvement of elastic net to achieve oracle properties. Unpublished Manuscript (2007)
10. Gross, M., Población, J.: Implications of model uncertainty for bank stress testing. J. Financ. Serv. Res. **55**, 31–58 (2017)
11. Hoerl, A.E.: Application of ridge analysis to regression problems. Chem. Eng. Prog. **58**, 54–59 (1962)
12. Hoerl, A.E., Kennard, R.W.: Ridge regression: biased estimation for nonorthogonal problems. Technometrics **12**(1), 55–67 (1970)
13. Hu, Q., Zeng, P., Lin, L.: The dual and degrees of freedom of linearly constrained generalized Lasso. Comput. Stat. Data Anal. **86**, 13–26 (2015)
14. Hurvich, C.M., Tsai, C.-L.: The impact of model selection on inference in linear regression. Am. Stat. **44**(3), 214–217 (1990)
15. James, G.M., Paulson, C., Rusmevichientong, P.: Penalized and constrained optimization: an application to high-dimensional website advertising. J. Am. Stat. Assoc. **115**(529), 107–22 (2020)
16. James, G., Witten, D., Hastie, T., Tibshirani, R.: An Introduction to Statistical Learning – With Applications in R, 7th edn. Springer, New York (2017). https://doi.org/10.1007/978-1-4614-7138-7
17. Kolb, S.M.: Learning constraints and optimization criteria. In: Workshops at the Thirtieth AAAI Conference on Artificial Intelligence (2016)
18. Meinshausen, N.: Sign-constrained least squares estimation for high-dimensional regression. Electron. J. Stat. **7**, 1607–1631 (2013)
19. Meinshausen, N., Buehlmann, P.: High-dimensional graphs and variable selection with the Lasso. Ann. Stat. **34**(3), 1436–1462 (2006)
20. Murphy, K.P.: Machine Learning: A Probabilistic Perspective. Adaptive Computation and Machine Learning Series. MIT Press, Cambridge (2012)
21. Roecker, E.B.: Prediction error and its estimation for subset-selected models. Technometrics **33**(4), 459–469 (1991)
22. Tibshirani, R.: Regression shrinkage and selection via the Lasso. J. Roy. Stat. Soc. Ser. B **73**, 273–282 (1996)

23. Tibshirani, R., Suo, X.: An ordered lasso and sparse time-lagged regression. Technometrics **58**(4), 415–423 (2016)
24. Wang, H., Leng, C.: A note on adaptive group Lasso. Comput. Stat. Data Anal. **52**(12), 5277–5286 (2008)
25. Wu, X., Liang, R., Yang, H.: Penalized and constrained LAD estimation in fixed and high dimension. In: Statistical Papers (2021)
26. Xiao, N., Xu, Q.-S.: Multi-step adaptive elastic-net: reducing false positives in high-dimensional variable selection. J. Stat. Comput. Simul. **85**(18), 3755–3765 (2015)
27. Yuan, M., Lin, Y.: Model selection and estimation in regression with grouped variables. J. R. Stat. Soc. Ser. B Stat. Methodol. **68**(1), 49–67 (2006)
28. Zhou, H., Lange, K.: A path algorithm for constrained estimation. J. Comput. Graph. Stat. **22**(2), 261–283 (2013)
29. Zou, H.: The adaptive Lasso and its oracle properties. J. Am. Stat. Assoc. **101**(476), 1418–1429 (2006)
30. Zou, H., Hastie, T.: Regularization and variable selection via the elastic net. J. R. Stat. Soc. Ser. B Stat Methodol. **67**(2), 301–320 (2005)
31. Zou, H., Zhang, H.H.: On the adaptive elastic-net with a diverging number of parameters. Ann. Stat. **37**(4), 1733–1751 (2009)

Fetal Heart Rate Classification with Convolutional Neural Networks and the Effect of Gap Imputation on Their Performance

Daniel Asfaw[1,2](✉), Ivan Jordanov[1], Lawrence Impey[2], Ana Namburete[3], Raymond Lee[1], and Antoniya Georgieva[2]

[1] University of Portsmouth, Portsmouth PO1 2UP, UK
daniel.asfaw@port.ac.uk
[2] Nuffield Department of Women's & Reproductive Health, John Radcliffe Hospital OX3 9DU, University of Oxford, Oxford, UK
[3] University of Oxford, Oxford OX1 3QG, UK

Abstract. Cardiotocography (CTG) is widely used to monitor fetal heart rate (FHR) during labor and assess the wellbeing of the baby. Visual interpretation of the CTG signals is challenging and computer-based methods have been developed to detect abnormal CTG patterns. More recently, data-driven approaches using deep learning methods have shown promising performance in CTG classification. However, gaps that occur due to signal noise and loss severely affect both visual and automated CTG interpretations, resulting in missed opportunities to prevent harm as well as leading to unnecessary interventions. This study utilises routinely collected CTGs from 51,449 births at term to investigate the performance of time series gap imputation techniques (GIT) when applied to FHR: Linear interpolation; Gaussian processes; and Autoregressive modelling. The implemented GITs are compared by studying their impact on the convolutional neural network (CNN) classification accuracy, as well as on their ability to correctly recover artificially introduced gaps.

The Autoregressive model has been shown to be more reliable in the classification and recovery of artificial gaps when compared to the Linear and Gaussian interpolation. However, the improvement in the classification accuracy is relatively modest and does not reach statistical significance. The median (interquartile range) of sensitivity at 0.95 specificity is 0.17 (0.14,0.18) and 0.16 (0.13, 0.17) for the Autoregressive model and the zero imputations (baseline method) respectively (Mann-Whitney U = 69, P = 0.16). Future work include investigation and evaluation of other gap imputation methods to improve the classification performance of CNN on larger dataset.

Keywords: CTG · FHR · Gap imputation · Deep learning · CNN

G. Nicosia et al. (Eds.): LOD 2022, LNCS 13810, pp. 459–469, 2023.
https://doi.org/10.1007/978-3-031-25599-1_34

1 Introduction

Continuous electronic fetal monitoring by Cardiotocography (CTG) is commonly performed during or preceding labor to assess the wellbeing of the fetus. CTG monitors fetal heart rate (FHR) and uterine contraction (UC) signals simultaneously. Abnormal CTG patterns may indicate a fetus at risk of or already injured due to lack of oxygen in utero. In a clinical setting, the abnormal CTG is defined based on the FHR baseline, presence of accelerations and decelerations in the signal, short- and long-term variability. However, visual interpretation of the FHR is problematic even for experienced clinicians [1], leading to an increase in unnecessary interventions and obstetrical litigation costs [2]. Existing automated CTG analysis methods are also unreliable with low sensitivity and specificity. One of the challenges of both visual or automated CTG interpretation is the poor quality of FHR signals produced, resulting in both short (few seconds) and long (many minutes) gaps in the signal [3].

Most of the noise and gaps in the FHR during labor are not isolated, and are due to many factors: poor sensor placement, movement of mother and fetus, common and sudden changes of the signal during decelerations and contractions, known technology limitations such as erroneous doubling/halving of values or FHR measures being masked by maternal heart rate [4,5]. Reliable gap imputation is important in order to improve the performance of downstream data analysis tasks, such as pre-processing and signal classification. Several methods have been proposed to impute the gaps in the FHR signal, such as linear interpolation [2], cubic spline interpolation [7], sparse representation with dictionaries [6], and Gaussian processes (GP) [8,9]. However, the performance of the gap imputation techniques has been evaluated only in small datasets and their effect on CTG classification task is unknown.

Generally, efficient gap imputation methods make the most of the available data while minimizing the loss in statistical terms and the bias inevitably brought by the inferred values of missing data. The techniques are usually divided into univariate and multivariate imputations, and their applicability and efficiency depend on the type of missingness (which is also categorized into three groups: missing completely at random; missing at random; and missing not at random) [12]. Commonly, gap imputation techniques are compared based on their ability to use few available data to impute gaps, especially long gaps, accurately and efficiently [13].

This work investigates performance of different FHR gap imputation techniques on two distinct tasks. First, ability of gaps imputation methods on recovering artificially introduced gaps. This is assessed by introducing artificial gaps with varying gap lengths at various locations in the signal. Second, the impact of gap imputation methods on the performance of a convolutional neural networks (CNNs), which is widely applied in classification of CTGs recently [11,15]. The CNN is trained to classify births with and without severe compromise on a large dataset (n = 51,449).

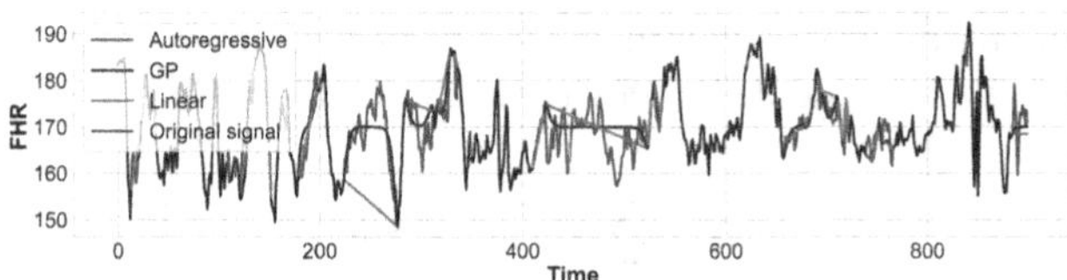

Fig. 1. Gap imputation of a sample FHR signal using Autoregressive, GP, and, Linear interpolation methods.

2 Dataset and Pre-processing

2.1 Dataset

The dataset was routinely collected as per standard clinical care in term births between 1993 and 2012 at Oxford (n= 51,449). The clinical protocol was to administer CTG during labour only to pregnancies deemed at 'high-risk'. We include in the study: births at gestation $\geq$ 36 weeks, with CTGs longer than 15 min, and ending within three hours of birth. Of the included, 452 are births with severe compromise - a composite outcome of intrapartum stillbirth, neonatal death, neonatal encephalopathy, seizures, and resuscitation followed by over 48 h in the neonatal intensive care unit. The rest of the cohort samples are labelled as normal. Here, we analyse the first 20 min CTG recordings (sample shown in Fig. 1a), under some circumstances called 'admission CTG' [16].

2.2 Noise Removal

The CTG datasets were collected as per standard fetal monitors, 4 Hz sampling rate for the FHR 2 Hz for the uterine contraction signal. Consistently with the works of [15], the FHR signals are downsampled to 0.25 Hz. A bespoke algorithm is applied to remove artefacts from the CTG signal, for example, erroneous maternal heart rate captureand and extreme outliers (FHR measurements $\geq$ 230 bpm or $\leq$ 50 bpm). The start time of the signal is adjusted to avoid significant gaps around the beginning and only those signals with signal loss less than 50% are included.

2.3 Gap Imputation

We investigate three time series gap imputation methods: Linear interpolation, Gaussian process (GP), and Autoregressive (AR) interpolation (Fig. 1b). Linear interpolation is performed by fitting a Linear regression on the available data points, and the fitted line is used to estimate the values of the missing points in the gaps [18].

The AR model is a multiple linear regression modelling technique that predicts the signal at a given time step n based on the signal values from the previous sample steps. The number of prior samples used for the prediction determines the order of the model (r), shown in (1). The optimal order of the autoregressive model is selected using Akaike's information criterion [19].

$$y_n = \sum_{k=0}^{r} \beta_k y_{n-r} + u_t \tag{1}$$

where u_t denotes additive white noise. GP is a collection of random variables, any finite number of which have a joint multivariate Gaussian distribution [20]. We use a combination of Matern and Exponential kernels to capture both rapidly and slowly changing components in FHR signals [20], given in (2). Additive Gaussian noise is also added to the kernels to account for the possible measurement error of the sensor

$$K = \sigma_1^2(1 + \frac{\sqrt{3}r}{\lambda_1})e^{-\frac{\sqrt{3}r}{\lambda_1}} + \sigma_2^2 e^{-\frac{r^2}{\lambda_2^2}} + \sigma_y^2 I \tag{2}$$

where $r = \sqrt{\sum_{i=1}^{n}(x_i - x_i)}$, λ_1 and σ_1^2, λ_2 and σ_2^2 are the length-scale and variance hyperparameters for the $matern_{\frac{3}{2}}$ and radial basis function (RBF) covariance functions, respectively. The GP model is implemented in Python, using the GPflow package [17].

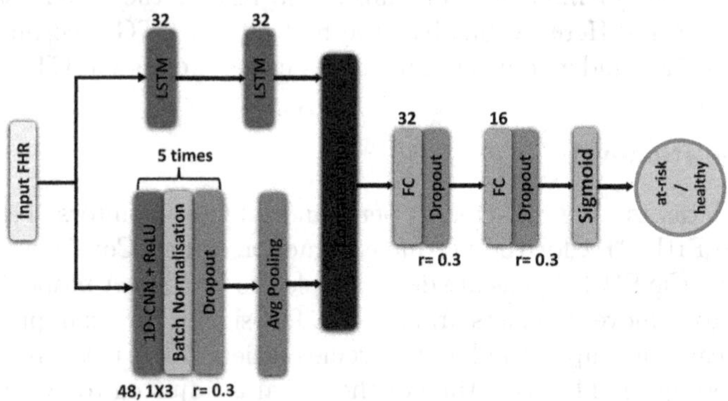

Fig. 2. Architecture of CNN-LSTM parallel network

3 Methods

3.1 Artificial Introduced Gaps

Artificial gaps are introduced in selected 20-minute FHR signals of good quality (0% signal loss). In total, 15,700 clean 20 min segments are extracted from the Oxford datasets. Different gap lengths are considered following the gap statistics of the dataset. The median (interquartile range [IQR]) gap length in the Oxford dataset's first 20-minute FHR signal is 8 (8,12) seconds, corresponding to gap lengths of 2 (2,3) in the dataset (FHR is sampled at 0.25Hz), while the median

(IQR) maximum gap length is 36 (24, 60) seconds, corresponding to gap lengths of 9 (6, 15) in our FHR dataset. Therefore, we considered gap lengths of 2, 6, 9 and 15 for the experiment to simulate short, moderate and long gaps. These gaps were introduced at three locations (start, middle, and, end) to investigate the effect of both gap length and location on the performance of gap imputation techniques.

3.2 Deep Learning Model Training Evaluation

Since the dataset (training data) has significantly fewer samples with severe compromise (n = 384) than without severe compromise (n = 43,296), we augment more positive samples to mitigate overfitting [21]. We extract additional 20-minute FHR segments from the first 1-hour FHR data with 50% overlap, increasing the number of severe samples by a factor of 4, and we further oversampled severe samples by a factor of 2, which led to an overall 8-fold increase of the training dataset severe samples.

Deep Learning Architecture. The proposed architecture constitutes a 1D-CNN and long-short-term memory (LSTM) cells: a five-layer 1D-CNN and two-layer LSTM networks, connected in parallel, followed by two fully-connected (FC) layers. Batch normalization and drop out layers are also applied at the end of each convolutional and FC layers. The topology of the model is shown in Fig. 2. The input data is prepared as (B, T, F), where B, T and F represent the batch size, length of times slices, and signal dimension, respectively. Each FHR sample has 300 data points (T = 300, F =1) since it is a 20-minute long and is sampled at 0.25Hz. A ReLU and a sigmoid activation functions are used at the end of each convolutional layer and output layer respectively.

Training Procedure. We split the dataset randomly into training (85%) and test (15%) sets while maintaining the class ratio in each subset. To mitigate overfitting, 10-fold cross-validation is adopted,where during each fold, the model is trained for a maximum of 400 epochs with early stopping (with a window size of 50 epochs). After the optimal model parameters are obtained, the model is evaluated on the test set. We report the average performance of the ten models on the test set. The binary cross-entropy loss (reweighted to account for the class frequencies) is used as a training objective. The network is trained for 400 epochs, using Adam optimiser with an initial learning rate of 0.001 and decayed by a factor of 2 every 50 epochs, and batch size of 512. Early stopping based on the partial area under the receiver operating characteristic (ROC) curve (PAUC) is used to train the model. The model is implemented using TensorFlow on an NVIDIA GTX 2080 Ti 12GB GPU machine.

Performance Metrics. The ability of gap imputation methods to recover artificially missing data is measured using three matrices: the root mean-squared error (RMSE) (3), mean absolute percentage error (MAPE) (4), and Pearson correlation coefficient (PCC) (5).

$$RMSE = \sqrt{\frac{1}{N}\sum_{i=1}^{N}(x_i - y_i)^2} \tag{3}$$

$$MAPE = \frac{1}{N}\sum_{i=1}^{N} \mid \frac{x_i - y_i}{x_i} \mid \times 100\% \tag{4}$$

$$PCC = \frac{\sum_{i=1}^{N}(x_i - \bar{x}_i)(y_i - \bar{y}_i)}{\sum_{i=1}^{N}(x_i - \bar{x}_i)^2 \sum_{i=1}^{N}(y_i - \bar{y}_i)^2} \times 100\% \tag{5}$$

where x_i is missing observation y_i is the imputed value at the location of the missing value, $\bar{x}$ is the overall average of the actual observed values, and $\bar{y}$ is the overall average of the imputed values at the location of a missing values. The classification performance of the model is evaluated using the PAUC and the true positive rate (TPR) at a 5% false-positive rate (FPR). The AUC measures the area underneath the entire ROC curve from (0, 0) to (1, 1). The PAUC is the AUC between 0 and 10% false-positive rates (FPR).

Table 1. Results of statistical comparison of the gap imputation methods.

Method	Gap length	RMSE (BPM)				MAPE (%)				PCC			
		Gap length				Gap length				Gap length			
		2	6	9	15	2	6	9	15	2	6	9	15
Linear	Start	5.15	11.34	15.31	23.62	3.79	8.33	14.87	33.83	−0.08	-0.04	−0.05	−0.04
	Middle	2.12	3.14	3.60	4.27	1.42	1.95	2.19	2.55	0.38	0.32	0.31	0.29
	End	5.20	11.28	15.32	23.89	3.63	14.38	18.38	38.35	−0.10	−0.05	−0.05	−0.05
GP	Start	3.88	5.91	7.25	9.86	2.61	3.75	4.54	6.10	−0.01	0.0	0.0	−0.01
	Middle	2.35	3.38	3.91	4.76	1.59	2.11	2.42	2.90	0.27	0.26	0.25	0.23
	End	3.68	5.81	7.07	9.45	2.50	3.72	4.47	5.89	−0.03	−0.01	-0.02	−0.01
Autoregressive	Start	4.44	5.91	6.40	6.97	3.00	3.72	3.97	4.23	0.08	0.07	0.06	0.07
	Middle	3.01	4.25	4.76	5.41	2.04	2.66	2.92	3.25	0.23	0.19	0.18	0.17
	End	4.25	5.75	6.22	6.82	2.86	3.63	3.85	4.14	0.07	0.07	0.07	0.07

4 Results

Table 1 shows the performance of the three gap imputations in recovering the artificially introduced gaps within the clean signals extracted from the dataset. The performance of AR methods is consistent, achieving the smallest RMSE and MAPE and highers PCC values across all gap locations and sizes. The performances of the Linear and GP interpolation techniques are comparable despite the Linear interpolation achieving superior performance in terms of PCC values. The results also show that when the gap length is small (regardless of the location of the gap), the three methods perform comparably. On the other hand, on moderate and long gap lengths the AR interpolation outperformed the other two methods.

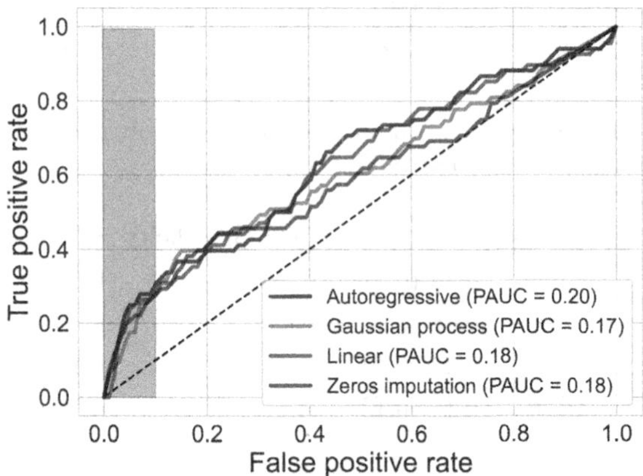

Fig. 3. ROC of the best performing models from the 10-fold cross-validation evaluated on the Oxford dataset's test set (68 severe compromises and 7703 without severe compromise).

Table 2. Effect of the maximum gap length on the CNN classifier.

Max gap length (seconds)	Total cases (severe)	Sensitivity at 0.95 Specificity		*PAUC*	
		AR	Zeros imputation	AR	Zeros imputation
<20	2506 (21)	0.20	0.19	0.19	0.19
20 – 48	2845 (25)	0.12	0.04	0.10	0.04
48	2420 (22)	0.41	0.27	0.32	0.23
All	7771 (68)	0.25	0.20	0.20	0.17

Effect of Imputation on CNN Classification. Table 2 shows the performance of the three interpolation methods on the classification task. The AR is best compared to the other two methods and the zeros imputation. in terms of PAUC and sensitivity values. Contrary to our expectation, the overall performance of both Linear and GP interpolation methods is comparable to the zero-imputation. The AR outperforms the other techniques and produces a decent separation between the two classes, especially at higher specificity values (bottom left corner of the plot, the grey shaded area in Fig. 3). However, the performance differences are modest. For example, based on the sensitivity at 0.95 specificity,

Table 3. Effect of the percentage signal loss on the CNN classifier.

Signal loss (%)	Total cases (severe)	Sensitivity at 0.95 Specificity		PAUC	
		AR	Zeros imputation	*AR*	Zeros imputation
<5%	2495(22)	0.14	0.19	0.14	0.15
5 – 15%	2961(25)	0.29	0.05	0.24	0.11
15%	2315(21)	0.29	0.27	0.23	0.20
All	7771(68)	0.25	0.21	0.20	0.17

Table 4. Effect of the three gap imputation methods on performance of the CNN (median (IQR) of the 10-fold cross validation).

Imputation method	Sensitivity at 0.95 Specificity	PAUC
Linear	0.13 (0.11, 0.16)	0.15 (0.13, 0.16)
GP	0.16 (0.13, 0.17)	0.15 (0.12, 0.16)
AR	0.17 (0.14, 0.18)	0.17 (0.15, 0.18)
Zeros imputation	0.16 (0.13, 0.17)	0.16 (0.14, 0.17)

values from 10 fold cross-validation, Mann-Whitney U tests (two tailed) shows no statistically significant difference between AR and zero-imputation U = 69.0, P = 0.16.

Effect of Gap Length and Signal Loss on CNN Classification. We perform post hoc analysis to evaluate the effect of percentage signal loss (proportion of the total gaps to the length of the signal), size of the gaps, and gap location on classification performance. Signal loss and gap lengths are computed from the initial pre-processed dataset (before gaps are imputed). Since small gaps are relatively easier to fill in and most common in the FHR signal, we look at the effect of the maximum gap lengths (the longest gaps that occur in the FHR) using the best performing AR and zeros imputation models from the 10-fold cross-validation. The results show that the CNN (when imputed using AR or zeros) performed better when the maximum gap length is higher (Table 3). For example, the sensitivity at 0.95 specificity is higher when the maximum gap length is greater than 48 s compared to when it is 20-48 s (0.41 vs 0.12 and 0.27 vs 0.04, when the gaps are imputed using AR and with zeros, respectively).

We further investigate whether the location of the maximum gaps in a signal affects the performance of the CNN. The results shows that when the maximum gaps are located at the very start, both the AR and zero imputation models struggle to separate between the two classes (sensitivity at 0.95 specificity is about 0.05). However this has less likely to do with the gap imputation as qualitative assessment of the imputed gaps shows the AR imputed the gaps decently regardless of the location as shown in Fig. 1.

In terms of the percentage signal loss, the performance of the AR model is less dependent on the percentage signal loss (Table 3). In contrast, the CNN trained with zero imputations perform well when the percentage of signal loss is higher. For example, the sensitivity at 0.95 specificity is higher when the percentage signal loss is greater than 15% compared to when it is 5-15% (0.27 vs 0.05).

5 Discussion

This work evaluates the gap imputation performance of Linear, Gaussian process (GP), and Autoregressive (AR) methods. These are evaluated on two separate tasks: the ability to recover artificially introduced gaps accurately and improve the performance of a CNN classifier.The comparison is performed on the initial 20 min FHR traces of a dataset with over 51,000 births [2]. In the classification task, the performance of the Linear and GP methods is comparable to the baseline (zeros imputation). Post hoc analysis of the effect of gap length, percentage of signal loss, and location of the gaps on the deep learning classifier indicate that the performance of the AR model is less dependent on the percentage signal loss. Still, it achieved higher performance when the maximum gap length in the signal is larger than 48 s or when there is a clean signal around the start of the 20-minute recording. We also observed poor classification performance when the signal gaps are imputed with AR and gaps occur at the beginning. This is likely due to the nature of the signal than the limitations with the gap imputation method. The performance of the AR imputation is superior in both signal recovery and classification tasks compared to the Linear and GP techniques. This result is significantly different from the finding of Feng et al. [9], where they reported AR performance worse than both GP and Linear interpolation. However, they compared different gap imputation methods using a single CTG segment from the CTU-CBH dataset [22] with a gap length of 10 data points. Feng et al. [9], also reported that spline interpolation achieved better performance than Linear and autoregressive methods in their signal recovery task. However, our initial experiments show that the performance of spline interpolation is significantly worse than the other three gap imputation methods in both signal recovering and classification tasks.

To the best of our knowledge, this is the first work to evaluate the performance of FHR gap imputation methods on a large dataset (using over 50,000 births). The proposed deep learning approach achieved superior accuracy when implementing the AR gap imputation. There is further ongoing work to increase the overall sensitivity by analysing and incorporating the full FHR trace (outside the scope of this work).

Finally, further work is considered to determine whether the gap imputation methods improve the classification performance when incorporating into the classification model more information: longer FHR segments; uterine contraction signals; and clinical risk factors (such as maternal age, fetal gestation, maternal co-morbidities, known to have a significant impact on the pre-labour risk of adverse outcome). Although the proposed deep learning methods achieved better performance when the gaps are imputed using AR, the performance gain is modest. Therefore, future work is needed to investigate other possibilities (for example, using deep learning based gap imputation methods) to model the signal gaps in conjunction with deep learning architectures on a larger dataset.

6 Conclusion

We compared the performance of Linear, GP, AR and Zero-valued gap imputation techniques for fetal heart rate processing and classification using FHR tracings from the onset of labour. Focusing on the 'admission' or early 20-minute segment, we investigated the potential of providing very early warning and triage women into high or low risk groups for further monitoring and/or review.

Our results showed that AR achieved superior performance in recovering artificially induced gaps than Linear and GP. The AR imputation also improved the performance of the proposed CNN classifier compared to the other gap imputation techniques. The results are clinically encouraging, given that majority of compromised babies are not expected to have problems, and if any, they are challenging to detect at the onset of labour. It is important to note that there is room to improve the classification performance of the CNN by analysing and incorporating the full FHR trace.

Despite the fact that the AR achieved superior performance compared to linear and GP in recovering gaps, it only improved the performance CNN classifier marginally compared to the other imputation methods and zeros imputation (baseline). Therefore, future work is needed to explore other gap imputation methods, including deep learning based approaches, to increase the accuracy of the CNN classifier.

References

1. Farquhar, C., Armstrong, S., Masson, V., Thompson, J., Sadler, L.: Clinician identification of birth asphyxia using Intrapartum Cardiotocography among neonates with and without encephalopathy in New Zealand. JAMA Netw. Open **3**, e1921363 (2020)
2. Georgieva, A., Redman, C., Papageorghiou, A.: Computerized data-driven interpretation of the intrapartum cardiotocogram: a cohort study. Acta Obstet. Gynecol. Scand. **96**, 883–891 (2017)
3. Hamelmann, P., et al.: Doppler ultrasound technology for fetal heart rate monitoring: a review. IEEE Trans. Ultrason. Ferroelectr. Frequency Control. **67**, 226–238 (2020)
4. Bakker, P., Colenbrander, G., Verstraeten, A., Van Geijn, H.: The quality of intrapartum fetal heart rate monitoring. Europ. J. Obstetrics Gynecol. Reprod. Biol. **116**, 22–27 (2004)
5. Kiely, D., Oppenheimer, L., Dornan, J.: Unrecognized maternal heart rate artefact in cases of perinatal mortality reported to the United States Food and Drug Administration from 2009 to 2019: a critical patient safety issue. BMC Pregn. Childbirth **19**, 501 (2019)

6. Barzideh, F., et al.: Estimation of missing data in fetal heart rate signals using shift-invariant dictionary. In: 2018 26th European Signal Processing Conference (EUSIPCO), pp. 762–766 (2018)
7. Spilka, J., Georgoulas, G., Karvelis, P., Chudáček, V., Stylios, C., Lhotská, L.: Discriminating normal from "abnormal" pregnancy cases using an automated FHR evaluation method. In: Hellenic Conference On Artificial Intelligence, pp. 521–531 (2014)
8. Feng, G., Quirk, J., Djurić, P.: Recovery of missing samples in fetal heart rate recordings with Gaussian processes. In: 2017 25th European Signal Processing Conference (EUSIPCO), pp. 261–265 (2017)
9. Feng, G., Quirk, J., Heiselman, C., Djurić, P.: Estimation of consecutively missed samples in fetal heart rate recordings. 2020 28th European Signal Processing Conference (EUSIPCO), pp. 1080–1084 (2021)
10. Zhao, Z., Zhang, Y., Comert, Z., Deng, Y.: Computer-aided diagnosis system of fetal hypoxia incorporating recurrence plot with convolutional neural network. Front. Physiol. **10**, 255 (2019). https://www.frontiersin.org/article/10.3389/fphys.2019.00255/full
11. Baghel, N., Burget, R., Dutta, M.: 1D-FHRNet: automatic diagnosis of fetal acidosis from fetal heart rate signals. Biomed. Sig. Process. Control **71**, 102794 (2022)
12. Enders, C.: Applied missing data analysis
13. Beveridge, S.: Least squates estimation of missing values in time series. Commun. Statist. Theory Methods **21**, 3479–3496 (1992)
14. Blix, E.: The admission CTG: is there any evidence for still using the test? Acta Obstetricia Et Gynecologica Scandinavica **92**, 613–619 (2013)
15. Petrozziello, A., Redman, C., Papageorghiou, A., Jordanov, I., Georgieva, A.: Multimodal convolutional neural networks to detect fetal compromise during labor and delivery. IEEE Access **7**, 112026–112036 (2019)
16. Blix, E.: The admission CTG: is there any evidence for still using the test? Acta Obstetricia Et Gynecologica Scandinavica **92**, 613–619 (2013). http://doi.wiley.com/10.1111/aogs.12091
17. Matthews, A., et al.: GPflow: a Gaussian Process Library using TensorFlow. J. Mach. Learn. Res. **18**, 1–6 (2017)
18. Boyd, D.: Systems analysis and modeling: a macro-to-micro approach with multidisciplinary applications
19. Hurvich, C., Tsai, C.: A corrected akaike information criterion for vector autoregressive model selection. J. Time Ser. Anal. **14**, 271–279 (1993)
20. Rasmussen, C., Williams, C.: Gaussian processes for machine learning. MIT Press (2006). OCLC: ocm61285753
21. Chawla, N., Bowyer, K., Hall, L., Kegelmeyer, W.: SMOTE: synthetic minority over-sampling technique. J. Artif. Intell. Res. **16**, 321–357 (2002)
22. Chudáček, V., et al.: Open access intrapartum CTG database. BMC Pregn. Childbirth **14**, 16 (2014)

Parallel Bayesian Optimization of Agent-Based Transportation Simulation

Kiran Chhatre[1(✉)], Sidney Feygin[2], Colin Sheppard[1,2], and Rashid Waraich[1,2]

[1] Lawrence Berkeley National Laboratory, Berkeley, CA 94720, USA
kiranchhatre3@gmail.com, {colin.sheppard,rwaraich}@lbl.gov
[2] Marain Inc., Palo Alto, CA 94306, USA

Abstract. MATSim (Multi-Agent Transport Simulation Toolkit) is an open source large-scale agent-based transportation planning project applied to various areas like road transport, public transport, freight transport, regional evacuation, etc. BEAM (Behavior, Energy, Autonomy, and Mobility) framework extends MATSim to enable powerful and scalable analysis of urban transportation systems. The agents from the BEAM simulation exhibit 'mode choice' behavior based on multinomial logit model. In our study, we consider eight mode choices viz. bike, car, walk, ride hail, driving to transit, walking to transit, ride hail to transit, and ride hail pooling. The 'alternative specific constants' for each mode choice are critical hyperparameters in a configuration file related to a particular scenario under experimentation. We use the 'Urbansim-10k' BEAM scenario (with 10,000 population size) for all our experiments. Since these hyperparameters affect the simulation in complex ways, manual calibration methods are time consuming. We present a parallel Bayesian optimization method with early stopping rule to achieve fast convergence for the given multi-in-multi-out problem to its optimal configurations. Our model is based on an open source HpBandSter package. This approach combines hierarchy of several 1D Kernel Density Estimators (KDE) with a cheap evaluator (Hyperband, a single multidimensional KDE). Our model has also incorporated extrapolation based early stopping rule. With our model, we could achieve a 25% L1 norm for a large-scale BEAM simulation in fully autonomous manner. To the best of our knowledge, our work is the first of its kind applied to large-scale multi-agent transportation simulations. This work can be useful for surrogate modeling of scenarios with very large populations.

Keywords: Bayesian optimization · Multiagent simulations · Traffic dynamics

1 Introduction

In order to get from location 'A' to location 'B' we all choose a convenient mode of transportation. According to US Census Bureau [1] study in 2017, people from New York metropolitan area spend 35.9 mins whereas people from the San Francisco bay area spend 32.1 mins in a typical one-way commute. Despite all the work transportation engineers put into making those journeys go smoothly, things go wrong more often than

K. Chhatre—The work was performed as a Berkeley Lab affiliate.

G. Nicosia et al. (Eds.): LOD 2022, LNCS 13810, pp. 470–484, 2023.
https://doi.org/10.1007/978-3-031-25599-1_35

they should. While the weather plays an important role, one of the biggest reasons that transportation systems run into issues is that they have to deal with another difficult element, which is human behavior. In order to design a good transportation system, it is vital to understand the people who are going to use it.

Many people get around with vehicles like cars, ride-hail services, and bikes. According to Transport Statistics Great Britain study of 2017, 83% [3] of all journeys were taken by car, van or taxi. Potential collisions on roads aren't confined to known junctions, moreover all back and forth lane switching causes congestion. Therefore, highway engineers tackle problems like this by developing traffic simulators. The forecasting of passenger travel requires multiple key elements that comprises the traffic simulator model. These key elements can be the estimation of trip generation (number of purposeful trips), trip distribution (destination choice), transportation mode choice, route assignment, trip chaining (the decision to link individual trips together in a tour), and several other aspects of traveler decisions [2].

The forecasting accounts for the prediction of specific transportation facility people will use in the future [4]. This forecast is dependent on the study of how people use transport. The factors affecting people's behavior are broad and are based on activities and learning how people allocate their time during an average day. Forecasting begins with the collection of data on current traffic, population, employment, trip rates, travel costs [5]. Using this data a traffic demand model is developed, which returns estimate of the future traffic by feeding in predicted data like future population and employment. Such future traffic prediction is useful to calculate the capacity of infrastructure, to estimate financial and social feasibility of the projects, and calculate environmental impacts.

In our study, we propose algorithms for optimization of the "Mode Choice Analysis" in the traffic demand model. Mode choice analysis determines what mode of transport will be used and which modes of transport will be shared by the people.

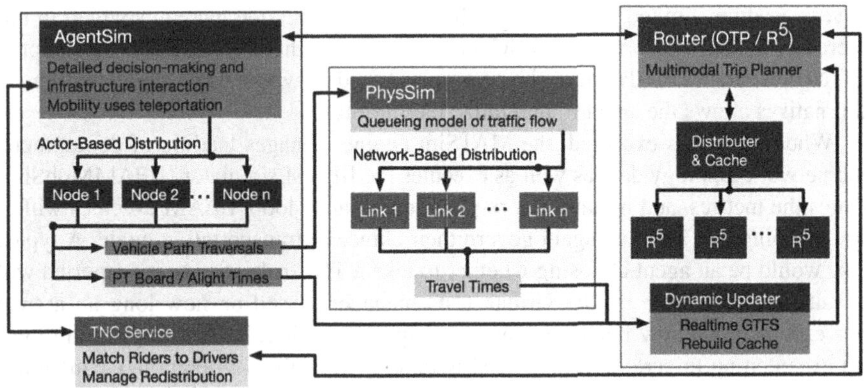

Fig. 1. BEAM infrastructure [6]

The large-scale transportation planning model of a particular whole city is executed through agent-based software modules. Complex systems have large number of interacting components that evolves over time. It's difficult to predict the emergent behavior of such system at a macro level. On the other hand, even if we possess the knowledge of the macro structure, it's difficult to find the microstructure that generates it. With the aid of computer-based modeling languages, we simulate complex patterns and understand more about how they arise in nature and society. In an agent-based model (ABM), the world is modeled using agents, the environment, and a description of agent-agent and agent-environment interactions, while an agent is an indivisible element with distinct properties and actions. Every agent is characterized as bounded rational and acts according to their perceived interests using decision-making rules in a predefined interaction topology. ABM abstracts the representation of a world that sometimes exaggerates certain aspects at the expense of others. However once the world is satisfactorily designed, the simulated experiments allows us to find out how agents interact with other agents that defines the patterns of behavior that are not defined at the level of any individual agent. With the help of ABM simulations we are able to explore how the individual's micro decisions lead to macro patterns of the world's behavior.

As shown in Fig. 1, BEAM [27] integrates MATSim [13] toolkit and therefore possess an agent-based modeling approach. Each agent in BEAM employ reinforcement learning across successive simulated days to maximize their personal utility through plan mutation (exploration) and selecting between previously executed plans (exploitation). One of the primary objectives of the BEAM framework is to model and find an equilibrium point for all limited resource markets like road capacity, vehicle seating, transportation network companies (TNC) fleet availability, and refueling infrastructure that have limited supplies. Once the resources are utilized by travelers, no other traveler can simultaneously use the same resource. TNCs are modeled as a fleet of taxis controlled by a centralized manager that responds to requests from customers and dispatches vehicles accordingly. The degree to which an agent uses a resource depends on the resource availability and the agent's behavior. In a particular case if supply of TNC drivers becomes limited, the wait time for hailing a TNC ride increases which in turn decreases the utility score of the TNC mode choice, and therefore reduced consumption of TNC resource. This dynamic choice process based on within day evaluation of modal alternatives allows the agent to maximize their utility.

When BEAM is executed, the MATSim engine manages loading of demographic and network topology data as well as executes the BEAM simulator (BEAMMobSim), scores the metrics, and initiates the re-planning iterative loop. BEAM executes within-day planning that helps an agent govern their choice of transportation mode. A typical case would be an agent choosing whether to take a TNC ride after being reported with a wait time for a ride by the virtual TNC manager. Based on how long an activity was executed and how much time an agent spent traveling, BEAM re-planning loops are executed up to certain user-defined iterations (typically 15 iterations). This utility maximization through the re-planning process, unlike dynamic choice process based on within days evaluation, occurs outside the simulation day. Eventually, running BEAM over successive iterations balances the trade-offs between all resources in the system.

BEAMMobSim is composed of the AgentSim and the PhySim. AgentSim executes the daily plan of the population by allowing the agents to dynamically resolve the limited transportation resources. Whereas PhySim is a vehicle movement simulator. BEAM uses the R5 [7] routing engine to accomplish multi-modal routing. In the simulator, PhySim and AgentSim run serially. The agents receive the routing calculation (route and travel time) from the R5 router and choose between alternative routes, and later this information is used by PhySim to simulate traffic flow, resolve congestion, and update travel times back in the router. In the subsequent BEAM iterations, the agents move (teleport) according to travel times that are consistent with previous iteration network congestion.

The Metropolitan Transportation Commission [8] (MTC) is the government agency responsible for regional transportation planning and financing in the San Francisco Bay Area. MTC coordinates transportation services in Bay area's Alameda, Contra Costa, Marin, Napa, San Francisco, San Mateo, Santa Clara, Solano, and Sonoma counties. Vital Signs [9] is a data-driven web product developed by MTC that compiles data from a variety of national, state and regional data sources and makes it available to the public, agency staff, and policymakers to understand and track progress on key regional issues.

San Francisco Municipal Transportation Agency (SFMTA) undertook a Mode Share Survey within the San Francisco Bay Area using a survey conducted as a telephone study among 841 Bay Area residents aged 18 and older between May and August 2019 [10]. Mode share percentages based off the total number of trips (n = 10,437) for all respondents were determined by collecting trip level information for all respondents. In order to align with the scope of our study, "urbanism-10k" scenario was used. The urbanism-10k scenario is based on the City of San Francisco, including the SF Municipal public transit service and a sample population of 10,000 agents.

The estimated mode share compiled by SFMTA and Vital Signs was adjusted to be suitable for the BEAM ubransim-10k mode choice model. The adjusted estimated mode share for the City of San Francisco (benchmark) is as Table 1.

The intercept value for each mode choice affects the BEAM simulation in complex ways. Each intercept vector consists of eight distinct values that allows the AgentSim to reproduce realistic mode split travel behavior that is consistent with the data compiled by Vital Signs web service. Nonetheless, evaluating the accurate intercept vector is challenging due to inadequate prior knowledge of a bounded region that restricts the search of the mode choice model extrema. This concludes that it is essential to modify the intercept values in such a way that the simulation output converges. This process of manipulating the intercepts is termed as the calibration of the mode choice model. In that respect, BISTRO [14] platform provides collaborative planning and evaluation of various optimization algorithms on agent-based modeling and simulation framework in response to different transportation policy strategies.

BEAM follows the MATSim convention for most of the input requirements to run a simulation and R5 convention for the road network and transit system inputs. Based on these external package requirements, BEAM requires population attributes, household population attributes, personal vehicle fleet, vehicle types for personal vehicles and the public transit fleet, R5 network and transit data, open street map network [11], transportation analysis zone data, hereafter called "TAZ" [2], and General Transit Feed Specification (GTFS) [12] data for each transit agency.

Table 1. Mode share for the City of San Francisco [10]

Mode	Benchmark (%)
Bike	2
Car	49
Drive Transit	4
Ride Hail	3
Ride Hail Pooled	2
Ride Hail Transit	1
Walk	22
Walk Transit	17

2 State of the Art

BEAM's mode choice model is based on a multinomial logit choice model (MNL) in which agents select modal strategies (e.g. "car" versus "walk to transit" versus "TNC") for each tour prior to the simulation day, but resolve the outcome of these strategies within the day (e.g. route selection, standard TNC versus pooled, etc.).

2.1 Multinomial Logit Model

In the conventional four-step transportation forecasting model, mode choice analysis comes the third after the trip generation step and the trip distribution step and is followed by the route assignment step. Trip distribution's zonal analysis yields a set of origin destination tables describing where the trips will be made. Mode choice analysis allows the agents in the forecasting model to choose the mode of transport. The multinomial logit model (MNL) forms the backbone of the BEAM's mode choice model. MNL proposes on making a choice of the heavier object from two objects with weights and greater the difference in weight, greater the probability of choosing correctly, where the perceived weight (u) is

$$u = v + e \quad \forall \quad e \in \mathcal{N}(\mu, \sigma^2) \tag{1}$$

And additionally, v is the real weight and e is a random variable independently and identically, normally distributed (having mean μ and variance σ^2) with v.

The economical root of MNL considers the additional characteristic of the object whose weight is being compared, converting the existing equation to

$$u(x) = v(x) + e(x) \quad \text{and} \quad \log\left(\frac{P_i}{1 - P_i}\right) = v(x_i) \tag{2}$$

where the second term is the MNL which is a log ratio of the probability of choosing a mode to the probability of not choosing the mode. The probability of choosing the mode depends on the real weight, in a general sense, the utility function of the mode choice. The higher the utility function, higher is the probability of choosing the mode choice. After a few algebraic manipulations to above equation, the probability value can be represented as

$$P_i = \frac{e^{v(x_i)}}{1 + e^{v(x_i)}} \tag{3}$$

and the utility function can be represented as

$$v(x_i) = \beta_i + \beta_{cost} \cdot \text{cost} + \beta_{time} \cdot \text{time} + \beta_{transfers} \cdot \text{total transfers} \tag{4}$$

where,

$\beta_{cost,time,transfers}$ = utility variable specific to the characteristic of the mode choice (i.e., cost, time and transfers).

β_i = alternative specific constant (decision variable), it represents the characteristic that is not related to mode choice that includes additional nudge to make the random part of the equation independently normally distributed (NID).

Cost, time, and total transfers = these are the mode choice specific parameters which are fixed values for every mode of transportation.

The term β_i is a mathematical value that primarily serves the purpose of transforming the random part of the above equation into an NID. However, the physical interpretation of the term can possibly denote an individual's feeling of being politically correct by choosing a mode (e.g. Ride hail, with positive β_i) over other (e.g. Car, with negative β_i). This may additionally represent the individual's justifiable decision with respect to their comfort, cost, and various other preferences.

The utility *v(x)* is not an observable parameter in the simulation. Rather the simulation output only specifies an observable response of whether a particular mode of transport was chosen (measured as 0 or 1). Rewriting the economical expression of choosing a mode for all available eight choices by isolating the exogenous variables (from the previous example, 'real weight' on an object) and their coefficients (a vector of estimable parameters) at an observation *t*,

$$\boldsymbol{U}_t = \boldsymbol{V}_t \cdot \boldsymbol{\beta} + e_t \tag{5}$$

where $\boldsymbol{U}_t = [0, 1]$ is a discrete response and coefficients vector are $\beta_i^j (i \in \{cost, time, transfers\}, j \in \{1, 2, \cdots, 8\})$

Therefore, the conditional probability $\Pr(U_t|V_t)$ measures the chance that the observed outcome for choosing a particular mode of transportation is a noteworthy possible outcome for given exogenous variables, where $\boldsymbol{\beta}$ describes the relationship between the observed and the real outcome. In this case the conditional probability takes the logistic form for two cases, first when the mode is selected and second when the mode is not selected:

$$\Pr(U_t = 1|V_t) = \frac{e^{V_t \cdot \beta}}{1 + e^{V_t \cdot \beta}} \quad \text{and} \quad \Pr(U_t = 0|V_t) = \frac{1}{1 + e^{V_t \cdot \beta}} \tag{6}$$

With this expression, we can calculate the estimator vector [15] β by maximizing the log-likelihood function $\ln \mathcal{L}(\beta)$.

$$\hat{\beta} = \underset{\beta}{\arg\max}[\ln \mathcal{L}(\beta)] \tag{7}$$

$$\hat{\beta} = \underset{\beta}{\arg\max}\left[\sum_t \left(U_t \ln\left(\frac{e^{V_t \cdot \beta}}{1 + e^{V_t \cdot \beta}}\right) + (1 - U_t)\ln\left(\frac{1}{1 + e^{V_t \cdot \beta}}\right)\right)\right] \tag{8}$$

2.2 Bayesian Optimization

Using Bayesian linear regression, we can infer the value of the function at point x given the value of the function at point x'. For a multivariate Gaussian prior distribution on vector $[f(x), f(x')]$ with a mean function μ and a covariance function ('kernel') Σ_0, in the absence of enough data, we can compute the likelihood of the value of vector $[f(x), f(x')]$ for a particular observation [22]. Due to Bayesian prior distribution setup, the section always occurs to be a univariate Gaussian distribution, the confidence interval for which using its standard deviation is computed using

$$\bar{X} \pm Z\frac{\sigma}{\sqrt{n}} \tag{9}$$

where $\bar{X}$ is density mean, σ is standard deviation, n is sample size, and Z score for 95% confidence is computed from the Z table [16]. These computations when generalized over multiple dimensions, takes the form of Gaussian process regression as follows:

$$\begin{bmatrix} f(x_1) \\ \vdots \\ f(x_k) \end{bmatrix} \sim N\left(\begin{bmatrix} \mu(x_1) \\ \vdots \\ \mu(x_k) \end{bmatrix}, \begin{bmatrix} \Sigma_0(x_1, x_1) & \cdots & \Sigma_0(x_1, x_k) \\ \vdots & \ddots & \vdots \\ \Sigma_0(x_k, x_1) & \cdots & \Sigma_0(x_k, x_k) \end{bmatrix}\right) \tag{10}$$

After creating a statistical inference model, with the help of a widely used acquisition function 'Expected Improvement' (EI) we can decide where to sample (Bayesian estimate) next. In this scenario, if we report the final solution after *n+1* evaluations instead of *n* evaluations, we would gain an improvement over global loss from F^* to min(F^*,*F(x)*). The benefit in the reduction of loss is given as $EI(x)$,

$$EI(x) = E_n\left[F^* - min(F^*, F(x))\right] = E_n\left[min(F^*, F(x))^+\right] \tag{11}$$

In the above form since we don't know the *F(x)*, the acquisition function uses the Bayesian posterior distribution of the objective to compute the conditional expectation of improvement of this probability distribution of *F(x)* after 'n' evaluations. Therefore, the expected improvement $EI_n(x)$ in closed form is evaluated using posterior mean and standard deviation in the formula:

$$EI_n(x) = [\Delta_n(x)]^+ + \sigma_n(x)\varphi\left(\frac{\Delta_n(x)}{\sigma_n(x)}\right) - |\Delta_n(x)|\,\phi\left(-\frac{|\Delta_n(x)|}{\sigma_n(x)}\right) \tag{12}$$

where $\Delta_n(x) = F_n^* - \mu_n(x)$, φ = probability density function of posterior, and ϕ = cumulative distribution function of posterior.

Parallel Bayesian Optimization. If the problem's domain allows multiple simultaneous evaluations, the setup explained above can be easily parallelized by parallelizing the EI in q points, where each point can consist of a multidimensional input vector. With parallel mode of evaluation, the Bayesian statistical model leverages diversity that aids in rapid convergence. The expression for EI is replaced by an unbiased gradient estimator $\nabla EI(x_{1:q})$ [17]. By assuming Y to be a vector that consists of evaluations of the objective function from all q parallel workers, which also happens to be multivariate normal distribution under the Gaussian process posterior. In this setting, when we draw an independent standard normal vector Z, the expected improvement can be written as:

$$Y = [f(x_1), \cdots, f(x_q)] = m + CZ \tag{13}$$

where mean $m = E[Y]$ and covariance C = Cholesky($\Sigma_0[Y]$)

$$\therefore \quad EI_{x_{1:q}} = E\left[(F^* - min\{Y\})^+\right] \tag{14}$$

With this background, under satisfied regulatory considerations according to infinitesimal perturbation analysis theory [18], we can switch derivative and the expectation as follows:

$$\nabla EI_{x_{1:q}} = E\left[\nabla (F^* - min\{m + CZ\})^+\right] \tag{15}$$

Using the gradient estimator, we use a multi-start stochastic gradient method and iterate until global convergence. In the beginning, we start with random starting points equivalent to the number of parallel Bayesian workers, and using the gradient method,

$$(\boldsymbol{x}_1, \cdots, \boldsymbol{x}_q) \leftarrow (\boldsymbol{x}_1, \cdots, \boldsymbol{x}_q) + \alpha \nabla EI_{x_{1:q}} \quad \text{where } \alpha \text{ is decaying step-size} \tag{16}$$

With the decaying step size, the simulation progresses in an asynchronous and controlled fashion only in the direction where the largest gradient estimate is observed so that eventually the simulation converges to the stationary point of the ∇EI surface. The parallelized setup scales for large expensive-to-evaluate simulation for significantly large number of parallel workers each optimizing a input vector of up to twenty dimensions.

The prominent setback of Gaussian process with n evaluations is its cubic complexity $\mathcal{O}(n^3)$ that comes from the inversion and determinant of the covariance kernel matrix Σ_0. In contrast, Tree-structured Parzen Estimator approach (TPE) scales linearly in the number of evaluations. Unlike GP based Bayesian estimate $\Pr(Y|X)$ (integral of product of EI and posterior distribution), TPE estimate models $\Pr(X|Y)$ and $\Pr(Y)$ separately. TPE defines $\Pr(X|Y)$ using two densities [19]

$$\Pr(X|Y) = \begin{cases} l(x) & if\ y < y^* \\ g(x) & if\ y \geq y^* \end{cases} \tag{17}$$

where y denotes the objective loss value. Unlike GP based aggressive approach of choosing a single best served loss point, TPE chooses y^* that allows some points to be included in $l(x)$ such that

$$\Pr(y < y^*) = \gamma \quad \text{where } \gamma \text{ is some quantile of observed } y \tag{18}$$

Therefore, the EI is computed as follows:

$$EI_{y^*}(x) = \int_{-\infty}^{y*} (y^* - y) \Pr(y|x) dy \quad \text{and} \tag{19}$$

$$\Pr(x) = \int_R \Pr(x|y) \cdot \Pr(y)\, dy = \gamma \cdot l(x) + (1-\gamma) \cdot g(x) \quad \text{so that} \tag{20}$$

$$EI_{y^*}(x) \propto \left(\gamma + \frac{g(x)}{l(x)} \cdot (1-\gamma)\right)^{-1} \tag{21}$$

The proportionality expression proves that the EI for evaluations is greater by minimizing the ratio $\frac{g(x)}{l(x)}$.

2.3 Hyperband

While the objective function $f : \chi \rightarrow R$ is expensive-to-evaluate, in BEAM simulation it is possible to define the cheap-to-evaluate approximate version $\tilde{f}(\cdot, b)$ of $f(\cdot)$ that are parameterized by budget-iters $b \in [BEAM - lofi, BEAM - hifi]$. In our work, the budget b is used to encode the number of iterations of the BEAM simulations. Hyperband [20] is a parameter-free multi-armed bandit hyperparameter optimization algorithm that repeatedly calls the SuccessiveHalving (SH) [21] method and is parameterized by different budgets b. To identify the best configuration from n randomly sampled configurations, Hyperband balances many aggressive runs on a smaller budget and very few runs that yields best results on maximum budget. For provided upper and lower limits of the budgets with a downsampling rate, total number of geometrically spaced SH stages are computed. For each stage, random configurations are drawn in a descending order where each configuration is allowed to run for computed iterations in an ascending order. For small to medium total budgets, Hyperband usually outperforms full function random search and Bayesian optimization evaluation methods. However, convergence to a global optimum is not guaranteed due to randomly drawn configurations.

3 Proposed Method

Our proposed method, as shown in Fig. 2, parameterizes the optimization mechanism for BEAM such that it significantly reduces calibration times. Our method replaces current manual methods, that not only costs days for a subject matter expert to optimize, but also does not guarantee a desirable convergence. The optimization maximizes the accuracy of the model estimates and minimizes the cost associated with the parameterization. In this method, the algorithm approximates the real-world context of the BEAM

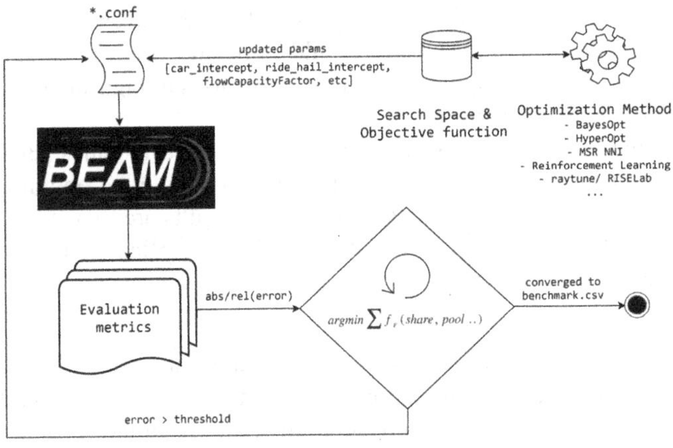

Fig. 2. BEAM optimization model

simulation at lower iterations to estimate the margin of error on an extrapolated output at higher iterations.

Algorithm 1: Bayesian Optimization with KDE and HyperBand

Input: BEAM observations Y, ConfigSpace Input Variables X, budget-iters = [BEAM-lofi, BEAM-hifi]
Output: x_{new}
Initiate Pyro Name Server
Initial random 9 BEAM runs: $D = \{(x_0, y_0), \cdots, (x_8, y_8)\}$
$s_{max} = log_{\eta}\left[\frac{BEAM-hifi}{BEAM-lofi}\right]$
for $s \in \{s_{max}, \cdots, 1, 0\}$ **do**
  Run HyperBand with $(\eta^s \cdot BEAM - hifi)$ as initial budget:
  Sample $x_{new} = \max\left(\frac{l(x)}{g(x)}\right) = \left(\frac{p(y<\alpha|x,D)}{p(y>\alpha|x,D)}\right)$
  Evaluate $y_{new} = BEAM(x_{new}) + \varepsilon$
  Refit $D \leftarrow D \cup (x_{new}, y_{new})$
end

Parallel Bayesian Optimization for Urbansim-10k Scenario. We used the Urbansim-10k scenario with a population of ten thousand agents for all optimization runs. The full BEAM run consists of fifteen iterations with an execution time of 3 h on a machine with minimum system requirements with 8Gb RAM. However, for a vanilla version of Bayesian optimization setup, the acquisition functions suggest only one sample location in the search space. Therefore, to be able to run multiple workers updating the same posterior distribution through newer observations in an accelerated manner, we chose a parallelized Bayesian optimization setup. Parallelizing allowed us to choose which trials can continue to execute a full run of fifteen iterations and which trials can be pruned because of a bad performance. Additionally, the HyperBand integration in the HpBandSter [23] package also provided us with another avenue to create a model-based early stop criterion. We use Gradle build automation tool for compilation of the BEAM code. Our Bayesian Optimization module focuses of high-fidelity optimization with an early stopping mechanism.

Based on the Sect. 2.2 and Sect. 2.3 the Python-based modules, BeamOptimizer and BeamWorker invoke the HpBandSter package to initiate the optimization task as described in Algorithm 1. Additionally, configs and results are two JSON objects that book-keep the optimization progress of the model. The HpBandSter project is forked from its GitHub project (Commit: 841db4b) [24].

BeamOptimizer. BeamOptimizer manages the parallelization of the trials, where each trial is evaluating different configurations. It imports all the components from BeamWorker script and the Bayesian optimization algorithm from the HpBandSter project. The Pyro Name Server allows tracking all the threads in the network. We have created the experiments with the number of parallel threads ranging between four and sixteen. The script allows to arbitrarily choose the scenario that we want to optimize. We tested the optimization of BEAM's "Beamville" and the "Urbanism-10k" scenario. Followed by a scenario selection, our model generates unique copies of the scenario's configuration file. Complete Bayesian optimization is run on a single Python process that initializes a Pyro [25] Name Server.

BeamWorker. BeamWorker is based on the ConfigSpace [26] package. ConfigSpace package manages and allows sampling of different types of hyperparameters. Each trial invoked by the BeamOptimizer is a copy of all procedures from the BeamWorker script. The optimizer algorithm executes the Successive Halving Algorithm (SH) (also known as HyperBand) during each iteration while simultaneously leveraging the new observation to update the posterior probability distribution. Each trial is associated with the iteration of the optimization algorithm, the budget of the current iteration, and the integer index of current trial. The most compute expensive budget consists of 21 BEAM iterations. The objective function is the L1 norm between last iteration and the benchmark which is summed to generate a one-dimensional scalar value. This scalar value is returned to the Bayesian optimizer as the loss incurred after evaluating the drawn decision variables. All eight decision variables as a standard continuous float value are drawn from a specified range using the ConfigSpace package.

The early stop mechanism included in the BeamWorker script restricts the trail who's L1 norm exceeds a predetermined value. The early-stopping module in our model determines the allowed highest L1 norm based on the total elapsed time since beginning the optimization experiment. The L1 threshold value decreases from 115% to 5% from the first elapsed 150 min until 750 min. This threshold value is dynamically imported in the BeamWorker script and is compared against the intermittent L1 norm at the end of third BEAM iteration.

Computing Platform. All experiments were conducted on Amazon Elastic Compute Cloud (EC2). For our project, we used a general purpose EC2 instance as these instances

Table 2. Parallel Bayesian Optimization on BEAM Urbansim-10k scenario. The results demonstrate L1 norm improvement for the BEAM's mode choice model on the specified configuration space. The configuration space is a hyperparameter for the optimization task.

Optimization-runs	BEAM iters	Parallel Worker	High L1 %	Low L1 %	ET hrs	ConfigSpace
10	21	8	183	60	6.5	[−20,20]
25	21	16	171	70	7	[−20,20]
40	21	4	≈180	61	–	[−20,20]
25	21	4	200	48	28	[−100,100]
50	21	5	40	**25**	51.5	Optimal ± 5%
100	21	16	160	**25 (40th iter)**	≈ 72	Optimal ± 20%

uses the resources in equal proportions for any code repository. We used a m5a.24xlarge instance for all experiments. This instance provided 96 vCPU (with 384 GB of computer memory), where each vCPU is a single physical CPU core on EC2 instance's operating system.

4 Experiments

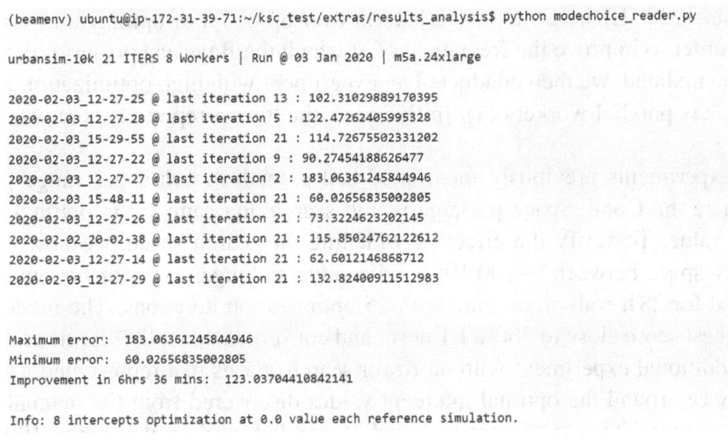

```
(beamenv) ubuntu@ip-172-31-39-71:~/ksc_test/extras/results_analysis$ python modechoice_reader.py

urbansim-10k 21 ITERS 8 Workers | Run @ 03 Jan 2020 | m5a.24xlarge

2020-02-03_12-27-25 @ last iteration 13 : 102.31029889170395
2020-02-03_12-27-28 @ last iteration 5 : 122.47262405995328
2020-02-03_15-29-55 @ last iteration 21 : 114.72675502331202
2020-02-03_12-27-22 @ last iteration 9 : 90.27454188626477
2020-02-03_12-27-27 @ last iteration 2 : 183.06361245844946
2020-02-03_15-48-11 @ last iteration 21 : 60.02656835002805
2020-02-03_12-27-26 @ last iteration 21 : 73.3224623202145
2020-02-02_20-02-38 @ last iteration 21 : 115.85024762122612
2020-02-03_12-27-14 @ last iteration 21 : 62.6012146868712
2020-02-03_12-27-29 @ last iteration 21 : 132.82400911512983


Maximum error:  183.06361245844946
Minimum error:  60.02656835002805
Improvement in 6hrs 36 mins:  123.03704410842141

Info: 8 intercepts optimization at 0.0 value each reference simulation.
```

Fig. 3. 8 parallel workers and 21 BEAM iterations

We conducted optimization experiments to validate the performance of our optimization routine on the BEAM Urbansim-10k scenario, a selective list is denoted in Table 2. The best score that we obtained through the Bayesian optimization model is 25% L1 norm. We varied the number of parallel workers and BEAM iterations to observe the optimization performance on the BEAM's mode-choice objective function. Figure 3 represents a sample output of one of our early experiments with objective loss of 60%. The total number of BEAM iterations for this run were 21 and the model executed

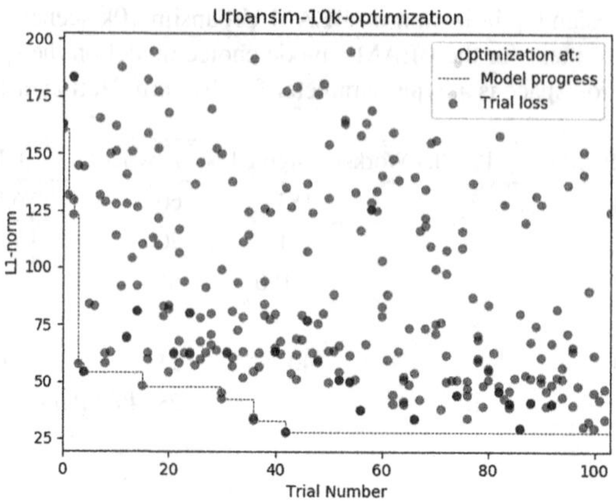

Fig. 4. 16 parallel workers and 100 optimization iterations

with eight parallel workers. The maximum L1 norm was observed to be 183% and the experiment ran up to 6.5 h wall clock time. We observed the early stop operation at fewer trials being executed later (fifth, nineth) than the specified third BEAM iteration.

Followed by this experiment, we increased the number of parallel workers and number of optimization iterations to 16 and 25 respectively. The model showed an improvement from 172% L1 norm to 70% L1 norm over up to 7 h of optimization wall clock time. In order to improve the frequency of at which the Bayesian posterior distribution was being updated, we then conducted an experiment with high optimization iterations (40) and less parallel workers (4). In this case, the lowest reported score was 61% L1 norm.

All experiments previously mentioned had a small search space range [−20,20] from where the ConfigSpace package would search and sample the optimal decision variable value. To verify the effect of widening the search space, we now expanded the search space between [−100,100] and conducted longer experiments out of which one lasted for 28 h wall clock time with 25 optimization iterations. The model started from highest score close to 200% L1 norm and converged up to 48% L1 norm. We conducted additional experiment with narrower search spaces that represented a region of 5% variance around the optimal intercept values discovered from the manual calibration experiment. This experiment included 16 parallel workers that led to 100% AWS EC2 CPU utilization with a maximum of 16 GB of memory usage per BEAM trial. We conducted total of 50 optimization iterations that lasted for 51.5 h of wall clock time. Due to a narrow search space, the model started off well at 40% L1 norm but could improve up to 25% L1 norm by the end of the experiment.

Moreover, we conducted very large scaled optimization experiments consisting of 100 optimization iterations with a maximum of 21 BEAM iterations per trial and 16 parallel workers as shown in Fig. 4. The experiment constantly utilized 100% EC2 CPU memory and lasted for more than three consecutive days. The search space was not manipulated around the ideal intercept values and had a variance of 20% around 0-mean value. This experiment was recorded to deliver the best performance of the Bayesian optimization model that costed about 2500 BEAM Urbansim-10k iterations. The experiment yielded an improvement from 160% L1 norm to 25% L1 norm at close to 40th optimization iteration. The optimization improvement tend to level out after the 45th optimization iteration.

5 Conclusion and Future Directions

We present an optimization model at scale to calibrate a multi-agent transportation Urbansim-10k BEAM simulation. Given a particular distributional forecast of demand patterns tuned to the particular time period, the BEAM simulation estimates the real time assignment decisions for agents to their mode-choice requests. In our study, we have focused on optimizing the intercepts for car, walk, walk transit, drive transit, ride hail transit, ride hail, ride hail pooled, walk transit, and bike parameters. All these parameters in totality contribute towards the mode choice model's convergence. In order to accelerate the convergence of such simulator, we developed a parallel Bayesian optimization model. Using our model, with a 20% variance search space around the 0-mean optimal value, we executed about 2500 BEAM iterations to achieve an L1 norm improvement from 160% to 25% using a fully automated procedure. The measured performance was independent of the chosen scenario and the BEAM's "*lastIteration*" value.

The ultimate usage of this optimization tool is planned to be made on the full San Francisco Bay Area population with all objectives applied. Alternative research questions can be formulated on this front. Research questions with broader context like multi-task learning and warm-start can be sought while considering all the input parameters. Extending the learned prior distribution of an experiment to geographically dissimilar scenario is also a problem of increasing importance.

Acknowledgements. The authors would like to thank Artavazd Balaian, Haitam Laarabi, Nikolay Ilin, Jessica Lazarus, Zachary Needell, and Rajnikant Sharma for many insightful conversations and technical support. The research was funded by Berkeley Lab's fellowship and the German National Scholarship provided by the Hans Hermann Voss Foundation.

References

1. Average One-Way Commuting Time by Metropolitan Areas – census.gov. https://www.census.gov/library/visualizations/interactive/travel-time.html. Accessed 02 May 2022
2. NCHRP Synthesis 406: Advanced Practices in Travel Forecasting. https://onlinepubs.trb.org/onlinepubs/nchrp/nchrp_syn_406.pdf. Accessed 02 May 2022
3. Transport Statistics Great Britain 2018 Report. https://assets.publishing.service.gov.uk/government/uploads/system/uploads/attachment_data/file/787488/tsgb-2018-report-summaries.pdf. Accessed 02 May 2022
4. TUMI's Road Space Requirements Representation. https://www.transformative-mobility.org/assets/publications/Road-Space-Requirements.pdf. Accessed 02 May 2022
5. TUMI's Vicious Cycle of Predict and Provide Representation. https://www.transformative-mobility.org/assets/publications/The-Vicious-Cycle-of-Predict-and-Provide.pdf. Accessed 02 May 2022

6. BEAM Documentation. https://beam.readthedocs.io/. Accessed 02 May 2022
7. R5: Rapid Realistic Routing on Real-world and Reimagined networks. https://github.com/conveyal/r5. Accessed 02 May 2022
8. Metropolitan Transportation Commission Website. https://mtc.ca.gov/. Accessed 02 May 2022
9. Vitalsigns Website. https://www.vitalsigns.mtc.ca.gov/. Accessed 02 May 2022
10. San Francisco Municipal Transportation Decision Survey 2019. https://www.sfmta.com/sites/default/files/reports-and-documents/2020/01/sfmta_travel_decision_survey_2019.pdf. Accessed 02 May 2022
11. OpenStreet Platform Website. https://www.openstreetmap.org/. Accessed 02 May 2022
12. General Transit Feed Specification Platform Website. https://gtfs.org/. Accessed 02 May 2022
13. Horni, A., et al.: The Multi-Agent Transport Simulation MATSim (2016)
14. Feygin, S.A., et al.: BISTRO: Berkeley Integrated System for Transportation Optimization. ArXiv abs/1908.03821 (2020): n. pag
15. Tuma, N.B.: Book Review: Econometric Analysis of Count Data. 3d edition, revised and enlarged. By Rainer Winkelmann. Berlin: Spring-Verlag, 2000. 282 pp. Sociological Methods & Research 31 (2003): 427 - 430
16. Confidence Z Score. http://www.z-table.com/. Accessed 02 May 2022
17. Wu, J., Frazier, P.: The Parallel Knowledge Gradient Method for Batch Bayesian Optimization. NIPS (2016)
18. Ho, Y.-C., Hu, J.: An infinitesimal perturbation analysis algorithm for a multiclass G/G/1 queue. Oper. Res. Lett. **9**, 35–44 (1990)
19. Bergstra, J., et al.: Algorithms for Hyper-Parameter Optimization. NIPS (2011)
20. Li, L., et al.: Hyperband: a novel bandit-based approach to hyperparameter optimization. J. Mach. Learn. Res. **18**, 185:1–185:52 (2017)
21. Jamieson, K.G., Talwalkar, A.S.: Non-stochastic Best Arm Identification and Hyperparameter Optimization. ArXiv abs/1502.07943 (2016): n. pag
22. Frazier, P.: A Tutorial on Bayesian Optimization. ArXiv abs/1807.02811 (2018): n. pag
23. Falkner, S., Klein, A., Hutter, F.: BOHB: robust and efficient hyperparameter optimization at scale. ICML (2018)
24. Forked HpBandSter Project at Commit: 841db4b. https://github.com/kschhatre/HpBandSter/. Accessed 02 May 2022
25. Pyro Nameserver Project Website. https://pyro4.readthedocs.io/en/stable/nameserver.html. Accessed 02 May 2022
26. ConfigSpace Project Website. https://automl.github.io/ConfigSpace/master/. Accessed 02 May 2022
27. Gopal, A.R., et al.: Modeling plug-in electric vehicle charging demand with BEAM, the framework for behavior energy autonomy mobility (2017)

Detecting Anomalies in Marine Data: A Framework for Time Series Analysis

Nicoletta Del Buono[1,2], Flavia Esposito[1,2](✉), Grazia Gargano[2,3], Laura Selicato[1,2], Nicolò Taggio[4], Giulio Ceriola[4], and Daniela Iasillo[4]

[1] Department of Mathematics, University of Bari Aldo Moro, via E. Orabona 4, Bari, Italy
{nicoletta.delbuono,flavia.esposito,laura.selicato}@uniba.it
[2] Members of INDAM-GNCS Research Group, Rome, Italy
[3] Istituto Oncologico Giovanni Paolo II, Bari, Italy
grazia.gargano@oncologico.bari.it
[4] Planetek Italia, Bari, Italy
{taggio,ceriola,iasillo}@planetek.it

Abstract. An ensemble framework for the analysis of time series from marine backgrounds is proposed to finally identify and classify anomalies in data time series collected from European Union's Earth Observation Programme Copernicus and Marine-EO project. The framework aims to estimate a prediction model for anomalies detection when new records are explored and then rank the magnitude of the anomalies eventually detected in some biogeochemical parameters of marine and ocean waters, such as chlorophyll-a concentrations, surface temperature profiles and dissolved oxygen.

Keywords: Anomaly · Outlier · Anomalies detection · Statistical models · Data pre-processing

1 Introduction

Marine environment observations are crucial to acquiring information about water bodies' environmental quality. Knowledge extracted from these data validly supports planning territorial interventions for their restoration. For this reason, it is necessary to obtain periodically and systematically reference information on the physical and biogeochemical state, the variability and dynamics of the ocean, and marine ecosystems. A helpful source of information on these processes is the European Union's Earth Observation Programme Copernicus [4] which monitors our planet by collecting data coming from dedicated satellites, called *Sentinels*, from dozens of other satellites, the so-called *participating missions* and *in situ* sensors. Copernicus marine environment monitoring service is helped by the service Nucleus for European Modelling of the Ocean model (NEMO), which provides a physically consistent description of all relevant ocean variables forecasting their evolution. These services support all marine applications. In particular, using data from satellites and *in situ* sensors, Copernicus

G. Nicosia et al. (Eds.): LOD 2022, LNCS 13810, pp. 485–500, 2023.
https://doi.org/10.1007/978-3-031-25599-1_36

Marine Services [5] provide daily water quality analyses and forecasts which offer an unprecedented capability to observe, understand and anticipate events in the marine environment. It contributes to the protection and sustainable management of marine biological resources, particularly for aquaculture, sustainable fisheries management, or the decision-making process of regional fisheries organizations. It thus enables a wide range of marine and coastal environmental applications.

These data are often time series that can be affected by anomalies. Anomalies or outliers –in the statistical sense– are (rare) points that differ significantly from the expected behavior of other data points. Identifying and detecting anomalies in marine physical and biochemical data can be a challenging task that successfully solving would bring added value to marine studies. Advantages can be obtained to check water quality, control pollution behavior, and climate changes, or assess coastal erosion. Moreover, when sea surface temperatures are under consideration, anomaly detection can help forecast problems in marine ecosystems or the occurrence of tropical cyclones.

The Marine-EO project is part of this context; it is devoted to collecting information about ocean parameters variability in time and space and proposes a set of support services to integrate better the Earth Observation (EO) and Copernicus-enabled resources. In particular, it aims to establish EO-based services, covering the sea-basins of the Mediterranean, North Atlantic, and the Arctic, by adapting Copernicus data and information to provide tempestive and reliable information on the Marine Environment in the thematic areas of Marine monitoring (environmental assessment and fish farming) and Maritime Security (support to navigation in Arctic Sea).

This paper proposes a workflow for the analysis of time series from marine backgrounds. In particular, starting from samples constituting data time series, a predictive model is estimated to adhere to these observations. Moreover, a proper comparison between expected and actual observations helps recognize whenever parameter anomalies are present.

The paper is organized as follows. The proposed new framework for marine anomalies detection is detailed in Sects. 2 and 4, together with some descriptions of data and marine areas which have been considered and inspected (Sect. 3). Some experimental results on physical and biogeochemical parameters of the waters of some interest areas are also reported in Sect. 4. A final Sect. 5 of conclusion and discussions closes the paper.

2 The Proposed Framework for Anomalies Detection

The framework focuses on hydrological time series and aims to estimate a prediction model for anomaly detection when new records are explored. The novelty of the proposal stays in its ability to rank the magnitude of the anomalies in the particular context of the water quality time series provided by the Marine-EO project. The complete workflow has been developed in R environment [11]; it integrates some already loaded packages with new designed functions for anomaly classification. All related codes have been stored on Github[1]. Figure 1 sketches the main steps of the proposed framework.

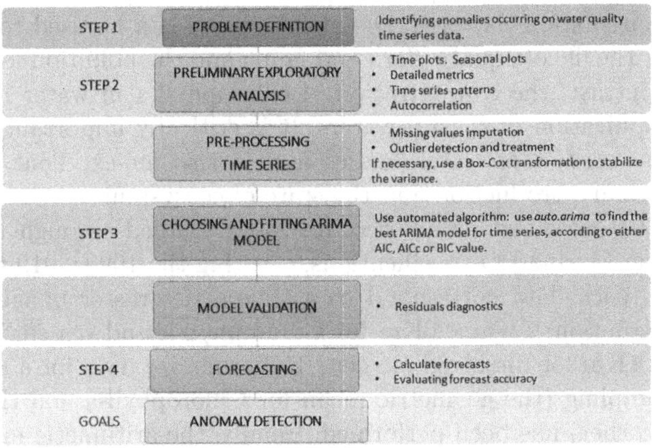

STEP 1	PROBLEM DEFINITION	Identifying anomalies occurring on water quality time series data.
STEP 2	PRELIMINARY EXPLORATORY ANALYSIS	• Time plots. Seasonal plots • Detailed metrics • Time series patterns • Autocorrelation
	PRE-PROCESSING TIME SERIES	• Missing values imputation • Outlier detection and treatment If necessary, use a Box-Cox transformation to stabilize the variance.
STEP 3	CHOOSING AND FITTING ARIMA MODEL	Use automated algorithm: use *auto.arima* to find the best ARIMA model for time series, according to either AIC, AICc or BIC value.
	MODEL VALIDATION	• Residuals diagnostics
STEP 4	FORECASTING	• Calculate forecasts • Evaluating forecast accuracy
GOALS	ANOMALY DETECTION	

Fig. 1. Workflow of the proposed approach

A preliminary Exploratory and Preprocessing phase is performed to provide a qualitative study of the time series and data cleaning operations managing missing values and outliers. This phase is followed by estimating the data best fit model; after a check of the working hypotheses, a forecasting procedure is applied according to different levels of prediction. The last workflow phase is devoted to detecting and then classifying the anomalies. This step distinguishes different levels of anomalies, from moderate to severe, which vary according to the parameter studied. A detailed description of the anomalies detection procedure is reported in the following sections.

It should be observed that since hydrological data are taken into study, due to their characteristics, the adoption of distribution-free methods is preferred for as long as possible.

3 Data Description

Data from two marine protected areas, the island of Zakynthos in the Ionian Sea, the Condor mouth, and an area of high interest for aquaculture, the Ribeira Quente in the Azores of the Atlantic Ocean, are considered. The Mediterranean Sea, with its peculiar characteristics and the presence of physical processes common to other oceans, can be considered a miniature ocean system. Many techniques that are fundamental to the general circulation of the global ocean occur –identically or similarly– within the Mediterranean basin.

[1] https://github.com/flaespo/AnomaliesMarineData.

Among the physical and biogeochemical parameters of marine and ocean waters, surface temperature profiles and dissolved oxygen and chlorophyll-a concentrations are analyzed.

Sea surface temperature anomalies are an important predictor of atmospheric and oceanic circulation patterns. Dissolved oxygen can be used to characterize the fitness for the life of aquatic flora and fauna and the pollution level in a water system. In contrast, the concentration of chlorophyll-a in water highlights the level of eutrophication in coastal waters. It is critically important for assessing trophic characteristics of water bodies and is also an excellent indicator for evaluating primary production and the state of ecosystems.

Each of the examined parameters has been studied through its data time series following Marine-EO specifications, covering the 1999–2019 time window. In particular, each data correspond to a time series raster image with a specific space resolution, $1\,\text{Km} \times 1\,\text{Km}$ for Chlorophyll-a and sea surface temperature, $4\,\text{Km} \times 4\,\text{Km}$ for dissolved oxygen. To harmonize data for a selected area, a monthly sampling (the geometric mean for Chlorophyll-a and the arithmetic mean for the other) has been performed. Finally, the arithmetic mean has been computed for all pixels within the same area. In the following, a complete description of the proposed framework is illustrated when the level of Chlorophyll-a in the Ionian sea on Zakynthos Greek island is considered. The results on other parameters and regions of interest are summarized for space reasons in Appendix A.

4 Methods and Results

This section describes in detail the main phases of the proposed framework for anomalies detection. The methods and the techniques included in it are described referring to the results obtained when data on the level of Chlorophyll-a in Zakynthos island sea are analyzed.

4.1 Preliminary Exploratory Analysis Phase

This initial phase in the framework aims to qualitative study and visually understand the general time-series behavior and its possible quirks (i.e., missing data, abrupt changes in time, or variables relationships). This phase provides plotting of the time course of the studied parameters measured monthly. Seasonal plots are also adopted to detect the presence of seasonal fluctuations: yearly reference period is plotted and scanned into monthly sub-periods as well. Figure 2 shows the behavior of Chlorophyll-a monthly acquired in the Ionian sea. A visual inspection of the qualitative graphs reported by the preliminary data analysis phase immediately reveals some interesting features: data present missing values, some observations that significantly differ from the others, and an upward shift in this parameter value has been observed since January 2018.

Time series internal dynamics can also be investigated using the tools present in this phase of the framework. Generally speaking, time series are characterized by systematic patterns, which need to be identified and separated from possible

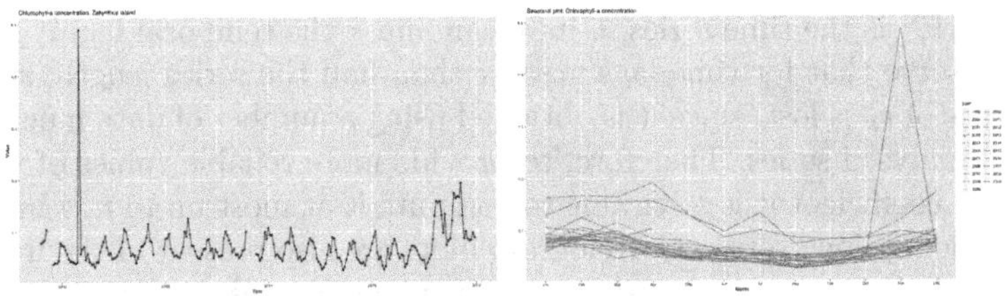

Fig. 2. *Time plot* and *seasonal plots*: Chlorophyll-a concentration in Zakynthos island, Greece.

accidental oscillations (erratic components) by equalization or smoothing methods. These two patterns are trend and seasonality. The trend is the underlying monotonic movement, which highlights a structural evolution of the phenomenon due to systematic causes over a long period. On the other hand, seasonality is constituted by variations having similar intensity each year but different magnitude during the same year. Then a residual component includes factors that influence the phenomenon randomly.

Different time series decomposition methods exist in the literature. The proposed framework includes Seasonal and Trend decomposition with the Local Weighted Regression (STL with LOESS) approach [3]. STL acts by averaging, with some weights, the time series for each month for the seasonal component and then smoothing the remaining part with LOESS for the trend component. This non-parametric methodology provides better flexibility in describing seasonal trends compared to other analytical approaches. It should be observed that for computations, no temporal continuity is required, allowing the presence of missing values, which almost will affect the robustness of the local month average. By the way, the presence of the smoothing procedure reduces the effect of the small number of observations used to calculate the monthly averages. It follows that missing values, even large numbers, do not affect a sufficiently reliable definition of the seasonal components.

The behavior of sea surface temperature is typically influenced by climate change [6] but, in this context, due to the "short time period" (relative to a climate scale) and that the areas are in a coastal zone (apart from the Condor Mouth) it is possible to assume independence from climate change.

Besides the general trend of the time series, it is also interesting to analyze whether the behavior of past data influences the current one. Each time series value can affect its previous one, reflected in the autocorrelation. This phenomenon defines the degree of linear dependence between data in a time series; it has been measured with the autocorrelation coefficient:

$$r_k = \frac{\sum_{t=k+1}^{n} (y_t - \bar{y})(y_{t-k} - \bar{y})}{\sum_{t=1}^{n} (y_t - \bar{y})^2}, \tag{1}$$

where $y_t \in \mathbb{R}^n$ is the time series, $\bar{y}$ its mean and k the temporal lag. It is important to observe that for time lags greater than half the series length, statistical significance of r_k is lost, since it is obtained using a number of data much smaller than the analysed series. Therefore, to provide more reliable values of the autocorrelation coefficient it is advisable to compute it at most up to $n/4$ in order to have sufficiently long series. Correlogram plots are also present in the framework to graphically inspect autocorrelation. This plot provides a clear representation of the time dependency structure, allowing to visually capture some general physical properties of the analysed hydrological data (such as seasonality, temporal linkage in the short and long term, non-stationarity). The correlogram plot of concentration of Chlorophyll-a in Zakynthos water is reported in Fig. 3.

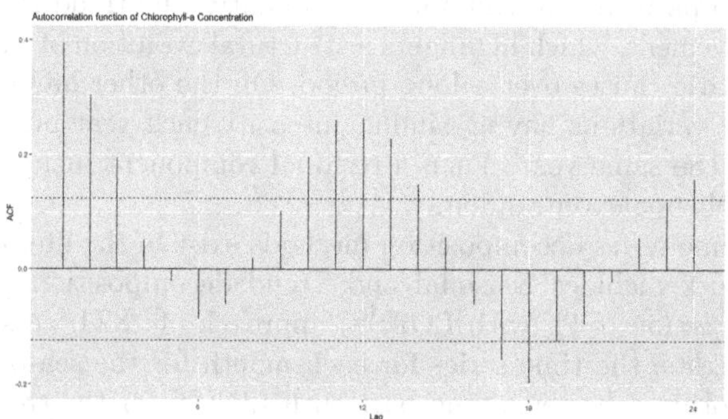

Fig. 3. Correlogram plot of concentration of Chlorophyll-a in Zakynthos water.

Even if the autocorrelation coefficient presents a some kind of variation, it is positive and with maximum values in correspondence to k values such as to configure an annual periodicity (e.g. for k equal to 12 or its multiples). While they are lower or negative for other values of k (e.g. for k equal to 6 and 18). This reflects strong correlation with values of the same periods of previous years. This behaviour characterizes a phenomenon which varies during each year and in a similar way from one year to the other. The same behaviour highlighted in Fig. 3 was found in most of the time series examined, confirming the prevalence of a seasonal component in analyzed data.

4.2 Pre-processing Time Series Phase

Entering the preprocessing phase requires to split data series with an hold out of 80-20. The first part will be preprocessed and used for fitting the model, whereas the latter for testing and evaluating the accuracy of the constructed model.

Measurement instrument malfunction or interruption of the survey service for maintenance is a very frequent problem in hydrological data time series acquisition; these occurrences cause the presence of missing data and outliers embedded in the time series which can lead to meaningless, highly biased, and/or even incorrect analyses. Building a prediction model by neglecting the presence of missing data and values that deviate significantly from the general trend in parameter values could cause distortions in the prediction model.

The proposed framework is equipped with appropriate tools able to imputing the missing values and replacing the outliers with estimates that are as consistent as possible with the data trend.

Missing data were linearly imputed from STL deseasoned series. Once the series is completed the seasonal component is added. Figure 4a shows the distribution of the missing data for the Chlorophyll-a concentration in Zakynthos waters.

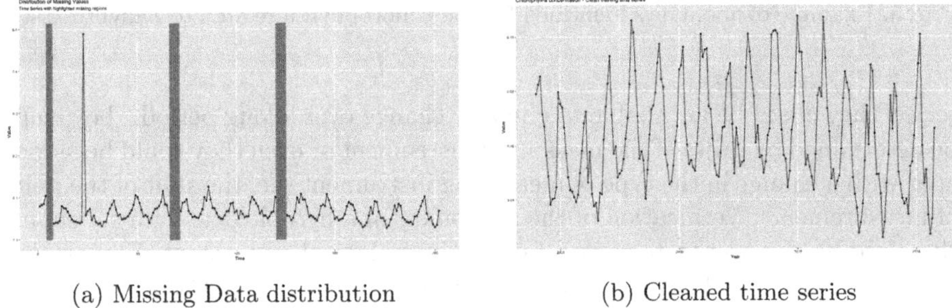

(a) Missing Data distribution (b) Cleaned time series

Fig. 4. Analysis of missing data on the parameter Chlorophyll-a concentration in Zakynthos sea

Similarly, anomalies are found from residual of STL decomposition, by looking for values outside the default range $\pm 2(q_{0.9} - q_{0.1})$, where q_p is the $p-$th residual quantile, estimated taking into account the behavior of the parameter. Figure 5 shows an example of anomalies for the Chlorophyll-a in Zakynthos water on its STL decomposition in which one anomaly is founded and corrected.

Once the time series data have been completed, the framework also checks if appropriate transformations are additionally required to finally prepare the series and then use it to construct the anomalies detection model. The Box and Cox transformation [1] described by the following equation is eventually adopted to stabilise the variance of a given time series $\{y_t\}_{t=t_0,\ldots,tfin}$

$$w_t = \begin{cases} \log(y_t) & \text{se } \lambda = 0, \\ (y_t^\lambda - 1)/\lambda & \text{se } \lambda \neq 0. \end{cases} \quad (2)$$

Equation 2 depends on the parameter λ that gives good results when the seasonal variation remains constant. In the proposed framework λ is estimated according to the proposal in [7]. It is worthy to note that at the end of the process, output needs to be back-transformed to obtain predictions readable on the original scale.

Time series can be stabilised to be stationary, this is an assumption most prediction models are based on. Different reasons can cause non-stationary data time

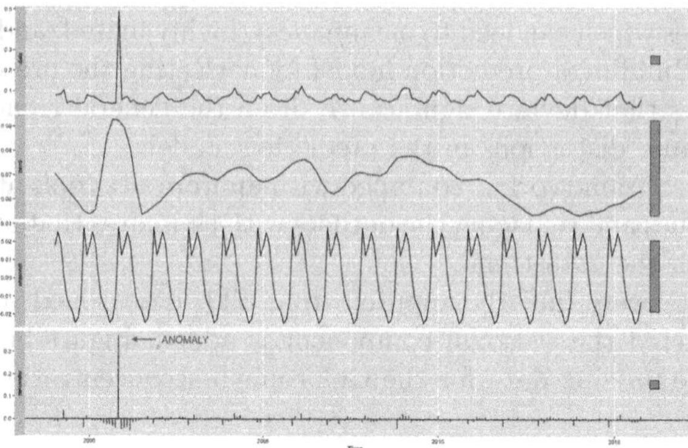

Fig. 5. Example of anomalies identified for the Chlorophyll-a related to Zakynthos sea.

series: they could be ascribable to climatic change over a long period, they could be determined by drifts of the measuring instrument or even they could be associated with a change in the type of measuring instrument (i.e. the shift of the point of measurement). Verification of the stationary of a hydrological data set should therefore be the preparatory step for any hydrological assessment. Exploratory and visual data analysis could provide information on non-stationary characteristics, at least for the most evident ones. Time and seasonal plots, such as correlogram, can be useful to identify non-stationary time series. Regarding this latter, the autocorrelation function of a stationary time series, tends to zero rather quickly, on the contrary of the slow decrease for non-stationary data. In this context, in addition to graphical tools, it is therefore particularly important to have statistical methodologies able to verify the stationary for a long series of data, and in lacking of the basics hypothesis, it is necessary to make corrections. These approaches can be useful to understand and evaluate the state of the ocean ecosystem but also to implement timely action to restore it. One way to make a time series stationary is the so-called differencing, which is about calculating the differences between consecutive observations. By using the lag operator B on the time series y_t, it can be defined the d-order differencing as $y_t^d = (1-B)^d y_t$, and the seasonal differencing as $y_t'' = (1-B)(1-B^m)y_t$ where m is the number of seasons. The first difference represents changes between two consecutive observations, whereas the latter changes between years. One way to statistically determine whether differentiation is necessary, is to use a unit root test. In our workflow, it is automatically included into the model the well known Kwiatkowski-Phillips-Schmidt-Shin (KPSS) test [9], that checks for null hypothesis regarding stationary data. Several KPSS allow to determine the optimal number needed to make stationary the time series.

Figure 4b shows the output from the Pre-processing time series phases for the Chlorophyll-a concentration in the Ionian sea. Other delays in differencing are unlikely to make interpretative sense and should, therefore, be avoided. Moreover, care should be taken not to apply more lags than necessary, as this may induce false dynamics or autocorrelations that do not really exist in the evaluated time series.

4.3 Fitting and Evaluation Model Generation Phase

Once the preliminary exploratory and preprocessing phases end, it enters the model fitting step. This phase is devoted to find the appropriate prediction model best fitting the hydrological time series. This will be subsequently adopted to produce predictions necessary for the anomalies detection. This latter phase, in fact, is performed on the basis of comparisons between the expected and the actual values to recognise whenever the presence of anomalous parameter occurs.

Since the seasonal nature of the analysed data, we used as fitting mathematical model SARIMA (Seasonal AutoRegressive Moving Average) method. It assumes stationarity of the time series and it is composed by two main parts: the AutoRegressive (AR) and the Moving Average (MA) models. The AR part allows to incorporate the effect of past values while the MA one represents a part of the time series not explained by trend or seasonality, permitting to set the model error as a linear combination of the error values observed in the past. An integrated part is added to the union of these two models, which allows to manage the differentiation of a series. It comes from ARIMA model that, depending on some parameters (p, d, q), can be written as:

$$\underset{\text{AR}(\,p)}{\underset{\uparrow}{(1-\phi_1 B-\cdots-\phi_p B^p)}}\ \underset{d\text{ differences}}{\underset{\uparrow}{(1-B)^d y_t}} = c + \underset{\text{MA}(\,q\,)}{\underset{\uparrow}{(1+\theta_1 B+\cdots+\theta_q B^q)\varepsilon_t}}, \quad (3)$$

where μ is the mean of $(1-B)^d y_t$, ε_t is the white noise and ϕ and θ the model coefficients in $\mathbb{R}$. The optimal values of the model hyperparameters (p, d, q) are estimated using a variation of the algorithm proposed in [8] and combining test for unit roots and the minimization of both AICc (*corrected Akaike's Information Criterion*) and MLE (*Maximum Likelihood Estimation*). These processes find the order of the model (p, d, q), estimate the parameters c, $\phi_1, \ldots, \phi_p$, $\theta_1, \ldots, \theta_q$ and then compute the best model fitting given data. SARIMA includes seasonal terms, similar to the not seasonal one, but with a back shift of the seasonal period, that is $ARIMA((p, d, q), (P, D, Q)_m)$, being P, D, Q and m (the number of time steps for a single seasonal period) four seasonal elements that are not part of ARIMA and must be configured. To check the correctness of trained SARIMA model in capturing right information embedded in data, a residuals analysis is also performed. In fact, a good forecasting method produces uncorrelated and zero mean residuals, while the presence of correlations between the residuals indicates information left in it that should be used in the forecasting calculation. If the residuals have a non-zero mean, then the forecasts are biased. Graphical analysis based on correlogram and the Ljung-Box[2] statistical test [10] were used together with a T-test for the hypothesis of zero mean residuals.

[2] LB test computes the statistics $T_{L\&B} = n(n+2)\sum_{k=1}^{h}(n-k)^{-1}r_k^2 \sim \chi(h)$, with h degrees of freedom usually chosen as $h = min(2m, n/5)$ [8].

Figure 6 reports the residual analysis performed on the estimated model for the Chlorophyll-a concentration in Zakynthos sea water.

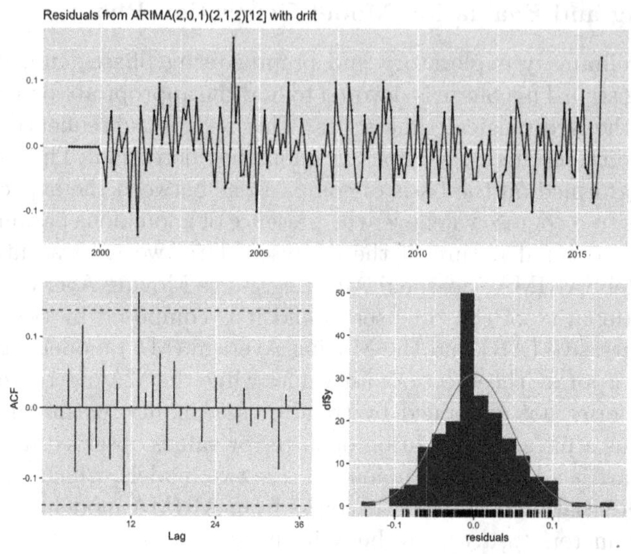

Fig. 6. Residual behaviour of the estimated model for the Chlorophyll-a concentration in Zakynthos sea water.

As it can be observed the series presents a little but significant peak, still coherent with white noise assumption. Moreover, statistical tests confirm the absence of residual correlation and zero mean (p-values equals to 0.1288 and 0.8660, for LB and t-test, respectively).

It should be noted that since real hydrological data are considered, it may happen that the lack of correlation between the residuals is not verified. In this case the model can still be used to make forecasts, but the forecast intervals may not be accurate due to the correlated residuals.

4.4 Forecasting and Model Evaluation

After the model of time series data is estimated and the white noise assumption on residuals checked, the framework proceedes with the forecasting phase using as confidence interval 80% and 95% as suggested in literature [2]. As an example, Fig. 7 reports the estimated model and the forecasting for the Chlorophyll-a concentration on the Zakynthos sea water.

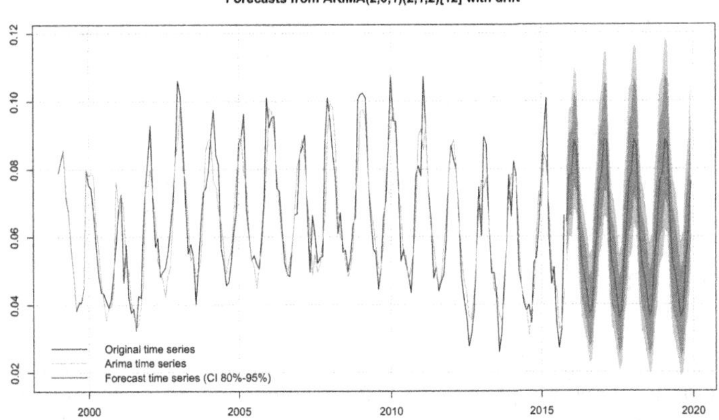

Fig. 7. Estimated model and the forecasting for the Chlorophyll-a concentration on the Zakynthos sea water *forecasting.*

Table 1. Performance values of the proposed predictive model for Chlorophyll-a concentration in the Zakhyntos waters.

	RMSE	MAE	MPE	MAPE	MASE
Training set	0.0079	0.0058	−1.1484	9.4642	0.6455
Test set	0.0472	0.0337	18.3203	30.9848	3.7262

The error[3] model is $e_{T+h} = y_{T+h} - \hat{y}_{T+h|T}$, where $\{y_1, \ldots, y_T\}$ is the training set and $\{y_{T+1}, y_{T+2}, \ldots\}$ the test set. For seasonal time series of length n and seasonal period m, errors are measured using: i) Standard and Root Mean Squared Error (MSE/RMSE); ii) Mean Absolute Percentage Error (MAPE) and iii)Standard and Scaled Mean Absolute Error (MAE/MASE).

Table 1 reports the performances of predictive model for Chlorophyll-a concentration in the Zakhyntos waters. All the measures confirm the goodness of the estimated model.

4.5 Anomalies Detection

To detect if a sample in the data time series is an anomaly, the framework compares the new value with the forecast results constructed as previously described. All values which fall outside the confidence interval of 80% were labelled as anomalous points. The anomalies detection phase of the proposed framework is also able to classify the levels of anomalies from moderate to severe (as illustrated

[3] This differ from residual, computed on training set, that are actual errors from the fitted and real values $e_t = y_t - \hat{y}_t$.

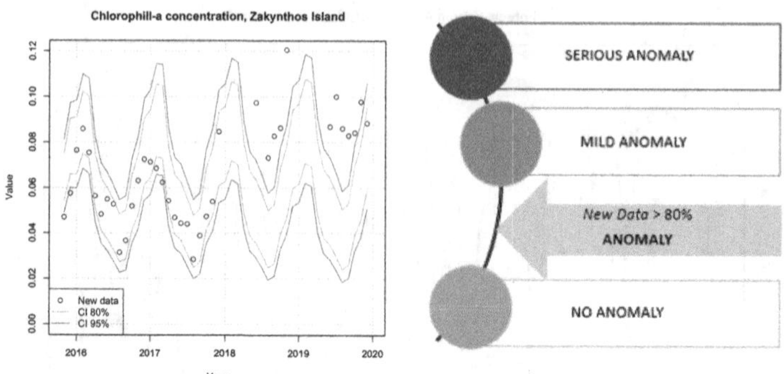

Fig. 8. Anomalies detection phase: new data are depicted as black circle points, while regions of serious and mild anomalies are bounded by red and orange solid lines, respectively (Color figure online)

in Fig. 8 where new data are depicted as black circle points, while regions of serious and mild anomalies are bounded by red and orange solid lines, respectively). These levels strictly depend on the nature of the parameter under studies. In the context described in the paper, for Chlorophyll-a concentration, if the value is below the prediction it is never labelled as severe as shown in Fig. 8. Appendix A described, with the help of visual panels, the analysis conducted for other parameters under study. From these it can be seen the levels of severity depends on the expert of domain. When surface temperature is taken into account, the severity degree of the anomaly occurs either the anomaly is above or below the prediction value, whereas for dissolved oxygen concentration if the value is above the prediction the anomaly is never severe.

Identification of anomalies is of great importance in order to evaluate impact of human activities on coastal environment and vice versa, in a framework of integrated coastal management. In Marine-EO the focus was put on Marine Protected Areas (MPA) and on aquaculture farms. For what concerns the former, to estimate the number and type of anomalies in a given time frame is important in order to evaluate the eventual impact of external factors to the health of the MPA and possibly to identify them by crossing with other information (e.g. changes occurring in surrounding areas or implementation of access policies). This will allow then to define actions to mitigate such impact on the coastal environment. Concerning aquaculture farms, it is important for the farmers to be rapidly informed of specific anomalies in order to be able to take measures to avoid production losses. In this case they are interested in those specific anomalies that can impact the fishes: different type of anomalies can have different impact. For example lower oxygen concentration is extremely dangerous, because it can result in fishes death, while higher oxygen concentration usually has not a relevant impact. A "positive" sea surface temperature anomaly, together with similar chlorophyll anomaly, can lead to dangerous phenomena like algal bloom or foster disease spreading. While low temperature can result in a decrease of growth rate.

5 Discussions

This paper briefly illustrates a novel framework for anomalies detection in data time series related to some physical and biogeochemical parameters of marine and oceanic waters, such as surface temperature profiles, dissolved oxygen and Chlorophyll-a concentrations of the waters of some areas of interest. The peculiarity of the approach is the possibility of classify a new value as an anomaly on the basis of a comparison with appropriate forecasting trained model and also indicating the level of detected anomaly point from severe to mild according to some threshold defined in agreement with the domain experts.

Acknowledgments. This work was supported in part by the GNCS-INDAM (Gruppo Nazionale per il Calcolo Scientifico of Istituto Nazionale di Alta Matematica) Francesco Severi, P. le Aldo Moro, Roma, Italy. The author F.E. was funded by REFIN Project, grant number 363BB1F4, Reference project idea UNIBA027 "Un modello numerico-matematico basato su metodologie di algebra lineare e multilineare per l'analisi di dati genomici".

A Appendix

The proposed framework was also applied to time series recording dissolved oxygen and sea surface temperature values in some marine zones of the archipelago of the Azores. Figures 9 and 10 report the results for the anomalies detection of the dissolved oxygen in the Condor Mouth (Azores) and of the Sea Surface Temperature in the Ribeira Quente (Azores), respectively.

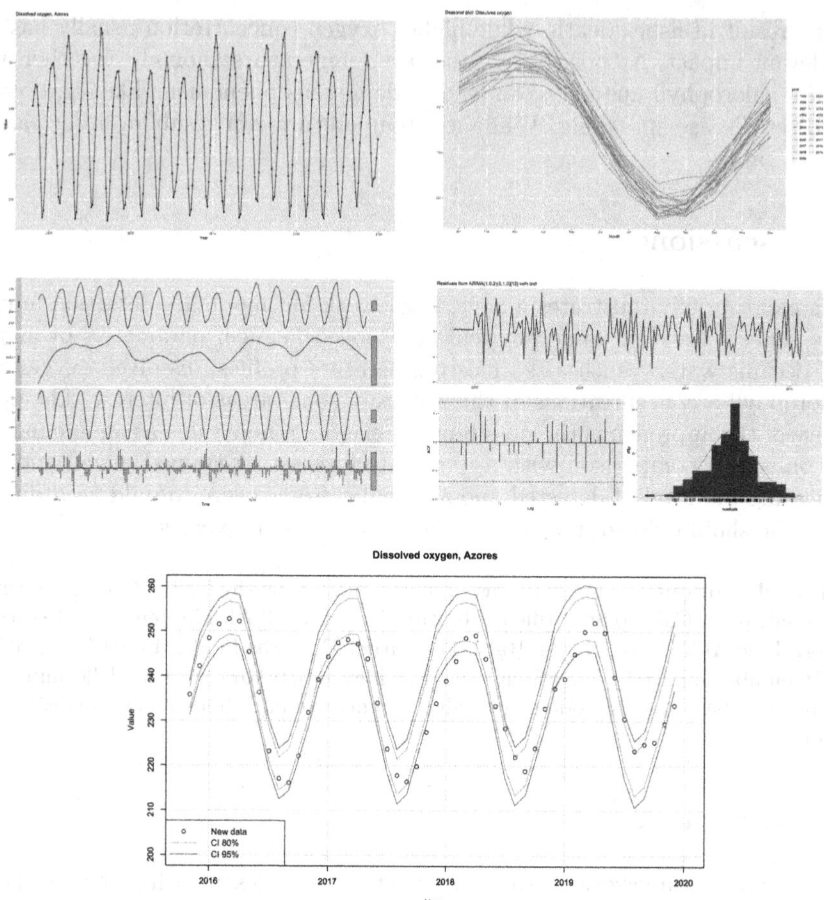

Fig. 9. Panel for the analysis of dissolved oxygen time series in the Condor Mouth, Azores. From left to right and from top to bottom, the five plots illustrate: time and seasonal plots of dissolved oxygen; the time series decomposition; the residual behaviour of the estimated model; the anomalies detection on new data (black circle points) through the estimate model.

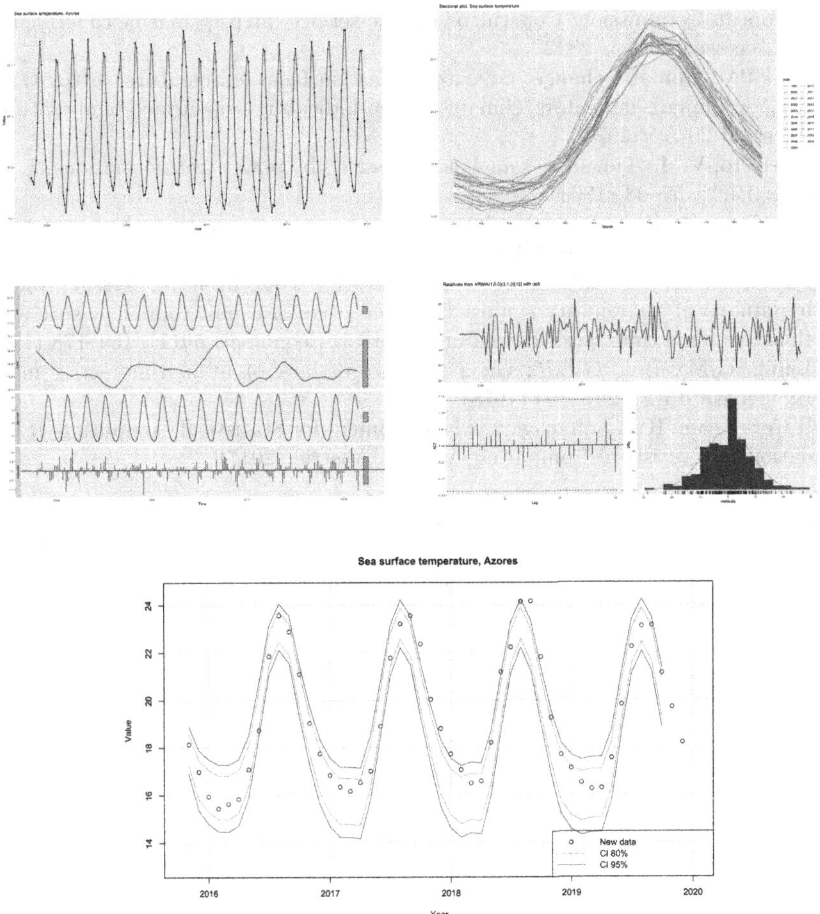

Fig. 10. Panel for the analysis of Sea Surface Temperature time series archipelago of the Azores. From left to right and from top to bottom, the five plots illustrate: time and seasonal plots of sea surface temperature; the decomposition of the time series; the residual behaviour of the estimated model; the anomalies detection on new data (black circle points) through the estimate model.

References

1. Box, G.E.P., Cox, D.R.: An analysis of transformations. J. R. Stat. Soc. Series. B Stat. Methodol. **26**(2), 211–252 (1964)
2. Brockwell, P.J., Davis, R.A.: Introduction to Time Series and Forecasting. STS, Springer, Cham (2016). https://doi.org/10.1007/978-3-319-29854-2
3. Cleveland, R.B., Cleveland, W.S., McRae, J.E., Terpenning, I.: STL: a seasonal-trend decomposition procedure based on loess. J. Offic. Stat. **6**(1), 3–73 (1990)
4. European Commission: Copernicus Europe's eyes on earth. https://www.copernicus.eu/en. Accessed 20 Apr 2022

5. European Commission: Copernicus marine service. https://marine.copernicus.eu/en. Accessed 20 Apr 2022
6. U. EPA: Climate change indicators: Sea surface temperature. https://www.epa.gov/climate-indicators/climate-change-indicators-sea-surface-temperature. Accessed 20 Apr 2022
7. Guerrero, V.M.: Time-series analysis supported by power transformations. J. Forecast. **12**(1), 37–48 (1993)
8. Hyndman, R., Khandakar, Y.: Automatic time series forecasting: the forecast package for R. J. Stat. Softw. **26** (2008)
9. Kwiatkowski, D., Phillips, P.C., Schmidt, P., Yongcheol, S.: Testing the null hypothesis of stationarity against the alternative of a unit root: how sure are we that economic time series have a unit root? J. Economet. **54**(1), 159–178 (1992)
10. Ljung, G.M., Box, G.E.P.: On a measure of lack of fit in time series models. Biometrika **65**(2), 297–303 (1978)
11. R Core Team: R: A Language and Environment for Statistical Computing. R Foundation for Statistical Computing, Vienna, Austria (2017)

Does Catastrophic Forgetting Negatively Affect Financial Predictions?

Alberto Zurli[1,2(✉)], Alessia Bertugli[3], and Jacopo Credi[2]

[1] Università di Modena e Reggio Emilia, Modena, Italy
alberto.zurli@unimore.it
[2] Axyon AI, Modena, Italy
{alberto.zurli,jacopo.credi}@axyon.ai
[3] Università di Trento, Trento, Italy
alessia.bertugli@unitn.it

Abstract. Nowadays, financial markets produce a large amount of data, in the form of historical time series, which quantitative researchers have recently attempted at predicting with deep learning models. These models are constantly updated with new incoming data in an online fashion. However, artificial neural networks tend to exhibit poor adaptability, fitting the last seen trends, without keeping the information from the previous ones. Continual learning studies this problem, called catastrophic forgetting, to preserve the knowledge acquired in the past and exploiting it for learning new trends. This paper evaluates and highlights continual learning techniques applied to financial historical time series in a context of binary classification (upward or downward trend). The main state-of-the-art algorithms have been evaluated with data derived from a practical scenario, highlighting how the application of continual learning techniques allows for better performance in the financial field against conventional online approaches (Code is available at https://github.com/albertozurli/cl_timeseries.).

1 Introduction

The financial market is a worldwide virtual place meant for the exchange of financial instruments, such as shares, contracts, stocks, or commodities. In the past, investments were made in person by a small circle of domain experts, basing their decisions on experience and a small amount of data. Nowadays, the actors and the dynamics involved have radically changed. The strong availability of real-time data and the great computational capabilities of computers brought that most of the investments are not made only by human traders. Therefore, they are assisted by an ever-increasing number of equipped "intelligent machines" able to understand the best timing for carrying out financial transactions. Consequently, the concept of algorithmic trading has taken hold. The algorithms are characterized by a set of rules defined to perform certain actions based on the state of the market. These rules are written by human traders who study particular techniques for market choices. The natural evolution of these algorithms is the use of

G. Nicosia et al. (Eds.): LOD 2022, LNCS 13810, pp. 501–515, 2023.
https://doi.org/10.1007/978-3-031-25599-1_37

cutting-edge machine learning and deep learning techniques to develop predictive models. This way, human intervention is no longer required to define steady rules, and decisions support tools can rely only on data and their patterns over time, through the use of tailored neural networks. The problem of using machine and deep learning techniques with time series is the need to periodically retrain the model to allow for the updating and acquisition of knowledge of the most recent data. This leads neural networks to suffer from catastrophic forgetting, meaning that their weights are overwritten in favor of last seen data, losing their predictive power over the older data. It has been shown that market trends over time are qualitatively similar, and therefore learning the behavior of the time series in the past could be useful for predicting its future trend. Continual learning is a deep learning technique developed to face the catastrophic forgetting problem, and that has recently shown promising and significant results even in image classification areas. The study and application of these techniques could be relevant within the financial market, due to its cyclical nature.

Exploring the potential of continual learning is a novel topic in the financial world and therefore, to the best of our knowledge, this is the first work that analyzes and evaluates the potential of continual learning in-depth. In particular, we analyzed whether catastrophic forgetting could lead to a bad predictive performance in financial time series classification, comparing the classical online learning paradigm with continual learning techniques designed to overcome catastrophic forgetting. Specifically, a problem of binary classification of time series has been studied, where each time series consists of a vector of consecutive daily samplings. The goal is to predict whether this will have an increasing or decreasing trend in the future, leading to a "buy" or "sell" decision. We define continual learning tasks as periods, which may have different lengths, but specific behavior. For example, the first task can be represented by a stationary prices trend, the second one by a sudden increase of prices, the third one by a slow drop. Each task is time bounded by a change of these regimes. Approaching the problem this way, we place it in a Domain-Incremental Learning (Domain-IL) scenario [20], where the distribution of the classes remains unchanged while the change of task is defined by a variation of the distribution of input data. Several state-of-the-art continual learning methods have been developed and deeply analyzed. Eventually, the experimental results achieved suggest that CL techniques, alleviating the forgetting phenomenon, exhibit better performance than online learning.

2 Related Work

2.1 Financial Market Predictions

Time series research has experienced strong growth in several sectors in the last few years: from the prediction of pedestrian and car trajectories for video surveillance [1,13] to the prediction of machinery failures in industries. Among them, financial market predictions have been deeply investigated leading to the development of increasingly sophisticated algorithms capable of predicting the trend of the financial market. Machine learning and deep learning techniques have been applied to financial time series, to make these algorithms as automatic as possible to facilitate the traders' decisions. However, due to the unpredictability of the market is still hard to design machine learning algorithms that can properly work on financial time series [14]. In fact, deep learning models, like 1-D convolutional neural networks, multi-layer perceptrons, temporal transformers [19]), that achieve the State Of The Art on other tasks, do not always exhibit satisfactory performance on financial problems. For this reason, deep learning for finance is at the cutting-edge of research and in the last few years, several methods have been designed or adapted specifically to financial time series [6,10,15].

2.2 Continual Learning Techniques

Real-world computational systems are exposed to continuous data flow, and they have to learn and remember multiple tasks. The traditional optimization technique applied to machine and deep learning models is not suited to tackle continuous data flow due to their nature to forget and overwrite the previously learned knowledge. Neural networks tend to overwrite the previously acquired knowledge by updating the network parameters when trained with data from a new task or distribution. This phenomenon is known as catastrophic forgetting [12]. It typically leads to a sudden drop in performance or, in the worst case, the total overwriting of the old task against the new one. If you train on sequential tasks, the performance of traditional neural networks decreases in past tasks as the number of tasks increases. Continual learning techniques try to solve this issue, through specifically designed algorithms that alleviate catastrophic forgetting. Several continual learning methods have been studied and presented in the literature in the last few years. However, no consensus has been reached about a globally valid continual learning algorithm. We can therefore distinguish three categories: replay-based, regularization-based and parameter isolation methods. *Replay-based methods* [2,3,16], also known as rehearsal, save examples from previous tasks. These examples are sampled from a memory buffer and used as input when training the current task. These previous task samples contribute to the loss function to prevent forgetting and interference. The rehearsal methods use only a subset typically limited by the size of the memory used to save the examples. In the absence of previous task data, pseudo-rehearsal techniques collect information regarding the distribution of data and, through a generative model, generate fictitious data but similar to the distribution of the original dataset. These methods generally scale very well but, for a large number of tasks, they can run into problems of saturation of the available memory. On the other hand, *regularization-based methods* [9,21] act on the parameters of the model, discouraging the updating of neurons or layers deemed relevant for the individual tasks. The various methods differ on the type of penalty to be applied and how this is calculated. This allows consolidating the knowledge of the previous tasks, leaving the model the possibility to observe and learn data coming from the current task. The regularization methods, with rare exceptions, are not particularly computational expansive but they lack scalability. To preserve past knowledge, the neural network weights are "frozen". So, when the number of tasks increases

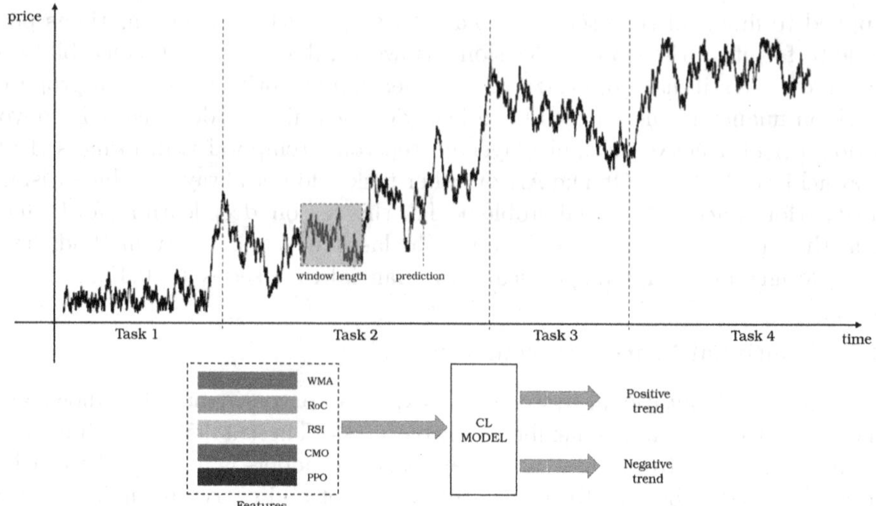

Fig. 1. Data flow of a financial time series in our setting. The prices time series is subdivided into tasks by a change-point detector (dashed red lines). Raw prices, relative to a certain period (determined by a window length), are turned into financial indicators, that become the inputs of the neural network. The problem consists of a binary classification to predict prices trend (positive or negative) at N time step later. (Color figure online)

over time, the neural network tends to saturate, bringing to low performance on the last tasks. Finally, *parameter isolation or architectural methods* [17] dedicate different parts of the model or make a different copy of if to each task exclusively. In the absence of constraints on the architecture, a new model's branch, with exclusive parameters, is instantiated for each new task. The final layers of the model involved in the classification can be exclusive or shared. The architectural methods are extremely expensive from the point of view of the memory and computational power required since there are N models for N tasks, but they have potentially infinite scalability.

3 Problem Formulation

A great advantage of trading using algorithms is the ability to process a large amount of data in real-time to extract as much information as possible. It is necessary to provide the model with data that best reflects the state of the market in each timestep to make the most appropriate investment choice. In this section, we present the data, their structure, and the features generated by the models.

3.1 Financial Time Series and Indicators

The input provided to the model consist of fixed-length sequences representing the trend of financial time series (daily close values) over a period of one month (i.e. 20 business days) of observations. Sequences are constructed by taking all possible consecutive 20-day windows in the available data. Each series is used to obtain the features and time sequences by building a different dataset, giving rise to an analysis of multivariate historical series for each financial series. Even if a neural network can extrapolate the information it deems relevant in a completely autonomous way, using only time series does not carry out any learning despite copying the label of the previous example in output. This fact demonstrates a dependence between time series even where they have no provable relationships. Consequently, it was necessary to manipulate the series by carrying out various engineering operations to obtain features. The first step is carried out to eliminate copying between outputs, a phenomenon that is not uncommon in time series and which is usually solved by creating a series obtained as the difference between data points of two consecutive instants. The input has doubled its dimensionality from a vector to a matrix. The increase in dimensionality is an aspect to be taken into account in a problem of this type: if, on the one hand, only the raw series is insufficient, on the other hand, using too many features could push the model to focus on unnecessary or worse misleading information. This aspect was taken into account in choosing which financial indicators to use as features that could provide helpful information. Thanks to moving averages, we can catch the real market trends, removing irrelevant fluctuations in the original series. Using *Weighted Moving Average*, we put a specific weight which decades overtime to any timestep, giving a higher impact on more recent timesteps. *Rate of Change* provides a first qualitative analysis regarding possible changepoints computing the percentual difference between the current timestep and another n step in the past. A minor variation means that current data do not differ from previous ones, while a bigger one indicates a possible change in the distribution. To measure variation speed and magnitude in time series, we used *Relative Strength Index*. This momentum oscillator reports the strength of the current time series trend, highlighting periods of excessive overconfidence and underconfidence of stocks. *Chande Momentum Oscillator* provides more fine-grain values in respect to RSI, enabling the model to confirm or deny results of the previous indicator while at the same time finding out new changepoints. Finally, we can keep track of market reversals with the *Percentage Price Oscillator*, another momentum oscillator able to compare moving averages with different temporal horizons. An exciting feature of the chosen indicators is that they collect information about the past and mediate it with the present. This allows forming features that represent the current state and a set of past timesteps used to constitute the feature itself. Many indicators are characterized by an observation period overtime in the series. Those listed below have been generated for 5 to 20 days to extract informative content both in the short and long term. Shorter periods could capture information that is too little specific for the sequence, just as long periods would provide information of a greater window than the observation window of the model itself.

3.2 Definition of Domain Regimes

The classic problems of continual learning do not deal with time series, but rather images. Consequently, it is necessary to understand which algorithms could be applied, taking into account that input data of a different nature lead to different assumptions. The first is the temporal aspect of the financial series, where the time series itself defines the order of tasks and examples. Namely, the only way to learn from these series is respecting the temporal order, hence excluding data shuffling or any kind of offline learning. However, this assumption does not indicate a limitation in the study of the problem since the training of the tasks is sequential. Continual learning is defined for different scenarios (task, domain, and class-IL) and the next step was to identify in which of these the problem treated is found. Since all the tasks come from the same time series, the data distribution of each task will also be the same; therefore, we can discard the class-IL scenario. Considering the problem of predicting the market by classification, for each task, the model receives as input a sequence equal to one month of data updated every day at the close of the markets, for a sequence length equivalent to 20 days. The reason for this choice is dictated above all by the opening and closing of the world market. The second reason for choosing a window of this size is given by the assumption that in a month of daily observations it is possible to collect a sufficient amount of data to make a reliable analysis. There is also a label associated with each sequence. There will be $y = 1$ if the value of the time series 20 days after the end of the sequence will be greater than the last data of the same, zero otherwise. This corresponds to one month of observation and prediction of the following month. The labels predicted by the model indicate how to act: a positive prediction suggests an increase in market value, and therefore it is generally advisable to buy. In contrast, a negative prediction invites to sell. For each task, the possible outputs will belong to the same domain and consequently, we are faced with a domain-IL scenario. In this scenario, there is no information regarding a task identifier, leading to discard architectural or parameter isolation methods. Figure 1 shows the data flow in our setting. Dashed red lines indicate a change of regime (task in the continual learning problem) in the raw time series; the network takes as input a series of financial indicators, derived from the raw prices series, relative to a period, defined by a window on the timeline; the neural networks operate a binary classification defining if the series exhibits a positive or a negative trend.

3.3 Continual Learning Techniques

In this section, all the examined algorithms will be explored, describing the tricks we sometimes employed to adapt these methods to the problem. In fact, not all of the state-of-the-art methods are suitable for the problem of classifying financial time series or do not offer the desired performance.

Gradient Episodic Memory. The early learning problems of multiple sequential tasks were based on Empirical Risk Minimization (ERM) [18], which defines the theoretical limits of the learning algorithm performance. This limitation comes from the inability to know the data distribution on which the algorithm will work. Gradient Episodic Memory (GEM) [11] was proposed as a learning method disconnected from data distribution and focused on an example by example observation. In particular, the classic example-label pair (x, y) is abandoned in favor of a triplet (x, y, t), where t is a task descriptor. Applied to financial time series, the task descriptor can be the task-id, as done in this work, or a complex structure describing the distribution to which the data belong. The main feature of this method is the episodic memory M_t capable of saving a subset of each task t. With T tasks and total available memory M, each task will have an exclusive memory equal to M/T. If the total number of tasks is not known a priori, it is possible to gradually reduce the number of examples for each task as the number of tasks increases. The goal of this method is to sequentially train a model on T tasks, preventing overwriting in future tasks with the constraint that training a task should not lead to worse performance in previous tasks. Given a triplet (x, y, t), the optimization problem to be solved is the following:

$$\begin{aligned} &\text{minimize}_\theta \quad \ell(f_\theta(x,t), y) \\ &\text{subject to} \quad \ell(f_\theta, \mathcal{M}_k) \leq \ell(f_\theta^{t-1}, \mathcal{M}_k) \text{ for all } k < t, \end{aligned} \tag{1}$$

The problem is complex to solve with this formulation, but it is possible to make two observations. The first one concerns the conservation of the parameters of each task: if the constraint on the loss is maintained, it is no longer necessary to save the state of the network at the end of each task. The second and, more important, allow us to represent the variation of the loss between two updates through the angle between the two gradient vectors if the function is locally linear, assumption valid between two gradient steps. The second observation allows us to rewrite the optimization constraints:

$$\langle g, g_k \rangle := \left\langle \frac{\partial \ell(f_\theta(x,t), y)}{\partial \theta}, \frac{\partial \ell(f_\theta, \mathcal{M}_k)}{\partial \theta} \right\rangle \geq 0, \text{ for all } k < t. \tag{2}$$

For every training step, there is, therefore, a system of k inequalities to resolve. It is easy to guess that this operation becomes more onerous as the number of tasks increases. In case of at least one constraint is not met, a gradient step in a new direction is required. This makes the optimization problem a QP-complete problem for that specific training step. Only approximations of the optimal solution are valid, and the authors of the method proposed a valid one using the dual problem. Buffer size plays a relevant role in the performance evaluation: if we use more memory, we could expect better performances. But, in time series problems, this becomes tricky. A too big memory should allow saving too much data, going to a pseudo-parallel training of more tasks, and, in a scenario where temporal order is fundamental, this aspect must be avoided.

Averaged Gradient Episodic Memory. Averaged Gradient Episodic Memory (A-GEM) [5] has been proposed as an optimization of the forerunner. To alleviate the weight of the computation, the authors opted for a relaxation of the constraints, going from loss reduction on examples of each of past tasks to an average on all the episodic memory. While the objective function to minimize remains the same, the constraints collapse to a single one valid for all past tasks.

As before, we can reformulate the objective function and the constraints regarding previous observation of the loss variation and gradient vectors:

$$\text{minimize}_{\tilde{g}} \quad \frac{1}{2}||g - \tilde{g}||_2^2 \quad \text{s.t.} \quad \tilde{g}^\top g_{ref} \geq 0 \tag{3}$$

where g_{ref} indicates gradient computed from a random batch obtained by the episodic memory from all previous tasks; in other words, A-GEM replaces $t-1$ inequalities with only one. However, it remains possible that the unique constraint is not met. In this scenario, there is no particular problem or approximation to compute but the solution is given by the projected gradient method:

$$\tilde{g} = g - \frac{g^\top g_{ref}}{g_{ref}^\top g_{ref}} g_{ref} \tag{4}$$

As GEM, this method exploits different algorithms to fill the buffer. Since in A-GEM, we got a single batch sampled from the whole buffer, the data distribution of the batch could not reflect the original distribution between tasks. We opted for a different strategy to fill the buffer to maintain the correct distribution, equal as far as possible to the stream one, using the reservoir sampling algorithm.

Synaptic Intelligence. A significant limitation in the development of neural networks capable of learning multiple sequential tasks lies in the one-dimensional structure of the neuron, leading a network to catastrophic forgetting. Defining which neurons are most responsible for learning is necessary to consolidate acquired knowledge on a task. The best way to assess how significant a neuron is for a task is to calculate its contribution to the global loss of the current task. In this way, at the end of each task, it will be possible to determine which neurons contribute most to the current task's learning and prevent their update in the future, maintaining knowledge of past tasks, thus avoiding forgetting. Synaptic Intelligence (SI) [21] does not require external memories or architecture variation. Still, it acts only on network parameters, defining an additional loss related to the state of the neurons themselves. When training on a new task, changes to important parameters are penalized to prevent old memories from being overwritten. We can compute weight importance ω_k^μ for each neuron θ_k related to task μ. During the training of a task, a learning trajectory $\theta(t)$ is described in the network parameter space. This trajectory will come as close as possible to a minimum for the loss function for each task. We can now consider an update $\delta(t)$ at time t, leading to a variation on the loss of the current task. This variation can be approximated by the gradient g_k and in such case the relation

$$\ell(\theta(t) + \delta(t)) - \ell(\theta(t)) \approx \sum_k g_k(t) \cdot \delta_k(t) \tag{5}$$

can be considered valid. The variation $\delta_k(t) = \theta'_k(t) = \frac{\partial \theta_k}{\partial t}$ therefore contributes to the variation of the global loss. If we want to compute the variation over the entire trajectory, we must sum up all small updates. This amounts to computing the path integral of the gradient vector along the parameter trajectory. Since the gradient is a conservative field, the value of the integral is equal to the difference in loss between the end point t_{end} and start point t_{start}. In addition, the integral can be decomposed as the sum of the impact of the importance ω_k^μ on loss variation. In practice, ω_k^μ is the online approximation of the running sum of the product of the gradient $g_k(t) = \frac{\partial L}{\partial \theta_k}$ with the update θ'_k. In a sequential tasks scenario, the model will have only a loss $\mathcal{L}_\mu$ available on the current task μ. Catastrophic forgetting occurs when minimizing $\mathcal{L}_\mu$ there is a significant increase of the loss $\mathcal{L}_v$ of past tasks $v < \mu$. In this context, the importance of parameters θ_k is determined by: 1) how much the parameter contributes to a loss drop and 2) the difference $\theta_k(t^\mu) - \theta_k(t^{\mu-1})$. To avoid a significant variation in these parameters, a modified loss has been proposed:

$$\widetilde{\mathcal{L}_\mu} = \mathcal{L}_\mu + c \sum_k \Omega_k^\mu (\widetilde{\theta_k} - \theta_k)^2 \tag{6}$$

with $\widetilde{\theta_k} = \theta_k(t^{\mu-1})$ and parameter c to mange regularization. Coefficient Ω_k^μ determines the regularization strenght of each parameter:

$$\Omega_k^\mu = \sum_{v<\mu} \frac{\omega_k^v}{(\Delta_k^v)^2 + \epsilon} \tag{7}$$

with $\Delta_k^v = \theta_k(t^v) - \theta_k(t^{v-1})$.

Elastic Weight Consolidation. Like the previous one, this method presented by Kirkpatrick *et al.* [9] is based on the possibility of determining a coefficient of importance for each neuron to be used in the computing of the global loss. Given task A, multiple valid network weights configurations θ_A ensure the same performances. In this way, when task B occurs, the model will maintain the performance binding model parameters in a solution space with low error for the previous task while maximizing performance on the new task B. Elastic Weight Consolidation (EWC) does not aim to find the optimal solution for each task but focuses on finding an intersection of low error solution space. To find which parameters are the most significant for a task, the authors addressed the problem from a probabilistic point of view. Optimizing the parameters of a network given the training set $\mathcal{D}$ is equal to probability $p(\theta|\mathcal{D})$. In presence of two independent tasks A($\mathcal{D}_A$) and B($\mathcal{D}_B$), with the Bayes' law we can write:

$$\log p(\theta|\mathcal{D}) = \log p(\mathcal{D}_B|\theta) + \log p(\theta|\mathcal{D}_A) - \log p(\mathcal{D}_B) \tag{8}$$

where the right-hand side depends only on the loss on task B $\log p(\mathcal{D}_B|\theta)$. Simultaneously, all the information regarding task A $\log p(\theta|\mathcal{D}_A)$ is soaked up by the posterior probability. Suppose the true posterior probability cannot be computed. In that case, we can obtain a good approximation from a Gaussian distribution with mean given by parameters θ_A and angular precision from the diagonal of the Fisher Information Matrix F. This matrix has three key properties: 1) it is equivalent to the second derivative of the loss near a minimum, 2) it can be computed from first-order derivates, and 3) it is guaranteed to be positive semi-definite. Given this approximation, the loss function $\mathcal{L}$ to minimize:

$$\mathcal{L}(\theta) = \mathcal{L}_B(\theta) + \sum_i \frac{\lambda}{2} F_i(\theta_i - \theta^*_{A,i})^2 \tag{9}$$

where $\mathcal{L}_B(\theta)$ is the loss for task B, λ sets the importance of old task compared to the new one. When moving to a third task C, EWC will try to keep the model parameters close to the learned parameters on both task A and B, where this can be enforced either with two separate penalties.

Experience Replay. The consolidation of acquired knowledge can occur in several ways in the human brain. One consists of periodic observation to consolidate the knowledge acquired but potentially overwritten over time. Experience Replay (ER) [16] uses an external memory buffer to save data from previous tasks, as example-label couple (x, y), without any reference about the task that the data belong to. During each training step, in addition to the batch of examples of the current task, another batch composed of examples from past tasks is sampled from the buffer, and the loss is computed on both of its, driven by hyperparameters α and β:

$$\mathcal{L} = \alpha \cdot \mathbb{E}_{(x,y)\sim D_t}[l(y, f(x))] + \beta \cdot \mathbb{E}_{(x,y)\sim M}[l(y, f(x))] \tag{10}$$

where $\mathcal{D}_t$ denotes training set of the current task and $\mathcal{M}$ the external memory.

Dark Experience Replay. Several CL algorithms have been proposed as improvements of ER. Dark Experience Replay (DER) [3] is one of these, and it relies on *dark knwowledge* for distilling past experiences, sampled over the entire training trajectory. Differently from the other rehearsal-based methods, this method retains the network's logits $z \triangleq h_{\theta_t}(x)$, instead of the ground truth labels y. This stratagem allows avoiding the loss of information due to the compression made by the final activation function. The corresponding loss function results:

$$\mathbb{E}_{(x,y)\sim D_t}[l(y, f(x))] + \alpha \mathbb{E}_{(x,z)\sim \mathcal{M}}\left[\, \|z - h_\theta(x)\|_2^2 \,\right] \tag{11}$$

This approach is related to *Knowledge Distillation* [8], a paradigm that allows the transfer of knowledge from a teacher to a student model. In particular, DER exploits a variant of this, known as self-distillation [7], in which transfer occurs between the same architecture. In this scenario, by saving logits of previous task examples, the model transfers knowledge to a version of itself in the future. Moreover, *Dark Experience Replay ++* has been proposed that equips Eq. 11 with an additional term on buffer datapoints, promoting higher conditional likelihood concerning their ground labels.

4 Experiments

4.1 Architecture

All the methods under examination were evaluated with the same architecture, using a fully-connected network composed of three hidden layers with respectively 100, 50, and 25 neurons each and with LeakyReLu as an activation function. All methods were tested using Stochastic Gradient Descent (SGD) with momentum as an optimizer for a total of 480000 training steps for each task. For rehearsal methods, we set the buffer size to 500 samples.

4.2 Evaluation Metrics

To properly assess learning quality at training time, it is mandatory to consider both single tasks as the whole training process. In other words, a CL algorithm should be evaluated both on the *past* and the *present* tasks to reflect in its behavior on the *future* unseen tasks. It is crucial to assess the ability to transfer knowledge across tasks to achieve this, along with average accuracy (ACC). More specifically, we would like to measure *Forward Transfer* (FWT) and *Backward Transfer* (BWT) [11] (*Forgetting* (FRG) [4] has been omitted because it is equal to BT except for the sign). The first one assesses the influence that learning a task t has on the performance on a future task $k > t$, whereas the second and third ones measure the performance degradation in subsequent tasks. FWT is computed as the difference between the accuracy before starting training on a given task and the random-initialized network, then averaged across all tasks. FRG and BWT compute the difference between the current accuracy and its best value for each task, presumably at the end of the training of the task itself. Except for FRG, the larger these metrics, the better the model. If two models have similar ACC, the preferable one is the one with larger BWT and FWT. While BWT measures the influence of a task on the previous ones and FWT the influence on the following ones, it is meaningless to discuss backward for the first task or forward for the last one.

4.3 Dataset

For the experimental analysis, two datasets have been employed. We used Brent Oil dataset[1], the historical series of the oil prices on a daily basis. In particular, we used 9282 time steps, collected between 02/01/1986 and 31/07/2021[2]. We also employed the copper dataset note1, consisting in 8500 time steps, taken from 02/01/1989 and 31/07/2021 note2. An example contains twenty consecutive daily observations. The next example is obtained by shifting the time window by one timestep. To provide more refined information to the model, it was decided to proceed towards an engineering of the features using some of

[1] https://datahub.io/core/oil-prices, https://help.yahoo.com/kb/SLN2311.html.

[2] Dates are reported as DD/MM/YY.

Table 1. Results of tested methods on Brent Oil dataset. For accuracy, backward and forward transfer bigger is better.

	Online	SI	EWC	ER	GEM	A-GEM	DER	DER++
Accuracy	65.04	70.34	71.64	72.29	65.25	65.47	72.59	**73.37**
Backward transfer	–	−6.06	−4.15	**−3.72**	−11.29	−5.74	−4.96	−4.11
Forward transfer	–	25.28	26.02	24.05	12.21	25.97	23.39	**27.24**

Table 2. Results of tested methods on copper dataset. For accuracy, backward and forward transfer bigger is better.

	Online	SI	EWC	ER	GEM	A-GEM	DER	DER++
Accuracy	58.28	65.46	68.01	68.00	54.23	58.86	64.09	**68.77**
Backward transfer	–	−5.08	−3.28	**−0.85**	−8.94	−6.74	−4.22	−3.11
Forward transfer	–	14.45	11.64	17.92	7.02	13.39	**14.32**	12.28

the most famous financial and statistical indicators, as explained in Sect. 3.1. Finally, for the definition of the various tasks within the time series, Bayesian Online Changepoint Detection (BOCD) was used, an online algorithm for the detection of changepoints, i.e. moments in which a significant change occurs in the data distribution. To verify the validity of this technique, we asked a financial expert to manually find out the change-points on the time series. The results almost completely coincide with those found by the algorithm, with an occasional variation of maximum 1 or 2 time steps. This allows the whole continual learning process to work without the need for further human intervention to detect regime changes, allowing us to use public datasets without further processing. Within each task, we split the data into two different sets: train and evaluation set. Between the two sets, we leave a gap excluding any sample whose evaluation time is posterior to the earliest prediction time in the validation set. This ensures that predictions on the validation set are free of look-ahead bias.

4.4 Quantitative Results

This section will discuss the results of each Continual Learning method, comparing them with the sequential training of each task without any continual technique, called *online learning*. Each method is evaluated not only with accuracy across all tasks but also with ad hoc Continual Learning metrics: *forward transfer* and *backward transfer*. In Table 1 and Table 2 are reported performances measured at the end of the whole training, respectively on the Brent oil dataset and on the copper dataset. To obtain less noisy performance estimates, values are reported as averages of three runs with different runs with different initialization. In the regularization methods, SI and EWC, we found an attenuation of the average forgetting across all tasks, even if the task nature heavily influences the accuracy of past tasks. This aspect is emphasized in SI, while EWC demonstrates good stability. Figure 2(a) shows the evolution of the accuracy of these

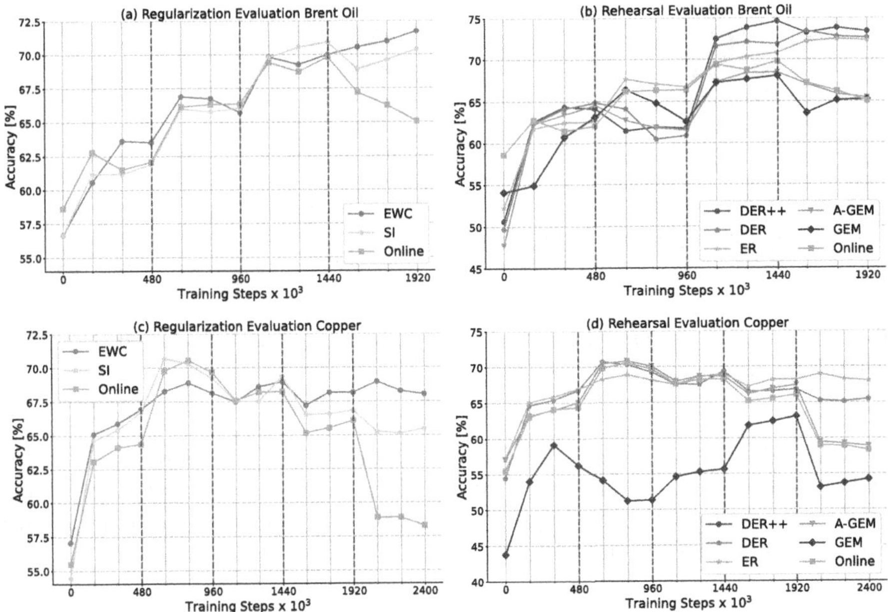

Fig. 2. Performance results. Dashed lines indicate task change: regularization methods performance on Brent Oil dataset (a), rehearsal method performance on Brent Oil dataset (b), regularization methods performance on the copper dataset (c), rehearsal method performance on the copper dataset (d).

methods during the training, compared to online learning on Brent oil dataset, while Fig. 2(c) on the copper dataset. SI and EWC experience similar accuracy on both datasets, while in the continual metrics, EWC performs slightly better. In replay-based methods (Fig. 2(b) on the Bernt oil dataset and Fig. 2(d) on the copper dataset), Gem and A-GEM are complex and hard to evaluate algorithms. GEM experiences the worst performance across all continual methods tested, especially on the copper dataset; probably due to the constraints violations introduced by the technique, event more and more frequently as the number of tasks increases. Although A-GEM significantly relaxes the constraints, accuracy remains aligned with baseline online learning while continual metrics are comparable to other methods. Vice versa, ER, DER and, DER++ experience the best performances. DER++ combines the features of the other two methods taking advantage of its: replaying past data from the buffer as ER, it uses logits information context as done in DER to improve further the predictions, bringing it to be the most performing method. In Fig. 3(a) and (b) are reported the training times of each method, respectively on Brent oil dataset and on the copper dataset. Again, the behavior is similar on both datasets, demonstrating the robustness of these algorithms on financial time series. SI demonstrates to be the quickest to be trained thanks to online estimate models weights. Simultaneously, the other regularization method, EWC, is the second most time-consuming

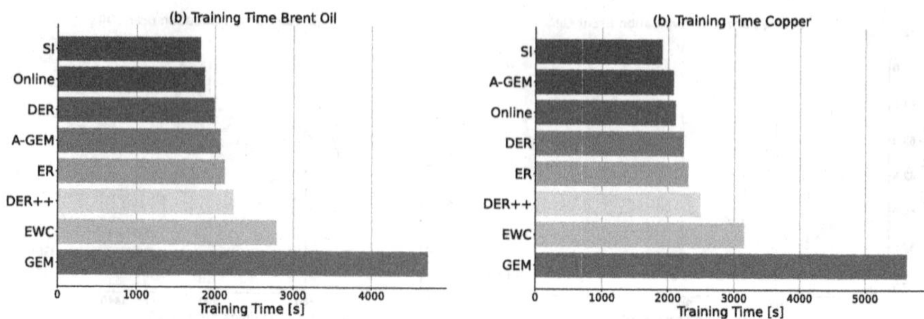

Fig. 3. Training time comparison on Brent Oil dataset (a) on the left and the copper dataset (b).

algorithm due to the necessity to compute the Fisher Information Matrix at the end of each task. GEM constraints, finally, do not make up a complexity element only from a computational perspective, but they also make this method the most time-consuming for the model training.

5 Conclusion

In this paper, an experimental analysis of continual learning algorithms on market predictions has been conducted, highlighting their significant contribution in the field of artificial intelligence applied to finance. A deep investigation of the most promising state-of-the-art continual learning algorithms has been made, discovering that not all of them are suitable to financial time series. Furthermore, we found that other factors such as training time, computational complexity, and memory requirements can be decisive in defining the scenario and choosing the most appropriate algorithm to apply. The formulation adopted represents only an exemplifying model of more complex dynamics, but the development of this tool could be a concrete help for professional traders in the future. As future work, a specific continual learning algorithm for financial time series could be designed, taking into account the variety and complexity of the markets.

Acknowledgement. Work supported by FF4EuroHPC: HPC Innovation for European SMEs, Project Call 1. FF4EuroHPC has received funding from the European High-Performance Computing Joint Undertaking (JU) under grant agreement No 951745.

References

1. Bertugli, A., Calderara, S., Coscia, P., Ballan, L., Cucchiara, R.: AC-VRNN: attentive conditional-vrnn for multi-future trajectory prediction. Comput. Vis. Image Underst. **210**, 103245 (2021)
2. Bertugli, A., Vincenzi, S., Calderara, S., Passerini, A.: Few-shot unsupervised continual learning through meta-examples. In: Neural Information Processing Systems Workshops (2020)

3. Buzzega, P., Boschini, M., Porrello, A., Abati, D., Calderara, S.: Dark experience for general continual learning: a strong, simple baseline. In: Neural Information Processing Systems (2020)
4. Chaudhry, A., Dokania, P.K., Ajanthan, T., Torr, P.H.S.: Riemannian walk for incremental learning: understanding forgetting and intransigence. In: Ferrari, V., Hebert, M., Sminchisescu, C., Weiss, Y. (eds.) ECCV 2018. LNCS, vol. 11215, pp. 556–572. Springer, Cham (2018). https://doi.org/10.1007/978-3-030-01252-6_33
5. Chaudhry, A., Ranzato, M., Rohrbach, M., Elhoseiny, M.: Efficient lifelong learning with A-GEM. In: International Conference on Learning Representations (2019)
6. Dixon, M.F., Halperin, I., Bilokon, P.: Machine Learning in Finance, 1st edn. Springer, Cham (2020). https://doi.org/10.1007/978-3-030-41068-1
7. Furlanello, T., Lipton, Z., Tschannen, M., Itti, L., Anandkumar, A.: Born again neural networks. In: International Conference on Machine Learning, pp. 1607–1616. PMLR (2018)
8. Hinton, G., Vinyals, O., Dean, J., et al.: Distilling the knowledge in a neural network. arXiv preprint arXiv:1503.02531 2(7) (2015)
9. Kirkpatrick, J.N., et al.: Overcoming catastrophic forgetting in neural networks. In: Proceedings of the National Academy of Sciences of the United States of America, vol. 114, no. 13, pp. 3521–3526 (2016)
10. Liu, X.Y., et al.: FINRL: a deep reinforcement learning library for automated stock trading in quantitative finance. In: Neural Information Processing Systems Workshops (2020)
11. Lopez-Paz, D., Ranzato, M.A.: Gradient episodic memory for continual learning. In: Advances in Neural Information Processing Systems (2017)
12. McCloskey, M., Cohen, N.J.: Catastrophic interference in connectionist networks: the sequential learning problem. Psychol. Learn. Motiv. **24**, 109–165 (1989)
13. Monti, A., Bertugli, A., Calderara, S., Cucchiara, R.: Dag-net: double attentive graph neural network for trajectory forecasting. In: International Conference on Pattern Recognition (2020)
14. de Prado, M.L.: Advances in Financial Machine Learning, 1st edn. Wiley Publishing, Hoboken (2018)
15. López de Prado, M.M.: Machine Learning for Asset Managers. Cambridge University Press, Cambridge (2020)
16. Rolnick, D., Ahuja, A., Schwarz, J., Lillicrap, T., Wayne, G.: Experience replay for continual learning. In: Neural Information Processing Systems (2019)
17. Schwarz, J., et al.: Progress & compress: a scalable framework for continual learning. In: International Conference on Machine Learning (2018)
18. Vapnik, V.: Principles of risk minimization for learning theory. In: Advances in Neural Information Processing Systems (1991)
19. Vaswani, A., et al.: Attention is all you need. In: Advances in Neural Information Processing Systems (2017)
20. van de Ven, G.M., Tolias, A.S.: Three scenarios for continual learning. In: Neural Information Processing Systems Workshops (2018)
21. Zenke, F., Poole, B., Ganguli, S.: Continual learning through synaptic intelligence. In: International Conference on Machine Learning (2017)

Mesoscale Events Classification in Sea Surface Temperature Imagery

Marco Reggiannini[1(✉)], João Janeiro[2,3], Flávio Martins[3,4], Oscar Papini[1], and Gabriele Pieri[1]

[1] Institute of Information Science and Technologies (ISTI), National Research Council of Italy, Pisa, Italy
{marco.reggiannini,oscar.papini,gabriele.pieri}@isti.cnr.it

[2] S2AQUA, Laboratório Colaborativo, Associação para uma Aquacultura Sustentável e Inteligente, Olhão, Portugal
joao.janeiro@s2aquacolab.pt

[3] Centre for Marine and Environmental Research (CIMA), University of Algarve,Faro, Portugal

[4] Higher Institute of Engineering (ISE), University of Algarve, Faro, Portugal
fmartins@ualg.pt

Abstract. Sea observation through remote sensing technologies plays an essential role in understanding the health status of marine fauna species and their future behaviour. Accurate knowledge of the marine habitat and the factors affecting faunal variations allows to perform predictions and adopt proper decisions. This is even more relevant nowadays, with policymakers needing increased environmental awareness, aiming to implement sustainable policies. There is a connection between the biogeochemical and physical processes taking place within a biological system and the variations observed in its faunal populations. Mesoscale phenomena, such as upwelling, countercurrents and filaments, are essential processes to analyse because their arousal entails, among other things, variations in the density of nutrient substances, in turn affecting the biological parameters of the habitat. This paper concerns the proposal of a classification system devoted to recognising marine mesoscale events. These phenomena are studied and monitored by analysing Sea Surface Temperature images captured by satellite missions, such as Metop and MODIS Terra/Aqua. Classification of such images is pursued through dedicated algorithms that extract temporal and spatial features from the data and apply a set of rules to the extracted features, in order to discriminate between different observed scenarios. The results presented in this work have been obtained by applying the proposed approach to images captured over the south-western region of the Iberian Peninsula.

Keywords: Image processing · Remote sensing · Mesoscale patterns · Sea surface temperature · Machine learning · Climate change

G. Nicosia et al. (Eds.): LOD 2022, LNCS 13810, pp. 516–527, 2023.
https://doi.org/10.1007/978-3-031-25599-1_38

1 Introduction

Evaluating the impact of climate change on coastal marine ecosystems may be a challenging task: near the coast, global drivers are modified by topography and by local atmospheric and oceanographic circulation patterns.

In particular, Ekman dynamics and large-scale thermocline processes control the coastal upwelling occurring at the Eastern Boundary Upwelling Ecosystems (EBUEs) [1,2]; winds directed towards the Equator drive upwelling, which transports deeper, colder and nutrient-rich waters to the surface. As a result, these areas host the most productive ecosystems in the global ocean [3], playing a major role in the marine primary production and the worldwide fisheries. Moreover, it was recently shown that upwelled water's low long-term warming rates may provide thermal refugia, stabilize changes in species distributions and enhance local biodiversity [4].

According to related literature, more than 71% of coastal zones are experiencing a net heat gain due to global warming [5]. Yet, both positive and negative trends were observed in different upwelling ecosystems [6]. Therefore, it is surmised that every upwelling ecosystem reacts differently to the changing climate.

Among the world's EBUEs, the Iberia/Canary Current System (ICCS) is one of the least studied [7]. Despite a general circulation similar to other EBUEs, in ICCS the discontinuity imposed by the Mediterranean Sea, combined with the seasonality of the large-scale atmospheric circulation, has a profound impact on the regional oceanography. The region's continental shelf is characterized by a large number of topographical features, such as prominent capes, promontories and submarine canyons, whose spatial scales are tens to hundreds of kilometers [8]. All the above highlight the importance of sub-seasonal temporal scales and sub-basin spatial scales, which explain the observed oceanographic patterns.

The identification and cataloguing of upwelling regimes occurring in an EBUE are important achievements towards the characterization of the system. Traditionally this task has been performed subjectively by experts, analysing Sea Surface Temperature (SST) maps of the area of interest. This procedure is manageable if few tens or even hundreds of images are used, but it turns into an unfeasible task as the number of scenes approaches thousands of images, that is the typical order of magnitude when the purpose is to investigate climate-related changes.

Nowadays, with the growing amount of remote sensing observations, automated techniques have been gaining momentum, using tools such as two-dimensional wavelet transforms [9], neural networks [10], or edge detection algorithms [11]. A complete automation encompasses three main challenges: (1) the presence of noise, mainly due to clouds and other atmospheric phenomena; (2) the fact that gradients are weak and provide an excess of information, generating several unconnected borderlines and complicating the structure identification task due to the fine edge detail; (3) the substantial morphological variation that prevents from an accurate geometric representation and the absence of a valid analytical model for the structures [12].

The main objective of this work is to design and develop automatic methods capable of accepting massive datasets of oceanographic SST imagery as input and returning a classification of the images according to the different regimes of observable upwelling patterns. The identification of a specific temperature pattern is based on the extraction of quantitative features from the SST maps. Indeed the emergence of a certain pattern is usually highly correlated with peculiarities in the temperature spatial arrangement at time fixed (e.g. the presence of abrupt variations in the temperature values within a certain neighbourhood), as well as with the observation of specific temperature trends at fixed locations, providing insights about the flowing of water masses between points at different temperature values. In a previous work dedicated to this topic [13], a custom visualisation tool was developed to extract and visualise the time series of the SST signals related to a given number of fixed locations within an area of interest. Based on this result, a novel step is introduced in the pipeline, with the objective of processing the mentioned signal series to extract quantitative descriptors of the signals trend. The computed quantities are finally used to fulfil the classification task by implementing a set of rules that assign each set of time series to a specific class according to the numerical values of the computed features.

The proposed method will be applied to the South Iberian region, contributing to understanding the formation of upwelling filaments and the effects of climate change in this particular EBUE. In its current form, the metrics used (e.g. the signal variation rate and its deviation from the mean value) are able to identify different types of mesoscale features.

The paper is arranged as follows: Sect. 2 concerns a detailed description of the employed dataset and the related ground truth classification; Sect. 3 thoroughly reports on the developed processing pipeline and describes a relevant use case; Sect. 4 concludes the paper by discussing the outcomes of this work and providing a few considerations about future perspectives.

2 Materials

2.1 SST Satellite Data

The identification and classification of upwelling events in a marine ecosystem have been performed by processing SST maps of an area of interest. These maps are compiled using data coming from satellite missions. In particular satellite data from the years 2009 to 2017 has been retrieved from two sources: EUMETSAT's *Metop* programme [14] and NASA's *Aqua* satellite [15].

Satellites of the Metop programme gather data through the Advanced Very High Resolution Radiometer (AVHRR); the information is processed at level L2P and binned in a single netCDF-4 file every 3 min, for a total of 480 images per day covering the entire globe, with a spatial resolution of 1km at nadir and an accuracy of 0.01 °C in the temperature measurement. We used data from the satellite Metop-A for the period 2009–2016 and from Metop-B for 2017.

The satellite Aqua uses the Moderate Resolution Imaging Spectroradiometer (MODIS) to gather data; the information is processed at level L2P and binned in a single netCDF-4 file every 5 min, for a total of 288 images per day covering the entire globe, with a spatial resolution of 1km at nadir and a temperature accuracy of 0.005 °C.

For both sources, only data covering the region of interest were downloaded. In particular, points with latitude between 35 ° and 40 ° N and longitude between 12 ° and 6 ° W were considered, resulting in 2–3 images per day at most. Moreover, since the data capture task is often prone to failures in registering SST values for areas that appear opaque (e.g. due to atmospheric events), a further selection has been performed by discarding images containing less than the 15% of the expected amount of data.

2.2 Types of Patterns

By looking at the SST images, four upwelling patterns have been identified:

1. a cold water filament going westwards, originating from the southward upwelling jet that runs along the western coast of Portugal;
2. a cold water filament going southwards, extending over Cape St. Vincent the upwelling jet mentioned above;
3. a clear stream of cool water running along the southern Iberian coast;
4. a warm countercurrent originating in the Gulf of Cádiz and running along the southern Iberian coast, eventually reaching Cape St. Vincent and turning northwards.

A more detailed description can be found in [13]. These four patterns will be called E1, E2, E3 and E4 respectively. Pattern E3 may be further divided into E3i, when the thermal gradient between the cool stream and the water in the Gulf of Cádiz is small, mostly occurring during winter; and E3u, with a more significant gradient. For our analysis we do not distinguish between these two subpatterns. Figure 1 shows some examples of the described patterns.

Based on this classification, a dataset with labels E1–E4 assigned to the corresponding SST maps has been provided by expert oceanographers. This represents the ground truth for the subsequent SST analysis.

3 SST Analysis

The main goal of the method proposed for the analysis of SST time series is to classify upwelling events exploiting the *dynamic* information contained in the temperature patterns, observed over a given time window. To this aim, the multiple SST signal sequences related to a given geographical area are extracted from the corresponding netCDF files and arranged in a single 2D plot, namely a *spaghetti plot* (see Fig. 2d). The resulting visualization allows for a direct and clearer interpretation of the ensemble of the SST trends in the considered area of interest. The software dedicated to this analysis has been developed within a Python framework.

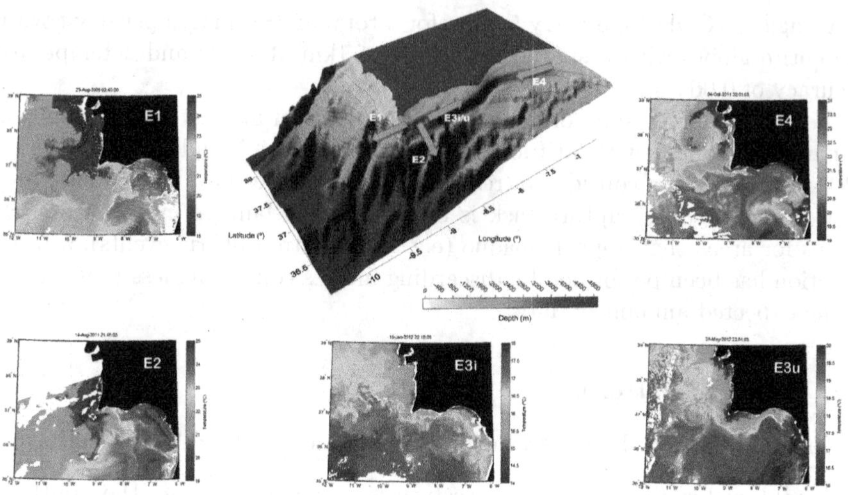

Fig. 1. Mesoscale patterns in the south-western Iberian Peninsula (from [13]).

3.1 Spaghetti Plot Generation

The steps fulfilled to generate a spaghetti plot for a given geographical area A are described below:

1. the rectangular area A selected on the SST map is split into a grid of N_A disjoint squares a_j, each with fixed size (typically between 0.01 and 0.25 degrees in latitude/longitude):

$$A = \bigcup_{j=1}^{N_A} a_j;$$

2. at a given time t (recorded in the netCDF file) the mean value of the SST signal is estimated by averaging the available n_j SST values, located within the corresponding a_j:

$$\mu_j(t) = \frac{1}{n_j} \sum_{i=1}^{n_j} \mathrm{SST}_i(t);$$

3. the previous step is repeated for each a_j and for each t within the considered time window, eventually returning N_A time series of the averaged SST signal;
4. the spaghetti plot is finally generated by simultaneously plotting all the N_A signals within the same coordinate system.

Each square a_j, and the corresponding averaged signal $\mu_j(t)$, is colour-coded so that the differences in the signal trends observed in different squares can be easily recognised by visual inspection. The reader can refer to [13,16] for further information about the generation of the spaghetti plots.

Figure 2 shows the case of an event classified as E3 in the ground truth, together with the corresponding spaghetti plot. This will be exploited in the following as a case study to describe step by step the implemented processing stages and the corresponding output. The spaghetti plot has been computed dividing the region of interest in squares of size 0.25 ° and considering a 15 days window for the time range, with the last observation coinciding with the classified event. The specified resolution values represent input parameters to the spaghetti plot generation task. They have been set up through empirical considerations after testing alternative values, and assessing that the mentioned choice achieves a better agreement with the ground truth. Indeed, it can be noticed that the temperature curves diverge and decrease, starting at times t beyond 8 September, in agreement with the temperature trends of an E3 pattern.

3.2 Features Extraction

Based upon the ground truth dataset, the corresponding set of spaghetti plots has been generated. The objective of the task described in this section is to investigate the discriminating properties of a number of features extracted from the signals. Generally speaking these features can be computed based on one or more SST sequences belonging to the same spaghetti plot. The goal is to define a set of rules to be applied to the extracted features, with the purpose of identifying the class of the observed mesoscale pattern.

As mentioned before, a single curve p_j in a spaghetti plot represents the spatially averaged trend of the signal in the j-th square of the grid. It is represented by the following array of n pairs, where n is the number of time samples:

$$p_j = \{(t_m, \mu_j(t_m)) \mid m = 1, \ldots, n\}.$$

Notice that n is not a constant value: it depends on the quantity and quality of the data actually exploitable in the netCDF file. Therefore, n can be considered as a reliability index, reflecting the confidence level associated with the estimators described hereafter.

The following statistics are computed:

1. the temporal *mean* of p_j,

$$\mu(p_j) = \frac{1}{n} \sum_{m=1}^{n} \mu_j(t_m);$$

2. the *standard deviation* of p_j,

$$\sigma(p_j) = \sqrt{\frac{1}{n} \sum_{m=1}^{n} (\mu_j(t_m) - \mu(p_j))^2};$$

3. the *linear regression coefficient* $\theta(p_j)$, defined as the slope of the straight line that better fits the curve p_j.

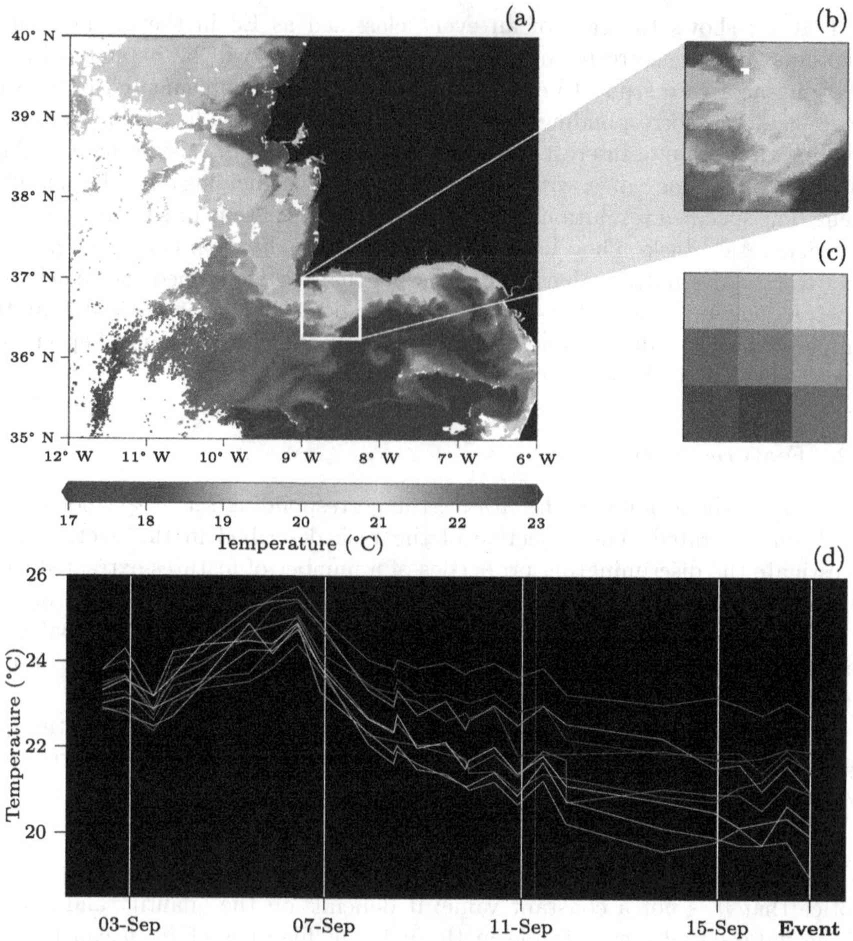

Fig. 2. Event of 16 September 2016 at around 21:35 UTC. (a) SST map at the date of the event; (b) detail of the SST in the reference area for the spaghetti plot (latitude between 36.25 ° and 37 ° N, longitude between 9 ° and 8.25 ° W, resolution 0.25 °); (c) reference grid; (d) generated spaghetti plot.

The three statistics are meant to describe in first approximation the local behaviour of the SST in the square a_j. Additional parameters, such as the coefficients of the quadratic regression for the curve p_j, have been taken into consideration, but finally discarded since they don't seem to capture the SST trend, as their integration in the pipeline didn't improve the classification performance. Notice that the choice of the linear regression coefficient $\theta(p_j)$ *does not imply* that we assume a linear correlation between SST and time during an upwelling event: that value is interpreted as a descriptor of the SST trend in the square a_j.

Figure 3 shows the statistics computed for the case study under investigation.

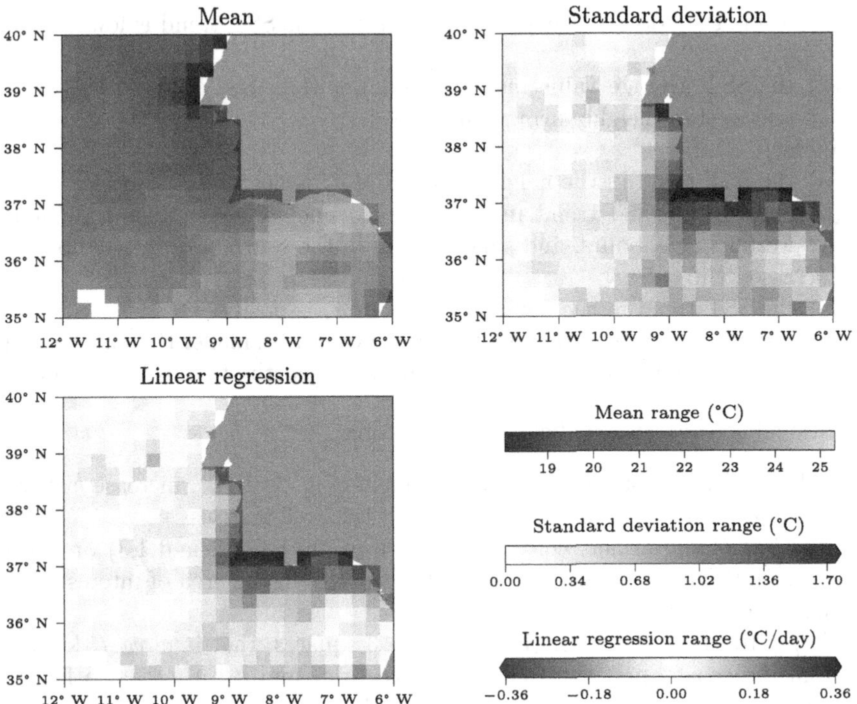

Fig. 3. Computed statistics for the period between 2 and 16 September 2016.

3.3 Classification Rules

Considering a square a_j in the area of interest, a score array (e_1, e_2, e_3, e_4), with values e_i normalized in $[0, 1]$, is defined. The value e_i represents a belief index for the corresponding event Ei to have occurred inside a_j at the end of the considered time range. Each e_i is obtained by applying a set of conditional rules to the statistics described in Sect. 3.2, computed inside both a_j and the neighbouring squares. The rules are modelled on the *a priori* knowledge of the oceanographic patterns, so that the score e_i is increased (by a fixed amount) only if the behaviour of the features μ, σ and θ, inside and in the neighbourhood of the square a, matches the one observed in the case of an Ei pattern. A qualitative description of these rules is reported below.

1. Increase e_1 if:
   (a) the SST trend $\theta(p_j)$ inside a_j is negative;
   (b) inside the eastern neighbouring squares, the SST trend is lower than in a_j;
   (c) the SST average value $\mu(p_j)$ inside a_j is lower than SST in both the northern and southern neighbouring squares.
2. Increase e_2 if:
   (1) the SST trend inside a_j is negative;

  (2) inside the northern neighbouring squares, the SST trend is lower than in a_j;
  (3) the SST average value inside a_j is lower than SST in both the eastern and western neighbouring squares.
3. Increase e_3 if:
   (a) the SST trend in the neighbourhood of a_j is negative;
   (b) inside a_j the SST trend and its average value are larger than in the north-western neighbours, and smaller than in the south-eastern neighbours.
4. Increase e_4 if:
   (a) the SST trend in the neighbourhood of a_j is positive;
   (b) inside a_j the SST trend and its average value are smaller than in the south-eastern neighbours, and larger than in the north-western neighbours.

Additional considerations affect the final scores:

- if the SST variation $\sigma(p_j)$ is large (namely $\sigma(p_j) \geq 1\,°\text{C}$), increase either e_1, e_2 and e_3 (if the SST decreases) or e_4 (if the SST increases);
- if a_j is globally either "cold" (in case of events E1, E2 and E3) or "warm" (event E4) with respect to all the other squares in the area of interest, boost the corresponding scores;
- if a_j is too near the coast (i.e. less than 3 squares, that is *circa* 75 km away from the coast), penalize (halve) the scores e_1 and e_2; if a_j is too far from the coast (i.e. more than 3 squares), penalize e_3 and e_4.

The application of the described classification rules to the case study is displayed in Fig. 4, where the four scores have been computed for each square in the grid.

In order to classify a_j, the maximum score $e_m = \max\{e_1, e_2, e_3, e_4\}$ is considered and, in case it is larger than a certain threshold (empirically decided), the square is marked with the corresponding "Em" colour label. If none of the scores exceeds the threshold, no label is assigned. Referring to our case study, the final outcome of the classifier is shown in Fig. 5, representing a heatmap where each square is labeled according to this rule, together with the percentage of the available SST data.

4 Discussion and Conclusion

In this work, a methodology for the analysis of SST time series has been proposed, with the objective of automating the classification of upwelling events by exploiting the dynamic information observed in the SST patterns. This ongoing study involves the analysis of large imagery datasets, using expert knowledge to contour the positions of the mesoscale feature, and eventually aiming at its characterisation.

The current results are promising and show patterns of differentiation among different mesoscale events occurring in the analysed area. Throughout Sect. 3, we

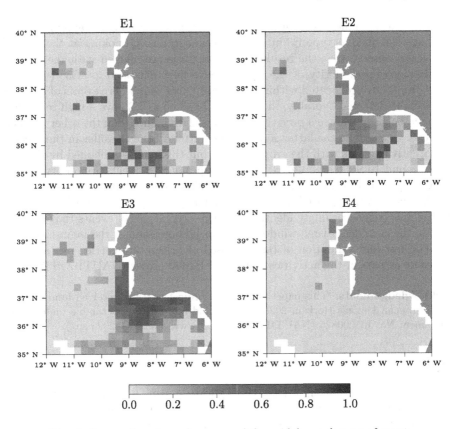

Fig. 4. Scores given to each square of the grid for each type of event.

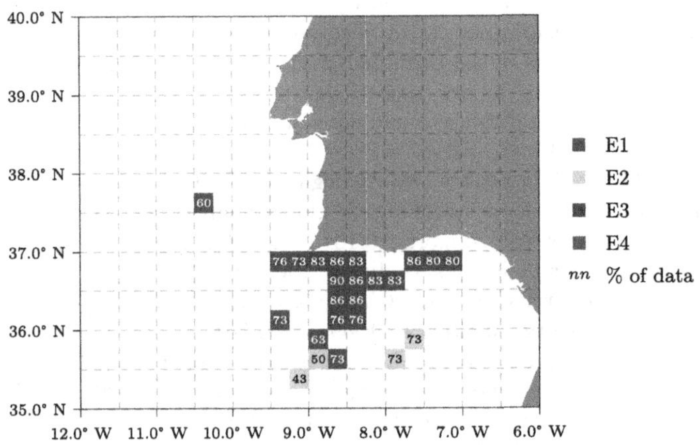

Fig. 5. Labels given to each square of the grid, depending on their scores.

presented a case study of an upwelling event that occurred on 16 September 2016 and was classified as type E3 by expert oceanographers. Looking at the map of Fig. 5, we observe that the results of our analysis are aligned with the ground truth, since the majority of the squares have been classified as "E3" (the blue ones), and they are located within the geographical zone where events of type E3 are usually detected. Notably, the squares classified as E3 have a high reliability value n, i.e. they feature a percentage of valid data higher than the other labelled squares. Concerning the latter squares, the output of the classifier in those cases apparently conflicts with the ground truth, but this can be explained by the fact that around 16 September 2016, different types of SST events are observed (e.g. an E1-type event occurs on 17 September). This indeed confirms the correctness of the classification procedure that, employing a "time series based" approach to the analysis, accordingly identifies multiple typologies of events that develop within the considered time window in the neighbourhood of the main event.

The test and validation of the proposed algorithm are carried out and will continue as part of the activities of the EU H2020 project NAUTILOS [17].

Acknowledgements. This paper is part of a project that has received funding from the European Union's Horizon 2020 research and innovation programme under grant agreement No. 101000825 (NAUTILOS[1];https://www.nautilos-h2020.eu/).

References

1. Messié, M., Ledesma, J., Kolber, D.D., Michisaki, R.P., Foley, D.G., Chavez, F.P.: Potential new production estimates in four eastern boundary upwelling ecosystems. Prog. Oceanogr. **83**(1), 151–158 (2009). https://doi.org/10.1016/j.pocean.2009.07.018
2. Ramajo, L., et al.: Upwelling intensity modulates the fitness and physiological performance of coastal species: Implications for the aquaculture of the scallop *Argopecten purpuratus* in the Humboldt Current System. Sci. Total Environ. **745**, 140949 (2020). https://doi.org/10.1016/j.scitotenv.2020.140949
3. FAO: The state of world fisheries and aquaculture 2018, meeting sustainable development goals (2018)
4. Varela, R., Lima, F.P., Seabra, R., Meneghesso, C., Gómez-Gesteira, M.: Coastal warming and wind-driven upwelling: A global analysis. Sci. Total Environ. **639**, 1501–1511 (2018). https://doi.org/10.1016/j.scitotenv.2018.05.273
5. IPCC: Summary for policymakers. In: Pörtner, H.-O. (eds.) IPCC Special Report on the Ocean and Cryosphere in a Changing Climate. Cambridge University Press (2019). https://doi.org/10.1017/9781009157964.001
6. Varela, R., Álvarez, I., Santos, F., de Castro, M.T., Gómez-Gesteira, M.: Has upwelling strengthened along worldwide coasts over 1982–2010? Scientific Rep. **5**, 10016 (2015). https://doi.org/10.1038/srep10016
7. Chavez, F.P., Messié, M.: A comparison of eastern boundary upwelling ecosystems. Prog. Oceanogr. **83**(1), 80–96 (2009). https://doi.org/10.1016/j.pocean.2009.07.032
8. Relvas, P., et al.: Physical oceanography of the western Iberia ecosystem: latest views and challenges. Progress Oceanography **74**(2), 149–173 (2007). https://doi.org/10.1016/j.pocean.2007.04.021

9. Liu, A.k., Peng, C.Y., Chang, S.Y.-S.: Wavelet analysis of satellite images for coastal watch. IEEE J. Oceanic Eng. **22**(1), 9–17 (1997). https://doi.org/10.1109/48.557535
10. Kriebel, S.K.T., Brauer, W., Eifler, W,: Coastal upwelling prediction with a mixture of neural networks. IEEE Trans. Geosci. Remote Sens. **36**(5), 1508–1518 (1998). https://doi.org/10.1109/36.718854
11. Simpson, J.J.: On the accurate detection and enhancement of oceanic features observed in satellite data. Remote Sens. Environ. **33**(1), 17–33 (1990). https://doi.org/10.1016/0034-4257(90)90052-N
12. Lea, S.M., Lybanon, M.: Automated boundary delineation in infrared ocean images. IEEE Trans. Geosci. Remote Sens. **31**(6), 1256–1260 (1993). https://doi.org/10.1109/36.317437
13. Reggiannini, M., Janeiro, J., Martins, F., Papini, O., Pieri, G.: Mesoscale patterns identification through SST image processing. In: Proceedings of the 2nd International Conference on Robotics, Computer Vision and Intelligent Systems – ROBOVIS, pp. 165–172. SciTePress (2021). https://doi.org/10.5220/0010714600003061
14. OSI SAF: Full resolution L2P AVHRR Sea Surface Temperature MetaGRanules (GHRSST) - Metop (2011)
15. NASA/JPL: GHRSST Level 2P Global Sea Surface Skin Temperature from the Moderate Resolution Imaging Spectroradiometer (MODIS) on the NASA Aqua satellite (GDS2) (2020)
16. Papini, O., Reggiannini, M., Pieri, G.: SST image processing for mesoscale patterns identification. Eng. Proceed. **8**, 5 (2021). https://doi.org/10.3390/engproc2021008005
17. Pieri, G., et al.: New technology improves our understanding of changes in the marine environment. In: Proceedings of the 9th EuroGOOS International Conference. EuroGOOS (2021)

Adapting to Complexity: Deep Learnable Architecture for Protein-protein Interaction Predictions

Junzheng Wu[1], Eric Paquet[2], Herna L. Viktor[1](✉), and Wojtek Michalowski[3]

[1] School of Electrical Engineering and Computer Science, University of Ottawa, Ottawa, ON, Canada
hviktor@uottawa.ca
[2] Digital Technologies Research Centre, National Research Council, Ottawa, ON, Canada
[3] Telfer School of Management, University of Ottawa, Ottawa, ON, Canada

Abstract. Protein-protein interactions play an important role in the development of new therapeutic treatments and prophylactic vaccines. For instance, the efficacy of a vaccine strongly depends to what extent an antibody may form a stable bond with an antigen. In-laboratory experiments are both time-consuming and expensive, which limits their scope to only the most relevant interactions. Computational experiments, on the other hand, have the potential to explore and screen a vast number of possibilities, thus providing experimentalists with the most promising cases. Protein-protein interactions may be learned by deep learning networks. Nonetheless, the training of these networks requires a large number of instances which may not be readily available, as in the case, of rare or new diseases. Furthermore, the learning process is made more complex by the scarcity of data about non-interacting proteins, making the network prone to overfitting. These two shortcomings are addressed in this paper. A new learnable pyramid network architecture is proposed in which the depth and the complexity of the network are directly learned from the data, which makes the network suitable for both small and large datasets. Our network outperforms state-of-the-art neural networks for protein-protein interaction predictions and is capable of adapting its architecture to datasets whose size varies from a few thousand to seventeen million. A classification accuracy of over 96% is achieved for all datasets.

Keywords: Protein-protein interaction · Deep learning · Adaptive learning

1 Introduction

The interaction between proteins is the origin of a multitude of cellular functions, such as signal transduction, cellular organization, and cell cycle progression [1]. Protein-protein interactions (PPIs) are paramount in understanding biological

G. Nicosia et al. (Eds.): LOD 2022, LNCS 13810, pp. 528–542, 2023.
https://doi.org/10.1007/978-3-031-25599-1_39

processes, and in designing therapeutic and prophylactic (preventive) proteins [15]. To investigate their behavior, high-throughput experimental approaches have been designed, such as chip-based techniques [25], microarrays [16], and interactomics [21]. Unfortunately, in vitro methods are time consuming and expensive as opposed to in silico approaches. Computational methods, especially those based on machine learning, have been successfully applied to PPI.

Most deep learning techniques employed in PPI originate from other fields: convolutional neural networks were introduced in computer vision, whereas LSTM neural networks were initially intended for time series. In contrast to the datasets used in these fields, which are large, PPI datasets are usually quite small. For instance, the ImageNet dataset [5] consists of 1,281,167 images, whereas Guo's dataset, which is the most commonly employed dataset in PPI, consists of only 60,671 protein pairs. A small dataset imposes strong limitations on deep learning algorithms, as there is a serious risk of overfitting when a complex model is employed. As a result, models were intentionally kept simple and the number of parameters was kept small to prevent overfitting. Depending on the dataset and the application, the number of protein pairs may vary from small to large and, irrespectively of the dataset size, the amino acidic complexity may range from low to high. This issue must be addressed by selecting the right architecture and depth for the neural network. This is usually done by inspection, which is a time-consuming and ad hoc approach. In addition, most protein interaction datasets, especially large ones, lack negative examples, i.e., non-interacting proteins. This is an important issue, as the algorithm must not only learn if there is an interaction, but also be able to detect the nonexistence of such an interaction. Indeed, for therapeutic or prophylactic vaccines, an antibody that interacts with an antigen may constitute a good candidate for a vaccine whereas a non-interacting antibody is not.

In this paper, we introduce a novel shallow-deep protein-protein interaction (SDPPI) deep learning architecture, which dynamically learns its architecture and depth directly from the dataset. In our work, the network learns the best network, or the best combination of networks (from shallow to deep) that minimizes both the loss function and the overfitting.

The paper is organized as followed. Our SDPPI network is presented in Section II. Our experimental results are then reported in Section III, including the network adaptability to complexity and datasets. Section IV concludes the paper.

2 Shallow-deep Protein-Protein Interaction Network Architecture

2.1 General Architecture

The architecture of our SDPPI neural network is reported in Fig. 1. Our network belongs to the Siamese network category, as both the top and the bottom branch share the same parameters and the same architecture. However, it differs from traditional Siamese networks in a number of ways.

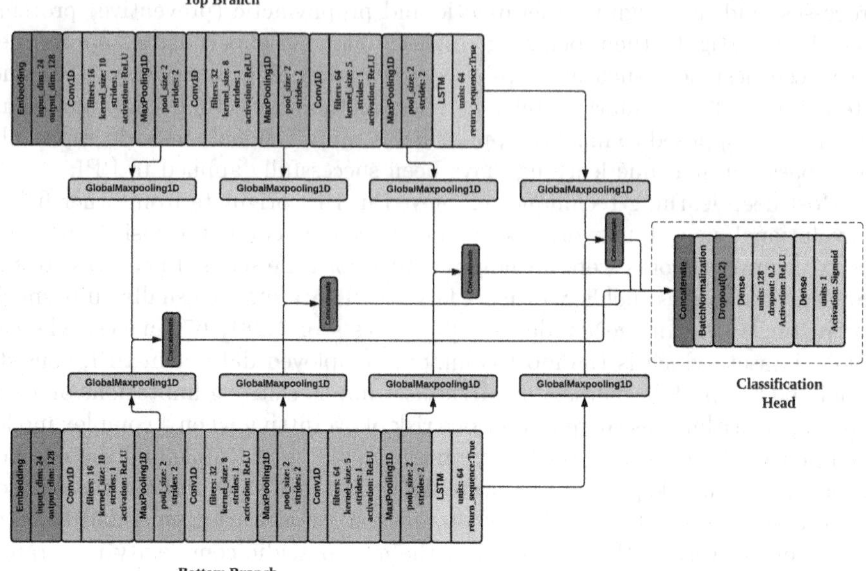

Fig. 1. Architecture of our shallow deep protein-protein interaction (SDPPI) neural network.

That is, as opposed to Siamese networks, both branches interact periodically. Such a network is called a pyramidal network. These couplings allow the network to analyze the incoming information at various complexity levels. For each complexity level i, the feature maps from both the top L_i (ligand) and the bottom R_i (receptor) branches are combined or concatenated into a single tensor V_i, as described by:

$$V_i = L_i \oplus R_i, \quad i = 0, 1, \ldots, n \tag{1}$$

Given a complexity level i, the subnetwork associated with this particular level may be assimilated to a nonlinear function f_i, which acts recursively on the previous levels $L_i = f_i(L_{i-1})$. As a result, the resulting concatenated tensor at complexity level j is given by:

$$V_j = \bigoplus_{i=0}^{j-1} V_i = \bigoplus_{i=0}^{j-1} [L_i \oplus R_i], \quad j = 1, 2, \ldots, n \tag{2}$$

Therefore, the feature map tensors are stacked together from shallow to deep levels. *Shallow neural networks process features directly, whereas deep networks extract features automatically during the training* [2]. The resulting concatenated feature maps are processed by the classification head, which consists of a batch normalization layer, a dropout layer, and dense layers. The normalization as well as the randomness introduced by the dropout layer serve as a regularization mechanism that aims to minimize overfitting. These layers do not affect the order in which the various complexity levels are concatenated.

The first dense layer, which is called the selecting dense layer, contains a layer of neurons that is activated using a ReLU function to obtain positive weights for the weighting of the various layers. If a shallow network is learned, the weights corresponding to the deeper layers will be small, whereas if a deeper network is learned, most weights will be non-negligible. The last layer consists of a single neuron having a sigmoid activation function that predicts the binding probability. The sigmoid function is employed to generate a probability between zero and one. The latter is converted to a binary classification: the proteins are binding if the probability is greater than 0.5, and non-binding otherwise. The selecting dense layer adaptively attributes weights to the concatenated layers according to:

$$Z_j = \mathrm{ReLU}\left(W_j \cdot V_j + b_j\right) \tag{3}$$

In other words, the hierarchical and periodical interaction between the two branches allows the selecting dense layer to discard certain layers, if deemed appropriate by the loss function, thus resulting in a shallower network. Furthermore, the selecting dense layer also constrains the gradient of the loss function. Indeed, when the weights associated with a particular complexity level increase as a result of the action of the selecting dense layer, their contribution to the back propagation gradient increases as well. Therefore, the selecting dense layer attributes the most weight to the most accurate levels. For instance, if the deepest level (V_4) has the highest accuracy, the weights attributed to the shallower layers are reduced by the selecting dense layer accordingly, along with their gradients (remember that the deepest level includes all the previous results, as specified by (2).

In general, if the number of instances is small, a shallower network should be chosen by the selecting dense layer. Indeed, a shallower network performs better than a deep network because the latter has a large number of parameters, which makes it prone to overfitting [2]. Comparatively, if the number of instances is large, a deeper network is selected, which is more suitable for complex datasets due to its large latent space. As pointed out by [24], a deep learning model with p parameters trained with instances of dimension d is quite likely to be subject to overfitting if the number of instances is greater than $(p-d)/2$. Therefore, the complexity and the depth of the model is strongly constrained by the size of the underlying dataset. This is why the model must adapt its architecture accordingly.

Therefore, our SDPPI employs a shallower network when the dataset is too small to train a deeper network properly. The selecting dense layer assigns greater importance to the shallower networks by increasing their weights while reducing the weights associated with the deeper networks. As a result, most of the gradient originates from the shallower networks. On the other hand, a deeper network is employed by the selecting dense layer if the number of instances is large. The selection is performed by attributing more weight to the deeper networks and less to the shallower networks, as well as their respective gradients. Therefore, SDPPI may simulate networks of various depth and determines the network architectures that are the most suitable for a given dataset irrespective of its size and without recourse to arbitrary rules or human intervention. According to the complexity of the dataset, a superposition of shallow and deep networks may be generated.

The action of selecting the dense layer may be explained mathematically as follows. Let us consider an SDPPI network with interaction levels. The input of the selecting dense layer is given by (2) Therefore, its input is the direct sum of each interaction level. The direct sum may be expressed as:

$$V = [V_0, V_1, \ldots V_n] \tag{4}$$

It follows that the weight matrix, associated with the dense selecting layer, may also be partitioned into submatrices; one for each level:

$$W = [W_0, W_1, \ldots, W_n] \tag{5}$$

As a result, the output of the selecting dense layer becomes:

$$O = \eta\,(V \cdot W) = \mathrm{ReLU}\left(\sum_{i=o}^{n} V_i W_i + b_i\right) \tag{6}$$

It follows that the depth of an SDPPI network is determined by the weight of its respective submatrices$\{W_i\}_{i=0}^{n}$: all their elements become extremely small when certain levels are not required by the network. When backpropagation is performed, each submatrix W_i is updated according to:

$$\begin{aligned} W_i &\leftarrow W_i - \alpha \frac{\partial \mathcal{L}}{\partial O} \frac{\partial O}{\partial\,(V \cdot W)} \frac{\partial\,(V \cdot W)}{\partial W_i} \\ &= W_i - \alpha \frac{\partial \mathcal{L}}{\partial O} \frac{\partial O}{\partial\,(V \cdot W)} V_i \end{aligned} \tag{7}$$

where α is the learning rate and $\mathcal{L}$ is the cost or loss function. Therefore, it is apparent from (7), that the weights at a given level are determined by the input of the selecting dense layer at the very same level V_i. The weights are updated in accordance with their effect on the cost function: they are increased if they reduce the cost function and decreased otherwise. This, in turn, determines which levels and depth are chosen by the selecting dense layer, thereby allowing the best architecture for the training dataset to be learned. As a result, the gradient assigned to each level becomes:

$$\delta V_i = \frac{\partial \mathcal{L}}{\partial O} \frac{\partial O}{\partial\,(V \cdot W)} \frac{\partial\,(V \cdot W)}{\partial V_i} \tag{8}$$

where the last member of (8) is given by:

$$\frac{\partial\,(V \cdot W)}{\partial V_i} = W_i + \frac{\partial V_{i+1}}{\partial V_i} W_{i+1} + \ldots + \frac{\partial V_n}{\partial V_i} W_n \tag{9}$$

Let us define:

$$\begin{aligned} L_i &\triangleq L(V_i) \\ R_i &\triangleq R(V_i) \end{aligned} \tag{10}$$

and the recursive function:

$$V_{i+1} = g(V_i) = f_i(L(V_i)) \oplus f_i(R(V_i)) \tag{11}$$

Then, (9) may be written as:

$$\frac{\partial (V \cdot W)}{\partial V_i} = W_i + \sum_{j=i+1}^{n} \frac{\partial g_{j-1}(\ldots g_i(V_i))}{\partial V_i} W_j \tag{12}$$

which means that greater gradient is attributed to V_i, implying that the network tends toward learning an architecture of depth i.

The ability of SDPPI to prevent overfitting may be understood by taking inspiration from ResNet [8]. Indeed, as stated earlier, each layer may be assimilated to a function. As a result, the output of a given branch at level j is:

$$\begin{aligned} L_j &= f_{j-1}(\ldots f_1(f_0(L_0))\ldots) \\ R_j &= f_{j-1}(\ldots f_1(f_0(R_0))\ldots) \end{aligned} \tag{13}$$

Then, the concatenated tensor becomes:

$$\begin{aligned} V_j &= \bigoplus_{i=0}^{j-1} [L_i \oplus R_i] \\ &= \bigoplus_{i=0}^{j-1} [f_i(\ldots f_0(L_0)\ldots) \oplus f_i(\ldots f_0(R_0)\ldots)] \end{aligned} \tag{14}$$

This means that the network input(L_0, R_0) appears at each complexity level, which implies that the gradient propagates through multiple paths, one for each complexity level. The periodical connections between the two branches may be assimilated to skip connections, as skip connections are extra connections between nodes in different layers of a neural network that skip one of more layers of nonlinear processing [17].

These connections constitute additional paths for the gradient to propagate along. As a result, error accumulation is alleviated, convergence is accelerated, and the number of parameters is reduced, thus mitigating overfitting. The skip connections improve feature reusability while facilitating the training process (parameter optimization). Indeed, operations such as maximum pooling and convolutions discard information as a result of downsampling. As it is lost, the information cannot be propagated to subsequent layers. The skip connections present in SDPPI allow re-inputting the discarded information into the network at each resolution level. The space associated with the loss function has a very large number of dimensions, which is equal to the number of parameters of the network. These spaces tend to have multiple local minima that make optimization of the network more difficult. Recently, Li demonstrated *et al.* [11] that skip connections contribute to the elimination of these detrimental minima, which makes the search space smoother, thus facilitating the optimization process. This is illustrated in Fig. 2.

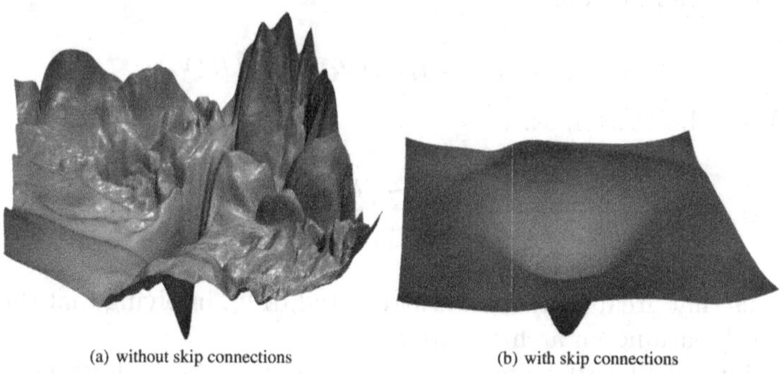

Fig. 2. Loss function parameter space with and without skip connections (with permission of the author [11])

2.2 Architecture of the Various Components

Our network is constructed of six components: the input and embedding layers, the convolutional and maximum pooling layers, the LSTM layer, the global maximum pooling layer, and the dense layers.

Input and Embedding Layers: There are 23 standard amino acids, namely alanine, arginine, asparagine, aspartic acid, cysteine, glutamine, glutamic acid, glycine, histidine, isoleucine, leucine, lysine, methionine, phenylalanine, proline, serine, threonine, tryptophan, tyrosine, valine, selenocysteine, and pyrrolysine. Each categorical amino acid is encoded with a number between 1 and 23, whereas ambiguous or unknown amino acids are attributed a value of 24. As mentioned earlier, sequences are restricted to 1,200 amino acids. Smaller amino acid sequences are padded with zeros. Therefore, the dimension of the input vector is fixed at 1,200. Naturally, the names (categories) neither reflect the similarity between amino acids nor their physicochemical properties. To solve this problem, a trainable embedding layer is added. The categorical input is first transformed to a set of one-hot vectors. Given a vocabulary of n amino acids and an embedding size of m, the embedding layer learns a $n \times m$ projection matrix. This layer learns, from the amino acids, a more informative and dense latent representation, which is better suited for machine learning.

Convolutional and Maximum Pooling Layers: Convolutional networks were initially introduced in computer vision [6]. They are known for their ability to extract local features and for providing translation invariance. As opposed to dense layers, they involve far fewer parameters, and therefore are less subject to overfitting. They evaluate the convolution between an input and a set of filters (also known as sliding windows or kernels). For each position of the filter, the outcome of the convolution operation is summarized or downsampled using a maximum pooling layer, which extracts the maximum value from the convolutional window. As illustrated in Fig. 1, each branch of SDPPI consists of three convolutional networks. The number of filters as well as their sizes were determined, for each layer, by inspection. The first convolutional network consists of 16 filters of size 10, the second one is formed of 32 filters of size 8, and the last one has 64 filters of size 5. The filters perform a multiresolution analysis from low to high resolution in the forward direction. The first convolutional network performs a low-resolution analysis, as its filters are larger and less numerous, whereas the last convolutional network performs a high-resolution analysis, as the filters are smaller and more numerous. This approach is reminiscent of the wavelet transform [14] in the sense that a multiresolution analysis is performed, noise is reduced, and feature maps are compressed.

Long Short-Term Memory Network: The LSTM is a recurrent neural network with long and short-term memory that solves the gradient vanishing problem by adding control gates. The LSTM is located just after the last convolutional network. Its role is essentially to introduce a long and short-term memory mechanism within the network after the multiresolution analysis is performed by the three convolutional neural networks and to reduce the number of output parameters.

Global Maximum Pooling Layer: In addition to the local maximum pooling layers, our SDPPI network employs multiple global maximum pooling layers, as illustrated in Fig. 1. These layers are involved in the periodical and hierarchical interaction between the two branches. The global layer determines the maximum for the entire feature map at each resolution level, each resolution level being determined by the corresponding convolutional networks. Therefore, the strength of the interaction between the two branches is based on the most salient feature-that is, the global maximum at a particular level. This interaction is repeated three times, once for each resolution level.

Batch Normalization and Dropout Layers: The batch normalization layer performs a batch normalization, both in origin and scale, which improves the training convergence while mitigating overfitting [9]. The dropout layer randomly sets to zero a certain number of neurons. In addition to introducing some randomness into the network, the dropout layer regularizes the loss function. Therefore, the occurrence of over-trusted neurons is less likely, thus contributing to the overall network robustness.

Dense Layer: Our network consists of two dense networks, both of which are located in the classification head, as illustrated in Fig. 1. The first network, which is called the selecting dense layer, consists of a single layer that is activated by a ReLU function.

As explained earlier, the selecting dense layer role is essentially to determine the depth of the network according to the size and the complexity of the training set. The last layer, which is a decision layer, consists of a single neuron with a sigmoid activation function that evaluates the binding probability.

3 Experimental Evaluation

Our experimental evaluation consists of three sets of experiments. In the first set, our results are compared against the state of the art for the benchmark datasets described below. In the second set, the performances and the scalability of our system are evaluated against a very large dataset, namely B-STRING. The last set provides an empirical demonstration of the ability of our network to adapt its architecture to the size and complexity of the dataset. The calculations were performed with TensorFlow on a Compute Canada Beluga supercomputer with two 2.4 GHz Intel Gold 6148 Skylake CPUs, 186 GB of RAM, and four NVIDIA 16 GB V100 SXM2 GPUs.

3.1 Datasets

Five datasets were employed in this study, namely Guo's benchmark dataset, the DIP dataset, the HIPPIE dataset, the INWEB dataset, and the B-STRING dataset. These datasets are described in the following subsections, and their characteristics are reported in Table 1. Our research focuses on Homo sapiens proteins, as they are the most suitable for therapeutic treatments [13] [4].

Table 1. Characteristics of the datasets.

Dataset	Guo	B-STRING	DIP	HIPPIE HQ	HIPPIE LQ	INWEB HQ	INWEB LQ
Positive Pairs	29,071	9,366,352	3,547	34,152	18,315	141,354	380,318
Negative Pairs	31,496	8,225,481	–	–	–	–	–
Total Pairs	61,197	17,591,832	3,547	34,152	18,315	141,354	380,318

Guo's Benchmark Dataset. This dataset was employed by Guo *et al.* in their study [7]. It consists of 20,971 interactive protein pairs (positive examples) and 31,496 non-interacting pairs (negative examples), in which each protein consists of, at most, 1,200 amino acids. A validation set, consisting of 2,943 interacting pairs and 3,057 non-interacting pairs, was randomly selected, and the remaining pairs, 26,128 positive examples and 28,439 negative examples, were employed in five-fold cross-validation. Consequently, the dataset was essentially balanced.

DIP Dataset. The Database of Interacting Proteins (DIP) dataset was first introduced by Xenerios *et al.* in 2002 [22]. Multiple releases have followed since. Version 20,160,430 was employed for this study. In order for the results to be comparable between datasets, and to avoid information leakage (a protein being in both the training and the testing set), proteins with more than 1,200 amino acids were excluded, as well as proteins already appearing in Guo's dataset. Indeed, the DIP dataset was employed for testing, whereas Guo's dataset was employed both for training and testing. This is the reason proteins appearing in Guo's dataset were not considered-to guarantee that the results achieved were the result of a prediction and not of a regression. The filtering procedure resulted in 3,242 protein pairs.

HIPPIE Dataset. The Human Integrated Protein-Protein Interaction Reference (HIPPIE) dataset [18] consists of 52,467 protein pairs. A confidence score is associated with each one of them. The dataset was divided into two subsets: a high-quality (HQ) dataset, for which the score was greater than 0.73, and the remaining pairs were attributed to a low-quality (LQ) dataset. To compare the results between datasets and to prevent information leakage, proteins with more than 1,200 amino acids, as well as proteins appearing in Guo's dataset were excluded.

INWEB Dataset. This dataset was employed in Li's study [12], and it is one of the largest PPI datasets available. As with the HIPPIE dataset, the pairs are divided between low quality and high quality; the quality threshold was fixed at one. The high-quality dataset consisted of 8,672 distinct proteins, whereas the low-quality dataset contained 15,288 proteins, resulting in 141,354 PPI for the former and 380,318 interactions for the latter. The filtering process was similar to the process employed for the DIP and HIPPIE datasets.

B-String Dataset. The STRING database [20] contains the 19,356 Homo sapiens proteins known to participate in 5,879,727 PPIs (positive examples). We constructed the B-String dataset that contains these known PPIs together with 5,879,727 negative examples (non-interacting proteins) in order to create a balanced dataset. The negative examples were generated as follows. A graph was constructed in which the nodes refer to the proteins, and the edges correspond to their binding strength: the shorter the length of an edge, the stronger the bond. Non-interacting proteins correspond to the most distant ones in terms of their geodesic distance in the graph.

It is impossible to visualize the entire graph because of the large number of proteins involved. Therefore, as an illustration, we create a graph from the first 30 proteins as shown in Fig. 3. Because there are 30 proteins, there are 30 vertices in the graph. The interactions are represented by edges; the stronger the interaction, the shorter the edge. For instance, the interaction between proteins 14 and 15 is much stronger than between proteins 3 and 29. The resultant balanced B-String dataset consists of nearly twelve million instances was thus created from the STRING database, called B-STRING.

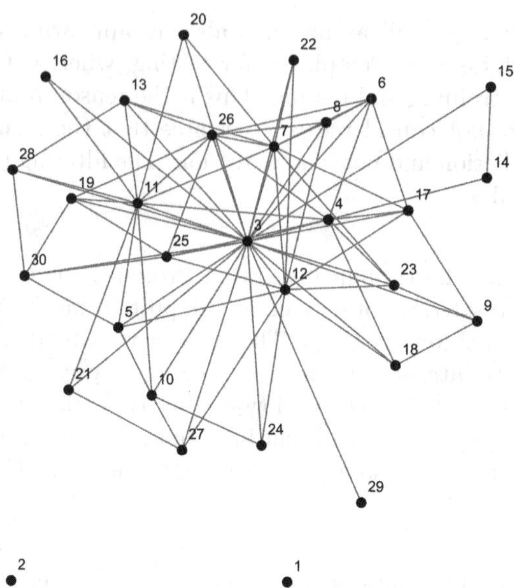

Fig. 3. Interaction graph for the first thirty proteins of STRING Homo sapiens.

3.2 Comparison Against the State of the Art for Protein-Protein Interactions

The performances of our SDPPI network were evaluated against Guo's dataset and the DIP, HIPPIE, and INWEB datasets. Recall that the accuracy of our network, against Guo's benchmark hold-out training set, was evaluating using five-fold cross-validation. The accuracy obtained for each fold, as well as the average five-fold accuracy are reported in Table 2. Our results were further compared against the state of the art for PPI, namely Li's [10] and Sun's [19], using the best model obtained from five-fold cross-validation (fold 2). In addition, following Li [10], we compare our work to three machine learning algorithms, namely Random Forests (RF), support vector machines (SVMs) and Gradient Boosting decision trees (GBDTs), using QuaLitative Characteristic (QLC) and QuaNtitative Characteristic (QNC) features extraction methods. The classification accuracy, precision, recall, and F1-scores are reported in Table 3 and the respective accuracies are shown in Table 4. The results confirm the value of our adaptive architecture, and indicates that we were able to construct accurate models against all datasets. Our network outperformed the state-of-the-art networks on all datasets except for HIPPIE HQ, for which it was *ex aequo* with Li [10].

3.3 Large-Scale Experiment

The performances of SDPPI against the B-STRING dataset were evaluated again using five-fold cross-validation. Recall that this dataset consists of 19,356

Table 2. Five-fold cross-validation result for Guo's benchmark dataset.

Fold	0	1	2	3	4	Average
Accuracy	98.95%	98.80%	98.95%	98.94%	98.94%	98.92%

Table 3. Comparison in terms of accuracy, precision, recall and F1-score for Guo's hold-out testing set.

Model		Accuracy	Precision	Recall	F1-score
SDPPI		**98.81%**	98.81%	**98.78%**	**98.79%**
DNN-PPI		98.35%	98.80%	97.83%	98.31%
SAE		96.82%	–	–	–
QLC	SVM	91.85%	93.98%	89.09%	91.47%
	RF	97.80%	98.65%	96.84%	97.74%
	GBDT	98.58%	98.94%	98.17%	98.55%
QNC	SVM	97.00%	98.29%	95.55%	96.90%
	RF	97.48%	98.57%	96.25%	97.40%
	GBDT	98.41%	98.83%	97.92%	98.38%
QNC + QLC	SVM	97.15%	98.36%	95.79%	97.06%
	RF	97.85%	98.65%	96.94%	97.79%
	GBDT	98.47%	**98.97%**	97.89%	98.43%

Table 4. Comparison of SDPPI to the state of the art for six benchmark datasets (accuracy).

Approach		DIP	HIPPIE-HQ	HIPPIE-LQ	inWeb-HQ	inWeb-LQ	Guo Hold-out
SDPPI		97.27%	**95.95%**	**95.33%**	**95.29%**	**95.91%**	**98.81%**
DNN-PPI (Li)		93.56%	94.28%	92.93%	93.29%	92.76%	98.38%
SAE (Sun)		93.77%	92.24%	87.04%	91.14%	87.99%	96.82%
RF	QNC-QLC	97.48%	93.05%	91.22%	93.49%	90.84%	97.85%
	QNC	**97.78%**	93.09%	91.52%	93.46%	90.99%	97.48%
	QLC	96.87%	92.46%	89.98%	92.86%	89.98%	97.78%
GDBT	QNC-QLC	97.46%	93.30%	91.29%	93.61%	90.58%	98.47%
	QNC	97.49%	93.60%	92.03%	93.53%	89.00%	98.42%
	QLC	97.40%	93.08%	90.66%	93.08%	90.88%	98.58%
SVM	QNC-QLC	93.23%	91.12%	88.30%	90.65%	84.23%	97.15%
	QNC	89.98%	90.69%	87.91%	89.49%	83.69%	97.00%
	QLC	84.55%	81.15%	75.12%	81.03%	72.85%	91.85%

Homo sapiens proteins participating in 5,879,727 PPIs and nearly twelve million instances were employed to train and test the network.

The results are reported in Table.5. The results were consistent from fold to fold, and an average accuracy of 96.08% was achieved despite the large number of instances. These results clearly demonstrate the ability of our network to adapt to large datasets through the selecting dense layer, in addition to demonstrating the scalability of our approach. Indeed, a deeper network was automatically selected by the selecting dense layer to reflect the complexity and the size of the dataset. They further illustrate the performance of our system.

Table 5. Five-fold cross-validation accuracies for the B-STRING dataset.

Fold	0	1	2	3	5	Average
	95.88%	96.15%	96.11%	96.15%	96.12%	96.08%

3.4 Selecting Dense Layer: An Empirical Analysis

As explained in Section III, the depth of the network is automatically determined according to the size and complexity of the dataset. The outcome may be a shallower network, a deeper network, or any combination thereof; the architecture of the resulting network is determined by the selecting dense layer.

Fig. 4. Gradient associated with the selecting dense layer for Guo's dataset.

Fig. 5. Gradient associated with the selecting dense layer for the B-STRING dataset.

In this section, we propose an empirical analysis of this mechanism. Small and large datasets were selected, namely Guo's dataset and the B-STRING dataset. For each dataset, the element-wise gradient with respect to the selecting dense layer was evaluated. These gradients are represented as heat maps: a cold color corresponds to a low gradient and a warm color corresponds to a higher one. Our results are reported in Figs. 4 and 5, for Guo's dataset and B-STRING, respectively. The left side of these figures corresponds to the shallower networks, while the right side corresponds to deeper networks. In Fig. 4, more weight and gradient is attributed to the shallower layers (warmer colors) whereas in Fig. 5, the deeper layers clearly dominate the shallower layers. Therefore, a shallower network was automatically created for the small dataset, and a denser network was generated for the larger and more complex one.

4 Conclusions

Large repositories of PPIs mostly contain interacting proteins. Consequently, they are highly unbalanced and unsuitable for learning. Therefore, a new approach was proposed for determining non-interacting pairs. A graph was constructed in which the nodes refer to the proteins, and the edges correspond to their binding strength: the shorter the length of an edge, the stronger the bond. Non-interacting proteins correspond to the most distant ones in terms of their geodesic distance in the graph. A large balanced dataset was thus created from the STRING database, called B-STRING. A new deep pyramidal network, which learns both its architecture and depth from the dataset, was proposed. This allows the network to adapt to the structural complexity and size of the dataset while preventing overfitting. This is of great importance as, for rare or new diseases, the size of the corresponding PPI datasets may be small, whereas for common and older diseases, their size is much larger. It also considerably reduces the number of parameters that must be determined by inspection, thereby streamlining the design of the network while removing arbitrariness and biases that result from human intervention. Our future work will focus on interaction type prediction and binding affinity estimation [3,23].

References

1. Alberts, B.: The cell as a collection of protein machines: preparing the next generation of molecular biologists. Cell **92**(3), 291–294 (1998)
2. Bejani, M.M., Ghatee, M.: A systematic review on overfitting control in shallow and deep neural networks. Artif. Intell. Rev. **54**(8), 6391–6438 (2021). https://doi.org/10.1007/s10462-021-09975-1
3. Chen, M., Ju, C., Zhou, G., Chen, X., Zhang, T., Chang, K., Zaniolo, C., Wang, W.: Multifaceted protein-protein interaction prediction based on siamese residual RCNN. Bioinformatics **35**(14), i305–i314 (2019)
4. Chin, M., Marks, C., Deane, C.M.: Humanization of antibodies using a machine learning approach on large-scale repertoire data. bioRxiv (2021)
5. Deng, J., Dong, W., Socher, R., Li, L., Li, K., Fei-Fei, L.: ImageNet: a large-scale hierarchical image database. In: Proceedings of the IEEE Conference on Computer Vision and Pattern Recognition, pp. 248–255. Miami, FL, USA (2009)
6. Fukushima, K., Miyake, S.: Neocognitron: a self-organizing neural network model for a mechanism of visual pattern recognition. In: Competition and Cooperation in Neural Nets, pp. 267–285. Berlin, Heidelberg (1982)
7. Guo, Y., et al.: PRED_PPI: a server for predicting protein-protein interactions based on sequence data with probability assignment. BMC. Res. Notes **3**(1), 1–7 (2010)
8. He, K., Zhang, X., Ren, S., Sun, J.: Deep residual learning for image recognition. In: Proceedings of the IEEE Conference on Computer Vision and Pattern Recognition, pp. 770–778. Las Vegas, NV, USA (2016)
9. Ioffe, S., Szegedy, C.: Batch Normalization: accelerating deep network training by reducing internal covariate shift. In: International Conference on Machine Learning, vol. 37, pp. 448–456. Lille, France (2015)

10. Li, H., Gong, X.J., Yu, H., Zhou, C.: Deep neural network based predictions of protein interactions using primary sequences. Molecules **23**(8), 1923–1939 (2018)
11. Li, H., Xu, Z., Taylor, G., Studer, C., Goldstein, T.: Visualizing the loss landscape of neural nets. arXiv preprint arXiv:1712.09913 (2017)
12. Li, T., et al.: A scored human protein-protein interaction network to catalyze genomic interpretation. Nat. Methods **14**(1), 61–78 (2017)
13. Liang, S., Zhang, C.: Prediction of immunogenicity for humanized and full human therapeutic antibodies. PLoS ONE **15**, 1–14 (2020)
14. Meyer, Y.: Wavelets and Operators, vol. 1. Cambridge University Press, Cambridge (1992)
15. Neek, M., Kim, T.I., Wang, S.W.: Protein-based nanoparticles in cancer vaccine development. Nanomed. Nanotechnol. Biol. Med. **15**(1), 164–174 (2019)
16. Neiswinger, J., et al.: Protein microarrays: flexible tools for scientific innovation. Cold Spring Harbor Protocols 2016(10), pdb-top081471 (2016)
17. Orhan, E., Pitkow, X.: Skip connections eliminate singularities. In: International Conference on Learning Representations. Vancouver, Canada (2018)
18. Schaefer, M.H., Fontaine, J.F., Vinayagam, A., Porras, P., Wanker, E.E., Andrade-Navarro, M.A.: HIPPIE: integrating protein interaction networks with experiment based quality scores. PLoS ONE **7**(2), 1–8 (2012)
19. Sun, T., Zhou, B., Lai, L., Pei, J.: Sequence-based prediction of protein protein interaction using a deep-learning algorithm. BMC Bioinf. **18**(1), 1–8 (2017)
20. Szklarczyk, D., et al.: STRING v11: protein-protein association networks with increased coverage, supporting functional discovery in genome-wide experimental datasets. Nucleic Acids Res. **47**(D1), D607–D613 (2019)
21. Titeca, K., Lemmens, I., Tavernier, J., Eyckerman, S.: Discovering cellular protein-protein interactions: technological strategies and opportunities. Mass spectrometry Rev. **38**(1), 79–111 (2019)
22. Xenarios, I., Salwinski, L., Duan, X.J., Higney, P., Kim, S.M., Eisenberg, D.: DIP, the database of interacting proteins: a research tool for studying cellular networks of protein interactions. Nucleic Acids Res. **30**(1), 303–305 (2002)
23. Zeng, M., Zhang, F., Wu, F., Li, Y., Wang, J., Li, M.: Protein-protein interaction site prediction through combining local and global features with deep neural networks. Bioinformatics **36**(4), 1114–1120 (2020)
24. Zhang, C., Bengio, S., Hardt, M., Recht, B., Vinyals, O.: Understanding deep learning (still) requires rethinking generalization. Commun. ACM **64**(3), 107–115 (2021)
25. Zhu, H., et al.: Global analysis of protein activities using proteome chips. Science **293**(5537), 2101–2105 (2001)

Capturing Self Fulfilling Beliefs in a Rare Event: Local Outlier Factor

Iulia Igescu(✉)

National Bank of Romania, 030031 Bucharest, Romania
iulia.igescu@bnro.ro

Abstract. The global Virus Crisis of 2020–2021 was a Rare Event when governments generated incomplete markets. An episode of such large scale market incompleteness is a unique platform to study beliefs role in orienting "animal spirit" - defined by Keynes as "a spontaneous urge to action". Beliefs become components of the output equilibrium stabilization mechanism of an economy. As proxy for beliefs this paper uses survey data (soft data). In this Rare Event flawed statistics affect hard data (industrial output), feed into flawed beliefs (panic, confidence loss), mirror into soft data, and anew into hard data. Beliefs become *self-fulfilling*. This gives rise to idiosyncratic outliers: soft and hard data have different *outlierness degrees*. By deploying Local Outlier Factor to detect outlier patterns in soft indicators, reported faster than hard ones, a policymaker can timely capture self fulfilling beliefs and departures from stability.

Keywords: Location shifts · Equilibrium change · Rare events · Animal spirit · Outliers · Local outlier factor · Learning with less data

1 Introduction

The goal of monetary policy is economic stability. During rare events economies are prone to a stability loss. As monetary policy decisions have become data driven, they depend on reliable data. A major challenge for a policy maker is learning equilibrium changes from a few data points. This paper uses Virus Crisis of 2020/2021 as a Rare Event to investigate a way of learning from less data to better understand how an economy is moving away from its equilibrium. Theoretically [9] shows that a confidence shock could push an economy into an entirely different equilibrium because its "stabilization mechanism is broken". One reason has to do with Keynes idea that a *position of equilibrium at full employment might not exist*. Friedman interpretation of Keynes statement as unemployment being a result of some market rigidities gave rise to the concept of *natural rate of unemployment* currently used in monetary policy. A more recent interpretation offers [12] where full employment does not exist because

The views expressed in this paper belong to the author only. This paper does not involve the official view of National Bank of Romania.

G. Nicosia et al. (Eds.): LOD 2022, LNCS 13810, pp. 543–557, 2023.
https://doi.org/10.1007/978-3-031-25599-1_40

it is an unstable equilibrium. Therefore a policymaker could instead pursue a policy to stabilize output using what the author calls *bonding*, through special purpose long term bonds. Another interpretation comes from [9] where economic equilibrium is Pareto inefficient most of the time.

There is therefore scope for policy intervention of a new type. As a result of a previous Rare Event, the Great Recession of 2008/2009, central banks have adopted monetary policy such as quantitative easing (QE), an increase in the size of the central bank balance sheet, or qualitative easing (QualE), a change in the central bank balance sheet risk composition. Such instruments find grounds in [16] and [2] where in an excess reserve equilibrium debt becomes perfect substitute for currency. In such equilibria monetary policy is lacking real effects because it has only one-asset, currency. As 2021 inflationary pressures mounted, central banks faced increased calls to stop QE/QualE, perceived as *inflationary*. However, little is understood in practice how long shall such a policy be employed. One conclusion of above-mentioned scholars would be that this policy should continue as long as such an equilibrium persists. Moreover, as [10] show, QualE has real effects when markets are incomplete. In their model, incompleteness arrives because markets open before agents are born. In that case randomness separated from fundamentals (capital, labor, technology), "sunspots" such as beliefs will serve as a way of coordinating markets to select an equilibrium. Beliefs become "self-fulfilling". They become part of the stabilization mechanism of an equilibrium. The work of [5] and [1] lay the theoretical foundations of sunspot equilibrium formation, while [8] find evidence of sunspots in a lab experiment.

Virus Crisis is in fact a Rare Event when governments generated incomplete markets: governments paid some markets to stop trading and become public goods (for e.g. hotels and restaurants). With incomplete markets hard economic indicators measure now only partly their original economic content. Hard data issues are easier to diagnose in the industry sector where a decrease in the number of hours worked translates into lower output level. Seasonal adjustment should use in this case hours rather than days worked, otherwise lower output is misinterpreted as Keynesian spillover effects. Using data on hours worked and wages [4] estimate labor demand and supply shocks during Virus Crisis in the United States. Labor supply shocks account for a larger share of hours decline.

Econometric models often rely on variables with outcome known in advance called leading indicators. Finance sector is relying on survey data as leading indicators, available *two months faster* than hard data. Soft data (survey data) quantify in fact agents *beliefs about a hard indicator*. This paper imparts that extremely negative yet faulty data on major economic indicators reported during the whole period of the Virus Crisis acted as a sunspot: an additional confidence shock reinforcing an initial panic episode, further propagating beliefs to a Confidence Crisis level. That happened despite a fast and successful monetary and fiscal policy intervention at the onset of the Virus Crisis.

As more than one soft variable describes a hard indicator, one first issue is how to select the most important indicators. Methods include screening, penalized likelihood, lasso, or boosting. An approach often used in econometrics is a composite leading index (CLI), a combination of leading indicators. As [7] show in this case variables do not systematically lead. One second issue is more subtle. While hard data are flawed statistically, flawed beliefs (panic, confidence loss) plague soft data. Hard data feed into soft data giving way to *circular relationships between hard and soft data*. In that case one could turn to unsupervised machine learning, for example impurity-based feature (i.e. variable) importance method of random forests. In [11] feature-based importance of random forests inflates the importance of numerical features. Authors recommend permutation importance (PI) with hierarchical clustering on Spearman rank-order correlations and keeping a single feature (indicator) from each cluster.

Formal econometric tests identify major structural breaks in the selected hard and soft variables. A structural break means that a variable moves to a new "regime" - a range of values with different dynamical properties. It is this change in dynamics that conveys some first information about a change in the equilibrium state; for example data show as novelties or outliers and move to a different "location". Usually outliers cannot form clusters, as each data point is unique. In this Rare Event data outliers have been accumulating for more than one year and they retain some information about the real output, therefore about the original equilibrium. Local Outlier Factor (LOF) introduced in [3] uses the principle that being outlying is not a binary property anymore. Authors quantify the abnormality degree of observations which is the *degree the object is being outlying*. They are outlying relative to their local neighborhood densities. The advantage is that when several abnormal observations tend to form a "pattern" observations change their local density. LOF detects patterns within non-clean data sets because it takes both local and global properties of data into account, for example the current location versus original location (original equilibrium). Moreover, soft data show that the economy started from a stable equilibrium, or to be more precise, one stabilized through previous QE/QualE interventions. That in turn allows LOF to measure departures from the global equilibrium.

Such large scale market incompleteness offers therefore a unique glimpse into how beliefs orient "animal spirit" as part of a stabilization mechanism. LOF in fact identifies peculiar patterns in soft indicator outliers. One pattern is overshooting. Another one is when soft indicators stagnate at a low level, despite hard data moving upward. One more pattern is with one stagnating dimension and improvements on the other one (usually both dimensions change). These patterns stand for contradictory beliefs. They reflect in the end in contradictory developments in hard data. Neither do beliefs properly orient animal spirit to allow the economy to go back to the old equilibrium, nor do they orient it to a new (feared lower) equilibrium level. Even though this Confidence Crisis has not changed equilibrium output, it hints at a broken stabilization mechanism of the economy. Furthermore, it has weakened output showing that *beliefs are self-fulfilling*. That in turn would support monetary policy to continue additional QualE. Fears of inflation due to QE/QualE are therefore unjustified. Incomplete markets, implying a lack of trade today or in the future, are in fact deflationary. A confidence crisis weakens demand, dampens output with a risk of a recession and threatens further equilibrium stability.

Data are on Romanian economy, as a proxy for an open economy with industry as main production sector and driven by superstar firms in Europe. Leading indicators are monthly industry survey data. The rest of the paper is as follows. First, unsupervised machine learning is a starting point in selecting the two most important soft indicators for monitoring purposes. Second, one should look at data issues in different "regimes" - for example ordinary observations in the Original Location Regime (the original equilibrium before the Virus Crisis) versus outliers in the Rare Event Regime (Virus Crisis). Third, Local Outlier Factor detects and monitors soft data agglomerations to understand how beliefs orient animal spirit in the Virus Crisis/Outlier Regime. Last section concludes.

2 Selecting Soft Indicators with Machine Learning

2.1 Defining Data and Their Regimes

The target hard indicator of this paper is the monthly *industry production index* (Industry Hard). One could look at soft data as quantifying monthly agent beliefs regarding some aspect of a hard indicator. For example the eight survey indicators in the industry sector are attributes along eight dimensions of this sector. Even though numeric, survey data remain qualitative data. A list of soft indicators is included in Appendix A.

One first issue is to define the "regimes" of an indicator. Data on each regime express changes in time series dynamics. A regime starts with a structural break in the hard indicator, in this case Industry Hard. Soft indicators often undergo cascading breaks during a hard indicator regime.

Using semi-supervised learning [13] finds three regimes. A first one includes data from April 2018 to May 2019 (i.e. 14 data points). This is the Original Location Regime or Stable Regime of so-called ordinary observations. Around this stable equilibrium gravitates a location shift from June 2019 up to February 2020 (i.e. 9 data points). This is due to a process of digitization and robotization along the value chain in industry started at the center of superstar firms (West Europe). It generates data points classified as novelties in machine learning. This is the Location Shift Regime or Novelty Regime. The Stable and Novelty locations will be referred to as the Restructuring Regime. The last Regime is a Rare Event location from March 2020 to May 2021 (i.e. 15 data points) that produced data classified as outliers. This is the Rare Event (Virus Crisis) Regime or Outlier Regime. Data on the Novelty and Outlier regimes are normalized with the mean and variance of Stable Regime, to make them comparable.

2.2 Permutation Importance versus Factor Importance - Selecting Leading Indicators

With few data points in each regime and to avoid overfitting, a second issue is selecting leading indicators. Multicollinearity affects the choice of indicators. The nature of data also changes with the regime (ordinary-novelty-outlier), highlighting the importance of a correct regime delimitation. One way is to use impurity-based feature (i.e. indicator) importance method of machine learning, for example of random forests. As [11] show that feature-based importance of random forests inflates the importance of numerical features (i.e. of numerical variables as it is the case here), a possible alternative is permutation importance (PI). This method often fails if there are multicollinear features. Therefore authors start with hierarchical clustering of soft variables with Spearman rank-order correlations, keeping a single feature (indicator) from each cluster.

In line with [11], this paper considers Feature Importance of Random Forest Regression (RFR). Feature Importance of RFR rates how important each indicator is for a decision tree. This method indeed inflates the importance of Confidence Index and Stocks, as these two factors predict ca. 50% of the target, see Fig. 1[1] (left). PI finds Stocks as the most important feature. In this case PI fails, as Fig. 1 (right) shows. Both Stocks and Confidence Index boxplots show their wide distribution, while Trends Soft have outliers. Permuting Stocks feature drops accuracy by 40% for example.

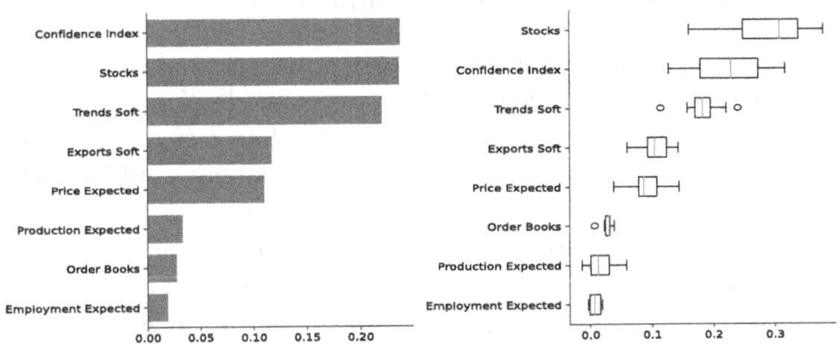

Fig. 1. At a first glance both Factor Importance of Random Forest Regression (left) and PI (right) order indicators in the same way. Multicollinearity affects results.

Figure 2 shows hierarchical clustering based on Spearman correlation of soft data. Here are two clusters. For example, Production Expected, Employment Expected and Stocks are highly correlated, as expected in economics (see lower right Spearman cluster on Fig. 2, right). Unsupervised RFR cluster hierarchy selects the following factors: Exports Soft, Confidence Index, Trends Soft, Production Expected, and Stocks as most important.

For monitoring purposes this paper needs to select two dimensions and prefers the following decision. From the first cluster keep Exports Soft, as highly correlated to Confidence Index, and as having a branch of its own. From the second cluster take instead Trends Soft with a branch of its own. Exports and Trends are negatively correlated. In fact these two factors (indicators) are important for policymakers. The later is a proxy for domestic, the former for external factors.

[1] Machine learning figures in this paper use as source scikit-learn in [17].

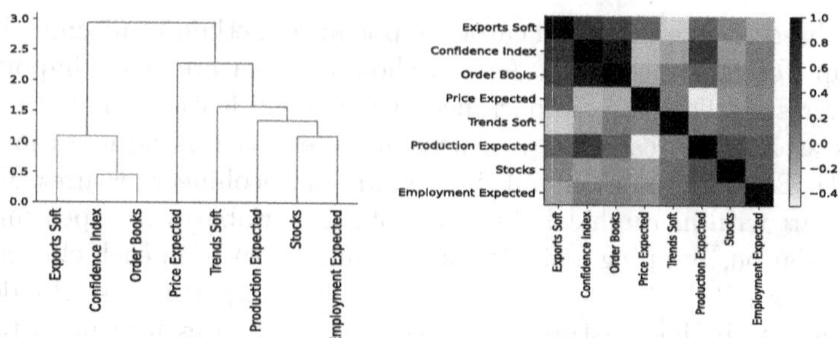

Fig. 2. There are two major clusters on the tree diagram (left). In the lower right corner of Spearman diagram (right), Production Expected is correlated with Stocks, Trends Soft, Employment Expected. In the upper left corner of Spearman a second cluster is made of Exports Soft, Confidence Index and Order Books, also correlated. Price Expected has a branch of its own.

2.3 Data Issues: Soft versus Hard Indicators

Figure 3 concentrates on location changes during crises. For example during Great Recession there was a clear shift to lower levels of production and persistently pessimistic *agent beliefs* regarding *production trends observed in recent months*, i.e. Trends Soft. There was a threat the economy would shift to a lower equilibrium level. Soft data correctly respond to changes in hard data (i.e. fundamentals). One could therefore learn from soft data anomalies about hard data anomalies, and indirectly about the stability of an equilibrium.

The industrial restructuring driven by superstar firms in 2019 has manifested in a Location Shift, an agglomeration below the Original Location. As industry output dropped gradually, it reflected into a slight yet persistent worsening in agents beliefs. If one takes into account also data from the last break before the Original Location, called here a pre-Restructuring location (March 2016 up to March 2018), these three locations follow a unique "circle-like" arrangement. There are no "squeezes" between them to indicate structural breaks, making these locations even harder to identify. During Virus Crisis hard and soft data are rather erratic at first. Slowly a new agglomeration forms right below the Novelty Location: one where industrial production is moving closer to previous Location Shift levels (pre-Virus Crisis levels) while beliefs about the state of the economy are dropping further. Beliefs seem therefore "wrong" i.e. not confident.

One important question is what allows for multiple agglomerations to form around this Original Location. It is a notably different development compared to Great Recession where both hard and soft data shifted to a distincter and lower equilibrium level. One plausible explanation is that the starting equilibrium during Great Recession was unstable, while during Virus Crisis it was a stable one. Romania did not pursue QE during the Great Recession, neither during the Location Shift. The magnitude of the Location Shift of 2019 - which allowed for such a big departure from equilibrium - raises the possibility of an equilibrium that superstar firms stabilized with help from QE/QualE of the European Central Bank at the center. Virus Crisis gives us a unique glimpse into how far and for how long could output depart from a stabilized equilibrium. It is hinting at an economy unable to reach its previous state without further intervention. A stabilized equilibrium means in fact a broken stabilization mechanism.

One first challenge is to separate locations, for e.g. Stable Regime versus Novelty Regime versus Outlier Regime, and classify values on the Novelty Regime. [13] addresses this issue using econometric tests and then uses classification with One-Class Support Vector Machines (O-C SVM) as in [15].

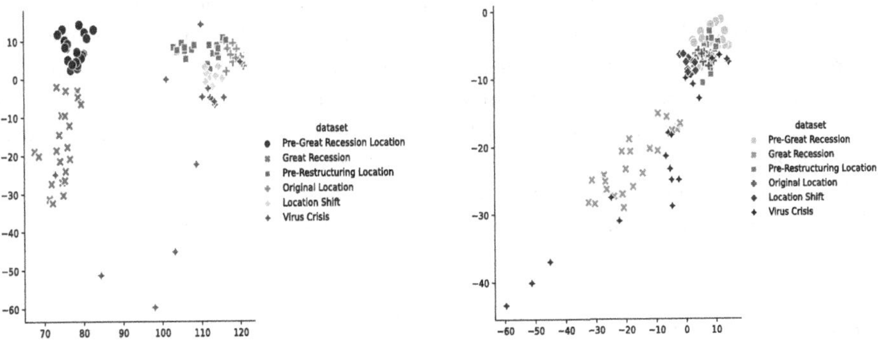

Fig. 3. Soft versus Hard Data during crises. During Great Recession both hard and soft data moved to a lower state. During Virus Crisis, after overshooting, soft data formed an agglomeration below the Location Shift Regime. Industry Hard is on x-axis. Trends Soft is on y-axis.

Fig. 4. Soft versus Soft Data during crises. During Great Recession soft data moved to a lower equilibrium. During Virus Crisis, after overshooting, soft data formed several agglomerations below the Location Shift Regime. Exports Soft is on x-axis. Trends Soft is on y-axis.

Note on Fig. 4 where Exports Soft on horizontal axis is plotted against Trends Soft on vertical axis that during the Virus Crisis soft data do not seem to have the usual "cloud pattern" of location shifts, unlike during the Great Recession. The Location Shift is also "glued" to the Original Location. Some observations of the Virus Crisis are engulfing these two locations. Therefore one second challenge is to find possible patterns that could act as an early warning of location (equilibrium) changes. Classifying agglomerations on the Virus Regime is the subject of next sections.

2.4 Data Issues: Measurement Issues

If one looks at the Virus Crisis in terms of market incompleteness, one has to account for non-fundamental uncertainty. This paper claims that one such source was the inability of statistics to keep up with rapidly changing markets. [6] points out measurement issues of inflation in the USA during Virus Crisis. The same applies to output statistics, for e.g. when seasonal adjustment fails to account for fewer hours worked or for changes in the structure of output when many goods and services became public goods.

As seen in Table 1 (parentheses imply negative numbers), statistical discrepancy has played a key role in the Virus Crisis. In fact most negative and positive "growth" was in fact statistical - statistical discrepancy holds sometimes even a 340% positive share of total growth or a 50% share of total loss. Statistical discrepancy is mainly a result of seasonal adjustment procedures and of aggregating procedures from micro to macro hard data. Rather than a measure of "loss", this paper embraces the view that negative output numbers during Virus Crisis measure *the degree of incompleteness of an economy.* In that case higher positive numbers measure increases in its degree of completeness. Output values (GDP values)[2] become in fact outliers that only partly measure "true" output levels.

Table 1. Real GDP, SA: share of main components. Change to previous quarter.

Time	GDP mil.RON	Industry %	Sales %	Statistical discrepancy
2020Q1	226	(310)%	(53)%	340%
2020Q2	(4682)	(29)%	(44)%	8%
2020Q3	1854	60%	59%	(47)%
2020Q4	1613	29%	34%	(36)%
2021Q1	1045	(3)%	36%	73%

During Virus Crisis hard data due to "wrong" statistics and soft data due to "wrong" beliefs translate into different *degrees of outlierness.*

3 Learning from Outliers

To learn from outliers, this paper relies on Local Outlier Factor (LOF) introduced in [3]. For comparisons purposes Appendix B includes also outlier classification using Isolation Forest (iForest) as in [14]. It becomes fast clear that iForest is not the appropriate method for the issue at hand because this method is searching for outliers that are "few and different." LOF is searching for outliers that are "many and similar" while still being "different."

[2] Data from Romanian National Institute of Statistics.

LOF method as unsupervised machine learning needs around 10 neighbors to stabilize the classifier. As more observations with different densities become available, they interact with the Restructuring Location (including both the Original Location and the Novelty Regime, therefore 23 neighbors) and with their closest neighbors. If new neighborhoods depart more from the Restructuring Location, LOF takes first the closest ones. One has to re-adjust neighborhood numbers over time by including some from the Restructuring Location.

As the number of outliers increases, sometimes LOF redefines a previous outlier as "ordinary observation" (a decrease in the degree of outlierness). They still remain outliers on the Outlier Regime; only their degree of outlierness measured by LOF scores differs from that of a new agglomeration. It turns out that a new "outlier" signals a transition from one agglomeration to another, translating most likely into a change from one set of beliefs to another.

3.1 April–July 2020: Activating Animal Spirit

Virus Crisis started in the middle of March 2020 when the government "shut down" the economy (the economy reached its lowest completeness level). Panic struck financial markets. Monetary policy effectively intervened through QE. In April 2020 the economy was at its lowest completeness level for a whole month. Government "re-opened" the industry sector completely in the middle of May and partially services in June. One would therefore expect the worst output data in April only and a recovery after that.

Even though in May 2020 the economy had already regained its industry completeness, soft industry indicators continued to wildly drop in May and June. In contrast to that, industrial production index returned equally wildly to growth in May. In June its levels were back to March 2020 levels, as shown in graph 5. Hard data become available with a two month delay. In May 2020 first quarter GDP data are made official. Even though GDP shows "surprisingly" positive growth, agents seem to *discard positive macro data* - as implied by worsening monthly soft data in May - *remaining under the influence of April panic* shock. The very negative number of industrial production index for April 2020 is published in June 2020 only. Soft data move further down.

For April-May period only soft data are available to a policymaker and their values show that March-April panic shock continues to shape agents beliefs.

Starting April 2020 consider LOF with 11 k-neighbors. In this case there are the two neighbors for the new outliers (March and April 2020) and nine neighbors from the Restructuring Location (Original Location). If one adds May and June data points, LOF classifies April-June 2020 as outliers, see Fig. 6. All LOF figures have Trends Soft on x-axis and Exports Soft on y-axis.

In July industrial output posts May data and they are positive for the first time. July soft data recover, but remain well below the level of the panic shock of April as Fig. 7 shows. LOF reclassifies April as ordinary observation and *correctly identifies May–July as local outliers*, underlining a first discrepancy

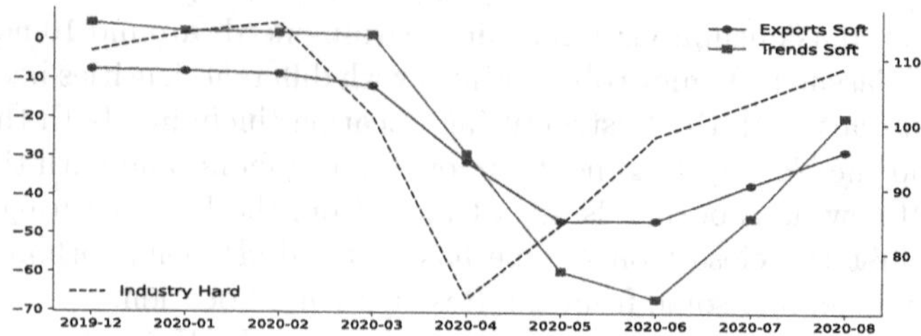

Fig. 5. Even though in May industrial production starts a recovery, soft indicators dive to lower levels till June 2020, a first consequence of a panic shock in April. In July 2020 the hard indicator weakens even though soft indicators improve. Hard data are on the right vertical axis.

between micro hard data (posting robust growth) and soft data (under a confidence shock). Despite a successful monetary policy intervention allowing for a fast recovery as measured by hard data, the *recovery measured by soft data* is *perceived as irrelevant.*

LOF April classification corresponds to an initial panic shock; monetary policy expects panic in such cases (therefore the initial QE response). LOF becomes in this case a useful tool to timely (note: soft data become available in the last week of each month) identify that beliefs remain under a confidence shock in May–July 2020. As a confidence shock is expected to activate animal spirit into weakening demand, monetary policy should continue QE.

3.2 October 2020 – April 2021: A Confidence Crisis

From July to August 2020 Industry Hard improves slightly above March 2020 levels and then stagnates. GDP data for the second quarter of 2020 published in August show extremely negative sales and industrial output growth, *"confirming"* agents beliefs of a poor economic state. A first puzzle is that extremely weak macro data in the second quarter are out of sync with strong micro data on both industry output and retail sales in the third quarter. A second one is that despite negative macro data, soft data continue to improve in August even though only to April 2020 levels - panic levels - to stagnate at this level in September. LOF classifies August-September soft values as "ordinary observations". Therefore August-September together with April form a new agglomeration of *low confidence around panic levels*, see Fig. 9. It is if as *agents have convinced themselves* that the economy is "doing poorly". Any additional shock could push the economy into a confidence crisis. This shock comes: further shut-downs in October 2020, a result of a second COVID-19 wave.

LOF promptly classifies October value as an outlier and wrongly re-classifies July as ordinary. With 11 k-neighbors as before, there are now 8 neighbors from the new densities and only 3 from the original density. LOF learns from

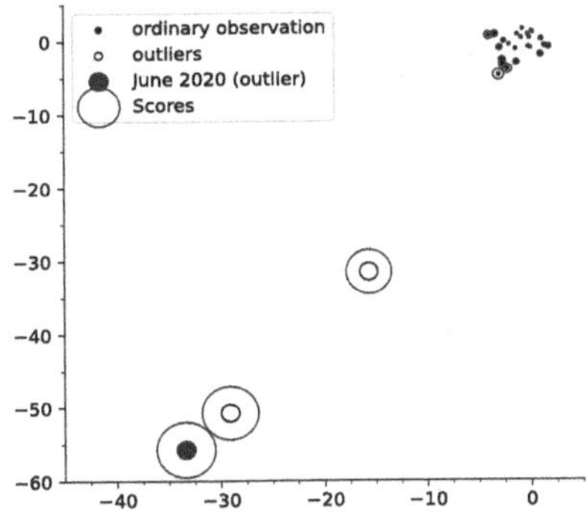

Fig. 6. In June 2020 LOF outlier reaches a low, despite Industry Hard being close to March 2020 levels. LOF classifies April–June soft data as outliers.

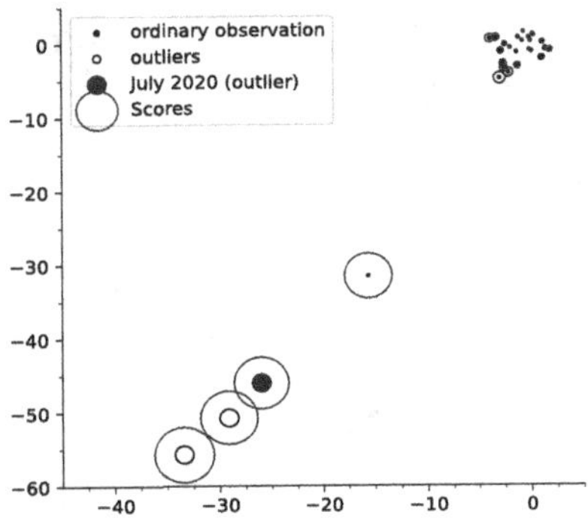

Fig. 7. In July 2020, April is reclassified as ordinary observation. Despite a first improvement, July soft values are classified below April (panic) levels and remain an outlier.

local neighbors more than from the original density. This is depicted in Fig. 8. To keep the same classification as before, one must force the classifier to add one more neighbor from the original density (Restructuring Regime). With 12 k-neighbors LOF classifies October as an ordinary observation in the range of April panic levels (as shown in Fig. 9) and July remains an outlier. To correctly classify in November and again in December, LOF must add one more k-neighbor every month forcing LOF to keep 4 values from the Restructuring Regime. In November 2020 with 13 k-neighbors LOF re-classifies October as outlier and November as ordinary, see Fig. 10.

October and November re-classifications signal a new emerging agglomeration: a "peculiar column", a vertical agglomeration with stagnant Trends Soft and upward movements in Exports Soft. This extends until March 2021, see Fig. 11. In this phase beliefs about the state of domestic economy indicate stagnation (correctly aligned to micro data), while those about external factors see permanent improvements (not correctly aligned to additional closures in Euro Area, therefore prolonged incompleteness translated into weaker "growth"). Any signs of hard data weakness are reinforcing low confidence beliefs, for example the fact that monthly industrial production index remains below its February 2020 level (pre-crisis level). Signs of improvement are discarded, for example strongly positive GDP growth of the third and fourth quarter in 2020. These contradictory beliefs are all indicators of a Confidence Crisis. Animal spirit is expected to act through increased savings, weakening demand. Indeed, retail sales have equally contradictory developments. Weak micro hard data are now out of sync with the very strong growth rates at macro level, as GDP components show in Table 1. Even though in January-March 2021 soft data remain

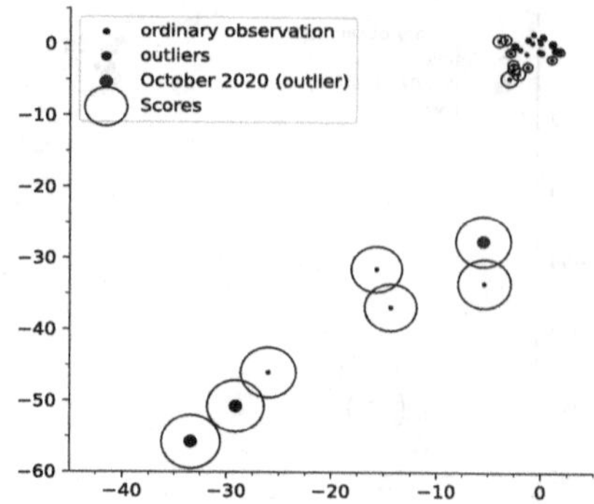

Fig. 8. With 11 k-neighbors, LOF classifies October 2020 as outlier. July is reclassified as ordinary, it decreases its outlierness level.

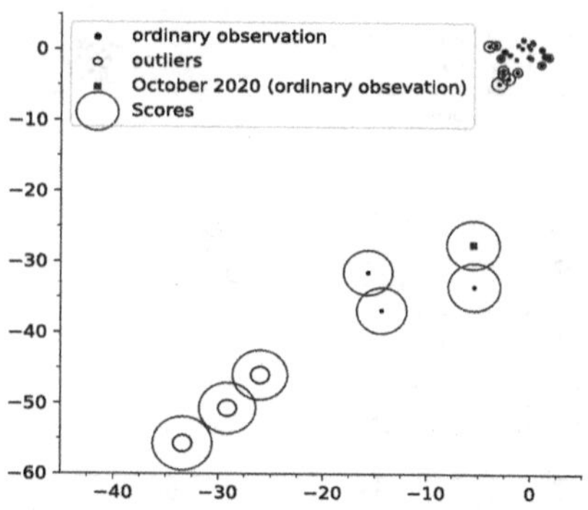

Fig. 9. With 12 k-neighbors, LOF keeps July as an outlier and includes October 2020 in the new August-September agglomeration of low confidence levels.

on the Confidence Crisis agglomeration, GDP real makes a full "recovery" and it is back to pre-Virus levels at the end of the first quarter of 2021. Statistical discrepancy holds a share of 73% of this total "growth". On the other hand industrial output at macro level posts negative growth. This supports the view that *even though output seems to be expanding, this is mostly due to an increase in the level of completeness of the economy.*

LOF timely identifies and visualizes a Confidence Crisis, signaling monetary policy to continue QualE, as strong GDP "growth" hides in fact a weakening in demand through animal spirit. That in turn could lead to a recession. This development is even more dangerous, as it seeds the idea of "too fast of a recovery" with "excess demand" - raising the prospect of inflationary pressures and preparing for a tight monetary policy. That would be the last step in cementing *self-fulfilling beliefs.*

In April 2021 the government announces a full "re-opening" of the economy in May 2021. The economy regains almost full completeness. LOF shows that soft data leave panic levels. In May values move close to original density. The fact that this Confidence Crisis ends "over night" could be an indicator of the right decision of monetary policy to continue supporting the economy, despite its apparent *growth* and full *recovery.*

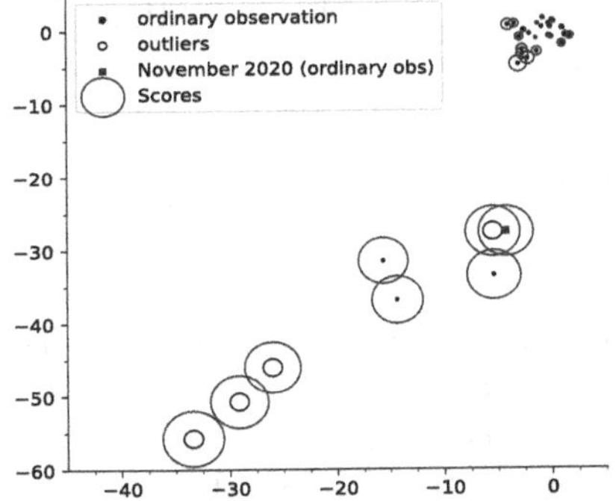

Fig. 10. In November 2020, with 13 k-neighbors LOF re-classifies October as outlier and November as ordinary. A peculiar agglomeration is developing, with beliefs of stagnant domestic trends and improved foreign trends. This is despite additional closures in West Europe. These are signs of an ongoing Confidence Crisis.

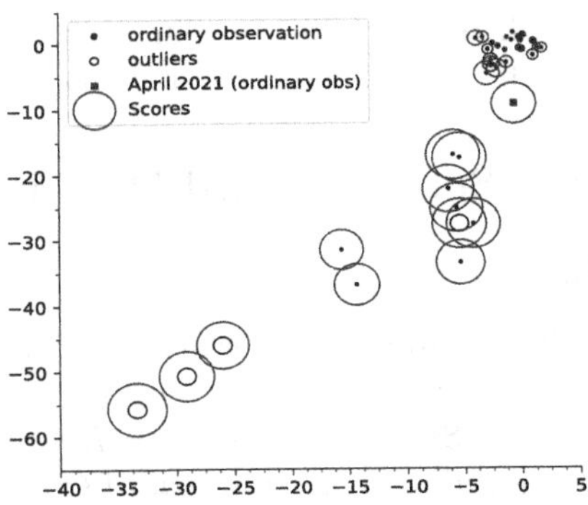

Fig. 11. April 2021 leaves the Confidence Crisis agglomeration whose start is marked by the October outlier. Both domestic and external factors are perceived as more favorable. LOF does not classify April as outlier, it includes it into the agglomeration pertaining to the original equilibrium. Still soft values remain below March 2020 values.

4 Conclusion

Virus Crisis of 2020–2021 and its large scale incompleteness offers a unique platform to understand how beliefs orient animal spirit in a Confidence Crisis. There are two features unique to this Rare Event.

First, there is a discrepancy between extremely negative macro data and strong micro data in the first half of 2020, followed by extremely positive macro data and weak micro data till the first quarter of 2021. A possible explanation advanced in this paper is that output does not measure loss or growth anymore. GDP has become a measure of the degree of incompleteness of an economy. The "loss" of the second quarter of 2020 measures the depth of incompleteness. "Growth" up to the first quarter of 2021 measures in fact increasing completeness. Second, both soft and hard data are plagued by anomalies with idiosyncratic properties: they are outliers that still convey some valid information about fundamentals dynamics. They have different degrees of outlierness.

The goal of this paper is to understand how far away from its state can in fact output move during a Rare Event. One would like to early detect data patterns pointing at shocks that an initial policy intervention could not account for. By deploying Local Outlier Factor (LOF), machine learning detects indeed "peculiar" soft data patterns, taken as a sign of "inconsistent" beliefs. LOF timely identifies a Confidence Crisis showing that beliefs fail to orient animal spirit in returning the economy back to its pre-Virus Crisis level. As the stabilization mechanism of the equilibrium appears broken, a Confidence Crisis could threaten an economy into a low output/high unemployment state, a poverty trap. Beliefs can therefore be self-fulfilling. In that case monetary policy should continue with QualE until the economy regains full completeness.

A Soft and Hard Data Indicators

Industry

Soft indicators are monthly survey data in various sectors of the economy freely available from [19]. They include *Production trend observed in recent months* (Trends Soft), *Assessment of Export Order-book Levels* (Exports Soft), *Confidence Indicator*, *Assessment of Order-book Levels* (Order Books), *Assessment of Stocks of Finished Products* (Stocks), *Production Expectations for the Months Ahead* (Production Expected), *Selling Price Expectations for the Months Ahead* (Price Expected), *Employment Expectations for the Months Ahead* (Employment Expected). Hard indicators is *industry production index* (Industry Hard) freely available monthly from [18].

B Classification of Outliers with Isolation Forest

For comparison purposes, Figs. 12 and 13 below use Isolation Forest as in [14]. Isolation Forest (iForest) randomly selects a variable and a split value between the maximum and minimum values of the selected variable. The number of splittings required to isolate an observation is equivalent to the path length from the root node to the terminal node of the tree. This is repeated over a forest of random trees. Outliers are therefore "few and different." There are only two parameters in this method: the number of trees to build and the sub-sampling size. This paper keeps the standard calibration. It turns out that this method is not appropriate for the problem at hand, which is searching for outliers that are "many and similar" while still being "different".

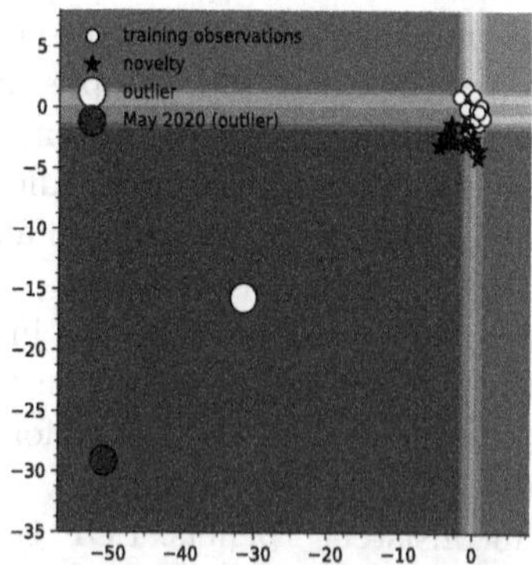

Fig. 12. iForest classifies April and May 2020 as outliers. Trends Soft is on x-axis.

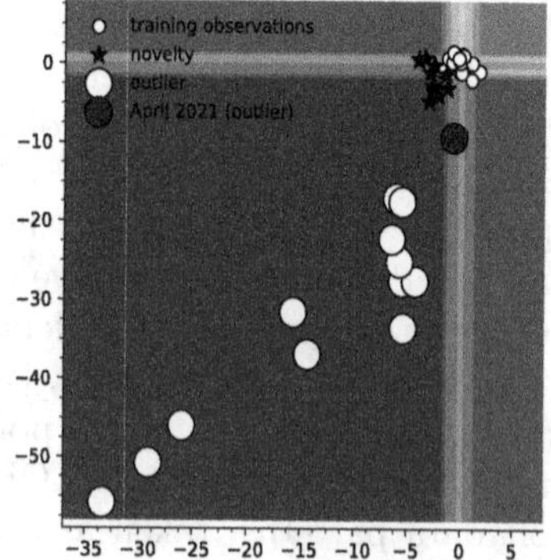

Fig. 13. iForest cannot distinguish between changes in densities. Trends Soft is on x-axis.

References

1. Azariadis, C.: Self-fulfilling prophecies. J. Econ. Theory **25**(3), 380–396 (1981)
2. Azariadis, C., Farmer, R.E.A.: Fractional Reserve Banking. University of Pennsylvania, Mimeo (1987)
3. Breunig, M.M., Kriegel, H.P., Ng, R.T., Sander, J.: LOF: identifying density-based local outliers. ACM SIGMOD Rec. **29**(2), 93–104 (2000). https://doi.org/10.1145/335191.335388
4. Brinca, P., Duarte, J.B., Faria-e-Castro, M.: Measuring labor supply and demand shocks during COVID-19. In: Working Paper 2020-011, Federal Reserve Bank of St. Louis, St. Louis, Missouri (2020)
5. Cass, D., Shell, K.: Do Sunspots matter? J. Polit. Econ. **91**(2), 193–227 (1983)
6. Cavallo, A.: Inflation with Covid Consumption Baskets. NBER WP 27352 (2020). https://doi.org/10.3386/w27352
7. Clements, M.P., Hendry, D.F.: Forecasting Economic Time Series. Cambridge University Press, Cambridge (1998)
8. Duffy, J., Fisher, E.O'N.: Sunspots in the laboratory. Am. Econ. Rev. **95**(3), 510–529 (2005)
9. Farmer, R.E.A.: Expectations, Employment, and Prices. Oxford University Press, Oxford (2010)
10. Farmer, R.E.A., Zabczyk, P.: A sunspot-based theory of unconventional monetary policy. Macroecon. Dyn. (2020). https://doi.org/10.1017/S1365100520000127
11. Geurts, P., Ernst, D., Wehenkel, L.: Extremely randomized trees. Mach. Learn. **63**(1), 3–42 (2006)
12. Igescu, I.: Bonding and dynamical changes. Eur. Econ. Rev. **50**(6), 1387–1402 (2006)
13. Igescu, I.: The role of animal spirit in monitoring location shifts with SVM: novelties versus outliers. In: Nicosia, G., et al. (eds.) LOD 2020. LNCS, vol. 12565, pp. 72–82. Springer, Cham (2020). https://doi.org/10.1007/978-3-030-64583-0_8
14. Liu, F.T., Ting, K.M., Zhou, Z.: Isolation forest. In: 2008 Eighth IEEE International Conference on Data Mining, pp. 413–422 (2008). https://doi.org/10.1109/ICDM.2008.17
15. Schoelkopf, B., Williamson, R., Smola, A., Shawe-Taylor, J., Platt, J.: Support vector method for novelty detection. In: Advances in Neural Information Processing Systems. NIPS 1999, vol. 12 (2000). https://papers.nips.cc/paper/1723-support-vector-method-for-novelty-detection.pdf
16. Wallace, N.: A legal restriction theory on the demand of "money" and the role of monetary policy. Federal Reserve Bank Minneapolis Q. Rev. **7**, 1–7 (1983)
17. Pedregosa, F., et al.: Scikit-learn: machine learning in Python. J. Mach. Learn. Res. **12**, 2825–2830 (2011)
18. Industry data from Romanian National Institute of Statistics. https://insse.ro
19. Survey data from European Commission. https://ec.europa.eu/info/business-economy-euro/indicators-statistics/economic-databases/business-and-consumer-surveys/download-business-and-consumer-survey-data-en

Hybrid Human-AI Forecasting for Task Duration Estimation in Ship Refit

Jiye Li(✉) and Daniel Lafond

Thales Research and Technology Canada, Québec, QC, Canada
{jiye.li,daniel.lafond}@thalesgroup.com

Abstract. Task duration estimation is an important element of scheduling and optimization in various task domains. However, work completion times may vary based on endogenous and exogenous factors. Such estimations are typically made by human experts and often entail assumptions and uncertainty. Such imprecision causes either longer or shorter than expected task durations leading to scheduling conflicts or inefficiencies. As an example, in the domain of ship refits, certain tasks do not take place very often. When there is little historical data available, task scheduling and planning can be very challenging and require ongoing replanning effort to compensate for estimation errors. While human experts can provide synthetic cases to train forecasting tools, it is not obvious how one can integrate human forecast with historical cases for decision support purposes. A new forecasting method is introduced in this paper that integrates human experts' inputs with historical data to create a hybrid model for forecasting and scheduling purposes. We demonstrate through experiments that the proposed hybrid model increases prediction accuracy by 5 to 10% compared to forecasting with only historical data.

Keywords: Cognitive science · Hybrid AI · Expert modeling · Ship refit · Task duration · Forecasting · Scheduling

1 Introduction

Forecasting task durations is of key importance in the scheduling domain [1]. A well-planned schedule can save significant time and budget for projects in different domains such as manufacturing, pharmaceutical warehouses, health care and food industry [2,3]. Data driven modeling often requires significant amounts of historical data to train accurate predictive models. However, in certain use cases, accessing historical data is very difficult. New tasks might not have any historical data available, and existing tasks may only have a very limited number of recorded historical cases.

As an example, refitting has become one of the most important activities inside a shipyard. Ship refit activities include repairing, fixing, restoring, renewing, mending, and renovating an old vessel. A study for overtime budget predictions

Supported by Thales Digital Solutions, Canada.

G. Nicosia et al. (Eds.): LOD 2022, LNCS 13810, pp. 558–572, 2023.
https://doi.org/10.1007/978-3-031-25599-1_41

for naval fleet maintenance facilities was discussed in [4]. The authors stressed the importance of accurate estimations for task scheduling and management. In [5], statistical and machine learning based models were explored on predicting task hours for fleet maintenance operational data. In this domain, historical data for task durations is very scarce. Task durations are estimated manually by human experts most of the time. Approaches on how to aggregate expert judgements were surveyed in [6] to combine expert judgments into forecasting models to form an augmented single predictive model.

The present study focuses on how to elicit and model expert forecasts to compensate for the lack of historical data, and how to augment the data driven method by integrating human forecasts to create a hybrid approach. We expect to improve optimization under uncertainty by forecasting task durations more accurately, resulting in more reliable schedules.

The remainder of this paper is organized as follows. First, we review existing methods on how to enrich a data set by incorporating domain experts' knowledge from two research fields, cognitive science and artificial intelligence in Sect. 2. Next, we introduce the synthetic data set created with a known ground truth for controlled experimental purposes in Sect. 3. Details on data generation to create a hybrid model are discussed in Sect. 4. Experimental design and results analysis are discussed in Sect. 5 and 6, respectively. We summarize this study in Sect. 7 and discuss directions for future work in Sect. 8.

2 Review of Existing Methods

The focus of this paper is to investigate how to increase data resources for task duration predictions in the ship refit domain. We reviewed relevant methods from two broad domains. Relevant approaches identified from the human judgment domain include the Delphi method, cross impact analysis, and policy capturing. Relevant approaches identified from the artificial intelligence and data mining domain include multi-criteria decision analysis, expert-augmented machine learning as well as the traditional non-hybrid supervised machine learning approaches.

2.1 Delphi Method

The Delphi method [7] was proposed to provide a collaborative human input resolution strategy. There are multiple phases on how the agreement could be reached among multiple users. Each user is asked for a prediction, then a resolution strategy is applied to make sure all the users reach the consensus, multiple rounds of re-voting could take place in order to reach a final agreement for all the users. This approach has been used in military and business domains for forecasting. However, a facilitator is required to control the session to allow the anonymous interactions among users, and make sure the users reach the consensus; also, the process may take several rounds with human input before reaching an agreement. The Delphi approach has a few limitations [8], such as not taking into consideration when events are dependent and interrelated. The occurrence of one event is assumed to have no impact on other events. Cross-impact analysis was proposed to circumvent this limitation.

2.2 Cross-Impact Analysis

The initial position paper for cross-impact analysis (CIA) [9] proposed an approach to analyse all factors interacting each other which impact the events in a subjective way. A matrix is designed with multiple factors as rows and columns, and each cell stands for the co-occurrence probabilities of two events. The experts are supposed to fill in such matrix to present subjective opinions on the impacts of the events. Several limitations are stated by the authors, such as the overburden of the experts for estimations in terms of larger events sets, ways to converge subjective estimations from multiple experts, not being able to consider the time dependence among the events and so on.

In [10], a Cross-Impact Analysis approach is used to estimate a task scheduling model in the scenarios of an emergency event, "dirty bomb attack" scenario. The authors claim that domain experts define multiple scenarios for a given problem, but it is difficult to analyze these events into manageable tasks in order to plan for the future tasks/predictions. CIA-ISM (Cross-Impact Analysis - Interpretative Structural Modelling) approach is used to generate and analyze the events, as well as the interactions between these events, to reduce uncertainty in the future. The goal of Cross-Impact Analysis is to forecast events by considering multiple events as dependent.

A qualitative-quantitative-qualitative approach is introduced in [11]. The authors demonstrated that hybrid approaches such as mixing qualitative with quantitative analysis increases the credibility of the findings. 28 experts were interviewed, and each interview lasted 45 min. A software "Szeno-Plan" is suggested to facilitate cross-impact analysis with 7 or more variables. The study shows that adding a quantitative approach in between the qualitative approaches increased the trust and confidence from the domain experts on the findings.

In [12], a causal cross-impact analysis approach is demonstrated in gaming and planning. A causal loop diagram was introduced to explain how interrelated variables affect each other. Arrows with "+" or "-" signs associate to a node in the diagram indicating positive or negative impact from one variable to the other. Different variables are defined: trend variables, event variables, actor and action variables. The interrelation values are between 0 and 1 for positive and negative values. A cross-impact matrix was also utilized to illustrate the model. The authors claimed that such approach could be applied to any strategic planning task.

2.3 Policy Capturing

The use of policy capturing systems has been proposed as a way to improve decision making (and forecasting) using human-machine teaming. We are interested in reviewing the background work in this field to evaluate the approaches on effectively using policy capturing techniques for decision making.

Previous work [13] has claimed that policy capturing (also called judgmental bootstrapping) had great advantages on incorporating experts' opinions in cases of limited availability of historical data or low quality of historical data.

Policy capturing is a judgment analysis method that typically uses linear statistical models such as multiple regression to model domain expert's judgment. In [14], different approaches of policy capturing, both linear based and non-linear based classification approaches, are demonstrated to assist effectively human-decision making process. The experiments demonstrated that judgment analysis helps one understand the psychological processes of real-time human machine interaction and decision making.

An online policy capturing system [15] was demonstrated for predicting human decisions based on small data set. This recent research work demonstrated that policy capturing is promising for integrating experts' estimations together with limited historical data for predictions.

Bootstrapping models have been shown to produce better decisions than human decision makers [16]. The achievement index was introduced to evaluate the success of bootstrapping approach. Human expertise intervention improved bootstrapping effectiveness.

2.4 Multicriteria Decision Analysis

Multi-criteria decision analysis (MCDA) is known to propose approaches for expert modeling based on several decision criteria. It is discussed in [17] that when the problem involves assessing a product based on multiple criteria, the MCDA can be used to provide recommendations considering all such conditions. Myriad is a software to help such analysis by building hierarchies for multi-criteria decision making problems [18]. It is used as a preference (or forecasting) model for multi-criteria analysis [19] and a leading tool for MCDA. Myriad was used to design the problem hierarchy, with the root of the tree representing the overall aggregation, and the leaves representing the attributes. Myriad also provides the abilities to visualize the aggregated results. In [20] Myriad was used as a decision support tool to derive a model of the problem, derive aggregation rules, evaluate and compare options with supporting visualizations and explanations.

2.5 Expert Augmented Machine Learning

An approach involving integrating expert evaluations for prediction was introduced in [21]. The study was based on a set of publicly available clinical data for mortality predictions. The authors used a random forest classifier to predict categories of mortality, as well as analyzing the feature importance. Human experts (clinicians) were asked for evaluations on the selected rules containing significant features without exposing the decisions. This hybrid approach of using the rules both experts and machine learning identified together showed better results than the machine learning approach alone. The approach has two advantages: 1) to identify the errors from historical data, and 2) to filter inconsistent rules (between historical data and expert estimated data), and use better quality rules as training data.

2.6 Data Mining and Data Analytics

The modern AI and Data Mining fields have emphasized greatly on statistical approaches rather than expert systems in recent years. We intend to incorporate more expert input into these statistical methods. A machine learning approach to estimate task durations is introduced in [22]. The data set contains large email data including texts. The purpose of this study was to estimate incoming tasks based on the calendar/email data from multiple users using machine learning and deep learning models, in other words, using the mainstream "data mining" approach. This paper demonstrated the challenges of estimating task durations, however there was no human input involved in the data collections. Also, the data covered not only numeric data, but also text data as well as other types of features.

The following two research papers used fleet maintenance historical data, showing that automatically forecasting for refit task durations and costs has been gaining more and more interest. The first one was a technical paper on predictive analytics for the Royal Canadian Navy Fleet Maintenance Facilities [5]. Two data sets of 62MB and 44MB each were introduced as the test data sets in this paper, which included around 132,292 individual ship maintenance tasks in 43,731 unique orders. The data ranges from February 5, 2004 to July 6, 2016. A linear trend has been discussed for one data set, plotting from data between 2008 to 2016; whereas for the other data set, the author claimed no trend on plotting data from 2013 to 2016, which indicates an unavailability of data during the same comparison period. This is a relatively rich data set, featuring individual ship repair tasks. Task duration predictions on individual ship level are performed, and overall results indicated that a regression tree (tree-based prediction model) was suitable for predicting task durations. However, the data is not publicly available.

Secondly, a machine learning model for overtime budget estimation was discussed in [4]. The authors mentioned that overtime happened when the tasks required more hours than planned to be completed, due to multiple factors such as age, types of the vessels, maintenance policies, supply chain constraints and etc. H2O auto ML software [23] was used for data pre-processing, feature selection and model building. The data set was collected over 7 years of vessel maintenance tasks.

3 Data Description

3.1 Problem Statement

In this study, we consider one of the ship refit tasks - deck painting task - as the scenario. Multiple factors could affect this task, such as the weather, the size of the deck, the conditions of the deck, the age of the ship and so on. The purpose for this study is to implement and compare different methods for estimating the painting task durations. We assume that 1) one or multiple experts are available to produce forecasts (but that this availability is limited and costly) and 2) there is limited historical data available (assumed to be correct data without errors or noise). Given the sparsity of the data, the research problem is to examine how to best make use of historical data and expert forecasts by comparing each method alone and investigating how to best combine these methods to create a hybrid forecasting capability.

Note that this problem could be considered either as a regression problem (estimating time needed for each task given task attributes) or a classification problem (assign a discrete duration category to each task). We formulated the problem as a classification problem for two reasons, 1) this study was intended for an initial model for task estimation, a finer granularity could be investigated in a follow-up study; 2) since the approach involves eliciting domain experts' inputs, it is deemed easier to request categorical rather than real-valued estimates. Therefore, we consider the task duration forecasting problem as multi-class classification problem.

3.2 Features

We consider the age of the ship, the previously traveled distance, the painting temperature, whether the ship has traveled in salt water in the most recent mission, the painting area, the size of the ship, and the deck material as features that could impact the duration of this painting task. We describe the features and values of our synthetic data sets in the following Table 1.

4 Synthetic Data Generation

We use the following *sklearn* function to generate the synthetic data set.

$$sklearn.datasets.make\_friedman1(n\_samples = 200000, n\_features = 7, \\ noise = 0.0, random\_state = 11) \quad (1)$$

The *make_friedman1()* function [24] is designed to create synthetic data set for regression problems, as shown in Eq 1. At least 5 features are required as input to this function, and the features are supposed to be independent variables uniformly distributed in the interval between 0 and 1. Since our features do not have values in-between 0 and 1, we transform each feature value ranges into the defined range as described in Table 1.

For example, for feature "age", the ship ages range from 1 to 30. We use *numpy.random.randint()* function to generate values from 1 to 30, and replace the original values from *make_friedman1()* by the randomly generated integers. For "painting_temperature", we consider 4 values to represent the 4 seasons. Therefore, *numpy.random.randint()* function is also used to randomly generate

Table 1. Synthetic data descriptions for ship painting task

ID	Feature	Description	Type	Value Range
1	Age	Age of the ship	Integer	[1, 30)
2	Last_traveled_dist	Traveled distance from the Most recent mission	Float	[1,30000)
3	Painting_temperature	Seasons (spring, summer, fall, winter) under which the current painting task is performed	Categorical	[1,2,3,4] for [spring, summer, fall, winter]
4	Traveled_in_sale_water	Whether the ship had traveled in salty water in the most recent mission	Boolean	[0,1] for [No, Yes]
5	Paint_area	Areas of deck to be painted (in square meters)	Float	(0, 500)
6	Ship_size	Size of the ship	Float	(0, 40000)
7	Deck_material	Material types of the deck (wood or metal)	Boolean	[0,1] for [wood, metal]
Class	Duration	Task durations (short, average and long)	Categorical	[1-short, 2-avg, 3-long]

integers between 1 to 4, and replace the original feature values with these integers. For "last_traveled_distance" feature, we multiply each value by 30000 to obtain the required value range; for "paint_area", we multiply the values by 500; and for "ship_size", we multiple the values by 40000. For "traveled_in_salt_water" and "deck_material", we discretize the value into two bins of 0 or 1.

4.1 Human-Based Decisions for Class Distributions

The above synthetic data generation results in a uniformly distributed dataset. We would like to consider a human decision rule based synthetic data set to have a more precise prediction performance. In this synthetic data generation, the decision label "Class" is created using the following human-based decision rules as shown in Fig. 1.

As an example, for a given painting task, if the painting area is less than 180 square meters, the ship has previously traveled in salt water, the ship is more than 18 years old, and the temperature is winter, then this painting task will have a long task duration.

We assume that in general, most of the painting tasks will have an average duration, with some having short or long durations. Therefore, naturally this synthetic data set is not perfectly balanced among the three classes. The class distributions are: short duration 24.7%, average duration 47.4%, and long duration 27.9%. We assume there are no missing attribute values in this data set.

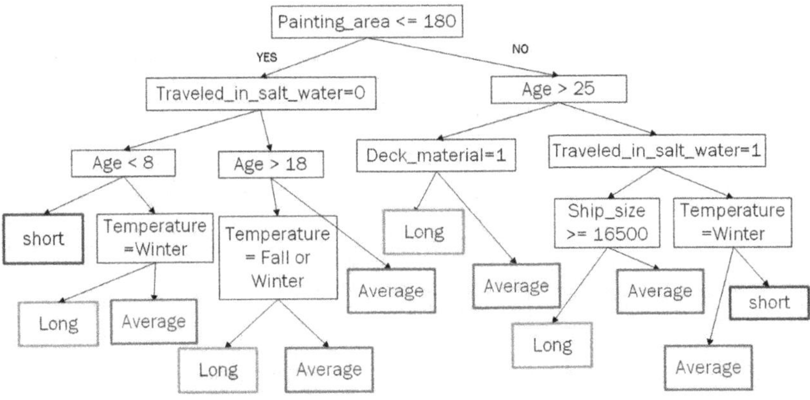

Fig. 1. Human decision tree for class generation of the synthetic data set

5 Experiment

We investigate the prediction performance of forecasting models based on the historical data set, expert data set, and hybrid data set through the following experiments. Since the task durations are categorical data, we use sklearn decision tree classification algorithms to train the prediction models. Note that there exist multiple classification algorithms and any algorithm would fit the purpose of this experiment, since the purpose is to compare the relative performance on the three data sets.

5.1 Experimental Design

A synthetic data set with 200k instances has been first generated from the *make_friedman1* function. 100k of these synthetic data is reserved for expert data pool, 80k data is used as testing data, and the rest of around 18k data is used as historical data pool. We randomly shuffle the test data first, then split into 10 different testing data. For each experiment, we test on these 10 test data sets and average the prediction accuracy. The following Fig. 2 demonstrates this experimental design. We compare the following three models:

- Historical data model for task duration forecast (randomly selected sample size of 50)
- Expert input model on the forecasting of task durations (randomly selected sample size ranges from 50 to 350)
- Hybrid model of historical data and human expert, and its impact on task duration forecast (sample size is the sum of historical data and expert data).

Therefore, in this initial study, we train 3 types of forecasting models for one task, through testing on multiple batches of 80000 test data, we compare which model would perform better than others.

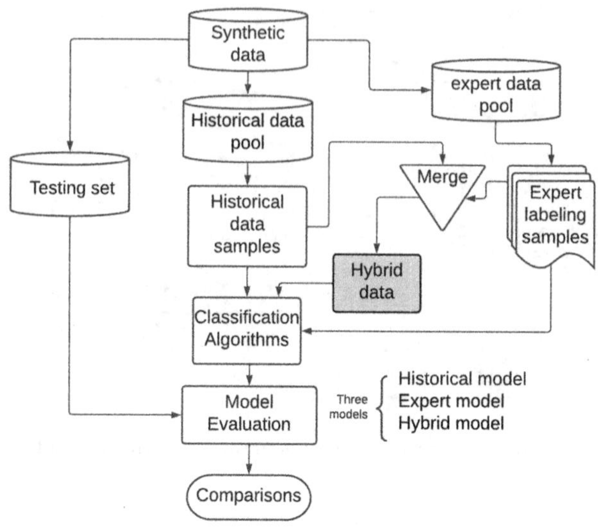

Fig. 2. Experimental design

5.2 Baseline Model

The historical data model for task duration forecast is considered as the baseline model. This model is trained with 50 randomly selected samples from the historical data pool.

5.3 Expert Data with Simulated Errors

To simulate human errors, we created expert data with random errors. We use *np.random.randint()* function to randomly generate the index of the dataframe (after converting expert csv data into dataframe), and alternated the labels to the next closest label. For example, to create an expert data with 10% errors for 60 data set which indicates 2 errors per categories (3 categories of 6 errors in total), we randomly selected 2 samples in each category, and changed their labels, either from short to average, or average to long, or long to short durations. We assume that errors estimated by the human experts are never too significant, such as estimating a short duration task to long duration task or vice versa.

Expert data was selected from the expert data pool of 50 to 350 samples. Expert data had the same category distributions (short, average and long durations) as the historical data set.

5.4 Experimental Results

Figure 3 shows prediction accuracy comparisons of hybrid models composed by a randomly selected 50 historical data set, and expert data sets with different size ranging from 50 to 350 with different errors of 0% to 40%.

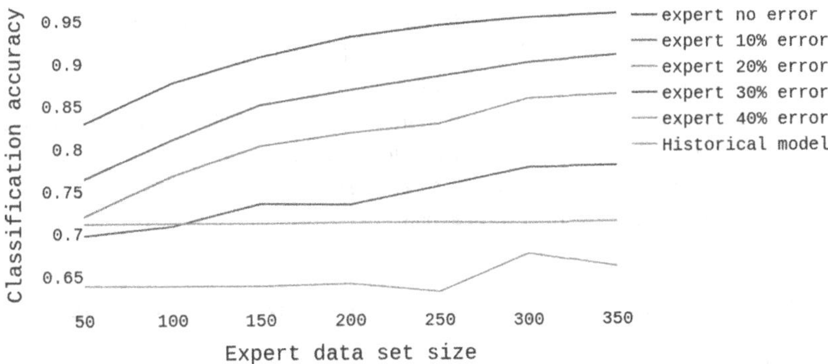

Fig. 3. Hybrid model comparisons for 50 historical samples and 50–350 expert data at different human error rate

For each prediction using the decision tree classification algorithm, the hyperparameters tuning process is applied with corresponding input data[1]. Each experiment result comes from the average of 100 random sampling of historical and expert data set tested against 10 different test data sets, to reduce the randomness and obtain the averaged model performance.

Table 2 shows classification accuracy for 50 historical samples with different variations of expert data size and error rate.

Table 2. Sample experimental results for synthetic data comparing historical model with hybrid model

Historical data size	Expert data size	Expert error rate	Accuracy from historical model	Accuracy from expert model	Accuracy from hybrid model
50	50	0.0	0.711420	0.718131	**0.828875**
50	50	0.1	0.711977	0.596376	**0.763460**
50	50	0.2	0.711078	0.488937	**0.719633**
50	50	0.3	0.711221	0.466485	0.697083
50	50	0.4	0.711757	0.428220	0.638376
50	100	0.0	0.711601	0.827130	**0.876010**
50	100	0.1	0.711482	0.703705	**0.809417**
50	100	0.2	0.710881	0.610328	**0.766529**
50	100	0.3	0.710786	0.528387	0.707698
50	100	0.4	0.711940	0.454457	0.637439

[1] The hyperparameter search space is defined as such: cv=10, scoring = 'accuracy', max_depth=list(range(3, 10)), min_samples_split=[2, 3].

We observe that the more expert data is integrated into the hybrid model, the better the model predicts. For error rates lower than 30%, the hybrid model always outperforms the historical model.

These results demonstrate that hybrid model (of both historical and expert data) performs better than historical models which serve as the baseline model. For historical data and expert data of error rate 10% and 20%, we observed a 5% average increase in prediction accuracy.

6 Data Sampling Approaches for Expert Querying

We investigated several data sampling techniques to be used during the expert data selection process. The purpose for this experiment was to investigate, in situations where we have access to multiple labeled data samples, how to select data that best represents the overall problem space. We were also interested to know whether certain data sampling approaches outperform random sampling in terms of their impact on the predictive accuracy of hybrid models. There exist several statistical sampling approaches, such as Latin Hypercube sampling, Halton sampling, Hammersley Sampling, Centroidal voronoi tessellation (CVT) sampling and etc. Each sampling approach has its own advantages and disadvantages, however, the choice depends on the use cases and the experimental results from the sampling methods.

In our experiment, we consider the following two sampling approaches from the pysmo library[2] The CVT sampling is also of interest, however, due to its slow converging time, we do not consider this approach in our use case.

- Latin Hypercube Sampling (LHS). Sampling space is firstly stratified, then data points are randomly selected from each partition. The goal is to obtain a set of more distributed sampling points in the parameter space.
- Halton sampling. This sampling approach is based on the Halton sequence, constructed by a set of co-prime bases. It is suitable for low-dimensional data, i.e., when there are no more than 10 features.

The following Table 3 shows part of the results from the hybrid model prediction accuracy comparisons on random sampling, LHS and Halton sampling on expert data. Hybrid model from Halton sampling for expert data always performs better than hybrid model with random sampling. LHS sampling is comparable to Halton sampling.

Considering the error rate of 10% and 20%, the following Figs. 4 and 5 describe the hybrid model prediction accuracy comparisons from random sampling, LHS and Halton sampling on expert data.

For both graphs, when the expert data size is no more than 100 samples, Halton sampling approach provides the best hybrid model performance, over both LHS and random sampling. For expert size large than 150 samples, random sampling seems to perform better than other sampling approaches. We also

[2] https://idaes-pse.readthedocs.io/en/1.5.1/surrogate/pysmo/index.html.

Table 3. Predictive accuracy for different sampling methods, human error rates and sample sizes

Historical data	Expert data	Expert error rate	Accuracy historical	Accuracy hybrid random	Accuracy hybrid LHS	Accuracy hybrid Halton
50	50	0.1	0.71	0.76	0.81	**0.85**
50	50	0.2	0.71	0.72	0.74	**0.76**
50	100	0.1	0.71	0.81	0.83	**0.85**
50	100	0.2	0.71	0.77	0.76	**0.78**

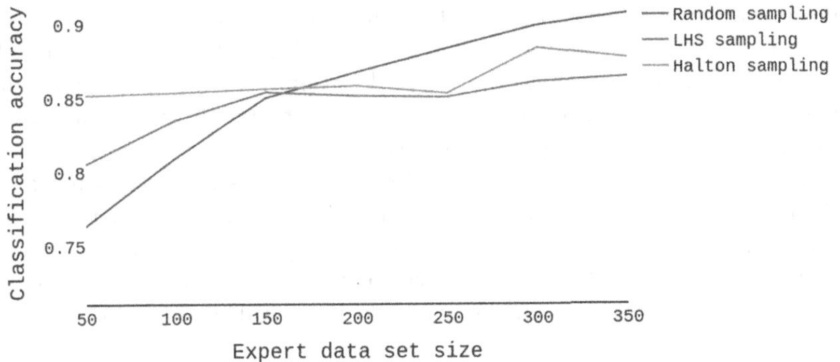

Fig. 4. Hybrid model comparisons with different sampling approaches on expert data with 10% errors

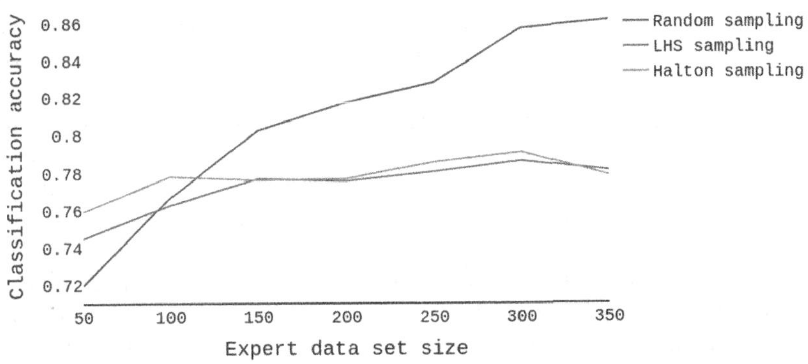

Fig. 5. Hybrid model comparisons with different sampling approaches on expert data with 20% errors

observe that Halton sampling always outperforms random sampling for error rate of 0%. The average precision increase for Halton sampling over historical model is 10%.

7 Conclusion

Herein, we explored approaches to integrate expert and historical data into a hybrid forecasting model to solve the scarce data problem. Our contributions are that such hybrid approach performs from 5% to 10% better than historical only data models (when assuming that experts are reasonably good estimators with 80–90% accuracy). Our objectives of this research focused on how to elicit and model expert forecasts to compensate for the lack of historical data, and how to augment the data driven method by integrating human forecasts to create a hybrid approach. In the context of scheduling problems, combining data driven forecasting and expert forecasts to improve planning under uncertainty is of particular interest because more accurate task duration estimates can lead to improved plan optimization in terms of duration, cost and risk management. Benefits of this approach are likely generalizable to many other domains. For example, in the context of pharmaceutical supply chains, a hybrid approach will enrich the prediction model [25] to increase forecasting precision for back-order products or newly available products. Experiments demonstrate that a hybrid method can outperform historical data models in terms of prediction accuracy. Potentially, it will be necessary to evaluate how accurate the experts are with their estimations with the available historical data before integrating their inputs.

8 Future Work

In the future, we first aim to develop and test a new two-stage hybrid forecasting method. In the first stage, we create three models: a model based on historical data, a separate model based on human expert forecasts, and a third hybrid model. Then in the second stage we train a meta-model on the same dataset while adding in the feature list for each case the prediction of each stage one model. The hypothesis is that the meta-model will have an added flexibility to learn under which circumstances it should rely on each of the three stage one models.

We are interested next to collaborate with real experts on integrating their inputs into the forecasting model. We are also interested to use multi-criteria modeling software such as Myriad to model expert's forecasts. It will be interesting to use Myriad to gather expert's input on multiple features as well as to design utility functions in order to aggregate features values into decisions.

It will also be worthwhile to consider process tracing [26] as a feature selection and data augmentation method for expert forecast modeling, which could be used to improve the hybrid model's performance.

We are also planning to adapt active learning as an intelligent querying method with continuously updated models to improve model accuracy through better case selection when collecting forecasts with experts [27].

Finally, multi-expert voting methods should be further investigated as well as multi-output forecasts. Indeed, an expert may be more confident to give a range of estimates, such as the most likely duration time, the pessimistic duration time and the optimistic duration time, instead of a precise number. Thus, three models per expert would need to be trained and integrated into future hybrid human-AI forecasting models uniquely modulating forecasts derived from human expertise with correlations learned from discrepancies with the historical data.

References

1. Lafond, D., Couture, D., Delaney, J., Cahill, J., Corbett, C., Lamontagne, G.: Multi-objective schedule optimization for ship refit projects: toward geospatial constraints management. In: Ahram, T., Taiar, R., Groff, F. (eds.) IHIET-AI 2021. AISC, vol. 1378, pp. 662–669. Springer, Cham (2021). https://doi.org/10.1007/978-3-030-74009-2_84
2. Torres, I.C., Armas-Aguirre, J.: Technological solution to improve outpatient medical care services using routing techniques and medical appointment scheduling. 2021 IEEE 1st International Conference on Advanced Learning Technologies on Education & Research (ICALTER), pp. 1–4 (2021)
3. Yeung, W., Choi, T., Cheng, T.C.E.: Optimal scheduling of a single-supplier single-manufacturer supply chain with common due windows. IEEE Trans. Autom. Control **55**(12), 2767–2777 (2010)
4. Eisler, C., Holmes, M.: Applying automated machine learning to improve budget estimates for a naval fleet maintenance facility. In: International Conference on Pattern Recognition Applications and Methods (2021)
5. Maybury, D.: Predictive analytics for the royal Canadian navy fleet maintenance facilities. DRDC - Centre for Operational Research and Analysis. Reference Document, DRDC-RDDC-2018-R150 December (2018)
6. McAndrew, T., Wattanachit, N., Gibson, GC., Reich, N.G.: Aggregating predictions from experts: a review of statistical methods, experiments, and applications. Wiley Interdiscip Rev Comput Stat. 13(2), e1514 Mar-Apr (2021)
7. Linstone, H.A., Turoff, M.: The Delphi Method: techniques and applications (Eds.), Addison-Wesley, Reading, MA (1975)
8. Bañuls, V.A., Turoff, M.: Scenario construction via Delphi and cross-impact analysis. Technol. Forecast. Soc. Change **78**(9), 1579–1602 (2011)
9. Porter, A., Xu, H.: Cross impact analysis. Project Appraisal **5**(3), 186–188 (1990)
10. Bañuls, V.A., Turoff, M., Hiltz, S.R.: Supporting collaborative scenario analysis through cross-impact. ISCRAM (2012)
11. Muskat, M., Blackman, D., Muskat, B.: Mixed methods: combining expert interviews, cross-impact analysis and scenario development. SSRN Electron. J. **10**(1), 9–21 (2012)
12. Duin, H.: Causal cross-impact analysis as a gaming tool for strategic decision making. In: Multidisciplinary Research on New Methods for Learning and Innovation in Enterprise Networks: Proceeding from the 11th Workshop of the Special Interest Group on Experimental Interactive Learning in Industrial Management, pp. 79–93 (2007)
13. Armstrong, J.S.: Judgmental Bootstrapping: inferring experts' rules for forecasting. In: Armstrong J.S., (eds) Principles of Forecasting. International Series in Operations Research & Management Science, vol 30. Springer, Boston, MA (2001)
14. Lafond, D., Roberge-Vallières, B., Vachon, F., Tremblay, S.: Judgment analysis in a dynamic multitask environment: capturing nonlinear policies using decision trees. J. Cognitive Eng. Decis. Making. **11**(2), 122–135 (2017)

15. Lafond, D., Labonté, K., Hunter, A., Neyedli, H.F., Tremblay, S.: Judgment analysis for real-time decision support using the cognitive shadow policy-capturing system. Human Interaction and Emerging Technologies, Volume 1018 (2020)
16. Zellner, M., Abbas, A.E., Budescu, D.V., Galstyan, A.: A survey of human judgement and quantitative forecasting methods. R. Soc. Open. sci. 8201187201187
17. Labreuche, C., Le Huédé, F.: Myriad: a tool for preference modeling application to multi-objective optimization. In: 7th International workshop on Preferences and soft constraints, October 1, Spain (2005)
18. Le Huédé, F., Grabisch, M., Labreuche, C., et al.: Integration and propagation of a multi-criteria decision making model in constraint programming. J. Heuristics **12**, 329–346 (2006)
19. Lafond, D., Gagnon, J., Tremblay, S., Derbentseva, D., Lizotte, M.: Multi-criteria assessment of a whole-of-government planning methodology using MYRIAD. In: IEEE International Multi-Disciplinary Conference on Cognitive Methods in Situation Awareness and Decision, pp. 49–55 (2015)
20. Barbaresco, F., Deltour, J.C., Desodt, G., Durand, G., Guenais, T., Labreuche, C.: Intelligent M3R radar time resources management: advanced cognition, agility & autonomy capabilities. In: 2009 International Radar Conference "Surveillance for a Safer World" (RADAR 2009), pp. 1–6 (2009)
21. Gennatas, E.D., et al.: Expert-augmented machine learning. In: Proceedings of the National Academy of Sciences, vol. 117, issue 9, pp. 4571–4577 Mar (2020)
22. White, R.W., Awaadallah, A.H.: Task duration estimation. In: WSDM '19: Proceeding of the Twelfth ACM International Conference on Web Search and Data Mining, Jan (2019)
23. LeDell, E., Poirier, S.: H2O AutoML: scalable automatic machine learning. In: 7th ICML Workshop on Automated Machine Learning (AutoML), July (2020)
24. Friedman, J.: Multivariate adaptive regression splines. Ann. Stat. **19**(1), 1–67 (1991)
25. Almentero, B.K., Li, J., Besse, C.: Forecasting pharmacy purchases orders. In: IEEE 24th International Conference on Information Fusion (FUSION), pp. 1–8 (2021)
26. Labonté, K., et al.: Combining process tracing and policy capturing techniques for judgment analysis in an anti-submarine warfare simulation. In: 65th International Annual Meeting of the Human Factors and Ergonomics Society, pp. 1557–1561 (2021)
27. Chatelais, B., Lafond, D., Hains, A., Gagné, C.: Improving policy-capturing with active learning for real-time decision support. In: Ahram, T., Karwowski, W., Vergnano, A., Leali, F., Taiar, R. (eds.) IHSI 2020. AISC, vol. 1131, pp. 177–182. Springer, Cham (2020). https://doi.org/10.1007/978-3-030-39512-4_28

Adaptive Zeroth-Order Optimisation of Nonconvex Composite Objectives

Weijia Shao(✉) and Sahin Albayrak

Technische Universität Berlin, Ernst-Reuter-Platz 7, 10587 Berlin, Germany
weijia.shao@campus.tu-berlin.de

Abstract. In this paper, we propose and analyse algorithms for zeroth-order optimisation of non-convex composite objectives, focusing on reducing the complexity dependence on dimensionality. This is achieved by exploiting the low dimensional structure of the decision set using the stochastic mirror descent method with an entropy alike function, which performs gradient descent in the space equipped with the maximum norm. To improve the gradient estimation, we replace the classic Gaussian smoothing method with a sampling method based on the Rademacher distribution and show that the mini-batch method copes with the non-Euclidean geometry. To avoid tuning hyperparameters, we analyse the adaptive stepsizes for the general stochastic mirror descent and show that the adaptive version of the proposed algorithm converges without requiring prior knowledge about the problem.

Keywords: Zeroth-order optimisation · Non-convexity · High dimensionality · Composite objective

1 Introduction

In this work, we study the following stochastic optimisation problem

$$\min_{x\in\mathcal{K}}\{F(x):=f(x)+h(x)=\mathbb{E}_{\xi}[f(x;\xi)+h(x)]\},\tag{1}$$

where f is a black-box, smooth, possibly nonconvex function, h is a white box convex function, and $\mathcal{K}\subseteq\mathbb{R}^d$ is a closed convex set. In many real-world applications, h and $\mathcal{K}$ are sparsity promoting, such as the black-box adversarial attack [3], model agnostic methods for explaining machine learning models [30] and sparse cox regression [27]. Despite the low dimensional structure restricted by h and $\mathcal{K}$, standard stochastic mirror descent methods [21] and the conditional gradient methods [14] have oracle complexity depending linearly on d and are not optimal for high dimensional problems.

The gradient descent algorithm is dimensionality independent when the first-order information is available [31]. For black-box objective functions, stronger dependence of the oracle complexity on dimensionality is caused by the biased gradient estimation [16]. In [39], the authors have proposed a LASSO-based gradient estimator for zeroth-order optimisation of unconstrained convex objective functions. Under the assumption of sparse gradients, the standard stochastic gradient descent with a LASSO-based gradient estimator has a weaker complexity dependence on dimensionality. The sparsity

G. Nicosia et al. (Eds.): LOD 2022, LNCS 13810, pp. 573–595, 2023.
https://doi.org/10.1007/978-3-031-25599-1_42

assumption has been further examined for nonconvex problems in [1], which proves a similar oracle complexity of the zeroth-order stochastic gradient method with Gaussian smoothing.

The critical issue of the algorithms mentioned above is the requirement of sparse gradients, which can not be expected in every application. We wish to improve the dependence on dimensionality by exploiting the low dimensional structure defined by the objective function and constraints. For convex problems, this can be achieved by employing the mirror descent method with distance generating functions that are strongly convex w.r.t. $\|\cdot\|_1$, such as the exponentiated gradient [18,40] or the p-norm algorithm [8]. However, a few problems arise if we apply these methods directly to optimising nonconvex functions. First, since these methods are essentially the gradient descent in $(\mathbb{R}^d, \|\cdot\|_\infty)$, the convergence of the mirror descent algorithm requires variance reduction techniques in that space. Existing variance reduction techniques [5,20,36] are developed for the standard Euclidean space, and deriving convergence from the equivalence of the norms in $\mathbb{R}^d$ introduces additional complexity depending on d [11]. Secondly, the exponentiated gradient [18] method and its extensions [40] work only for decision sets in the form of a simplex or cross-polytope with a known radius. Therefore, they can hardly be applied to general cases. The p-norm algorithm is more flexible and has an efficient implementation for ℓ_1 regularised problems [37]. However, handling ℓ_2 regularised problems with the p-norm algorithm is challenging.

The primary contribution of this paper is the introduction and analysis of algorithms for zeroth-order optimisation of nonconvex composite objective functions. To reduce the complexity dependence on dimensionality without assuming sparse gradients, we employ an entropy alike distance generating function in the stochastic mirror descent method (ZO-ExpMD), which performs gradient descent in $(\mathbb{R}^d, \|\cdot\|_\infty)$. To improve the gradient estimation in that space, we use the mini-batch approach [12] and show that the additional complexity introduced by switching the norms depends on $\ln d$ instead of d. Furthermore, we replace the gradient estimation methods applied in [1] and [38] with a smoothing method based on the Rademacher distribution. Our analysis shows that the total number of oracle calls required by ZO-ExpMD for finding an ϵ-stationary point is bounded by $\mathcal{O}(\frac{\ln d}{\epsilon^4})$, which improves the complexity bound $\mathcal{O}(\frac{d}{\epsilon^4})$ attained by proximal stochastic gradient descent (ZO-PSGD) [21]. To avoid tuning parameters, we extend and analyse the adaptive stepsizes [7,25] for constrained problems with composite objectives. Then we apply the adaptive stepsizes to ZO-ExpMD and show that the same complexity upper bound can be obtained without knowing the smoothness of f. In addition to the theoretical analysis, we also demonstrate the performance of the developed algorithms in experiments on generating contrastive explanations of deep neural networks [6].

The rest of the paper is organised as follows. Section 2 reviews related work. In Sect. 3, we present and analyse our algorithms. Section 4 demonstrates the empirical performance of the proposed algorithms. Finally, we conclude our work with some future research directions in Sect. 5.

2 Related Work

Zeroth-order optimisation of nonconvex objective functions has many applications in machine learning, and signal processing [28]. Algorithms for unconstrained nonconvex problems have been studied in [10,26,32] and further enhanced with variance reduction techniques [17,29]. The high dimensional setting has been discussed in [1,39], in which algorithms with weaker complexity dependence on dimensionality are proposed. In practice, weaker dependence on dimensionality can also be achieved by applying the sparse perturbation techniques introduced in [33].

It is popular to solve constrained problems with zeroth-order Frank-Wolfe algorithms [1,2,14], which require the smoothness of the objective functions. We are motivated by the applications of adversarial attack and explanation methods based on the ℓ_1 and ℓ_2 regularisation [3,6,30], for which the objective functions contain non-smooth components. Our work is based on exploiting the low dimensional structure of the decision set, which has been discussed in [9,18,22,37,40] for online and stochastic optimization of convex functions and further extended for zeroth-order convex optimization in [8,38]. To efficiently implement both ℓ_1 and ℓ_2 regularised problems, we used an entropy alike function as the distance-generating function in the stochastic composite mirror descent method. Similar versions of the entropy alike function have previously been applied to unconstrained online convex optimisation [4,34]. We combine it with the algorithmic ideas of mini-batch [12] and adaptive stepsizes [7,25] to solve nonconvex optimisation problems.

3 Algorithms and Analysis

We start the theoretical analysis by introducing some important results of zeroth-order stochastic methods in a finite-dimensional vector space $\mathbb{X}$ equipped with an inner product $\langle \cdot, \cdot \rangle$ and some norm $\|\cdot\|$. Based on them, we then construct and analyze our algorithms in $\mathbb{R}^d$.

3.1 Adaptive Stochastic Composite Mirror Descent

Similar to the previous works on stochastic nonconvex optimisation [21], the following standard properties of the objective function f are assumed.

Assumption 1. *For any realisation ξ, $f(\cdot;\xi)$ is G-Lipschitz and has L-Lipschitz continuous gradients with respect to $\|\cdot\|$, i.e.*

$$\|\nabla f(x;\xi) - \nabla f(y;\xi)\|_* \leq L\|x-y\|,$$

for all $x, y \in \mathbb{X}$, which implies

$$|f(y;\xi) - f(x;\xi) - \langle \nabla f(x;\xi), y-x\rangle| \leq \frac{L}{2}\|x-y\|^2.$$

Assumption 2. *For any $x \in \mathbb{X}$, the stochastic gradient at x is unbiased, i.e.*

$$\mathbb{E}[\nabla f(x;\xi)] = \nabla f(x).$$

Assumption 1 and 2 imply the G-smoothness and L-smoothness of f due to the inequalities

$$|f(x) - f(y)| \leq \mathbb{E}[|f(x;\xi) - f(y;\xi)|] \leq G\|x-y\|,$$

and
$$\|\nabla f(x) - \nabla f(y)\|_* \leq \mathbb{E}[\|\nabla f(x;\xi) - \nabla f(y;\xi)\|_*] \leq L\|x - y\|.$$

Our idea is based on the stochastic composite mirror descent (SCMD), which iteratively updates the decision variable following the rule given by

$$x_{t+1} = \arg\min_{x \in \mathcal{K}} \langle g_t, x\rangle + h(x) + \eta_t \mathcal{B}_\phi(x, x_t), \tag{2}$$

where g_t is an estimation of the gradient $\nabla f(x_t)$ and ϕ is a distance generating function, i.e. 1-strongly convex w.r.t. $\|\cdot\|$. Define the generalised projection operator

$$\mathcal{P}_\mathcal{K}(x, g, \eta) = \arg\min_{y \in \mathcal{K}} \langle g_t, y\rangle + h(y) + \eta \mathcal{B}_\phi(y, x) \tag{3}$$

and the generalised gradient map

$$\mathcal{G}_\mathcal{K}(x, g, \eta) = \eta(x - \mathcal{P}_\mathcal{K}(x, g, \eta)). \tag{4}$$

Following the literature on the stochastic optimisation [1,21], our goal is to find an ϵ-stationary point x_R, i.e. $\mathbb{E}[\|\mathcal{G}_\mathcal{K}(x_R, \nabla f(x_R), \eta_R)\|^2] \leq \epsilon^2$. Given a sequence of estimated gradients, the convergence of SCMD is upper bounded by the following proposition, the proof of which can be found in the appendix.

Proposition 1. *Let $g_1, \ldots, g_T$ be any sequence in $\mathbb{X}$, $x_1, \ldots, x_T$ be the sequence generated by* (2) *with a distance generating function ϕ. Then, for any f satisfying Assumption 1 and 2, we have*

$$\begin{aligned}
&\mathbb{E}[\frac{1}{T}\sum_{t=1}^{T}\|\mathcal{G}_\mathcal{K}(x_t, \nabla f(x_t), \eta_t)\|^2] \\
\leq &\frac{6}{T}\sum_{t=1}^{T}\mathbb{E}[\sigma_t^2] + \frac{4}{T}\mathbb{E}[\sum_{t=1}^{T}\eta_t(F(x_t) - F(x_{t+1}))] \\
&+ \frac{1}{T}\mathbb{E}[\sum_{t=1}^{T}\eta_t(2L - \eta_t)\|x_{t+1} - x_t\|^2],
\end{aligned} \tag{5}$$

where we denote by $\mathbb{E}[\sigma_t^2] = \mathbb{E}[\|g_t - \nabla f(x_t)\|_*^2]$ the variance of the gradient estimation.

Setting $\eta_1, \ldots, \eta_T = 2L$, the convergence of SCMD depends on the convergence of the variance terms $\{\sigma_t^2\}$, which requires variance reduction techniques.

In practice, it is difficult to obtain prior knowledge about L. To avoid the expensive tuning, we propose an adaptive algorithm with a similar convergence guarantee. The idea is similar to the adaptive stepsizes for unconstrained stochastic optimisation [25], which sets $\eta_t = \sqrt{\sum_{s=1}^{t-1}\|g_s\|_*^2 + \beta}$ for some $\beta > 0$ to control the last term in (5). For composite objectives, $\|x_{t+1} - x_t\|^2$ depends not only on g_t but also on $\nabla h(x_{t+1})$, for which we set $\eta_t \propto \sqrt{\sum_{s=1}^{t-1}\|\mathcal{G}_\mathcal{K}(x_t, g_t, \eta_t)\|^2 + 1}$. To analyse the proposed method, we assume that the feasible decision set is contained in a closed ball.

Assumption 3. *There is some $D > 0$ such that $\|\mathcal{P}_{\mathcal{K}}(x, g, \eta)\| \leq D$ holds for all $\eta > 0$, $x \in \mathcal{K}$ and $g \in \mathbb{X}$.*

Assumption 3 is typical in many composite optimisation problems with regularisation terms in their objective functions. In the following lemma, we propose and analyse the adaptive SCMD. Due to the compactness of the decision set, we can also assume that the objective function takes values from $[0, B]$.

Assumption 4. *There is some $B > 0$ such that $F(x) \in [0, B]$ holds for all $x \in \mathcal{K}$.*

Lemma 1. *Assume 1, 2, 3 and 4. Define sequence of stepsizes*

$$\begin{aligned} \alpha_t &= (\sum_{s=1}^{t-1} \lambda_s^2 \alpha_s^2 \|x_s - x_{s+1}\|^2 + 1)^{\frac{1}{2}} \\ \eta_t &= \lambda \alpha_t. \end{aligned} \tag{6}$$

for some $0 < \lambda \leq \lambda_t \leq \kappa$. Furthermore we assume $D\lambda \geq 1$. Then we have

$$\mathbb{E}[\frac{1}{T}\sum_{t=1}^{T} \|\mathcal{G}_{\mathcal{K}}(x_t, \nabla f(x_t), \eta_t)\|^2] \leq \frac{13}{T}\sum_{t=1}^{T} \mathbb{E}[\sigma_t^2] + \frac{C}{T}. \tag{7}$$

where we define $C = 33\kappa^2 B^2 + \frac{16\sqrt{2}L^2 D}{\lambda}(1 + 2D\lambda)$.

Sketch of the proof The proof starts with the direct application of Proposition 1. The focus is then to control the term $\sum_{t=1}^{T} \eta_t(2L - \eta_t)\|x_{t+1} - x_t\|^2$. Since the sequence $\{\eta_t\}$ is increasing, we assume that $\eta_t > L$ starting from some index t_0. Then we only need to consider those stepsizes $\eta_1, \ldots, \eta_{t_0-1}$. Adding up $\sum_{t=1}^{t_0-2} \|x_{t+1} - x_t\|^2$ yields a value proportional to η_{t_0-1}. Thus, the whole term is upper bounded by a constant. The complete proof can be found in the appendix.

The adaptive SCMD does not require any prior information about the problem, including the assumed radius of the feasible decision set. Similar to SCMD with constant stepsizes, its convergence rate depends on the sequence of $\{\sigma_t^2\}$, which will be discussed in the next subsections.

3.2 Two Points Gradient Estimation

In [1], the authors have proposed the two points estimation with Gaussian smoothing for estimating the gradient, the variance of which depends on $(\ln d)^2$. We argue that the logarithmic dependence on d can be avoided. Our argument starts with reviewing the two points gradient estimation in the general setting. Given a smoothing parameter $\nu > 0$, some constant $\delta > 0$ and a random vector $u \in \mathbb{X}$, we consider the two points estimation of the gradient given by

$$\nabla f_\nu(x) = \mathbb{E}_u[\frac{\delta}{\nu}(f(x + \nu u) - f(x))u]. \tag{8}$$

To derive a general bound on the variance without specifying the distribution of u, we make the following assumption.

Assumption 5. *Let $\mathcal{D}$ be a distribution with* $\mathrm{supp}(\mathcal{D}) \subseteq \mathbb{X}$. *For $u \sim \mathcal{D}$, there is some $\delta > 0$ such that*

$$\mathbb{E}_u[\langle g, u\rangle u] = \frac{g}{\delta}.$$

Given the existence of $\nabla f(x,\xi)$, Assumption 5 implies $\mathbb{E}_u[\langle \nabla f(x,\xi), u\rangle \delta u] = \nabla f(x,\xi)$. Together with the smoothness of $f(\cdot,\xi)$, we obtain an estimation of $\nabla f(\cdot,\xi)$ with a controlled variance, which is described in the following lemma. Its proof can be found in the appendix.

Lemma 2. *Let C be the constant such that $\|x\| \le C\|x\|_*$ holds for all $x \in \mathbb{X}$. Then the follows inequalities hold for all $x \in \mathbb{X}$ and f satisfying Assumptions 1, 2 and 5.*

a) $\|\nabla f_\nu(x) - \nabla f(x)\|_* \le \frac{\delta\nu C^2 L}{2}\mathbb{E}_u[\|u\|_*^3]$.
b) $\mathbb{E}_u[\|\nabla f_\nu(x;\xi)\|_*^2] \le \frac{C^4 L^2 \delta^2 \nu^2}{2}\mathbb{E}_u[\|u\|_*^6] + 2\delta^2\mathbb{E}_u[\langle \nabla f(x;\xi), u\rangle^2 \|u\|_*^2]$.

For a realisation ξ and a fixed decision variable x_t, $\mathbb{E}[\sigma_t^2]$ can be upper bounded by combining the inequalities in Lemma 2. While most terms of the upper bound can be easily controlled by manipulating the smoothing parameter ν, it is difficult to deal with the term $\delta^2\mathbb{E}_u[\langle \nabla f(x;\xi), u\rangle^2 \|u\|_*^2]$. Intuitively, if we draw $u_1, \ldots, u_d$ from i.i.d. random variables with zero mean, δ^{-2} is related the variance. However, small $\mathbb{E}[\|u\|_*^k]$ indicates that u_i must be centred around 0, i.e. δ has to be large. Therefore, it is natural to consider drawing u from a distribution over the unit ball with controlled variance. For the case $\|\cdot\|_* = \|\cdot\|_\infty$, this can be achieved by drawing $u_1 \ldots, u_d$ from i.i.d. Rademacher random variables.

3.3 Mini-Batch Composite Mirror Descent for Non-Euclidean Geometry

With the results in Subsects. 3.1 and 3.2, we can construct an algorithm in $\mathbb{R}^d$, starting with analyzing the gradient estimation based on the Rademacher distribution.

Lemma 3. *Suppose that f is L-smooth w.r.t. $\|\cdot\|_2$ and $\mathbb{E}[\|\nabla f(x) - \nabla f(x,\xi)\|_2^2] \le \sigma^2$ for all $x \in \mathcal{K}$. Let $u_1, \ldots, u_d$ be independently sampled from the Rademacher distribution and*

$$g_\nu(x;\xi) = \frac{1}{\nu}(f(x+\nu u;\xi)) - f(x;\xi))u \tag{9}$$

be an estimation of $\nabla f(x)$. Then we have

$$\mathbb{E}[\|g_\nu(x;\xi) - \nabla f_\nu(x)\|_\infty^2] \le \frac{3\nu^2 d^2 L^2}{2} + 10\|\nabla f(x)\|_2^2 + 8\sigma^2. \tag{10}$$

The dependence on d^2 in the first term of (10) can be removed by choosing $v \propto \frac{1}{d}$, while the rest depends only on the variance of the stochastic gradient and the squared ℓ_2 norm of the gradient. The upper bound in (10) is better than the bound $(\ln d)^2(\|\nabla f(x)\|_1^2 + \|\nabla f(x;\xi) - \nabla f(x)\|_1^2)$ attained by Gaussian smoothing [1]. Note that $g_\nu(x;\xi)$ is an unbiased estimator of $\nabla f_\nu(x)$. Averaging $g_\nu(x;\xi)$ over a mini-batch can significantly reduce the variance alike quantity in $(\mathbb{R}^d, \|\cdot\|_\infty)$, which is proved in the next lemma.

Lemma 4. *Let $X_1, \dots, X_m$ be independent random vectors in $\mathbb{R}^d$ such that $\mathbb{E}[X_i] = \mu$ and $\mathbb{E}[\|X_i - \mu\|_\infty^2] \le \sigma^2$ hold for all $i = 1, \dots, m$. For $d \ge e$, we have*

$$\mathbb{E}[\|\frac{1}{m}\sum_{i=1}^{m} X_i - \mu\|_\infty^2] \le \frac{e(2\ln d - 1)\sigma^2}{m}. \tag{11}$$

Proof. We first prove the inequality $\mathbb{E}[\|\sum_{i=1}^m X_i - \mu\|_p^2] \le me(\ln d - 1)\sigma^2$ for $p = 2\ln d$. From the assumption $d \ge e$ and $p = 2\ln d$, it follows that the squared p norm is $2p - 2$ strongly smooth [35], i.e.

$$\|x + y\|_p^2 \le \|x\|_p^2 + \langle g_x, y\rangle + (p-1)\|y\|_p^2, \tag{12}$$

for all $x, y \in \mathbb{R}^d$ and $g_x \in \partial\|\cdot\|_p^2(x)$. Using the definition of p-norm, we obtain

$$\begin{aligned}(p-1)\|y\|_p^2 =& (p-1)(\sum_{i=1}^{d} |y_i|^p)^{\frac{2}{p}} \\ \le& (p-1)d^{\frac{2}{p}}\|y\|_\infty^2 \\ \le& e(2\ln d - 1)\|y\|_\infty^2.\end{aligned} \tag{13}$$

Combining (12) and (13), we have

$$\|x + y\|_p^2 \le \|x\|_p^2 + \langle g_x, y\rangle + e(2\ln d - 1)\|y\|_\infty^2. \tag{14}$$

Next, let X and Y be independent random vectors in $\mathbb{R}^d$ with $\mathbb{E}[X] = \mathbb{E}[Y] = 0$. Using (14), we have

$$\begin{aligned}\mathbb{E}[\|X + Y\|_p^2] \le& \mathbb{E}[\|X\|_p^2] + \mathbb{E}[\langle g_X, Y\rangle] + e(2\ln d - 1)\mathbb{E}[\|Y\|_\infty^2] \\ =& \mathbb{E}[\|X\|_p^2] + \langle \mathbb{E}[g_X], \mathbb{E}[Y]\rangle + e(2\ln d - 1)\mathbb{E}[\|Y\|_\infty^2] \\ =& \mathbb{E}[\|X\|_p^2] + e(2\ln d - 1)\mathbb{E}[\|Y\|_\infty^2],\end{aligned} \tag{15}$$

Note that $X_1 - \mu, \dots, X_m - \mu$ are i.i.d. random variable with zero mean. Combining (15) with a simple induction, we obtain

$$\mathbb{E}[\|\sum_{i=1}^{m}(X_i - \mu)\|_\infty^2] \le \mathbb{E}[\|\sum_{i=1}^{m}(X_i - \mu)\|_p^2] \le me(2\ln d - 1)\sigma^2. \tag{16}$$

The desired result is obtained by dividing both sides by m^2. □

Our main algorithm, which is described in Algorithm 1, uses an average of estimated gradient vectors

$$g_t = \frac{1}{m\nu}\sum_{j=1}^{m}(f(x_t + \nu u_{t,j}; \xi_{t,j}) - f(x_t; \xi_{t,j}))u_{t,j}, \tag{17}$$

and the potential function given by

$$\phi : \mathbb{R}^d \to \mathbb{R}, x \mapsto \sum_{i=1}^{d}((|x_i| + \frac{1}{d})\ln(d|x_i| + 1) - |x_i|) \tag{18}$$

to update x_{t+1} at iteration t. The next lemma proves its strict convexity.

Algorithm 1. Zeroth-Order Exponentiated Mirror Descent

Require: $m > 0, \nu > 0, x_1$ arbitrary and a sequence of positive values $\{\eta_t\}$
Define $\phi : \mathbb{R}^d \to \mathbb{R}, x \mapsto \sum_{i=1}^d ((|x_i| + \frac{1}{d}) \ln(d|x_i| + 1) - |x_i|)$
for $t = 1, \dots, T$ **do**
  Sample $u_{t,j,i}$ from Rademacher distribution for $j = 1, \dots m$ and $i = 1, \dots d$
  $g_t := \frac{1}{m\nu} \sum_{j=1}^m (f(x_t + \nu u_{t,j}; \xi_{t,j}) - f(x_t; \xi_{t,j})) u_{t,j}$
  $x_{t+1} = \arg\min_{x \in \mathcal{K}} \langle g_t, x \rangle + h(x) + \eta_t \mathcal{B}_\phi(x, x_t)$
end for
Sample R from uniform distribution over $\{1, \dots, T\}$.
Return x_R

Lemma 5. *For all $x, y \in \mathbb{R}^d$, we have*

$$\phi(y) - \phi(x) \geq \langle \nabla\phi(x), y - x \rangle + \frac{1}{2\max\{\|x\|_1, \|y\|_1\} + 2} \|y - x\|_1^2$$

The Proof of Lemma 5 can be found in the appendix. If the feasible decision set is contained in an ℓ_1 ball with radius D, then the function ϕ defined in (18) is $\frac{1}{D+1}$-strongly convex w.r.t $\|\cdot\|_1$. With ϕ, update (2) is equivalent to mirror descent with stepsize $\frac{\eta_t}{D+1}$ and the distance-generating function $(D + 1)\phi$. The performance of Algorithm 1 is described in the following theorem.

Theorem 1. *Assume 1, 2 for $\|\cdot\| = \|\cdot\|_2$, 3 for $\|\cdot\| = \|\cdot\|_1$ and 4. Furthermore, let f be G-Lipschitz continuous w.r.t. $\|\cdot\|_2$. Then running Algorithm 1 with $m = 2Te(2\ln d - 1)$, $\nu = \frac{1}{d\sqrt{T}}$ and $\eta_1 =, \dots, = \eta_T = 2L(D + 1)$ guarantees*

$$\mathbb{E}[\|\mathcal{G}_\mathcal{K}(x_R, \nabla f(x_R), 2L)\|_1^2] \leq \sqrt{\frac{2e(2\ln d - 1)}{mT}} (6V + 4LB), \tag{19}$$

where we define $V = \sqrt{10G^2 + 8\sigma^2 + 2L^2}$. Furthermore, setting

$$\lambda_t = \frac{1}{\max\{\|x_s\|_1, \|x_{s+1}\|_1\} + 1}$$

$$\alpha_t = (\sum_{s=1}^{t-1} \lambda_s^2 \alpha_s^2 \|x_{s+1} - x_s\|_1^2 + 1)^{\frac{1}{2}},$$

we have

$$\mathbb{E}[\frac{1}{T} \sum_{t=1}^T \|\mathcal{G}_\mathcal{K}(x_t, \nabla f(x_t), \eta_t)\|_1^2] \leq 13V \sqrt{\frac{2e(2\ln d - 1)}{mT}} + \frac{C}{T}.$$

where we define $C = 132B^2 + 40\sqrt{2}L^2 D(D + 1)$.

Proof. First, we bound the variance of the mini-batch gradient estimation. Define

$$g_{t,i} = \frac{1}{\nu} (f(x_t + \nu u_{t,j}; \xi_{t,j}) - f(x_t; \xi_{t,j})) u_{t,j}.$$

Since $g_{t,1}, \ldots, g_{t,m}$ are unbiased estimation of $\nabla f_\nu(x_t)$, we have

$$\mathbb{E}[\|g_t - \nabla f_\nu(x_t)\|_\infty^2] \leq \frac{e(2\ln d - 1)}{m}(\frac{3\nu^2 d^2 L^2}{2} + 10\|\nabla f(x_t)\|_2^2 + 8\sigma^2).$$

Using Lemma 3 and the distribution of u, we obtain

$$\|\nabla f_\nu(x_t) - \nabla f(x_t)\|_\infty^2 \leq \frac{\nu^2 d^2 L^2}{4}.$$

For $m \geq 2e(2\ln d - 1)$, we have

$$\begin{aligned}\mathbb{E}[\|g_t - \nabla f(x_t)\|_\infty^2] &\leq 2\mathbb{E}[\|g_t - \nabla f_\nu(x_t)\|_\infty^2] + 2\mathbb{E}[\|\nabla f_\nu(x_t) - \nabla f(x_t)\|_\infty^2] \\ &\leq \frac{e(2\ln d - 1)}{m}(20\|\nabla f(x_t)\|_2^2 + 16\sigma^2) + 2\nu^2 d^2 L^2 \\ &\leq \frac{2e(2\ln d - 1)}{m}(10G^2 + 8\sigma^2) + \frac{2L^2}{T} \\ &\leq \sqrt{\frac{2e(2\ln d - 1)}{mT}}(10G^2 + 8\sigma^2 + 2L^2).\end{aligned} \tag{20}$$

where the last inequality follows from $m = 2Te(2\ln d - 1)$.

Next, we analyse constant stepsizes. Note that the potential function defined in (18) is $\frac{1}{D+1}$ strongly convex w.r.t. to $\|\cdot\|_1$. Our algorithm can be considered as an mirror descent with distance generating function given by $(D+1)\phi$, stepsizes $\frac{\eta_t}{D+1} = 2L$. Applying Proposition 1 with stepsizes $2L$, we have

$$\begin{aligned}\mathbb{E}[\|\mathcal{G}_\mathcal{K}(x_R, \nabla f(x_R), 2L)\|_1^2] &\leq \frac{6}{T}\sum_{t=1}^{T}\mathbb{E}[\sigma_t^2] + \frac{4L}{T}(F(x_1) - F^*) \\ &\leq \sqrt{\frac{2e(2\ln d - 1)}{mT}}(6V + 4LB),\end{aligned} \tag{21}$$

where we define $V = 10G^2 + 8\sigma^2 + 2L^2$. To analyze the adaptive stepsizes, Lemma 1 can be applied with distance generating function $(D+1)\phi$, stepsizes $\frac{\alpha_t}{D+1}$ and

$$\lambda_t = \frac{1}{\max\{\|x_t\|_1, \|x_{t+1}\|_1\} + 1}.$$

It holds clearly $0 < \lambda = \frac{1}{D+1} \leq \lambda_t \leq 1 = \kappa$. W.l.o.g., we assume $2 \geq \lambda D \geq 1$. Then we obtain

$$\mathbb{E}[\frac{1}{T}\sum_{t=1}^{T}\|\mathcal{G}_\mathcal{K}(x_t, \nabla f(x_t), \eta_t)\|_1^2] \leq 13V\sqrt{\frac{2e(2\ln d - 1)}{mT}} + \frac{C}{T}. \tag{22}$$

where we define $C = 132B^2 + 40\sqrt{2}L^2 D(D+1)$. □

The total number of oracle calls for finding an ϵ-stationary point with constant is upper bounded by $\mathcal{O}(\frac{\ln d}{\epsilon^4})$, which has a weaker dependence on dimensionality compared to

$\mathcal{O}(\frac{d}{\epsilon^4})$ achieved by ZO-PSGD [21]. The adaptive stepsize has slightly worse oracle complexity than the well-tuned constant stepsize. Note that a similar result can be obtained by using distance generating function $\frac{1}{2(p-1)}\|\cdot\|_p^2$ for $p = 1 + \frac{1}{2\ln d}$. Since the mirror map at x depends on $\|x\|_p$, it is difficult to handle the popular ℓ_2 regulariser. Our algorithm has an efficient implementation for Elastic Net regularisation, which is described in the appendix.

4 Experiments

We examine the performance of our algorithms for generating the contrastive explanation of machine learning models [6], which consists of a set of positive pertinent (PP) features and a set of pertinent negative (PN) features[1]. For a given sample $x_0 \in \mathcal{X}$ and machine learning model $f : \mathcal{X} \to \mathbb{R}^K$, the contrastive explanation can be found by solving the following optimisation problem [6]

$$\min_{x \in \mathcal{K}} \quad l_{x_0}(x) + \gamma_1\|x\|_1 + \frac{\gamma_2}{2}\|x\|_2^2.$$

Define $k_0 = \arg\max_i f(x_0)_i$ the prediction of x_0.The loss function for finding PP is given by

$$l_{x_0}(x) = \max\{\max_{i \neq k_0} f(x)_i - f(x)_{k_0}, -\kappa\},$$

and PN is modelled by the following loss function

$$l_{x_0}(x) = \max\{f(x_0 + x)_{k_0} - \max_{i \neq k_0} f(x_0 + x)_i, -\kappa\},$$

where κ is some constant controlling the lower bound of the loss. In the experiment, we first train a LeNet model [23] on the MNIST dataset [23] and a ResNet20 model [13] on the CIFAR-10 dataset [19], which attains a test accuracy of 96%, 91%, respectively. For each class of the images, we randomly pick 20 correctly classified images from the test dataset and generate PP and PN for them. We set $\gamma_1 = \gamma_2 = 0.1$ for MNIST dataset, and choose $\{x \in \mathbb{R}^d | 0 \leq x_i \leq x_{0,i}\}$ and $\{x \in \mathbb{R}^d | x_i \geq 0, x_i + x_{0,i} \leq 1\}$ as the decision set for PP and PN, respectively. For CIFAR-10 dataset, we set $\gamma_1 = \gamma_2 = 0.5$. ResNet20 takes normalized data as input, and images in CIFAR-10 do not have an obvious background colour. Therefore, we choose $\{x \in \mathbb{R}^d | \min\{0, x_{0,i}\} \leq x_i \leq \max\{0, x_{0,i}\}\}$ and $\{x \in \mathbb{R}^d | 0 \leq (x_i + x_{0,i})\nu_i + \mu_i \leq 1\}$, where ν_i and μ_i are the mean and variance of the dimension i of the training data, as the decision set for PP and PN, respectively. The search for PP and PN starts from x_0 and the center of the decision set, respectively.

Our baseline method is ZO-PSGD with Gaussian smoothing, the update rule of which is given by

$$x_{t+1} = \arg\min_{x \in \mathcal{K}} \langle g_t, x \rangle + h(x) + \eta_t \|x - x_t\|_2^2.$$

[1] The source code is available at https://github.com/VergiliusShao/highdimzo.

We fix the mini-batch size $m = 200$ for all candidate algorithms to conduct a fair comparison study. Following the analysis of [21, Corollary 6.10], the optimal oracle complexity $\mathcal{O}(\frac{d}{\epsilon^2})$ of ZO-PSGD is obtained by setting $m = dT$ and $\nu = T^{-\frac{1}{2}}d^{-1} = m^{-\frac{1}{2}}d^{-\frac{1}{2}}$. The smoothing parameters for ZO-ExpMD and ZO-AdaExpMD are set to

$$\nu = m^{-\frac{1}{2}}(2e(2\ln d - 1))^{\frac{1}{2}}d^{-1}$$

according to Theorem 1. For ZO-PSGD, ZO-ExpMD, multiple constant stepsizes $\eta_t \in \{10^i | 1 \leq i \leq 5\}$ are tested. Figure 1 plots the convergence behaviour of the candidate algorithms with the best choice of stepsizes, averaging over 200 images from the MNIST dataset. Our algorithms have clear advantages in the first 50 iterations and achieve the best overall performance for PN. For PP, the loss attained by ZO-PSGD is slightly better than ExpMD with fixed stepsizes, however, it is worse than its adaptive version. Figure 2 plots the convergence behaviour of candidate algorithms averaging over 200 images from the CIFAR-10 dataset, which has higher dimensionality than the MNIST dataset. As can be observed, the advantage of our algorithms becomes more significant. Furthermore, choices of stepsizes have a clear impact on the performances of both ZO-ExpMD and ZO-PSGD, which can be observed in Fig. 3, 4 and Fig. 5, 6 in the appendix. Notably, ZO-AdaExpMD converges as fast as ZO-ExpMD with well-tuned stepsizes.

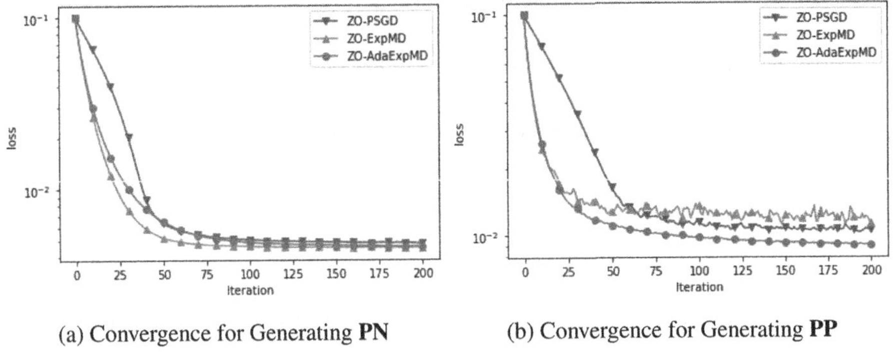

(a) Convergence for Generating **PN** (b) Convergence for Generating **PP**

Fig. 1. Black box contrastive explanations on MNIST

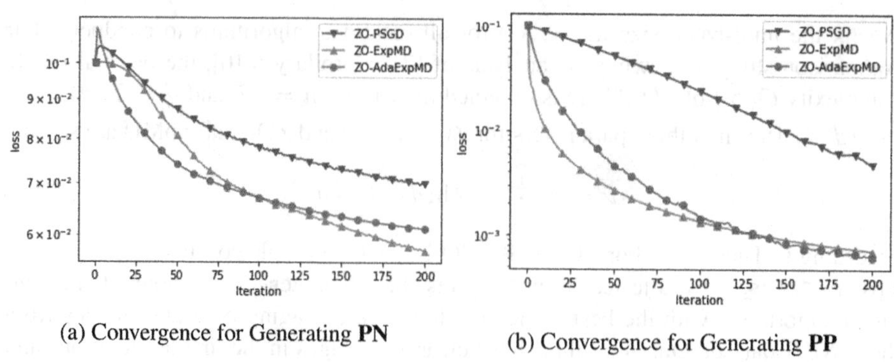

(a) Convergence for Generating **PN**

(b) Convergence for Generating **PP**

Fig. 2. Black box contrastive explanations on CIFAR-10

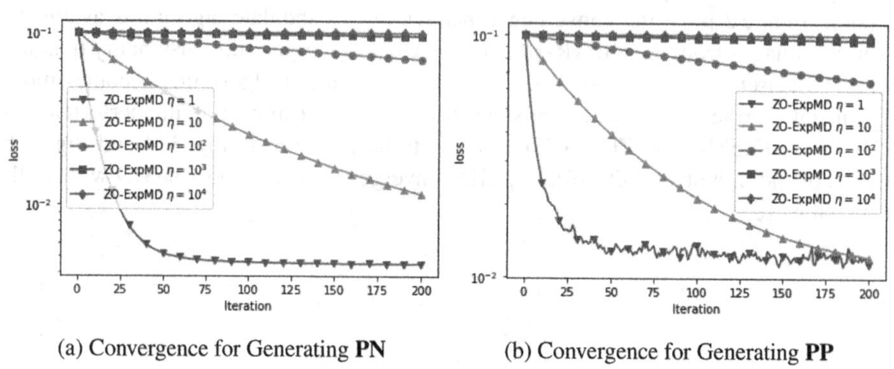

(a) Convergence for Generating **PN**

(b) Convergence for Generating **PP**

Fig. 3. Impact of step size on ZO-ExpMD on MNIST

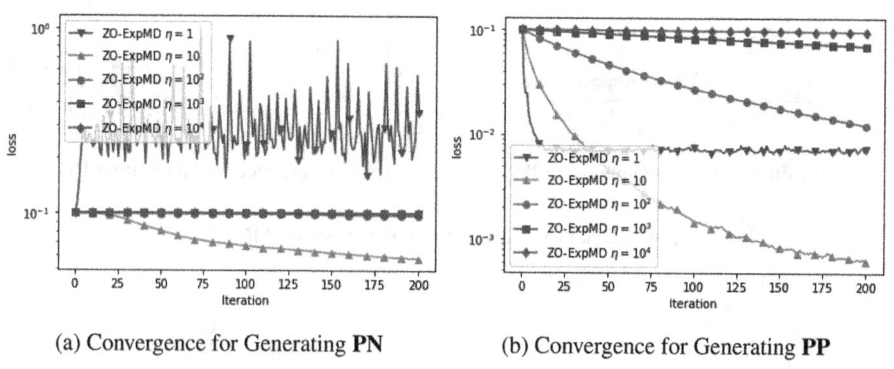

(a) Convergence for Generating **PN**

(b) Convergence for Generating **PP**

Fig. 4. Impact of step size on ZO-ExpMD on CIFAR-10

5 Conclusion

Motivated by applications in black-box adversarial attack and generating model agnostic explanations of machine learning models, we propose and analyse algorithms for zeroth-order optimisation of nonconvex objective functions. Combining several algorithmic ideas such as the entropy-like distance generating function, the sampling method based on the Rademacher distribution and the mini-batch method for non-Euclidean geometry, our algorithm has an oracle complexity depending logarithmically on dimensionality. With the adaptive stepsizes, the same oracle complexity can be achieved without prior knowledge about the problem. The performance of our algorithms is firmly backed by theoretical analysis and examined in experiments using real-world data.

Our algorithms can be further enhanced by the acceleration and variance reduction techniques. In the future, we plan to analyse the accelerated version of the proposed algorithms together with variance reduction techniques and draw a systematic comparison with the accelerated or momentum-based zeroth-order optimisation algorithms.

Acknowledgements. The research leading to these results received funding from the German Federal Ministry for Economic Affairs and Climate Action under Grant Agreement No. 01MK20002C.

A Missing Proofs

A.1 Proof of Proposition 1

Proof (Proof of Proposition 1). First of all, we have

$$\begin{aligned}
&F(x_{t+1}) - F(x_t)\\
\leq &\langle \nabla f(x_t) + \nabla h(x_{t+1}), x_{t+1} - x_t\rangle + \frac{L}{2}\|x_{t+1} - x_t\|^2\\
\leq &\langle \eta_t \nabla\phi(x_{t+1}) - \eta_t\nabla\phi(x_t), x_t - x_{t+1}\rangle\\
&+ \langle \nabla f(x_t) - g_t, x_{t+1} - x_t\rangle + \frac{L}{2}\|x_{t+1} - x_t\|^2\\
\leq & - \eta_t\|x_{t+1} - x_t\|^2 + \langle \nabla f(x_t) - g_t, x_{t+1} - x_t\rangle + \frac{L}{2}\|x_{t+1} - x_t\|^2 \qquad (23)\\
\leq & - \eta_t\|x_{t+1} - x_t\|^2 + \frac{1}{\eta_t}\sigma_t^2 + \frac{\eta_t\|x_t - x_{t+1}\|^2}{4} + \frac{L}{2}\|x_{t+1} - x_t\|^2\\
= & - \frac{\eta_t}{2}\|x_{t+1} - x_t\|^2 + \frac{1}{\eta_t}\sigma_t^2 + (\frac{L}{2} - \frac{\eta_t}{4})\|x_{t+1} - x_t\|^2\\
= & - \frac{1}{2\eta_t}\|\mathcal{G}_\mathcal{K}(x_t, g_t, \eta_t)\|^2 + \frac{1}{\eta_t}\sigma_t^2 + (\frac{L}{2} - \frac{\eta_t}{4})\|x_{t+1} - x_t\|^2,
\end{aligned}$$

where the first inequality uses the L-smoothness of f and the convexity of h, the second inequality follows from the optimality condition of the update rule, the third inequality is obtained from the strongly convexity of ϕ and the fourth line follows from the definition of dual norm. It follows from the $\frac{1}{\eta_t}$ Lipschitz continuity [21, Lemma 6.4] of $\mathcal{P}_\mathcal{K}(x_t, \cdot, \eta_t)$ that $\mathcal{G}_\mathcal{K}(x_t, \cdot, \eta_t)$ is 1-Lipschitz. Thus, we obtain

$$\begin{aligned}
&\|\mathcal{G}_{\mathcal{K}}(x_t, \nabla f(x_t), \eta_t)\|^2 \\
\leq& 2\|\mathcal{G}_{\mathcal{K}}(x_t, \nabla f(x_t), \eta_t) - \mathcal{G}_{\mathcal{K}}(x_t, g_t, \eta_t)\|^2 + 2\|\mathcal{G}_{\mathcal{K}}(x_t, g_t, \eta_t)\|^2 \\
\leq& 2\sigma_t^2 + 2\|\mathcal{G}_{\mathcal{K}}(x_t, g_t, \eta_t)\|^2 \\
\leq& 6\sigma_t^2 + 4\eta_t(F(x_t) - F(x_{t+1})) + \eta_t(2L - \eta_t)\|x_{t+1} - x_t\|^2.
\end{aligned} \tag{24}$$

Averaging from 1 to T and taking expectation, we have

$$\begin{aligned}
&\mathbb{E}[\frac{1}{T}\sum_{t=1}^{T}\|\mathcal{G}_{\mathcal{K}}(x_t, \nabla f(x_t), \eta_t)\|^2] \\
\leq& \frac{6}{T}\sum_{t=1}^{T}\mathbb{E}[\sigma_t^2] + \frac{4}{T}\mathbb{E}[\sum_{t=1}^{T}\eta_t(F(x_t) - F(x_{t+1}))] \\
&+ \frac{1}{T}\mathbb{E}[\sum_{t=1}^{T}\eta_t(2L - \eta_t)\|x_{t+1} - x_t\|^2],
\end{aligned} \tag{25}$$

which is the claimed result.

A.2 Proof of Lemma 1

Proof (Proof of Lemma 1). Applying proposition 1, we obtain

$$\begin{aligned}
&\mathbb{E}[\frac{1}{T}\sum_{t=1}^{T}\|\mathcal{G}_{\mathcal{K}}(x_t, \nabla f(x_t), \eta_t)\|^2] \\
\leq& \frac{6}{T}\sum_{t=1}^{T}\mathbb{E}[\sigma_t^2] + \frac{4}{T}\mathbb{E}[\sum_{t=1}^{T}\eta_t(F(x_t) - F(x_{t+1}))] \\
&+ \frac{1}{T}\mathbb{E}[\sum_{t=1}^{T}\eta_t(2L - \eta_t)\|x_{t+1} - x_t\|^2].
\end{aligned} \tag{26}$$

W.l.o.g., we can assume $F(x_0) = 0$, since it is an artefact in the analysis. The second term of the upper bound above can be rewritten into

$$\begin{aligned}
&\sum_{t=1}^{T}\eta_t(F(x_t) - F(x_{t+1})) \\
=& \eta_1 F(x_0) - \eta_T F(x_{T+1}) + \sum_{t=1}^{T}(\eta_t - \eta_{t-1})F(x_t) \\
\leq& B\sum_{t=1}^{T}(\eta_t - \eta_{t-1}) \\
\leq& B\eta_T \\
\leq& 4\kappa^2 B^2 + \frac{1}{16\kappa^2}\eta_T^2
\end{aligned} \tag{27}$$

where the first inequality follows from $F(x_0) = 0$ and $F(x_{T+1}) \geq 0$ and the last line uses the Hölder's inequality. Using the definition of η_T, we have

$$\begin{aligned}\frac{1}{16\kappa^2}\eta_T^2 &\leq \frac{\lambda^2}{16\kappa^2}\sum_{t=1}^{T}\lambda_s^2\alpha_s^2\|x_{s+1}-x_s\|^2+\frac{1}{16}\\ &\leq \frac{1}{16}\sum_{t=1}^{T}\|\mathcal{G}_{\mathcal{K}}(x_t,g_t,\eta_t)\|+\frac{1}{16}\\ &\leq \frac{1}{8}\sum_{t=1}^{T}\|\mathcal{G}_{\mathcal{K}}(x_t,\nabla f(x_t),\eta_t)\|+\frac{1}{8}\sum_{t=1}^{T}\sigma_t^2+\frac{1}{16}\end{aligned} \tag{28}$$

Next, define

$$t_0 = \begin{cases}\min\{1 \leq t \leq T | \eta_t > L\}, & \text{if } \{1 \leq t \leq T | \eta_t > 2L\} \neq \emptyset\\ T, & \text{otherwise.}\end{cases}$$

Then, the third term in (26) can be bounded by

$$\begin{aligned}&\sum_{t=1}^{T}\eta_t(2L-\eta_t)\|x_{t+1}-x_t\|^2\\ =&\sum_{t=1}^{t_0-1}\eta_t(2L-\eta_t)\|x_{t+1}-x_t\|^2+\sum_{t=t_0}^{T}\eta_t(2L-\eta_t)\|x_{t+1}-x_t\|^2\\ =&2L\sum_{t=1}^{t_0-1}\frac{\alpha_t\eta_t\|x_{t+1}-x_t\|^2}{\alpha_t}\\ =&\frac{2\sqrt{2}L}{\lambda}\sum_{t=1}^{t_0-1}\frac{\eta_t^2\|x_{t+1}-x_t\|^2}{\sqrt{2\sum_{s=1}^{t-1}\lambda_s^2\alpha_s^2\|x_{s+1}-x_s\|^2+2}}\\ \leq&4\sqrt{2}LD\sum_{t=1}^{t_0-1}\frac{\eta_t^2\|x_{t+1}-x_t\|^2}{\sqrt{\sum_{s=1}^{t-1}\lambda_s^2\alpha_s^2\|x_{s+1}-x_s\|^2+4\lambda^2\alpha_t^2D^2}}\\ \leq&4\sqrt{2}LD\sum_{t=1}^{t_0-1}\frac{\eta_t^2\|x_{t+1}-x_t\|^2}{\sqrt{\sum_{s=1}^{t}\lambda^2\alpha_s^2\|x_{s+1}-x_s\|^2}}\\ \leq&8\sqrt{2}LD\sqrt{\sum_{t=1}^{t_0-1}\lambda^2\alpha_t^2\|x_{t+1}-x_t\|^2}\\ \leq&8\sqrt{2}LD(\alpha_{t_0-1}+2\lambda D\alpha_{t_0-1})\\ \leq&\frac{8\sqrt{2}LD}{\lambda}(1+2D\lambda)\eta_{t_0-1}\\ \leq&\frac{16\sqrt{2}L^2D}{\lambda}(1+2D\lambda),\end{aligned} \tag{29}$$

where we used the assumption $\lambda D \geq 1$ for the first inequality, lemma 6 for the third inequality and the rest inequalities follow from the assumptions on λ_t,D and η_{t_0-1}. Combining (26), (27), (28) and (29), we have

$$\begin{aligned}
&\mathbb{E}[\frac{1}{T}\sum_{t=1}^{T}\|\mathcal{G}_{\mathcal{K}}(x_t, \nabla f(x_t), \eta_t)\|^2] \\
\leq &\frac{13}{T}\sum_{t=1}^{T}\mathbb{E}[\sigma_t^2] + \frac{1}{T}(\frac{1}{2} + 32\kappa^2 B^2) + \frac{16\sqrt{2}L^2 D}{\lambda T}(1 + 2D\lambda).
\end{aligned} \tag{30}$$

For simplicity and w.l.o.g., we can assume $\kappa^2 B^2 \geq \frac{1}{2}$. Define $C = 33\kappa^2 B^2 + \frac{16\sqrt{2}L^2 D}{\lambda}(1 + 2D\lambda)$, we obtain the claimed result.

□

A.3 Proof of Lemma 2

Proof (Proof of Lemma 2). Let $\nabla f_\nu(x)$ be as defined in (8), then we have

$$\begin{aligned}
&\|\nabla f_\nu(x) - \nabla f(x)\|_* \\
= &\|\mathbb{E}_u[\frac{\delta}{\nu}(f(x + \nu u) - f(x))u] - \nabla f(x)\|_* \\
= &\frac{\delta}{\nu}\|\mathbb{E}_u[(f(x + \nu u) - f(x) - \langle\nabla f(x), \nu u\rangle)u]\|_* \\
\leq &\frac{\delta}{\nu}\mathbb{E}_u[(f(x + \nu u) - f(x) - \langle\nabla f(x), \nu u\rangle)\|u\|_*] \\
\leq &\frac{\delta\nu C^2 L}{2}\mathbb{E}_u[\|u\|_*^3]
\end{aligned} \tag{31}$$

where the second equality follows from the Assumption 5, the third line uses the Jensen's inequality, and the last line follows the L smoothness of f. Next, we have

$$\begin{aligned}
&\mathbb{E}_u[\|\nabla f_\nu(x;\xi)\|_*^2] \\
= &\mathbb{E}_u[\frac{\delta^2}{\nu^2}|f(x + \nu u;\xi) - f(x;\xi)|^2\|u\|_*^2] \\
= &\frac{\delta^2}{\nu^2}\mathbb{E}_u[(f(x + \nu u;\xi) - f(x;\xi) - \langle\nabla f(x;\xi), \nu u\rangle + \langle\nabla f(x;\xi), \nu u\rangle)^2\|u\|_*^2] \\
\leq &\frac{2\delta^2}{\nu^2}\mathbb{E}_u[(f(x + \nu u;\xi) - f(x;\xi) - \langle\nabla f(x;\xi), \nu u\rangle)^2\|u\|_*^2] \\
&+ \frac{2\delta^2}{\nu^2}\mathbb{E}_u[\langle\nabla f(x;\xi), \nu u\rangle^2\|u\|_*^2] \\
\leq &\frac{C^4 L^2\delta^2\nu^2}{2}\mathbb{E}_u[\|u\|_*^6] + 2\delta^2\mathbb{E}_u[\langle\nabla f(x;\xi), u\rangle^2\|u\|_*^2],
\end{aligned} \tag{32}$$

which is the claimed result.

A.4 Proof of Lemma 3

Proof (Proof of Lemma 3). We clearly have $\mathbb{E}[uu^\top] = I$. From lemma 2 with the constant $C = \sqrt{d}$ and $\delta = 1$, it follows

$$\begin{aligned}
\mathbb{E}[\|g_\nu(x;\xi)\|_\infty^2] \leq& \mathbb{E}[\frac{d^2L^2\nu^2}{2}\mathbb{E}_u[\|u\|_\infty^6] + 2\mathbb{E}_u[\langle\nabla f(x;\xi), u\rangle^2\|u\|_\infty^2]] \\
\leq& \mathbb{E}[\frac{d^2L^2\nu^2}{2} + 2\mathbb{E}_u[\langle\nabla f(x;\xi), u\rangle^2]] \\
\leq& \frac{d^2L^2\nu^2}{2} + 2\mathbb{E}[\|\nabla f(x;\xi)\|_2^2] \\
\leq& \frac{d^2L^2\nu^2}{2} + 4\mathbb{E}[\|\nabla f(x) - \nabla f(x;\xi)\|_2^2] + 4\|\nabla f(x)\|_2^2 \\
\leq& \frac{d^2L^2\nu^2}{2} + 4\sigma^2 + 4\|\nabla f(x)\|_2^2
\end{aligned} \tag{33}$$

where the second inequality uses the fact the $\|u\|_\infty \leq 1$ and the third inequality follows from the Khintchine inequality. The variance is controlled by

$$\begin{aligned}
&\mathbb{E}[\|g_\nu(x;\xi) - \nabla f_\nu(x)\|_\infty^2] \\
\leq& 2\mathbb{E}[\|g_\nu(x;\xi)\|_\infty^2] + 2\|\nabla f_\nu(x)\|_\infty^2 \\
\leq& \nu^2d^2L^2 + 8(\|\nabla f(x)\|_2^2 + \sigma^2) + 2\|\nabla f(x)\|_\infty^2 + 2\|\nabla f(x) - \nabla f_\nu(x)\|_\infty^2 \\
\leq& \nu^2d^2L^2 + 8(\|\nabla f(x)\|_2^2 + \sigma^2) + 2\|\nabla f(x)\|_\infty^2 + \frac{\nu^2d^2L^2}{2} \\
\leq& \frac{3\nu^2d^2L^2}{2} + 10\|\nabla f(x)\|_2^2 + 8\sigma^2,
\end{aligned} \tag{34}$$

which is the claimed result.

A.5 Proof of Lemma 5

Proof (Proof of Lemma 5). We first show that each component of ϕ is twice continues differentiable. Define $\psi : \mathbb{R} \mapsto \mathbb{R} : x \mapsto (|x| + \frac{1}{d})\ln(d|x|+1) - |x|$. It is straightforward that ψ is differentiable at $x \neq 0$ with

$$\psi'(x) = \ln(d|x| + 1)\,\mathrm{sgn}(x).$$

For any $h \in \mathbb{R}$, we have

$$\begin{aligned}
\psi(0 + h) - \psi(0) =& (|h| + \frac{1}{d})\ln(d|h| + 1) - |h| \\
\leq& (|h| + \frac{1}{d})d|h| - |h| \\
=& dh^2,
\end{aligned}$$

where the first inequality uses the fact $\ln x \leq x - 1$. Furthermore, we have

$$\begin{aligned}\psi(0+h) - \psi(0) &= (|h| + \frac{1}{d})\ln(d|h|+1) - |h| \\ &\geq (|h| + \frac{1}{d})(\frac{|h|}{|h|+\frac{1}{d}}) - |h| \\ &\geq 0,\end{aligned}$$

where the first inequality uses the farc $\ln x \geq 1 - \frac{1}{x}$. Thus, we have

$$0 \leq \frac{\psi(0+h) - \psi(0)}{h} \leq dh$$

for $h > 0$ and

$$dh \leq \frac{\psi(0+h) - \psi(0)}{h} \leq 0$$

for $h < 0$, from which it follows $\lim_{h\to 0} \frac{\psi(0+h)-\psi(0)}{h} = 0$. Similarly, we have for $x \neq 0$

$$\psi''(x) = \frac{1}{|x| + \frac{1}{d}}.$$

Let $h \neq 0$, then we have

$$\frac{\psi'(0+h) - \psi'(0)}{h} = \frac{\ln(d|h|+1)\operatorname{sgn}(h)}{h} = \frac{\ln(d|h|+1)}{|h|}.$$

From the inequalities of the logarithm, it follows

$$\frac{1}{|h| + \frac{1}{d}} \leq \frac{\psi'(0+h) - \psi'(0)}{h} \leq d.$$

Thus, we obtain $\psi''(0) = d$. Since ψ is twice continuously differentiable with $\psi''(x) > 0$ for all $x \in \mathbb{R}$, ϕ is strictly convex, and we have, for all $x, y \in \mathbb{R}^d$, there is a $c \in [0, 1]$ such that

$$\phi(y) - \phi(x) = \nabla\phi(x)(y - x) + \frac{1}{2}\sum_{i=1}^{d} \frac{1}{|cx_i + (1-c)y_i| + \frac{1}{d}}(x_i - y_i)^2. \quad (35)$$

For all $v \in \mathbb{R}^d$, we have

$$\begin{aligned}
&\sum_{i=1}^{d} \frac{v_i^2}{|cx_i + (1-c)y_i| + \frac{1}{d}} \\
=&\sum_{i=1}^{d} \frac{v_i^2}{|cx_i + (1-c)y_i| + \frac{1}{d}} \frac{\sum_{i=1}^{d}(|cx_i + (1-c)y_i| + \frac{1}{d})}{\sum_{i=1}^{d}(|cx_i + (1-c)y_i| + \frac{1}{d})} \\
\geq&\frac{1}{\sum_{i=1}^{d}(|cx_i + (1-c)y_i| + \frac{1}{d})}(\sum_{i=1}^{d}|v_i|)^2 \\
\geq&\frac{1}{c\|x\|_1 + (1-c)\|y\|_1 + 1}(\sum_{i=1}^{d}|v_i|)^2 \\
=&\frac{1}{\max\{\|x\|_1, \|y\|_1\} + 1}\|v\|_1^2,
\end{aligned} \tag{36}$$

where the first inequality follows from the Cauchy-Schwarz inequality. Combining (35) and (36), we obtain the claimed result. □

B Efficient Implementation for Elastic Net Regularization

We consider the following updating rule

$$\begin{aligned}
y_{t+1} &= \nabla\phi^*(\nabla\phi(x_t) - \frac{g_t}{\eta_t}) \\
x_{t+1} &= \arg\min_{x\in\mathcal{K}} h(x) + \eta_t \mathcal{B}_\phi(x, y_{t+1}).
\end{aligned} \tag{37}$$

It is easy to verify

$$(\nabla\phi^*(\theta))_i = (\frac{1}{d}\exp(|\theta_i|) - \frac{1}{d})\operatorname{sgn}(\theta_i).$$

Furthermore, (37) is equivalent to the mirror descent update (2) due to the relation

$$\begin{aligned}
x_{t+1} &= \arg\min_{x\in\mathcal{K}} h(x) + \eta_t \mathcal{B}_\phi(x, y_{t+1}) \\
&= \arg\min_{x\in\mathcal{K}} h(x) + \eta_t\phi(x) - \langle \eta_t \nabla\phi(y_{t+1}), x\rangle \\
&= \arg\min_{x\in\mathcal{K}} h(x) + \eta_t\phi(x) - \langle \eta_t \nabla\phi(x_t) - g_t, x\rangle \\
&= \arg\min_{x\in\mathcal{K}} \langle g_t, x\rangle + h(x) + \eta_t \mathcal{B}_\phi(x, x_t).
\end{aligned}$$

Next, We consider the setting of $\mathcal{K} = \mathbb{R}^d$ and $h(x) = \gamma_1\|x\|_1 + \frac{\gamma_2}{2}\|x\|_2^2$. The minimiser of

$$h(x) + \eta_t \mathcal{B}_\phi(x, y_{t+1})$$

in $\mathbb{R}^d$ can be simply obtained by setting the subgradient to 0. For $\ln(d|y_{i,t+1}| + 1) \leq \frac{\gamma_1}{\eta_{t+1}}$, we set $x_{i,t+1} = 0$. Otherwise, the 0 subgradient implies $\mathrm{sgn}(x_{i,t+1}) = \mathrm{sgn}(y_{i,t+1})$ and $|x_{i,t+1}|$ given by the root of

$$\ln(d|y_{i,t+1}| + 1) = \ln(d|x_{i,t+1}| + 1) + \frac{\gamma_1}{\eta_t} + \frac{\gamma_2}{\eta_t}|x_{i,t+1}|$$

for $i = 1, \dots, d$. For simplicity, we set $a = \frac{1}{d}$, $b = \frac{\gamma_2}{\eta_t}$ and $c = \frac{\gamma_1}{\eta_t} - \ln(d|y_{i,t+1}| + 1)$. It can be verified that $|x_{i,t+1}|$ is given by

$$|x_{i,t+1}| = \frac{1}{b} W_0(ab \exp(ab - c)) - a, \tag{38}$$

where W_0 is the principle branch of the *Lambert function* and can be well approximated [15]. For $\gamma_2 = 0$, i.e. the ℓ_1 regularised problem, $|x_{i,t+1}|$ has the closed form solution

$$|x_{i,t+1}| = \frac{1}{d} \exp(\ln(d|y_{i,t+1}| + 1) - \frac{\gamma_1}{\eta_t}) - \frac{1}{d}. \tag{39}$$

The implementation is described in Algorithm 2.

Algorithm 2. Solving $\min_{x \in \mathbb{R}^d} \langle g_t, x \rangle + h(x) + \eta_t \mathcal{B}_\phi(x, x_t)$

```
for i = 1, ..., d do
    z_{i,t+1} = ln(d|x_{i,t}| + 1) sgn(x_{i,t}) - g_{i,t}/η_t
    y_{i,t+1} = (1/d exp(|z_{i,t+1}|) - 1/d) sgn(z_{i,t+1})
    if ln(d|y_{i,t+1}| + 1) ≤ γ_1/η_t then
        x_{t+1,i} ← 0
    else
        a ← β
        b ← γ_2/η_t
        c ← γ_1/η_t - ln(d|y_{t+1,i}| + 1)
        x_{t+1,i} ← (1/b) W_0(ab exp(ab - c)) - a
    end if
end for
Return x_{t+1}
```

B.1 Impact of the Choice of Stepsizes of PGD

Lemma 6. *For positive values $a_1, \dots, a_n$ the following holds:*

1.

$$\sum_{i=1}^{n} \frac{a_i}{\sum_{k=1}^{i} a_k + 1} \leq \log(\sum_{i=1}^{n} a_i + 1)$$

2.

$$\sqrt{\sum_{i=1}^{n} a_i} \leq \sum_{i=1}^{n} \frac{a_i}{\sqrt{\sum_{j=1}^{i} a_j^2}} \leq 2\sqrt{\sum_{i=1}^{n} a_i}.$$

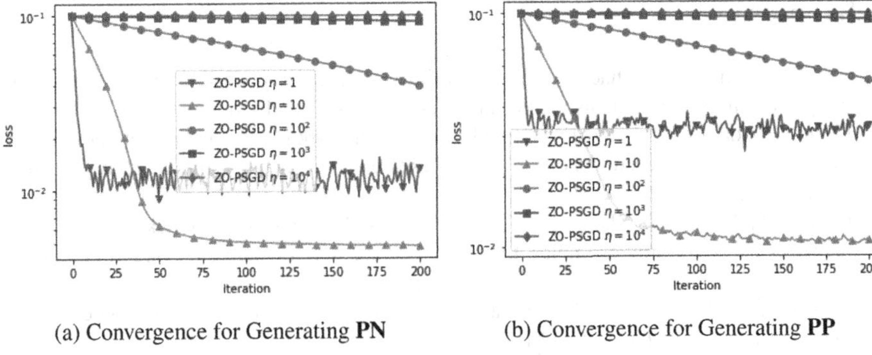

(a) Convergence for Generating **PN** (b) Convergence for Generating **PP**

Fig. 5. Impact of step size on ZO-PSGD on MNIST

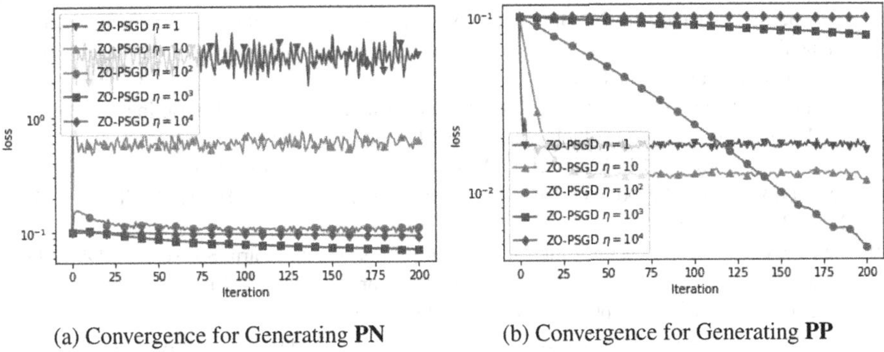

(a) Convergence for Generating **PN** (b) Convergence for Generating **PP**

Fig. 6. Impact of step size on ZO-PSGD on CIFAR-10

Proof. The proof of (1) can be found in Lemma A.2 in [24] For (2), we define $A_0 = 1$ and $A_i = \sum_{k=1}^{i} a_i + 1$ for $i > 0$. Then we have

$$\begin{aligned}
\sum_{i=1}^{n} \frac{a_i}{\sum_{k=1}^{i} a_k + 1} &= \sum_{i=1}^{n} \frac{A_i - A_{i-1}}{A_i} \\
&= \sum_{i=1}^{n} (1 - \frac{A_{i-1}}{A_i}) \\
&\leq \sum_{i=1}^{n} \ln \frac{A_i}{A_{i-1}} \\
&= \ln A_n - \ln A_0 \\
&= \ln \sum_{i=1}^{n} (a_i + 1),
\end{aligned}$$

where the inequality follows from the concavity of log.

References

1. Balasubramanian, K., Ghadimi, S.: Zeroth-order nonconvex stochastic optimization: handling constraints, high dimensionality, and saddle points. Found. Comput. Math. 1–42 (2021)
2. Chen, J., Zhou, D., Yi, J., Gu, Q.: A Frank-Wolfe framework for efficient and effective adversarial attacks. In: Proceedings of the AAAI Conference on Artificial Intelligence, pp. 3486–3494 (2020)
3. Chen, P.Y., Sharma, Y., Zhang, H., Yi, J., Hsieh, C.J.: Ead: elastic-net attacks to deep neural networks via adversarial examples. In: Thirty-Second AAAI Conference on Artificial Intelligence (2018)
4. Cutkosky, A., Boahen, K.: Online learning without prior information. In: Conference on Learning Theory, pp. 643–677. PMLR (2017)
5. Cutkosky, A., Orabona, F.: Momentum-based variance reduction in non-convex SGD. In: Advances in Neural Information Processing Systems, vol. 32 (2019)
6. Dhurandhar, A., et al.: Explanations based on the missing: towards contrastive explanations with pertinent negatives. In: Bengio, S., Wallach, H., Larochelle, H., Grauman, K., Cesa-Bianchi, N., Garnett, R. (eds.) Advances in Neural Information Processing Systems, vol. 31. Curran Associates, Inc. (2018)
7. Duchi, J., Hazan, E., Singer, Y.: Adaptive subgradient methods for online learning and stochastic optimization. J. Mach. Learn. Res. **12**(Jul), 2121–2159 (2011)
8. Duchi, J.C., Jordan, M.I., Wainwright, M.J., Wibisono, A.: Optimal rates for zero-order convex optimization: the power of two function evaluations. IEEE Trans. Inf. Theory **61**(5), 2788–2806 (2015)
9. Gentile, C.: The robustness of the p-norm algorithms. Mach. Learn. **53**(3), 265–299 (2003)
10. Ghadimi, S., Lan, G.: Stochastic first-and zeroth-order methods for nonconvex stochastic programming. SIAM J. Optim. **23**(4), 2341–2368 (2013)
11. Ghadimi, S., Lan, G.: Accelerated gradient methods for nonconvex nonlinear and stochastic programming. Math. Program. **156**(1–2), 59–99 (2016)
12. Ghadimi, S., Lan, G., Zhang, H.: Mini-batch stochastic approximation methods for nonconvex stochastic composite optimization. Math. Program. **155**(1), 267–305 (2016)
13. He, K., Zhang, X., Ren, S., Sun, J.: Deep residual learning for image recognition. In: 2016 IEEE Conference on Computer Vision and Pattern Recognition (CVPR), pp. 770–778 (2016). https://doi.org/10.1109/CVPR.2016.90
14. Huang, F., Tao, L., Chen, S.: Accelerated stochastic gradient-free and projection-free methods. In: International Conference on Machine Learning, pp. 4519–4530. PMLR (2020)
15. Iacono, R., Boyd, J.P.: New approximations to the principal real-valued branch of the lambert w-function. Adv. Comput. Math. **43**(6), 1403–1436 (2017)
16. Jamieson, K.G., Nowak, R., Recht, B.: Query complexity of derivative-free optimization. In: Advances in Neural Information Processing Systems, vol. 25 (2012)
17. Ji, K., Wang, Z., Zhou, Y., Liang, Y.: Improved zeroth-order variance reduced algorithms and analysis for nonconvex optimization. In: International Conference on Machine Learning, pp. 3100–3109. PMLR (2019)
18. Kivinen, J., Warmuth, M.K.: Exponentiated gradient versus gradient descent for linear predictors. Inf. Comput. **132**(1), 1–63 (1997)
19. Krizhevsky, A.: Learning multiple layers of features from tiny images. Master's thesis, University of Tront (2009)
20. Lan, G.: An optimal method for stochastic composite optimization. Math. Program. **133**(1–2), 365–397 (2012)
21. Lan, G.: First-Order and Stochastic Optimization Methods for Machine Learning. Springer, Cham (2020). https://doi.org/10.1007/978-3-030-39568-1

22. Langford, J., Li, L., Zhang, T.: Sparse online learning via truncated gradient. J. Mach. Learn. Res. **10**(3) (2009)
23. LeCun, Y., et al.: Handwritten digit recognition with a back-propagation network. In: Advances in Neural Information Processing Systems, vol. 2 (1989)
24. Levy, Y.K., Yurtsever, A., Cevher, V.: Online adaptive methods, universality and acceleration. In: Advances in Neural Information Processing Systems, pp. 6500–6509 (2018)
25. Li, X., Orabona, F.: On the convergence of stochastic gradient descent with adaptive stepsizes. In: The 22nd International Conference on Artificial Intelligence and Statistics, pp. 983–992. PMLR (2019)
26. Lian, X., Zhang, H., Hsieh, C.J., Huang, Y., Liu, J.: A comprehensive linear speedup analysis for asynchronous stochastic parallel optimization from zeroth-order to first-order. In: Advances in Neural Information Processing Systems, vol. 29 (2016)
27. Liu, S., Chen, J., Chen, P.Y., Hero, A.: Zeroth-order online alternating direction method of multipliers: convergence analysis and applications. In: International Conference on Artificial Intelligence and Statistics, pp. 288–297. PMLR (2018)
28. Liu, S., Chen, P.Y., Kailkhura, B., Zhang, G., Hero, A.O., III., Varshney, P.K.: A primer on zeroth-order optimization in signal processing and machine learning: principals, recent advances, and applications. IEEE Sig. Process. Mag. **37**(5), 43–54 (2020)
29. Liu, S., Kailkhura, B., Chen, P.Y., Ting, P., Chang, S., Amini, L.: Zeroth-order stochastic variance reduction for nonconvex optimization. In: Advances in Neural Information Processing Systems, vol. 31 (2018)
30. Natesan Ramamurthy, K., Vinzamuri, B., Zhang, Y., Dhurandhar, A.: Model agnostic multilevel explanations. In: Advances in Neural Information Processing Systems, vol. 33, pp. 5968–5979 (2020)
31. Nesterov, Y.: Introductory Lectures on Convex Optimization: A Basic Course, vol. 87. Springer, New York (2003). https://doi.org/10.1007/978-1-4419-8853-9
32. Nesterov, Y., Spokoiny, V.: Random gradient-free minimization of convex functions. Found. Comput. Math. **17**(2), 527–566 (2017)
33. Ohta, M., Berger, N., Sokolov, A., Riezler, S.: Sparse perturbations for improved convergence in stochastic zeroth-order optimization. In: Nicosia, G., et al. (eds.) LOD 2020. LNCS, vol. 12566, pp. 39–64. Springer, Cham (2020). https://doi.org/10.1007/978-3-030-64580-9_5
34. Orabona, F.: Dimension-free exponentiated gradient. In: NIPS, pp. 1806–1814 (2013)
35. Orabona, F., Crammer, K., Cesa-Bianchi, N.: A generalized online mirror descent with applications to classification and regression. Mach. Learn. **99**(3), 411–435 (2015)
36. Pham, N.H., Nguyen, L.M., Phan, D.T., Tran-Dinh, Q.: Proxsarah: an efficient algorithmic framework for stochastic composite nonconvex optimization. J. Mach. Learn. Res. **21**(110), 1–48 (2020)
37. Shalev-Shwartz, S., Tewari, A.: Stochastic methods for l 1-regularized loss minimization. J. Mach. Learn. Res. **12**, 1865–1892 (2011)
38. Shamir, O.: An optimal algorithm for bandit and zero-order convex optimization with two-point feedback. J. Mach. Learn. Res. **18**(1), 1703–1713 (2017)
39. Wang, Y., Du, S., Balakrishnan, S., Singh, A.: Stochastic zeroth-order optimization in high dimensions. In: International Conference on Artificial Intelligence and Statistics, pp. 1356–1365. PMLR (2018)
40. Warmuth, M.K.: Winnowing subspaces. In: Proceedings of the 24th International Conference on Machine Learning, pp. 999–1006 (2007)

Inferring Pathological Metabolic Patterns in Breast Cancer Tissue from Genome-Scale Models

Matteo N. Amaradio[1], Giorgio Jansen[1,2], Varun Ojha[3], Jole Costanza[4], Giuseppe Di Fatta[5], and Giuseppe Nicosia[1(✉)]

[1] Department of Biomedical and Biotechnological Sciences, University of Catania, Catania, Italy
giuseppe.nicosia@unict.it
[2] Department of Biochemistry, University of Cambridge, Cambridge, UK
[3] School of Computing, Newcastle University, Newcastle Upon Tyne, UK
[4] Fondazione Instituto Nazionale Di Genetica Molecolare, Milan, Italy
[5] Free University of Bozen-Bolzano, Bolzano, Italy

Abstract. We will consider genome-scale metabolic models that attempt to describe the metabolism of human cells focusing on breast cells. The model has two versions related to the presence or absence of a specific breast tumor. The aim will be to mine these genome-scale models as a multi-objective optimization problem in order to *maximize biomass* production and *minimize the reactions* whose enzymes that catalyze them may have undergone mutations (oncometabolite), causing cancer cells to proliferate. This study discovered characteristic pathological patterns for the breast cancer genome-scale model used. This work presents an in silico BioCAD methodology to investigate and compare the metabolic pathways of breast tissue in the presence of a tumor in contrast to those of healthy tissue. A large number of genome-scale metabolic model simulations have been carried out to explore the solution spaces of genetic configurations and metabolic reactions. An evolutionary algorithm is employed to guide the search for possible solutions, and a multi-objective optimization principle is used to identify the best candidate solutions.

Keywords: Metabolic engineering · Genome-scale metabolic models · Pathological patterns · Breast cancer · Oncometabolite · Multi-objective optimization

1 Introduction

Metabolism is the structure and behavior of the network of chemical reactions that occur within living organisms and serve to ensure their life. Dysfunctions in metabolism could be associated with many diseases. To understand metabolism, there has been a desire to include it in a broader model to simulate and predict the behavior of a given organism. Two types of computational pathway models have been developed to accomplish this

G. Nicosia et al. (Eds.): LOD 2022, LNCS 13810, pp. 596–612, 2023.
https://doi.org/10.1007/978-3-031-25599-1_43

task (Umeton et al. 2012): the first is kinetic, which uses differential equations (Van Rosmalen et al. 2021), while the second is based on constraints that are imposed on the system according to in-depth knowledge of metabolic networks (Orth et al. 2010).

The constraint-based model considers the stoichiometry of the reactions of metabolism, and thus through this methodology, metabolic networks can be easily converted into mathematical models to which a constraint-based analysis can then be applied (Angione et al. 2013). In this approach, model predictions depend on constraints on reactions and the definition of an objective function. Metabolic reactions can occur with enzymes or otherwise spontaneously or catalyzed by small molecules, implying that no gene is necessary for their catalysis (Keller et al. 2015). Most chemical reactions require enzymatic catalysis to occur. Therefore, for a reaction to take place, it is necessary for the enzyme that catalyzes it to be present and especially well-functioning. In the simplest case, a reaction is catalyzed by a single enzyme, which in turn is translated by a specific gene. The expression and translation of this gene imply the eligibility of a reaction. More complicated cases involve multiple genes and protein complexes whose relationships are described through Boolean logic.

In fact, if we want to turn off a particular reaction, we will have to silence all the enzymes that catalyze it, but even more convenient would be to knock-out the nucleic sequences that code for these proteins, that is, the genes. This is because, from a single gene, it is possible to get thousands of copies of the protein for which it translates, so by knocking out the gene and then acting upstream with a single knock-out, we would stop the production of many copies of a specific enzyme.

The knock-outs are modeled using the Gene-Protein Reaction (GPR) mapping, which links a group of a gene to a set of reactions. The knock-out cost (kc) is defined according to the Boolean relationships between genes. If two genes are linked by an 'and,' the knock-out cost of this gene set is 1, and this means that if we would delete the reaction catalyzed by these genes, we can delete one of these genes because both genes are necessary to catalyze the reaction (the two-protein subunit form a complex that catalyzes the reaction). Instead, if a gene set is composed of two genes linked by an 'or', then the kc of the gene set is 2 (this means that these two proteins are isoenzymes) (Patanè et al. 2015); so, if we would silence the reaction catalyzed from the enzyme translated by these genes, we must knock-out both. In this paper, we will focus on the analysis of metabolism.

We will consider genome-scale metabolic models (Patané et al. 2018) that attempt to describe the metabolism of human cells, focusing on breast cells. The model has two versions related to the presence or absence of a tumor. The aim will be to interpret these models as a system of constraints (Nicosia and Stracquadanio 2007) for an optimization problem to maximize biomass production, maximizing the flux of that reaction while minimizing other reactions. Obviously, by maximizing biomass, we want the cell to continue to function; at the same time, we will pose other objective functions, thus obtaining a multi-objective optimization problem (Amaradio et al. 2022).

In our case, such functions will be a minimization of certain reactions whose enzymes that catalyze them may have undergone mutations (oncometabolite), causing cancer cells to proliferate. Breast cancer is one of the most common global malignancies and the leading cause of cancer deaths (Katsura et al. 2022). Breast cancer is a heterogeneous disease in which genetic and environmental factors are involved. In 2018, 268670 new cases of breast cancer were reported in the United States (Barzaman et al. 2020). Based on both molecular and histological evidence, breast cancer could be categorized into three groups: breast cancer expressing hormone receptor (estrogen receptor (ER+) or progesterone receptor (PR+)), breast cancer expressing human epidermal receptor 2 (HER2+) and triple-negative breast cancer (TNBC) (ER−, PR−, HER2−). The treatment approaches should be based on breast cancer molecular characteristics (Barzaman et al. 2020).

2 Key Enzymes and Oncometabolites

Metabolism is intrinsically linked with many other cellular functions, and dysfunctions in metabolism are often identified as the causes of several diseases. Over the past decades, the list of reactions that make up an organism's metabolism has been cataloged in detail. This process has focused on characterizing individual reactions in detail. With the advancement in the understanding of metabolism, there has been a need to include it in broader models to simulate, predict and in some way understand the behavior of an organism. Constraint-based models that are based on genome-scale reconstructions of metabolism, on the other hand, seek to include all known reactions for a given organism through the integration of gene information and knowledge about biochemical reactions.

Reactions are defined simply through their stoichiometry, and metabolic networks can be easily converted into simple mathematical models, to which constraint-based analysis can then be applied. In this approach, the model predictions depend on constraints on reaction floors and a target (Rana et al. 2020). The publication of large network databases has simplified the reconstruction of metabolic models, and this has obviously encouraged the emergence of an increasing number of them. Several widely used and manually curated models have been published using a bottom-up approach, i.e., component-by-component. This required the creation of databases of metabolic reconstructions; one of the most important is the BiGG (Norsigian et al. 2020). Tumors acquire definitive changes in metabolism during disease development and may become dependent on metabolites (oncometabolites) (Vander Heiden 2011).

Oncometabolites, under physiological conditions, can have vital metabolic functions in cells. However, they can support malignant transformation through various mechanisms under pathological conditions. In addition to the oncometabolites presented by us in this research work, several oncometabolites such as 27-hydroxycholesterol (27HC), Lysophosphatidic acid (LPA), Quinurenine (AHR) were found in the case of breast cancer (Mishra P. et al. 2015). Instead, among the proteins identified as tumor suppressors, we can include the P53 protein, the PTEN, and BRCA 1/2 (Liu et al. 2015).

The Krebs Cycle and the 3 Groups of Enzymes Considered. The enzymes that were considered are part of the Krebs cycle. The Krebs cycle is used by all aerobic organisms to generate energy through the oxidation of the obtained acetyl-CoA. The Krebs cycle consists of eight intermediate steps, which are regulated by the following enzymes: Citrate synthase; Aconitase; Isocitrate Dehydrogenase; Alpha-ketoglutarate dehydrogenase; Succinyl-CoA synthetase; Succinate dehydrogenase; Fumarase.

Isocitrate Dehydrogenase. The enzyme Isocitrate Dehydrogenase, also referred to as IDH, catalyzes within the Krebs cycle the following reaction:

$$Isocitrate + NAD^{+} \leftrightarrows \alpha - ketoglutarate + CO_2 + NADH$$

There are three human IDH isoforms, that is, the closely related homodimeric IDH1 and IDH2 (70% homology) and the more distantly related heterotetrameric (2α, 1β, 1γ) IDH3. IDH1 localizes to the cytoplasm and peroxisomes; IDH2 and IDH3 localize to mitochondria (Liu et al. 2020). *IDH1* and *IDH2* undergo mutations correlating with >80% of low-grade glioma (LGG) and ~20% of acute myeloid leukemia (AML) cases (Liu et al. 2020). By contrast, no tumor-associated *IDH3* mutations are reported. IDH3 catalyzes the NAD+-dependent oxidative decarboxylation of d-isocitrate, giving 2-oxoglutarate (2OG) in the TCA cycle and a reaction reported to be irreversible under physiological conditions. IDH1 and IDH2 catalyze the reversible oxidized nicotinamide adenine dinucleotide phosphate (NADP+)-dependent oxidative decarboxylation of d-isocitrate to 2OG in a manner regulating isocitrate and 2OG levels and which provides reduced nicotinamide adenine dinucleotide phosphate (NADPH) (Liu et al. 2020). Cancer-associated substitutions in IDH1 and IDH2 impair wild-type (WT) activity–producing 2OG by promoting a 'neomorphic' reaction that converts 2OG to d-2-hydroxyglutarate (D-2-HG), using NADPH as a substrate (Liu et al. 2020). D-2-HG is one of the most well-characterized oncometabolite that is associated with pathogenic IDH mutations (Chou et al. 2021). A decade of research has provided a detailed description of the effects of 2HG on the acceleration of oncogenesis, such as affecting epigenetics by hypermethylation *via* inhibitions of 2-oxoglutarate- (2OG-) dependent dioxygenases, blocking DNA and histone demethylation (Ježek 2020). The neomorphic activity of mutated IDH1 or IDH2 enzymes causes a dramatic elevation of 2HG levels, which themselves are sufficient to promote glioma genesis or leukemo genesis in hematopoietic cells through the maintenance of dedifferentiation and increased proliferation (Ježek 2020). Accumulation of 2HG leads to aberrations in DNA and histone methylation that cause reversible dedifferentiation into a stem cell-like phenotype (Mishra and Ambs, 2015). Also, a key component of the hypoxia-inducible factor (HIF) pathway, the enzyme prolyl hydroxylase domain-2 (PHD2/EglN1), has been found to be activated by r-2HG (Ježek 2020).

Succinate Dehydrogenase. This enzyme is also referred to as SDH and catalyzes the following reaction:

$$Succinate + Ubiquinone \leftrightarrows Fumarate + ubiquinol$$

Succinate dehydrogenase (SDH), also known as mitochondrial complex II, is a mitochondrial enzyme with the unique property to participate in both the citric acid cycle, where its oxidases succinate to fumarate, and the electron transport chain, where it reduces ubiquinone to ubiquinol. SDH is composed of six subunits encoded by SDHA, SDHB, SDHC, SDHD, SDHAF1, and SDHF2, the last two coding for associated accessory factors. The complex is embedded in the inner mitochondrial membrane and exhibits a matrix-facing and a membrane-integrated domain performing the two distinct chemical reactions that produce fumarate, released into the soluble mitochondrial matrix, and ubiquinol, released into the inner membrane (Dalla Pozza et al. 2020). Mutations of SDH have been identified in several types of cancer and have been shown to contribute to an abnormal accumulation of succinate in the cytosol of tumoral cells and in the extracellular fluids of the patients. Following these observations confirming its role in the genesis of some tumors, SDH has been defined as a *tumor suppressor* and succinate an important *oncometabolite* (Dalla Pozza et al. 2020).

Fumarase. Fumarate hydratase or Fumarase (FH) is an enzyme that catalyzes a reversible hydration reaction, that is, the addition of a molecule of H2O, resulting in the transformation of fumarate to malate.

$$Fumarate + H_2O \leftrightarrows (S) - Malate$$

The reaction involved in the Krebs cycle occurs at the mitochondrial level. The enzyme FH exists in two isoforms encoded by the same FH gene, which is localized in the mitochondrion or the cytosol, depending on the activation of the mitochondrial localization signal. The mitochondrial isoform is part of the Krebs or citrate cycle, whereas the cytosolic isoform metabolizes fumarate produced by the urea cycle and the metabolism of amino acids and purines. Due to the essential role of FH in energy production. (Peetsold et al. 2021). Mutations of FH have been described in the literature and have been implicated in the pathogenesis of various diseases. For instance, the homozygous germline loss of *FH* is the cause of an autosomal recessive metabolic disease called fumaric aciduria (Schmidt et al. 2020). Heterozygous germline mutations of *FH* predispose to Hereditary Leiomyomatosis and Renal Cell Cancer (HLRCC), a cancer syndrome characterized by cutaneous, uterine leiomyomas, and renal cancer (Peetsold et al. 2021). These findings hint at a key role of *FH* loss in human cancers.

3 Designing a Genome-Scale Metabolic Network

A chemical reaction can be broken down into several conceptual elements: the reactants, the products, the actual reaction, stoichiometry, and the kinetic laws governing the reaction. To analyze or simulate a network of reactions, however, it is necessary to add other information, such as the compartments in which they occur, the units of measurement that describe the amount of reactant needed to react, and how much product will be formed from a generic reaction. Defining a model in SBML involves creating a list of one or more components. *Compartments:* A container of defined volume for well-mixed substances where certain reactions take place. In the example of metabolism, such compartments may be mitochondria, Golgi apparatus, and ribosomes. *Species:* A substance that plays a role in a reaction. Thus, they can be simple ions or molecules. *Parameters*: A quantity with a symbolic name. There can be global parameters or restricted to a single reaction. *Unit of measurement:* The unit with which a given quantity is expressed. Appropriate software can read the information from an SBML model to convert it to another format and then use it through a method, as in our case. Going then to look at the XML text file that expresses a model, it will have a structure typical of the XML standard, a text file in which each element consists of the corresponding start and end tag characters.

Compartments. A compartment in an SBML file identifies a finite volume in which species are located. Usually, they refer to actual parts of the cell, which refers to the compartments present in a model cell in breast cancer. For each compartment, we give some parameters: the name, which is the unique name it will have within the model, the spatial dimensions of the compartment, and the constant parameter indicating whether the size in volume will remain constant; in this case, the compartments do not have constant size.

Species. Species represent the elements, such as ions or molecules, that participate in the reactions. For species, we have some parameters that express the name and the compartment to which they refer. In addition, we have the '*OnlySubstanceUnits*' value, which refers to the interpretation of the value of the substance in a mathematical formula. In this case, when the dimension of the compartment changes, one must recalculate the total amount of the substance, expressing the values as units. The values of *'boundaryCondition'* and constant, equal in this case to false, stand for the fact that the quantity of the substance is not determined. The opposite case could, for example, be that of a simulation in which the level of a species is kept constant by an external mechanism during a reaction. In the two parameters more *"internal"*, finally, we indicate in *FORMULA* the chemical formula, and in *CHARGE* the electrical charge possessed, as in the case of an ion.

Reactions. As the last tag, we then have the list of chemical reactions. For each one, we indicate in the required parameter name the name of the reaction itself, as well as an ID value expressing a shorter identifier. Both values must be unique, i.e., there cannot be multiple reactions with the same name. It is then indicated via the logical parameter reversible whether the reaction is reversible or not. In the next 'notes' tag, in addition to other reaction identifiers, the most important property is given by the value GENE_ASSOCIATION, which expresses the association with certain genes. This information will then be fundamental. Such as drawing up the GPR map, as mentioned in the introduction. There are then the lists of reagents and products, reported respectively in <listOfReactants> and <list0fProducts>, in which all species are involved in the process. For each of them, the stoichiometric coefficient is highlighted, which will then be read and stored in the S matrix as a stoichiometry parameter. Finally, we have the tag <kineticLaw>, that is, related to the kinetic equation. In it, it is initially indicated that the law to be considered is given by the flow values. Then the parameters of a lower bound, upper bound, initial flux, and coefficient of the objective function are defined. This coefficient, for the models considered, is almost always zero. The only reactions for which there is a positive value are related to biomass and, in some models, related to healthy tissue cells to ATP production.

3.1 The FBA Model

Each XML file was interpreted through MATLAB to make it compatible with flux balance analysis standards using libSBML. A special function of the SBML library in MATLAB was exploited for this purpose:

Funtion model = TranslateSBML ('filename' optional), validateFlag (optional), verboseFlag (optional)

This function returns a structure containing the model data, which can be used, for example, by the COBRA Toolbox in MATLAB. The model must be interpreted and modified appropriately by a script, for example, by renaming some of the fields to make it usable by the algorithm and by the function glpk (Patanè et al. 2015). Each model thus obtained is a MATLAB structure consisting of 15 subfields that define the properties of the network and, thus, the problem of FBA optimization. The fields are given in Table 1:

Table 1. FBA fields.

Fields	Description
nrxn	It is the number of reactions present in the model. It, therefore, corresponds to the elements present in the listOfReactions of the initial XML file
nmetab	The number of metabolites involved in reactions. It corresponds to the number of species
S	Numerical matrix of size nmetab × nrxn, representing the stoichiometric matrix, that is, the relationships between reactants and metabolites produced, defining which metabolites and in what proportions are needed to generate a given product
rxns	Vector of dimension nrxn that contains the name of the reactions that identify the model
metabs	Vector of dimension nmetab containing the name of metabolites considered in the model
f	Numerical vector of dimension nrxn that contains the coefficients of the natural objective function. Through it we then determine, in our case, the flows of the reactions to be maximized. In the simplest example, there will be a single element equal to 1, corresponding to the function of biomass
g	Numerical vector of dimension nrxn containing the coefficients of the synthetic objectives. At each coefficient, as indeed in the case of f, each coefficient is related to the flux of a reaction
vmin	Numerical vector of dimension nrxn containing the lower bounds corresponding to all reaction flows in the model
vmax	Numerical vector of dimension nrxn containing the upper bounds corresponding to all reaction flows in the model. This vector (and the previous one) defines the constraints of the system
present	Logical vector of dimension nrxn, whose values (1 or 0) indicate the presence or absence of the corresponding reaction in the model (initially all the elements are equal to 1)
nbin	The number of genes in the model
pts	Vector of size nbin that contains the description of genes or sets of them

(*continued*)

Table 1. (*continued*)

Fields	Description
G	Matrix of size nbin × nrxn representing GPR (Gene-Protein-Reaction) mapping. In it the generic element $G_{l,j}$ is equal to 1 if the l-th gene manipulation maps onto reaction j, and 0 otherwise. GPR succeeds in capturing, through this Boolean approach, the complexity of the biological reactions
ko_cost	Vector of size nbin representing the knock-out cost of the individual genes (determined by the relationships between genes)

Table 2. Characteristics of Genome-scale metabolic models.

Models	Number of reactions (*nrxn*)	Number of metabolites (*nmetab*)	Number of genes (*nbin*)
Breast cancer	6602	4782	984
Breast normal (control)	6712	4820	1059

To carry out the optimization and simulations, initially, the models of human cell metabolism of breast tissue in the presence of a tumor affecting this tissue. For the model, there is also a corresponding model of cell metabolism in the case of healthy tissue. Thus, in total, there are 2 metabolic models (Table 2), each of which is thus assumed to be able to model in a plausible way the behavior, that is, the fluxes of reactions affecting the cell in human tissues. Such metabolic networks, used by Nam et al. (Nam et al. 2014), and taken from medical databases" 'The cancer genome Atlas' (TCGA), 'Cancer Cell Line Encyclopedia (CCLE), 'NCBI gene expression omnibus 'and are encoded with the SBML standard, as XML files.

4 Multi-objective Optimization for Genome-Scale Metabolic Models

Nowadays, most engineering problems require dealing with multiple conflicting objective functions instead of a single objective function. The multi-objective optimization (Biondi et al. 2006) is an efficient technique for finding a set of solutions that define the best tradeoff between competing objective functions while satisfying several constraints (Sharifi et al. 2021). Compared with a single optimization problem, in multi-objective optimization, we have more than one function to be considered as the objective of the optimization procedure. Obviously, one does not have, unless in very special cases, a single solution since there is no immediate comparison between two solutions, since in evaluating them, the values of the objective functions may, from time to time, be better in one or the other. The definition of such optimums is related to Multi-objective Optimization. A first intuitive definition might be that of a solution that cannot be improved in all the objectives required in optimization. Equivalently, if a solution is a Pareto optimum, and another solution improves the value of one of the objective functions, then necessarily there must be at least another objective function in which the Pareto optimum has a better value. This definition can be formalized, of course, by introducing the concept of dominance (Patané et al. 2018). Let it, therefore, be given a problem of multi-objective optimization, which we can generally denote by

$$\min F(x_1, \ldots, x_n) s.t. (x_1, \ldots, x_n) \in P \tag{4.1}$$

where F: $\Omega \rightarrow R^h$ is the vector defined by the objective functions, i.e., $F(x) = (f_1(x), f_2(x), \ldots, f_h(x))$, with $f_i(x)$, $i \in \{1, \ldots, h\}$ real-valued functions. The function F then associates with each point in the space Ω the respective values of the objective functions. $P \subset \Omega$ is called the admissible region of the problem (4.1). The points $x \in \Omega$ are candidate solutions, while $x \in P$ are the admissible solutions or points.

Optimal Fluxes. Table 3 below shows the optimal values identified through the Flux Balance Analysis. These values are, therefore, in the case of the single optimization problem, and the only objective function considered is the generic biomass function or biomass and ATP production. This value will serve as a comparison term for all subsequent tests. The objective functions considered will therefore be:

$$f_{obj} = fl_{bio} \tag{4.2}$$

in the case of the cancer model, and

$$f_{obj} = 0.0010 \cdot fl_{ATP} + 10.9109 \cdot fl_{bio} \tag{4.3}$$

in the case of normal metabolism models. The biomass reaction is a very complicated reaction, has about 40 reactants, which we do not report, and 3 products: adp_c, h_c, and pi_c. The flux of ATP is related to the reaction identified as ATPS4m, which has the following kinetic law (or kinetic equation/reaction):

$$4.0h_c + adp_m + pi_m \rightarrow atp_m + H_2O_m + 3.0h_m$$

Table 3. Optimal flux values for the Biomass reaction alone or for the metabolic objective function.

Disease/Tissue	Pathological case (biomass flux)	Healthy case (objective function flux)
Breast	0.091651635797115	1.999999999998552

5 Results

Finally, we report in this last part a brief description of the models examined and the computational results related to them obtained through the multi-objective evolutionary algorithm (Amaradio et al. 2022). It is possible to visualize the Pareto Front obtained in the last generation of the multi-objective evolutionary algorithm (Cutello et al. 2010). The purpose of the simulations is to minimize the production of certain specific metabolites, or rather to minimize the flow of certain reactions linked to genes that are related to their production (King et al. 2016). The reactions considered are the same for each model, and they are Isocitrate Dehydrogenase (NADP), which is bound to the enzyme IDH1. This irreversible reaction is localized in the cytoplasm ("_c") and has a kinetic reaction:

$$icit_c + nadp_c \rightarrow akg_c + CO_{2c} + nadph_c$$

Isocitrate Dehydrogenase NADP, bound to the enzyme IDH2. A difference of the former, the compartment in which it takes place is the mitochondrion ("_m"), and the reaction is reversible. The kinetic reaction is

$$icit_m + nadp_m \rightleftarrows akg_m + CO_{2m} + nadph_m$$

Succinate Dehydrogenase, corresponding to the SDHA, SDHB, SDHC, SDHD genes, and the compartment is still the mitochondrion ("_m"); the kinetic reaction is

$$Fad_m + succ_m \rightleftarrows Fadh2_m + Fum_m$$

Fumarase Mitochondrial, related to the FH gene, is a reversible reaction localized in the mitochondrion ("_m"); the kinetic reaction is

$$Fum_m + H_2O_m \rightleftarrows L - mal_m$$

The purpose of the simulation is to minimize a function of the flows of these 4 reactions, which then constitutes the synthetic objective function, while maximizing the natural objective function, which in the tumor and healthy case models is equal to (4.2) and (4.3), respectively. The synthetic objective functions considered are the following:

- The sum of the *absolute values* of the flows associated with the 4 reactions.
- The *algebraic sum* of the floors associated with the 4 reactions.
- A test was also performed requiring *maximization of the minor flows* (i.e., the negative flows) and, similarly, the *minimization of the major flows* (i.e., the positive flows) of the 4 reactions. With these objective functions, by maximizing negative flows, we want to force those flows to reach zero. Similarly, by minimizing positive flows, we want to force those flows to reach zero.

The number of generations of the multi-objective evolutionary algorithm (Amaradio et al. 2022) is fixed for all simulations at 1000 (Cutello et al. 2006). We see a graphical representation (Fig. 6) of the S matrix, in which each point corresponds to values of the stoichiometric matrix other than 0. The algorithm will have as its initial generation the one corresponding to several points equal to pop. In that generation, all genes are present, and as mentioned, subsequent generations will vary in the gene knock-outs considered. We report in Fig. 1 an initial control test. In it, maximization of biomass and minimization of all reactions in the model. As can be seen, the algorithm tends to find points that best satisfy the conditions of the minimization and maximization set. In the x-axis, we have the value of biomass, expressed as difference from the initial case. The y-axis, on the other hand, is shown as the sum of the flows of the reactions. The Pareto Front shows values found in the more advanced generations. Initially, the control test was repeated, with the results shown in Fig. 7. The other case considered, on the other hand, consists of the minimization of the sum of the flows (Fig. 8).

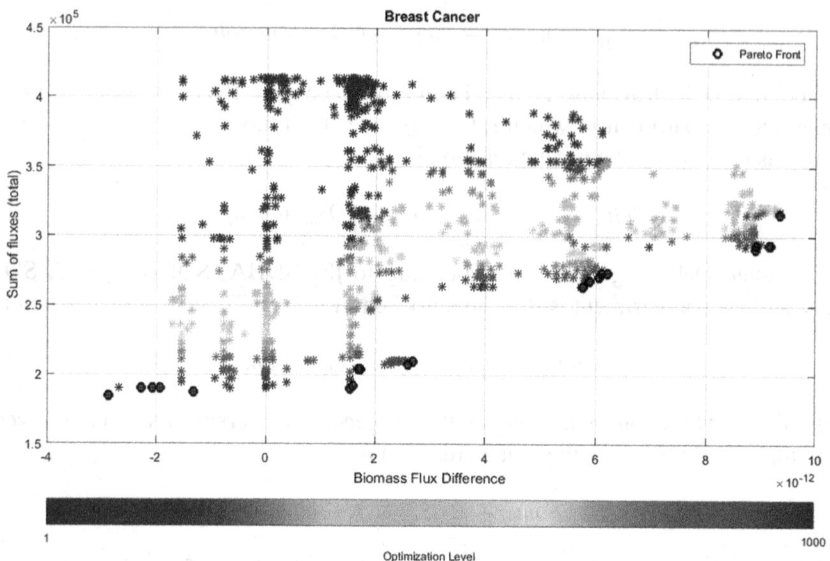

Fig. 1. Control test. The fluxes of all reactions are minimized while maximizing the biomass (breast cancer model). Circles indicate the non-dominated points of the Pareto front; all other candidate solutions are feasible points. The closer we get to the Pareto front, the more the global sum of flows decreases; conversely, the further we move away from the Pareto front, the more the global sum of flows increases (i.e., the level of optimization decreases).

Next, we then proceed to the 4 reactions considered in the model. In Fig. 2, we have the results obtained by setting as an objective the minimization of the absolute value of the sum of the flows of the reactions. The results are grouped along the value of flux equal to zero. It is, therefore, appropriate to see the graphic in more detail, as shown in Fig. 3.

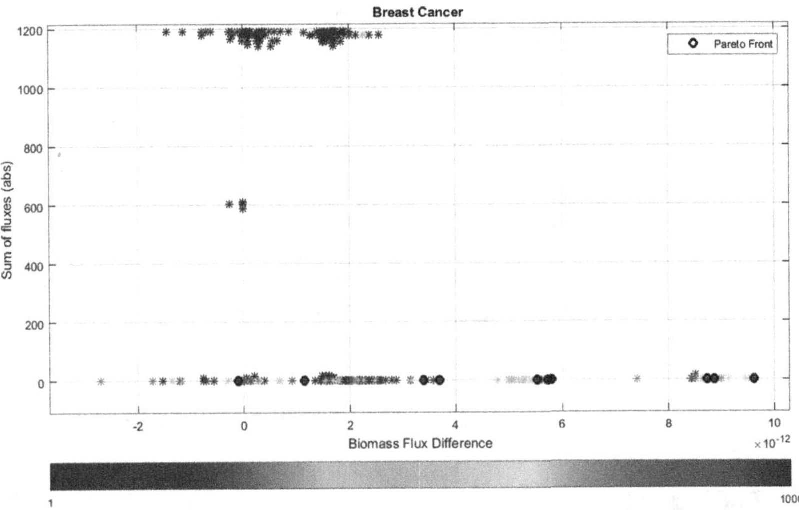

Fig. 2. Three operational patterns. Minimization of the absolute value of the sum of reactions and maximization of biomass. Circles indicate the non-dominated points of the Pareto front; all other candidate solutions are the feasible points. In this case, we have three distinct patterns clustered at three specific values of the overall sum of flows (about 0, 600 and 1200). The pattern at the bottom of the figure (the one with a sum of fluxes about 0) is shown in Fig. 3. The middle and top patterns show only poorly optimized feasible points.

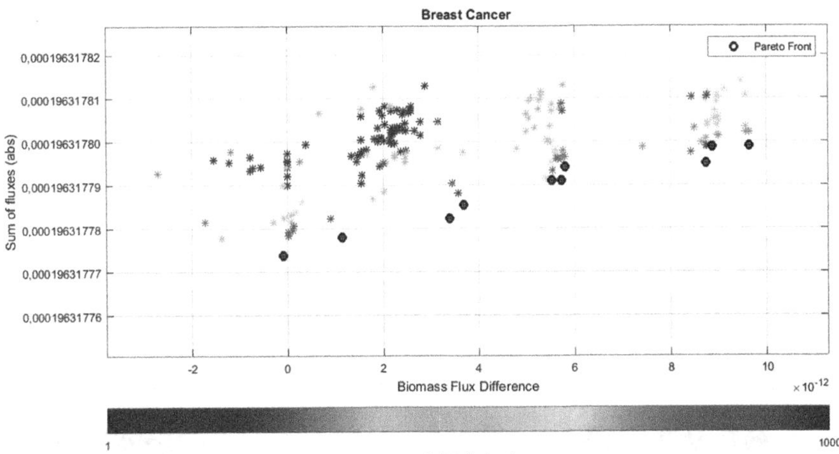

Fig. 3. Zoom of the pattern at the bottom of Fig. 2. Results were obtained by setting the minimization of the absolute value of the sum of reaction fluxes as the objective function. Similar to the findings in Fig. 1, we observe that the closer we get to the Pareto front, the more the global sum of flows decreases (i.e., the level of optimization increases); conversely, the further we move away from the Pareto front, the more the global sum of flows increases.

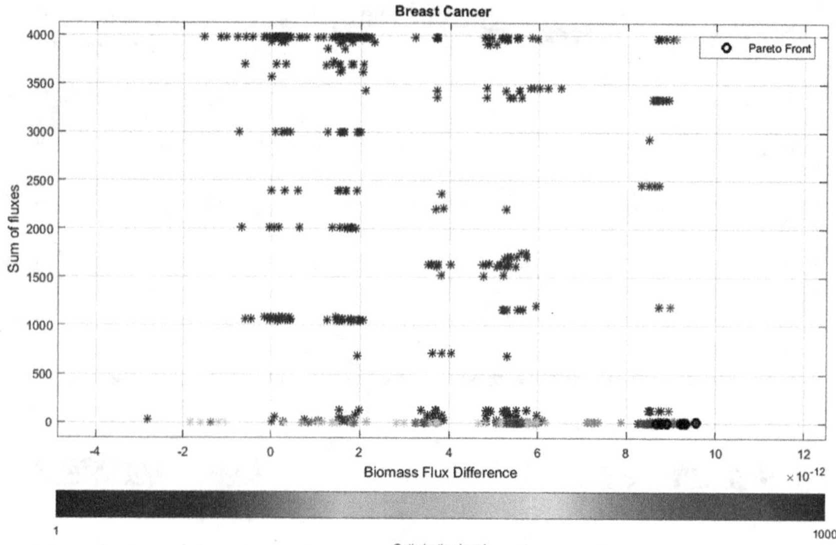

Fig. 4. Breast cancer genome-scale metabolic model. Minimization of the absolute value of the sum of reactions and maximization of biomass. Circles indicate the non-dominated points of the Pareto front; all other candidate solutions are the feasible points.

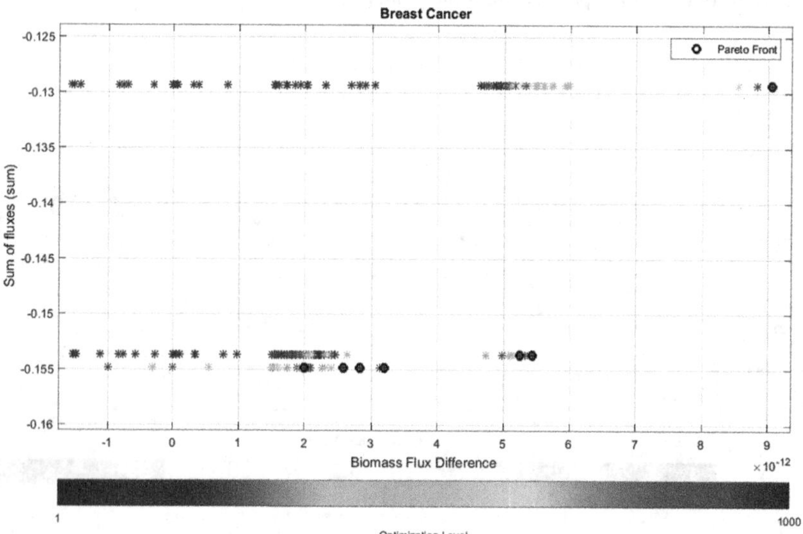

Fig. 5. Breast cancer genome-scale metabolic model. Minimization of the sum of reactions and maximization of biomass. Circles indicate the non-dominated points of the Pareto front; all other candidate solutions are the feasible points.

In contrast, the second test carried out consists of a minimization of the flows that in the initial test have maximum value, and similarly a maximization of the flows that have a minimum value (Fig. 4). Again, it is appropriate to see in more detail the obtained fronts shown in Fig. 5.

The last simulation presented is related to the minimization of the sum of the floors of the reactions considered. As can be seen, the results obtained are similar but not the same as the case of minimization of the absolute value.

Similar simulations were also carried out considering the breast normal model. We, therefore, also report here the graphical results obtained by the multi-objective evolutionary algorithm in a parallel manner with respect to the presence of the tumor then. The biochemical purpose of such a simulation is obviously to be considered different. Note that the biomass values, as mentioned several times above, correspond here to a composite reaction, in which the production of biomass and of ATP. Again, we report an initial test carried out by varying the fluxes randomly.

6 Conclusions

The *in silico* metabolism simulations carried out in this research work show the effectiveness of the methodology used for the biochemical interpretation of the results and the analysis of the individual gene in the modified metabolic network found. In fact, each individual network corresponds to a series of knock-outs performed or not on individual genes in the genome-scale metabolic model, represented with an n-tuple of Boolean variables, where n is the total number of defined genes. The next step in research would then be to compare the *in silico* results with studies performed *in vitro*, or analyses of patients and real human cells. It is also evident that these kinds of simulations are a powerful tool in the hands of researchers, as they allow numerous tests to be carried out in a relatively short time, allowing them to select the most interesting results to be compared and further investigated. BioCAD methodologies of this type are useful for these models and maybe a new frontier in the study of problems that are much more realistic and usable even in practice than was possible to imagine at the time of their conception. The striking thing about this approach is its great versatility, which allows a method such as the one described in this paper to be used in a wide variety of biological circuit design problems.

Appendix

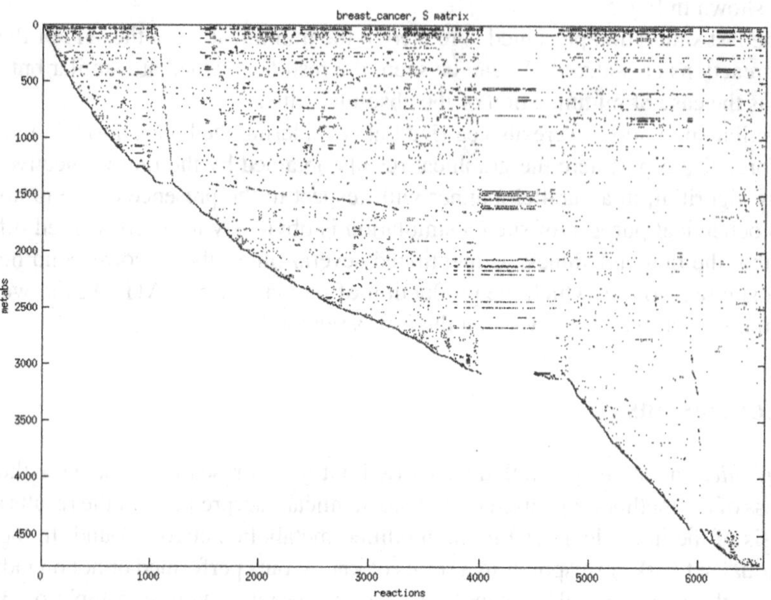

Fig. 6. Genome-scale metabolic model of breast cancer. A graphical representation of the S-matrix is presented, in which on the x-axis are the reactions, and on the y-axis are the metabolites. Each point on the S matrix represents values of S other than 0.

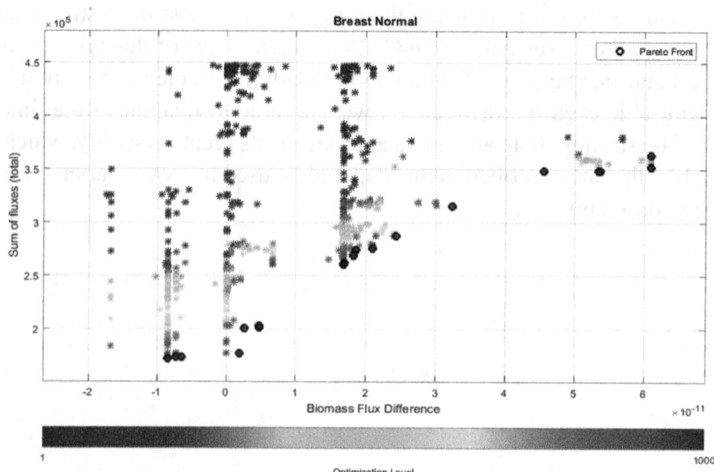

Fig. 7. Control test for the breast genome-scale metabolic model. The fluxes of all reactions are minimized while maximizing the biomass. Circles indicate the non-dominated points of the Pareto front; all other candidate solutions are the feasible points.

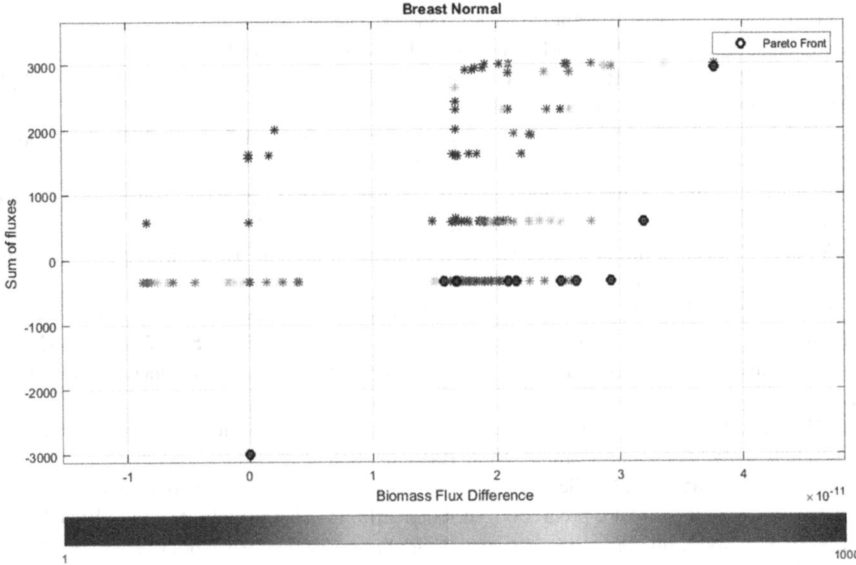

Fig. 8. Breast genome-scale metabolic model. Minimization of the sum of flows for the reactions considered while maximizing the biomass. Circles indicate the non-dominated points of the Pareto front; all other candidate solutions are the feasible points.

References

Amaradio, M.N., Ojha, V., Jansen, G., Gulisano, M., Costanza, J., Nicosia, G.: Pareto optimal metabolic engineering for the growth-coupled overproduction of sustainable chemicals. Biotechnol. Bioeng. **119**(7), 1890–1902 (2022)

Angione, C., Costanza, J., Carapezza, G., Lió, P., Nicosia, G.: A design automation framework for computational bioenergetics in biological networks. Mol. BioSyst. **9**(10), 2554–2564 (2013)

Barzaman, K., et al.: Breast cancer: biology, biomarkers, and treatments. Int. Immunopharmacol. **84**, 106535 (2020)

Biondi, T., Ciccazzo, A., Cutello, V., D'Antona, S., Nicosia, G., Spinella, S.: Multi-objective evolutionary algorithms and pattern search methods for circuit design problems. J. Univers. Comput. Sci. **12**(4), 432–449 (2006)

Chou, F.J., Liu, Y., Lang, F., Yang, C.: D-2-hydroxyglutarate in glioma biology. Cells **10**(9), 2345 (2021)

Cutello, V., Lee, D., Leone, S., Nicosia, G., Pavone, M.: Clonal selection algorithm with dynamic population size for bimodal search spaces. In: Jiao, L., Wang, L., Gao, X.-b, Liu, J., Wu, F. (eds.) ICNC 2006. LNCS, vol. 4221, pp. 949–958. Springer, Heidelberg (2006). https://doi.org/10.1007/11881070_125

Cutello, V., Nicosia, G., Pavone, M., Stracquadanio, G.: An information-theoretic approach for clonal selection algorithms. In: Hart, E., McEwan, C., Timmis, J., Hone, A. (eds.) ICARIS 2010. LNCS, vol. 6209, pp. 144–157. Springer, Heidelberg (2010). https://doi.org/10.1007/978-3-642-14547-6_12

DallaPozza, E., et al.: Regulation of succinate dehydrogenase and role of succinate in cancer. Semin. Cell Dev. Biol. **98**, 4–14 (2020)

Ježek, P.: 2-Hydroxyglutarate in cancer cells. Antioxid. Redox Signal. **33**(13), 903–926 (2020)

Katsura, C., Ogunmwonyi, I., Kankam, H.K., Saha, S.: Breast cancer: presentation, investigation, and management. Br. J. Hosp. Med. (London, England) **83**(2), 1–7 (2005)

Keller, M.A., Piedrafita, G., Ralser, M.: The widespread role of non-enzymatic reactions in cellular metabolism. Curr. Opin. Biotechnol. **34**, 153–161 (2015)

King, Z.A., et al.: BiGG models: a platform for integrating, standardizing and sharing genome-scale models. Nucleic Acids Res. **44**(D1), D515–D522 (2016)

Hucka, M., et al.: The systems biology markup language (SBML): language specification for level 3 version 2 Core release 2. J. Integr. Bioinform. **16**(2), 20190021 (2019)

Liu, Y., et al.: Targeting tumor suppressor genes for cancer therapy. BioEssays: News Rev. Mol. Cell. Dev. Biol. **37**(12), 1277–1286 (2015)

Liu, S., Cadoux-Hudson, T., Schofield, C.J.: Isocitrate dehydrogenase variants in cancer - cellular consequences and therapeutic opportunities. Curr. Opin. Chem. Biol. **57**, 122–134 (2020)

Mishra, P., Ambs, S.: Metabolic signatures of human breast cancer. Mol. Cell. Oncol. **2**(3), e992217 (2015)

Nam, H., et al.: A systems approach to predict oncometabolites via context-specific genome-scale metabolic networks. PLoS Comput. Biol. **10**(9), e1003837 (2014)

Nicosia, G., Stracquadanio, G.: Generalized pattern search and mesh adaptive direct search algorithms for protein structure prediction. In: Giancarlo, R., Hannenhalli, S. (eds.) WABI 2007. LNCS, vol. 4645, pp. 183–193. Springer, Heidelberg (2007). https://doi.org/10.1007/978-3-540-74126-8_17

Norsigian, C.J., et al.: BiGG models 2020: multi-strain genome-scale models and expansion across the phylogenetic tree. Nucleic Acids Res. **48**(D1), D402–D406 (2020)

Orth, J., Thiele, I., Palsson, B.: What is flux balance analysis? Nat. Biotechnol. **28**, 245–248 (2010)

Patanè, A., Santoro, A., Costanza, J., Carapezza, G., Nicosia, G.: Pareto optimal design for synthetic biology. IEEE Trans. Biomed. Circuits Syst. **9**(4), 555–571 (2015)

Patané, A., Jansen, G., Conca, P., Carapezza, G., Costanza, J., Nicosia, G.: Multi-objective optimization of genome-scale metabolic models: the case of ethanol production. Ann. Oper. Res. **276**(1–2), 211–227 (2018). https://doi.org/10.1007/s10479-018-2865-4

Peetsold, M., et al.: fumarase deficiency: a case with a new pathogenic mutation and a review of the literature. J. Child Neurol. **36**(4), 310–323 (2021)

Rana, P., Berry, C., Ghosh, P., Fong, S.S.: Recent advances on constraint-based models by integrating machine learning. Curr. Opin. Biotechnol. **64**, 85–91 (2020)

Sharifi, M.R., Akbarifard, S., Qaderi, K., Madadi, M.R.: A new optimization algorithm to solve multi-objective problems. Sci. Rep. **11**(1), 20326 (2021)

Schmidt, C., Sciacovelli, M., Frezza, C.: Fumarate hydratase in cancer: a multifaceted tumor suppressor. Semin. Cell Dev. Biol. **98**, 15–25 (2020)

Umeton, R., Nicosia, G., Dewey, C.F.: OREMPdb: a semantic dictionary of computational pathway models. BMC Bioinform. **13**(4), 1–9 (2012)

Van Rosmalen, R.P., Smith, R.W., Martins Dos Santos, V.A.P., Fleck, C., Suarez-Diez, M.: Model reduction of genome-scale metabolic models as a basis for targeted kinetic models. Metab. Eng. **64**, 74–84 (2021)

Vander Heiden, M.G.: Targeting cancer metabolism: a therapeutic window opens. Nat. Rev. Drug Discov. **10**(9), 671–684 (2011)

Author Index

G. Nicosia et al. (Eds.): LOD 2022, LNCS 13810, pp. 613–616, 2023.
https://doi.org/10.1007/978-3-031-25599-1

T

U

V

W

X

Z

d in the United States
er & Taylor Publisher Services

Print
by Ba